Homogeneous and Heterogeneous Photocatalysis

NATO ASI Series

Advanced Science Institutes Series

A series presenting the results of activities sponsored by the NATO Science Committee, which aims at the dissemination of advanced scientific and technological knowledge, with a view to strengthening links between scientific communities.

The series is published by an international board of publishers in conjunction with the NATO Scientific Affairs Division

A Life Sciences B Physics	Plenum Publishing Corporation London and New York
C Mathematical and Physical Sciences	D. Reidel Publishing Company Dordrecht, Boston, Lancaster and Tokyo
D Behavioural and Social Sciences E Engineering and Materials Sciences	Martinus Nijhoff Publishers The Hague, Boston and Lancaster
F Computer and Systems Sciences G Ecological Sciences	Springer-Verlag Berlin, Heidelberg, New York and Tokyo

Series C: Mathematical and Physical Sciences Vol. 174

Homogeneous and Heterogeneous Photocatalysis

edited by

Ezio Pelizzetti

Department of Analytical Chemistry, University of Turin,
Turin, Italy

and

Nick Serpone

Department of Chemistry, Concordia University,
Montreal, Canada

D. Reidel Publishing Company

Dordrecht / Boston / Lancaster / Tokyo

Published in cooperation with NATO Scientific Affairs Division

Proceedings of the NATO Advanced Research Workshop on
Homogeneous and Heterogeneous Photocatalysis
Maratea, Potenza, Italy
September 1-7, 1985

Library of Congress Cataloging in Publication Data

NATO Advanced Research Workshop on Homogeneous and Heterogeneous Photocatalysis (1985 : Maratea, Italy)

Homogeneous and heterogeneous photocatalysis.
(NATO ASI series C, Mathematical and physical sciences; vol. 174)
"Proceedings of the NATO Advanced Research Workshop on Homogeneous and Heterogeneous Photocatalysis, Maratea, Potenza, Italy, September 1–7, 1985."
Includes indexes.
1. Photochemistry—Congresses. 2. Catalysis—Congresses. I. Pelizzetti, Ezio, 1944– , II. Serpone, Nick, 1939– . III. Series.
QD701.N34 1985 541.3′5 86–3152
ISBN 90–277–2221–8

Published by D. Reidel Publishing Company
P.O. Box 17, 3300 AA Dordrecht, Holland

Sold and distributed in the U.S.A. and Canada
by Kluwer Academic Publishers,
190 Old Derby Street, Hingham, MA 02043, U.S.A.

In all other countries, sold and distributed
by Kluwer Academic Publishers Group,
P.O. Box 322, 3300 AH Dordrecht, Holland

D. Reidel Publishing Company is a member of the Kluwer Academic Publishers Group

Printed in The Netherlands

TABLE OF CONTENTS

FOREWORD

Ever since the oil crisis of 1973, researchers in various fields of chemistry have proposed various schemes to conserve energy, as well to convert the sun's abundant and limitless supply of energy to produce chemical fuels (e.g., hydrogen from water, ...). The enthusiasm had no previous parallel in the mid-1970's. Unfortunately, despite the several good proposals, the results have proven - in retrospect - somewhat disappointing from an economic viable point of view. The reasons for the meagre results are manyfold not the least of which are the experimental difficulties encountered in storage systems. Moreover, the lack of a concerted, well orchestrated interdisciplinary approach has been significant. By contrast, the chemical advances made in the understanding of the processes involved in such schemes have been phenomenal. A recent book on this issue (M. Gratzel, Energy Resources through Photochemistry and Catalysis, 1983) is witness to the various efforts and approaches taken by researchers. In the recent years, many more groups have joined in these efforts, and the number of papers in the literature is staggering !

One of the motives for organizing this NATO Advanced Research Workshop stemmed from our view that it was time to take stock of the accomplishments and rather than propose new schemes, it was time to consider seriously avenues that are most promising. For this very purpose, our approach, as we viewed it, was to bring together experts from the various chemical fields (Photochemistry, Electrochemistry, Catalysis, Colloid, Semiconductor, and Biochemistry) - an interdisciplinary approach - and from various NATO (and non-NATO) countries to one meeting place such that we could fully and informally discuss recent achievements and problems.

Our hope from this Workshop was that the participants would put forward and point out the important problems yet to resolve towards achieving a viable system for energy conversion and energy storage. Additionally, and by no means least, we saw the Workshop participants tackling other important issues regarding photocatalysis in the synthesis of chemically useful products, and in the environmental pollution area.

The Workshop would not have been the same without the dedicated help of the staff of the Hotel Villa del Mare, to whom we owe the creation of a very warm and pleasant atmosphere, conducive to scientific and social interactions among the participants.

Finally, we thank the Directors of the Scientific Affairs Division of NATO and of the Programme on Selective Activation of Molecules for granting us the financial support for a very fruitful Advanced Research Workshop.

Ezio Pelizzetti
Università di Torino
Torino - ITALY

Nick Serpone
Concordia University
Montreal - CANADA

September 10, 1985

ORGANIZING COMMITTEE

Prof. E. Pelizzetti
Dipartimento di Chimica Analitica
Università di Torino
via Pietro Giuria 5 - 10126 TORINO - ITALY

Prof. N. Serpone
Dept. of Chemistry
Concordia University
1455 de Maisonneuve - Blvd. West - MONTREAL - QUEBEC H3G 1M8 - CANADA

Dr. F. Cecchini
C.N.R.
via Nizza 128 - 00198 ROMA - ITALY

Dr. A. Harriman
The Royal Institution
21, Albemarle Street - LONDON W1X 4BS - U.K.

Prof. G. McLendon
Dept. of Chemistry
University of Rochester
River Station - ROCHESTER - N.Y. 14627 - USA

Prof. F. Scandola
Dipartimento di Chimica
Università di Ferrara
via Borsari 46 - 44100 FERRARA - ITALY

LIST OF LECTURERS

Prof. M.A. Fox
Dept. of Chemistry
University of Texas
AUSTIN TX 78712-1167 - USA

Prof. M. Gratzel
Institut de Chimie Physique
E.P.F.L.
Ecublens - 1015 LAUSANNE - CH

Dr. A. Heller
AT&T Bell Laboratories
600 Mountain Avenue
Murray Hill - NEW JERSEY 07974 - USA

Dr. B.G. Oliver
Environmental Contaminants Division
National Water Research Institute - Canada Centre for Inland Waters
P.O.Box 5050 - BURLINGTON - ONTARIO L7R 4A6 - CANADA

Dr. P. Pichat
C.N.R.S. - Ecole Centrale de Lyon
Bat. Chimie
B.P. 163 - 69131 ECULLY CEDEX - FRANCE

Prof. G.N. Schrauzer
Dept. of Chemistry
University of California - San Diego
Revelle College - LA JOLLA - CALIF. 92093 - USA

Prof. G.A. Somorjai
Dept. of Chemistry
University of California - Berkeley
BERKELEY - CALIFORNIA 94720 - USA

Dr. H. Van Damme
C.N.R.S. - Centre de Recerche sur le Solides a Organisation Cristalline Imparfaite
45045 ORLEANS CEDEX - FRANCE

LIST OF CHAIRMEN AND PARTICIPANTS

CANADA

Dr. D. Belanger — Institut Nat. de la Recherche Scientifique
Case Postale 1020
Varennes - Quebec JOL 2 PO

Dr. E. Borgarello — Dept. of Chemistry - Concordia University
1455 de Maisonneuve Blvd. West
Montreal - Quebec H3G 1M8

Prof. J. Bolton — The University of Western Ontario
Dept. of Chemistry
London N 6A 5B7

Dr.J.P. Dodelet — Institut Nat. de la Recherche Scientifique
Case Postale 1020
Varennes - Quebec JOL 2 PO

FRANCE

Dr. E. Amouyal — Laboratoire de Physico-Chimie des Rayonnements
Université Paris-Sud
91405 ORSAY CEDEX

Dr. C. Giannotti C.N.R.S.
Institut de Chimie des Substances Naturelles
91190 GIF-SUR-YVELFE

Prof. S.J.Teichner Laboratoire de Thermodynamique et Cinetique
Chimiques - Université Claude Bernard
43, Boulevard du 11 Novembre 1918
69622 VILLEURBANNE CEDEX

F.R.GERMANY

Prof. H. Kisch Institut fur Anorganische Chemie der Universitat
Erlangen-Nurnberg
Egerlandstr. 1
D-8520 ERLANGEN

Prof. R. Memming Institut f. Physikalische Chemie
Universitat Hamburg
Laufgraben 24
D-2000 HAMBURG 13

Dr. R. Millini Institut fur Anorganische Chemie der Universitat
Erlangen-Nurnberg
Egerlandstr. 1
D-8520 ERLANGEN

GREECE

Dr. E. Papaconstantinou Chemistry Dept.
N.R.C. Demokritos
ATHENS

ISRAEL

Dr. M. Halmann — Isotope Dept.
Weizmann Institute of Science
REHOVOT 76100

Dr. J. Manassen — The Weizmann Institute of Science
REHOVOT 76100

Dr. L.A.Rajbenbach — Soreq Nuclear Research Center
YAVNE 70600

ITALY

Prof. G.G.Aloisi — Dip. di Chimica - Università di Perugia
via Elce di Sotto 8
06100 PERUGIA

Dr. R.Amadelli — Centro di Fotochimica CNR
via Borsari 46
44100 FERRARA

Prof. V.Augugliaro — Istituto di Ingegneria Chimica
Università di Palermo - Viale delle Scienze
90128 PALERMO

Prof. V.Balzani — Istituto Chimico G.Ciamician
Università di Bologna - via Selmi 2
40126 BOLOGNA

Dr. M.Barbeni Dip. di Chimica Analitica - Università di Torino
via Pietro Giuria 5
10125 TORINO

Prof. C.A.Bignozzi Dip. di Chimica - Università di Ferrara
via Borsari 46
44100 FERRARA

Dr. C.Chiorboli Centro di Fotochimica CNR
via Borsari 46
44100 FERRARA

Dr. M.Ciano FRAE CNR
via de Castagnoli 1
40126 BOLOGNA

Dr. A.di Domenico Istituto Superiore di Sanità
Dept. of Comparative Toxicology and Ecotoxicology
viale Regina Elena 299
00161 ROMA

Prof. M.T.Indelli Dip. di Chimica - Università di Ferrara
via Borsari 46
44100 FERRARA

Dr. A.Maldotti Dip. di Chimica - Università di Ferrara
via Borsari 46
44100 FERRARA

Dr. C.Minero Dip. di Chimica Analitica - Università di Torino
via Pietro Giuria 5
10125 TORINO

Dr. L.Palmisano — Istituto di Ingegneria Chimica
Università di Palermo - Viale delle Scienze
90128 PALERMO

Dr. E.Passalacqua — Istituto CNR di Metodi e Processi Chimici per la trasformazione e l'accumulo dell'Energia
via S.Lucia Sopra Contesse 39
98013 PISTUNINA (MESSINA)

Prof. E.Pramauro — Dip. di Chimica Analitica - Università di Torino
via Pietro Giuria 5
10125 TORINO

Prof. M.A.Rampi — Dip. di Chimica - Università di Ferrara
via Borsari 46
44100 FERRARA

Prof. A.Sclafani — Istituto di Ingegneria Chimica
Università di Palermo - Viale delle Scienze
90128 PALERMO

Prof. L.Stradella — Istituto di Chimica Generale e Inorganica
Facoltà di Farmacia - Università di Torino
via Pietro Giuria 9
10125 TORINO

Dr. M.L.Tosato — Istituto Superiore di Sanità
Dept. of Comparative Toxicology and Ecotoxicology
viale Regina Elena 299
00161 ROMA

JAPAN

Dr. T.Sakata Institute for Molecular Science
MYODAIJI OKAZAKI 444

Prof. H.Tsubomura Laboratory for Chemical Conversion of Solar Energy
and Dept. of Chemistry - Faculty of Engineering
Science - Osaka University
Toyonaka , OSAKA 560

PORTUGAL

Dr. S.Costa Centro de Quimica Estrutural , Complexo I
Instituto Superior Técnico - Avenida Rovisco Pais
1096 LISBOA Codex

SPAIN

Prof. G.Munuera Departamento de Quimica General
Facultad de Quimica - Universidad de Sevilla
SEVILLA

Dr. J.A.Navio Santos Departamento de Quimica General
Facultad de Quimica - Universidad de Sevilla
SEVILLA

SWITZERLAND

Dr. J.Kiwi — Institut de Chimie Physique
E.P.F.L. - Ecublens
1015 LAUSANNE

TURKEY

Prof. F.Baykut — Dept. of Chemical Engineering
Faculty of Engineering - University of Istanbul
Laleli - ISTANBUL

Dr. G.Baykut — Dept. of Chemical Engineering
Faculty of Engineering - University of Istanbul
Laleli - ISTANBUL

U.K.

Prof. M.Archer — Dept. of Physical Chemistry
University of Cambridge
Sensfield Road
CAMBRIDGE CB2 1EP

Prof. R.Bickley — Scool of Studies in Chemistry
University of Bradford
BRADFORD BD7 1DP

USA

Dr. L.Brus AT&T Bell Laboratories
600 Mountain Avenue
Murray Hill - New Jersey 07974

Prof. J.H.Fendler Dept. of Chemistry
and Institute of Colloid and Surface Science
Clarkson University
POTSDAM - New York 13676

Dr. A.J.Frank Solar Energy Research Institute
1617 Cole Boulevard
GOLDEN CO 80401

Dr. M.E.Gress Dept. of Energy
Division of Chemical Sciences
WASHINGTON DC 20545

Dr. K.Krist Gas Research Institute
8600 W. Bryn Mawr Ave.
CHICAGO IL 60631

Prof. N.S.Lewis Dept. of Chemistry
Stanford University
STANFORD CA 94305

Dr. D.Meisel Argonne National Laboratory
Chemistry Division
ARGONNE IL 60439

Dr. A.J.Nozik Solar Energy Research Institute
1617 Cole Boulevard
GOLDEN CO 80401

Prof. D.F.Ollis
Chemical Engineering Dept.
North Caroline State University
RAYLEGH NC 27695-7905

Dr. S.Tunesi
University of Wisconsin
Dept. of Agricoltural Sciences
MADISON - Wisconsin 53706

Prof. D.G.Whitten
Dept. of Chemistry
University of Rochester
ROCHESTER - New York 14627

YUGOSLAVIA

Dr. O.I.Micic
Boris Kidric Institute of Nuclear Sciences
Vinca 11001 - BEOGRAD

PHOTOINDUCED CHARGE SEPARATION: REQUIREMENTS NEEDED FOR IDEAL RELAYS AND PHOTOSENSITIZERS

Vincenzo Balzani, Alberto Juris
Istituto Chimico "G. Ciamician" dell'Universita' and Istituto FRAE-CNR, Bologna, Italy

Franco Scandola
Istituto Chimico dell'Universita' and Centro di Fotochimica del CNR, Ferrara, Italy

ABSTRACT. The requirements needed for ideal relay and photosensitizer species to be used in photoinduced electron transfer processes in homogeneous solution are examined and discussed. Transition metal complexes appear to be the most appropriate class of chemical molecules to take into consideration in the attempt to find out or to design useful relays and photosensitizers. Some guidelines that should be followed in this research are illustrated, with particular reference to the requirements of a high turnover number for a relay and of a high turnover number and a long excited state lifetime for a photosensitizer. The problem of charge separation is discussed and an attempt is made to identify the factors that should be controlled to optimize the cage escape efficiency.

1. INTRODUCTION

In the last few years the important role played by photosensitizer (P) and relay (R) species in a variety of electron transfer processes, including the conversion of solar energy into fuels, has clearly emerged (1-10). Progress has been very rapid and, to some extent, we have now reached the point where we need to evaluate and consolidate our achievements before undertaking further investigations. In this article we will try to formulate the requirements needed for ideal photosensitizers and relays and to discuss the problems of charge separation in homogeneous solution.

2. PHOTOSENSITIZED ELECTRON TRANSFER PROCESSES

For the sake of simplicity we will make reference to a homogeneous photosensitized electron transfer reaction (which could be, for example, the widely discussed photosensitized water splitting process (5,11-15)) where both a photosensitizer and a relay are employed. Most

E. Pelizzetti and N. Serpone (eds.), Homogeneous and Heterogeneous Photocatalysis, 1–27.

likely such homogeneous systems will be less useful than heterogeneous ones for practical applications. Nevertheless they are quite interesting because, especially in their "sacrificial" versions (vide infra), they offer the opportunity to probe specific properties of photosensitizer and relay species and to get information that may also be useful for the improvement of heterogeneous systems (6).

The role of a photosensitizer in a photoinduced electron transfer process is to absorb light so as to make available the energy needed to overcome the kinetic and/or thermodynamic barriers of the desired reaction. In the simplest case (Fig. 1), the photosensitizer P absorbs a photon and is promoted to an excited state, $^*P$, which is a reductant strong enough to reduce the substrate A to A^-; the oxidized form of the photosensitizer, P^+, obtained in this way can then oxidize the other substrate B to B^+ with regeneration of P and completion of the desired net reaction. (Note that here and in the following we make reference to a system where the excited state of the photosensitizer acts as a reductant; of course, a parallel discussion can be made for the case in which the excited state of the photosensitizer acts as an oxidant.) In practice, however, such simple systems do not work because the reaction between the excited photosensitizer and the substrate A is always too slow compared with the extremely fast decay rate of the excited state (Fig. 1, dotted arrow). In other words, there is something like a short-circuit which immediately dissipates the absorbed light energy and prevents the occurrence of the desired net reaction. In order to overcome this difficulty, one should try to prolong the excited state lifetime and/or to catalyze the reaction between $^*P$ and A. Both these possibilities, however, are severely hampered by intrinsic factors, as it will be shown later. Therefore, it is generally necessary to use a relay, R, i.e. a species which is able to undergo a very fast electron transfer reaction with the excited state before it undergoes deactivation (Fig. 2). The reduced form of the relay, R^-, obtained from the excited state reaction should be a reductant strong enough to reduce A to A^-, with regeneration of R. (Having chosen the example of a reducing excited state, here and in the following only a reducible relay is considered.)

It should now be noticed that the oxidized form of the photosensitizer and the reduced form of relay, produced in the reaction between the excited photosensitizer and the relay, may undergo the so-called back electron transfer reaction (Fig. 2, dashed arrows). Such a reaction is usually very fast and prevents the occurrence of the desired process involving the substrate species A and B. Minimization of the effects of the back electron transfer reaction is a very important goal which can be pursued facilitating cage escape of the products of the excited state reaction (vide infra) and speeding up, by appropriate catalysts, the reactions of P^+ and R^- with the two substrates (16-19).

For the purpose of studying single reactions and optimizing single reactants of the overall cycle, one can use the so-called "sacrificial" systems, where B is replaced by a strong reductant (or A by a strong oxidant) so as to involve P^+ (or R^-) in a fast irreversible reaction that competes successfully with the back electron transfer reaction. The best known example of these sacrificial cyclic processes is the

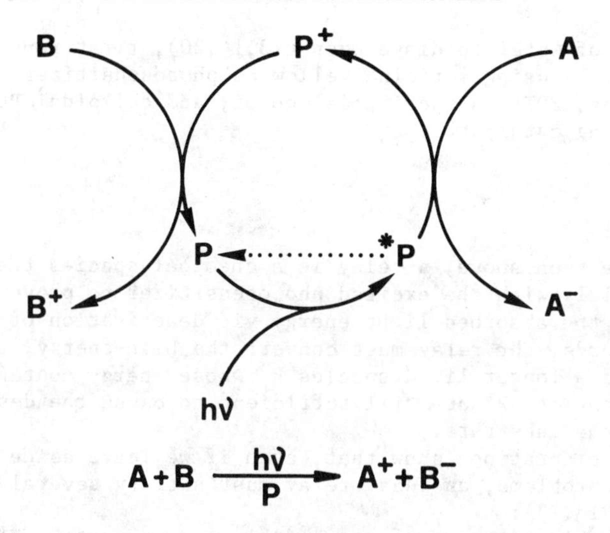

Fig. 1. Schematic representation of the simplest system for electron transfer photosensitization.

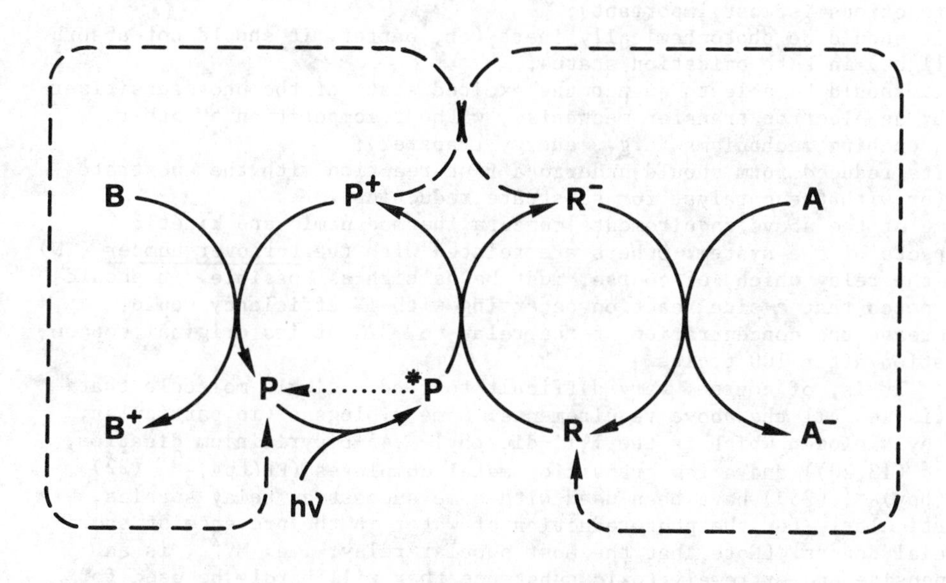

Fig. 2. Schematic representation of a photoinduced electron transfer reaction involving a photosensitizer and a relay.

photoreduction of water to dihydrogen (13,14,20), first reported by Shilov et al. (21) using acridine yellow as photosensitizer, methylviologen as a relay, EDTA as sacrificial donor, and colloidal Pt as hydrogen evolving catalyst.

3. RELAYS

As we have seen above, a relay is a chemical species that must react very rapidly with the excited photosensitizer to prevent the dissipation of the absorbed light energy via deactivation of $^*P$ (Fig. 2). In other words, the relay must convert the high-energy, short-lived $^*P$ species into a longer lived species R^- whose energy content is smaller than that of $^*P$ but still sufficient to cause the desired reaction with the substrate.

Simple considerations show that, even if we leave aside economic and ecological problems, an ideal relay must satisfy several specific requirements (Fig. 3):

- it should be able to undergo a reversible redox reaction at a suitable potential;
- it should be thermally inert in both oxidation states (for the water splitting reaction, stability toward hydrogenation and oxygenation reactions is most important);
- it should be photochemically inert (or, better, it should not absorb light) in both oxidation states;
- it should be able to quench the excited state of the photosensitizer by an electron transfer mechanism, without competition by other quenching mechanisms (e.g., energy transfer);
- its reduced form should undergo a fast reaction with the substrate (or with the catalyst for substrate reduction).

Some of the above requirements concern thermodynamic and kinetic aspects of the system. Others are related with the turnover number (TN) of the relay which, of course, must be as high as possible. It should be noted that a side reaction occurring with 1% efficiency would decrease the concentration of the relay to ≃37% of its original concentration after 100 cycles.

It is, of course, very difficult to find a single molecule that satisfies all the above requirements. Some viologens (in particular, methylviologen which is the 1,1'-dimethyl-4,4'-bipyridinium dication, MV^{2+} (13,20)) and a few transition metal complexes ($Rh(bpy)_3^{3+}$ (22), $Co(bpy)_3^{2+}$ (23)) have been used with some success as relay species, particularly for the photoreduction of water in the presence of sacrificial donors. (Note that the most popular relay, i.e. MV^{2+}, is an expensive and extremely toxic substance that will hardly be used for large scale applications).

In the search for and for the design of good relays, the following considerations may be useful. Organic molecules do not seem to be suitable candidates since saturated compounds are not likely to be reducible at reasonable potentials and unsaturated ones undergo thermal and/or photochemical reactions very easily. (A main drawback of MV^{2+} is an irreversible side reaction (24)). As far as inorganic compounds are

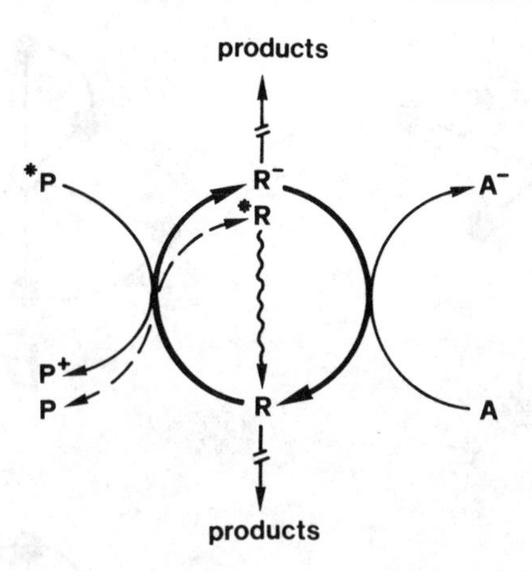

Fig. 3. Schematic illustration of the requirements needed for an ideal relay (see text).

concerned, the involvement of transition metal ions seems to be necessary in order to keep the reduction potentials within reasonable limits. Inorganic or fully saturated organic ligands should be preferred for the reasons seen above as well as to avoid light absorption (vide infra). Along these lines, metal aquo ions (e.g., Eu^{3+}_{aq}, V^{3+}_{aq}) and amine complexes (e.g., $Co(NH_3)_6^{3+}$, $Cr(NH_3)_6^{3+}$) could be considered as suitable candidates. Unfortunately, most of these species (especially the aquo ions) exhibit very bad kinetic properties for electron transfer processes because of high intrinsic barriers and/or nonadiabaticity problems (25). Furthermore, most coordination compounds are thermally and/or photochemically labile towards ligand dissociation in the oxidized and/or reduced form (26,27). Such ligand dissociation reactions, except for the special case of aquo ions, cause the depletion of the relay and the consequent failure of the whole system.

A major advance in the solution of these problems has occurred in the last few years with the design and the synthesis of cage-type ligands and cage-type complexes (28-31). The general strategy (Fig. 4) is that of replacing the unidentate ligands of a metal complex with fewer multidentate ligands, up to the limit of a single multidentate cage-type ligand which encapsulates the metal ion. In this way, ligand dissociation is prevented and the complex can be cycled between two oxidation states a great number of times (32). Interestingly, in several cases encapsulation also improves the kinetic properties (33-35).

Very promising relay species are the macrobicyclic polyazacryptates of cobalt prepared and characterized by Sargeson and co-workers (29,32,36-41). The prototype of these complexes is $Co(sep)^{3+}$, where sep = sepulchrate is a trivial name for 1,3,6,8,10,13,17,19-octaazabicyclo-[6.6.6]eicosane (Fig. 5). As discussed in detail elsewhere (42), this caged complex may be viewed as a perturbed version of $Co(NH_3)_6^{3+}$, in the sense that it mantains the "static" properties of $Co(NH_3)_6^{3+}$ that are related to the presence of 6 nitrogen atoms in the first coordina-

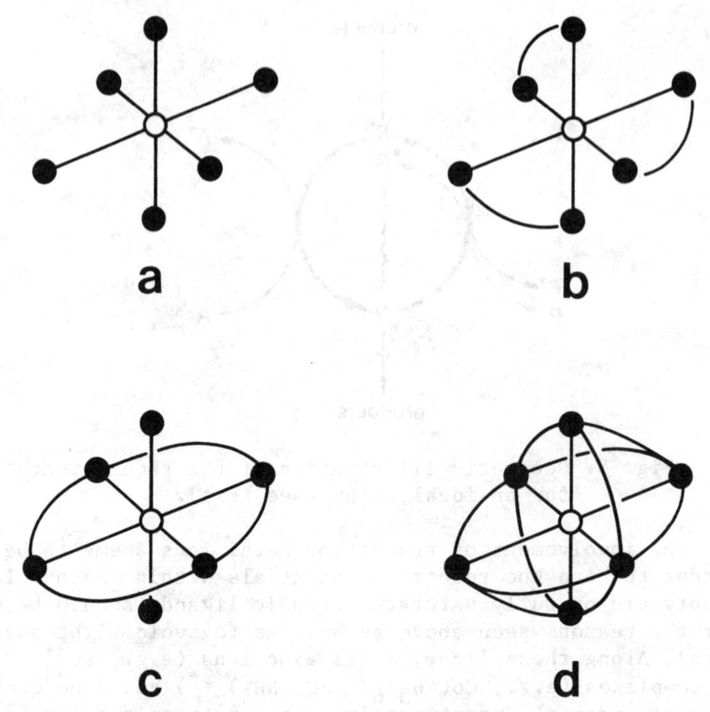

Fig. 4. Schematic representation of the transformation of a simple complex containing six unidentate ligands (a) into a cage-type complex (b).

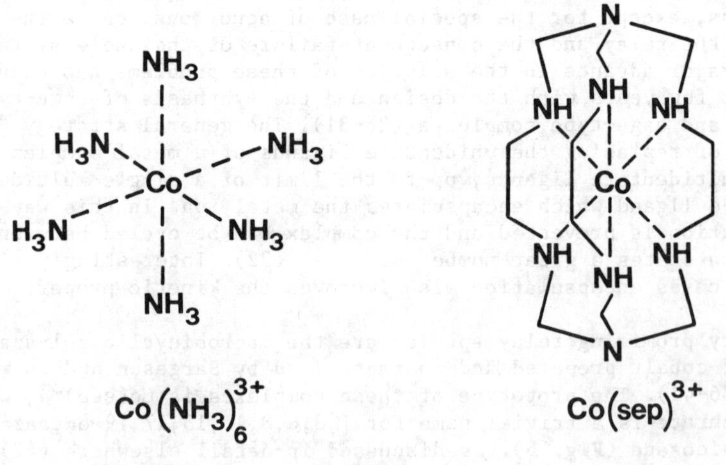

Fig. 5. Structural formulae of $Co(NH_3)_6^{3+}$ and $Co(sep)^{3+}$.

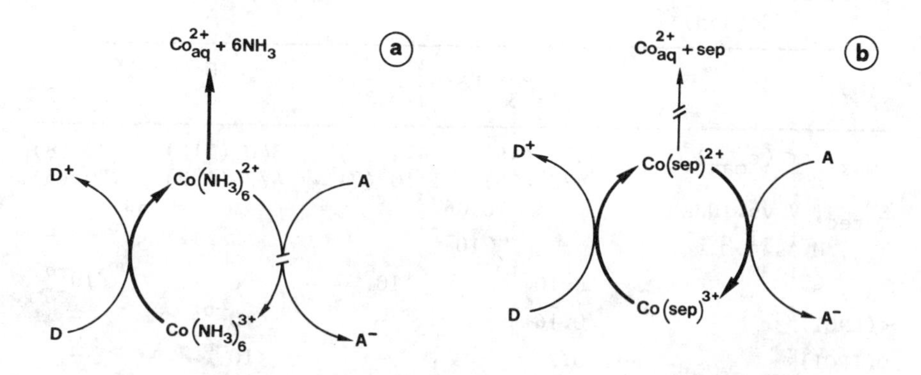

Fig. 6. $Co(NH_3)_6^{3+}$ is useless as a relay because its reduced form rapidly decomposes (a). In contrast, the reduced form of $Co(sep)^{3+}$ is inert and thus this complex is a good relay (b).

tion sphere (e.g., the absorption spectrum), but exhibits a completely different behavior when large amplitude nuclear motions come into play (i.e., in chemical reactions) because the 6 nitrogen atoms are linked together by covalent bonds. Thus, while the reduction at the metal center of $Co(NH_3)_6^{3+}$ causes the decomposition of the complex because the presence of a $\sigma^*(e_g)$ antibonding electron makes the reduced form labile towards ligand dissociation (43) (Fig. 6a), the corresponding process for $Co(sep)^{3+}$ does not cause any decomposition (Fig. 6b) because there is no simple ligand to eject but a single polydentate ligand structure which encapsulates the metal ion. A comparison between some properties of $Co(NH_3)_6^{3+}$ and $Co(sep)^{3+}$ is shown in Table I.

There are now available a large variety of encapsulated cage complexes of cobalt (29,32,36-41) which are substitution inert in both the Co(III) and Co(II) oxidation states. From long term irradiation experiments on the sacrificial photoreduction of water a turnover number higher than 5000 has been estimated for $Co(sep)^{3+}$, a figure that is at least two orders of magnitude higher than that found for the turnover number of MV^{2+} (50). Other nice properties of $Co(sep)^{3+}$ and related complexes are the high yield of separated electron transfer products (when coupled with Ru(II)-polypyridine photosensitizers

TABLE I. Comparison between some properties of $Co(NH_3)_6^{z+}$ and $Co(sep)^{z+}$.

	$Co(NH_3)_6^{z+}$		$Co(sep)^{z+}$	
	z = 3	z = 2	z = 3	z = 2
λ_{max}, nm (ε_{max})[a]	339 (46) 472 (56)	474 (8) 1110 (3)	340 (116) 472 (109)	470 (8) 920 (5)
E^o_{red}, V vs. NHE[b]	+0.06		-0.30	
k_{ex}, $M^{-1}s^{-1}$ [c]	10^{-7}		5	
k_{aq}, s^{-1} [d]	$6x10^{-12}$	$>10^6$	e	$<10^{-6}$
Φ(LMCT)[f]	0.16	-	$<10^{-6}$	-
Φ(IPCT)[g]	0.2	-	$<10^{-6}$	-

a) from refs. 32, 44, 45; b) from ref. 46; c) rate of the self-exchange electron transfer reaction, from ref. 46; d) rate of the ligand aquation reaction, from refs. 36, 39, 47; e) stable even in boiling 12 M HCl, ref. 32; f) quantum yield of photodecomposition upon ligand-to-metal charge transfer excitation, from refs. 45, 48; g) quantum yield of photodecomposition upon excitation of the ion-pair charge-transfer band of the ion pair with I^-, from refs. 48, 49.

(41,50-52)) and the very low overpotential for hydrogen production with colloidal Pt catalyst (41). It is also important to note that this family of cobalt cage complexes is a rather flexible one: (i) the redox potential of the Co(III)/Co(II) couple can be controlled by changing apical ligand substituents; (ii) the rate of the electron self-exchange reaction of the Co(III)/Co(II) couple can be changed by altering the structure of the cage ligand; (iii) the charge of the complexes can be controlled by the introduction of charged substituents; (iv) steric factors may be altered by appropriate substituents. Further flexibility can be expected by the replacement of nitrogen atoms in the first coordination sphere: cobalt cage complexes containing sulphur atoms in the first coordination sphere have already been prepared (33).

Thus, the cobalt cage complexes can satisfy several of the requirements needed (vide supra) for an ideal relay (50,51,53,54). Their major drawbacks are the occurrence of competing energy transfer processes caused by the presence of low energy excited states in the metal ion (41,50,52), and the absorption of light in competition with the photosensitizer (50). Both these deficiencies can in principle be overcome replacing Co with II- and III-row transition metal ions which, having a higher ligand field strength, should not absorb in the visible and should not possess low energy excited states. Such replacement, however, may cause the appearance of other drawbacks. For example, the Rh(III) cage complexes, which do not absorb in the visible region and do not have low energy excited states, are so difficult to reduce (55) to be unable to quench the excited $Ru(bpy)_3^{2+}$ complex. Furthermore, the reduced forms of the Rh(III) caged complexes so far prepared are not as stable as those of the cobalt complexes, presumably because the ligand cavities do not have the right size to host Rh^{2+}. Further synthetic work is needed to arrive at fully satisfactory relay systems.

Another problem concernig relays is worth mentioning. Since most of the desired net reactions (including H_2 evolution from water) imply a multi-electron reduction process, simple one-electron relays like $Co(sep)^{3+}$ must be coupled with an electron charge-storage catalyst (56) which allows one to overcome the kinetic barriers related to the formation of intermediate radical species. Alternatively, the relay must be able to give rise to a multi-electron transfer process (two electrons in the case of H_2 generation) that requires either a change of more than one unit in the oxidation state of the metal ion or the presence of two or more one-electron active sites (e.g., two metal ions both of which change their oxidation state by one unit as a consequence of the reaction with the photosensitizer). These possibilities have not yet been explored with sufficient effort although some research groups (22,23,57-60) have reported on the use of relays that also play the role of catalysts, leading to H_2 evolution without the involvement of heterogeneous species.

In conclusion, the following guidelines can be suggested for the design of an ideal relay:

- to choose compounds that contain a transition metal ion, in order to ensure the presence of a redox site at a suitable potential;
- to prefer metal ions of the II or III transition row, in order to avoid the presence of low energy metal centered excited states that could cause light absorption and quenching via energy transfer;
- to select ligands that are not easy to oxidize or to reduce, in order to avoid interferences in the redox process and the presence of low energy charge transfer excited states that could cause light absorption and quenching via energy transfer;
- to link together the unidentate ligands to make a single multidentate cage-type ligand, in order to prevent thermal and photochemical ligand dissociation reactions and to ensure a reversible redox behavior and good kinetic properties.

4. PHOTOSENSITIZERS

As we have seen above, a photosensitizer is a chemical species which must be able to use light energy to induce an electron transfer reaction between two substrate species (Figs. 1 and 2). To do that, a photosensitizer must be involved in excitation and redox processes without being consumed. It is immediately clear that the requirements needed for an ideal photosensitizer are more numerous than those needed for an ideal relay since the latter is only involved in a ground state redox process. The properties of an ideal photosensitizer are as follows (61) (Fig. 7):

- reversible redox behavior;
- suitable ground and excited state redox potentials;
- stability towards thermal and photochemical decomposition reactions;

- sufficiently large extinction coefficient (ε) at suitable wavelengths (λ) (e.g., for solar energy conversion, high absorption in the visible region);
- small energy gap between the reactive excited state and the excited states that can be populated by light absorption;
- high efficiency (η) of population of the reactive excited state;
- suitable lifetime (τ) of the reactive excited state;
- high energy content of the reactive excited state;
- good kinetic factors (i.e., high self-exchange rate constant, k_{ex}) for ground and excited state electron transfer reactions.

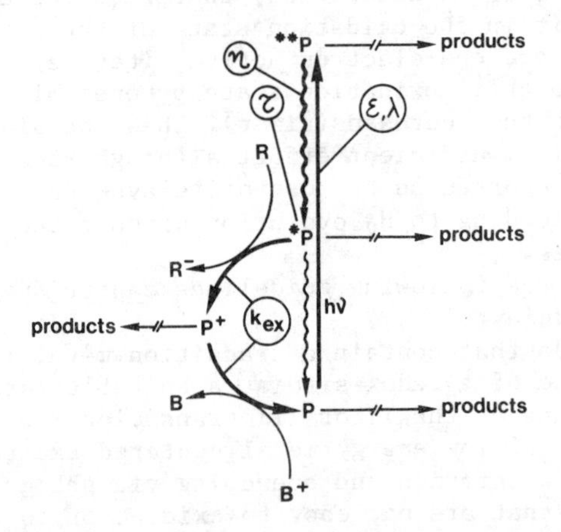

Fig. 7. Schematic illustration of some requirements needed for an ideal photosensitizer (see text).

These kinetic, thermodynamic, spectroscopic, and excited state requirements are very difficult to be met by a single molecular species. The search for and the design of valuable photosensitizers is therefore a very difficult task. For the ground state redox-related properties, the discussion made before for relays remains valid. As to the spectroscopic and excited state properties, some points which are worth mentioning are as follows.

The reactive excited state must have a lifetime sufficiently long (>10 ns) to be involved in the desired bimolecular reaction with the relay. This limit excludes (i) practically all the excited states of any molecule except the lowest spin-forbidden excited state, and (ii) even the lowest spin-forbidden excited state when its energy content is too small or when the excited state geometry is substantially different from the ground state geometry (vide infra). Furthermore, the excited state lifetime should not be too long (<1 ms), to avoid complications from bimolecular side reactions (annihilation reactions or quenching by impurities). It should also be noted that the lowest spin-forbidden excited state is spectroscopically inaccessible and must be populated via absorption to upper lying, spin-allowed excited states. This introduces the requirements of a small energy gap between the excited states that can be populated by light absorption and the reactive excited state, and a high efficiency of population of the reactive excited state, in order to avoid problems related to short wavelength excitation and loss of efficiency.

A rapid survey of chemical compounds shows that entire classes of molecules can be discarded as potential photosensitizers. Saturated organic molecules have in general bad absorption properties and too short excited state lifetimes (62). Aromatic hydrocarbons may have large singlet-triplet separation, low efficiency of population of the reactive excited state and a too long triplet lifetime (63). Simple inorganic anions and non-transition metal cations do not exhibit intense absorption bands in the visible and near U.V. spectral regions and a sufficiently long excited state lifetime. Thus, taking also into account the previously discussed limitations concerning the redox behavior, we remain again with transition metal complexes as the most appropriate candidates to the role of photosensitizers (64).

The guidelines to find out or to design transition metal complexes that may satisfy the numerous requirements needed for a photosensitizer are now reasonably well known. This, of course, does not mean that the task is an easy one.

In transition metal complexes there are basically three orbital types of excited states, namely metal centered (MC), charge transfer (either ligand-to-metal, LMCT, or metal-to-ligand, MLCT), and ligand centered (LC) (65). The energies of such states are respectively related to the ligand field strength, the redox properties of metals and ligands, and intrinsic spectroscopic properties of the ligands. The MC and LMCT excited states are most often obtained by promoting an electron into the σ^* metal-ligand antibonding orbitals and, therefore, they are strongly distorted with respect to the ground state geometry. This facilitates fast radiationless transitions (including ligand dissociation reactions): as a consequence, when the lowest excited state of a complex is MC or LMCT in nature, its lifetime is very short and the complex cannot play the role of a photosensitizer.

The properties of the LC excited states of a transition metal complex depend on the specific nature of the ligand. For most transition metal complexes the LC excited states lie very high in energy. In some cases, however, it may happen that the lowest triplet LC excited state is the lowest excited state of the complex (66-68). In such cases the properties of the LC excited state are usually more satisfactory than those of the lowest excited state of the free ligand. In particular, the presence of the (heavy) metal atom facilitates the intersystem crossing process, which increases the efficiency of population of the reactive excited state and brings the excited state lifetime into a more suitable time scale.

For various reasons that we are going to discuss, the MLCT excited states are the most suitable ones as reactive excited states of photosensitizers. When a MLCT excited state is the lowest excited state of a complex, it must involve the promotion of a metal electron into a relatively low energy ligand orbital, which can only be a delocalized π^* orbital. As a consequence, the MLCT excited state will not be distorted and, therefore, it will be reasonably long lived. Furthermore, the absorption spectrum of the complex will exhibit high intensity MLCT bands at relatively low energies that, of course, are not shown by the simple metal ion or by the free ligand (65). For MLCT excited states, the singlet-triplet energy gap is small because of the charge transfer nature of the transition, and the efficiency of population of the reactive triplet excited state from the upper singlet is high because of the small energy gap and the efficient spin-orbit coupling induced by the (heavy) metal ion. The presence of a MLCT state at low energy also implies that the ground state of the complex should not be too difficult to oxidize and/or to reduce. There is, in fact, an obvious (although not simple (69-72)) correlation between the oxidation (at the metal) and reduction (at the ligands) potentials and the energy of the MLCT transition. When this set of favorable spectroscopic and excited state properties is coupled with chemical and photochemical stability (vide infra), transition metal complexes are suitable candidates as photosensitizers.

The $Ru(bpy)_3^{2+}$ complex, where bpy = 2,2'-bipyridine, is presently the best known and most widely used electron transfer photosensitizer (1,8,9,13,73-76). Some relevant properties of this complex are shown in Fig. 8. The lowest excited state is MLCT in nature. It is interesting to note that this excited state can be used both as a reductant and as an oxidant: thus, it allows us to envisage both oxidative (Fig. 7) and reductive schemes for photosensitization. As one can see from Fig. 8, this molecule satisfies reasonably well most of the requirements needed for a photosensitizer. Besides the properties illustrated in Fig. 8, it should be recalled that $Ru(bpy)_3^{2+}$ behaves reversibly in redox processes (77,78) and exhibits very good electron transfer kinetic parameters (the rate constants of the self exchange reactions involving the excited state and the various ground state species are $>10^6$ $M^{-1}s^{-1}$ (79,80)). When $Ru(bpy)_3^{2+}$ is coupled with the MV^{2+} relay in sacrificial systems for hydrogen evolution, its turnover number is estimated to be from 300 (81,82) to 10000 (50). Even the highest figure, however, cannot be considered satisfactory because of the high cost of the complex. It should also be recalled that in the natural photosynthetic process each chlorophyll molecule processes at least 10^5 photons in its lifetime in a leaf (56). Other not completely satisfactory properties, especially for applications in the field of solar energy conversion, are (i) the position of the absorption band in the visible, which does not extend enough toward the red to match the emission spectrum of the sun, and (ii) the relatively high threshold energy, 2.12 eV (the maximum thermodynamic efficiency, 32%, of solar energy conversion devices is obtained with a threshold of 1.48 eV (56)).

Of the hundreds of polypyridine complexes of Ru(II) (72,83-89) and of other metal ions (90-94) synthesized and characterized in the last few years, none is a <u>substantially</u> better photosensitizer than $Ru(bpy)_3^{2+}$. This huge amount of work, however, has not been useless inasmuch as it has allowed us to elucidate the various factors that govern the spectroscopic, excited state, and redox properties of coordination compounds. Furthermore, this synthetic work has made available a large series of complexes having variable excited state energy and excited state redox potentials, which are most useful for

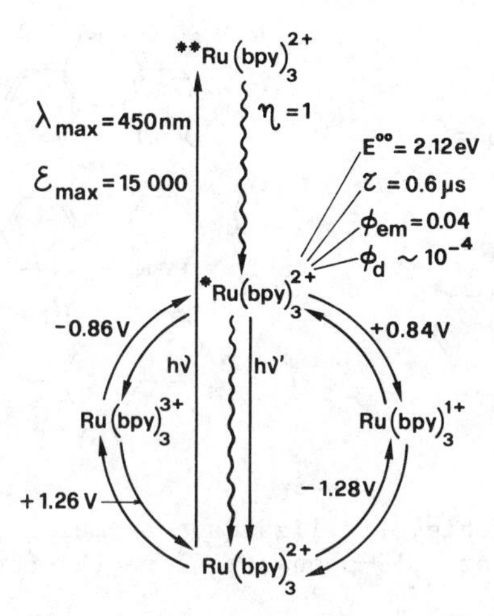

Fig. 8. Properties of the $Ru(bpy)_3^{2+}$ photosensitizer.

systematic studies in the field of energy (95) and electron (80,96-98) transfer kinetics.

Rather than discussing in detail the various factors which govern the photochemical, photophysical, and electrochemical behavior of polypyridine complexes, we will only mention briefly some of the lessons learned from the study of this family of compounds.

Concerning the problem of the cost, an obvious solution would be to replace Fe(II) for Ru(II) in $Ru(bpy)_3^{2+}$. The iron complex, however, is useless because Fe(II), as all the first row metals, has a lower ligand field strength than Ru(II). As a consequence, the lowest excited state of Fe(II)-polypyridine complexes is a triplet MC excited state which is strongly distorted and therefore very short lived (τ = 0.8 ns (91)).

The displacement of the absorption spectrum to the red and the related decrease in the excited state energy can be easily obtained by replacing one of the bpy ligands of $Ru(bpy)_3^{2+}$ with another ligand that is easier to reduce. The mixed ligand $Ru(bpy)_2(biq)^{2+}$ complex, where biq = 2,2'-biquinoline, shows an absorption maximum at 547 nm and is able to collect about twice solar energy compared with $Ru(bpy)_3^{2+}$ (99). At the same time, the threshold energy (i.e. the zero-zero energy of the lowest energy reactive excited state) goes down to 1.73 eV (from 2.12 eV of $Ru(bpy)_3^{2+}$) which allows a maximum thermodynamic efficiency of solar energy conversion of 29% (compared with 21% of $Ru(bpy)_3^{2+}$ (99)). However, the excited state lifetime of $Ru(bpy)_2(biq)^{2+}$ is shorter and the turnover number is not known.

The consumption of $Ru(bpy)_3^{2+}$ in these cyclic systems is known to be due, at least in part, to a ligand dissociation photoreaction (100,101). This reaction is thought to occur as a consequence of an activated ($\Delta E \simeq 4000\ cm^{-1}$) surface crossing from the lowest ^{3}MLCT

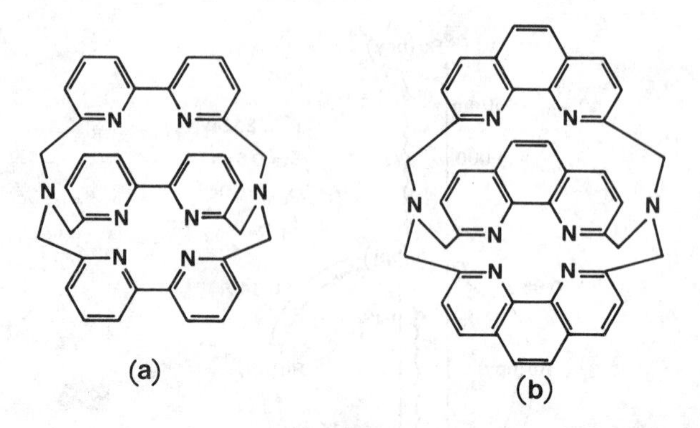

Fig. 9. Macrobicyclic ligands containing 2,2'-bipyridine (a) or 1,10-phenanthroline (b) (from ref. 105).

excited state to a distorted ^{3}MC excited state. This process controls the excited state lifetime at room temperature and also accounts for the strong decrease in the luminescence intensity and luminescence lifetime with increasing temperature (101-104). The obvious remedy to the ligand dissociation process would be to link together the three bpy ligands so as to form a single cage-type ligand which encapsulates the metal ion. A first success in this direction has been recently reported by Lehn and co-workers (105) who have synthetized the polypyridine and polyphenanthroline cage-type ligands shown in Fig. 9. The Ru(II) complexes of such ligands, however, have not yet been prepared and there might also be problems concerning the right shape and size of the cage. If the Ru-N bonds were too long or the bite angles were unfavorable, the ligand field strength would be small and the decrease in energy of the lowest ^{3}MC excited state would compromise the lifetime of the ^{3}MLCT excited state. This actually happens for $Ru(trpy)_2^{2+}$ (trpy = 2,2',2''-terpyridine) whose lifetime at room temperature is less than 5 ns (106,107).

As mentioned above, the lifetime of ^{3}MLCT at high temperature is controlled by a thermally activated surface crossing to ^{3}MC (108). To increase the excited state lifetime at room temperature, one should increase the energy gap between ^{3}MC and ^{3}MLCT. This can be done by replacing the bpy ligands with other ligands exhibiting higher ligand field strength (which would raise ^{3}MC) and/or higher electronic affinity (which would lower ^{3}MLCT) (70,88). This, however, is not so easy to do. For example, the previously mentioned 2,2'-biquinoline ligand has a higher electronic affinity but also a smaller ligand field strength (104). Replacement of bpy with $phpy^-$ (phenylpyridine anion (109-111)), not yet reported for Ru(II), would have the opposite, but presumably still unsatisfactory, result. Partial replacement of the bpy ligands may yield a useful compromise (88).

Polypyridine complexes of other transition metal ions may also exhibit some interesting properties as photosensitizers. In all cases, however, there are drawbacks which make such complexes worse than the ruthenium ones. For example, the Cr(III) complexes have long lived excited states (ms range (93)), but they can only act as excited state oxidants and give rise to an extremely labile reduced form (TN_3 = 4 (112)). As another example, the osmium complexes have a large ^3MC - ^3MLCT energy gap, but yet a shorter lifetime than the ruthenium complexes (by a factor of about 20) because of faster radiative and nonradiative rate constants for the direct decay to the ground state (113).

In conclusion, $Ru(bpy)_3^{2+}$ possesses several suitable properties to play the role of an electron transfer photosensitizer, but it cannot be considered as an ideal photosensitizer for practical applications. It seems also unlikely that a much better photosensitizer than $Ru(bpy)_3^{2+}$ will be found in the family of metal polypyridine complexes. Another extremely interesting family of photosensitizers is constituted by porphyrins. The work on complexes of the latter family has recently been the object of excellent reviews (114-116) and will not be dealt with here. Even with porphyrins, however, there are severe drawbacks and problems to be solved. Studies on other series of transition metal compounds have only been carried out occasionally (68,117,118) and should be strongly encouraged. A few suggestions derived from the above discussion for the design of an ideal photosensitizer are as follows:

- to choose compounds that contain a transition metal ion, in order to ensure the presence of a redox site at a suitable potential;
- to prefer metal ions of the II or III transition row, in order to avoid that the lowest excited state is MC in nature, which would imply an extremely short lifetime;
- to select ligands which are easy to reduce, in order to have a MLCT excited state as the lowest one; this assures long lifetime, small S-T energy gap, high η_{isc}, intense absorption bands at low energy, and favorable electronic factor for oxidative quenching (vide infra);
- to link together the ligands so as to encapsulate the metal ion, in order to prevent thermal and photochemical decomposition.

5. CAGE ESCAPE

As mentioned above, a most important defect of photosensitized electron transfer processes in homogeneous solution is the occurrence of a fast back electron transfer reaction between P^+ and R^-, i.e. the products of the excited state electron transfer reaction (Figs. 2 and 10). In several cases, such a dissipative reaction takes place in the solvent cage in which P^+ and R^- are formed, before they have a chance to escape. For example, for $Ru(bpy)_3^{3+}$ and MV^+ the cage recombination efficiency is estimated to amount to about 75% (119).

According to the simple kinetic scheme of Fig. 10, the rate constants of the quenching reaction (k_q) and of the back electron transfer reaction (k_b) are given by the following equations:

$$k_q = \frac{k_d}{1 + (k_{-d}/k_e) + (k_{-d}k_{-e}/k_{-e}(g)k_e)} \qquad (1)$$

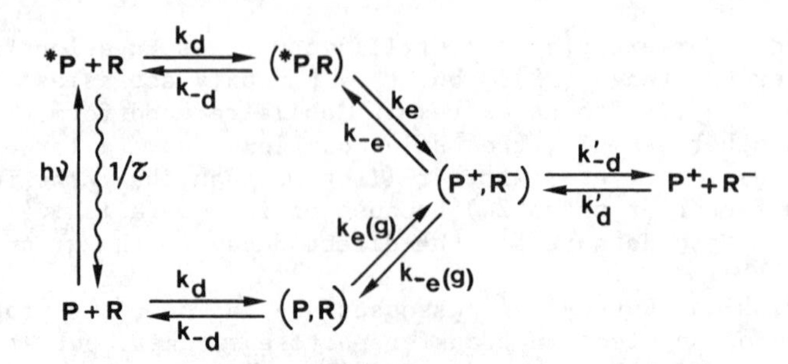

Fig. 10. Simple kinetic scheme for photoinduced electron transfer reactions in fluid solution.

$$k_b = \frac{k'_d}{1 + (k'_{-d}/k_{-e}(g)) + (k'_{-d}k_e(g)/k_{-d}k_{-e}(g))} \tag{2}$$

where k_d, k'_d, k_{-d}, and k'_{-d} are the diffusion or dissociation rate constants and k_e, k_{-e}, $k_e(g)$, and $k_{-e}(g)$ are the unimolecular electron transfer rate constants of the various steps. For energy conversion systems, the back electron transfer reaction must be exoergonic ($k_{-e}(g) >> k_e(g)$), so that eq. (2) reduces to

$$k_b = k'_d \frac{k_{-e}(g)}{k'_{-d} + k_{-e}(g)} \tag{3}$$

The cage escape efficiency, $\eta_{c.e.}$, is given by

$$\eta_{c.e.} = \frac{k'_{-d}}{k'_{-d} + k_{-e}(g)} \tag{4}$$

and is one of the most important parameters of any photosensitizer/relay couple. In the case of a system for solar energy conversion (97), the quenching reaction represents the process in which the light (or spectroscopic) energy is converted into chemical energy, the back electron transfer reaction represents a dissipative channel for the converted energy, and the cage escape efficiency is related to the maximum fraction of the converted energy that can be stored. According to the mechanism of Fig. 10, the cage escape efficiency is related to the rate of the observable back electron transfer reaction by the simple expression

$$k_b = k'_d (1 - \eta_{c.e.}) \tag{5}$$

Thus, from the experimental point of view $\eta_{c.e.}$ is expected to increase when k_b decreases. It should also be noted that any attempt to improve $\eta_{c.e.}$ should not introduce variants that decrease k_q, the rate constant of the excited state reaction which must succesfully compete with the excited state decay, $1/\tau$.

Measurements of product yields of photochemical electron transfer reactions are not numerous in the literature and a cursory examination of the available data gives the impression that yields are extremely variable and hardly fit within any conceptually simple model (120). Following eq. (4), $\eta_{c.e.}$ should be affected by all the factors that affect k'_{-d} or $k_{-e}(g)$.

The Eigen-Debye equation (121) gives the following expression for the rate constant of cage escape

$$k'_{-d}(\mu=0) = \frac{2kT}{\pi r^3 \eta} \; \frac{W_r(\mu=0)/RT}{1 - \exp(-W_r(\mu=0)/RT)} \tag{6}$$

where

$$W_r(\eta=0) = \frac{Z_{(P^+)} Z_{(R^-)} N e^2}{\varepsilon r} \tag{7}$$

This expression holds at zero ionic strength, and can be obtained from the relationship

$$k'_{-d} = k'_d / K'_a \tag{8}$$

if the Fuoss equation (122) and the Debye-Smoluchowsky equation (123) are used for K'_a and k'_d, respectively. The situation is not very clear concerning the dependence of k'_{-d} on the ionic strength of the medium. If, as it is often done (124), a Broensted-Debye ionic strength dependence is used for k'_d (which is the same ionic strength dependence holding for K'_a), k'_{-d} would turn out to be ionic strength independent. On the other hand, there have been proposals (120,125,126) of calculating k'_{-d} from eq. (6) by using an ionic strength dependent work term

$$W_r = \frac{Z_{(P^+)} Z_{(R^-)} N e^2}{\varepsilon r \, (1 + \beta r \sqrt{\mu}\,)} \tag{9}$$

where

$$\beta = \left(\frac{8 \, N^2 e^2}{1000 \, \varepsilon RT} \right)^{1/2} \tag{10}$$

In this approach, k'_{-d} would be expected to decrease or increase with increasing ionic strength, depending on whether P^+ and R^- have the same or different charge type. It can be shown (127) that the two points of view correspond to using different approximations in the integration of the distance dependence of the Debye-Smoluchowsky equation for k'_d. If no approximation is made, the Debye-Smoluchowsky equation for

k'_d is integrated numerically, and eq. (8) is used, k'_{-d} turns out to depend on ionic strength qualitatively as predicted by eqs. (6), (9), and (10). (Actually, eqs. (6), (9), and (10) tend to overestimate slightly the ionic strength dependence).

Summarizing, the rate constant of cage escape is dependent on the following parameters (eqs. (6), (9), and (10)): (i) electric charge product; (ii) ionic strength (if $Z_{(P^+)}Z_{(R^-)} \neq 0$); (iii) solvent dielectric constant, ε (if $Z_{(P^+)}Z_{(R^-)} \neq 0$); (iv) solvent viscosity, η; (v) temperature, T. Detailed investigations carried out in a few laboratories (1,120,128) have demonstrated that a change in the electric charges of the reactants can indeed control the cage escape efficiency. For example it has been shown that, using $Cr(4,4'-(COO)_2bpy)_3^{3-}$ as a photosensitizer, the efficiency of cage escape can be tuned by tuning the effective electric charge of the complex with changing pH (129). Using Fe^{2+}_{aq} as a quencher, high cage escape efficiency is obtained at pH = 0, when protonation of the carboxylic groups switches the electric charge of the photosensitizer from negative to positive, while undetectable cage escape efficiency is obtained at pH = 5. In contrast, quenching by $Fe(CN)_6^{4-}$ gives negligible efficiencies at pH = 0 and high efficiences at pH = 5. In these cases, of course, the increase in the cage escape efficiency is accompanied by a decrease of the quenching rate constant, k_q. In this context, it is interesting to note that if we compare the $^*Ru(bpy)_3^{2+}-MV^{2+}$ and $^*Ru(bpy)_3^{2+}-Co(sep)^{3+}$ photosensitizer-relay couples (41), we find that the latter has smaller values for k_q and k_b and a higher value ($\simeq 1$) for $\eta_{c.e.}$. This is in qualitative agreement with the expectations based on the different charge products of the two systems, even if intrinsic parameters of the two relays (e.g., degree of adiabaticity, vide infra) might also play a role. Although these and similar results can be ascribed to the change of k'_{-d} with changing electrostatic repulsion, we would like to emphasize that the charge product of the two reaction partners may also control the detailed nature of the encounter complex and thus it may affect to some extent the rate constant of the electron transfer step, $k_{-e}(g)$.

Effects of ionic strength (126) and dielectric constant (130) on the yields of separated products, as expected from the above simple model, have also been reported. Much study has also been devoted to the control of the cage escape efficiency (or, more generally, of the rate of the back electron transfer reaction) by addition of polyelectrolytes (131-134) or colloids (inorganic oxides, micelles, or vesicles) (135) on the basis of electrostatic and/or hydrophobic interactions (note that the latter is not explicitly considered in the above simple treatment). For example, using $Ru(bpy)_3^{2+}$ as photosensitizer and zwitterionic viologens as quenchers, the presence of colloidal silica was found to retard the back electron transfer reaction (135,136). In the system containing $Ru(bpy)_3^{2+}$ as photosensitizer and the zwitterionic viologen N,N'-bis(4-sulfonatotolyl)4,4'-bipyridinium as quencher, addition of the polyelectrolyte polyvinylsulphate was found to cause a significant retardation of the back electron transfer reaction (132); however, in the same system no effect was found on the cage escape efficiency. This result is interesting because it shows that, contrary to the prediction of eq. (5), a change in the rate of the experimental back electron transfer reaction may not be accompanied by a consequent change in the cage escape efficiency (vide infra).

The unimolecular electron transfer rate constant $k_{-e}(g)$ can be given by the expression (97,137)

$$k_{-e}(g) = \nu_n \underline{k} e^{-\Delta G^{\neq}/RT} \qquad (11)$$

where ν_n is the relevant frequency for nuclear motions, $\underline{k}$ is the electronic transition coefficient (which is equal to 1 for adiabatic reactions and lower than 1 for nonadiabatic ones) related to the interaction energy H_{if}, and $\Delta G^{\neq}$ is the classical free activation energy related to the free energy change ΔG and to the classical intrinsic barrier $\Delta G^{\neq}(0)$ by the equation

$$\Delta G^{\neq} = \Delta G^{\neq}(0) \left(1 + \Delta G/4\Delta G^{\neq}(0)\right)^2 \qquad (12)$$

In principle, we can thus increase the cage escape efficiency decreasing $k_{-e}(g)$ via changes in ΔG, $\Delta G^{\neq}(0)$, or $\underline{k}$.

A back electron transfer reaction is usually strongly exoergonic (in fact, it <u>must</u> be strongly exoergonic for solar energy conversion processes). This means that, for the back electron transfer step, reactant and product energy surfaces are imbedded and the process lies in the so-called inverted region. The onset of this region is at more negative ΔG values the higher is $\Delta G^{\neq}(0)$. In the inverted region, eqs. (11) and (12) would predict a quadratic decrease of the logarithm of the rate constant with increasing driving force. However, in this region it is more appropriate (8) to consider the unimolecular back electron transfer step as a nonradiative transition, for which the logarithm of the rate constant, according to the quantum mechanical treatment (138-140), should decrease linearly with increasing driving force (energy gap law). Thus, in both cases, theory predicts that $k_{-e}(g)$ should decrease with increasing driving force, which means that the cage escape efficiency should increase as $-\Delta G$ increases. As noted by Sutin and Creutz (1), this would be an extremely favorable prediction concerning systems for solar energy conversion: the more energy converted, the higher efficiency of the separation process which may allow energy storage. These theoretical expectations are fulfilled by nonradiative deactivations of excited states of aromatic molecules (140) and transition metal complexes (141,142), as well as for intramolecular electron transfer processes (143) and for electron transfer in frozen solutions (144). However, at most very tenuous "vestiges" (145) of a decrease in the rate constant with increasing exoergonicity has been found for bimolecular processes in fluid solution (97,98). The reasons for this behavior must be very subtle and are not yet fully understood. Experimentally, the failure to observe the inverted behavior has encouraged (97,146) the use of empirical free energy relationships (147,148) implying an asympthotic $\Delta G^{\neq} \dashrightarrow 0$ behavior in the exoergonic region, in contrast with the prediction of eq. (12).

As far as the electronic factor $\underline{k}$ is concerned, the ideal case for a photosensitizer-relay couple would be that of a system where the

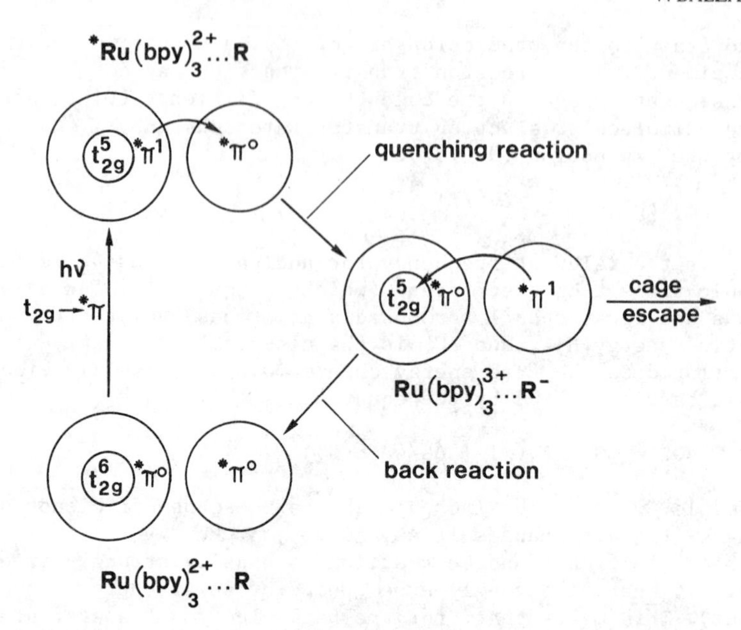

Fig. 11. Schematic representation of the orbitals involved in the quenching and back electron transfer reactions of the $Ru(bpy)_3^{2+}$ photosensitizer with an oxidizing quencher (see text).

excited state electron transfer step (k_e in Fig. 10) is adiabatic while the back electron transfer step ($k_{-e}(g)$ in Fig. 10) is slow because of nonadiabaticity reasons. Although at a first glance such a requirement could seem absurd, it is satisfied by the most popular photosensitizer, $Ru(bpy)_3^{2+}$. As we can see from Fig. 11, light excitation of $Ru(bpy)_3^{2+}$ promotes an electron from the t_{2g} metal orbital to the $^*\pi$ ligand orbital. Reduction of the relay, which is assumed to involve a $^*\pi$ LUMO orbital, is characterized by an excellent interaction energy because of the favorable overlap between the donor and acceptor orbitals in the encounter. The back electron transfer step between $Ru(bpy)_3^{3+}$ and R^-, however, requires the transfer of an electron from the $^*\pi$ orbital of the relay to the <u>inner</u> t_{2g} metal orbitals. The orbital overlap, of course, is smaller and there is good evidence of a nonadiabatic behavior of this reaction (80,149). It should be pointed out that this favorable condition derives from the MLCT nature of light excitation which promotes the reducing electron from the center to the periphery of the complex. LMCT excitation, of course, represents the opposite case where <u>oxidation</u> of the relay would offer the same type of kinetic advantages.

When the photosensitizer and/or the relay are nonspherical species but molecules having a complex structure with anisotropic distribution of electric charges, functional groups, and redox reactive centers, the kinetic scheme of Fig. 10 (and, as a consequence, eq. (5)) may be no

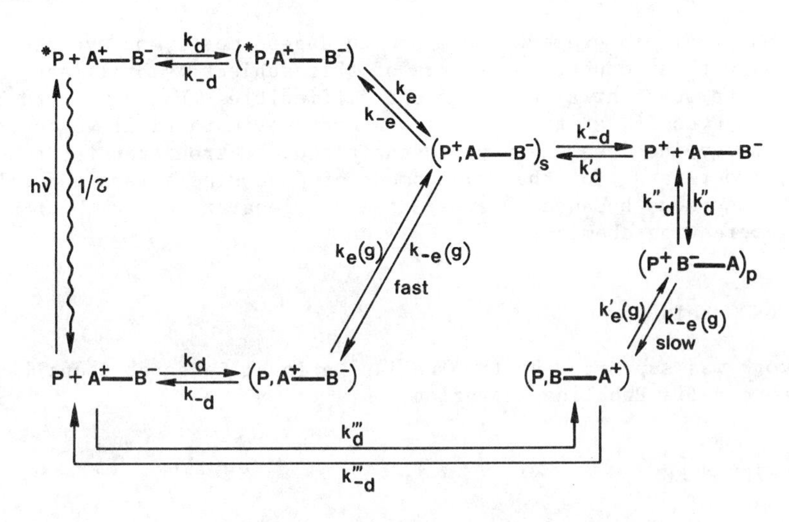

Fig. 12. A more complicated kinetic scheme (cf. Fig. 10) that can be needed when the relay and/or the photosensitizer are non-spherical molecules (see text).

longer appropriate. Consider, for example, the quenching of a spherical and neutral photosensitizer P by a nonspherical relay A^+-B^- having (i) charges of different sign localized on distinct regions of the molecular structure and (ii) the reactive redox site localized in the A^+ molecular region (Fig. 12). The quenching reaction will take place in a precursor complex ($^*P,A^+$-B^-), giving rise to a successor complex $(P^+,A$-$B^-)_s$ where the back electron transfer reaction can take place very rapidly because the two redox centers are close together and thus the electronic factor is very good. For the back electron transfer reaction taking place from the separated redox products, coulombic attraction leads to the formation of a precursor complex having the $(P^+,B^-$-$A)_p$ structure, where the P^+ and A redox centers are far away. Within this structure the back electron transfer reaction may be very slow because of a poor electronic factor. Evidence for the involvement of more than one distinct encounter complexes in bimolecular electron transfer reactions in homogeneous solution are clearly emerging from recent studies (150-152). Complexities of this type should be considered when designing relay-photosensitizer couples to be used in homogeneous solution.

6. CLOSING REMARKS

The above discussion has shown that it is very difficult to find out or to design molecules that satisfy to a reasonable degree the numerous requirements needed for an ideal relay or an ideal photosensitizer. However, many factors governing the desired properties are well known and real improvements seem now to depend mostly on synthetic efforts. Without doubt, the future of this subject is brilliant, despite the substantial problems and difficulties that remain to be overcome. Even if, at the end, it would prove impractical to convert solar energy into fuels via photosensitized electron transfer reactions, the rewards for the development of photosensitizers and relays will be enormous because these species can generate a wealth of novel, most interesting chemistry.

ACKNOWLEDGMENTS

This work was supported by the Italian National Research Council and Ministero della Pubblica Istruzione.

REFERENCES

1) Sutin, N; Creutz, C. Pure Appl. Chem., **52**, 2717 (1980).
2) Albini, A. Synthesis, 249 (1981).
3) Connolly, J.S. (Ed.) Photochemical Conversion and Storage of Solar Energy, Academic Press, London (1981).
4) Rabani, J. (Ed.) Photochemical Conversion and storage of Solar Energy 1982, Weizman, Jerusalem (1982).
5) Harriman, A.; West, M.A. (Eds.) Photogeneration of Hydrogen, Academic Press, London (1982).
6) Graetzel, M. (Ed.) Energy Resources through Photochemistry and Catalysis, Academic Press, London (1983).
7) Julliard, M.; Chanon, M. Chem. Rev., **83**, 425 (1983).
8) Meyer, T.J. Progr. Inorg. Chem., **30**, 389 (1983).
9) Balzani, V.; Bolletta, F.; Ciano, M.; Maestri, M. J. Chem. Educ., **60**, 447 (1983).
10) Chanon, M. Bull. Soc. Chim. France, 209 (1985).
11) Graetzel, M. in Photochemical Conversion and Storage of Solar Energy, Connolly, J.S. (Ed.), Academic Press, London (1981), p. 131.
12) Lehn, J.M. in Photochemical Conversion and Storage of Solar Energy, Connolly, J.S. (Ed.), Academic Press, London (1981), p. 161.
13) Kalyanasundaram, K. Coord. Chem. Revs., **46**, 159 (1982).
14) Porter, G. in Light, chemical change and life, Coyle, J.D.; Hill, R.R.; Roberts, D.R. (Eds.), Open University Press, Milton Keynes (1982), p. 362.
15) Harriman, A. J. Photochem., **25**, 33 (1984).
16) Kiwi, J.; Kalyanasundarwm, K.; Graetzel, M. Struct. Bonding (Berlin), **49**, 37 (1981).
17) Harriman, A.; Porter, G.; Richoux, M.C. in Photogeneration of Hydrogen, Harriman, A.; West, M.A. (Eds.), Academic Press, London (1982), p. 67.
18) McLendon, G. in Energy Resources through Photochemistry and Catalysis, Graetzel, M. (Ed.), Academic Press, London (1983), p. 99.

19) Zamaraev, K.I.; Parmon, V.N. in Energy Resources through Photochemistry and Catalysis, Graetzel, M. (Ed.), Academic Press, London (1983), p. 123.
20) Darwent, J.R. in Photogeneration of Hydrogen, Harriman, A.; West, M.A. (Eds.), Academic Press, London (1982), p. 23.
21) Koryakin, B.V.; Dzhabiev, T.S.; Shilov, A.E. Dokl. Akad. Nauk. SSSR, **233**, 620 (1977).
22) Lehn, J.M.; Sauvage, J.P. Nouv. J. Chim., **1**, 449 (1977).
23) Krishnan, C.V.; Brunschwig, B.S.; Creutz, C.; Sutin, N. J. Am. Chem. Soc., **107**, 2005 (1985).
24) Johansen, O.; Launikonis, A.; Loder, J.W.; Mau, A.W.-H.; Sasse, W.H.F.; Swift, J.D.; Wells, D. Aust. J. Chem., **34**, 981 (1981).
25) Cannon, R.D. Electron transfer reactions, Butterworths, London (1980).
26) Balzani, V.; Carassiti, V. Photochemistry of Coordination Compounds, Academic Press, London (1970).
27) Adamson, A.W.; Fleischauer, P.D. (Eds.) Concepts of Inorganic Photochemistry, Wiley-Interscience, New York (1975).
28) Lehn, J.M. Acc. Chem. Res., **11**, 49 (1978).
29) Sargeson, A.M. Chem. Br., **15**, 23 (1979).
30) Lehn, J.M. Pure Appl. Chem., **52**, 2441 (1980).
31) Lehn, J.M. Science, **227**, 849 (1985).
32) Creaser, I.I.; Geue, R.J.; Harrowfield, J.M.; Hertl, A.J.; Sargeson, A.M.; Snow, M.R.; Springborg, J. J. Am. Chem. Soc., **104**, 6016 (1982).
33) Dubs, R.V., Gahan, L.R.; Sargeson, A.M. Inorg. Chem., **22**, 2523 (1983).
34) Yee, E.L.; Hupp, J.T.; Weaver, M.J. Inorg. Chem., **22**, 3465 (1983).
35) Sabbatini, N.; Dellonte, S.; Bonazzi, A.; Ciano, M.; Balzani, V. Inorg. Chem., in press.
36) Creaser, I.I.; Harrowfield, J.M.; Hertl, A.J.; Sargeson, A.M.; Springborg, J.; Geue, R.J.; Snow, M.R. J. Am. Chem. Soc., **99**, 3181 (1977).
37) Gainsford, G.J.; Geue, R.J.; Sargeson, A.M.; J. Chem. Soc. Chem. Commun., 233 (1982).
38) Gahan, L.R.; Hambley, T.W.; Sargeson, A.M.; Snow, M.R. Inorg. Chem., **21**, 2699 (1982).
39) Curtis, N.J.; Lawrance, G.A.; Sargeson, A.M. Aust. J. Chem., **36**, 1327 (1983).
40) Geue, R.J.; Hambley, T.W.; Harrowfield, J.M.; Sargeson, A.M.; Snow, M.R. J. Am. Chem. Soc., **106**, 5478 (1984).
41) Creaser, I.I.; Gahan, L.R.; Geue, R.J.; Launikonis, A.; Lay, P.A.; Lydon, J.D.; McCarthy, M.G.; Mau, A.W.-H.; Sargeson, A.M.; Sasse, W.H.F. in press.
42) Balzani, V.; Sabbatini, N.; Scandola, F. Chem. Revs., submitted.
43) Basolo, F.; Pearson, R.G. Mechanism of Inorganic Reactions, John Wiley, New York (1967).
44) Jorgensen, C.K. Adv. Chem. Phys., **5**, 33 (1963).
45) Manfrin, M.F.; Varani, G.; Moggi, L.; Balzani, V. Mol. Photochem., **1**, 387 (1969).

46) Creaser, I.I.; Sargeson, A.M.; Zanella, A.W. Inorg. Chem., **22**, 4022 (1983).
47) Lilie, J.; Shinohara, N.; Simic, M.G. J. Am. Chem. Soc., **98**, 6516 (1976).
48) Pina, F.; Ciano, M.; Moggi, L.; Balzani, V. Inorg. Chem., **24**, 844 (1985).
49) Adamson, A.W.; Sporer, A.H. J. Am. Chem. Soc., **80**, 3865 (1958).
50) Lay, P.A.; Mau, A.W.H.; Sasse, W.H.F.; Creaser, I.I.; Gahan, L.R.; Sargeson, A.M. Inorg. Chem., **22**, 2347 (1983).
51) Houlding, V.; Geiger, T.; Koelle, U.; Graetzel, M. J. Chem. Soc. Chem. Commun., 681 (1982).
52) Mok, C.-Y.; Zanella, A.W.; Creutz, C.; Sutin, N. Inorg. Chem., **23**, 2891 (1984).
53) Geiger, T.; Nottenberg, R.; Pelaprat, M.-L.; Graetzel, M. Helv. Chim. Acta, **65**, 2507 (1982).
54) Rampi Scandola, M.A.; Scandola, F.; Indelli, A.; Balzani, V. Inorg. Chim. Acta, **76**, L67 (1983).
55) Harrowfield, J.M.; Hertl, A.J.; Lay, P.A.; Sargeson, A.M. J. Am. Chem. Soc., **105**, 5503 (1983).
56) Bolton, J.R. Science, **202**, 705 (1978).
57) Mulazzani, Q.G.; Venturi, M.; Hoffman, M.Z. J. Phys. Chem., **86**, 242 (1982).
58) Krishnan, C.V.; Creutz, C.; Mahajan, D.; Schwarz, H.A.; Sutin, N. Isr. J. Chem., **22**, 98 (1982).
59) Creutz, C.; Schwarz, H.A.; Sutin, N. J. Am. Chem. Soc., **106**, 3036 (1984).
60) Creutz, C.; Sutin, N. Coord. Chem. Rev., **64**, 321 (1985).
61) See also ref. 20.
62) See for example: Orlandi, G.; Flamigni, L.; Barigelletti, F.; Dellonte, S. Radiat. Phys. Chem., **21**, 113 (1983).
63) Birks, J.B. Photophysics of Aromatic Molecules, Wiley-Interscience, New York (1970).
64) Balzani, V.; Scandola, F. in Photochemical Conversion and Storage of Solar Energy, Connolly, J.S. (Ed.), Academic Press, New York (1981), p. 97.
65) Crosby, G.A. J. Chem. Educ., **60**, 791 (1983).
66) DeArmond, M.K.; Hillis, J.E. J. Chem. Phys., **54**, 2247 (1971).
67) Belser, P.; von Zelewsky, A.; Juris, A.; Barigelletti, F.; Tucci, A.; Balzani, V. Chem. Phys. Lett., **89**, 101 (1982).
68) Ballardini, R.; Varani, G.; Indelli, M.T.; Scandola, F. submitted.
69) Curtis, J.C.; Meyer, T.J. Inorg. Chem., **21**, 1562 (1982).
70) Caspar, J.V.; Meyer, T.J. Inorg. Chem., **22**, 2444 (1983).
71) Dodsworth, E.S.; Lever, A.B.P. Chem. Phys. Lett., **112**, 567 (1984).
72) Juris, A.; Belser, P.; Barigelletti, F.; von Zelewsky, A.; Balzani, V. Inorg. Chem., in press.
73) Balzani, V.; Bolletta, F.; Gandolfi, M.T.; Maestri, M. Top. Curr. Chem., **75**, 1 (1978).
74) Whitten, D.G. Acc. Chem. Res., **13**, 83 (1980).
75) Graetzel, M. Acc. Chem. Res., **14**, 376 (1981).
76) Watts, R.J. J. Chem. Educ., **60**, 834 (1983).

77) Saji, T.; Aoyagui, S. Electroanal. Chem., **58**, 401 (1975).
78) Tokel-Takvoryan, N.E.; Hemingway, R.E.; Bard, A.J. J. Am. Chem. Soc., **95**, 6582 (1973).
79) Sutin, N. Acc. Chem. Res., **15**, 275 (1982).
80) Sandrini, D.; Maestri, M.; Belser, P.; von Zelewsky, A.; Balzani, V. J. Phys. Chem., in press.
81) Moradpour, A.; Amouyal, E.; Keller, P.; Kagan, H. Nouv. J. Chim., **2**, 547 (1978).
82) Keller, P.; Moradpour, A.; Amouyal, E.; Kagan, H.B. Nouv. J. Chim., **4**, 377 (1980).
83) Crutchley, R.J.; Kress, N.; Lever, A.B.P. J. Am. Chem. Soc., **105**, 1170 (1983).
84) Kitamura, N.; Kawanishi, Y.; Tazuke, S. Chem. Phys. Lett., **97**, 103 (1983).
85) Allen, G.H.; White, R.P.; Rillema, D.P.; Meyer, T.J. J. Am. Chem. Soc., **106**, 2613 (1984).
86) Morris, D.E.; Ohsawa, Y.; Segers, D.P.; DeArmond, M.K.; Hanck, K.W. Inorg. Chem., **23**, 3010 (1984).
87) Skarda, V.; Cook, M.J.; Lewis, A.P.; McAuliffe, G.S.G.; Thomson, A.J.; Robbins, D.J. J. Chem. Soc. Perkin Trans. II, 1309 (1984).
88) Balzani, V.; Juris, A.; Barigelletti, F.; Belser, P.; von Zelewsky, A. Riken Q., **78**, 78 (1984).
89) Wacholtz, W.M.; Auerbach, R.A.; Schmehl, R.H.; Ollino, M.; Cherry, W.R. Inorg. Chem., **24**, 1758 (1985).
90) Watts, R.J.; Missimer, D. J. Am. Chem. Soc., **100**, 5350 (1978).
91) Creutz, C.; Chou, M.; Netzel, T.L.; Okumura, M.; Sutin, N. J. Am. Chem. Soc., **102**, 1309 (1980).
92) Bergkamp, M.A.; Guetlich, P.; Netzel, T.L.; Sutin, N. J. Phys. Chem., **87**, 3877 (1983).
93) Serpone, N.; Ponterini, G.; Jamieson, M.A.; Bolletta, F.; Maestri, M. Coord. Chem. Rev., **50**, 209 (1983).
94) Nishizawa, M.; Suzuki, T.M.; Sprouse, S.; Watts, R.J.; Ford, P.C. Inorg. Chem., **23**, 1837 (1984).
95) Scandola, F.; Balzani, V. J. Chem. Educ., **60**, 814 (1983).
96) Sutin, N.; Creutz, C. J. Chem. Educ., **60**, 809 (1983).
97) Balzani, V.; Scandola, F. in Energy Resources through Photochemistry and Catalysis, Graetzel, M. (Ed.), Academic Press, London (1983), p. 1.
98) Indelli, M.T.; Ballardini, R.; Scandola, F. J. Phys. Chem., **88**, 2547 (1984).
99) Juris, A.; Barigelletti, F.; Balzani, V.; Belser, P.; von Zelewsky, A. Isr. J. Chem., **22**, 87 (1982).
100) Van Houten, J.; Watts, R.J. Inorg. Chem., **17**, 3381 (1978).
101) Durham, B.; Caspar, J.V.; Nagle, J.K.; Meyer, T.J. J. Am. Chem. Soc., **104**, 4803 (1982).
102) Van Houten, J.; Watts, R.J. J. Am. Chem. Soc., **98**, 4853 (1976).
103) Kemp, T.J. Prog. React. Kinet., **10**, 301 (1980).
104) Barigelletti, F.; Juris, A.; Balzani, V.; Belser, P.; von Zelewsky, A. Inorg. Chem., **22**, 3335 (1983).
105) Rodriguez-Ubis, J.-C.; Alpha, B.; Plancherel, D.; Lehn, J.-M. **Helv. Chim. Acta, 67, 2264 (1984).**

106) Lin, C.T.; Boettcher, W.; Chou, M.; Creutz, C.; Sutin, N. J. Am. Chem. Soc., **98**, 6536 (1976).
107) Kirchhoff, J.R.; McMillin, D.R.; Marnot, P.A.; Sauvage, J.-P. J. Am. Chem. Soc., **107**, 1138 (1985).
108) Caspar, J.V.; Meyer, T.J. J. Am. Chem. Soc., **105**, 5583 (1983).
109) Chassot, L.; Mueller, E.; von Zelewsky, A. Inorg. Chem., **23**, 4249 (1984).
110) King, K.A.; Spellane, P.J.; Watts, R.J. J. Am. Chem. Soc., **107**, 1431 (1985).
111) Maestri, M.; Sandrini, D.; Balzani, V.; von Zelewsly, A.; Chassot, L. submitted.
112) Ballardini, R.; Juris, A.; Varani, G.; Balzani, V. Nouv. J. Chim., **4**, 563 (1980).
113) Kober, E.M.; Sullivan, B.P.; Meyer, T.J. Inorg. Chem., **23**, 2098 (1984).
114) Richoux, M.-C. in Photogeneration of Hydrogen, Harriman, A.; West, M.A. (Eds.), Academic Press, London (1982), p. 39.
115) Darwent, J.R.; Douglas, P.; Harriman, A.; Porter, G.; Richoux, M.-C. Coord. Chem. Rev., **44**, 83 (1982).
116) Harriman, A. in Energy Resources through Photochemistry and Catalysis, Graetzel, M. (Ed.), Academic Press, London (1983), p. 163.
117) Gray, H.B.; Maverick, A.W. Science, **214**, 1201 (1981).
118) Peterson, J.R.; Kalyanasundaram, K. J. Phys. Chem., **89**, 2486 (1985).
119) Chan, S.-F.; Chou, M.; Creutz, C.; Matsubara, T.; Sutin, N. J. Am. Chem. Soc., **103**, 369 (1981).
120) Scandola, F.; Ballardini, R.; Indelli, M.T. in Photochemical, Photoelectrochemical, and Photobiological Processes, Hall, D.O.; Palz, W. (Eds.), Reidel, Dordrecht (1981), p. 66.
121) Eigen, M. Z. Physik. Chem. (Frankfurt), **1**, 176 (1954).
122) Fuoss, R.M. J. Am. Chem. Soc., **80**, 5059 (1958).
123) Debye, P. Trans. Electrochem. Soc., **82**, 265 (1942).
124) See for example: Rybak, W.; Haim, A.; Netzel, T.L.; Sutin, N. J. Phys. Chem., **85**, 2856 (1981).
125) Balzani, V.; Scandola, F.; Orlandi, G.; Sabbatini, N.; Indelli, M.T. J. Am. Chem. Soc., **103**, 3370 (1981).
126) Gore, B.L.; Harriman, A.; Richoux, M.C. J. Photochem., **19**, 209 (1982).
127) Chiorboli, C.; Indelli, M.T.; Rampi, M.A.; Scandola, F. manuscript in preparation.
128) Harriman, A.; Porter, G.; Richoux, M.C. J. Chem. Soc. Faraday Trans. 2, **78**, 1955 (1982).
129) Indelli, M.T.; Scandola, F. manuscript in preparation.
130) Richoux, M.C. J. Photochem., **22**, 1 (1983).
131) Sassoon, R.E.; Gershuni, S.; Rabani, J. J. Phys. Chem., **89**, 1937 (1985).
132) Sassoon, R.E.; Aizenshtat, Z.; Rabani, J. J. Phys. Chem., **89**, 1182 (1985).
133) Meyerstein, D.; Rabani, J.; Matheson, M.S.; Meisel, D. J. Phys. Chem., **82**, 1879 (1978).

134) Sassoon, R.E.; Rabani, J. J. Phys. Chem., **84**, 1319 (1980).
135) Calvin, M.; Willner, I.; Laane, C.; Otvos, J.W. J. Photochem., **17**, 195 (1981).
136) Furlong, D.N.; Johansen, O.; Launikonis, A.; Loder, J.W.; Mau, A.W.-H.; Sasse, W.H.F. Aust. J. Chem., **38**, 363 (1985).
137) Sutin, N. Progr. Inorg. Chem., **30**, 441 (1983).
138) Englman, R.; Jortner, J. Mol. Phys., **18**, 145 (1970).
139) Freed, K.F.; Jortner, J. J. Chem. Phys., **52**, 6272 (1970).
140) Henry, B.R.; Siebrand, W. in Organic Molecular Photophysics, Birks, J.B. (Ed.), Wiley, New York, **1**, 153 (1973).
141) Kober, E.M.; Sullivan, B.P.; Dressick, W.J.; Caspar, J.V.; Meyer, T.J. J. Am. Chem. Soc., **102**, 7383 (1980).
142) Caspar, J.V.; Kober, E.M.; Sullivan, B.P.; Meyer, T.J. J. Am. Chem. Soc., **104**, 630 (1982).
143) Wasielewski, M.R.; Niemczyk, M.P.; Svec, W.A.; Pewitt, E.B. J. Am. Chem. Soc., **107**, 1080 (1985).
144) Miller, J.R.; Calcaterra, L.T.; Closs, G.L. J. Am. Chem. Soc., **106**, 3047 (1984).
145) Creutz, C.; Sutin, N. J. Am. Chem. Soc., **99**, 241 (1977).
146) Scandola, F.; Balzani, V. J. Am. Chem. Soc., **101**, 6140 (1979).
147) Rehm, D.; Weller, A. Isr. J. Chem., **8**, 259 (1970).
148) Agmon, N.; Levine, R.D. Chem. Phys. Lett., **52**, 197 (1977).
149) Creutz, C.; Keller, A.D.; Sutin, N.; Zipp, A.P. J. Am. Chem. Soc., **104**, 3618 (1982).
150) Simon, J.D.; Peters, K.S. J. Am. Chem. Soc., **103**, 6403 (1981).
151) Rougee, M.; Ebbesen, T.; Ghetti, F.; Bensasson, R.V. J. Phys. Chem., **86**, 4404 (1982).
152) Haim, A. Comments Inorg. Chem., **4**, 113 (1985).

PHOTOINDUCED CHARGE SEPARATION: TOWARDS THE DESIGN OF SUPERMOLECULAR SYSTEMS BASED ON TRANSITION METAL COMPLEXES

Franco Scandola, Carlo Alberto Bignozzi
Dipartimento di Chimica, Università di Ferrara and
Centro di Fotochimica CNR, Ferrara, Italy

Vincenzo Balzani
Istituto Chimico "G. Ciamician" dell'Università and
Istituto FRAE-CNR, Bologna, Italy

ABSTRACT. An approach to the design of supermolecular systems for photoinduced charge separation based on transition metal complexes is described. Specific aspects of electron transfer kinetics and intervalence transfer spectroscopy that are relevant to the problem are discussed. Studies in which several metal-containing moieties are linked via cyanide bridges to the $Ru(bpy)_2^{2+}$ chromophore are reviewed, and strategies for future work are outlined.

1. INTRODUCTION

The photosensitizer-relay couple in homogeneous solution discussed in the previous paper (1) has no simple counterpart in the natural photosyntetic systems, that make use of a photosensitizer and a series of relay species specifically organized within the framework of the photosyntetic membrane (2). It is this organization that is considered to be responsible for the high efficiency of the natural photosyntetic systems. As a number of in vitro studies have shown (3), a very specific combination of thermodynamic and kinetic factors and spatial organization causes the fast migration apart of the primary electron-hole pair generated in the photosensitizer by light absorption. The migration occurs in a subnanosecond time scale via adjacent donor and acceptor units (relays) until a spatially separated pair of oxidizing and reducing centers is reached which survives long enough (milliseconds) for slower and more complex energy-storing reactions to take place.

Any attempt to reproduce the complex organization of a natural photosyntetic reaction center would be presently a hopeless undertake. However, studies in model molecules for the photoreactive center incorporating a photosensitizer (usually a porphyrin) and a relay (most often a quinone) have received a great impetus in recent years (4). Very recently, molecular triads consisting of a photosensitizer covalently bound to both an acceptor and a donor subunit

E. Pelizzetti and N. Serpone (eds.), Homogeneous and Heterogeneous Photocatalysis, 29–49.

have been synthesized and studied (5). The studies in this field have made use almost exclusively of organic photosensitizer and relay subunits. This has been clearly dictated on one side by the aim of mimicking natural photosynthesis, and on the other side by the availability of synthetic strategies for assembling the organic subunits.

In spite of their more difficult synthetic handling, inorganic photosensitizers and relays have, as discussed in the previous paper (1), a number of important practical advantages over organic analogues. Thus, the synthesis and characterization of complex inorganic molecular species containing a photosensitizer and one or more acceptor or donor (relay) subunits seems to be a goal of considerable interest. As will be discussed in the following paragraphs, there are theoretical reasons which further justify the interest in inorganic sensitizer-relay supermolecular systems. In fact, models specifically developed for inorganic systems are available for rationalizing: (i) the kinetics of the intramolecular radiationless electron transfer steps expected to take place in these systems, and (ii) the absorption spectra originating from the optical transitions corresponding to these electron transfer steps.

In this paper, some general considerations relevant to the design of photosensitizer-relay supermolecular systems will be developed, and the results of some studies recently initiated in our laboratories towards the construction of such systems will be reviewed.

2. GENERAL CONSIDERATIONS

Let us first examine on very general grounds the types of charge separation which can be expected to occur in a supermolecular system. The simplest of such systems consists of a photosensitizer, P, and a relay, A, bound together via some kind of bridge, L. If appropriate thermodynamic and kinetic requirements (to be discussed later on) are satisfied, excitation of this molecule may lead to a charge separation step, k_e, as shown in Fig. 1 for the case of a reducing P and an oxidizing A (a parallel discussion could be made for the case of an oxidizing photosensitizer bound to a reducing relay D). In general, it is to be expected that a back electron transfer step, k_b, (electron-hole recombination) will follow the charge separation, leading the system back to the ground state. The possibility to utilize such charge separation process depends on the relative rates of the unimolecular back electron transfer step, k_b, and of any bimolecular process, k_s, by which an external scavenger (D, in Fig. 1, representing the particular case of a reducing scavenger) can intercept the charge-separated state. Although the study of such molecules will be interesting for many fundamental reasons (vide infra), it is clear that, from a practical point of view, these systems are not expected to offer substantial advantages over a conventional homogeneous photochemical cycle involving well designed free sensitizer-relay couples.

A logical extension of this model would be that of incorporating into the same molecule the photosensitizer P, the primary acceptor A,

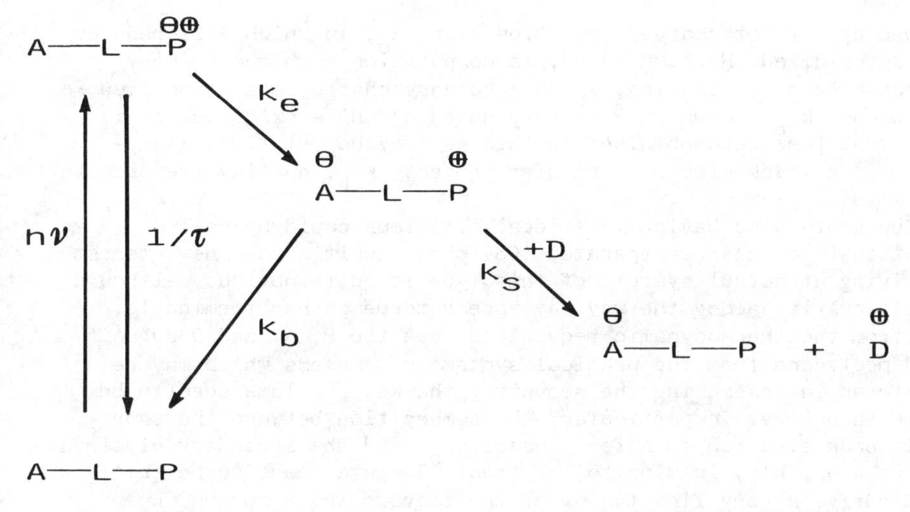

Fig. 1. Schematic representation of photoinduced charge separation in a binuclear supermolecule.

and the secondary donor D. This would lead to a sequence of possible events as shown in Fig. 2. In this example, excitation is again

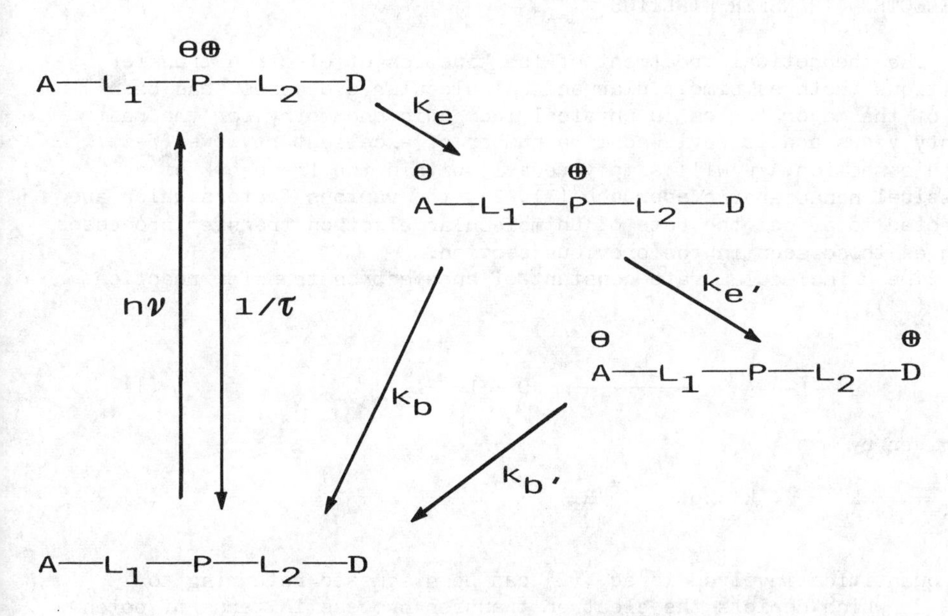

Fig. 2. Schematic representation of photoinduced charge separation in a trinuclear supermolecule (CS photosensitizer).

followed by a first charge separation step, k_e, in which A is reduced and P is oxidized. Here, however, in competition with the primary electron-hole recombination, k_b, a secondary charge separation step could occur, k'_e, in which P is re-reduced and D is oxidized. It is likely that the state obtained in this way may be relatively long-lived, as the back electron transfer process, k'_b, now involves remote redox sites.

Supermolecules having this ideal behaviour could be called "directional" or charge-separated (CS) photosensitizers. The interest in arriving at actual systems of this type is quite obvious, although the difficulties along the way may appear to be rather formidable. Aside from the thermodynamic requisites that the P, A, and D subunits should meet, and from the pratical synthetic problems which may be encountered in assembling the subunits, the key problems seem to be kinetic in nature. In particular, the competition between the short-circuit back electron transfer process, k_b, and the secondary electron transfer step, k'_e, leading to the final CS state seems to be critical. Clearly, a very fine tuning of the factors which govern the kinetics of these electron transfer processes would be necessary in order to achieve efficient charge separation with a system of this type.

3. ELECTRON TRANSFER KINETICS

The theoretical treatment of the kinetics of electron transfer reactions (both as bimolecular and unimolecular processes) has been one of the major topics in physical inorganic chemistry for the past twenty years and is reviewed in a number of excellent reviews (6-11). In this section, we will simply recall, within the framework of a classical nonadiabatic approach (11,12), the various factors which are expected to affect the rate of unimolecular electron transfer processes such as those seen in the previous section.

The unimolecular rate constant of an electron transfer reaction (eq. (1)),

$$\text{D—L—A} \xrightarrow{k_e} \overset{+}{\text{D}}\text{—L—}\overset{-}{\text{A}} \qquad (1)$$

is given by

$$k_e = \nu_n \underline{k} \exp(-\Delta G^{\neq}/RT) \qquad (2)$$

The quantities involved in eq. (2) can be discussed referring to Fig. 3, which depicts the electron transfer process in terms of potential energy surfaces for reactants and products. The ν_n term in the pre-expotential part of the rate constant is an effective nuclear frequency, which corresponds to the parabolic potential energy surfaces of Fig. 3 and can be expressed as a function of both the inner vibrational frequencies of the reactant and product species and the

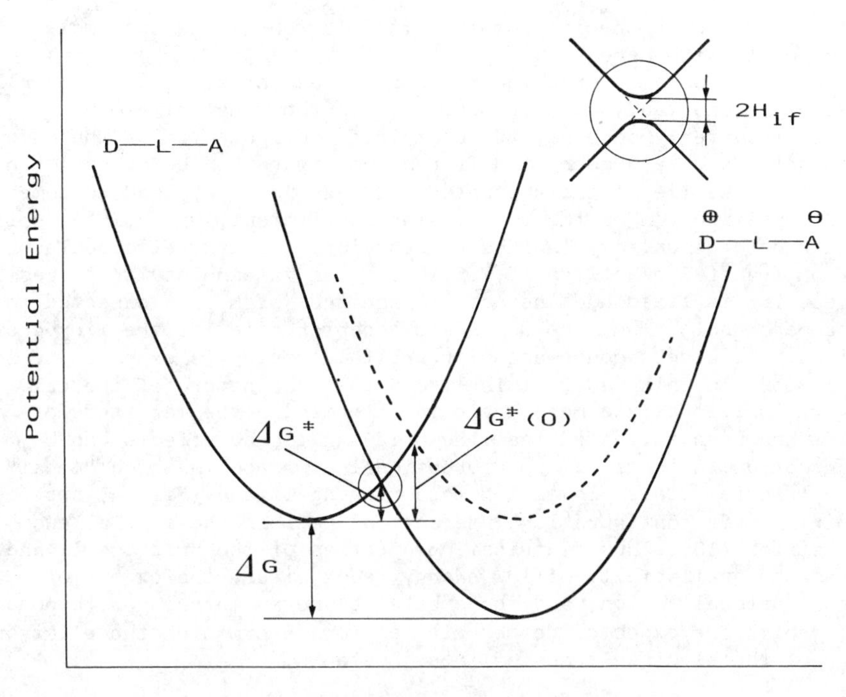

Fig. 3. Potential energy surfaces and kinetic parameters for a unimolecular electron transfer process.

outer solvent reorganizational frequency (11). The $\underline{k}$ term in the preexponential part of the rate constant is the trasmission coefficient, related to the probability of conversion from reactant to product in the avoided crossing region, which can be expressed as a function of the electronic coupling matrix element H_{if} between the initial and final states of the system (eqs. (3) and (4)),

$$\underline{k} = \frac{2\,[1-\exp(-\,\nu_{el}/2\nu_n)]}{2-\exp(-\,\nu_{el}/2\nu_n)} \tag{3}$$

$$\nu_{el} = \frac{2H_{if}^2}{h}\left(\frac{\pi^3}{4\Delta G^{\neq}(0)RT}\right)^{1/2} \tag{4}$$

where $\Delta G^{\neq}(0)$ is the so-called intrinsec barrier, to be discussed below. The activation free energy term, $\Delta G^{\neq}$ in eq. (2) is related to the energy required to reach the distorted nuclear configuration (crossing point) at which the Franck-Condon principle is satisfied for the electron transfer process. Within this general and common model, a few points deserve some special comments in relation to the supermolecular systems dealt with here.

The electronic interaction H_{if} depends essentially on the overlap between the donor and acceptor orbitals. In the usual bimolecular electron transfer processes, the geometric configuration of the encounter is not fixed, a number of different configuration being explored within the lifetime of the encounter complex. This may lead to some kind of optimization of the trasmission coefficient, which is actually considered to be unitary (adiabatic behavior) for many bimolecular electron transfer reactions (6,7,9,10). In an intramolecular process, the geometry is fixed and the actual magnitude of H_{if} is expected to play a major role. When the donor and acceptor moieties are linked by a bridge, the direct donor-acceptor orbital overlap is expected to be rather poor. In this case, it is likely that the nature of the bridging group will play a major role in determining the magnitude of H_{if}. It has been suggested that the bridging ligand effects can be accounted for in terms of a perturbational approach in which "local" charge transfer states involving the bridging ligand (Fig. 4) provide the mixing (via configuration interaction) between the initial and final states (13). Thus, intrinsic properties of the bridging ligand, such as the availability of low-energy redox sites, the extent of internal delocalization, and the orbital symmetry matching with donor and acceptor are expected to be critical for determining the electronic factor of the electron transfer rate constants.

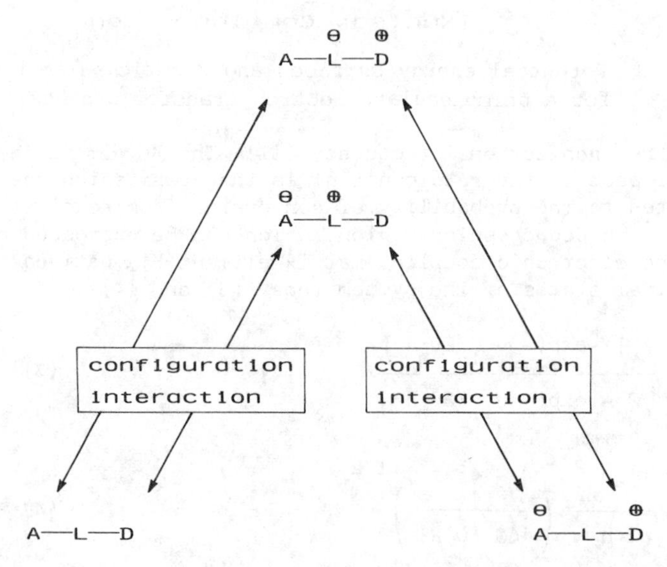

Fig. 4. Schematic representation of the through-bridge electronic interaction between the donor and acceptor subunits of a binuclear supermolecule.

As far as the exponential term of eq. (2) is concerned, the free energy of activation is classically related by means of a free energy relationship (eq. 5) to: (i) an "intrinsic barrier" parameter, $\Delta G^{\neq}(0)$,

$$\Delta G^{\neq} = \Delta G^{\neq}(0) \left[1 + \frac{\Delta G}{4\Delta G^{\neq}(0)} \right]^2 \qquad (5)$$

which represents the degree of distortion (horizontal displacement, in Fig. 3) between the reactant and product curves; (ii) the actual free energy change (vertical displacement, in Fig. 3) of the electron transfer process. For a given degree of distortion, the classical model predicts that $\Delta G^{\neq}$ initially decrease with increasing exergonicity. This is the so called "Marcus inverted region" of electron transfer reactions. According to more rigorous, but not more pratical, quantum mechanical models (14-21), an inverted behaviour (rates decreasing with increasing driving force) is still predicted to occur, although with some quantitative differences with respect to the classical model (linear instead of quadratic dependence, usually referred to as ""energy gap law"). For reasons which are not yet completely clear (22,23), common bimolecular electron transfer reactions in fluid solution do not behave experimentally as expected in this regard, since no conclusive example of inverted behavior has been reported to date for these systems. On the other hand, however, electron transfer processes in rigid matrices (24) or within covalently bound supermolecular systems (25,26), offer several experimental examples of the inverted behavior. Moreover, the energy gap law is also obeyed for a number of electronically excited states whose decay can be viewed as an intramolecular electron transfer (27-29). Since a number of the intramolecular electron transfer processes in Figs. 1 and 2 are likely to be highly exergonic for practical systems, the actual extent of inversion could be one of the important factors in determining the rate of each single step.

4. OPTICAL INTERVALENCE TRANSFER

In a molecule containing several potential redox sites (e.g., in a polynuclear coordination compound), electrons can be transferred from one site to another not only thermally, but also optically (30-32). These optical transitions, which are usually called intervalence transfer (IT) transitions, are responsible for the typical colors of many mixed-valence polynuclear inorganic systems. The classical theory of IT spectra, which has been mainly worked out by Hush, is reviewed in a number of comprehensive reviews (30-32). In Hush's theoretical model, the properties of an optical IT transition (energy, intensity, band shape) are intimately related to the kinetic and thermodynamic factors for the corresponding thermal electron transfer process. Thus, in constructing inorganic model systems for charge separation along the lines of Figs. 1 and 2, considerable attention should be payed to their IT spectra.

The Hush model can be discussed in terms of the same potential energy diagram (Fig. 5) as used for the thermal electron transfer processes. According to this model, the optical IT energy ("vertical"

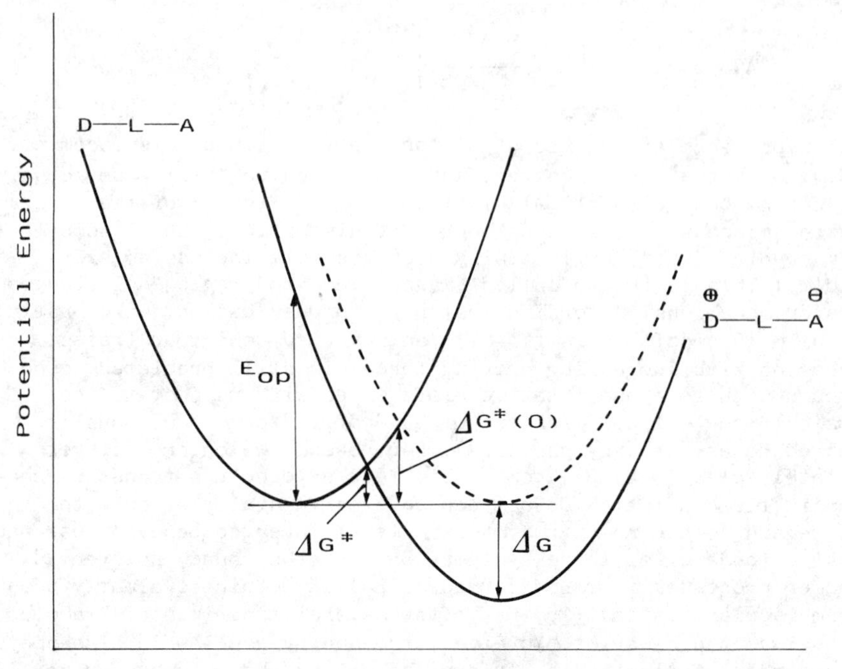

Fig. 5. Optical intervalence transfer and thermal electron transfer in a binuclear supermolecule.

energy or $\bar{\nu}_{max}$ for the IT band) E_{op}, is related by eq. (6) to both the intrinsic barrier parameter $\Delta G^{\neq}(0)$ and the thermodynamic driving force of the thermal reaction, ΔG, and, by virtue of eq. 5

$$E_{op} = 4\Delta G^{\neq}(0) + \Delta G \qquad (6)$$

also to the actual activation free energy, $\Delta G^{\neq}$, of the corresponding thermal process. On the other hand, the halfwidth of the IT band, $\Delta\bar{\nu}_{1/2}$, is related to the optical energy and to the thermodynamic driving force of the thermal process by the following equation:

$$\Delta\bar{\nu}_{1/2}(cm^{-1}) = 48.06(E_{op} - \Delta G)^{1/2} \qquad (7)$$

Finally the "integrated" IT intensity, $\varepsilon_{max}\,\Delta\bar{\nu}_{1/2}$, is related to the optical energy, to the distance, r, between the two redox centers involved, and to the electronic matrix element, H_{if}, between the initial and final states of the transition (or of the thermal electron transfer process):

$$\varepsilon_{max} \Delta\bar{\nu}_{1/2} = \frac{r(\mathring{A})^2}{4.20 \times 10^{-4} E_{op}} H_{if}^2 \qquad (8)$$

The above expressions clearly show that, at least in principle, the relevant kinetic parameters of a unimolecular thermal electron transfer process (i.e., the degree of adiabaticity, k, and the activation free energy, $\Delta G^{\neq}$) should be calculable from the spectroscopic parameters of the corresponding IT transition. More practically, a very quick estimate of the relative "degree of adiabaticity" of various unimolecular electron transfer paths in a complex molecule can be made by looking at the relative intensities of the corresponding IT spectral transitions.

It should be recalled that the Hush model was originally conceived for metal-to-metal charge transfer transitions in mixed-valence bi- or polynuclear complexes. It should be pointed out, however, that in principle this treatment should be extendible to any kind of transition between localized redox centers in a complex molecule, including ligands which can be reduced or oxidized at reasonable energies. As a matter of fact, it is entirely possible to discuss the usual metal-to-ligand charge transfer (MLCT) (29) and the less popular ligand-to-ligand charge transfer (LLCT) (33-35) bands in a mononuclear complex in terms of the Hush formalism for IT. transitions.

5. SUPERMOLECULAR SYSTEMS BASED ON $Ru(bpy)_2(CN)_2$

As a first experimental step towards the design of inorganic CS sensitizers, a number of bi- and trinuclear complexes containing $Ru(bpy)_2^{2+}$ as the photosensitizer P unit, CN^- as the L_1 and L_2 bridging ligands, and various attached metal-containing moieties have been recently sinthesized and studied.

The reasons for choosing $Ru(bpy)_2^{2+}$ as P can be traced back to a number of the properties which make $Ru(bpy)_3^{2+}$ a very good, if not ideal, photosensitizer (1). Among a number of possible candidates as bridging ligands, cyanide has the advantage of having a strong ligand field. This should ensure that ligand field states, which provide the main radiationless deactivation path in this type of complexes at room temperature (36-39), remain at substantially higher energy than the d-π^* state of the $Ru(bpy)_2^{2+}$ chromophore.

The photophysical properties of such a photosensitizer-bridge unit are known in some detail from the work of Demas and coworkers (40-43). This unit exhibits a prominent d-π^* absorption band in the visible and a typical long-lived d-π^* phosphorescence in fluid solution (Table I). The excited-state energy coupled with the ground-state redox potentials indicate that the excited state is a powerful reductant and a very mild oxidant (Table I). Contrary to what happens with its more famous $Ru(bpy)_3^{2+}$ congener, the excited-state properties of $Ru(bpy)_2(CN)_2$ depend to a remarkable extent on solvent (41,44,45).

The supermolecular systems studied are of the type shown in Fig. 6:

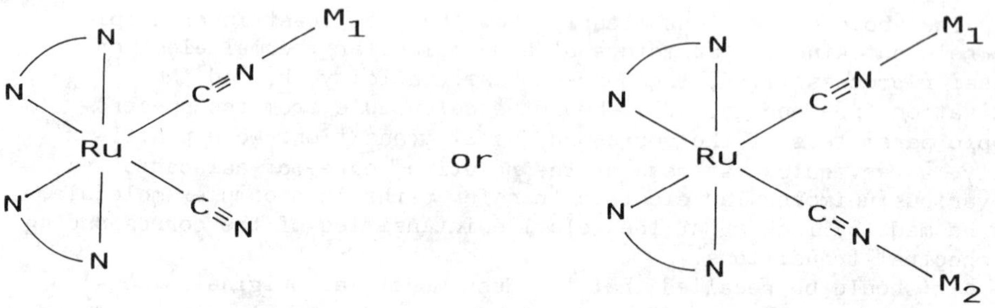

Fig. 6. Schematic structural representation of the binuclear and trinuclear supermolecular systems based on $Ru(bpy)_2(CN)_2$.

where M_1 and M_2 are metal complex moieties. These species can be subdivided into groups according to both the intrinsic properties of the M_1, M_2 moieties (excited-state energies, redox potentials) and the effect of these moieties on the properties (absorption and emission spectra, lifetimes) of the original $Ru(bpy)_2(CN)_2$ chromophore. A general discussion of such supermolecular complexes in terms of second sphere perturbation of the $Ru(bpy)_2(CN)_2$ chromophore has been given elsewhere (46).

5.1. M_1, M_2: Platinum(II) Complex Moieties

Several Pt(II) complex moieties have been shown to form supermolecular adducts with $Ru(bpy)_2(CN)_2$ <u>via</u> cyanide bridges (44,47). For a number of $PtClXY^{n+}$ moieties ($X = C_2H_4$, DMSO; Y = Cl, n = 0; Y = 4-Mepy, n = 1), equilibrium mixtures of free $Ru(bpy)_2(CN)_2$, binuclear, and trinuclear species were studied (47), while for $Pt(dien)^{2+}$ (dien = dietylenetriamine) stable 1:1 and 1:2 adducts were isolated and investigated in detail (44). The behavior observed was similar in all cases. Typical data for M_1, M_2 = $Pt(dien)^{2+}$ are given in Table I.

The consequences of adduct formation are essentially: (i) blue shifts in both adsorption and emission spectra; (ii) substantial lifetime enhancement (the figure for the 1:2 adduct in Table I appears to be exceptional among all the Pt(II) adducts studied); (iii) roughly constant excited-state reducing power; (iv) increasing excited-state oxidizing power. This behavior can be easily rationalized in terms of the stabilization of the d(π) (t_{2g} in O_h symmetry) ruthenium orbitals due to increased back bonding to bidentate cyanides. This shifts to higher energy the d-π^* excited states responsible for

Table I. Properties of $Ru(bpy)_2(CN)_2$ and of its Adducts with $Pt(dien)^{2+}$.[a]

	$Ru(bpy)_2(CN)_2$	$Ru(bpy)_2(CN)_2$–$[Pt(dien)]^{2+}$	$Ru(bpy)_2(CN)_2$–$[Pt(dien)]_2^{4+}$
λ_{max}(abs), nm	505	460	426
λ_{max}(em), nm	680	630	580
τ, ns	205	630	90
$^*E_{1/2}(S^+/S^*)$,[b] V	−1.32	−1.16	−1.45
$^+E_{1/2}(S^*/S^-)$,[b] V	0.37	0.57	0.81

a) Deaerated DMF solutions, room temperature. Data from ref.44.
b) Excited state redox potentials vs SCE; S represents the complex.

adsorption and emission. The parallel increase in the energy of the d-d states keeps the efficiency of the thermally activated radiationless deactivation pathway low, so that the main effect on the lifetime is given by the increased energy gap with respect to the ground state. The reducing power of the excited state is hardly affected by adduct formation because of the parallel increase in excited-state energy and stabilization of the d(π) redox orbitals. The excited-state oxidizing power increases since the increase in excited-state energy is not compensated by changes in energy of the redox orbitals (which, in this case, are π^*bpy orbitals (48)).

Very similar behavior is also observed when Zn^{2+} ions are linked to the $Ru(bpy)_2(CN)_2$ chromophore (46). The above described behavior is to be considered typical of all the supermolecular systems of this type involving M_1 and M_2 subunits which do not possess (i) low-energy excited states and (ii) low-energy redox sites. From the point of view of charge separation, such supermolecular systems obviously do not represent a forward step. At least, however, the study of these systems shows that metal coordination to the cyanides of $Ru(bpy)_2(CN)_2$ does not per se destroy the useful excited-state properties of the chromophore, and gives an estimate of the minimal consequences of such a process.

5.2. M_1, M_2 = Ni(II) or Co(II) Aquo ions.

These systems were investigated several years ago by Demas,et al. (41) in the context of a study on the quenching of excited

$Ru(bpy)_2(CN)_2$ by several transition metal complexes. Actually, no definite molecular species were isolated in that study, but rather association between $Ru(bpy)_2(CN)_2$ and Co^{2+}_{aq} or Ni^{2+}_{aq} was detected by observing a static quenching (49) component accompanying the usual dynamic quenching process. A kinetic scheme suitable to interpret the data of Demas, et al. (41) is shown in Fig. 6. Since both Co^{2+}_{aq} and Ni^{2+}_{aq} are redox inactive in the potential range spanned by the excited $Ru(bpy)_2(CN)_2$ chromophore (Table I), the quenching (both dynamic and static) is assumed to proceed by an energy transfer mechanism (41).

The quenching scheme shown in Fig. 7 offers a very interesting example of simultaneously operating "outer-sphere" and "inner-sphere" mechanism for energy transfer. These two types of mechanism have long been known in the field of electron transfer reactions, and this system gives further emphasis to the already noted (50,51) analogy between energy and electron transfer kinetics and mechanisms.

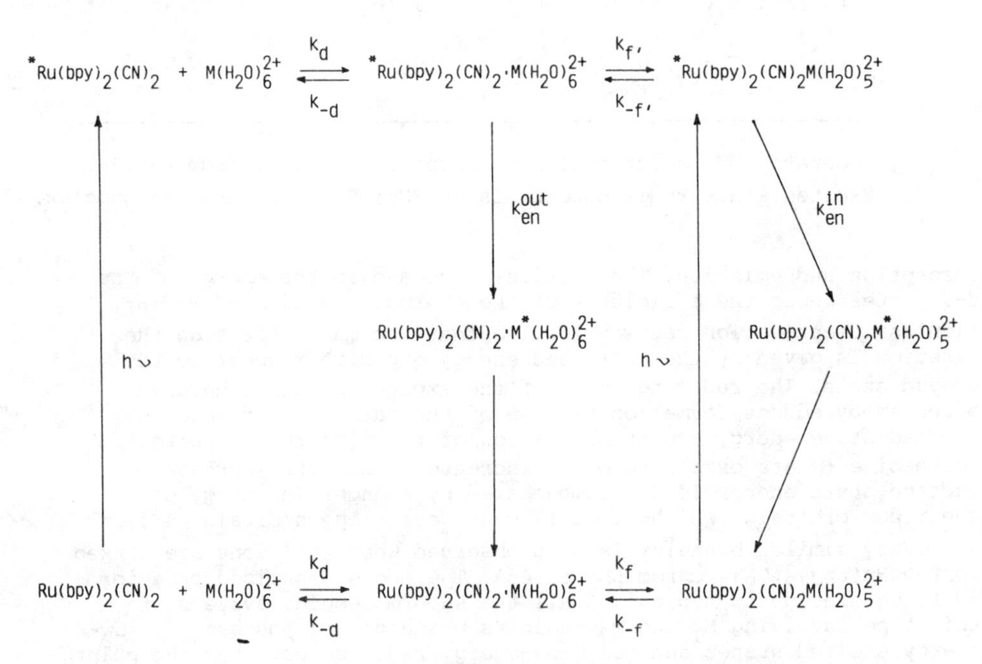

Fig. 7. Kinetic scheme for the static (inner-sphere) and dinamic (outer- sphere) quenching of $Ru(bpy)_2(CN)_2$ by $M(H_2O)_6^{2+}$ (M = Co, Ni).

The bimolecular rate constants for dynamic quenching, k_q, are 6.4×10^6 and 9×10^6 $M^{-1}s^{-1}$ for Co^{2+}_{aq} and Ni^{2+}_{aq}, respectively. These values are much lower than the diffusion controlled limit, which implies that $k_q \cong (k_d/k_{-d})(k_{en}^{out} + k_f)$. Since k_d/k_{-d} is roughly 1 M^{-1} and k_q is substantially higher than the rate constants for water exchange of the aquo metal ions (which is an upper limit for k_f) (52), the experimental k_q values (s^{-1}) should essentially correspond to the unimolecular outer-sphere energy transfer rate constant, k_{en}^{out}, values. Recent single-photon counting experiments (53), on the other hand, indicate that in both cases the rate constant for the inner-sphere energy transfer process, k_{en}^{in}, is higher than 1×10^9 s^{-1}. The comparison between these values emphasizes the large increase in energy transfer probability when the mechanism is changed from outer-sphere to inner-sphere. Since an exchange energy transfer process requires simultaneous HOMO-HOMO and LUMO-LUMO overlap (50), this increase is understable in terms of (i) the decrease in metal-to-metal distance and especially (ii) the more favorable symmetry requirements (particularly for t_{2g}-t_{2g} HOMO-HOMO overlap) in the inner-sphere process.

In conclusion, these systems show that intramolecular energy transfer quenching of the sensitizer subunit may be remarkably fast in supermolecules of this type. Thus, metal containing moieties having low-lying excited states should be avoided as far as possible in the design of supermolecular systems for light-induced charge separation.

5.3. M_1, M_2 = $Ru(NH_3)_5{}^{3+/2+}$, $Ru(NH_3)_4py^{3+/2+}$.

Altogether, thirteen supermolecules of the type shown in Fig. 6 have been synthesized whith these subunits (54,55) by controlling (i) the number of subunits bound(only M_1 or both M_1 and M_2), (ii) the nature of the M_1 and/or M_2 subunits; and (iii) the 2+ or 3+ oxidation state of the M_1 and/or M_2 subunits.

These subunits differ sharply from those discussed in the previous sections in that they are species that can be easily reduced or oxidized ($E°$ = +0.067 and +0.299 V for the $Ru(NH_3)_6{}^{3+/2}$ and $Ru(NH_3)_5py^{3+/2+}$ parent compounds (56)). Thus, a variety of intramolecular electron transfer processes should be thermodynamically feasible in these supermolecules following excitation of the $Ru(bpy)_2(CN)_2$ chromophore. The position of the lowest d-d states of these subunits is not known with certainty. Several indirect lines of evidence indicate that these states should be comparable in energy with the lowest excited state of the $Ru(bpy)_2(CN)_2$ chromophore. Therefore, although energy transfer quenching similar to that discussed in the previous section cannot be ruled out completely, this process is expected to be relatively slow due to small driving force.

Experimentally, the main effects of the introduction of these subunits are: (i) the emission of the $Ru(bpy)_2(CN)_2$ chromophore is completely quenched; (ii) the spectrum of the supermolecules is quite different from the sum of the spectra of the component subunits, exhibiting a number of new absorption bands. By a careful comparison of the spectra of different supermolecules of this class and by using standard theoretical models (Section 4), these bands can be assigned to the one-electron optical transitions interconnecting the various redox centers of the molecule. As an example, the spectrum of $[(NH_3)_5RuNCRu(bpy)_2CNRu(NH_3)_5]^{5+}$ is shown in Fig. 8.

Using the model outlined in Section 4, the energy and shape of the bands in the spectra of these supermolecules can be satisfactorily accounted for (54,55) in terms of the energetics of the various redox

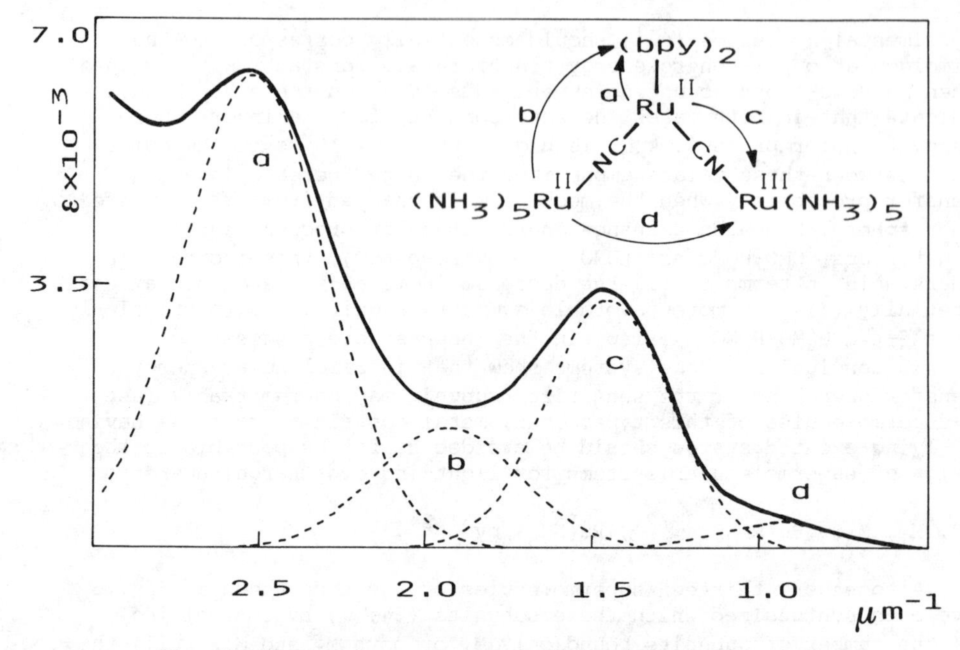

Fig. 8. Absorption spectrum and band assignement for $[(NH_3)_5RuNCRu(bpy)_2CNRu(NH_3)_5]^{5+}$.

sites (which can be generally inferred from the electrochemistry of the supermolecules (55,57) of the subunits, or of suitable model systems) and of the reorganizational barriers (which in these systems depend mainly on solvent and transfer distance). The band intensities (which are expected to depend primarily on the electronic matrix element H_{if} between the centers involved in the transition) generally follow (see, e.g., Fig. 8) the sequence: metal-to-bound ligand > metal-to-bridged metal > metal-to-remote ligand > metal-to-remote metal. This sequence corresponds quite plausibly to an ordering of the effective site-to-site interactions in these supermolecular systems.

The lack of the usual d-π* $Ru(bpy)_2(CN)_2$ emission in the supermolecular systems is as expected if the presence of low-energy redox centers in the attached moieties opens up new intramolecular electron transfer channels for the deactivation of the excited state. These channels are shown for the case of $[(NH_3)_5RuNCRu(bpy)_2CN-$ $-Ru(NH_3)_5]^{5+}$ in Fig. 9, in terms of simple one-electron configurations (i.e., ignoring spin considerations for the sake of simplicity). The observed emission quenching simply indicates that the primary intramolecular electron transfer processes, e.g., processes 1 or 2 in Fig. 9, are exceedingly fast in all these systems.

In all the supermolecular systems of this type investigated, the products of the primary intramolecular electron transfer steps

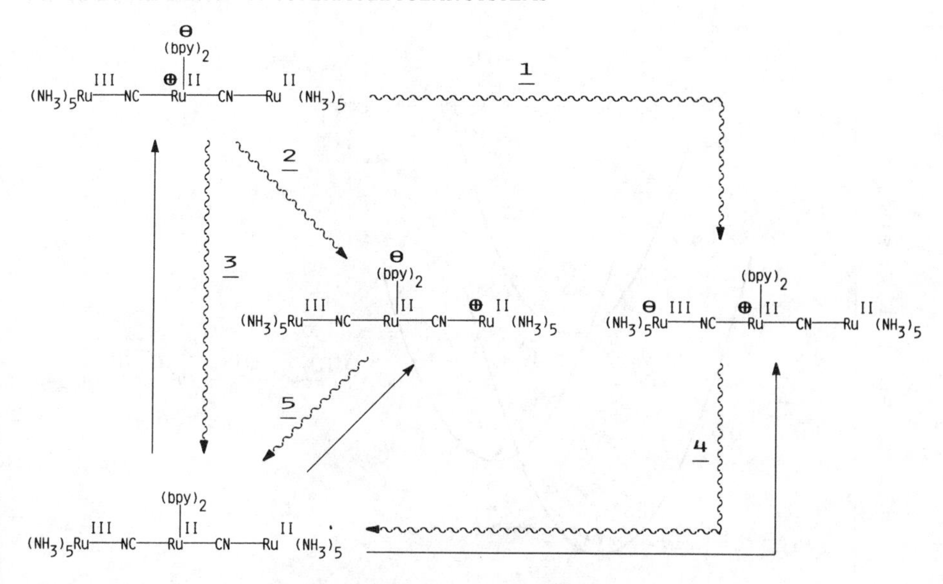

Fig. 9. Intramolecular electron transfer pathways for the deactivation of excited $[(NH_3)_5RuNCRu(bpy)_2CNRu(NH_3)_5]^{5+}$.

have escaped detection by laser photolysis, even upon lowering the temperature to 77 K (54,55). This implies that further deactivation steps (e.g., the back intramolecular electron transfer processes 4 or 5 in Fig. 8) are also fast on a nanosecond time scale. It should be noticed that, according to a scheme such as that of Fig. 9, a two-step electron transfer sequence involving remote redox sites (1 + 4 or 2 + 5 in Fig. 9) is found to be faster than the direct bpy-to-ruthenium step (3 in Fig. 9) which corresponds to the usual radiationless decay of the d-π* excited state of the $Ru(bpy)_2(CN)_2$ chromophore. Although a number of other factors may be influential (e.g., some relaxation of the spin forbiddness in the two-step sequence), a major reason for this behavior seems to lie in the "inverted" character of the direct deactivation, as opposed to the "normal" character of the processes involved in the two-step sequence. This difference is due to both thermodynamic factors (lower driving force for the latter processes) and intrinsic barriers (larger electron transfer distances, an thus larger solvent reorganizational energies, for the latter processes). This situation is sketched in terms of potential energy surfaces in Fig 10, for the 1 + 4 sequence of the above example.

If the system in Fig. 9 is compared with the hypotetical model of

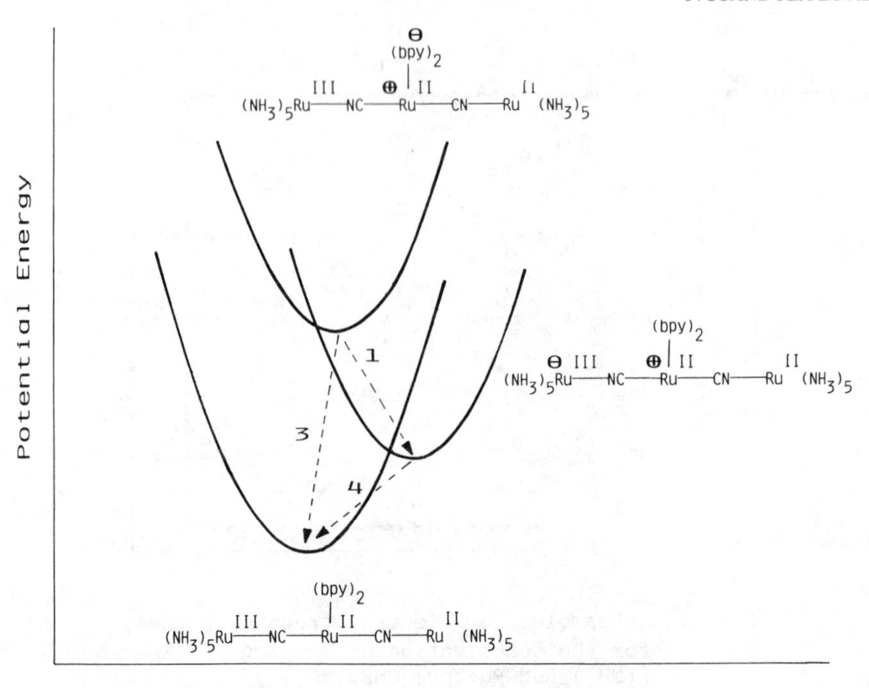

Fig. 10. Schematic representation in terms of potential energy surfaces of the deactivation pathways indicated as 1 + 4 in Fig. 9.

Fig. 2, process 1 corresponds to the primary k_e charge separation step, and process 4 corresponds to the primary back electron transfer step k_b. It should be noticed that the secondary charge separation step, k_e' in Fig. 2, would be meaningless in the system of Fig. 9, since it would coincide (except for an exchange of the two terminal redox sites) with the back electron transfer step 4. In principle, such a secondary charge separating step could be observable in an asymmetric analogue such as $[(NH_3)_5RuNCRu(bpy)_2CNRu(NH_3)_4py]^{5+}$ (Fig. 11). Laser experiments performed on this system, however, failed to give any evidence for the end-to-end CS intermediate (55). This may be due to one of the following reasons: (i) the back electron transfer step is much faster than the secondary charge separation step (process 4 faster than process 6, or process 5 faster than process 7) so that the CS intermediate is not substantially populated; (ii) the charge recombination process in the intermediate (process 8) is fast on a nanosecond timescale.

Although it might well be that the lifetime of the CS intermediate

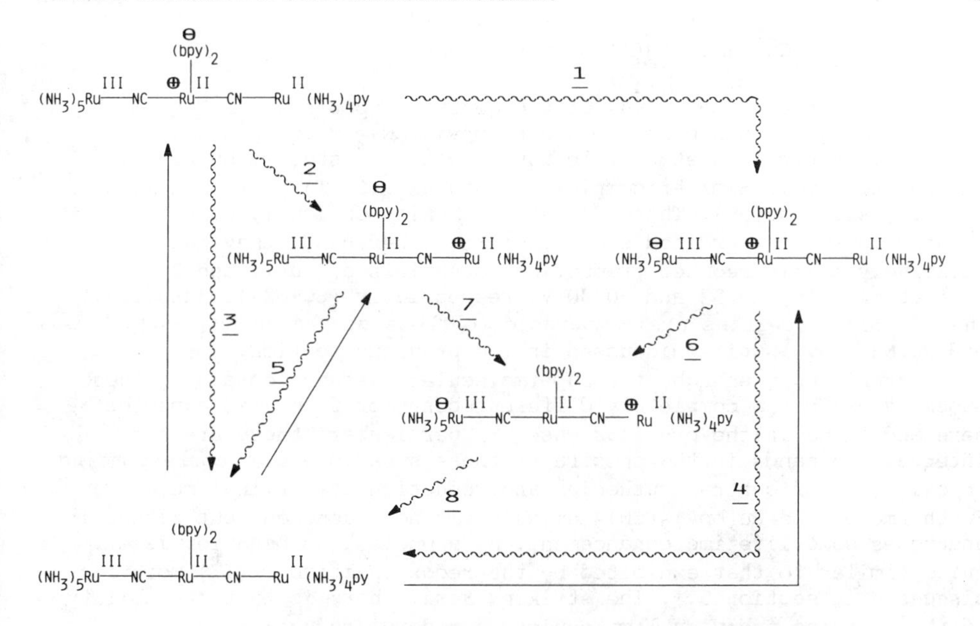

Fig. 11. Intramolecular electron transfer pathways for the deactivation of excited $[(NH_3)_5RuNCRu(bpy)_2CNRu(NH_3)_4py]^{5+}$.

is too short for observation (simple calculations using the Hush model and the available spectroscopic information give expected values in the 0.5-2 ns range), the most likely hypothesis seems to be that of inefficient population of the intermediate. As a matter of fact, if process 5 is compared with process 7 (or process 4 with process 6), the two processes are virtually identical as far as the electronic factors and the intrinsic barriers are concerned, while the former has a more favourable driving force. As both processes fall in the "normal" free energy region, the former is also expected to be the fastest one.

These studies seem to suggest that, with supermolecular systems containing these subunits, the chances to observe long-lived charge separation are bound to the possibility of changing at least one of the bridging ligands. In particular, the two-bridge system should be made asymmetric so as to orient the competition between the primary back electron transfer step and the secondary charge separation step against the driving force difference. Moreover, the electronic coupling between the two terminal groups should be decreased, in order to make the end-to-end charge recombination process substantially nonadiabatic. It should be noticed that no system exhibiting appreciably intense end-to-end intervalence transfer absorption could likely fulfill the nonadiabaticity condition.

5.4. M_1, M_2 = Cu(I) or Ag(I) Solvated ions.

Supermolecular complexes of the general type shown in Fig. 6, containing Ag(I) ions bound to the $Ru(bpy)_2(CN)_2$ chromophore in 1:1 or 1:2 ratio in acetonitrile have first been studied by Kinnaird and Whitten (58). Similar complexes containing Cu(I) have been more recently studied (59). These d^{10} metal subunits obviously do not possess low-energy excited states; on the other hand, they can be relatively easily reduced (reduction potentials of Cu(I) and Ag(I) in acetonitrile, -0.32 and +0.30 V, respectively (60-62)). Actually, their redox properties are comparable to those of the $Ru(NH_3)_5^{3+}$ and $Ru(NH_3)_4py^{3+}$ units discussed in the previous section.

Surprisingly enough, the supermolecular systems containing these subunits exhibit a completely different behavior from what expected here and found in the previous case. In particular, there are no intervalence bands in the spectra of these supermolecules corresponding to oxidation of central ruthenium and reduction of terminal metal. Furthermore, the $Ru(bpy)_2(CN)_2$ emission is not quenched, but rather it undergoes some lifetime enhancement. As a whole, the behavior is quite similar to that exhibited by the redox inactive Pt(II) moieties discussed in section 5.1. The striking result here is that the ability of the units as individual molecules to undergo reduction cannot be used, either optically or non-radiatively, once these units are incorporated into the supermolecular system. This must have to do with specific symmetry constraints to electron transfer in the supermolecular system. If one considers that these subunits only have s-type redox orbitals available which give zero overlap with the π system of the bridging ligands and central ruthenium, these unexpected observations can be rather easily rationalized. A related case could be that of the analogous complex containing Cu(II) aquo ions first investigated by Demas (49), which, in spite of the favourable redox properties of the attached unit, does not show the expected intervalence transfer absorption band (although quenching takes place). Here again the redox orbital is a σ-type e_g (O_h) orbital of copper. The importance of orbital symmetry in determining the intensity of intervalence absorption of mixed-valence metal cyanides was actually pointed out many years ago by Braterman (63).

The study of the supermolecular systems of this class gives the interesting suggestion that symmetry restrictions could be used to achieve charge separation. This requires, of course, that such restrictions can be made operative against the final charge recombination process, but not in the earlier charge separation steps. This is in principle possible, since the various electron transfer processes do not generally involve the same orbitals.

6. CONCLUDING REMARKS

Although the efforts towards the construction of an inorganic CS photosensitizer are still at a preliminary stage, the initial studies summarized in this article have enlightened a number of interesting facts taking place when a standard inorganic photosensitizer and one or more metal-containing subunits are assembled in a supermolecular complex. In particular, some information has been obtained on the interplay of thermodynamic and kinetic (both nuclear and electronic) factors that determine the efficiency of the intramolecular radiationless processes (energy and electron transfer) which take place following light absorption by the supermolecule. Part of this insight has been gained through the study of the rich intervalence transfer (in the generalized sense of site-to-site optical charge transfer) spectra of the supermolecules, although the very appearance of this type of bands probably indicates that these systems are too strongly coupled for achieving long-lived charge separation. Strategies for future work in this direction can be devised on the basis of these initial studies.

ACKNOWLEDGEMENTS

This work was supported by the Consiglio Nazionale delle Ricerche (Comitato Chimica and Progetto Finalizzato Chimica Fine e Secondaria, Tematica Ah1), and Ministero della Pubblica Istruzione.

REFERENCES

1) Balzani, V.; Juris, A.; Scandola, F. Previous article in this volume.
2) Dutton, P. L.; Prince, R. C.; Tiede, D. M. Tunneling in Biological Systems; Chance, B., De Vault, D. C., Frauenfelder, H., Marcus, R. A., Schriffer, J. R., Sutin, N., (Eds); Academic Press, New York, 1979, p. 319.
3) Jortner, J. J. Am. Chem. Soc., 102, 6676 (1980).
4) Joran, A. D.; Leland, B. A.; Geller, G. G.; Hopfield, J. J.; Dervan, P. B. J. Am. Chem. Soc., 106, 6090 (1984), and references therein.
5) Moore, T. A.; Gust, D.; Mathis, P.; Mialocq, J. C.; Chachaty, C.; Bensasson, R. V.; Land, E. J.; Doizi, D.; Liddell, P. A.; Lehaman, W. R.; Nemeth, G. A.; Moore, A. L. Nature, 307, 630 (1984).
6) Marcus, R. A., Annu. Rev. Phys. Chem., 15, 155 (1964).
7) Reynolds, W. R.; and Lumry, R. W., Mechanisms of Electron Transfer. Ronald Press, New York, 1966.
8) Hush, N. S. Electrochim. Acta, 13, 1005 (1968).
9) Sutin, N., in Inorganic Biochemistry, Eichorn, G. L. (Ed.), Elsevier, New York, 1973, p. 611.
10) Cannon, R. D., Electron Transfer Reactions. Butterworths, London, 1980.
11) Sutin, N. Prog. Inorg. Chem., 30, 441 (1983).
12) Balzani, V.; Scandola, F. in Energy Resources through Photochemistry and Catalysis, Gratzel, M. (Ed.); Academic Press, New York, 1983.

13) Day, P. Comments Inorg. Chem. 1, 155 (1981).
14) Ulstrup, J., Charge Transfer in Condensed Media. Springer-Verlag, Berlin, 1979.
15) Levich, V. G., Adv. Electrochem. Electrochem. Eng. 4, 249 (1966).
16) German, E.D.; Dvali, V. G.; Dogonadze, R. R., Kuznetsov, A. M., Elektrokhimiya, 12, 639 (1976).
17) Dogonadze, R. R., in Reactions of Molecules at Electrodes. Hush, N. S. (Ed.) Wiley, London, 1971, p. 135.
18) Kestner, N. R.; Logan, J.; Jortner, J.; J. Phys. Chem., 78, 2148 (1974).
19) Ulstrup, J.; Jortner, J. J. Chem. Phys., 63, 4358 (1975).
20) Efrima, S.; Bixon, M., Chem. Phys., 13, 447 (1976).
21) Van Duyne, R. P.; Fischer, S. F., Chem. Phys., 5, 183 (1974).
22) Marcus, R. A.; Siders, P., J. Phys. Chem., 86, 622 (1982).
23) Indelli, M. T.; Ballardini, R.; Scandola, F. J. Phys. Chem., 88, 2547 (1984).
24) Miller, J. R.; Beitz, J. V.; Huddleston, R. K. J. Am. Chem. Soc., 106, 5057 (1984).
25) Miller, J.R.; Calcaterra, L. T.; Closs, G. L., J. Am. Chem. Soc., 106, 3047 (1984).
26) Wasielewski, M. R.; Niemczyk, M. P.; Svec, W. A.; Pewitt, E. B. J. Am. Chem. Soc., 107, 1080 (1985).
27) Caspar, J. V.; Meyer, T. J. Inorg. Chem., 22, 2444 (1983).
28) Allen, G. H.; White, R. P.; Rillema, D. P.; Meyer, T. J. J. Am. Chem. Soc., 106, 2615 (1984).
29) Meyer, T. J. Prog. Inorg. Chem., 30, 389 (1983).
30) Hush, N. S. Prog. Inorg. Chem., 8, 391 (1967).
31) Brown, D. B., (Ed.) Mixed Valence Compounds, Reidel, Dordrecht, Holland, 1980.
32) Creutz, C. Prog. Inorg. Chem., 30, 1 (1983).
33) Vogler, A.; Kunkely, H. J. Am. Chem. Soc., 103, 1559 (1981).
34) Vogler, A.; Kunkely, H. Angew. Chem. Int. Ed. Engl., 21, 77 (1982).
35) Crosby, G. A.; Highland, R. G.; Truesdell, K. A. Coord. Chem. Rev., 64, 41 (1985).
36) Van Houten, J.; Watts, R. J. J. Am. Chem. Soc., 98, 4853 (1976).
37) Kemp, T. J. Prog. React. Kinet., 10, 301 (1980).
38) Durham, B.; Caspar, J. V.; Nagle, J. K.; Meyer, T.J. J. Am. Chem. soc., 104, 4803 (1982).
39) Barigelletti, F.; Juris, A.; Balzani, V.; Belser, P.; Von Zelewsky, A. Inorg. Chem., 22, 3335 (1983).
40) Peterson, S. H.; Demas, J. N. J. Am. Chem. Soc., 98, 7880 (1976).
41) Demas, J. N.; Addington, J. W.; Peterson, S. H.; Harris, E. W. J. Phys. Chem., 81, 1039 (1977).
42) Demas, J. N.; Harris, E. W.; Mc Bride, R. P. J. Am. Chem. Soc., 99, 3547 (1977).
43) Peterson, S. H.; Demas, J. N. J. Am. Chem. Soc., 101, 6571 (1979).
44) Bignozzi, C. A.; Scandola, F. Inorg. Chem., 23, 1540 (1984).
45) Belser, P.; Von Zelewski, A.; Juris, A.; Barigelletti, F.; Balzani, V. Gazz. Chim. Ital., In press.
46) Balzani, V.; Sabbatini, N.; Scandola, F. Chem. Rev., In press.

47) Bartocci, C.; Bignozzi, C.A.; Scandola, F.; Rumin, R.; Courtot, P. Inorg. Chim. Acta 76, L 119 (1983).
48) Roffia, S.; Ciano M. J. Electroanal. Chem. Interfacial Electrochem. 77, 349 (1977).
49) Balzani, V.; Moggi, L.; Manfrin, M. F.; Bolletta, F.; Laurence, G. S. Coord. Chem. Rev., 15, 321 (1975).
50) Balzani, V.; Bolletta, F.; Scandola, F. J. Am. Chem. Soc., 102, 2152 (1980).
51) Scandola, F.; Balzani, V. J. Chem. Educ.60, 814 (1983).
52) Basolo, F.; Pearson, R. G. Mechanism of Inorganic Reactions, Wiley, New York 1967, p. 156.
53) Chiorboli, C.; unpublished results.
54) Bignozzi, C.A.; Roffia, S.; Scandola F. J. Am. Chem. Soc., 107, 1644 (1985).
55) Bignozzi, C.A.; Roffia, S.; Paradisi, C.; Scandola, F., manuscript in preparation.
56) Brown, G. M.; Krentzien, H.J.; Abe, M.; Taube, H. Inorg. Chem. 18, 3374 (1979).
57) Roffia, S.; Paradisi, C.; Bignozzi, C. A. J. Electroanal. Chem. Interfacial Electrochem. In press.
58) Kinnaird, M.G.; Whitten, D.G. Chem. Phys. Lett. 88, 275 (1982).
59) Bignozzi, C. A.; unpublished results.
60) Farha, F. Jr.; Iwamoto, R. T. J. Electroanal.Chem. 8, 55 (1964). For Cu(I) more cathodic values have been reported by other authors (61,62). The discussion remains valid independently of the particular value adopted.
61) Kolthoff, I. M.; Coetzee, J. F. J. Am Chem. Soc., 79, 1852 (1957).
62) Senne, J. K.; Kratochvil, B. Anal. Chem., 43, 79 (1971).
63) Braterman, P. S. J. Chem. Soc., (A), 1471 (1966).

FUNDAMENTAL STUDIES INTO PRIMARY EVENTS IN PHOTOCATALYSIS EMPLOYING CdS AND TiO_2 SEMICONDUCTORS: PHOTOLUMINESCENCE, LASER FLASH PHOTOLYSIS AND PULSE RADIOLYSIS

Nick Serpone* and Ezio Pelizzetti**

*Department of Chemistry, Concordia University
1455 deMaisonneuve Blvd. West, Montreal,
Quebec, CANADA H3G 1M8

**Dipartimento di Chimica Analitica, Università di Torino
Via Pietro Giuria 5, Torino 10125, ITALIA

ABSTRACT. Two semiconductor materials have been the focus of much attention in photocatalytic processes as in the photocleavage of water, of hydrogen sulfide, in the photo-oxidation of alcohols and of carboxylate compounds, among others. These materials are the n-type CdS and TiO_2. Many of the processes are carried out with powders. Unfortunately, powder dispersion, where the particles have sizes $\gtrsim$ 2 μm, are not amenable to studies that might lead to an understanding of the primary events taking place on the particle surface or in the bulk of the particle. Suspensions in which the particle size is $\lesssim$ 100 A are transparent solutions which make possible the utilization of optical spectroscopic techniques to follow the events immediately after light excitation. This paper presents and reviews some of the recent attempts in this regard, and emphasizes the results from photoluminescence, laser flash photolysis and pulse radiolysis studies. With the above handles, we are beginning to unravel and understand the mechanisms of photocatalysis.

1.0 INTRODUCTION

Much recent interest has been focused on photo-induced processes at semiconductor particle surfaces. The relevant oxidizing and catalytic properties exhibited by semiconductors (n-type) under bandgap illumination have been exploited to achieve various types of photo-oxidations and photo-reductions. Two properties of semiconductors that have generated such attention are the high extinction coefficients and the fast carrier diffusion to the solid/electrolyte interface. These have led to large catalytic surfaces and to higher efficiencies. Two semiconductors that have been investigated extensively in photocatalytic processes (1) (the photo-cleavage of water (2) and hydrogen sulfide (3), the photo-oxidation of alcohols (4), the water-gas shift reaction (5), the photo-decarboxylation of carboxylic acids (6), the photodegradation of wastes (7), and the photodeposition and recovery of noble trace metals (8)) have been CdS (1,3,4f) and TiO_2 (1,2,4a-e,5-8). These studies have employed macrodispersoins in which the particle sizies $\geq$ 2 μm.

Absorption of light having energy greater than that of the bandgap,

E. Pelizzetti and N. Serpone (eds.), Homogeneous and Heterogeneous Photocatalysis, 51–89.

by semi-conductor dispersions leads to the generation of electron/hole pairs which, under the influence of the electric field, move (within the bandgap) into the conduction and valence band, respectively. It is the resulting non-equilibrium distribution of conduction band electrons (e^-_{cb}) and valence band holes (h^+_{vb}) that leads to either reduction or oxidation processes. The question in these photocatalytic processes rests on the nature and identity of the primary events that immediately follow light excitation. Unfortunately, powders are too opaque and consequently Raileigh scattering too severe for information on these primary events to be obtained by optical spectroscopic techniques. However, semiconductor dispersions in which the particles are of colloidal dimensions ($\lesssim$ 100-200A) provide a route whereby spectroscopic techniques, normally employed for homogeneous systems, can be used to probe these events inasmuch as Raileigh scattering is negligible under these conditions. The success, or in some cases the lack of success, of the CdS and TiO_2 semiconductors in photocatalytic processes underlines the need for a fundamental understanding of the basic properties of semiconductor particles.

This paper deals with the optical properties that have recently been the focus of investigations that utilized techniques heretofore not common in studies of semiconductor dispersions. The emphasis is placed on spectrofluorimetric, laser flash photolytic, and pulse radiolytic investigations.

2.0 CADMIUM SULFIDE DISPERSIONS

2.1 Photoluminescence

The presence of defects and impurities in semiconductor crystalline particles can lead to an alteration and creation of midgap electronic energy levels. Therefore the measurement of the time and wavelength dependence of the photo-emission should provide clues as to the chemical nature of these defects and impurities. Moreover, these will also influence electron transfer phenomena at the semiconductor/electrolyte interface via (i) the provision of midgap levels that can mediate electron transfer at the interface, or (ii) via interfering in the electron/hole recombination reaction that competes with the redox chemistry at the interface.

Cadmium sulfide is an important n-type semiconductor (bandgap ~2.4 eV) which has been investigated extensively in the last five decades owing to its luminescent characteristics (9) when excited by light photons ($E_{photons} > E_{bandgap}$) or by electron beams. Two principal emissions are observed: (a) a rather narrow, edge luminescence (often known as green luminescence) occurring at ~520 nm and (b) a broad red luminescence positioned at ~700-720 nm (10). The assignment of these emissions to a certain crystal defect or impurity has not been without controversy.

Kroger and Meyer (11) first suggested that the green emission resulted from the recombination of free holes and free electrons through excitonic states. Alternative assignments have been offered by others (12-14) who noted that the green luminescence results from the radiative recombination of a free carrier of one sign with a trapped carrier of the opposite sign. The work of Lambe and coworkers (14) suggested that the process involves at least one free carrier: a trapped electron recombines radiatively with a free hole. However, Collins (15) has claimed that the green emission results from the radiative recombination of a free electron with a trapped hole at a sulfur vacancy with the recombination level located at 2.41 eV below the conduction band; this suggestion stemmed (a) from quenching experiments of the emission by heat treatment in sulfur vapour, (b) from the introduction of green edge emission in CdS single crystals by 200-keV electron bombardment, and (c) from the time dependence of the emission lasting 10 μsec after excitation by a pulse of 1 MeV electrons. By contrast, Vuylsteke and Sihvonen (16) remarked that the green emission at 524 nm (2.38 eV) occurs through direct recombination of e_{cb}^- and h_{vb}^+ while the red luminescence at 718 nm is due to sulfur vacancies, this on the basis of variations in the luminescence intensity of the green and red bands and simultaneously in the variations of the photoconductivity of the crystals. Later, Kulp and Kelley (17) proposed another model for the CdS luminescence: a sulfur interstitial atom is the centre for edge emission and the sulfur vacancy is the centre for the red emission band. The field has been reviewed by Halsted and coworkers (18).

The matter of the photoluminescence of small CdS aggregates suspended in electrolyte solutions has been taken up recently by Ramsden and Gratzel (10) in order to determine the essential features of the emission and to compare it with measurements made on single crystals and polycrystalline layers. One feature of this particular work (10) was to determine how the preparation of CdS sols affects the photoluminescence. CdS sols were prepared from $Cd(NO_3)_2$ and $CdCl_2$ salts in the presence of $(NaPO_3)_6$ as the stabilizer; the respective emission spectra are depicted in Figures 1 and 2, respectively. Also, an understanding of the photoemissions should lead to an understanding of interfacial electron transfer at the surface of the particles, inasmuch as redox chemistry must compete with other processes occurring either at the surface or in the particle bulk.

The sol prepared from the nitrate salt showed both green (∿515 nm; ∿2.4 eV) and red (∿700 nm) luminescence (Figure 1). The former, weak edge luminescence was attributed (10) to direct recombination of free electrons and holes (reaction 1), in keeping with the earlier assignments of Vuylsteke and Sihvonen (16) on CdS bulk crystals. The red luminescence was ascribed to the

$$e_{cb}^- \text{ (CdS)} + h_{vb}^+ \text{ (CdS)} \rightarrow \text{CdS} + h\nu \text{ (∿515 nm)} \quad (1)$$

reaction of photogenerated holes with sulfur vacancies (V_S) (reaction 2).

$$h_{vb}^+ + V_S \rightarrow V_S^+ + h\nu \text{ (∿700 nm)} \quad (2)$$

Corroboration of this assignment to V_S obtained from addition of Cd^{2+} ions to the CdS sol that was found to increase the quantum yield of the red luminescence, Φ_{red} (Cd^{2+} ions increase the number of V_S on the particle surface); addition of S^{2-} led to smaller Φ_{red} because sulfide

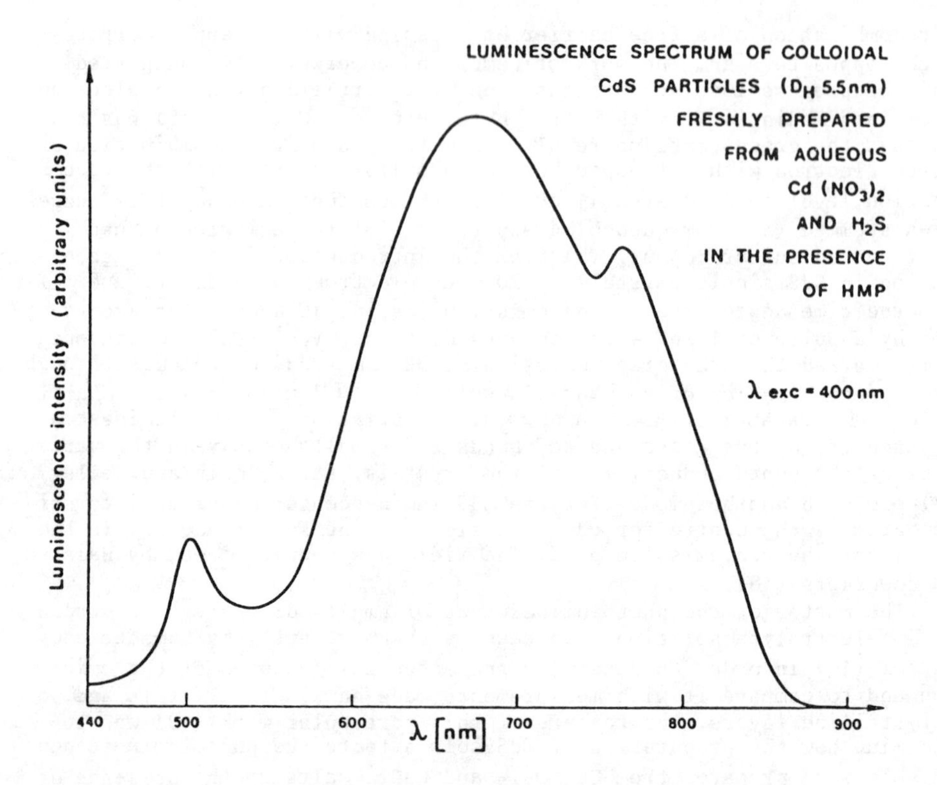

Figure 1.- Photoluminescence from freshly prepared CdS particles prepared from $Cd(NO_3)_2$, H_2S, and $(NaPO_3)_6$. From ref. 10.

ions increase the number of sulfur vacancies and also scavenge the h^+_{vb} (reaction 3) (10). A further interesting observation on this sol is that

$$h^+_{vb} \quad + \quad S^{2-} \rightarrow S^- \tag{3}$$

addition of methylviologen (MV^{2+}) totally quenches the red luminescence by either or both of two mechanisms: (1) reaction of MV^{2+} with conduction band electrons (reaction 4), and/or (2) charge transfer interaction of MV^{2+} with sulfur vacancies (reaction 5). More remarkable is that MV^{2+} induces a strong

$$e^-_{cb} \quad + \quad MV^{2+} \rightarrow MV^+ \tag{4}$$

$$V_S \quad + \quad MV^{2+} \rightleftarrows V_S^{\delta+} \cdots MV^{(2-\delta)+} \tag{5}$$

emission at 530 nm (Figure 3) (10), which neither of the two mechanisms above could explain. It was suggested that the red-shifted green emis-

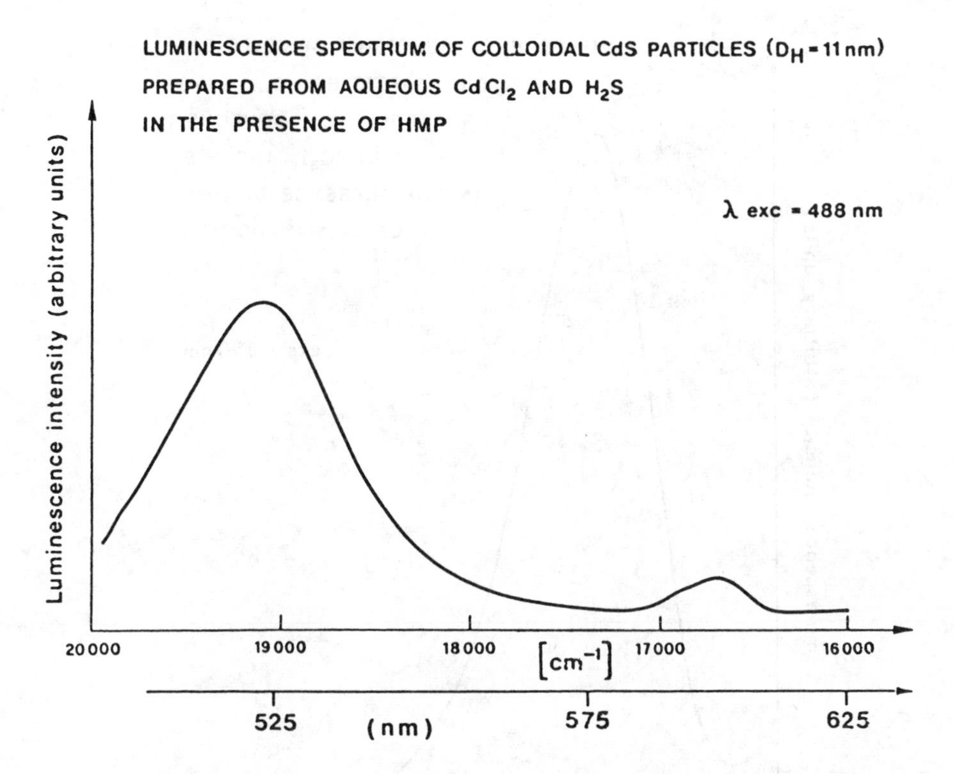

Figure 2.- Photoluminescence from CdS particles prepared from aqueous $CdCl_2$, H_2S, and $(NaPO_3)_6$. The feature at 16700 cm^{-1} is the Raman band of water. From ref. 10.

sion arises from cadmium vacancies, V_{Cd}^{2-}, formed by the presence of MV^{2+} adsorbed on the particle surface (reaction 6). To the extent that holes and V_{Cd}^{2-} are in

$$2(MV^{2+})_{ads} \rightleftarrows 2(MV^{+})_{ads} + V_{Cd}^{2-} + 2h^{+} + Cd_{aq}^{2+} \qquad (6)$$

equilibrium (equation 7), that is, V_{Cd}^{2-} is a trap for valence band holes, the green emission at 530 nm was believed to occur through interaction of e_{cb}^{-}

$$h_{vb}^{+} + V_{Cd}^{2-} \rightleftarrows V_{Cd}^{-} \qquad (7)$$

with V_{Cd}^{-} (equation 8).

$$V_{Cd}^{-} + e_{cb}^{-} \rightarrow V_{Cd}^{2-} + h\nu \quad (\sim 530 \text{ nm}) \qquad (8)$$

The CdS sol prepared from $CdCl_2$ showed a green luminescence centred

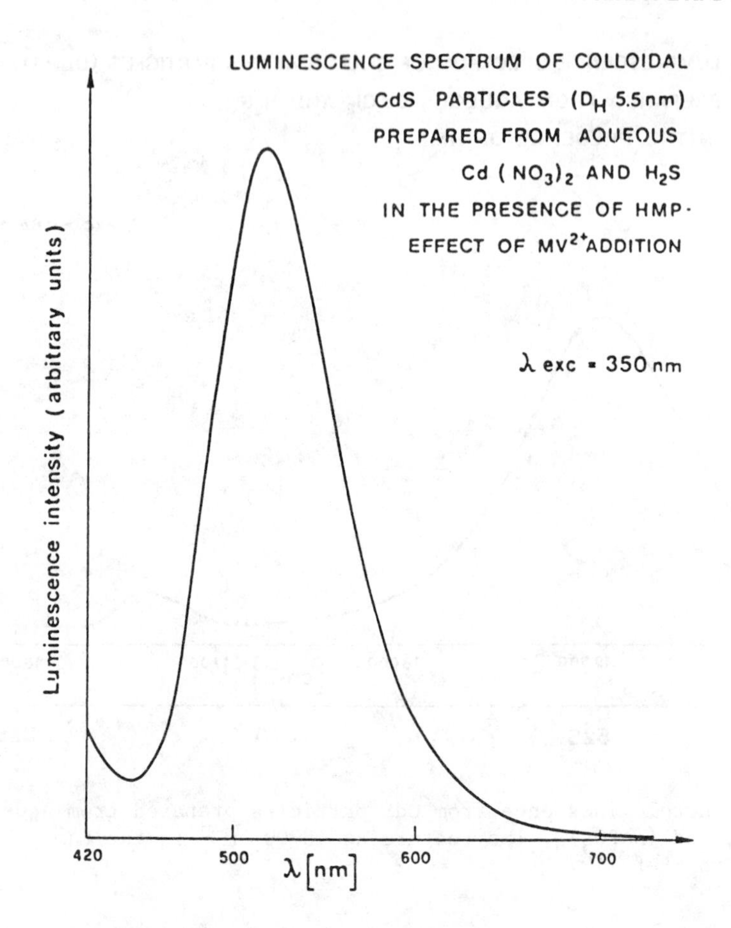

Figure 3.- Photoluminescence from CdS colloids prepared from aqueous $Cd(NO_3)_2$, H_2S, and $(NaPO_3)_6$ after MV^{2+} addition. From ref. 10.

at ∿522 nm which has been attributed to interstitial sulfur defects (I_S) formed through substitutional doping of CdS by Cl^- ions (10). Since e^-_{cb} and I_S are in equilibrium (10)(equation 9) this green emission is produced upon trapping of a photogenerated hole by I^-_S (reaction 10), consistent with the results of Kulp and Kelley (17).

$$I_S + e^-_{cb} \rightleftarrows I^-_S \tag{9}$$

$$h^+_{vb} + I^-_S \rightarrow I_S + h\nu \quad (\sim 522 \text{ nm}) \tag{10}$$

The photodegradation and the fluorescence of colloidal cadmium sulfide are influenced by dissolved substances at low concentrations (19). While O_2 does not quench the emission (∿620-660 nm), anions such as S^{2-}, Br^-, and I^-, the cations Tl^+, Ag^+, Pb^{2+}, and Cu^{2+} as well as nitrobenze' ne all quench the fluorescence when these are adsorbed on the CdS particle surface. In the case of anions, h^+_{vb}/anion pairs form at the surface of the colloidal particles before neutralization by e^-_{cb} takes place. Quenching by the cations and nitrobenzene occurs by scavenging of e^-_{cb}. Degradation of CdS by light occurs only in the presence of oxygen ($\Phi_{dec}\sim 0.04$) via oxidation of surface sulfides, S^{2-}_s, by h^+_{vb} (reaction 11) followed by reaction 12 (19). The $O_2^{\bar{\cdot}}$ radical anions mainly form H_2O_2 (20). The presence of Tl^+ or Pb^{2+} drastically promotes photodegra-

$$S^{2-}_s + h^+_{vb} \rightarrow S^{\bar{\cdot}}_s \qquad (11)$$

$$S^{\bar{\cdot}}_s + O_2 \rightarrow S_s + O_2^{\bar{\cdot}} \qquad (12)$$

dation ($\Phi_{dec}\sim 0.24$), while Cu^{2+} and Ag^+ inhibit degradation. These effects were ascribed (19) to electron scavenging by the adsorbed metal ions followed by re-oxidation of the reduced metal by oxygen or by h^+_{vb}.

Methylviologen also promotes photodegradation of CdS sols in aerated solution and both MV^{2+} and colloidal Pt efficiently quench the CdS fluorescence (20). In the presence of the former, photodegradation and emission have a common intermediate that is scavenged by MV^{2+} adsorbed (ads) at the surface of the colloidal particles. MV^{2+} scavenges photogenerated e^-_{cb} (reaction 13) and the resulting $(MV^+)_{ads}$ further transfers electrons to oxygen via reaction 14.

$$e^-_{cb} + (MV^{2+})_{ads} \rightarrow (MV^+)_{ads} \qquad (13)$$

$$(MV^+)_{ads} + O_2 \rightarrow (MV^{2+})_{ads} + O_2^{\bar{\cdot}} \qquad (14)$$

The result of these two reactions is an excess of h^+_{vb} which escape recombination and so more efficiently oxidize S^{2-}_s via reaction 11. The holes can also re-oxidize $(MV^+)_{ads}$ to $(MV^{2+})_{ads}$. Fluorescence quenching by platinum occurs through electron scavenging (equations 15 and 16) (20). Even when colloidal particles of CdS are embedded in a Nafion

$$H^+ + e^+_{cb} + Pt \rightarrow Pt\text{-}H \qquad (15)$$

$$Pt\text{-}H + O_2 \rightarrow O_2^{\bar{\cdot}} + H^+ + Pt \qquad (16)$$

polymer film (N125) the emission is statically quenched by MV^{2+} (21), and the luminescence lives for $\sim$1 μsec, at least three orders of magnitude longer than colloidal CdS (10). Water has a dramatic effect on the luminescence intensity of CdS in Nafion (21).

The effect of water on the CdS emission has been investigated for CdS particles dispersed in acetonitrile (22). Increasing the water content from 0.6 to 40% resulted in an increase of the average particle size from 15 to 23 Å with a concomitant decrease in their bandgap from 2.5 to 2.3 eV; both the absorption and the emission blue-shift with decreasing size of the particles consistent with the concepts of quantum size effects enunciated by Brus (23). Colloidal CdS particles, dispersed in acetonitrile/water solvent mixtures, show only the red luminescence ($\sim$660-740 nm depending on water content)(22). The origin of this luminescence was attributed to sulfur vacancy centres (equation 17) and the quantum yield of luminescence was ∿100 times greater in acetonitrile/10% water than in water indicating the higher concentration of radiative recombination centres in the former mixed solvent system (22). Henglein has also observed variations in absorption and emission spectra for different aggregates of different size in the same colloidal CdS sol (24).

$$h^{+}_{vb} + V^{+}_{S} \rightarrow V^{2+}_{S} + h\nu\ (\sim 660\text{-}740\ nm) \qquad (17)$$

Very small, colorless CdS colloidal particles (as small as 9 Å) have been prepared in 2-propanol solutions at -78°C and in aqueous media at ambient temperature with $(NaPO_3)_6$ present (24). These small particles possess little of the semiconductor properties (Q-state CdS) of the larger, macrocrystalline particles and the first excited state can be reached only by photon absorption in the ultraviolet (< 480 nm). The various absorption maxima in the spectrum were taken to arise from a size distribution of the colloids with preferential agglomeration numbers (24). These numbers represent integral multiples of a base number, m_o (about 8), which refers to the number of CdS 'molecules' in the basic colloid particle. Particles below a certain size showed only one broad fluorescence band at much longer wavelengths than the onset of absorption. Figure 4 depicts two fluorescence spectra of a colloidal CdS sol on excitation at two different wavelengths (305 nm and 410 nm);

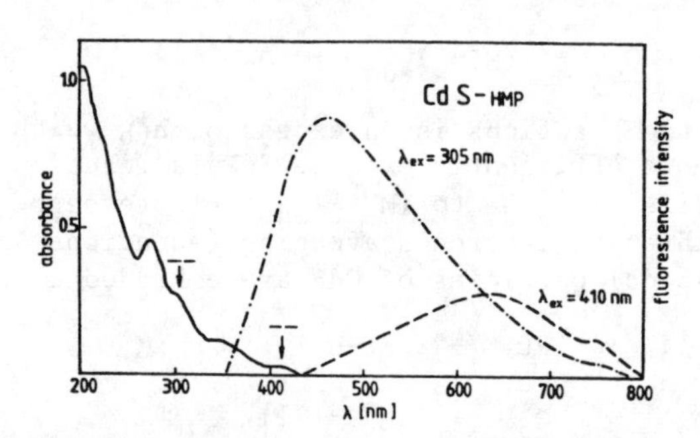

Figure 4.- Absorption spectrum and fluorescence spectra at two wavelengths of excitation for CdS particles prepared in the presence of $(NaPO_3)_6$. From ref. 24b.

the broad luminescence bathochromically shifts on increasing the wavelength of the exciting light. Aging favours an increase in intensity of the longer wavelength band; the onset of absorption also red-shifts. As the particle size increases, this fluorescence band is shifted to longer wavelengths and finally an additional, rather narrow band appears at ∿460 nm along with a broad band at ∿630 nm (24b). The multi-exponential luminescence decay in the narrow band occurred in <1 nsec, while the decay of the broad band took place from several nsec to ∿100 nsec. The 460-nm emission was attributed to direct fluorescence from the exciton-like state produced by light absorption, while the broad band from recombination of charge carriers trapped in defect states at the particle surface. Electron acceptors such as MV^{2+} only quench the broad emission band (24c).

The effect of the synthetic preparation of colloidal CdS on the photochemical behaviour of the sols has also been studied by Thomas and coworkers (25). Figure 5 illustrates the 77 K emission spectrum of a CdS aqueous powdered dispersion and of two colloids prepared in the presen-

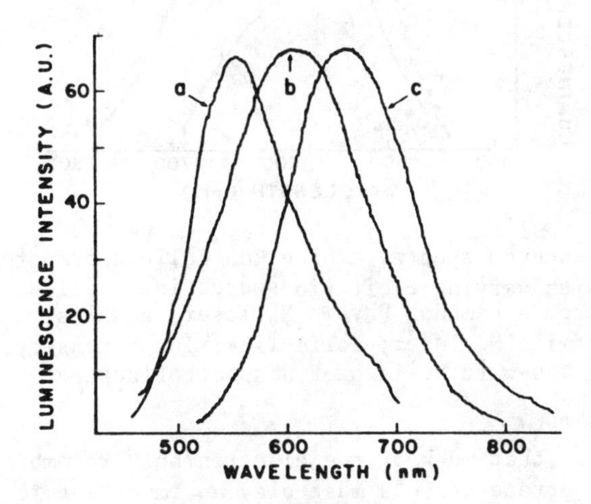

Figure 5.- Steady-state luminescence spectra 77 K: (a) CdS dispersion; (b) CdS colloid precipitated at ca. 0°C (ice bath); (c) CdS colloid precipitated at 70°C. From ref. 25.

ce of sodium dodecyl sulfate (SDS). The differences in the emission maxima indicate that emission originated from different midgap energy states owing to differences in the nature of the luminescent centres produced in the formation of the particles, under the different conditions used. Also, the luminescence lifetimes at 77 K are >0.5 μsec for the colloids but for the powder dispersion τ_{lum} <1 nsec; at room temperature, τ_{lum} of all the CdS samples was <1 nsec. Moreover, the emission wavelength maxima (at 77 K) for a particular CdS colloidal sample was dependent on the excitation light intensity (see Figure 6); no such effect was observed for the polycrystalline powder dispersion (25). Increasing the intensity of the excitation light resulted in a noticeable blue shift in the emission maximum. These differences are probably due to the large number of defect sites which can serve as either electron or hole traps whose energy levels are located within the bandgap such that recombination leads to emission at longer wavelengths than direct e^-_{cb}/h^+_{vb} recombination (edge emission). At the high excitation light intensities, the trapping sites are saturated and the major position of the luminescence subsequently arises from direct transition between the conduction band and valence band (25).

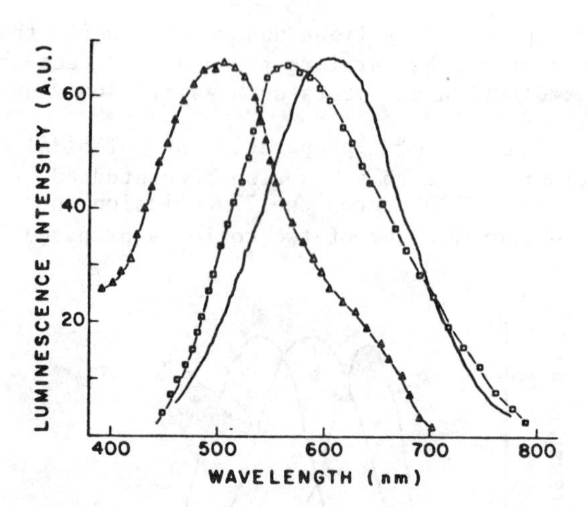

Figure 6.- Luminescence spectra of the CdS colloid prepared at 0°C obtained at 77 K with varying excitatio source intensities: triangles, high intensity from a Lambda Physik N_2 laser; squares, moderate intensity from a Nitromite N_2 laser; solid line, low intensity, steady state excitation froma 150-watt Xe lamp of a spectrofluorimeter. From ref. 25.

To the extent that much of the electron/hole recombination occurs on the surface, surface defects must play an important role in the photophysics of recombination (26). For an aqueous CdS powder dispersion, the emission maximum occurs at 560 nm while fluorescence maxima for CdS colloids, stabilized with SDS or cetyldimethylbenzylammonium chloride occur at 620 nm and 650 nm, respectively. The onset of the absorption at 520 nm (2.4 eV) of all the CdS samples was not affected by the surface effects. The effect of light intensity on the emission from CdS samples was also re-affirmed (26). The electron/hole pairs, formed by excitation with high light intensities, for a large part reside within the bulk of the CdS particle and these gave rise to the emission at 525 nm, while the electron/hole pairs at the surface recombine to give luminescence at >600 nm.

An important question must now be asked on the basis of the above findings. Does hight intensity, in fact, produce these defect sites (emission centres, surface states, ...) or are the sites already available initially on formation of the colloidal particles? The studies of Thomas and coworkers (25,26) would seem to suggest that the surface states (electron and/or hole traps) are available *a priori* and excitation by ultra-bandgap energies populates these surface states predominantly by low intensity light, and totally by high intensity light, after which bulk luminescent centres become populated that give rise to edge-type luminescence (see discussion above and ref. 10). Recent investigations on CdS single crystals by Brodin et al. (27-29), however, seem to indicate otherwise. Under two-photon excitation with a ruby laser pulse (power density 10-100 MW/cm^2) structure defects formed in the bulk of the single crystal (28,29). As well, interaction of the ruby laser radiation with a CdS single crystal (surface damage initiated with power densities of ∿70 MW/cm^2 produced structure defects (largely of the donor type) in a thin sub-surface layer (∿1 μm thick) that affect the optical properties of the crystal. It appears that laser irradiation induces the escape of the more volatile component, namely, the surface atoms from interstitial positions. The emission line at 488.5 nm (4.2 K) assigned to interstitial sulfur was quenched by laser illumination. The heavier cadmium atoms seemed to be less affected by laser radiation; it led to cadmium vacancies and enhancement of the 488.9 nm line (27). Thus, high light intensity irradiation can generate native point defects in a crystal thereby creating new recombination channels for the electrons and holes. This fact should not be precluded in similar experiments with quasi-spherical colloidal particles.

It is clear that the controversies have not ended. The many experimental variables (mode of preparation, size distribution, impurities, dopants like Cu^{2+}, etc...) and the resulting intrinsic properties of CdS, not to mention the various interpretations, have made the study of CdS semiconductor particles an exciting one, where much is being learned about the details of semiconductor photophysics.

2.2 Laser Flash Photolysis

CdS is an intrinsic n-type semiconductor with a large predominance of thermal electrons over holes. The linear dependence of the photoluminescence with incident light flux has shown that only one carrier has a transient concentration above the pre-existing thermal level (30). This carrier is the h^+; the excess holes have short lifetimes owing to their interaction with the sulfide of CdS (lattice dissolution) and/or reaction with excess sulfide ions on the surface. The electron/hole luminescence lifetime of aqueous crystallites of CdS is $\leq 10^{-11}$ sec (31-34). The notion has also been made that luminescence quenching by added species (electron or hole scavengers) implies recombination emission is governed by surface kinetics (30). This is one mode of quenching; equally true is quenching by destruction of luminescent centres in the subsurface (27).

2.2.1. Nanosecond Laser Flash Photolysis

We have noted earlier that in photocatalytic processes employing semiconductor powders it is difficult to probe the primary events occurring upon light excitation. The monodispersed semiconductor particles of small enough dimension yield optically transparent dispersions with negligible Railegh scattering such that kinetic analysis of interfacial electron transfer processes and primary events in semiconductor/electrolyte systems and in naked semiconductor can be carried out by laser photolysis techniques (35). We will now entertain some recent experiments that have attempted to answer some of the intriguig questions in photocatalysis.

Hydrated electrons form in the solution bulk by laser illumination

(347.1 nm; 3.57 eV) of a CdS colloid, and the yield is increased by excess sulfide anions; the quantum yield is 0.15 e^-_{aq}/photon absorbed in the presence of 0.002 M Na_2S (36). Photoelectron emission was a monophotonic process under the conditions of the experiment. The absorption signal observed immediately after the laser flash at 400-800 nm (spectrum of e^-_{aq}) decayed with a half life of ∿2-3 μsec. At the largest laser power possible, the absorption signal decayed in two steps; the half-life for the faster step was 0.3 μsec, and 2-3 μsec for the slower one (36). This longer lifetime was attributed to some reaction of the photoemitted electrons with a component or an impurity of the solution. Electron emission from CdS to the solution bulk with the 3.57 eV laser light excitation used is energy short by ∿1 eV, the process requiring $\geq$ 4.6. Henglein (36) suggested that with the positive holes scavenged by sulfide anions, so many electrons are being stored on the particle for a short time that a tremendous cathodic shift of all the electronic energy levels in the colloid (conduction and valence bands) takes place with respect to the solution bulk. At this point, some of the generated electrons are on sufficiently high negative potentials (3.57 eV is now sufficient energy) for injection into the aqueous phase to become possible. This excess of electrons on CdS can lead to the cathodic dissolution of CdS via reaction 18 (37).

$$CdS + 2e^- \rightarrow Cd + S^{2-} \qquad (18)$$

Using the single photon counting technique, Gratzel and coworkers (38) have determined the decay time of the luminescence from CdS colloids to be ∿0.3 ± 0.2 nsec; in fact it is much less than this value (see below). From such a short lifetime of the electron/hole pair in a CdS particle, the electrons and holes can only be involved in redox processes if the reducible of oxidizable species are adsorbed to the surface of the particle. Diffusion of species in the solution bulk to the semiconductor particle surface is simply too slow to compete with e^-_{cb}/h^+_{vb} recombination. These ideas were borne out by experiments involging MV^{2+} as the electron acceptor; MV^+ is promptly formed within the pulse and no growth of the 602-nm absorption of MV^+ was observed (38). The complementary valence band process was the photoanodic corrosion of CdS (reaction 19) in the absence of hole scavengers.

$$CdS + 2h^+ \rightarrow Cd^{2+} + S \qquad (19)$$

The yield of MV^+ increased rapidly in the presence of such hole scavengers as adsorbed sulfide anions, thereby increasing significantly the lifetime of the conduction band electrons and resulting in a more efficient interfacial electron transfer to adsorbed MV^{2+} (38).

Kuczynski and Thomas (26) also noted prompt formation of MV^+ (τ<5 nsec (19,20,39)) upon 337-nm or 490-nm pulsed laser excitation of a CdS/SDS/MV^{2+} colloid. Irradiation at high light intensities (0.1 J/pulse, λ 490 nm) and high MV^{2+} concentrations resulted in the formation of two species with absorption maxima at 610 and 540 nm and these species exhibited different kinetics (Figure 7A). Differences in the decay

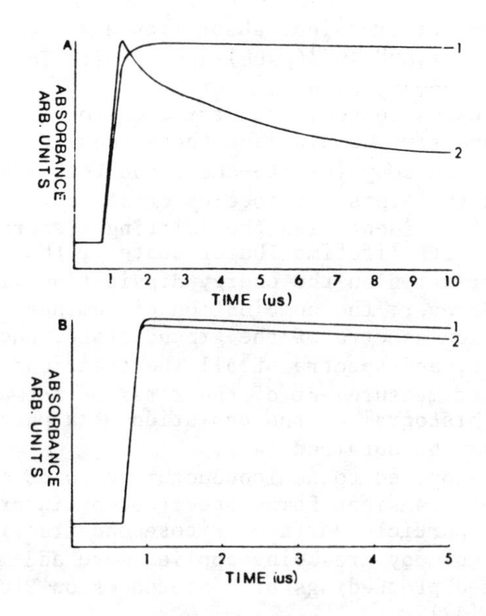

Figure 7.- Pulsed laser studies of MV^+ formation and decay in CdS/SDS colloids. (A) high intensity laser pulse (0.1 J/pulse, 490 nm): transient absorption observed at 540 nm (1), and at 610 nm (2). (B) measurement as in (A) at low laser intensity (<< 0.01 J/pulse) transient absorption observed at 540 nm (1) and at 610 nm (2). From ref. 26.

curves arise from two distinct species formed on the colloidal particles surface; the absorption of one of these decreases at λ 610 nm while that of the other species grows at 540 nm (26). The former is the characteristic absorption of MV^+ monomer whereas the absorption decay at 540 nm is due to the dimer $(MV^+)_2$ (40) (also see below). At low laser intensity, the decay of the 610-nm species decreases and the two decay curves become identical (Figure 7B). Formation of the dimer $(MV^+)_2$ was efficient only at high laser intensities. Moreover, the data suggested movement of MV^+ monomer on the CdS/SDS particle surface over periods of >260 μsec (26). Such long time may have consequence in MV^+ back-reacting with h^+_{vb}.

2.2.2. Picosecond Laser Flash Photolysis

Rentzepis (41) has recently pointed out that a single type of measurement - may it be quenching of some luminescent level or of some radicals with reactive species, or fluorescence or quantum yield measurements - cannot provide the necessary data needed to identify directly the electronic states involved in processes and the rates that govern their population and depletion. The advent of picosecond (10^{-12}sec; psec), mode-locked laser instrumentation has made possible detailed experimental studies by means of transient absorption and luminescence spectroscopy on a timescale (10^{-9}–10^{-12} sec) appropriate for the observation of the fundamental primary events.

Many relaxation processes in semiconductor/electrolyte systems occur in the sub-nanosecond time domain, and these require both luminescence and absorption spectroscopy (excite-and-probe technique) to identify the mechanism(s) and the transient species or states. Determination of the emission spectrum identifies the emitting electronic levels and affords an estimate of its lifetime. Unfortunately, the emission spectrum provides no information on the energy dissipation via nonradiative decay channels. However, the combination of luminescence data and time-resolved absorption spectra of the ground state, showing its depletion and repopulation, and spectra of all the transient species or states, affords a direct measurement of the rates of formation and decay such that a unique histogram of the evolution of a photochemical or photocatalytic event can be obtained (41).

This novel technique applied to semiconductor colloids complements well with the technique of transient Raman spectroscopy in probing events and kinetics on the particle surface. Picosecond transient emission and absorption spectroscopy are being applied more and more in this area. Recent published proceedings of conferences on Picosecond Phenomena attest to this (42).

High-density, nonequilibrium electrons and holes can be produced in semiconductors through interaction with psec laser pulses. The real-time observation of relaxation, thermalization, and recombination of these non-equilibrium carriers in direct gap semiconductors such as CdS necessitates psec time resolution. Suffice to note the recent studies of Saito and coworkers (43) on the dynamics of the free carrier-exciton system in a highly excited CdS crystal at 10 K and 70 K, and those of Huppert et al. (44,45) on the time-resolved photoluminescence on n-type CdS single crystal photo-electrodes. The latter studies are particularly significant in our understanding of the photophysics in single crystals of CdS, the results of which might lead to a better view of the photophysics and events on CdS colloid particles and subsequently to powder dispersions used in photocatalysis.

The work of Huppert and coworkers (44,45) demonstrated that the decay times of the band edge luminescence is very sensitive to the crystal surface conditions. Untreated CdS when excited by psec laser pulses (power densities ~ 1 GW/cm^2) produced a luminescence which decayed with $\tau < 10$ psec. Surface modification by mechanical polishing increased τ to ~ 50 psec. More dramatic changes resulted upon chemical etching of the crystal surface with acids; the emission decay was biphasic, $\tau_1 = 100 \pm 20$ psec and $\tau_2 = 700 \pm 100$ psec. At low temperatures, CdS crystals showed an additional emission (Figure 8) at 590 nm (2.1 eV) ascribed to interstitial, singly charged cadmium ions located close to some other defects that contribute to their formation (e.g., O^{2-} instead of S^{2-}). The decay rate of this sub-bandgap emission of a chemically etched CdS crystal at ambient temperature was 16 nsec, longer than the band luminescence decay rate by an order of magnitude ($\tau \sim 1$ nsec). The detection of such and similar hole traps within the bandgap is significant. In the present case it means that a significant concentration of bulk hole

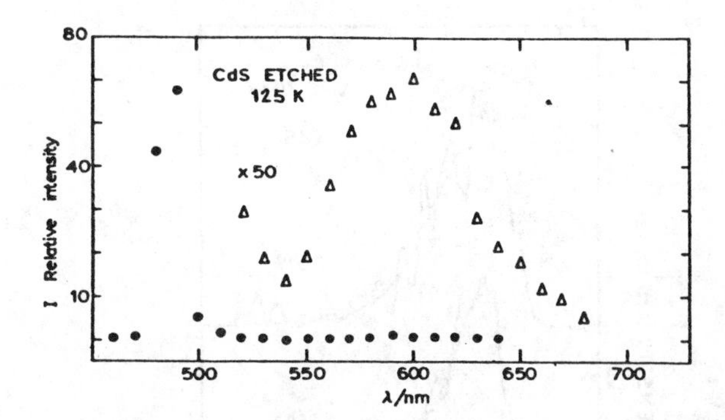

Figure 8.- Spectrum of the photoluminescence transients measured for a CdS single crystal in air at 125 K. From ref. 45.

traps existis in CdS samples that are distributed ∿0.3 eV above the valence band (45). That such a surface stata may be involved in hole transfer to hole acceptors in solution located energetically well above the valence band is worth considering (45). It is not clear, however, whether the laser pulse might have created the defect; the possibility was not discussed.

We noted above that nanosecond flash photolysis studies (38) of CdS sols containing MV^{2+}, adsorbed on the CdS surface, showed that the elementary process of reaction 13 was too fast to be resolved by this technique. By contrast, the kinetics of the process were resolved when CdS was replaced by TiO_2 sols. Our recent work (33,34) has addressed this question and in addition has attempted to time-resolve the fast electronic processes indicated in equation 20.

$$e^-_{cb} + h^+_{vb} \longrightarrow \left[\begin{array}{c}\text{green emission intermediate(s)}\\ \\ \text{red luminescence intermediate(s)}\end{array}\right] \longrightarrow h\nu \qquad (20)$$

Figure 9 illustrates the time-dependent evolution of the luminescence (λ >520 nm) of three CdS sols in Ar-purged aqueous/triethanolamine solutions. The risetime of the luminescence signal was pulse-limited and the decay time was estimated as $\leq$ 30 psec. In free colloids, the elementary processes can be described by reaction 21:

$$CdS \xrightarrow[355\ nm]{h\nu} e^-_{cb}(CdS) + h^+_{vb}(CdS) \xrightarrow{30\ psec} CdS + h\nu \qquad (21)$$

Alternatively, the fast emission decay could have arisen from recombination of the h^+_{vb} with electrons trapped by interstitial sulfur, I_S^-

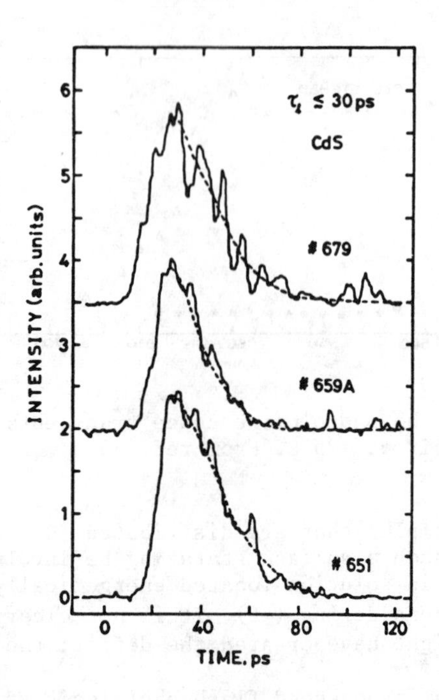

Figure 9.- Luminescence intensity vs. time (psec) plots showing formation and decay of the emission after photo-excitation of three CdS sols as those used in ref. 10. From ref. 34.

(reaction 10) or with electrons in sulfur vacancies, V_S (reaction 2). In Figure 10 is depicted the temporal course of the luminescence of CdS in the presence of MV^+/MV^{2+}. Four features are noteworthy: (i) the laser pulse, (ii) a fast decaying transient emission, τ = 0.9 nsec, (iii) a concomitant luminescence rise, with $\tau \sim 1$ nsec, and (iv) a slower decaying emission, τ = 3.2 nsec.

From the short lifetime of the electron/hole pair in CdS sols, the e^-_{cb} can only reduce species adsorbed on the particle surface inasmuch as diffusion displacement required for reaction with acceptor species in the solution bulk is too slow to effectively compete with e^-_{cb}/h^+_{vb} annihilation. In the presence of MV^+/MV^{2+}, the luminescence from argon-purged CdS sols should be that arising, directly or indirectly, from interaction between e^-_{cb} or h^+_{vb} with MV^{2+} or MV^+, respectively. We have identified (33,34) the fast decay component with interaction of a photogenerated hole with MV^+ (reaction 22)

$$h^+_{vb} + (MV^+)_{ads} \rightarrow (MV^{2+})_{ads} + \text{emission} \tag{22}$$

present in the sol under the experimental conditions used. In a sense, $(MV^+)_{ads}$ can be considered as an electron trapped by MV^{2+} adsorbed on

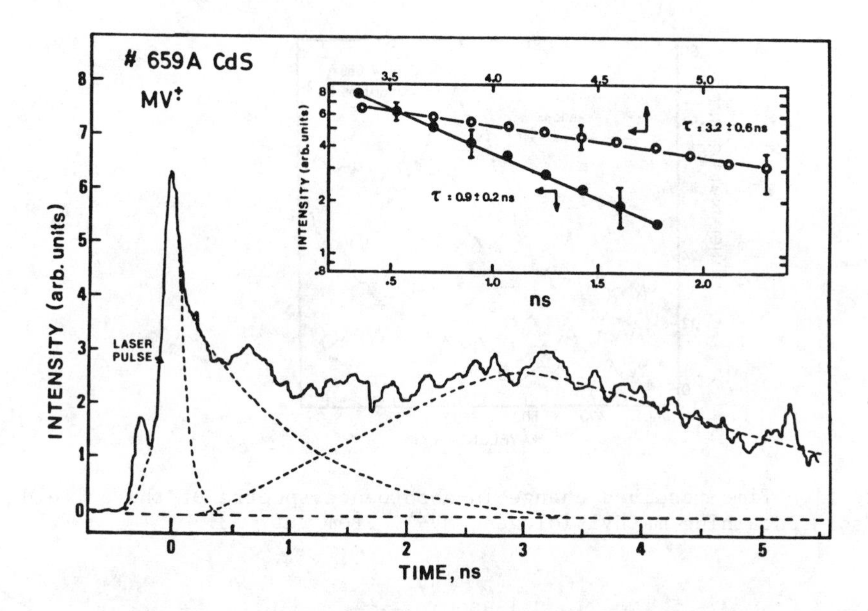

Figure 10.- Streak camera trace of the time-resolved luminescence intensity profile of a CdS colloid in the presence of MV^+. The inset shows the kinetic analysis of the two decay profiles. Solid curve, experimental data; dashed curve, deconvoluted data. From ref. 33.

the particle surface. The slower decaying emission component was attributed to reaction 8, where V_{Cd}^{2-} denotes a cadmium vacancy whose formation (10) is induced by MV^{2+} and V_{cd}^{+} denotes a trapped valence band hole by V_{Cd}^{2-}. To the extent that the transient emission decay of reaction 22 and the risetime of the luminescence in reaction 8 occurred in near identical times suggested that V_{Cd}^{-} is formed in ∿1 nsec through reaction 23 or its equivalent:

$$2(MV^{2+})_{ads} \rightleftarrows 2(MV^{+})_{ads} + V_{Cd}^{2-} + 2h^{+} + Cd_{aq}^{2+} \qquad (23)$$

Time-resolved change-in-absorbance spectra of CdS in the presence of MV^{2+} (Figure 11) showed a rapid increase in transient absorption with time and formation of a distinct absorption band that peaks at ∿530 nm. This absorption growth was seen at both 530 nm and 596 nm at 1 nsec. At 2 nsec, the band at 530 nm formed but transient decay became evident at 596 nm. The kinetics at the two wavelengths indicated formation of transients occur in 1 nsec. These transient absorptions were identified with formation of reduced methylviologen, MV^+ (λ_{max} ∿600 nm) and with dimer formation of MV^+, namely $(MV^+)_2$, according to reaction 24:

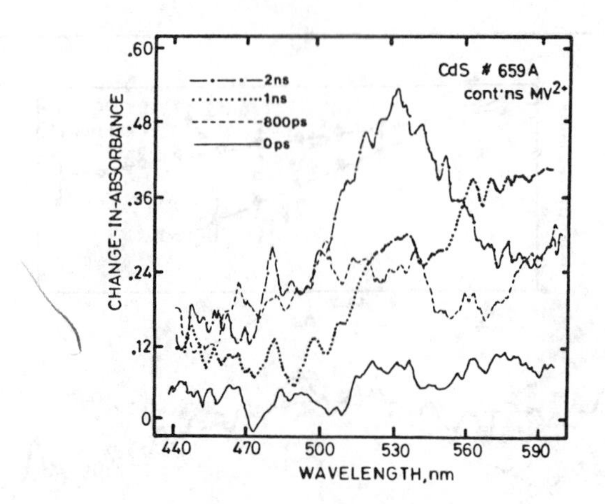

Figure 11.- Time-resolved change-in-absorbance spectra of the colloid 659A sol containing methylviologen, MV^{2+}. From ref. 33.

$$e^-_{cb} + (MV^{2+})_{ads} \xrightarrow{\tau_r = 1 \text{ nsec}} [(MV^+)_{ads} \rightleftarrows 1/2(MV^+)_2] \quad (24)$$

$$\qquad\qquad 600 \text{ nm} \qquad 530 \text{ nm}$$

To our knowledge, this was the first spectral kinetic evidence of the formation of the dimer, $(MV^+)_2$ (33,34).

3.0 TITANIUM DIOXIDE DISPERSIONS

Titanium dioxide is also an important n-type semiconductor that has received much attention in the photosplitting of water and photo-oxidation of alcohols. Unfortunately, the photophysics of TiO_2 sols and powders are not as well known as those of CdS. Nevertheless, much work has been undertaken lately and the chemistry of TiO_2 has been reviewed extensively (46).

Light excitation of TiO_2 (anatase, rutile, amorphous) aqueous dispersions with energies greater than bandgap ($\geq$ 3.2 eV) leads to charge separation of electrons and holes (equation 25) in a manner analogous to single

$$TiO_2 \xrightarrow[\geq 3.2 \text{ eV}]{h\nu} e^-_{cb}(TiO_2) + h^+_{vb}(TiO_2) \quad (25)$$

crystal semiconductor electrodes (47). The separated charge carriers at the particle surface can recombine radiatively and nonradiatively in the absence of adsorbed redox species. The nature and reactivity of the particle surface are important parameters in photocatalysis. One approach taken to probe surface events has been to study the photophysics of adsorbed fluorescent molecules on the surface. Below we present some recent, but by no means exhaustive, work on this approach (38,48-57). Moreover, we report on some work recently begun to probe the photophysics of "naked" TiO_2 colloidal particles (58a,b) and on electron transfer reactions at TiO_2 colloids followed by means of pulse-radiolysis technique (59).

3.1 Photoluminescence

Amorphous TiO_2 shows a green emission at 77 K when excited within the bandgap (60); no emission was observed at ambient temperature from aqueous suspension of amorphous TiO_2 (57). Chandrasekaran and Thomas (57) report that commercial TiO_2 and crystalline colloidal TiO_2 sols show a weak but broad emission (Φ_{em} ∿10^{-3}) centred at 375 nm (∿ bandgap 3.2 eV emission) and tailing to ∿700 nm (see Figure 12). The low

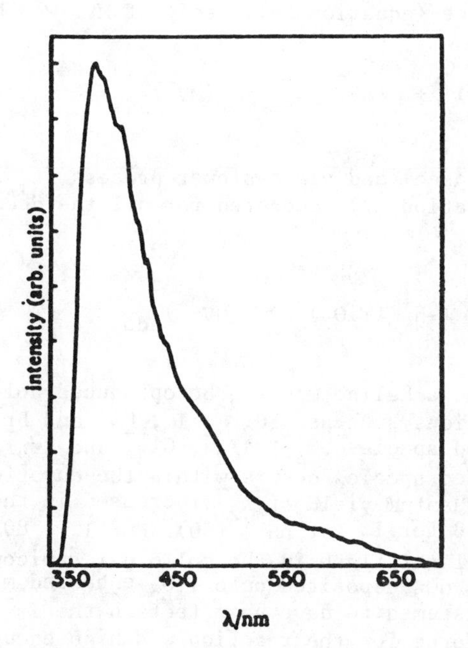

Figure 12.- Emission spectra of TiO_2 particles suspended in water; excitation wavelength was 320 nm (see text). From ref. 57.

energy tail of the luminescence was ascribed to sub-bandgaps caused by crystal defects. Interestingly, oxygen adsorbed on the particle surface appeared to have no effect on the emission and thus was not reactive with the hole/electron pair (57). We have also obtained similar spectra as that of Figure 12 from an aqueous colloidal sol of TiO_2 (6 g/L; particle size ∿100-200 A; ambient temperature) available from another study. <u>However</u> we cannot agree with the above interpretation inasmuch as, under identical conditions, both a milk colloidal solution and a diffuser plate gave identical spectra as that of Figure 12. We suggest that what has been taken as an emission spectrum of TiO_2 is in fact scattered light. This notwidthstanding, TiO_2 sols do emit as indicated by the luminescence decay upon exciting the sol with psec 355-nm laser pulses; the emission lifetime is ∿60-100 psec measured with a 550-nm cutoff filter and a 10-psec time resolution Streack Camera (61). In addition, O_2 is chemisorbed on the surface of TiO_2 colloidal particles and consumes part of the photogenerated electrons to form $O_2^{\cdot-}$ (38).

3.2 Nanosecond Laser Flash Photolysis

Photogenerated electrons by nsec laser pulse excitation of TiO_2 sols in the presence of methylviologen (MV^{2+}) and in the absence of any stabilizer led to the prompt formation of MV^+ (in < 6 nsec) (57). This radical cation is formed by electron transfer to MV^{2+} species adsorbed on the particle surface (equation 26); decay of MV^+ was biphasic proceeding via a fast

$$e^-_{cb}(TiO_2) \;+\; (MV^{2})_{ads} \;\rightarrow\; (MV^+)_{ads} \tag{26}$$

process ($t_{1/2}$ ∿60 nsec) and via a slower process ($t_{1/2}$ ∿1 μsec). The fast process (equation 27) accounted for all the decay in acidic pH but only for

$$(MV^+)_{ads} + h^+_{vb}(TiO_2) \;\rightarrow\; (MV^{2+})_{ads} \tag{27}$$

60% of the decay in alkaline media. Photoproduced holes can be scavenged by adsorbed anions such as $S_2O_8^{2-}$, I^-, Cl^- and Br^- to give the corresponding oxidized species SO_4^{-}, I_2^-, Cl_2^- and Br_2^- (50,57). Hole transfer to the adsorbed species occurs within the duration of the 10-nsec laser pulse; the quantum yield of X_2^- increased in the order Cl_2^- < Br_2^- < I_2^- attaining 0.8 for I_2^- at pH 1 (50). For I_2^-, 80% of the holes produced by the nsec laser flash in the colloidal semiconductor are scavenged by iodide. RuO_2 deposited onto TiO_2 enhanced markedly Cl_2^- and Br_2^- formation but seemed to have no effect on the I_2^- yield. In this case the driving force for the reaction was high enough for efficient hole transfer even in the absence of RuO_2 deposits (50).

Earlier, Duonghong and coworkers (38) had shown that pulsed laser excitation of colloidal TiO_2 leads to long-lived e^-_{cb} whose rate of reaction with MV^{2+} is strongly influenced by the MV^{2+} concentration (Figure 13) and by the pH of the aqueous suspension (Figure 14). The pH effect on the yield of MV^+ was exploited to derive the flat-band potential of the particle for which a value of E_{fb} = 0.130 - 0.059 (pH) V (vs. NHE) was obtained. The second-order rate constant for reduction of MV^{2+} by excited TiO_2 particles was $1.2 \times 10^7\ M^{-1}s^{-1}$ at pH 5 (38). Under the conditions of the experiments (PVA added as a stabilizer to colloidal TiO_2 sols at pH >3) (38), MV^+ formed rather slowly (inset of Figure 13

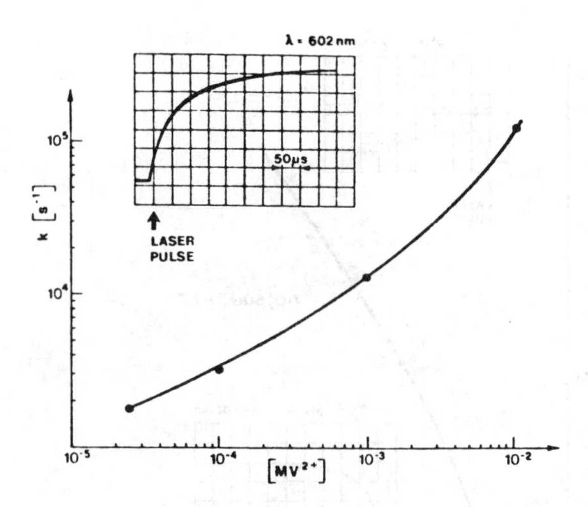

Figure 13.- 347-nm laser pyotolysis of colloidal TiO_2 (500 mg/L, pH 5). The observed rate constant for formation of MV^+ is plotted against $[MV^{2+}]$. Insert shows oscilloscope trace depicting time course of 602-nm absorption for $[MV^{2+}] = 10^{-3}$M. From ref. 38.

shows growth of MV^+ absorption occurs in the microsecond to millisecond time domain). This indicated that electron transfer occurs from TiO_2 to MV^{2+} in the solution bulk and diffusing towards the particle surface. In fact, in the absence of any stabilizer, only ~7% of MV^{2+} is adsorbed to the TiO_2 particle surface (57); obviously, the presence of PVA precludes MV^{2+} surface adsorption.

The effect of $[MV^{2+}]$ on the yield of MV^+ (Figure 15) has been used to determine the concentration of electron/hole pairs formed by the laser excitation (38). At relatively low electron acceptor concentration, $[MV^{2+}] \leq 2 \times 10^{-5}$ M, the concentration of e^-/h^+ pairs was greater than $[MV^{2+}]$ and all of the acceptor species were reduced. At night $[MV^{2+}]$ where $[MV^{2+}] >> [e^-/h^+]$, reduction was incomplete and a limit was reached at $[MV^+] \sim 2.6 \times 10^{-5}$M $= [e^-_{cb}]$. This number would be dubious (e^-/h^+ pair recombination can occur via other channels) were it not for the fact the quantum yield of MV^+ formation is approximately unity (38). One additional feature of Figures 13 and 14 is worth noting, and this is the lifetime of the conduction band electrons in the TiO_2 particle under the experimental conditions used. Inasmuch as the growth of MV^+ occurred over a period of at least several milliseconds, the inference might be made that the lifetime of the e^-_{cb}/h^+_{vb} pair might also be in the milliseconds. This cannot be correct because it neglects the possibility that the hole may be removed rapidly from the reaction sphere by either being trapped in some surface state or might be used in water oxidation. As noted later (58), the lifetime of the electron/hole pair is much shorter and lies in the sub-nanosecond time domain.

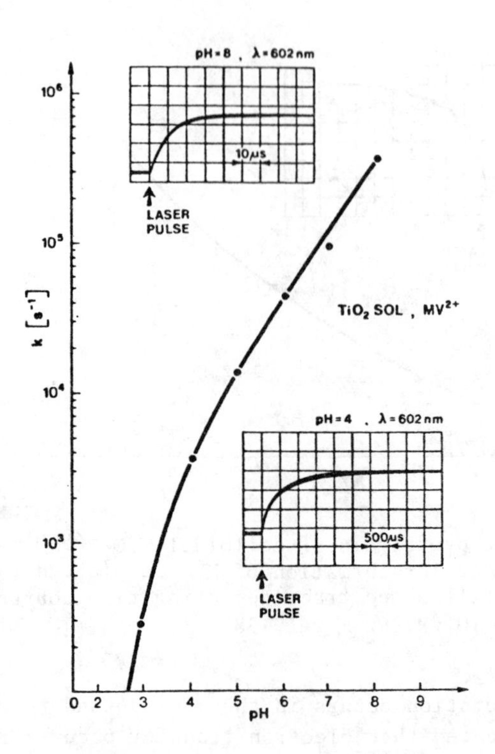

Figure 14.- Laser photolysis of TiO_2 colloids (500 mg/L) at various pH values. Effect of pH on the rate constant of electron transfer from the conduction band to MV^{2+}; $[MV^{2+}] = 10^{-3}$M. Insert shows oscilloscope traces illustrating growth of MV^{+} absorption at pH 8 and 4. From ref. 38.

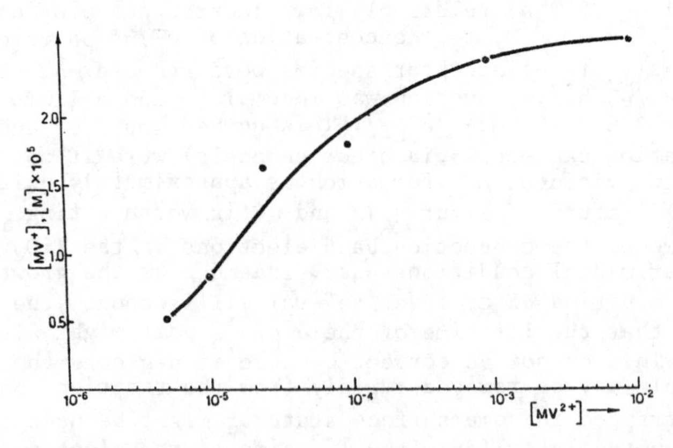

Figure 15.- Amount of MV^{+} generated after completion of e^-_{cb} reaction with MV^{2+}; $[TiO_2]$ = 500 mg/L, pH 5, $[MV^{2+}] = 10^{-3}$M. From ref. 38.

Nanosecond laser flash excitation of TiO_2 sols containing a hole scavenger (PVA, pH 10) produces excess electrons and the absorption of these is observed immediately after the flash (Figure 16) (62-64). The spectrum consists of a broad band in the visible region, the maximum occurring at ∿630 nm; the absorption signal was stable for seconds (insert of Figure 16). When the TiO_2 sol contained an adsorbed electron scavenger (Pt, pH 2.5), the absorption of excess positive holes could be traced immediately after the flash (Figure 17) (63). The broad ab-

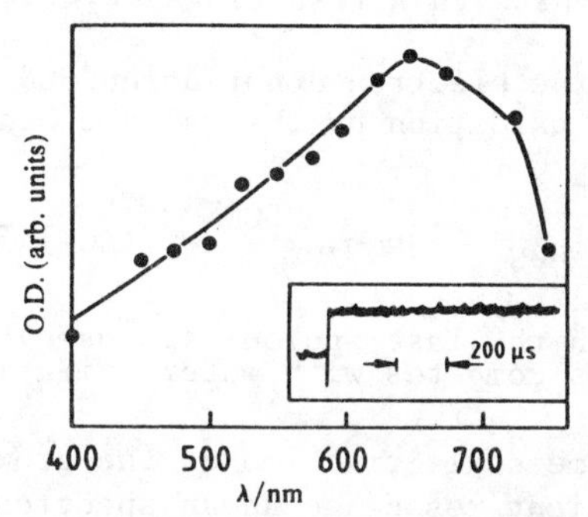

Figure 16.- Absorption spectrum of a laser-flashed solution of 6.3 x 10^{-3}M TiO_2 containing 5 x 10^{-3}M PVA and time profile of the absorption signal (inset) at pH 10. From ref. 63.

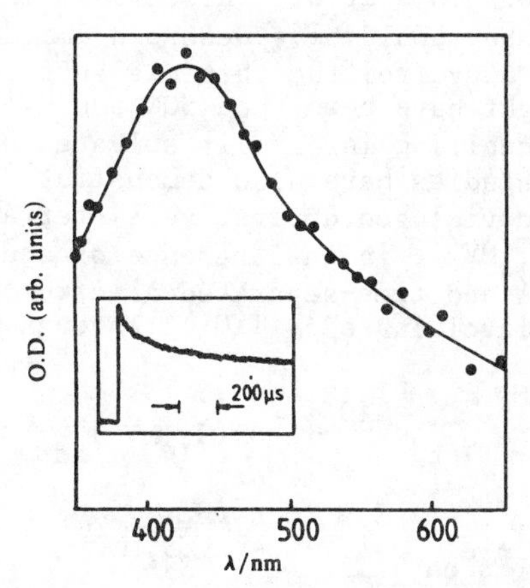

Figure 17.- Absorption spectrum of a laser-flashed TiO_2/Pt colloid (3.8 x 10^{-3}M TiO_2 and 1.6 x 10^{-5}M Pt) and decay of the absorption (inset) at pH 2.5. From ref. 63.

sorption at 430 nm decayed after the nsec flash in a multi-exponential manner indicating the participation of various first order processes of different lifetimes. O_2 had no effect on the h^+ spectrum bu oxidizable substrates led to faster absorption decay. Trapped holes were identified with $O^{\bullet -}$ radical anions which absorb in the ultraviolet in aqueous media; in TiO_2 particles, transitions into the conduction band may occur the energies of which lie in the visible region (63). The age of a colloidal solution continually exposed to light had an influence on the absorption intensity; after 30 min of its preparation, the absorption of the sol decayed to 70% and after 4 h to 50% of its original value. Further storage of the sol for 24 h showed no additional changes. Apparently, the original 'rough' surface of the particles become 'polished' on aging in sunlight with a resulting decrease in the number of hole traps (63).

Contrary to MV^{2+}, the electron donor anion SCN^- is adsorbed on the surface of TiO_2 and its oxidation by the valence band holes, via reaction 28, is

$$(SCN^-)_{ads} + h^+_{vb} \rightarrow SCN\bullet \xrightarrow{SCN^-} (SCN)_2^- \qquad (28)$$

prompt, occurring within the laser pulse (<15 nsec) (38). Hole transfer to adsorbed SCN^- species competes with water oxidation via adsorbed OH radicals.

The microsecond time scale risetime of the growth of MV^+ (38) has been confirmed by transient resonance Raman spectroscopy (56). Curiously, the resonance Raman spectrum of MV^+ and $(SCN)_2^-$ formed after excitation ($\Delta t \sim 10$ nsec) of aqueous colloidal TiO_2 suspensions were identical to the corresponding spectra of these species formed via other routes. In fact, the unshifted spectra suggested that $(SCN)_2^-$ and MV^+ are fully solvated by H_2O within a Δt of 5 nsec (56). If these products were adsorbed when created they could have desorbed immediately and diffused some tens of angstroms away from the particle surface at $\Delta t \sim 5$ nsec. An alternative pathway might have been that SCN and MV^+ were initially created by electron tunnelling in a fully solvated condition (56).

Laser photolysis studies have also demonstrated (54) that reduction of the amphiphilic viologen derivative (1-tetradecyl-1'-methyl--4,4'-bipyridinium), $C_{14}MV^{2+}$, in the presence of aqueous TiO_2 sols occurs via two sequential and time-separated electron transfer reactions (equations 29 and 30) involving e^-_{cb} (TiO_2). Through surface adsorption of this

$$(C_{14}MV^{2+})_{ads} + e^-_{cb} \xrightarrow{10^8 s^{-1}} (C_{14}MV^+)_{ads} \qquad (29)$$

$$(C_{14}MV^+)_{ads} + e^-_{cb} \xrightarrow{5 \times 10^4 s^{-1}} C_{14}MV^o \qquad (30)$$

amphiphilic electron relay, reactions 29 and 30 take place rapidly and are completed in <100 μsec at pH 11. Under the same conditions, the simple MV^{2+} did not undergo this two-electron reduction. Laser flash excitation at 347 nm of $C_{14}MV^+$ species yielded hydrated electrons through a monophotonic process (reaction 31), (53) assisted by TiO_2 particles. In the absence of the semiconductor particles, reaction 31 did not occur. Most likely, the negative surface charge of the TiO_2

$$(C_{14}MV^{+})_{ads} \xrightarrow{h\nu} (C_{14}MV^{2+})_{ads} + e^{-}_{aq} \quad (31)$$

particle enhanced escape of the photo-ejected electron from its parent ion into the bulk aqueous phase. Alternatively, inasmuch as $C_{14}MV^{+}$ was strongly adsorbed to the TiO_2 surface it can be considered as a filled surface state of the semiconductor (53). Light excitation could then promote electrons from these surface states to the conduction band continuum from which electron emission into water and hydration would occur (53). A cofacial dimeric viologen, DV^{4+}, was also reduced by e^{-}_{cb} (TiO_2) via a simultaneous two-electron transfer event to give DV^{2+} and DV^{3+} species. The processes involved are noted in reaction 32 and 33, respectively (53). The product of reaction 32 formed rapidly within 20 nsec, at least those DV^{4+} adsorbed on the particle surface. The rest of DV^{2+} formation occurred in the microsecond time domain following inte-

$$DV^{4+} + 2e^{-}_{cb} \rightarrow DV^{2+} \quad (32)$$

$$DV^{4+} + DV^{2+} \rightarrow 2DV^{3+} \quad (33)$$

raction of e^{-}_{cb} with bulk solution DV^{4+} species. By contrast, $Rh(bpy)_3^{3+}$ underwent one-electron reduction to $Rh(bpy)_3^{2+}$ by $e^{-}_{cb}(TiO_2)$; a subsequent dark reaction produced $Rh(bpy)_2^{+}$ (53). The implications of these reduced viologens and the rhodium complex to artificial photosynthetic systems have been dealt with by Moser and Gratzel (53,54). Thus, systems involving $C_{14}MV^{2+}$ and TiO_2/Pt species at pH 12 produced H_2 vigorously at 0.5 ml/h under the conditions of the experiment (54). Clearly, the fundamental studies carried out that have involved flash photolysis of transparent semiconductor sols bear significantly in our understanding of the processes in the water splitting reaction, on using electron relays and sacrificial donors or acceptors.

3.3 Pulse Radiolysis

The pulse radiolysis technique can be usefully applied to the characterization of the thermodynamic and kinetic properties of semiconductor colloids.

The study of the equilibrium between one-electron redox comples, V^{2+}/V^{+}, with colloidal TiO_2

$$V^{+} + (TiO_2)_c \rightleftarrows V^{2+} + (TiO_2^{-})_c \quad (34)$$

was performed by generating the viologen V^{+} radicals through radiolytic means; the decay of V^{+} was followed by kinetic spectrophotometry (59,65). The rate of the protonation reaction

$$(TiO_2^{-})_c + H^{+} \rightleftarrows H_{ads} + (TiO_2)_c \quad (35)$$

was determined conductometrically (59). When the radical is methyl viologen and the media are acidic (pH <5), charge injection takes place from MV^+ into the conduction band of TiO_2. The nature of the electrons in the semiconductor appears to depend on the different structures of the colloids. X-ray analysis of the colloids shows that the free carrier absorption is associated with a more crystalline structure, while colloids with localized electrons show mainly an amorphous structure (65).

In the same work (65), the electron transfer kinetics from MV^+ to TiO_2 colloids and α-Fe_2O_3 were investigated. For TiO_2 the rate constants depend on the particle dimensions and on the adsorbed radiation dose.

The determination of the flat-band potential was also carried out with different TiO_2 colloids.

The results were explained by the effects of adsorbed ionic charge on the semiconduction surface (66, 67).

Moreover, a detailed study on the electron transfer kinetics of such viologen radicals as the amphiphilic derivatives 1,1'-diheptyl--4,4'-bipyridinium,HV^{2+}, and $C_{14}MV^{2+}$ has also been carried out with amorphous TiO_2 colloids (70 A); the effect of Pt deposition on semiconductor catalyst was probed to obtain information on the crucial aspect of the hydrogen evolution process.

The rate constant for electron transfer from MV^+ to amorphous TiO_2 particles is virtually independent of the total dose (and thus of $[MV^+]$) and linearly dependent on $[TiO_2]$. Both these features are in contrast to the behaviour of MV^+ decay catalyzed by colloidal gold (68) or platinum (69). A value of $k = 2.7 \times 10^7$ dm^3 mol^{-1} (particles) s^{-1} was reported and is 10^3 slower than the diffusion-controlled rate constant.

The rate determing step is clearly the heterogenous electron-transfer step, which may occur to a localized surface state.

The rate of proton consumption in the MV^+/TiO_2 system (reaction 35) is exponential, as shown in Fig. 18, with $k_{obs} = 5.5$ s^{-1}.

Under the same conditions the decay of MV^+ measured spectrophometrically yelds $k_{obs} = 45$ s^{-1}. There is a mismatch of approximately an order of magnitude between the rates of reaction (34) and (35) under the experimental conditions quoted in the legend of Fig. 18 (pH 2.6).

The equilibrium situation (34), when V^+ is MV^+, has been investigated under different experimental conditions. Equilibrium (34) is achieved before an appreciable amount proceeds toward reaction (35).

Analysis of the process substantiates the argument of a lack of a depletion layer in the particles of the reported size (70).

In fact, the charge density of electrons in the particle following the electron-injection reaction is quite high reaching 10^{19}-10^{20} electron cm^{-3}.

On the other hand, the initial density of charge carriers can be calculated from the experimental data ca. 1×10^{18} cm^{-3} thus sustaining the above statement on the depletion layer.

The Fermi level is raised by ca. 130 mV at the highest charge density injected into the particles. This change could, however, be increased by increasing the dose/[particle] ratio or by using stronger reducing radicals. The effect of the former ratio on the Fermi level is similar to the effect of light intensity (71).

The effect of the loading with Pt, either as addition of a separate colloid or by photodeposition, has been found to affect dramatically

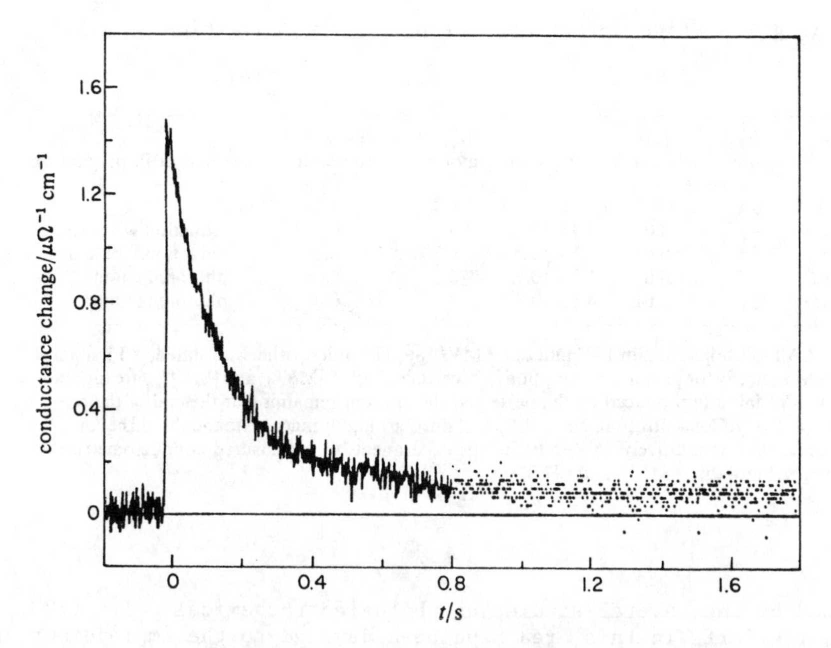

Figure 18.- Rate of consumption of protons measured by the conductivity technique: $5x10^{-4}$ mol dm^{-3} MV^{2+}, pH 2.6, 5 g dm^{-3} TiO_2; all solutions contain 0.17 mol dm^{-3} propan-2-ol and are Ar saturated.

both the equilibrium concentration of MV^+ as well as its rate of disappearance. This rate is substantially increased (compared with the rate in the absence of Pt), while the equilibrium level of MV^+ at the end of the reaction is practically reduced to zero in the acidic pH range.

Generally speaking metal-like behavior is observed once Pt is present and the dependence on reactant concentrations and dose resemble those observed for colloidal metals (68,69) rather than those observed for colloidal TiO_2.

Table 1 gives a comparative compilation of such experiments.

3.4 Charge Injection in Dye Sensitization

In the above section 3.2, recent examples have been given on the importance of basic investigations with pulsed lasers, and these examples made reference to conduction band electron acceptors. Next, we will delve into species (dyes) which are potential injectors of electrons into the conduction band of semiconductors. Much work has also been done in this area. Below we restrict the discussion to studies of charge injection in dye sensitization of wide bandgap semiconductors.

Photosensitization of electron transfer across a semiconductor/solution interface plays an important role in light energy conversion as

TABLE 1.- Effect of Pt and method of its deposition on the decay of MV^{+a}

	$[TiO_2]$ /g dm^{-3}	[Pt] /g dm^{-3}	$t_{1/2}^{b}$/s	$[MV^+]^c$ /μmol dm^{-3}	$[MV^+]_{eq}{}^c$ /μmol dm^{-3}	method of Pt preparation
(*a*)	0.5	—	0.11	5.8	1.9	—
(*b*)	—	0.01	3.9×10^{-5}	6.1	0.03	reduction with citrate
(*c*)	0.5	0.01	3.3×10^{-4}	6.3	0.21	reduction with citrate
(*d*)	0.5	0.01	1.7×10^{-5}	6.2	0.06	photoreduction
(*e*)	0.5	0.01	4.6×10^{-3}	19	0.05	photoreduction^d

[a] All solutions contain 10^{-3} mol dm^{-3} MV^{2+} at pH 2 unless otherwise stated. [b] First half-life of MV^+ decay [or proton consumption in experiment (*e*)]. [c] $[MV^+]_0$ and $[MV^+]_{eq}$ are the amount of MV^+ initially produced by the pulse and the amount remaining at the end of the reaction, respectively. Concentrations below 0.5 μmol dm^{-3} are highly inaccurate and should be considered only semi-quantitatively. [d] Results in this experiment were measured conductometrically at 5×10^{-4} mol dm^{-3} MV^{2+} and pH 3.2.

witnessed by the several studies on photoelectrochemical cells (72). The major efforts in this area have been devoted to the improvement of visible light response of wide bandgap semiconductors (e.g., ZnO, TiO_2, $SrTiO_3$,...). Sensitization can be achieved, among others, by adsorption of some suitable dye molecules onto the semiconductor surface which, upon light excitation, injects an electron into the conduction band of the semiconductor. While the overall performance of dye-sensitized semiconductor systems has been reported extensively (73), details of the electron injection process have been scarce. This is the result of difficulties that are encountered in the application of fast kinetic methods to studies of solid electrodes and powders. However, this is not the case of semiconductor colloidal sols (dimensions ∿20-200 A) that yield transparent solutions, and allow direct application of laser photolysis techniques to unravel the interfacial charge transfer processes (33-35,38,74).

In this regard, the excellent studies by Kamat and Fox (75) on the erythrosine sensitization of colloidal TiO_2 in CH_3CN, by Kiwi (52) on the electron transfer from the excited $Ru(bpy)_3^{2+}$ to colloidal TiO_2 at elevated temperatures, and the very recent extensive work by Moser and Gratzel (48) on the eosine-Y sensitized electron injection into TiO_2 are worth noting. In the latter study, the initial electron transfer event (48,76) and the fate of the injected electron were explored with particular emphasis on the back reaction of the electron with the eosine cation, EO^+. As well, the competing electron trapping process by noble metals deposited on the surface of the semiconductor colloidal particles was also examined (48).

Coulombic interactions between the TiO_2 surface ane eosin dianion play a vital role in the adsorption of EO onto colloidal TiO_2 particles, inasmuch as absorption spectral shifts were seen in the pH range (7.5 to 5) where the charge of the colloidal particles changes from negative to positive. Addition of TiO_2 to a EO solution at pH 4 also had a pronounced effect on the luminescence which red-shifted upon adsorption of EO to the TiO_2 surface indicating binding to the OH groups at the TiO_2 surface with relatively acidic character (48). The eosin fluorescence was strongly quenched by TiO_2 particles; the smallest emission intensity obtained for the highest mean occupancy of TiO_2 particles: 160 molecules of EO per particle (surface eosin concentration $\sim 2 \times 10^{13}$ cm^2; eosin-eosin distance on the surface $\sim$22 A). Efficient dipolar energy transfer is possible. Using a picosecond laser/streak camera system, fluorescence lifetimes (τ_F) in the range of 50-60 psec have been measured for eosin adsorbed on SnO_2 and In_2O_3 surfaces (77). These short lifetimes on the two semiconductors as well as on glass were interpreted as most likely due to energy transfer followed by trapping at defect sites. Liang et al. (77), did not however, consider electron injection for the decrease of τ_F from 1.4 nsec to $\sim$60 psec in the adsorbed state.

Addition of TiO_2 to a solution of eosin leads to dramatic changes in the photoredox behaviour of EO. The formation of EO^+ by photoexcitation of the ground state $EO(S_o)$ in colloidal TiO_2 solution was confirmed by Rossetti and Brus (78) by time-resolved laser Raman Spectroscopy. Differences in the Raman spectrum of EO^+ in water and aqueous TiO_2 solutions were attributed to protonation of EO^+ by surface hydroxyl groups. The quantum yield of EO^+ formation, Φ_{EO^+}, increased from 0.27 to 0.35 upon increasing TiO_2 from 0.1 to 0.5 g/L and reached a plateau at 0.38 at 3 g/L TiO_2 (77). In pure water, the photogeneration of EO^+ is a relatively slow ($4.2 \times 10^4 s^{-1}$) and inefficient process ($\Phi_{EO^+} \sim 0$) arising from dismutation of the triplet state (reaction 34). By contrast, in the presence of TiO_2, EO^+ was generated efficiently $\Phi_{EO^+} \sim 0.4$, pH 3) and at high rate ($k_{inj} \sim 8.5 \times 10^8 s^{-1}$), its formation being completed in <10 nsec (48). The mechanism of formation of EO^+ was different from that in water and involved electron injection from the lowest singlet excited

$$2\ EO(T_1) \rightarrow EO^+ + EO^- \quad (1.3 \times 10^9 M^{-1} s^{-1}) \qquad (36)$$

state of EO, $EO(S_1)$, to the conduction band of colloidal TiO_2 particles (equation 37) (79). It should be noted that charge injection was only

$$EO(S_1) + TiO_2 \xrightarrow{k_{inj}} EO^+ + e^-_{cb}(TiO_2) \qquad (37)$$

observed at pH $\leq$6, under conditions where eosin is associated with TiO_2 particles; close proximity of reactants is required for electron transfer to compete with the other channels of $EO(S_1)$ deactivation, viz., intersystem crossing (equation 38) and radiative and nonradiative decay (equation 39a and 39b, respectively). While charge injection from $EO(T_1)$ is thermodynamically possible, it did not occur; τ_{T_1} was the same in water as it was in TiO_2 aqueous solutions (equation 40).

$$EO(S_1) \xrightarrow{k_{isc}} EO(T_1) \quad (38)$$

$$EO(S_1) \xrightarrow{k_r} EO(S_0) + h\nu \quad (39a)$$

$$EO(S_1) \xrightarrow{k_{nr}} EO(S_0) \quad (39b)$$

$$EO(T_1) + TiO_2 \not\longrightarrow EO^+ + e^-_{cb}(TiO_2) \quad (40)$$

Back electron transfer between $e^-_{cb}(TiO_2)$ and EO^+ occurred *via* a rapid intraparticle reaction between $EO^+ \ldots e^-_{cb}(TiO_2)$ pairs associated with the same TiO_2 host aggregate and *via* a slower process involving bulk diffusion (48). The rate constant for intraparticle recombination is $2 \times 10^5 s^{-1}$, about 4000 times slower than that for electron injection. This enables light induced charge separation to be sustained on a colloidal TiO_2 particle for several microseconds, sufficient to trap the electron by a noble metal deposit on TiO_2.

The intimate processes of dye sensitization of a TiO_2 semiconductor particle and photosensitization of electron injection in noble metal loaded TiO_2 particles are illustrated in Figure 19 and Figure 20, respectively (48).

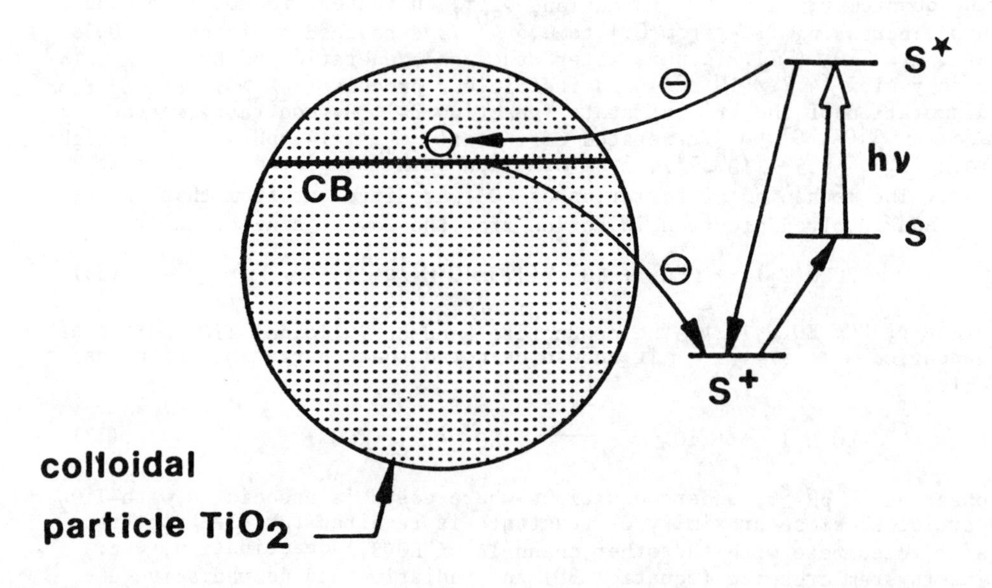

Figure 19.- Schematic illustration of charge injection and intra-particle back electron transfer in the photosensitization of a colloidal semiconductor particle without redox catalyst. From ref. 48.

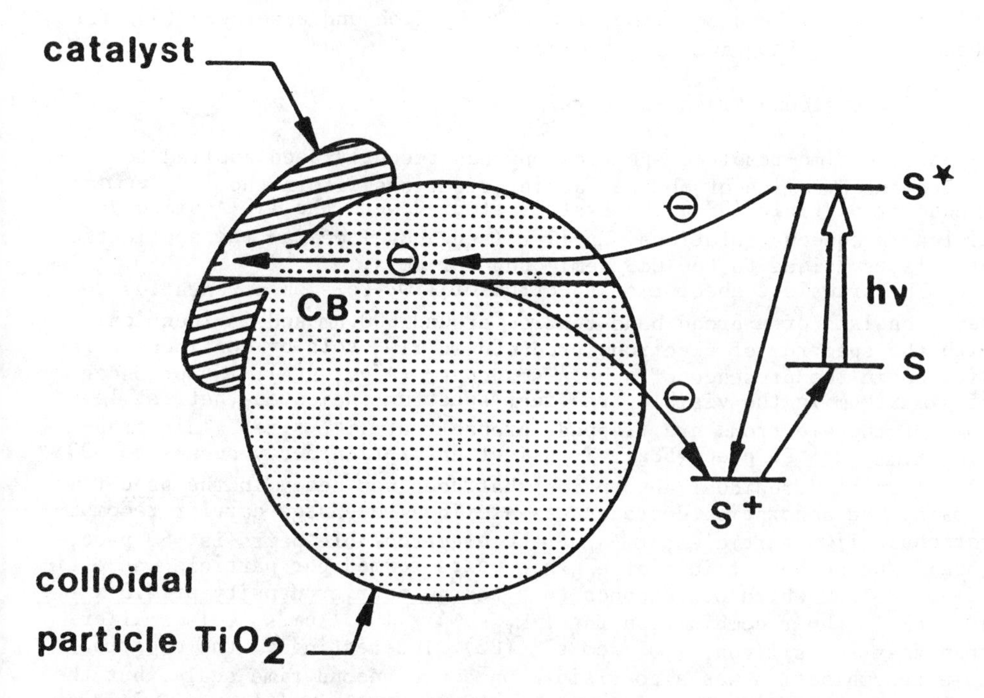

Figure 20.- Schematic illustration of charge injection and intra-particle back electron transfer in the photosensitization of a colloidal semiconductor particle loaded with a noble metal catalyst. From ref. 48.

In the former process, light excitation of EO to S_1 leads to injection of electrons to the conduction band of TiO_2 leaving EO^+ at the surface. In the second process, the role of the noble metal catalyst is to trap the photo-injected electron and to intercept the rapid back reaction between $EO^+ ... e^-_{cb}(TiO_2)$ pairs. This allows a larger fraction of all the EO^+ formed during the laser pulse to escape to the bulk of the solution with the subsequent back reaction being relatively slow.

On the basis of shifts of the absorption peak locations in the absorption spectrum of the dye, it appears that the dye sensitization process involves, initially, electron excitation to a dye excited state followed by electron injection into the semiconductor conduction band (80), rather than initial electron excitation directly into the conduction band (81). That is, dye sensitization proceeds *via* reaction (41) rather than through reaction (42).

$$S \xrightarrow{h\nu} S^* \xrightarrow{TiO_2} e^-_{cb}(TiO_2) + S^+ \quad (41)$$

$$S \xrightarrow{h\nu, TiO_2} e^-_{cb}(TiO_2) + S^+ \quad (42)$$

This notion has been confirmed recently by Itoh and coworkers (55) for rhodamine B on SnO_2 and TiO_2 semiconductors.

3.5 Picosecond Flash Photolysis

Picosecond time-resolved spectroscopy has recently been applied to elucidate the dynamics of charge carrier reactions within the TiO_2 semiconductor particle (58a). The systems studied were the 110-A sized particles in aqueous solutions. Earlier (see above) we saw the application of this technique to the CdS semiconductor sols.

The transient absorption spectrum of TiO_2 sols observed after 20 psec consists of a broad band centred around 600 nm and is identical with the spectrum of electrons produced in TiO_2 sols by continous irradiation in the presence of a hole scavenger (e.g., PVA). The presence of a maximum in the visible transient spectrum indicated that, at least some of the electrons are trapped, probably at Ti^{4+} sites. This trapping time is $\leq$ 20 psec (58a). Transient absorption decay occurs to 500 psec leaving a residual absorption that does not decay in the psec time domain. The absorption decrease was ascribed to charge carrier recombination in TiO_2 particles; $t_{1/2}$ of the electron/hole pairs is $\sim$60 psec. (58a). The number of initial e^-_{cb}/h^+_{vb} pairs formed per particle was estimated as $\sim$300 which corresponds to a charge carrier density of 4.5 x 10^{20} cm^{-3}; the recombination rate, k_{rec} $\sim$4 x 10^{-11} cm^3s^{-1}, is smaller than that for silicon, $\sim 10^{-9}$ cm^3s^{-1} (82). Interestingly, the electron/hole recombination was also visible on a nanosecond time scale, but the number of electron/hole pairs was $\sim$6 (58a). Henglein (64) has failed to observe this absorption signal when flashing the TiO_2 sol in the absence of an adsorbed scavenger for either electrons or holes in that particular TiO_2 sol, the recombination of the charge carriers must have occurred within the laser pulse.

In the time domain 500 psec to 10 nsec, the transient absorption spectra showed the growth of a new band in the 550-400 nm region. It has tentatively been assigned to the formation of peroxides at the TiO_2 surface following reaction 43 in which the $(Ti\text{-}O^-)_{surf}$ species subsequently dimerizes (58a).

$$(Ti\text{-}OH^-)_{surf} + h^+_{vb} \rightarrow (Ti\text{-}O^{\bar{\bullet}})_{surf} + H^+ \qquad (43)$$

We have also measured directly the rate of electron injection from the excited singlet state of eosin, $EO(S_1)$, to the conduction band of colloidal TiO_2 particles (83). The rate constant for interfacial electron transfer is (9.5 $\pm$ 1.4) x $10^8 s^{-1}$ at 298 K and pH 3; this is in very good agreement with the value estimated of 8.5 x $10^8 s^{-1}$ reported earlier from the nanosecond laser studies (see above) (48). Moreover, k_{isc} of reaction 38 is 5 x $10^8 s^{-1}$; the time constant for the growth of the EO^+ is 400 psec (reaction 37) which also represents the lifetime of $EO(S_1)$ (83). By contrast, the lifetime of $EO(S_1)$ in the absence of TiO_2 is 1.2 $\pm$ 0.4 nsec, consistent with the value of 1.4 nsec reported by Liang and coworkers (77). These results have confirmed that electron transfer from the adsorbed dye S_1 state to the conduction band of the semiconductor is a very rapid reaction (48).

The picosecond laser flash photolysis technique will find interesting applications in the future in the field of newly discovered surface derivatized TiO_2 particles which have been shown to be effective in the water cleavage by visible light (84).

4.0 ACKNOWLEDGEMENTS

Our work is supported by the Natural Science and Engineering Research Council of Canada, and by the Consiglio Nazionale delle Ricerche (Roma) through its 'Progetto Finalizzato di Chimica Fine e Secondaria'. We are grateful to these two agencies for their support. We also wish to thank all of our collaborators whose efforts have made possible some of the work reported here. We are also grateful to the American Chemical Society, the Chemical Society of London, Verlag Chemie GmbH, Elsevier Sequoia and the North Holland Physics Publ. Division for permission to reproduce figures from their respective journals.

5.0 REFERENCES

(1) a) M. Gratzel, Ed., "Energy Resources through Photochemistry and Catalysis", Academic Press, New York, 1983.

b) E. Pelizzetti, M. Barbeni, E. Pramauro, W. Erbs, E. Borgarello M.A. Jamieson, and N. Serpone, Quimica Nova (Brasil), in press.

(2) N. Serpone, E. Pelizzetti, and M. Gratzel, Coord. Chem. Rev., 64, 225 (1985).

(3) a) M. Barbeni, E. Pelizzetti, E. Borgarello, N. Serpone, M. Gratzel, L. Balducci, and M. Visca, Int. J. Hydrogen Energy, 10, 249 (1985).

b) N. Buhler, K. Meier, and J.F. Reber, J. Phys. Chem., 88, 3261 (1984).

c) E. Borgarello, N. Serpone, E. Pelizzetti, and M. Barbeni, J. Photochem., submitted.

(4) a) P. Pichat, J.M. Hermann, J. Disdier, H. Courbon, and M.N. Mozzanega, Nouv. J. Chim., 5, 627 (1981).

b) P. Pichat, M.N. Mozzanega, J. Disdier, and J.M. Hermann, Nouv. J. Chim., 6, 59 (1982).

c) E. Borgarello and E. Pelizzetti, Chim. & Ind. (Milano), 65, 474 (1983).

d) T. Kawai and T. Sakata, J. Chem. Soc. Chem. Commun., 694 (1980).

e) T. Sakata and T. Kawai, Chem. Phys. Lett., 80, 341 (1981).

f) N. Serpone, E. Borgarello, E. Pelizzetti, and M. Barbeni, Chim. & Ind. (Milano), 67, 318 (1985).

(5) a) S. Sato and J.M. White, J. Am. Chem. Soc., 102, 7206 (1980).

b) S.M. Fang, B.H. Chen, and J.M. White, J. Phys. Chem., 86, 3126 (1982).

(6) a) B. Kraeutler and A.J. Bard, J. Am.Chem. Soc., 100, 2239 (1978); ibid., 100, 5985 (1978).

b) H. Yoneyama, Y. Takao, H. Tamura, and A.J. Bard, J. Phys. Chem., 87, 1417 (1983).

(7) a) E. Pelizzetti, M. Barbeni, E. Pramauro, E. Borgarello, M.A. Jamieson, N. Serpone, and H. Hidaka, Chim. Ind. (Milano), in press.

b) H. Hidaka, H. Kubota, M. Gratzel, N. Serpone, and E. Pelizzetti, Nouv. J. Chim., 9, 67 (1985).

c) M. Barbeni, E. Pramauro, E. Pelizzetti, E. Borgarello, and N. Serpone, Chemosphere, 14, 195 (1985).

d) M. Barbeni, E. Pramauro, E. Pelizzetti, E. Borgarello, M. Gratzel, and N. Serpone, Nouv. J. Chim., 8, 547 (1984).

e) A.L. Pruden and D.F. Ollis, J. Catal., 82, 404 (1983).

f) A.L. Pruden and D.F. Ollis, Environ. Sci. Technol., 17, 628 (1983).

g) C.Y. Hsiao, C.L. Lee, and D.F. Ollis, J. Catal., 82, 418 (1983).

h) B.G. Oliver, E.G. Cosgrove, and J.H. Carey, Environ. Sci. Technol., 13, 1075 (1979).

(8) a) E. Borgarello, R. Harris, and N. Serpone, Nouv. J. Chim., submitted.

b) B. Kraeutler and A.J. Bard, J. Am. Chem. Soc., 100, 4317 (1978).

c) J.S. Curran, J. Domenich, N. Jaffreizic-Renault, and R. Philippe, J. Phys. Chem., 89, 957 (1985).

(9) F.A. Kroger, Physica, 7, 1 (1940).

(10) J.J. Ramsden and M. Gratzel, J. Chem. Soc. Faraday Trans., I, 80, 919 (1984).

(11) F.A. Kroger and H.J.G. Meyer, Physica, 20, 1149 (1954).

(12) M. Schon, Z. Physik, 119, 463 (1942).

(13) H.A. Klasens, Nature (London), 158, 306 (1946).
H.A. Klasens and M.E. Wise, J. Opt. Soc. Am., 38, 226 (1948).

(14) J.J. Lambe, C.C. Klick, and D.L. Dexter, Phys. Rev., 103, 1715 (1956).

(15) R.J. Collins, J. Appl. Phys., 30, 1135 (1959).

(16) A.A. Vuylsteke and Y.T. Sihvonen, Phys. Rev., 113, 40 (1959).

(17) B.A. Kulp and R.H. Kelley, J. Appl. Phys., 31, 1057 (1960).

(18) a) R.E. Halsted, M. Aven, and H.D. Coghill, J. Electrochem. Soc., 112, 177 (1965).

b) R.E. Halsted, D. Curie, and J.S. Prener, "Physics and Chemistry of II-VI Compounds", M. Aven and J.S. Prener, Eds., North Holland Publ., Amsterdam, 1967, chapter 8 and 9.

(19) A. Henglein, Ber. Bunsenges. Phys. Chem., 86, 301 (1982).

(20) A. Henglein, J. Phys. Chem., 86, 2291 (1982).

(21) J.P. Kuczynski, B.H. Milosavljevic, and J.K. Thomas, J. Phys. Chem., 88, 980 (1984).

(22) J.J. Ramsden, S.E. Weber, and M. Gratzel, submitted. (We thank Prof. M. Gratzel for a preprint of this article).

(23) L.E. Brus, J. Chem. Phys., 80, 4403 (1984).

(24) a) A. Henglein, personal communication, May 1985.

b) A. Fojtik, H. Weller, U. Koch, and A. Henglein, Ber. Bunsenges Phys. Chem., 88, 969 (1984).

c) A. Henglein, in "Modern Trends of Colloid Science in Chemistry and Biology", F. Eicke, Ed., Birkhauser Verlag, Basel, 1985.

(25) J.P. Kuczynski, B.H. Milosavljevic, and J.K. Thomas, J. Phys. Chem., 87, 3368 (1983).

(26) J.P. Kuczynski and J.K. Thomas, J. Phys. Chem., 87, 5498 (1983).

(27) M.S. Brodin, N.A. Davydova, and I. Yu. Shablii, Phys. Stat. Sol., 115, 641 (1983).

(28) M.S. Brodin, N.A. Davydova, and I. Yu. Shablii, Fiz. Tekh. Poluprov., 10, 625 (1976).

(29) M.S. Brodin, I. Ya. Gorodetskii, N.S. Korsunskaya, and I. Yu. Shablii, Ukr. Fiz. Zh., 24, 1539 (1974).

(30) R. Rossetti and L.E. Brus, J. Phys. Chem., 86, 4470 (1982).

(31) D. Huppert, P. Rentzepis, R. Rossetti, and L.E. Brus, unpublished results, quoted in ref. 32.

(32) R. Rossetti, S.M. Beck, and L.E. Brus, J. Am. Chem. Soc., 106, 980 (1984).

(33) N. Serpone, D.K. Sharma, M.A. Jamieson, M. Gratzel, and J.J. Ramsden, Chem. Phys. Lett., 115, 473 (1985).

(34) N. Serpone, in "Photoelectrochemistry, Photocatalysis and Photoreactors", M. Schiavello, Ed., D. Reidel Publ. Co., Dordrecht, 1985, pp. 351-372.

(35) M. Gratzel, Acc. Chem. Res., 14, 376 (1981).

(36) Z. Alfassi, D. Bahnemann, and A. Henglein, J. Phys. Chem., 86, 4656 (1982).

(37) M. Gutierrez and A. Henglein, Ber. Bunsenges. Phys. Chem., 87, 474 (1983).

(38) D. Dounghong, J.J. Ramsden, and M. Gratzel, J. Am. Chem. Soc., 104, 2977 (1982).

(39) Y. Nakato, A. Tsumura, and H. Tsubomura, Chem. Phys. Lett., 85, 387 (1982).

(40) E.M. Kosover and J.L. Cotter, J. Am. Chem. Soc., 86, 5524 (1964).

(41) P.M. Rentzepis, Science, 218, 1183 (1982).

(42) See for example, "Picosecond Phenomena I, II, and III", published by Springer-Verlag, Berlin, 1978, 1980, and 1982.

(43) H. Saito, W. Graudszus, and E.O. Gobel, in "Picosecond Phénomena III", K.B. Eisenthal, R.M. Hochstrasser, W. Kaiser, and A. Laubereau, Eds., Springer-Verlag Series in Chemical Physics, Berlin, vol. 23, 1982, p. 353.

(44) D. Huppert, Z. Harzion, S. Gottesfeld, and N. Croitoru, in ref. 43, p. 360.

(45) Z. Harzion, D. Huppert, S. Gottesfeld, and N. Croitoru, J. Electroanal. Chem., 150, 571 (1983).

(46) a) J.R. Ufford and N. Serpone, Coord. Chem. Rev., 57, 301 (1984).

b) N. Serpone, M.A. Jamieson, F. Disalvio, P.A. Takats, L. Yeretsian, and J.R. Ufford, Coord. Chem. Rev., 58, 87 (1984).

c) M.A. Jamieson, N. Serpone, and E. Pelizzetti, Coord. Chem. Rev. in press (1985).

(47) A.J. Bard, J. Photochem., 10, 59 (1979).

(48) J. Moser and M. Gratzel, J. Am. Chem. Soc., 106, 6557 (1984).

(49) M. Gratzel and A.J. Frank, J. Phys. Chem., 86, 2964 (1982).

(50) J. Moser and M. Gratzel, Helv. Chim. Acta, 65, 1436 (1982).

(51) R. Humphry-Baker, J. Lilie, and M. Gratzel, J. Am. Chem. Soc., 104, 422 (1982).

(52) J. Kiwi, Chem. Phys. Lett., 83, 594 (1981).

(53) J. Moser and M. Gratzel, J. am. Chem. Soc., 105, 6547 (1983).

(54) M. Gratzel and J. Moser, Proc. Natl. Acad. Sci. USA, 80, 3129 (1983).

(55) K. Itoh, Y. Chiyokawa, M. Nakao, and K. Honda, J. Am. Chem. Soc., 106, 1163 (1984).

(56) R. Rossetti, S.M. Beck, and L.E. Brus, J. Am. Chem. Soc., 104, 7322 (1982).

(57) K. Chandrasekaran and J.K. Thomas, J. Chem. Soc. Faraday I, 80, 1163 (1984).

(58) a) J. Moser, G. Rothenberger, M. Gratzel, D.K. Sharma, and N. Serpone, J. Am. Chem. Soc., submitted.

b) N. Serpone et al., work in progress.

(59) E. Borgarello, E. Pelizzetti, D. Meisel, and W.A. Mulac, J. Chem. Faraday I, 81, 143 (1985).

(60) S.K. Deb, Solid State Commun., 11, 713 (1972).

(61) D.K. Sharma and N. Serpone, unpublished results.

(62) D. Bahnemann, A. Henglein, J. Lilie, and L. Spanhel, J. Phys. Chem., 88, 709 (1984).

(63) D. Bahnemann, A. Henglein, and L. Spanhel, Faraday Discuss. Chem. Soc., 78, 151 (1984).

(64) A. Henglein, Pure & Appl. Chem., 56, 1215 (1984).

(65) M.M. Dimitrijeric, D. Savic, O.I. Micic, and A.J. Nozik, J. Phis. Chem., 88, 4278 (1984).

(66) A.J. Bard, F.-R. F. Fan, A.S. Gioda, G. Nagasubramanian, and H.S. White, Faraday Discuss., 70, 19 (1980).

(67) G. Cooper, J.A. Turner, and A.J. Nozik, J. Electrochem. Soc., 129, 1973 (1981).

(68) D. Meisel, W.A. Mulac, and M.S. Matheson, J. Phys. Chem., 85, 179 (1981).

(69) M.S. Matheson, P.C. Lee, D. Meisel, and E. Pelizzetti, J. Phys. Chem., 87, 394 (1983).

(70) A. Henglein, in "Photochemical Conversions and Storage of Solar Energy", Part A, J. Rabani ed., Weizmann Science Press, Jerusalem, 1982, p. 115.

(71) M. Ward, J. White, and A.J. Bard, J. Am. Chem. Soc., 105, 27 (1983).

(72) See for example, T. Iwasaki, A. Fujishima, and K. Honda, in ref.1a, chapter 11, p. 359.

(73) a) R. Memming, Philips Tech Rev., 38, 160 (1978).

b) K. Rajeshwar, P. Singh, and J. DuBow, Electrochim. Acta, 23, 111 (1978).

c) H. Gerischer, Ber. Bunsenges. Phys. Chem., 77, 771 (1973).

(74) D. Dounghong, E. Borgarello, and M. Gratzel, J. Am. Chem. Soc., 103, 4685 (1981).

(75) P.V. Kamat and M.A. Fox, Chem. Phys. Lett., 102, 379 (1983).

(76) J. Moser, M. Gratzel, D.K. Sharma, and N. Serpone, Helv. Chim. Acta, submitted.

(77) Y. Liang, A.M. Ponte-Goncalves, and D.K. Negus, J. Phys. Chem., 87, 1 (1983).

(78) R. Rossetti and L.E. Brus, J. Am. Chem. Soc., 104, 7321 (1982).

(79) V. Kasche and L. Lindquist, Photochem. Photobiol., 4, 923 (1965).

(80) D.A. Gulino and H.G. Drickamer, J. Phys. Chem., 88, 1173 (1984).

(81) a) S. Anderson, E.C. Constable, M.P. Dare-Edwards, J.B. Goodenough, A. Hamnet, K.R. Seddon, and R.D. Wright, Nature (London), 280, 571 (1979).

b) M.P. Dare-Edwards, J.B. Goodenough, A. Hamnet, K.R. Seddon, and R.D. Wright, Faraday Discuss. Chem. Soc., 70, 285 (1980).

c) J.B. Goodenough, A. Hamnet, M.P. Dare-Edwards, G. Campet, and R.D. Wright, Surf. Sci., 101, 531 (1980).

(82) P.T. Landsberg and G.S. Konsik, J. Appl. Phys., 56, 1696 (1984).

(83) J. Møser, M. Gratzel, D.K. Sharma, and N. Serpone, Helv. Chim. Acta, in press (1985).

(84) D. Duonghong, N. Serpone, and M. Gratzel, Helv. Chim. Acta, 67, 1012 (1984).

DYNAMICS OF INTERFACIAL ELECTRON TRANSFER REACTIONS IN COLLOIDAL SEMICONDUCTOR SYSTEMS AND WATER CLEAVAGE BY VISIBLE LIGHT

Michael Grätzel
Institut de chimie physique
Ecole Polytechnique Fédérale
Ch-1015 Lausanne
Switzerland

ABSTRACT. In this lecture I shall present recent results concerning the nature and reactivity of charge carriers in colloidal semiconductor particles. Using laser photolysis technique with picosecond time resolution, we have been able to monitor directly the dynamics of fundamental processes such as charge carrier trapping and recombination. Furthermore, it is possible, by using flash excitation of colloidal semiconductors, to determine the rate of interfacial charge transfer from the conduction and valence band of the particle to species present at the surface or in solution. In the second part of my lecture, I shall address the question of catalytic water cleavage by UV and visible light. Of particular importance in this context is the question of visible light sensitization of wide band semiconductors such as titanium oxide. A recent discovery in our laboratory has allowed to push the quantum yields for incident monochromatic visible light to current conversion to an unprecedented high level of more than 50%. Finally, I wish to present some surprising and very promising results concerning a new homogeneous and highly active catalyst for the oxidation of water to oxygen.

PICOSECOND TIME RESOLVED STUDIES OF CHARGE CARRIER TRAPPING AND RECOMBINATION IN COLLOIDAL SEMICONDUCTOR PARTICLES.

These studies were carried out with colloidal titanium dioxide (anatase) particles having a diameter of 120 Å. Irradiation of such colloidal solutions in the presence of a hole scavenger such as polyvinyl alcohol or formate ions results in the accumulation of electrons in the particles. As a result, the solution assumes a beautiful blue color under illumination. It was found that up to 300 electrons can be stored in one TiO_2 particle. (A TiO_2 particle of 120 Å size has about 3600 conduction band states. Therefore, at most 10% of the available states are occupied by electrons.) The absorption spectrum of these stored electrons is shown in Fig. 1:

E. Pelizzetti and N. Serpone (eds.), Homogeneous and Heterogeneous Photocatalysis, 91–110.

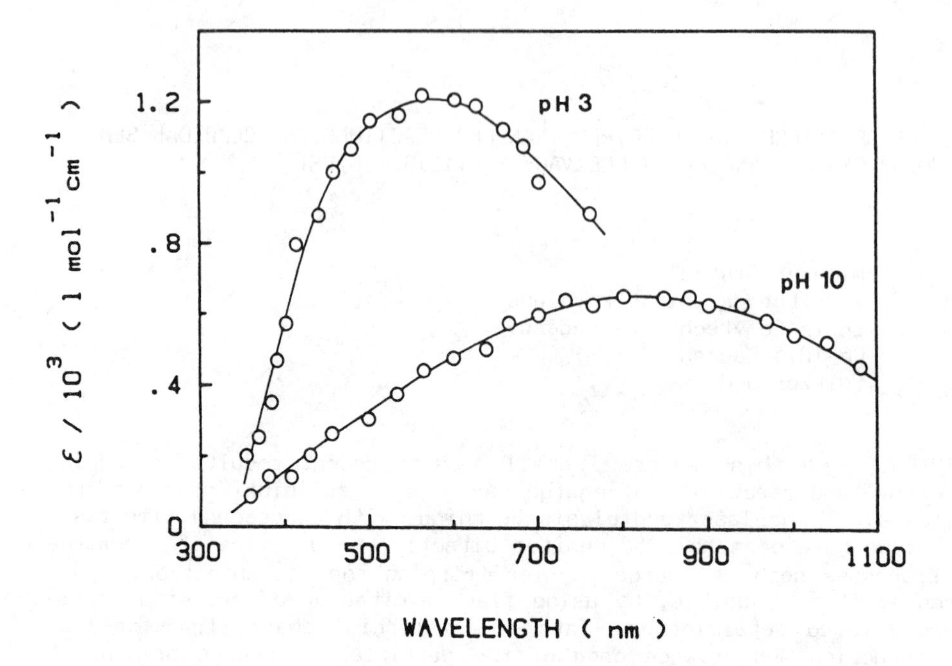

Fig. 1 Absorption spectrum of conduction band electrons in colloidal TiO_2 particles at pH 3 and pH 10.

The electron spectrum was found to be sensitive to the pH of the solution. Under alkaline conditions the electron absorption is very broad and has a maximum around 800 nm. Lowering the pH to 3 produced a pronounced blue shift in the spectrum which under these conditions shows a peak at 620 nm. The sensitivity of the electron absorption to the solution pH would indicate that they are located in the surface region of the particles. This has been confirmed by recent ESR experiments which show that under acidic conditions the electrons are trapped at the TiO_2 surface in the form of Ti^{3+} ions (1). Using redox titration, we have recently been able to determine the extinction coefficient of the trapped electrons (2). For the colloidal solutions of pH 3 the extinction coefficient at 600 nm is 1200 M^{-1} cm^{-1}.

Taking advantage of the characteristic optical absorption of trapped electrons in the colloidal TiO_2 particles, we have recorded their recombination with free and trapped holes in the picosecond to microsecond domain (3). Fig. 2 shows the temporal evolution of the transient spectrum after excitation of TiO_2 with a frequency tripled (353 nm) Nd laser pulse of ca. 40 ps duration.

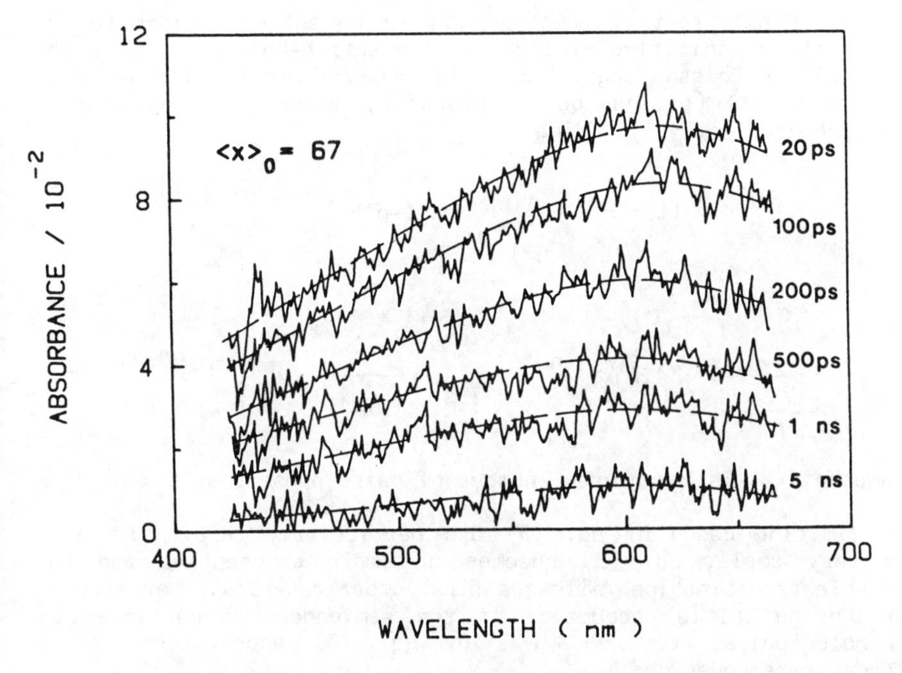

Fig. 2 Transient spectrum observed at various time intervals after picosecond excitation of colloidal TiO_2. Conditions: $[TiO_2]$ = 17 g/l, pH 2.7, Ar saturated solution, optical pathlength 0.2 cm. Average number of electron-hole pairs present initially in one TiO_2 particle is 67.

In Fig. 2 the spectrum of the trapped electron develops within the leading edge of the laser pulse indicating that the trapping time of the electron is less than 40 ps. Subsequently, the electron absorption decays due to recombination with valence band holes.

$$TiO_2\ (e^-_{tr} + h^+) \xrightarrow{k_r} TiO_2 \tag{1}$$

We have conceived a stochastic model to analyze the kinetics of this reaction. Since the recombination takes place between a restricted number of charge carriers restricted to the minute reaction space of a 120 Å sized colloidal TiO_2 particle, it cannot be treated by conventional homogeneous solution kinetics. The time differential of the probability that a particle contains x electron-hole pairs at time t is given by:

$$dP_x(t)/dt = k(x+1)^2 P_{x+1}(t) - kx^2 P_x(t) \tag{2}$$

where x = 0,1,2, ...

This system of differential equations is to be solved subject to the condition that the initial distribution of electron-hole pairs over the particles follows Poisson statistics. The average number of pairs present at time t, $\langle x \rangle (t)$, can be calculated by means of the generating function technique (3, 4) yielding:

$$\langle x \rangle (t) = \sum_{n=1}^{\infty} c_n \exp(-n^2 kt) \tag{3}$$

where

$$c_n = 2 \exp(-\langle x \rangle_o) (-1)^n \, n \sum_{i=n}^{\infty} \frac{\langle x \rangle_o^{\,i}}{(n+i)!} \prod_{j=1}^{n} (n-i-j) \tag{4}$$

The parameter $\langle x \rangle_o$ is the average number of pairs present at t = 0.

Two limiting cases of eq. (3) are particularly relevant: When $\langle x \rangle_o$ is very small, eq. (3) becomes a simple exponential and the electron-hole recombination follows a first order rate law. Conversely, at high average initial occupancy of the semiconductor particles by electron-hole pairs, i.e. $\langle x_o \rangle > 30$, eq. (3) approximates to a second order rate equation:

$$\langle x \rangle (t) = \frac{\langle x \rangle_o}{1+\langle x \rangle_o kt} \tag{5}$$

In Fig. 2 the initial concentration of electron-hole pairs was sufficiently high to allow for evaluation of the recombination process by the second order rate equation, eq. (5). This analysis gives for the recombination rate coefficient the value 3.2×10^{-11} cm^3 s^{-1} corresponding to a lifetime of 30 ns for an electron-hole pair in a colloidal TiO_2 particle with a size of 120 Å. Experiments were also carried out at low laser fluence where on the average less than one electron-hole pair was generated initially by the laser pulse. Under these conditions, hole trapping, presumably by surface hydroxyl groups, competes with recombination leading to a product whose reaction with trapped electrons is relatively slow. The trapping rate constant for the valence band hole was derived as 4×10^6 s^{-1}.

With regards to the use of colloidal TiO_2 as a photocatalyst for the light induced cleavage of water, the fact that recombination of electrons with free holes is about ten times faster than hole trapping is disadvantageous. It explains the experimental finding (vide infra) that deposition of highly active redox catalysts, such as Pt or RuO_2, on the surface of the colloidal TiO_2 particles is required to obtain good yields in the water cleavage process. The role of these catalytic deposits is to intercept electron-hole recombination by trapping the charge carrier and to accelerate their reaction with water leading to the formation of hydrogen and oxygen.

The characteristic optical absorption of electrons in TiO_2 particles can be used to monitor directly interfacial electron transfer reactions. This opens up the way to determine the heterogeneous rate constants for fast charge transfer from the conduction band of the semiconductor to acceptors present in solution. Thus, laser excitation of colloidal semiconductors combined with fast kinetic spectroscopy offers a very useful complement to the application of conventional electrochemical techniques which achieve only a relatively low time resolution to probe electron transfer events at the semiconductor/solution interface. In the following, we use the reduction of cobalticenium dicarboxylate ($Co(CpCOO)_2^-$) as an example to illustrate this procedure:

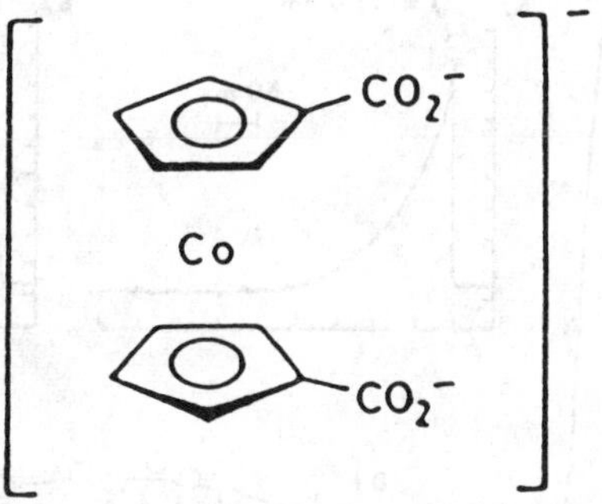

The kinetics of cobalticinium reduction were recorded by following the decay of the electron absorption at 780 nm and the growth of cobaltocene at 484 nm.

$$\underbrace{e^-_{cb}(TiO_2)}_{\lambda_{max}\ 780\ nm} + Co(CpCOO)_2^- \rightarrow \underbrace{Co(CpCOO)_2^{2-}}_{\lambda_{max}\ 484\ nm} \quad (6)$$

Fig. 3 shows transient absorption spectra obtained from the laser photolysis of colloidal TiO_2 in the presence of 5×10^{-4} M $Co(CpCOO)^-$. The solution contained also 0.1% polyvinylalcohol which acts as a hole scavenger. Immediately after the laser flash one obtains the spectrum of the conduction band electron with a maximum at 800 nm. Concomitantly with the decay of the electron absorption one observes the formation of the spectrum of the cobaltocenedicarboxylate. From the kinetic evaluation a second order rate constant of $k_6 = 4 \times 10^4\ m^{-1}\ s^{-1}$ is obtained.

This is much smaller than the rate of a diffusion controlled reaction which for 120 Å sized TiO_2 particles is expected to be about $5x10^{10}$ M^{-1} s^{-1}. From this one infers that the interfacial charge transfer is the rate determining step. In such as case, the heterogeneous rate constant is related to k_6 via (7):

$$k_{het} = k_6 \left[Co(CpCOO)_2^-\right] / (4\pi R^2) \quad (7)$$

where R is the semiconductor particle radius.

The value derived is k = $2x10^{-5}$ cm^{-1} s^{-1} indicating a relatively slow rate for the interfacial redox reaction at pH 10.

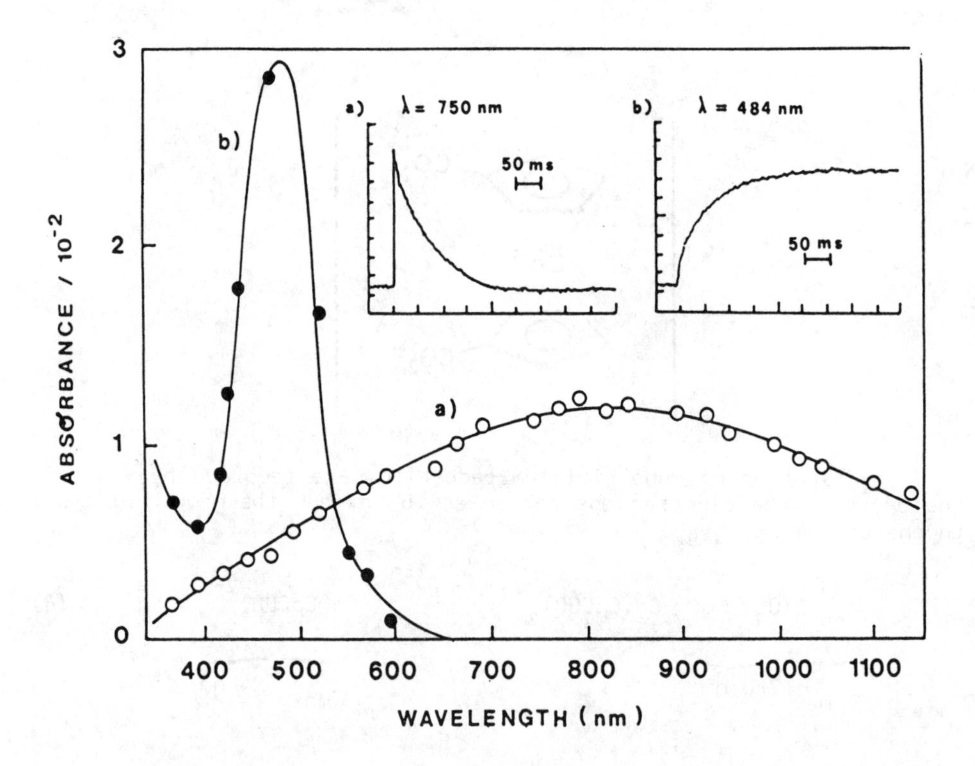

Fig. 3 Transient spectra obtained from the laser photolysis of alkaline (pH 10) solutions of colloidal TiO_2 (0.5 g/l) in the presence of $5x10^{-4}$ M $Co(CpCOO)_2^-$ (concentration of PVA is 0.5 g/l): o, spectrum ~ 10 μs after the laser pulse; •, spectrum 400 ms after the laser pulse. Insert shows temporal evolution of the absorbance at 750 and 484 nm.

WATER PHOTOLYSIS USING SEMICONDUCTOR DISPERSIONS.

The main thrust has been in the application of semiconductor-based microheterogeneous systems to split water photochemically into hydrogen and oxygen. UV-light induced water decomposition will be discussed first. Most of the work in this field has concentrated on wide bandgap semiconductors such as TiO_2 or $SrTiO_3$ loaded with noble metal catalysts. Excitation of the semiconductor by ultraviolet light generates electron hole pairs, eq. (8):

$$TiO_2 \xrightarrow{h\nu} TiO_2\,(e^-_{cb} + h^+) \qquad (8)$$

The electrons are trapped by the noble metal deposit, e.g. Pt, where hydrogen is subsequently evolved:

$$2e^-_{cb} + 2H^+ \xrightarrow{(Pt)} H_2 \qquad (9)$$

The concomitant hole reaction involves water oxidation to oxygen:

$$4h^+ + 2H_2O \rightarrow O_2 + 4H^+ \qquad (10)$$

The first study of this type was published by Bulatov and Khidekel (7) who reported that acidic aqueous dispersions of TiO_2 loaded with Pt decompose water under bandgap irradiation. Numerous other investigations followed and these have been reviewed (8).

In recent years, it became apparent that the simple sequence of reactions (8) to (10) could not account for all the experimental observations. In particular, during photolysis of water in closed systems, there is frequently only hydrogen evolution observed, and no or very little oxygen is released in the gas phase (9). This has led to the postulate that the hydrogen observed arises from the oxidation of Ti^{3+} ions present in the TiO_2 particles or that there is anodic oxidation of carbonaceous impurities. In our earlier investigations (10), we had attributed lack of oxygen generation to the photouptake of O_2 via reduction by conduction band electrons of TiO_2. Recent water photolysis experiments with TiO_2 particles loaded with Pt or Rh have provided ample confirmation of this hypothesis. Using redox indicators such as permanganate or o-diansidine, it could unambiguously be shown that peroxo- or superoxo species were formed during O_2-photouptake and also as the anodic reaction product during photolytic reduction of water to hydrogen. In contrast to materials such as ZnO, where photouptake of oxygen leads to the formation of free hydrogen peroxide in solution, in the case of TiO_2 the peroxide is associated with the solid. According to a recent study (11), the capacity for peroxide uptake by TiO_2 (anatase) is as high as 20 molecules H_2O_2/nm^2 surface area. It is worth mentioning that apart from oxygen reduction by conduction band electrons, peroxides can also be produced via the oxidation of surface OH-groups by valence band holes. Investigations are presently undertaken to determine the detailed mechanism of peroxide formation as well as the structure of the peroxo complex associated with the TiO_2 particles. Very likely, μ-peroxo titanium dimers are the prevailing species and their formation has been shown to occur via valence band hole pairing in other oxides such as MgO (12).

The fact that in titania based water cleavage systems peroxide can be produced in the anodic reaction instead of oxygen is advantageous, since it opens up a way to overcome the H_2/O_2 separation problem which is inherent to microheterogeneous energy conversion devices. However, there is also a disadvantage which arises from the inhibitory effect of the peroxide on the water cleavage activity of the catalyst during the photolysis, as H_2 and peroxide accumulate in the system. Fortunately, a striking improvement of the performance of the system is achieved (13) by addition of Ba^{2+} ions to the solution. We attribute the promoting effect of barium ions to the formation of insoluble barium peroxide. The solubility product of BaO_2 is so small ($K = 10^{-17.5}$) that Ba^{2+} can act as a peroxide scavenger. In this way, peroxide is removed from the systems allowing the catalyst to operate at higher efficiency.

Water cleavage systems that produce peroxides instead of oxygen would be without value for the solar generation of hydrogen if no methods could be found to decompose these peroxo-species in a simple and complete fashion. It was found earlier that prolonged flushing of the TiO_2 dispersion with an inert gas such as Ar or nitrogen affords already at least partial decomposition . However, once the hydrogen evolution curve approaches the plateau region, it is no longer possible to regenerate entirely the initial catalytic activity by this procedure. A very important observation was made in our laboratory during studies of the TiO_2/Pt dispersions containing barium ions. It was found that heating the catalyst for a few hours at 200-300°C under nitrogen restored entirely its initial water cleavage activity. Apparently, exposing the catalyst to elevated temperature leads to complete decomposition of the peroxo-species formed concomitantly with hydrogen. If the decomposition temperature could be lowered further, e.g. by using appropriate catalysts, the formation of peroxides could well be tolerable, and even desirable, in applied systems, where it would present a viable solution to the hydrogen-oxygen separation problem.

While peroxide is the likely oxidation product from water photolysis in TiO_2 particulate systems, concomitant formation of near stoichiometric amounts of oxygen and hydrogen has been observed in special cases. These comprise high temperature and low pressure conditions during irradiation, the use of NaOH covered TiO_2 particles and gaseous instead of liquid water (14), open systems where an inert gas is bubbled through the solution and the use of bifuncional catalysts (15, 16, 17). In the latter case, ruthenium dioxide is loaded in addition to Pt onto the TiO_2 particles. The role of RuO_2 has so far not been entirely clarified. On the one hand, it could promote transfer of holes from the valence band of the semiconductor to water, and this has unequivocally been established for visible light induced oxygen evolution on CdS particles (18). On the other hand, when TiO_2 is used as a support, RuO_2 could also act as a catalyst for the decomposition of peroxides, since hole trapping by surface OH-groups is a very rapid reaction.

Investigators that irradiate semiconductor dispersions in liquid water at room temperature in a closed flask should as a rule not expect any significant accumulation of O_2 in the gas phase since the catalyst that is employed for hydrogen generation, e.g. Pt, promotes oxygen reduction to peroxide at the same time. These considerations apply also to visible light induced water cleavage devices where apart from reduction at the hydrogen evolution catalyst, oxygen is also removed by reaction with electron relay species such as MV^+. In the event that the investigator, who performs water photolysis with a titania-based catalyst in a closed system, observes only appearance of H_2 but not O_2 in the gas phase, it is likely that peroxo- or superoxo-species are formed in the anodic part of the water splitting reaction. The alternatives that organic impurities or Ti^{3+}-ions serve as sacrificial electron donors in these systems can readily be excluded by determining the carbon and Ti^{3+} ion content of the catalyst. For the latter analysis, ESR and colorimetric techniques are available.

For solar application, it is crucial to shift the wavelength response of the light absorber into the visible. So far, three different strategies have been tested. The first involves the use of semiconductor particles with smaller bandgap than TiO_2. Apart from a recent report (16) on colloidal V_2O_5, Eg = 2.7 eV, these studies have concentrated on CdS particles as light harvesting units. In this case, it becomes mandatory to use a highly active oxygen evolution catalyst, such as RuO_2 or Rh_2O_3, to promote water oxidation by valence band holes and suppress photocorrosion. Photoinduced oxygen uptake is an undesirable side reaction in this system since it leads to oxidation of CdS to $CdSO_4$ (19). As was suggested by Harbour et. al. (20), this reaction is expected to occur also in water cleavage systems when the photolysis is performed in a closed vessel without intermittent removal of gaseous products, and it would lead to the destruction of the semiconductor. It is possible to avoid the photo-uptake of oxygen, which involves reduction of O_2 by conduction band electrons, by using CdS electrodes instead of particles.

Chromium doping of TiO_2 has been attempted to shift its wavelength response in the visible. The problem with these particle preparations so far has been long-term stability. Relatively high temperatures (at least 800°C) are required to introduce Cr ions in the TiO_2 lattice. Unfortunately, under these conditions, chromium catalyzes the transformation of anatase into rutile. Since TiO_2 (rutile) particles are unsuitable for water cleavage, one must avoid this transition by doping at lower temperature, i.e. 400°C. However, Cr-doped colloidal TiO_2 particles prepared in this fashion are not entirely stable and slowly release Cr ions in the aqueous solution thereby losing their visible light response.

The third approach to split water by visible light involves dye sensitization, and this has given the most promising results so far. In practically all cases $Ru(bipy)_3^{2+}$ or related chromophores served as sensitizers. These were used in conjunction with two redox catalysts, i.e. Pt and RuO_2, to promote water reduction and oxidation, respectively. An exception is the Prussian blue catalyzed system of Kaneko et al. (21). It is advantageous to use titania as a catalyst support since the TiO_2 particles are able to serve as oxygen carriers, as was discussed already above. This allows to sustain the water cleavage reaction in closed systems. However, clay-supported catalysts (again, a combination of Pt and RuO_2) have also been employed in conjunction with $Ru(bipy)_3^{2+}$ as a sensitizer (22).

Apart from particulate dispersions, membrane-based systems are becoming increasingly important. In this context, attention is drawn to a recent patent by Toshiba Corporation (23). This system uses a TiO_2 membrane loaded on one side with Pt and on the other with RuO_2. A $Ru(bipy)_3^{2+}$-derivative is used as a sensitizer and is attached to the RuO_2 side of the membrane. Light induced charge injection is followed by electron migration to the Pt side where hydrogen is evolved. The $Ru(bipy)_3^{3+}$ complex evolves oxygen from water which reaction is promoted by the colloidal RuO_2 catalyst (24). A similar approach has been published by Velasco (25) who employed a Nafion membrane to separate the same two catalysts. In his case, light was absorbed by a combination of two sensitizers $(Rubipy)_3^{2+}$ and $Ru(bipyrazyl)_3^{2+}$ and the electron was transported across the membrane by a viologen type carrier.

In our earlier studies (26), we used $Ru(bipy)_3^{2+}$ or a surfactant derivative as a sensitizer and added in some cases methyl viologen as electron acceptor. These were used in connection with an industrially developed catalyst consisting of amorphous oriented anatase particles loaded with Pt and RuO_2 (27). Initial results were promising, however, the reproducibility of catalyst preparation, in particular the method of loading with Pt, turned out to be a problem. Conventional exchange or impregnation procedures could not be applied since it was important to avoid exposing the TiO_2 support to high temperatures which would have decreased its surface area and degree of hydroxylation. Both properties were found to have a crucial influence on water cleavage activity. An attempt to reproduce the earlier results with a similar catalyst (28) yielded hydrogen and oxygen generation under UV-light excitation, while the quantum yield of H_2 evolved in the visible was very small ($\phi \sim 10^{-4}$).

Over the last few years, it has been possible to improve the reliability of systems that afford water cleavage by visible light. A new and exciting development in this area has been the sensitization of TiO_2 by surface derivatization with transition metal complexes (29). For ex. irradiation of acidic (pH 2) solutions of RuL_3^{2+} $2Cl^-$ (L = 2,2'-bipyridine-4,4'-di-carboxylic acid) in the presence of TiO_2 at 100°C leads to the loss of one bipyridyl ligand and the chemical fixation of the RuL_2^{2+} fragment at the surface of the TiO_2 particles through formation of Ru-O-Ti bonds. Recent investigations have shown that a redox mechanism is likely to be operative in the ligand loss process. The RuL_3^{2+} excited state undergoes oxidative quenching at the TiO_2 surface:

$$*RuL_3^{2+} \xrightarrow{TiO_2} RuL_3^{3+} + TiO_2(e^-_{cb}) \quad (11)$$

Subsequently, the Ru(III) complex releases a ligand forming the RuL_2 fragment:

$$RuL_3^{2+} \rightarrow RuL_2^{2+} + L \quad (12)$$

which is fixed at the TiO_2 surface.

These surface complexes are very stable and shift the absorption onset of TiO_2 beyond 600 nm. The reflectance spectrum shown in Fig. 4 exhibits apart from the bandgap transition of TiO_2 below 400 nm a pronounced absorption in the visible with a maximum at 480 nm and a tail extending beyond 600 nm. The features in the visible resemble closely those observed for cis-$Ru(bipy)_2(H_2O)_2$ adsorbed onto hectorite and are therefore attributed to cis-RuL_2^{2+}, chemically linked to the TiO_2 particles via one or two oxygen bridges.

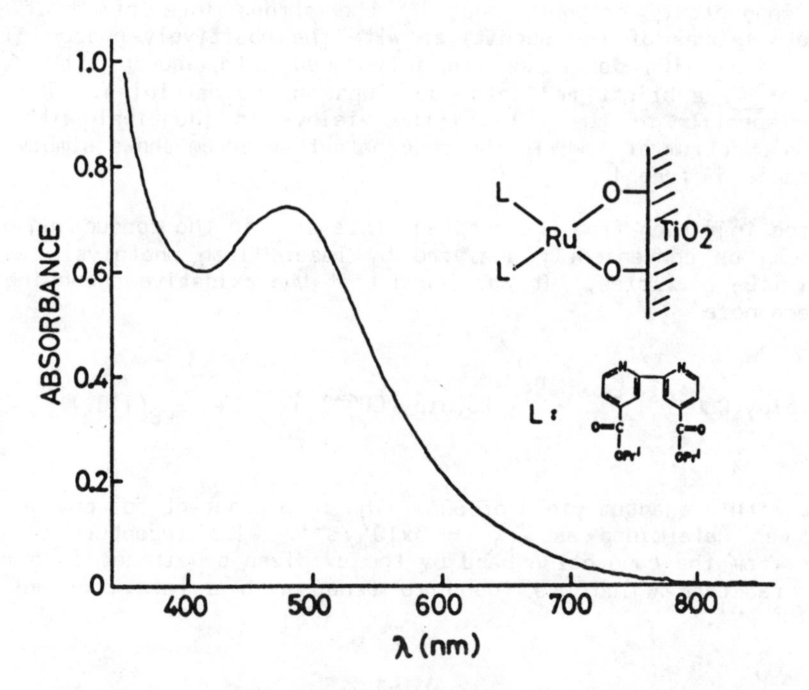

Fig. 4 Reflectance spectrum of RuL_2^{2+}-derivatized TiO_2 particles loaded simultaneously with 0.5% Pt and RuO_2. Absorption maximum in the visible is 480 nm.

RuL_2^{2+}-derivatized TiO_2 particles loaded with RuO_2 and Pt are active in producing hydrogen from water by visible light in the absence of sacrificial organic donor. Illumination at 100°C of 50 mg catalyst dissolved in 40 ml H_2O (pH 2, HCl) with $\lambda > 420$ nm light resulted in the formation of H_2 with an initial rate of 30 μl/h. Typically, 400 μl of H_2 were produced during 20 hours of irradiation. The catalyst maintained its activity over at least two weeks of photolysis at 100°C during which it was exposed to various pH conditions and repeatedly washed with water and centrifuged. The total amount of H_2 produced corresponded to a turnover number of 80 with respect to RuL_2^{3+}. We have also observed O_2 generation during photolysis at 100°C, and injection of 400 μl gas sampled after 12 hours of irradiation showed that 240 and 120 μl H_2 and O_2, respectively, were produced. Oxygen appearance in the gas phase was not consistent observed, however. In particular, at lower than boiling temperature only H_2 was found. We attribute this effect to photouptake of O_2 by the TiO_2 particles and have obtained evidence for the occurrence of such a process by peroxide analysis as described in ref. (16).

Further studies showed that that ruthenium tris (2,2'-bipyridil-4,4'-dicarboxylate), $Ru(bipy(COO^-)_2)_3^{4-}$, 1, in contrast to $Ru(bipy)_3^{2+}$ is a potent sensitizer of TiO_2. In acidic aqueous solution 1 is strongly adsorbed onto the surface of TiO_2. This chemisorption is undoubtedly brought about by the strong interaction of the carboxylate groups of the sensitizer with the positively charged TiO_2 surface. When TiO_2 particles are introduced into aqueous (pH < 5) solution of 1, a bright red color develops on the particles. The reflectance spectrum of the TiO_2 in the visible is identical with the absorption spectrum of 1 while the supernatant spectrum shows simply the disappearance of free 1.

Charge injection from the excited state of 1 in the conduction band of TiO_2 can be conveniently analyzed by laser flash photolysis using colloidal TiO_2 particles. It was found that the oxidative quenching of this chromophore

$$^{*}Ru(bipy(COO^-)_2)_3^{4-} \xrightarrow{k_{inj}} Ru(bipy(COO^-)_2)_3^{3-} + e^-_{CB}(TiO_2) \qquad (13)$$

occurred with a quantum yield of 60%. The rate constant for charge injection was determined as $k_{inj} = 3\times10^7\ s^{-1}$. The recapture of the electron from the conduction band by the oxidized sensitizer is a much slower reaction which was found to occur with a rate constant of $k_b = 2\times10^5\ s^{-1}$.

$$e^-_{CB}(TiO_2) + Ru(bipy(COO^-)_2)_3^{3-} \xrightarrow{k_b} Ru(bipy(COO^-)_3)_3^{4-} \qquad (14)$$

Thus, in this colloidal semiconductor/sensitizer system charge separation by light is sustained over at least several microseconds. This suffices to capture the electron by a suitable catalyst such as Pt deposited on the TiO_2 surface which affords water reduction to hydrogen. Fig. 5 gives a schematic outline of these elementary processes.

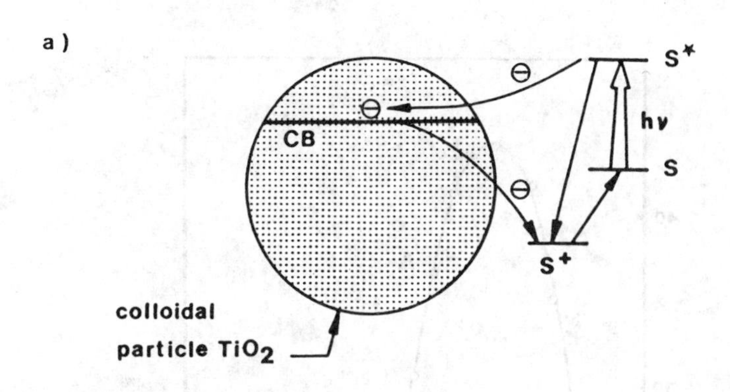

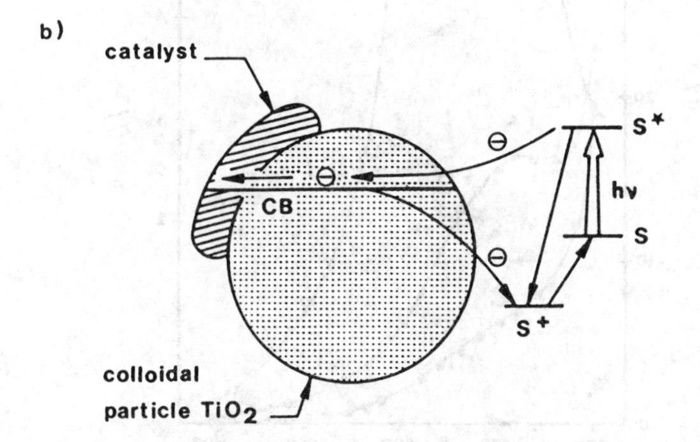

Fig. 5 Charge injection and intraparticle back electron transfer in the photosensitization of a semiconductor particle.
a) without redox catalyst;
b) particles loaded with catalyst.

Occurrence of efficient charge injection from 1 into the conduction band of TiO_2 was confirmed by photoelectrochemical investigations. These employed a polycrystalline anatase electrode (30). Loading of the electrode with sensitizer was carried out by dipping it for 30 min. in a 1.5×10^{-4} M solution of 1 (pH 4) and subsequent rinsing with water. Coloration of the TiO_2 surface by the chromophore is readily visible. Strikingly high photocurrents under visible light excitation were obtained with such electrodes. Monochromatic incident photon to current conversion efficiencies as high as 44% were obtained at the wavelength of maximum absorption in the presence of 10^{-3} M hydroquinone as supersensitizer, Fig. 6.

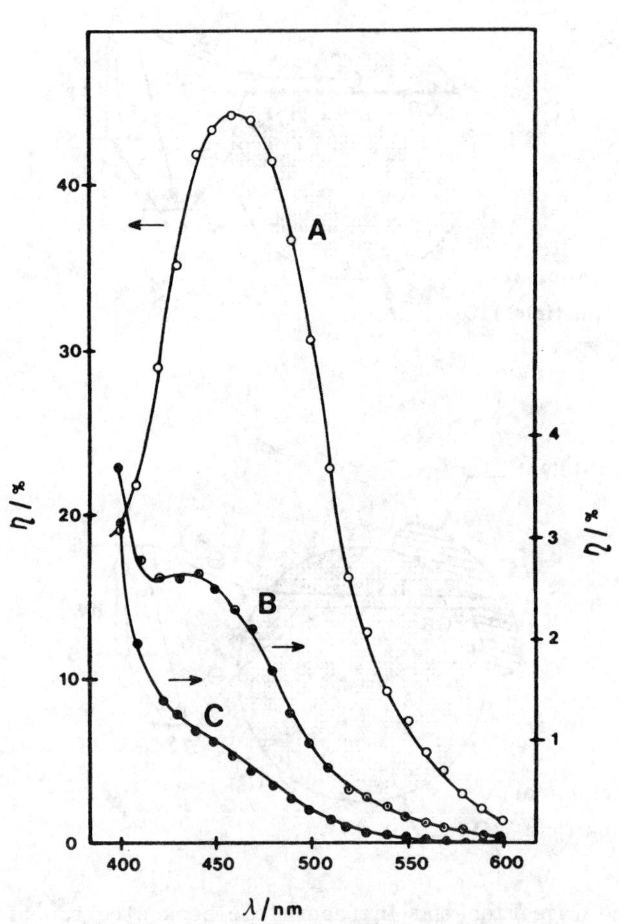

Fig. 6 Photocurrent action spectrum in the visible of the polycrystalline TiO_2 (anatase) electrode sensitized by 1, $Ru(bpy)_3^{2+}$ (B), and without sensitizer (C). Current efficiencies were calculated by dividing the photocurrent density by the incident monochromatic light flux. Solutions contain 10^{-2} M NaCl and 10^{-3} M hydroquinone. Higher hydroquinone concentrations did not improve efficiencies.
Conditions: (A) pH 2,6, adjusted with HCl, 0.0 V (SCE);
(B) 1.5×10^{-4} M $Ru(bpy)_3^{2+}$, pH 7, -0.1 V (SCE);
(C) no sensitizer, pH 7, -0.1 V (SCE).

Apart from strong adhesion and efficient charge injection, this effect must be attributed to the roughness of the electrode surface. On a smooth surface, 1 absorbs only ca. 1% of 470 nm light at monolayer coverage. However, the surface of the electrode employed here is rough and porous, the roughness factor being around 200, Fig. 7. Such an electrode adsorbs a considerably larger amount of sensitizer than a smooth surface. Combined with the fact that the pores act as a light trap inducing multiple reflection, this might lead to a practically total absorption of incident 470 nm light by the chromophore. The coupling of sensitization of TiO_2 by $Ru(bipy(COO^-)_2)_3^{4-}$ to water oxidation using heterogeneous (RuO_2) and homogeneous oxygen evolution catalysts is presently being investigated.

Fig. 7 Scanning electron micrograph of the electrode surface employed in the photochemical studies.

A NEW HOMOGENEOUS WATER OXIDATION CATALYST DERIVED FROM RUTHENIUM TRIS 2,2-BIPYRIDYL, 5,5'-DICARBOXYLIC ACID.

In view of the very interesting properties of carboxylated ruthenium tris bipyridyl complexes as sensitizers of titanium oxide particles and electrodes, we undertook a broad investigation of photoredox reactions of this class of chromophores in aqueous solution. These studies concerned first Ru(II) tris (2,2'-bipyridyl, 4,4'-dicarboxlate). At pH < 2 the carboxylate groups of this complex are fully protonated. The protonated form has a redox potential of 1.56 V (NHE) which is ca. 300 mV more positive than that of $Ru(bipy)_3^{2+}$. Oxygen evolution was observed when acidic solutions of this sensitizer were irradiated with visible light (31) in the presence of sacrificial electron acceptors such as $S_2O_8^{2-}$. This was attributed to the in situ formation of a molecular catalyst for the oxidation of water to oxygen. The stability of the catalyst was, however, poor and oxygen generation ceased after a few turnover numbers. A drastic enhancement of the stability of the catalyst was achieved by using the related isomer, i.e. ruthenium tris (2,2-bipyridyl-5,5'-dicarboxylate), as a starting compound in the photolysis. Turnover numbers exceeding 1000 were readily obtained during irradiation. At this time, a more detailed investigation of the chemical nature and the mechanism of operation of the homogeneous oxygen generation catalyst generated in this system was warranted.

Careful spectroscopic analysis of the product formed under photolysis of ruthenium tris (2,2'-bipyridyl-5,5'-dicarboxylate) in 1N H_2SO_4 alowed to identify the catalytically active species as the mixed valent, μ-oxygen bridged dimeric complex, i.e.:

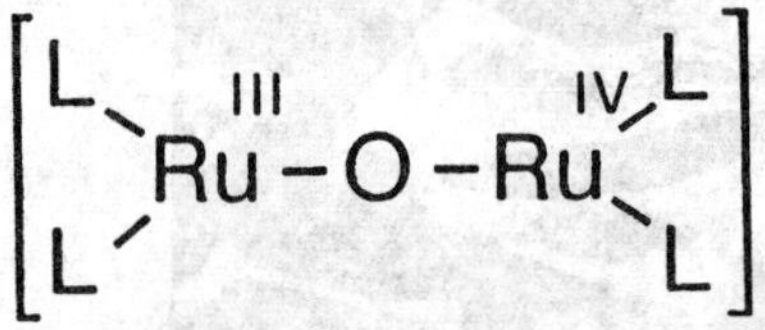

$$\left[L_2 \overset{III}{Ru} - O - \overset{IV}{Ru} L_2 \right]$$

where L stands for:

HOOC COOH

(2,2'-bipyridyl-5,5'-dicarboxylic acid structure: two pyridine rings, N N)

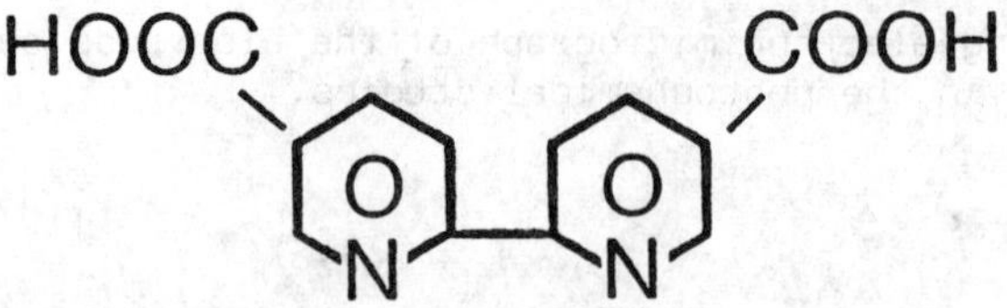

The absorption spectrum in Fig. 8 is similar to that of the simple bipyridyl complex (32) $(bipy)_2$ Ru(III)-O-Ru(IV) $(bipy)_2$. The μ-oxo dimer was identified to be the highly active homogeneous catalyst for the oxidation of water to oxygen.

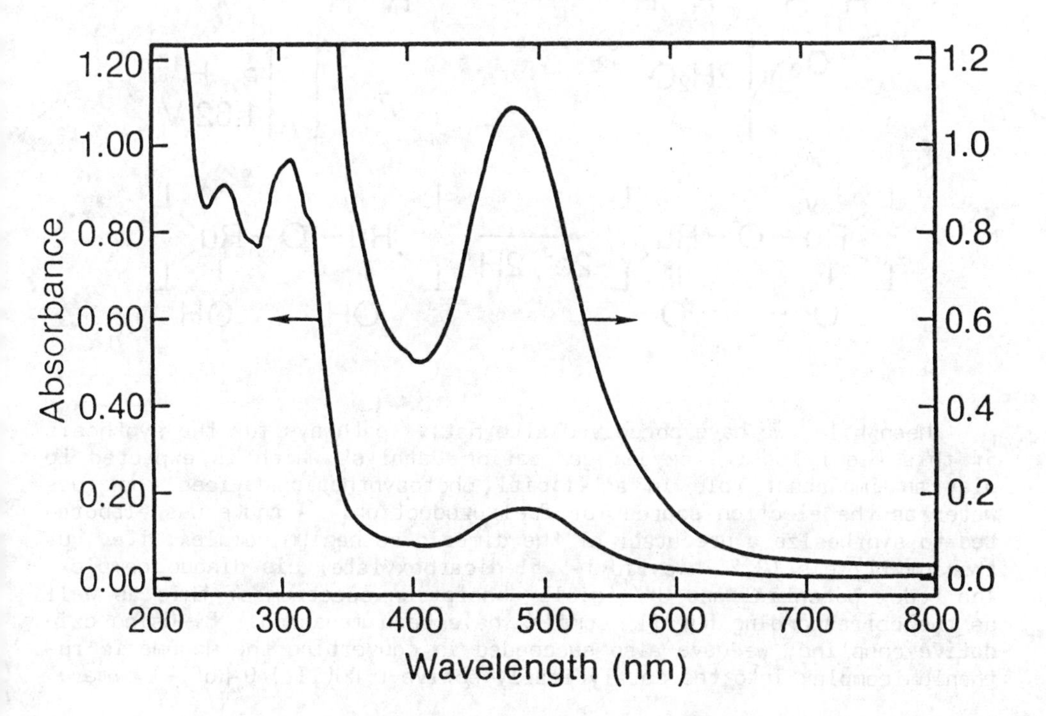

Fig. 8 Absorption spectrum of an aqueous solution (1N H_2SO_4) of L_2 Ru(III)-O-Ru(IV)L_2; the maximum in the visible is around 480 nm with an extinction coefficient of 10'000 M^{-1} cm^{-1}.

A study was undertaken to analyze the electrochemical properties of the μ-oxygen bridged ruthenium dimer. The presence of a pronouned catalytic wave in the cyclic voltammogram even at high scanning speeds provided strong evidence for the efficient mediation of oxygen evolution from water by this complex. This was fully confirmed by experiments in which oxygen was generated thermally and photochemically. Based on these results, the following scheme for the mechanism of intervention of the dimer in the water oxidation process was elaborated:

Water Decomposition Cycle

L Ru(III)–O–Ru(III) L (each with H_2O) ⇌ (0.66 V / e^-, H^+) L Ru(III)–O–Ru(IV) L (H_2O, OH)

↓↑ e^-, H^+ 1.32 V

L Ru(IV)–O–Ru(IV) L (OH, OH) → ($-2e^-$, $2H^+$) L Ru(V)–O–Ru(V) L (O⋯⋯O)

↑ $2H_2O$, ↖ O_2

Meanwhile, we have conceived alternative pathways for the synthesis of this highly active oxygen generation catalyst which is expected to play an important role in artificial photosynthetic devices that use water as the electron source for fuel production. A route was elaborated to synthesize a precursor of the dimeric ruthenium complex, i.e. ruthenium(II) bis (2,2'-bipyridyl-5,5'-dicarboxylate) cis diaquo complex. The redox potential and UV/visible absorption spectrum of this as well as the corresponding Ru(III) complex have been measured. By using oxidative coupling, we have also succeeded in converting the monomeric ruthenium complex into the catalytically active L Ru(III)-O-Ru(IV) dimer.

CONCLUSIONS.

The catalytic conversion of light into chemical energy is a relatively young field which has received a major impetus from the oil crisis in the beginning of the seventies. Given the short time that has elapsed, the progress made has been enormous. In particular, a wealth of knowledge has been acquired in the domain of light induced and thermal electron transfer processes which constitute a crucial step in the overall conversion process. Research in this area continues to advance very rapidly, a recent highlight being the design of chemically linked donor and acceptor molecules which allows to unravel the intrinsic features of intra-molecular electron transfer processes. The exploration of organized molecular assemblies reaction media of minute size to control kinetically the dynamics of photo-initiated redox events is another focal point in the field of artificial photosynthesis.

ACKNOWLEDGMENT.

This work was supported by a grant from the Gas Research Institute, Chicago, USA (subcontract by the Solar Energy Research Institute, Golden, Colorado, USA) and by the Swiss National Science Foundation.

REFERENCES.

1. R.F. Howe and M. Grätzel, J.Phys.Chem., 89, 4495 (1985).
2. U. Kölle, J. Moser and M. Grätzel, Inorg.Chem., 24, 2253 (1985).
3. G. Rothenberger, J. Moser, M. Grätzel, D.H. Sharma and N. Serpone, J.Am.Chem.Soc., in press.
4. D.A. McQuarrie, J.Appl.Prob., 4, 413 (1967).
5. W. Feller, "An Introduction to Probability Theory and its Applications", Vol. 1, Wiley & Sons, N.W. (1971).
6. J. Moser and M. Grätzel, J.Am.Chem.Soc., 105, 6547 (1983).
7. A.V. Bulatov and M.L. Khidekel, Izv.Akad.Nauk SSSR, Sci.Khim. 1902 (1976).
8. K. Kalyanasundaram, M. Grätzel and E. Pelizzetti, Coord.Chem.Rev., in press.
9. E. Yesodharan and M. Grätzel, Helv.Chim.Acta, 66, 2145 (1983).
10. D. Duonghong, E. Borgarello and M. Grätzel, J.Am.Chem.Soc., 103, 4685 (1981).
11. J.R. Harbour, J. Tromp and M.L. Hair, Can.J.Chem., 63, 204 (1985).
12. B.V. King and F. Freund, Phys.Rev. B, 29, 5814 (1984).
13. B. Gu, J. Kiwi and M. Grätzel, Nouv.J.Chim., 9, 539 (1985).

14. S. Sato and K. Yamaguti, Abstract A-40, Proceedings, Fifth Int'l Conference on Photochemical Conversion of Solar Energy, Osaka, Japan, August 26-31, 1984.
15. E. Borgarello, J. Kiwi, E. Pelizzetti, M. Visca and M. Grätzel, Nature, 284, 158 (1981); J.Am.Chem.Soc., 103, 6423 (1981).
16. S.A. Naman, S.M. Aliwi and K. Al-Emara, Abstract B-27, Proceedings Fifth Int'l Conference on Photochemical Conversion of Solar Energy, Osaka, Japan, August 26-31, 1984.
17. G. Blondeel, A. Harriman, D. Williams, Sol.Energy Mat., 9, 217 (1983).
18. N.M. Dimitrijevic, S. Li and M. Grätzel, J.Am.Chem.Soc., 106, 6565 (1984).
19. D. Meissner, R. Memming, S. Li, S. Yesodharan and M. Grätzel, Ber.Bunsenges.Phys.Chem., 89, 121 (1985).
20. J.R. Harbour, R. Wolkow and M.L. Hair, J.Phys.Chem., 85, 4026 (1981).
21. M. Kaneko, N. Takabayashi, A. Yamada, Chem.Lett., 1647 (1982). M. Kaneko, N. Takabayashi, Y. Yamamauchi, A. Yamada, Bull.Chem.Soc. Jpn, 57, 156 (1984).
22. H. Nijs, J.J. Fripiat, H. van Damme, J.Phys.Chem., 87, 1279 (1983).
23. "Photodecomposition of Water", Toshiba Corp., Japan, Kokai, Tokky-o Koho JP 58, 125 601.
24. J. Kiwi and M. Grätzel, Angew.Chem., Int'l Ed., 18, 624 (1979).
25. J. Gonzalez Velasco, Abstract A-57, Proceedings, Fifth Int'l Conference on Photochemical Conversion of Solar Energy, Osaka, Japan, August 26-31, 1984.
26. E. Borgarello, J. Kiwi, E. Pelizzetti, M. Visca and M. Grätzel, Nature, 284, 158 (1981); J.Am.Chem.Soc., 103, 6423 (1981).
27. European Patent 0043 251.
28. J. Rabani, private communication.
29. D. Duonghong, N. Serpone and M. Grätzel, Helv.Chim.Acta, 67, 1012 (1984).
30. M. Kondelka, A. Monnier, J. Sanchez and J. Augustynski, J.Mol.Cat., 25, 295 (1984).
31. J. Desilvestro, D. Duonghong, M. Kleijn and M. Grätzel, Chimia, 39, 102 (1985).
32. J.A. Gilbert, D.S. Eggleston, W.R. Murphy, D.A. Gesalowitz, S.W. Gersten, D.J. Hodgson and T.J. Meyer, J.Am.Chem.Soc., 107, 3855 (1985).

ELECTRONIC SPECTROSCOPY OF SEMICONDUCTOR CLUSTERS

Louis Brus
AT&T Bell Laboratories
Murray Hill, NJ 07974, USA

ABSTRACT. Semiconductor clusters (crystallites) of 20-30Å diameter are synthesized and stabilized using liquid phase arrested precipitation reactions. Their electronic states are characterized via *in situ* absorption and luminescence spectroscopy. The clusters, although crystalline, are too small to have the bulk semiconductor band gap. The physical nature of their electronic states is discussed.

1. INTRODUCTION

There are two important limiting cases in the electronic structure of crystalline materials. These two limits are very weak and very strong chemical bonding between the individual repeating components (atoms or molecules) in the crystal. In the weak bonding limit, typified by crystals of the rare gases and aromatic hydrocarbons (e.g., crystalline benzene), the crystal electronic states are very similar to the internal electronic states of the components. It is often possible to describe the crystal excited states as an excitation hopping from component to component. The width of the crystal electronic bands is narrow with respect to the spacing between component electronic states.

The strong bonding limit is exemplified, in reduced dimensionality, by systems such as polyacetylene and graphite, and in three dimensions by many inorganic metals, semiconductors, and insulators. In these systems electron delocalization is so extensive that the crystal properties are collective in nature and have no relationship to the electronic states of the components. The intercomponent chemical bonds are strong, with the well known example of diamond where the atomization energy is 7.4 eV per carbon atom in the covalently bonded network.

These strongly bonded materials, which traditionally are the subject of solid state physics, pose novel questions to chemists interested in structure, bonding, and reactivity. Strong chemical bonding and extensive delocalization imply that a small crystallite must achieve "moderate size" before full solid state electron delocalization is achieved. The term "cluster" in this context refers to the intermediate size of more than a few atoms but short of the solid state limit. Scientists interested in the transformation from molecular to bulk solid state behavior seek to define "moderate size" for each type of material and physical property.

E. Pelizzetti and N. Serpone (eds.), Homogeneous and Heterogeneous Photocatalysis, 111–121.

Such clusters represent entire new classes of large molecules, and have hybrid electronic properties not characteristic of either the small molecule or solid state limits. Our understanding of their physical and chemical properties is primitive, due to the experimental problem of reproducible preparation and characterization. These materials almost certainly will have important and presently unpredictable electronic and chemical applications.

Physicists and chemists of diverse backgrounds have began to focus on such clusters.[1] The "physical" approach for single element clusters typically involves gas phase atomic aggregation, followed by characterization of ionization potential, size distributions, and reactivity patterns.[2–4] Novel and intriguing observations have recently been reported. A systematic difficulty in these experiments is structural characterization of clusters with more than a few atoms.

We describe a "chemical" approach involving controlled or arrested cluster precipitation from a liquid phase.[5–14] Through control of reaction kinetics and solvent properties, it is now possible to make and characterize colloidal clusters with narrow size distributions. Such extremely dilute colloidal clusters can be stabilized against aggregation with each other, and are available for *in situ* spectroscopic examination on a time scale of days, in a liquid or frozen glass environment.

Colloidal metals and semiconductors of larger particle size have been employed for decades by chemists interested in surface (interface) photochemistry and catalysis. We describe now our efforts to extend colloidal synthetic methods to make and stablize the smallest possible homogeneous nucleation "seeds" (clusters). We describe in some detail physical characterization via electronic absorption, luminescence, and especially transmission electron microscopy. We also discuss the physical nature of the electronic wavefunctions in these large molecules.

2. EXPERIMENTAL OBSERVATIONS

2.1 Electronic Spectra

The large fraction of experimental work has concentrated on metal sulfide clusters for two principal reasons. First, precipitation from M^{++} and $S^{=}$ ions is straightforward and an extensive colloid literature exists for larger ($\geqslant$100Å) crystallites. Second, and of equal importance, is the fact that the solid state physics of these materials is understood in detail. Especially in the case of CdS, the crystal molecular orbitals (e.g., Bloch functions and band structure) are experimentally known at energies far above the fundamental gap as well as near the gap (E_g = 2.54 eV) itself. Thus an unambiguous comparison of cluster and bulk crystal properties is possible.

We will show in section 3 that the most sensitive indicator of cluster size is the optical spectra in the neighborhood of the band gap. We describe now the experimental observations. The thermally broadened room temperature absorption spectra of large CdS crystals is shown by the solid line in figure 1. This spectrum is featureless and rises sharply in the far ultraviolet. The observed spectrum for $\simeq$50Å diameter crystallites is close to this spectrum, with the exception of a ~0.2 eV blue shift in the apparent band gap.[6] At $\simeq$ 30Å diameter, however, there is a substantial change in the spectrum.[8] The apparent absorption edge shifts blue $\simeq$0.7 eV, and a partially resolved peak is apparent.

We will see that peak is in fact the lowest excited (and delocalized) electronic state ("exciton") of the crystallite.

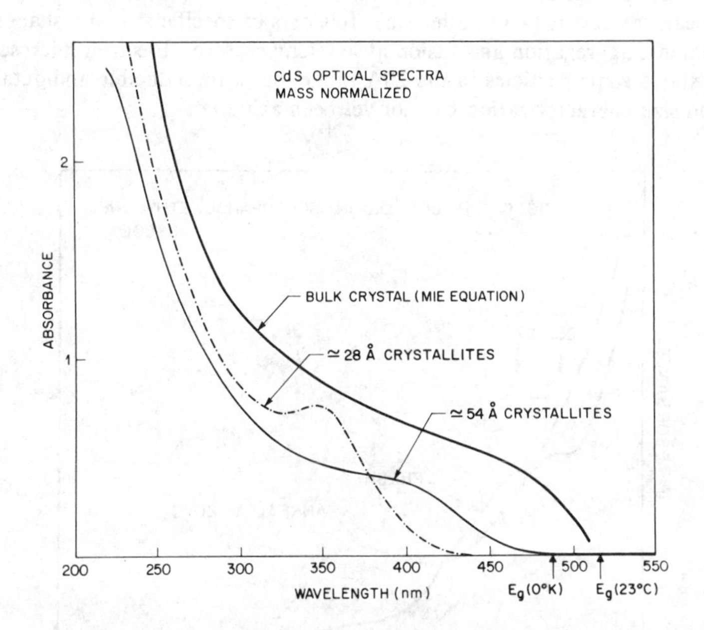

Figure 1. Comparison of optical absorption spectra of ≃54Å diameter crystallites and ≃28Å diameter crystallites with bulk crystal spectra. The ≃28Å crystallites were made by precipitation in methanol at −80°C.

These effects are *not* predicted in classical optics (e.g., Maxwell's equations).[8] It is true that, for larger crystallites with diameters on the same scale as the optical wavelength λ, crystallite spectra are size dependent. This occurs because of contributions from, and interference among, the various electromagnetic multipole scattering moments, each of which has a different size dependence. A well known example is the variation in color of colloidal gold. However, in the present case where we have diameters $<< \lambda$, only the electric dipole scattering term contributes, and the absorption cross section is simply proportional to the crystallite volume. The total crystalline mass, and not the size distribution, determines the optical density. The individual crystallites interact with the electromagnetic field in exactly the same fasion as molecules. We must conclude that these small crystallites are somehow chemically or electronically different than bulk CdS.

In general, we have found that the smallest possible "seeds" are precipitated and stabilized from non-aqueous solvents at the lowest possible temperature, typically just above the solvent melting point. We have avoided using stabilizing surfactant molecules in order to eliminate possible surfactant participation in the optical spectra.

An even larger blue shift occur is observed for synthesis at −80°C in an isopropanol based solvent, as compared with methanol in figure 1, as first reported by Henglein.[12] Figure 2 shows that in the initial −80°C spectra there are resolved peaks at 253 nm and 275 nm. Annealing this colloid at room temperature causes growth of the ≃30Å diameter peak near 350 nm, and disappearance of the 253 and 275 nm peaks. We believe that the two short λ peaks represent two smaller sizes (clusters of specific size and shape) that are only stable against aggregation and fusion at low temperature. Electron microscopic examination shows some particles in the 20Å range, but a reproducible and detailed structural and size characterization has not yet been achieved.

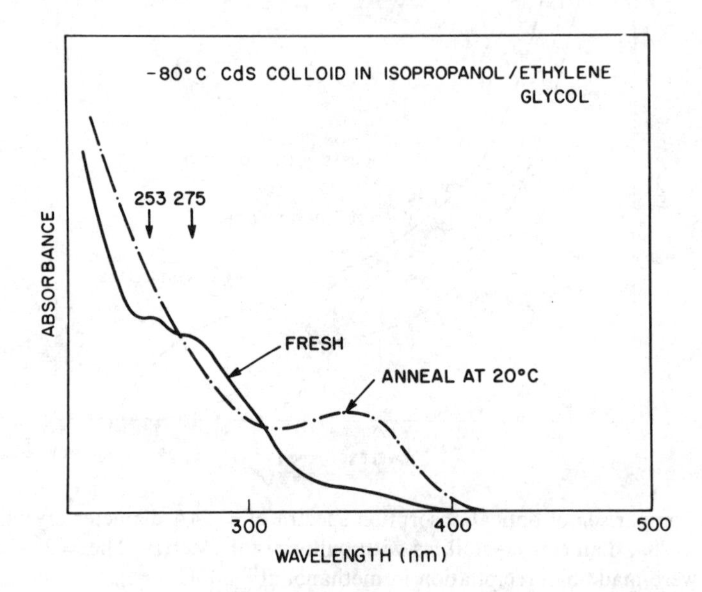

Figure 2. CdS cluster optical spectra at −80°C after precipitation in 8/1 isopropanol/ethylene glycol. The dot-dash curve shows the spectra of larger clusters made by small cluster fusion at 20°C.

An especially clear example of spectral change due to small size is observed in ZnS, a semiconductor with an ultraviolet band gap at 3.85 eV. Figure 3 shows the spectrum of colloidal ZnS crystallites with an average diameter of 20 ± 4Å.[9] The apparent band gap is shifted blue by .6 eV, and the peak at threshold is particularly well resolved. We emphase that this peak grows in relative intensity in small clusters. The figure also shows the temperature dependence observed upon cooling to 77 K in a organic glass forming 5/2 mixture of ethanol/methanol. The peak intensifies, shifts blue and narrows with a shoulder becoming visible at 250 nm. The temperature dependence indicates that the electronic transition is strongly coupled to vibrational degrees of freedom in the ZnS cluster. The peak at 250 nm maybe a higher lying excited state.

Somewhat different phenomena occur in PbS clusters. Bulk PbS is metallic in appearance with an infrared bandgap at 0.3 eV. As shown in figure 4, in ≃25Å clusters the far red and infrared optical absorption characteristic of the bulk material is absent.[10]

The absorption spectrum increases into the ultraviolet without resolved peaks at threshold, in contrast to CdS and ZnS.

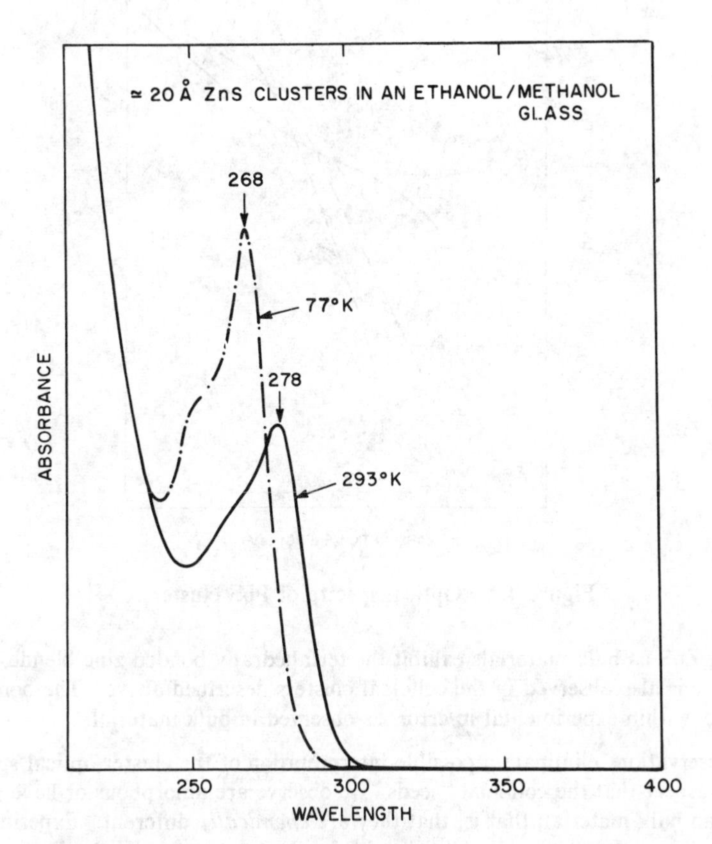

Figure 3. Optical spectra of ZnS clusters as a function of temperature in a glass forming ethanol/methanol solvent.

2.2 Transmission electron micrographs

Electron microscopic structural characterization is invaluable in understanding the cluster optical spectra. Figure 5 shows a micrograph of a ≃25Å PbS cluster at an experimental resolution of 2.5Å (point to point).[10] As bulk material, PbS shows simple cubic (rocksalt) structure in which each ion has 6 nearest neighbors counterions. These 25Å clusters show the same structure. The micrograph shows the stacked atomic planes of a simple cubic lattice. The plane to plane distance is 2.9Å in the cluster as well as in the bulk lattice. The figure demonstrates that direct TEM atomic imaging is a powerful structural tool for clusters.

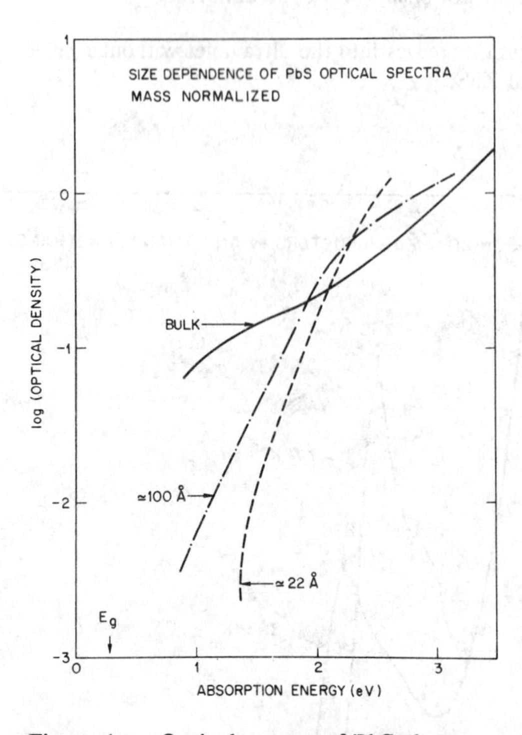

Figure 4. Optical spectra of PbS clusters.

CdS and ZnS as bulk materials exhibit the tetrahedrally bonded zinc blende lattice. This structure is also observed in the colloidal clusters described above. The bond lengths are the same, within experimental in error, as observed in bulk material.

These observations eliminate a possible interpretation of the cluster optical spectra. One might suspect that the colloidal "seeds" we observe are amorphous or have a different structure than bulk material; that is, that they are *chemically* different. Experimentally this is not the case; the same unit cell exists in the cluster as in the bulk. We conclude that these clusters are in fact *electronically* different than bulk CdS. That is, our "seeds" are true clusters as defined in the introduction. This conclusion[6] has been confirmed and extended by recent work in the laboratories of Henglein[12] and Nozik.[14]

While the clusters are essentially crystalline, some questions remain. Our sensitivity to the position of surface atoms is not high. There could be surface reconstruction, as is commonly observed in high vacuum surface physics. The micrographs are also not sensitive to single point lattice defects, such as the sulfur vacancies suggested by luminescence experiments in section 4. Additionally, our accuracy in determination of the absolute bond length is not high.

3. THE QUANTUM SIZE EFFECT IN SEMICONDUCTOR CLUSTERS

The bulk band gap corresponds to the threshold energy necessary to create an electron and hole pair, at rest with respect to the lattice and sufficiently far apart that the Coulomb interaction is unimportant. Above band gap photon absorption at shorter wavelength creates electrons and holes with additional kinetic energy.

In a small crystallite this scheme is modified. A photon creating an electron and hole pair must of necessity supply the kinetic energy necessary to spatially localize the two charges. This is a purely quantum effect increasing as R^{-2}, if R is the crystallite radius. The time average separation between hole and electron is also necessarily small. Thus, there is a Coulomb attraction between electron and hole varying as R^{-1}. The localization energy shifts the optical absorption threshold (i.e., crystallite lowest excited electronic state) to higher energy as R^{-2}, while the Coulomb energy shifts it to lower energy as R^{-1}. For sufficiently small sizes the localization term dominates, and the lowest excited state shifts blue of the bulk band gap. This is the qualitative effect seen in these semiconductor crystallites.

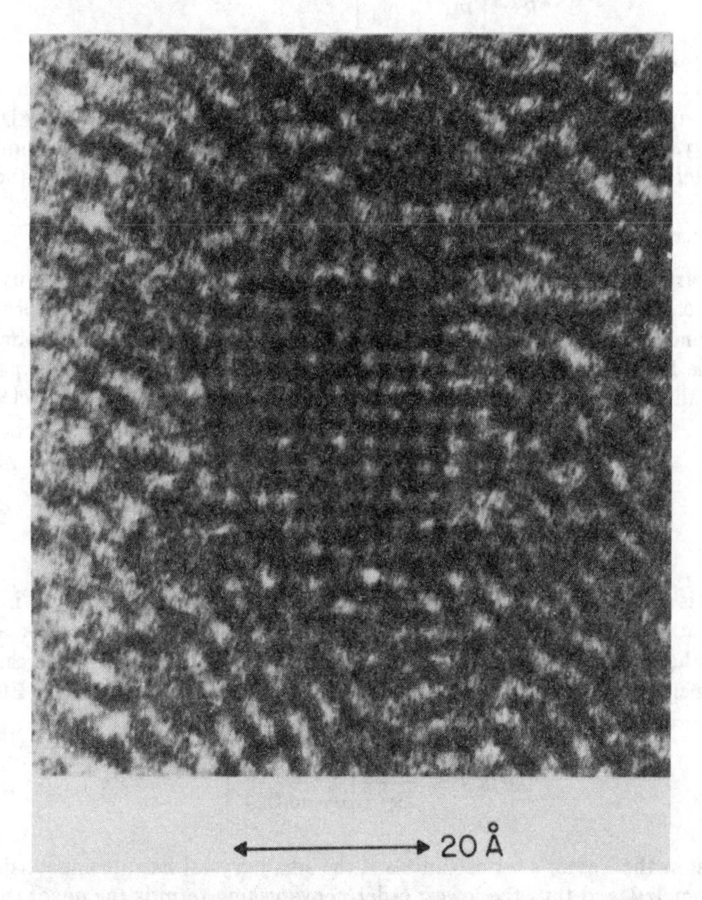

Figure 5. Transmission electron micrograph of a PbS cluster on an amorphous carbon substrate. The lines are planes of atoms viewed edgewise. The amorphous carbon contributes a background speckle.

The fact that the unit cell is the same in the cluster and the bulk material allows us to numerically calculate these two effects, under certain approximations, in the immediate neighborhood of the band gap E_g.[7] Electrons moving through the cluster interior diffract off the nuclei in the same way they diffract in bulk material. This diffraction makes them behave kinematically as if they have an effective mass m^*. If the crystallite surfaces are modelled as infinite potential barriers, then the localization energy is simply the "particle in a box" energy for a particle of effective mass m^*. This result is only valid at energies where m^* is independent of energy. Additionally, the Coulomb term can be estimated, using the high frequency dielectric constant of the bulk semiconductor, by incorporating the effect of the dielectric interface on the Coulomb energy. An approximate result for the energy of the lowest excited state as a function of R is

$$E = E_g + \frac{\hbar^2\pi^2}{2R^2}\left[\frac{1}{m_e} + \frac{1}{m_h}\right] - \frac{1.8e^2}{\epsilon R} + \text{smaller terms} \quad (1)$$

Here m_e and m_h are m^* for the electron and hole respectively. This equation shows that infrared band gap materials, such as GaAs, with small m^* and large ϵ will show large shifts to higher energy. In this case the localization term is far larger than the Coulomb term. The Coulomb term is more important for CdS and ZnS, and calculation shows the quantum localization and Coulomb terms numerically cancel for R = 30–40Å. For smaller R, E increases as shown in the figures.

For small particles the quantum localization term dominates. Equation 1 fails when E comes into regions where m_e or m_h is different than its band gap value. In order to handle this situation, and to consider the effect of spatial localization for molecular orbitals anywhere in the Brillouin zone, we have proposed a general scheme involving expansion of the small crystallite wavefunction Ψ in the infinite crystal, delocalized Bloch molecular orbitals Φ:[10]

$$\Psi_{i,\mu}(r) = \int_k f_i(\bar{k})\phi_{k,\mu}(\bar{r})d\bar{k} \quad (2)$$

where

$$\Phi_{k,\mu}(\bar{r}) = e^{ik\cdot r}\chi_{k,\mu}(r) \quad (3)$$

The basic idea is that one must form a wavepacket of pure k molecular orbitals in order to spatially localize Ψ_i inside the small crystallite. f_i is a Gaussian-like distribution around some central value k_o; the width of f_i in k space increases as R decreases. One can then calculate the energy shift $\Delta E(k_o)$ due to spatial localization at k_o by expanding E(k) in a Taylor's series around k_o. The result is[10]

$$\Delta E(k_o) \simeq \frac{\pi^2\hbar^2}{R^2}\left[\sum_{xyz}\frac{1}{m_i(k_o)}\right] \quad (4)$$

The linear term in the Taylor's series cancels if the small crystal has a compact shape with a center of symmetry, and thus the lowest order nonvanishing term is the quadratic (i.e., effective mass) term in equation 4.

This result indicates that, *for arbitrarily shaped bands*, it is the local effective mass at k_o that controls the quantum localization energy shift. The calculation also indicates that a direct bandgap at the Brillouin zone boundary, as in PbS, is equivalent to a direct gap at the zone center, as in ZnS and CdS. The absolute value of k does not matter. In small clusters the valence and conduction bands will narrow, towards their centers of gravity where $\frac{1}{m} \simeq 0$. The general effect is that the valence to conduction band optical spectra shift blue, and increase in intensity more abruptly with decreasing wavelength. This is the effect observed in PbS in figure 5.

PbS remains an enigma in that there is no resolved lowest excited state ("exciton") peak, as observed in ZnS and CdS. It may simply be that lifetime broadening of the exciton is extreme. An alternative possibility discussed elsewhere is that PbS behaves as an indirect gap semiconductor in 25Å crystallites, due to an unusually shaped valence band.[10]

4. CLUSTER LUMINESCENCE

Time resolved luminescence spectroscopy is a sensative and highly developed tool for study of structure and relaxation dynamics in both molecular chemistry and solid state physics. Room temperature luminescence from aqueous CdS crystallites has been reported, and utilized as a method of monitoring the surface redox chemistry of photogenerated electron and hole pairs.[5,11,13] The luminescence can be quenched extraordinarily efficiently by surface oxidation or reduction of a single adsorbed molecule.

Spectral and temporal resolution of CdS cluster luminescence, as a function of temperature, provides information about cluster "defects" that are not apparent in the optical absorption or TEM spectra.[15] The luminescence of CdS clusters having an absorption threshold at $\simeq$380 nm is a broad structureless band peaking near 600 nm. The luminescence decay is strongly multiexponential at every emission wavelength and temperature, down to 1.7°K as shown in figure 6. The decay lengthens radically at lower temperatures, and finally becomes independent of temperature below $\simeq$20°K. The major components below 20°K are in the 10-40 μsec range. At every temperature, the decay is faster on the high energy side than the low energy side of the emission band.

The long lifetimes and the fact that the emission occurs at far lower energy than the strongly absorbing excited state ("exciton") seen in absorption imply that low lying emitting state has a low oscillator strength and is not apparent in the absorption spectra. In solid state language, both hole and electron must be trapped, and not delocalized over the entire crystallite, to explain the long observed lifetimes. One carrier must be in a rather shallow trap to explain the $\sim$30 meV Arhennius activation energy seen for the longest decay component in figure 6. The other carrier must be in a deep trap to explain the large shift between exciton peak and emission peak. This deep trap is apparently strongly coupled to lattice phonons, as is commonly the case in solid state physics, explaining the broad nature and absence of resolved phonon structure in the emission band.

A possible mechanism is purely radiative tunnelling[16–18] (i.e., distant donor-acceptor pair emission) between a shallowly trapped hole and a deeply trapped electron at a sulfur vacancy. The involvement of sulfur vacancies has been indicated in the room temperature luminescence.[13] In this process close pairs emit at higher energy than distant pairs due to the Coulomb attraction. Close pairs also tunnel faster, and this explains the shorter decays

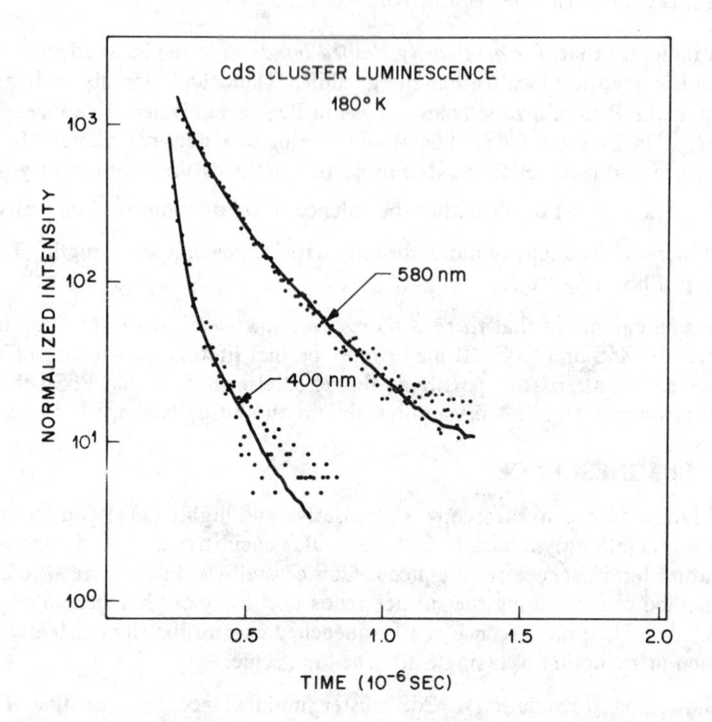

Figure 6. Time resolved luminescence decay of CdS clusters at the indicated emission wavelengths and temperature. The absorption spectrum is the initial −80°C spectrum in figure 2.

at higher energies. The low temperature lifetimes then provide information about the distance between localized electronic states (structural defects and surface states) in these clusters. This subject is under active current investigation.

ACKNOWLEDGEMENT: We have enjoyed fruitful collaboration with T. Harris in the luminescence studies, R. Rossetti in the synthetic efforts, and R. Hull, J. M. Gibson, and S. Nakahara in the TEM characterization

REFERENCES

1) J. Jortner, Ber. Bungenges Phys. Chem. *88*, 188 (1984).

2) R. L. Whetten, D. M. Cox, D. J. Trevor, and A. Kaldor Phys. Rev. Lett. *54*, 1494 (1985).

3) D. E. Powers *et al.*, J. Phys. Chem. *86*, 2556 (1982).

4) V. E. Bondybey and J. H. English, J. Chem. Phys. *76*, 2165 (1982).

5) R. Rossetti and L. Brus, J. Phys. Chem. *86*, 4470 (1982).

6) R. Rossetti, S. Nakahara, and L. E. Brus, J. Chem. Phys. *79*, 1086 (1983).

7) L. E. Brus, J. Chem. Phys. *80*, 4403 (1984).

8) R. Rossetti, J. L. Ellison, J. M. Gibson, and L. E. Brus J. Chem. Phys. *80*, 4464 (1984).

9) R. Rossetti, R. Hull, J. M. Gibson and L. E. Brus, J. Chem. Phys. *82*, 552 (1985).

10) R. Rossetti, R. Hull, J. M. Gibson, and L. E. Brus, "Hybrid Electronic Properties Between the Molecular and Solid State Limits: Lead Sulfide and Silver Halide Crystallites," J. Chem. Phys. (August 1985).

11) A. Henglein, Ber. Bungenges Phys. Chem. *86*, 301 (1982).

12) A. Fojtik, H. Weller, U. Koch and A. Henglein, Ber. Bungenges Phys. Chem. *88*, 969 (1984).

13) J. J. Ramsden and M. Grätzel, J. Chem. Soc. Faraday Trans. I, *80*, 919 (1984).

14) A. J. Nozik, F. Williams, M. T. Nenadovic, T. Rajh, and O. I. Micic J. Phys. Chem. *89*, 397 (1985).

15) T. Harris and L. E. Brus, unpublished data.

16) D. G. Thomas, J. J. Hopfield and W. M. Augustyniak, Phys. Rev. *140A*, 202 (1965).

17) K. Era, S. Shionoya, Y. Washizawa, and H. Ohmatsu J. Phys. Chem. Solids *29*, 1843 (1968).

18) K. Era, S. Shionoya, and Y. Washizawa, J. Phys. Chem. Solids *29*, 1827 (1968).

CATALYSED DECAY OF OXIDISING RADICALS IN WATER

A. Harriman
Davy Faraday Research Laboratory, The Royal Institution,
21 Albemarle Street, London, W1X 4BS, U.K.

P. Neta and M.C. Richoux
Center for Chemical Physics, National Bureau of Standards,
Gaithersburg, MD20899, U.S.A.

ABSTRACT

The pulse radiolytic technique has been used to generate a variety of oxidising radicals in aqueous solution. The oxidants include halide radicals, $bipy_3Ru^{3+}$, porphyrin π-radical cations and oxidised promethazine. Most of the radicals are unstable with respect to disproportionation, although the timescale for this deactivation varies enormously. Addition of colloidal $RuO_2.2H_2O$ or Prussian Blue, both reasonable O_2- evolving catalysts, can enhance the rate of decay of the oxidising radicals. With unstable radicals, the catalyst simply enhances disproportionation. Stable but weakly oxidising radicals can transfer charge to the catalyst until equilibrium is attained. This can involve storage of several hundred oxidising equivalents on each colloidal particle . With strong oxidising radicals that are also stable in water, the catalyst can cause water oxidation.

INTRODUCTION

Pulse radiolysis has been employed to monitor kinetic parameters associated with the catalysed reduction of water to H_2 (1 - 7). This work has been highly successful and many useful concepts have been identified. For example, it is known now that colloidal particles of Ag function as microelectrodes and store hundreds of electrons on one particle (1). Analysis of the results in terms of electrochemical parameters(3) has enabled calculation of the total charge stored on the particle. Discharge occurs via water reduction. With colloidal Pt particles, H_2 formation is much easier because of the more favourable overpotential (5) and few electrons are stored on the particle. In fact, it is believed that the electrons are stored on the catalyst surface in the form of hydrogen atoms or hydride ions (7).

Gold sols have also been used to catalyse the reduction of water to H_2 using methyl viologen as an electron relay (6). It was shown that the nature of the sol support had a pronounced effect upon the

E. Pelizzetti and N. Serpone (eds.), Homogeneous and Heterogeneous Photocatalysis, 123–145.

efficiency of H_2 generation. With a charged support, complexation and dimerisation of the relay became important whilst H_2 formation was the major deactivation route when neutral supports were used (6).

Many of these studies have shown that the charged particles live for reasonable times in aqueous solution. Discharge can be observed on the timescale of many seconds (3,6) for Ag and Au colloids, depending upon pH. Also, the importance of interparticle reactions has been noted (6).

Very little work has been reported on the kinetic parameters associated with O_2 evolution on colloidal particles. Some stopped-flow(8) and cyclic voltammetric (9) studies have been described in preliminary form but no detailed work has followed. Preliminary experiments have been made using pulse radiolysis to monitor interaction between oxidised zinc porphyrin and RuO_2 deposited upon colloidal TiO_2 (10,11). This work provided strong evidence to show that the metalloporphyrin π-radical cation interacted with the catalyst particle on a fast timescale. It was also shown from conductivity measurements that protons were ejected over the same time domain. Our own subsequent work (12) confirmed that the metalloporphyrin π-radical cation interacted with colloidal RuO_2 at the diffusional controlled rate limit.

Recent work by Henglein (13) has described oxidative processes on Pt colloids when attacked by strong oxidising radicals (e.g. $OH^{\cdot}$). Absorption changes consistent with oxide formation have been reported. Again, the reaction between oxidising radical and catalyst particle was diffusion controlled (13) but, in most cases, the fate of this reaction was not identified.

We have started experiments aimed at identifying the nature of the interaction between oxidised radicals and O_2-evolving catalysts. These experiments rely heavily upon the technique of pulse radiolysis and, normally, we have used optical absorption to monitor the fate of the radicals and/or catalyst species. As catalyst, we have used Prussian Blue or colloidal $RuO_2.2H_2O$. These catalysts are not necessarily the most effective O_2- evolving catalysts available and it is realised that, in particular, $RuO_2.2H_2O$ is prone towards anodic corrosion. Their usage has been controlled by their compatibility with the pulse radiolysis technique - especially light absorption and scattering. However, it has been demonstrated in independent experiments (14,15) that they are active O_2-evolving catalysts when Ce^{4+} (in 1N H_2SO_4) or tris (2,2'-bipyridyl)ruthe nium (III) ($2 < pH < 7$) ions are used as one electron oxidants.

EXPERIMENTAL

Zinc meso-tetrakis (4-sulphonatophenyl) porphine ($ZnTSPP^{4-}$, sodium salt) and the isomeric zinc meso-tetrakis (N-methyl-4-pyridyl) porphine ($ZnTMPyP^{4+}$, tosylate salts) were prepared and purified as described previously and full details regarding the pulse radiolytic oxidation of the compounds is given elsewhere (12). In brief, an aqueous solution containing metalloporphyrin ($1x10^{-4}$ M), KBr ($1x10^{-2}$ M) and buffer ($1x10^{-3}$ M) was saturated with N_2O and irradiated with a single 50 ns pulse of 2MeV electrons delivered with a Febetron 705 accelerator. The usual radiation dose was 500 rd, as measured by KSCN dosimetry, and the concentration of metalloporphyrin π-radical cation (ZnP+•), formed according to

$$ZnP + Br_2^{-\bullet} \rightarrow ZnP^{+\bullet} + 2Br^-$$

was monitored at 700 nm. Usually, the buildup of radical cation was complete with 100 μs and the final concentration was ca. 3×10^{-6} M. A fresh aliquot of solution was used for each pulse.

The rate of decay of the metalloporphyrin radical cations was measured spectrophotometrically at 700 nm in the presence of various concentrations of added catalyst. In the absence of catalyst, decay occurred by mixed kinetics and it was approximated to a first order decay using a computer "best fit" procedure. In the presence of catalyst, the decay remained mixed and it was only on rare occasions that a clean pseudo first order decay was found. Several traces were averaged for each point but, even so, the reported lifetimes are approximations. This is particularly so for the values observed in the absence of catalyst. Plots of first order rate constants vs concentration of catalyst were linear, at least over the limited concentration range studied, and the bimolecular rate constants (k_{cat}) were calculated from the slopes. These rate constants have an expected accuracy of ±25%.

For steady-state irradiations, a 150W quartz/halogen light source was filtered to remove IR and light of λ < 520 nm. A solution of the metalloporphyrin (8×10^{-5} M) in water containing $Na_2S_2O_8$ (5×10^{-3} M), Na_2SO_4 (0.46 M) and the appropriate buffer (1×10^{-3} M) was purged thoroughly with N_2. The solution was housed in a pyrex membrane polargraphic detector and the concentration of O_2 in the solution was monitored electrochemically and displayed on an x-t chart recorder (16).

Colloidal $RuO_2.2H_2O$ was prepared by dissolving $KRuO_4$ (82 mg) and poly (styrene/maleic anhydride 1/1) (50 mg) in water (50 cm^3). The pH was adjusted to 7 and, after stirring for 30 min, H_2O_2 (1 cm^3 3% aqueous solution) was added. After 30 mins, the solution was heated to ca. 70°C for a few mins to ensure destruction of any residual H_2O_2 and the pH was readjusted to 7 by addition of dilute HCl. In the absence of high concentrations of electrolyte, the colloid was stable against flocculation over several weeks standing and the average particle size was (10 ± 2) nm diameter, as measured by light scattering. Prussian blue (PB) was purchased from Aldrich Chemicals and used as received.

RESULTS and DISCUSSION

Pulse Radiolysis

Radiolysis of water results in formation of several primary radicals together with some molecular products:

$$H_2O \rightsquigarrow H^\bullet + OH^\bullet + e^-_{aq} + H_2 + H_2O_2 + H_3O^+$$

The total radiation yield of radicals is G = 6.0 where G refers to the number of radicals formed per 100 eV of energy absorbed. The solvated electron is a strong reducing agent but it can be transformed into hydroxyl radicals simply by saturating the solution with N_2O.

$$N_2O + e^-_{aq} + H_2O \longrightarrow N_2 + OH^{\bullet} + OH^-$$

Although the hydroxly radicals are powerful oxidising agents they tend to add to unsaturated bonds, forming long-lived adducts. In order to convert these primary radicals into more suitable one-electron oxidants it is usual to add a high concentration (ca. 10^{-2}M) of halide or pseudo-halide ions to the aqueous solution.

$$OH^{\bullet} + N_3^- \longrightarrow OH^- + N_3^{\bullet}$$

This gives rise to the radicals listed in Table 1. The redox potentials of these secondary radicals (E_{ox}) are provided in the table and it is seen that they are strong one-electron oxidants. Most of them are able to oxidise water-soluble metalloporphyrins to the corresponding π-radical cation.

$$MP + N_3^{\bullet} \longrightarrow MP^{\dot{+}} + N_3^-$$

In the presence of N_2O and halide ions, over 90% of the water radicals can be used to oxidise a metalloporphrin or similar substrate. The remainder being reducing $H^{\bullet}$.

Table 1 Halide and pseudo-halide radicals used in pulse radiolytic oxidations

Radical	E_{ox} (V vs NHE)
$OH^{\bullet}$	2.8
$Cl_2^{\dot{-}}$	2.3
$N_3^{\bullet}$	1.9
$Br_2^{\dot{-}}$	1.69
$(SCN)_2^{\dot{-}}$	1.5
$I_2^{\dot{-}}$	1.13

These secondary radicals absorb strongly in the UV region and normally decay by disproportionation. For example, $Cl_2^{\dot{-}}$ has an absorption maximum at 340 nm (ε = 1.2 x 10^4 M^{-1} cm^{-1}) and it disproportionates at the diffusion controlled rate limit (k_D = 4 x 10^9 M^{-1} s^{-1})(17). In a typical pulse radiolysis experiment, the radiation dose is ca.500 rads giving rise to 2 x 10^{-6} M radicals. Thus, the first half-life for $Cl_2^{\dot{-}}$ will be ca. 120 μs. Assuming favourable thermodynamics, $Cl_2^{\dot{-}}$ will oxidise water-soluble additives with a bimolecular rate constant in the range 10^8 - 10^{10} M^{-1} s^{-1} so that the concentration of additive must exceed 10^{-4} M for oxidation to compete favourably with disproportionation.

Radiolytic oxidation of water-soluble metalloporphyrins

Original experiments by Neta (18) showed that the pulse radiolytic technique afforded a very convenient method for formation of metalloporphyrin π-radical cations in aqueous solution. In particular, $Br_2^{\overline{\cdot}}$ was found to be a useful oxidant for zinc porphyrins and it was possible to obtain quantitative conversion of the oxidising radicals into porphyrin π-radical cations. These π-radical cations possess characteristic absorption bands in the near IR and their rate of formation can be followed easily at 700 nm where the ground state compounds do not absorb. By varying the concentration of porphyrin and measuring the pseudofirst order rate constant for absorption growth at 700nm bimolecular rate constants for formation of π-radical cation (k_{ox}) can be derived. Some values are collected in Table 2 for oxidation of the metalloporphyrins at pH7 with $Br_2^{\overline{\cdot}}$ at an ionic strength of 10^{-2}M.

Table 2 Oxidation of zinc porphyrins with $Br_2^{\overline{\cdot}}$ at pH7 (μ = 0.01 M)

Porphyrin	10^{-8} x k_{ox} (M^{-1} s^{-1})	ΔG^o (kJ mol^{-1})
$ZnTSPP^{4-}$	6	-76.2
$ZnTMPyP(2)^{4+}$	41	-49.0
$ZnTMPyP(3)^{4+}$	43	-67.0
$ZnTMPyP(4)^{4+}$	42	-49.2

These bimolecular rate constants are essentially diffusion controlled, taking into account coulombic forces, as expected from the thermodynamamic driving forces for oxidation. Many other metalloporphyrins, not only zinc porphyrins, are oxidised readily under such conditions and the other oxidising radicals listed in Table 1 can be used in place of $Br_2^{\overline{\cdot}}$.

Figure 1 gives the absorption spectrum of $ZnTMPyP(3)^{5+\cdot}$ as formed after pulse radiolytic oxidation of the ground state porphyrin with $Br_2^{\overline{\cdot}}$ at pH7. It is seen that there are two overlapping transient species observed during the course of the experiment. The rapidly decaying species (k_D = 5.7 x 10^3 s^{-1}) absorbs strongly around 400 nm and very weakly at 700 nm. This absorption is assigned to the adduct formed between porphyrin and H°. On longer timescales, the π-radical cation persists. Because of problems with the H° adduct, all kinetic measurements pertinent to the π-radical cation should be made at 700 nm if possible.

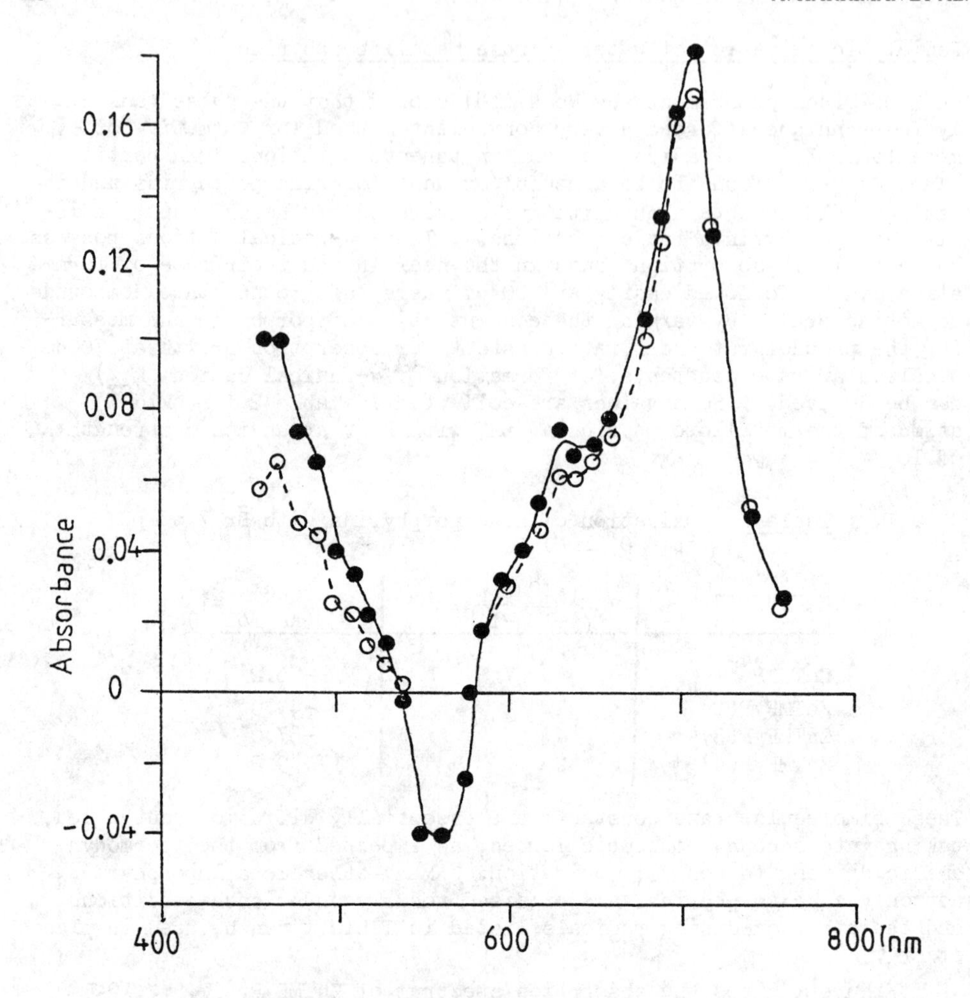

Fig. 1 Absorption spectra recorded after oxidation of $ZnTMPyP(3)^{4+}$ in water at pH 7 with Br_2^-; spectra recorded after 10 us (o) and 1 ms (•).

A further problem encountered with these experiments concerns the observation that the absorption spectra of $ZnTMPyP(4)^{5+\cdot}$ depend markedly upon the nature of the oxidising radical. This effect is shown in Figure 2.

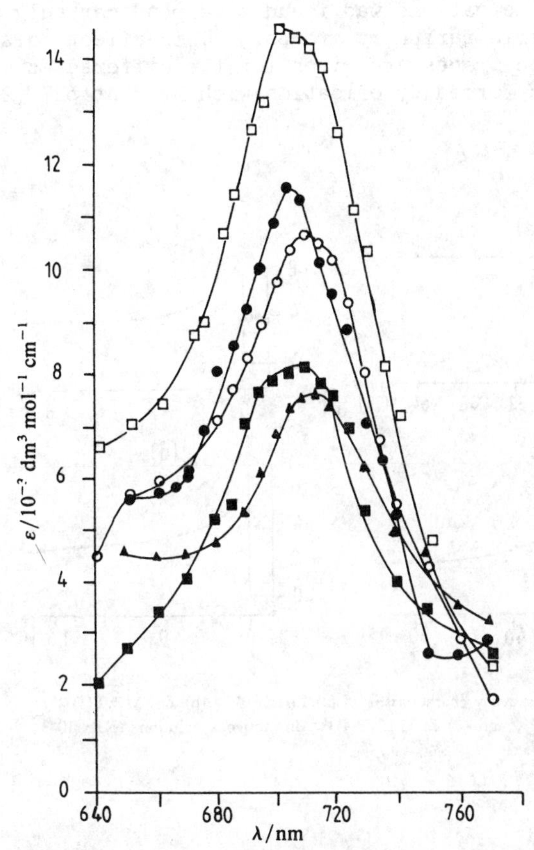

Fig. 2 Absorption spectra of $ZnTMPyP^{\cdot 5+}$ as formed by pulse-radiolytic oxidation of $ZnTMPyP^{4+}$ (3×10^{-4} mol dm^{-3}) in N_2O-saturated aqueous solution with various oxidants. The spectra were recorded after completion of the formation reaction (20–100 μs after the pulse). The solutions contained ■, $AgClO_4$ (5×10^{-3} mol dm^{-3}) at pH 3 ($HClO_4$); ●, KBr (10^{-2} mol dm^{-3}) at pH 7 (phosphate); ○, NaN_3 (10^{-2} mol dm^{-3}) at pH 7 (phosphate); □, KSCN (10^{-2} mol dm^{-3}) at pH 7 (phosphate) and ▲, KBr (10^{-2} mol dm^{-3}) at pH 12 (KOH).

It is clear that both the molar extinction coefficient and the absorption maximum vary according to the background electrolyte. This suggests complexation between the π-radical cation and anions present in solution (19). Such effects were not noted for the meta isomer nor for $ZnTSPP^{4-}$ and, consequently, all further work is restricted to these two compounds.

Decay of zinc porphyrin π-radical cations in water

Absorption spectra of the various zinc porphyrin π-radical cations are all quite similar to that given in Figure 2, although they do show some dependence upon the type of water-solubilising group (18). However, the stability of the cations was found to depend markedly upon the nature of the porphyrin periphery groups. This effect is shown in Figure 3 where kinetic traces are given for the different zinc porphyrin π-radical cations formed by oxidation with $Br_2^{\cdot -}$ at pH7 (20).

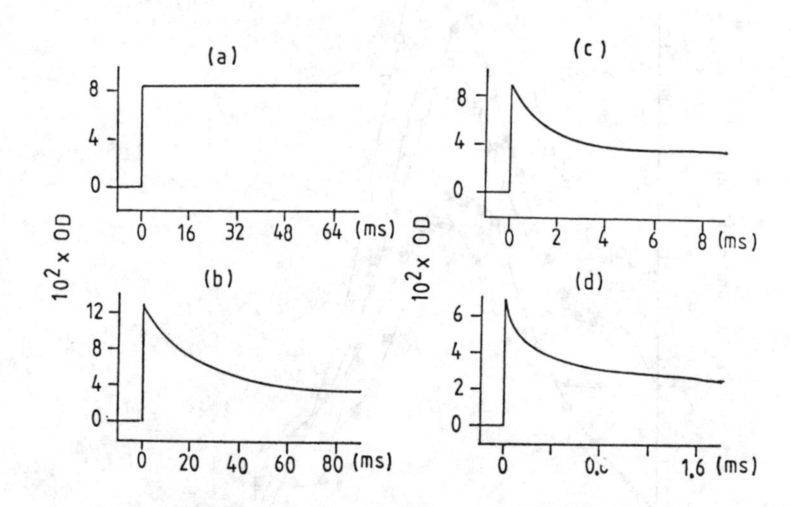

FIG. 3 Traces showing decay characteristic of (a) $ZnTSPP^{\cdot 3-}$, (b) $ZnTMPyP(3)^{\cdot 5+}$, (c) $ZnTMPyP(2)^{\cdot 5+}$, and (d) $ZnTMPyP(4)^{\cdot 5+}$ in aqueous solution (KBr 10^{-2} mol dm^{-3}) at pH7.

It is seen that $ZnTSPP^{3\cdot -}$ is stable on the time scale of the experiment and, on longer times, it was found that $ZnTSPP^{3\cdot -}$ decayed via an approximately first order process with a half-life of 6 s. The decay route possibly involves reduction by radiolytic products (e.g. H_2O_2). In contrast, the radicals derived from the isomeric $TMPyP^{4+}$ complexes decay rapidly. In all cases, there is a fast second order process that depletes ca. 50% of the absorbance at 700 nm. This decay occurs on the ms timescale and as shown in Figure 3, the relative stabilities of the various $TMPyP^{4+}$ radicals follow the order;

para < ortho < meta.

The residual absorbance decays via first order kinetics, as shown in Figure 4 for $ZnTMPyP(4)^{5\dot{+}}$.

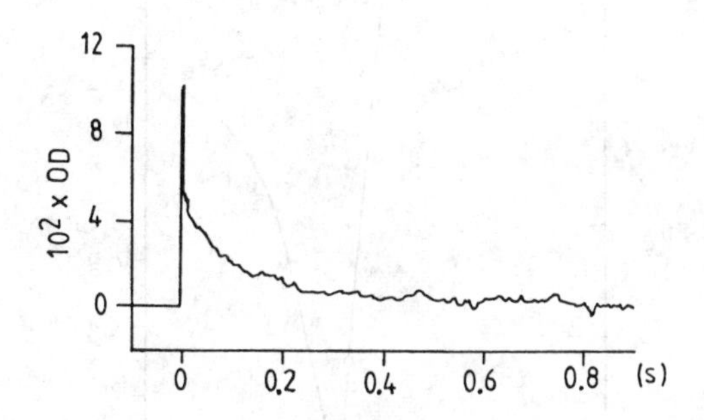

FIG. 4 Decay of $ZnTMPyP(4)^{\cdot 5+}$ in aqueous solution (KBr 10^{-2} M) at pH7 on a longer time scale.

From product analysis (21), together with chemical and photochemical studies, these decay processes can be interpreted in terms of the following scheme:

$$2\ ZnP^{\dot{+}} \rightleftharpoons (ZnP^{\dot{+}})_2 \rightleftharpoons ZnP + ZnP^{2+}$$
$$(ZnP^{\dot{+}})_2 \downarrow 2\ ZnP$$

Here, disproportionation of the π-radical cations involves transient formation of a dimer which decays, at least in part, to form a dication (ZnP^{2+}). This species is unstable with respect to attack by hydroxide ions and is rapidly converted into an isoporphyrin (21). The isoporphyrin is the long-lived transient species and it is slowly converted into rearranged products (21).

The lifetimes of the various radicals were found to depend upon pH, as shown in Figure 5. In all cases, decay increases with increasing pH. The effect is due to acid/base equilibria associated with an axially coordinated water molecule (19).

$$H_2O\text{–}(Zn)^{\dot{+}} \rightleftharpoons H^+ + OH\text{–}(Zn)^{\cdot}$$

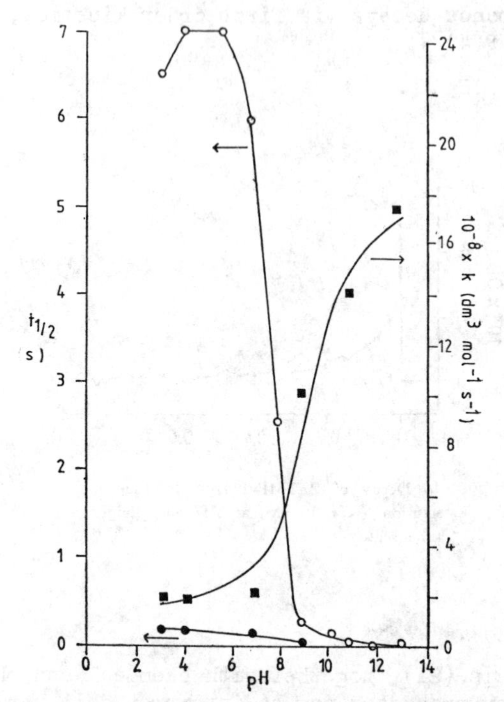

FIG. 5 Effect of pH on the half-lives for decay of $ZnTSPP\cdot^{3-}$ (○) and $ZnTMPyP(4)\cdot^{5+}$ (■) and the intermediate formed from $ZnTMPyP(4)\cdot^{5+}$ (●); aqueous solution (KBr 10^{-2} M).

Clearly, the $TSPP^{4-}$ complex offers the most stable π-radical cation of the zinc porphyrins studied here. Replacing the central zinc ion with other cations has a large effect upon the radical stability as shown by Table 3.

Table 3 Decay rate constants of metal $TSPP^{4-}$ π-radical cations in water

metal ion	$10^{-7} \times k_D$ $(M^{-1}s^{-1})$	Z
Mg(II)	0	1.0
Zn(II)	0	1.5
Pb(II)	1.6	2.0
2H(I)	2.9	-
Fe(III)	130	3.2
Co(III)	210	3.2
In(III)	130	3.5

It is seen that the second order rate constants for decay of the π-radical cations depend upon the induction parameter Z (22), which is a measure of the electronegativity of the central metal ion. Overall, any effect that pushes electron density onto the porphyrin ring (i.e. electron donating substituents on the periphery or electronegative cations) will raise the stability of the π-radical cation and vice versa.

To demonstrate this effect more clearly, the π-radical cations derived from the isomeric tetrakis (hydroxyphenyl) porphyrins were studied. It was found (23) that in neutral solution all the π-radical cations decayed via disproportionation with $k_D = (2.0 \pm 0.3) \times 10^8\ M^{-1}s^{-1}$. In alkaline solution, the hydroxyl groups are deprotonated but the π-radical cations derived from the ortho and meta isomers still decay fairly quickly via disproportionation ($k \sim 6 \times 10^7\ M^{-1}\ s^{-1}$). In contrast, the para isomer does not decay in basic solution. For this π-radical cation, resonance structures such as

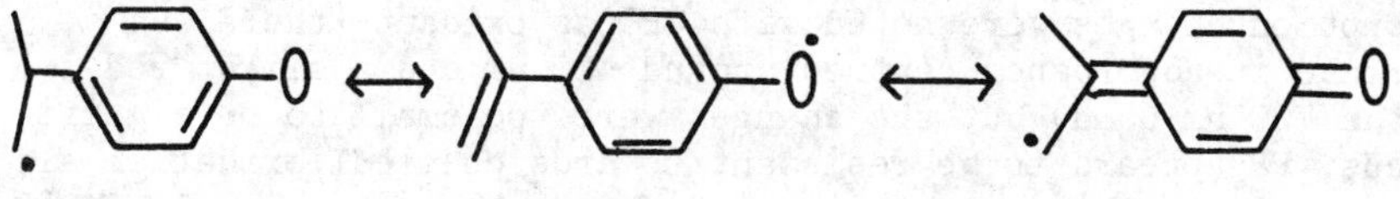

O ⟷ Ȯ ⟷ =O

can be drawn which allow the radical character to be removed entirely from the porphyrin ring. These resonance forms cannot be drawn for the ortho and meta isomers.

Catalysts

Two types of O_2-evolving catalyst have been used in this work, namely colloidal $RuO_2.2H_2O$ and Prussian Blue (PB). The colloidal oxide was studied because we have found it to be one of the best O_2-evolving catalysts in photochemical systems (24). It is stable over a wide pH range and, because of its low light scattering, it can be used in experiments relying upon optical absorption measurements. However, it does absorb light throughout the visible region and it precipitates at high electrolyte concentrations. PB was studied because it possesses an intense intervalence charge transfer absorption band, centre at 720 nm, that might allow kinetic measurements to be made on the catalyst itself.

Insoluble PB ($Fe_4(FeCN_6)_3$) disperses in water to form a three dimensional polymeric structure consisting of alternate ferric and ferrous ions on cubic lattice sites. The ferric ions are coordinated to the N atoms and the ferrous ions to the C atoms of bridging cyanide groups and the structure can interchelate K^+ or Na^+ ions. The open structure of PB is somewhat akin to a zeolite and it can accommodate large cations and water molecules. Dispersions of PB in water are completely stable throughout the range $1 < pH < 12$ although precipitation occurs at high ionic strength and the colour is bleached at $pH > 12$. Light scattering from the colloid is minimal at concentrations below 10^{-4} M. The compound is photostable, even upon prolonged irradiation with visible light, and the addition of strong oxidants (e.g. Cl_2, Br_2, $S_2O_8^{2-}$) had no effect upon the intensity or position of of the intervalence band. Similar oxidation studies performed under pulse radiolytic conditions showed that the long wave-length absorption band of PB was unaffected by the presence of $Br_2^{\cdot-}$, $SO_4^{\cdot-}$, $N_3^{\cdot}$ or Tl^{2+}. Small absorption changes were noted with $OH^{\cdot}$ as oxidant (there was a small increase in absorbance centred around 420 nm and a slight red shift of the 720 nm band) but the changes were too small to be significant. Thus, PB appears to be resistant towards chemical oxidation although recent electrochemical studies (25) have shown that insoluble PB adsorbed onto electrodes has a reversible wave at 1.11V vs NHE and an irreversible wave at about 1.5V vs NHE. Electrochemical oxidation at 1.3V resulted in absorption changes similar to those noted for oxidaation with $OH^{\cdot}$ but they were much clearer. Electrochemical oxidation at > 1.6V oxidised the cyanide ligands and resulted in complete bleaching of the compound. The absorption changes noted in the electrochemical experiments depend upon the extent of oxidation and it is possible that the absence of absorption changes in the pulse radiolysis studies is due to the low degree of oxidation that would occur under such conditions.

Hydrous RuO_2 has not been well characterised. It has an average particle radius of 5 nm which corresponds to an aggregate number of 3700. With neutral reactants, the bimolecular rate constant for diffusional encounter with such a particle is 8×10^9 $M^{-1}s^{-1}$. The surface area is $3 \times 10^{-12} cm^2$ and the percentage $RuO_2.2H_2O$ on the surface is ca. 20%. The colloid is protected with poly(styrene/maleic anhydride) copolymer which, of course, is associated with the surface. There will also be oxyanions (RuO_4^-, RuO_4^{2-}) in contact with the surface so that, overall the material is quite complex.

The hydrous oxide is amorphous and it exhibits no X-ray diffraction patterns. Upon heating, the material loses two molecules of water in successive steps; one water is lost at ca. 120°C and the second at ca. 180° C. Cyclic voltammograms recorded with the colloid in IN H_2SO_4 using a C paste working electrode showed a small reversible peak centred at 0.8V which decreased by 59mV per pH unit. This probably corresponds to charge storage via :

$$2\,RuO_2.2H_2O + 2e \rightleftharpoons Ru_2O_3.3H_2O + 2\,OH^- \qquad (E^o = 0.80V)$$

At higher potentials, a reversible wave occurs at 1.05V, independent of pH, which most probably corresponds to oxidation of surface bound perruthenate ions.

$$RuO_4 + e \rightleftharpoons RuO_4^- \qquad (E^o = 1.05V)$$

At still higher potentials, irreversible waves are found that almost certainly correspond to anodic corrosion processes.

$$RuO_4 + 4H^+ + 4e \rightleftharpoons RuO_2.2H_2O \qquad (E^o = 1.49\ V)$$

$$RuO_4^- + e \rightleftharpoons RuO_4^{2-} \qquad (E^o = 1.40\ V)$$

In basic solution, ruthenate ions are stable and may participate in the redox chemistry.

$$RuO_4^{2-} + 4H_2O + 2e \rightleftharpoons RuO_2.2H_2O + 4OH^- \qquad (E^o = 1.05\ V)$$

Thus, the electrochemistry of the colloid is quite complicated and there are a number of processes that can compete with water oxidation. In addition, the colloid surface will be negatively charged (pzc = 3) so that complexation with cations might be expected.

Effects of catalysts on secondary radicals

As mentioned earlier, we could not observe any direct reaction between PB and secondary radicals such as $Cl_2^{-\cdot}$, $Br_2^{-\cdot}$ or $N_3^{\cdot}$. Because this catalyst absorbs strongly in the near UV region it was not possible to monitor decay of the radicals in the presence of high concentrations of PB. Such experiments can be made with colloidal $RuO_2.2H_2O$ although light absorption by the catalyst restricts its concentration to less than 5×10^{-4} M.

Experiments were performed with $Cl_2^{-\cdot}$ as oxidant at pH 2.8 in 0.1 M NaCl. A low radiation dose was used (250 rad) to slow down the rate of disproportionation of the radical

$$2\ Cl_2^{-\cdot} \rightleftharpoons Cl_2 + 2\ Cl^-$$

$$Cl_2 + H_2O \rightleftharpoons HOCl + HCl$$

but, even so, the average first half-life of $Cl_2^{-\cdot}$ was only 220 μs. Addition of colloidal $RuO_2.2H_2O$ increased the decay of $Cl_2^{-\cdot}$, as monitored at 340 nm, and at a catalyst concentration greater than 10^{-4} M the decay profile could be satisfactorily approximated to pseudofirst order kinetics. At lower colloid concentrations, the decay profile was mixed first and second order but the pseudofirst order rate constant could still be extracted (26). Figure 6 shows the derived plot for the pseudofirst order rate constant (k) vs the concentration of added $RuO_2.2H_2O$. From the gradient, the bimolecular rate constant for interaction between $Cl_2^{-\cdot}$ and $RuO_2.2H_2O$ is $(1.2 \pm 0.2) \times 10^7\ M^{-1}s^{-1}$.

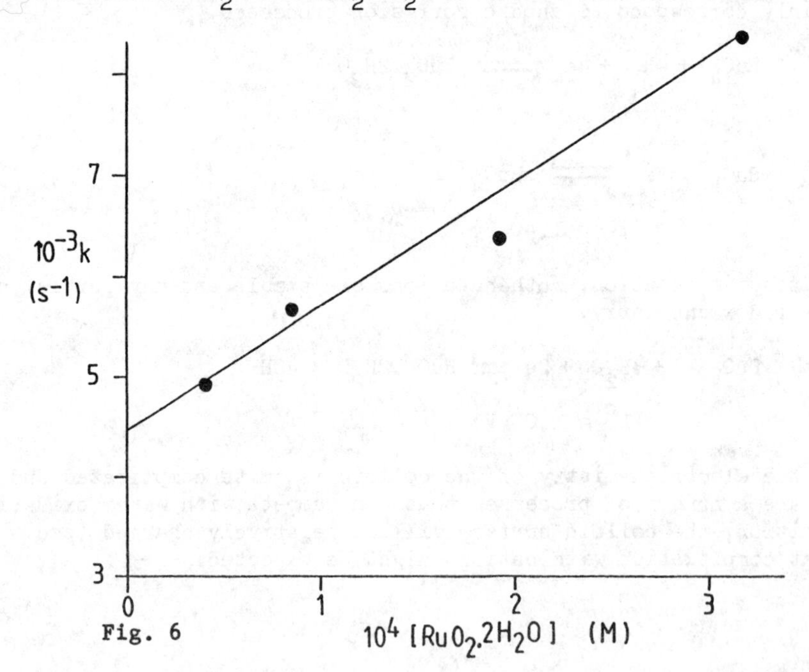

Fig. 6

In these experiments, decay of the $Cl_2^{-\cdot}$ absorbance at 340 nm was complete to the baseline in a single step. Correcting the bimolecular rate constant for the particulate nature of the colloid gives an encounter rate constant of (1.2 x 10^7 x 3700) 4.4 x 10^{10} $M^{-1}s^{-1}$. A similar bimolecular rate constant was found for reaction between the colloid and $Br_2^{-\cdot}$ at pH7. Clearly these rate constants are diffusion controlled and they serve to demonstrate the high efficiency with which the colloid can react.

The outcome of reaction between $Cl_2^{-\cdot}$ and colloidal $RuO_2.2H_2O$ was not studied. Probably, a multitude of processes occur including oxidation of surface bound RuO_4^- and Cl^- ions, oxidation of the colloid to RuO_4 and disproportionation of $Cl_2^{-\cdot}$. The high redox potential of the radical should allow water oxidation to proceed but, in the presence of high concentrations of Cl^-, this is unlikely to occur.

Reaction between PB and $bipy_3Ru^{3+}$

This reaction was studied in order to obtain kinetic information about any direct interaction between an oxidising species and PB. Oxidation of $bipy_3Ru^{2+}$ with $Cl_2^{-\cdot}$ or $Br_2^{-\cdot}$ under pulse radiolytic conditions results in quantitative formation of $bipy_3Ru^{3+}$; the rate of decay of which increases with increasing pH. At pH 3.2, the half-life of $bipy_3Ru^{3+}$ was ca.0.1s, as monitored at 610 nm. Addition of PB caused an enhancement in the rate of decay of $bipy_3Ru^{3+}$. By measuring the pseudofirst order rate constant for decay of $bipy_3Ru^{3+}$ as a function of the concentration of PB, the bimolecular rate constant for the reaction

$$bipy_3Ru^{3+} + PB \longrightarrow bipy_3Ru^{2+} + PB^+$$

$$\Delta G^o = -25.1 \text{ kJ mol}^{-1}$$

was found to be (2.1 $\pm$ 0.5) x 10^5 $M^{-1}s^{-1}$ at an ionic strength of 0.01 M. Since the extinction coefficient of $bipy_3Ru^{3+}$ at 610 nm is low, the derived rate constant is rather inaccurate.

In separate experiments, it was shown the PB catalysed the oxidation of water to O_2 using $bipy_3Ru^{3+}$ as the oxidant. Oxygen formation was observed in both photochemical systems, using persulphate as sacrificial acceptor (21), and in chemical systems in which solid $bipy_3Ru^{3+}$ was added to buffered water containing a suspension of PB. The optimum pH for O_2 formation was (7.0 $\pm$ 0.5) where the yield of O_2 was about 40% of that expected from the stoichiometric oxidation of water. Under these conditions there is some complexation between oxidant and PB which might account for some of the inefficiency. Also, $bipy_3Ru^{3+}$ is fairly unstable in water at pH7 so that there are kinetic problems associated with O_2 production.

Reaction between promethazine and colloidal $RuO_2.2H_2O$

This system was studied because oxidation of promethazine (Pz) at pH <5 gives rise to a stable π-radical cation ($t_{\frac{1}{2}}$ > 10 s). This is considerably more stable than $bipy_3Ru^{3+}$ and oxidation product absorbs strongly at around 500 nm. Pz was oxidised with $Br_2^{-\cdot}$ (KBr = 0.03M) at pH 4.4 and the pseudofirst order rate constant for decay of $Pz^{\cdot+}$ was measured as a function of the concentration of added colloidal $RuO_2.2H_2O$. Again a linear relationship was observed (Figure 7) from which the bimolecular rate constant for interaction was calculated to be (5.0 ± 0.2) x 10^5 $M^{-1}s^{-1}$. Correction for the aggregation number of the colloid gives an encounter rate constant of (5 x 3700 x 10^5) 1.85 x 10^9 M^{-1} s^{-1}. This value approaches the diffusion controlled limit.

In these experiments, decay of $Pz^{\cdot+}$ was incomplete and the amount of residual $Pz^{\cdot+}$ (%R) decreased with increased concentration of colloid. This effect is also shown in Figure 7.

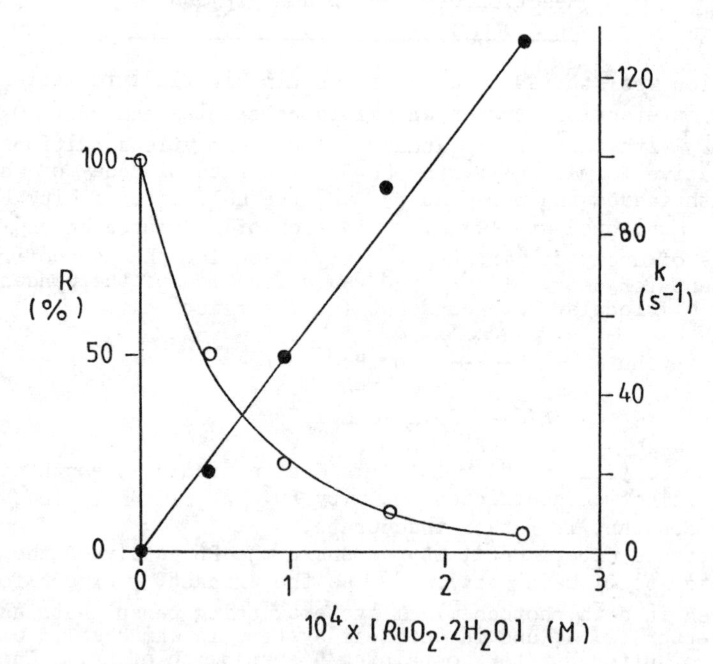

Fig. 7 Effect of colloidal $RuO_2.2H_2O$ upon the first order decay rate constant (•) and the percentage residual oxidised form (o) for promethazine at pH 4.4.

Thus, it appears that some kind of an equilibrium is established. From the redox potential of the $Pz^{\cdot+}$/Pz couple (E° = 0.80V) it is clear that water oxidation is improbable at pH < 7 so that the observed reaction with the colloid probably involves charge storage. If this is the case the Nernst equation can be used to calculate the specific capacity of the colloid.

Thus, the concentration of $Pz^{+\cdot}$ present immediately after completion of the formation reaction was 3.6×10^{-6} M and the initial concentration of Pz was 5×10^{-4} M. Assuming equilibrium between $Pz^{+\cdot}$ and the colloid is attained, the voltage on the colloid will be equal to

$$E = 0.80 + 0.059 \log ([Pz^{+\cdot}]/[Pz])$$

and the total amount of charge transferred to the colloid is given by:

$$Q = F\left(3.6 \times 10^{-6} \times \left(\frac{100 - \%R}{100}\right)\right)$$

The capacitance of the catalyst can be expressed in the form

$$C = Q/E$$

and, using the surface area of the colloid particles, the specific capacity of the catalyst becomes $8 \mu F/cm^2$. Compared to compact RuO_2 electrodes, this is a very low value but it is known that the capacitance of RuO_2 depends markedly upon the state of hydration.

The number of oxidising equivalents transferred to a single particle can be expressed as

$$N = Q/F\,[\text{Particles}].$$

At the lowest concentration of $RuO_2.2H_2O$ used in these experiments, N equals 133 and corresponds to about 18% of the total surface RuO_2 molecules. Thus, the surface has not been saturated and, consequently, the charging reaction can still occur at the diffusion controlled limit. It is of interest to perform experiments in which the number of oxidising equivalents greatly exceeded the total concentration of surface $RuO_2.2H_2O$ but this has not been possible as yet.

Catalysed decay of metal $TMPyP^{4+}$ complexes

As described above, the metal $TMPyP^{4+}$ complexes readily undergo one electron oxidations but the resultant π-radical cations disproportionate rapidly. The nature of the central metal ion has a pronounced effect upon the rate constant for disproportionation but even in the most favourable case this is still very fast. We have studied the effect of colloidal $RuO_2.2H_2O$ upon the decay of many different metal $TMPyP^{4+}$ π-radical cations in water but the results are very similar in each case. Generally, the colloid catalyses disproportionation of the π-radical cation, forming the isoporphyrin, without any involvement of water oxidation. For some systems, the colloid also enhanced conversion of the isoporphyrin into ring-opened products.

Recently, we described in detail the influence of both colloidal $RuO_2.2H_2O$ and PB on the decay of the π-radical cation derived from $ZnTMPyP(3)^{4+}$ (21). The two catalysts exhibited essentially identical behaviour. Disproportionation was enhanced significantly by the presence of the catalyst. For both systems, deprotonation of the axially bound water molecule caused a considerable increase in the rate of the disproportionation step, probably due to lowering coulombic repulsion forces. Careful monitoring with a Clark membrane electrode showed that there was no O_2 formed upon contact between π-radical cation and catalyst.

These π-radical cations, which are inherently unstable, seem to have no capability to oxidise water to O_2 in the presence of such catalysts. This is simply because the catalyst seems able to enhance the disproportionation reaction without directing any of the oxidising equivalents towards water oxidation. Perhaps this has something to do with the charge on the reactants. Both catalysts present a negative surface charge that might complex the positively charged porphyrins. Thus, there will always be some $TMPyP^{4+}$ in contact with the catalyst. Oxidising this surface bound porphyrin might be a way to discharge any oxidising equivalents stored on the catalysts. If this is the case, disproportionation will always be favoured over water oxidation. In this respect, it is important to note that we have repeatedly failed to induce O_2 generation from such systems.

Catalysed decay of metal $TSPP^{4-}$ complexes

Oxidation of metal $TSPP^{4-}$ complexes can give rise to stable π-radical cations. Again, the stability depends upon the state of protonation of any axially-bound water molecule with the radical being less stable in alkaline solution. In general, the oxidation potentials of the $TSPP^{4-}$ complexes tend to be about 0.32V lower than the corresponding $TMPyP^{4+}$ complexes, but their electrochemistry is much more reversible. We have studied the effect of colloidal $RuO_2 2H_2O$ and PB on the rate of decay of several metal $TSPP^{4-}$-π-radical cations and, below, we summarise our findings with $ZnTSPP^{4-}$.

Decay of the π-radical cation derived from $ZnTSPP^{4-}$, obtained by pulse radiolytic oxidation with $Br_2^{\bar{\cdot}}$ at pH 9 and 11, was not affected by the presence of PB. We calculate that the rate constant for reaction

$$ZnTSPP^{3\bar{\cdot}} + PB \longrightarrow ZnTSPP^{4-} + PB^{+}$$

$$\Delta G^{o} = 20.3 \text{ kJ mol}^{-1}$$

is less than 200 M^{-1} s^{-1} (μ = 0.01 M). Under identical experimental conditions, the rate constant for oxidation of ferrocyanide by the π-radical cation

$$ZnTSPP^{3\bar{\cdot}} + Fe(CN)_6^{4-} \longrightarrow ZnTSPP^{4-} + Fe(CN)_6^{3-}$$

$$\Delta G^{o} = -52.1 \text{ kJ mol}^{-1}$$

was found to be (3.7 ± 0.3) x 10^7 M^{-1} s^{-1}. On the basis of these experiments, we conclude that if PB exists in aqueous solution in equilibrium with free ferrocyanide then the degree of dissociation must be less than 10^{-2}%. Water oxidation does not occur in any of these systems.

The rate of decay of $ZnTSPP^{3-\bullet}$ was considerably enhanced by the presence of colloidal $RuO_2.2H_2O$ (12). As before, by measuring the pseudofirst order rate constant for decay of the radical as a function of the concentration of added colloid, bimolecular encounter rate constants (k_{cat}) could be calculated. It was noted this rate constant depended upon pH as shown by Figure 8.

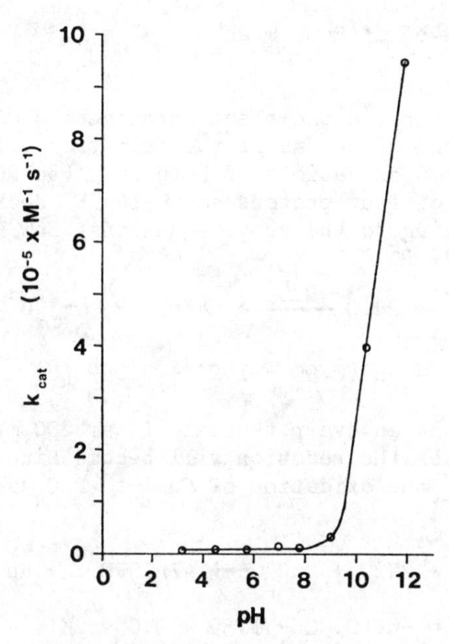

FIG. 8 Effect of pH upon the derived k_{CAT} for $ZnTSPP^{\cdot 3-}$.

Thus in alkaline solution k_{cat} approaches the diffusion controlled rate limit but it is very slow below pH9.

The redox potentials for the relevant couples are given below (vs NHE):

$$ZnTSPP^{\cdot 3-} + e \rightleftharpoons ZnTSPP^{4-} \qquad E_{\frac{1}{2}} = 0.90\ V$$

$$RuO_4 + e \rightleftharpoons RuO_4^- \qquad E^o = 0.95\ V$$

$$O_2 + 4H^+ + 4e \rightleftharpoons 2H_2O \qquad E^o = 1.23\ V$$

$$RuO_4 + 4H^+ + 4e \rightleftharpoons RuO_2 + 2H_2O \qquad E^o = 1.39\ V$$

Half-wave potentials for the one-electron oxidation of the metalloporphyrins are independent of pH, as is the oxidation of RuO_4^- ions. However, the four electron oxidations of both RuO_2 (to RuO_4) and water(to O_2) involve transfer of four protons so that E^o decreases by 59 mV per pH unit. According to the above equations, $ZnTSPP^{\cdot 3-}$ is able to oxidise water to O_2 at pH > 6:

$$4\ ZnTSPP^{\cdot 3-} + 2H_2O \xrightarrow{RuO_2} 4ZnTSPP^{4-} + 4H^+ + O_2$$

$$\Delta G^o = -nF(0.90 - 1.23 + 0.059\ pH)$$

but since there will be an overpotential of ca. 200 mV for this process on the catalyst surface the reaction will become kinetically fast only at pH > ca.9. Also, the oxidation of RuO_2 to RuO_4 should become important at around pH9:

$$4ZnTSPP^{\cdot 3-} + 2H_2O + RuO_2 \longrightarrow 4ZnTSPP^{4-} + RuO_4 + 4H^+$$

$$\Delta G^o = -nF(0.90 - 1.39 + 0.059\ pH)$$

so that the observed enhancement in k_{cat} at pH > 9 probably reflects the onset of the oxidation of water and RuO_2. Below this pH, water oxidation should be quite slow and the major electron transfer processes are probably oxidation of surface-bound RuO_4^- ions.

$$ZnTSPP^{3-\cdot} + RuO_4^- \longrightarrow ZnTSPP^{4-} + RuO_4$$

$$\Delta G^o = 5\ kJ\ mol^{-1}$$

$$k = 1.6 \times 10^6\ M^{-1}\ s^{-1}\ (pH\ 11)$$

and the normal charging of the colloid as outlined for promethazine.

In separate experiments (12), it was found that the $ZnTSPP^{3-\cdot}$ species was able to oxidise water to O_2 in the presence of colloidal $RuO_2 \cdot 2H_2O$. Both the rate and yield of O_2 generation depended upon pH in the same manner as Fig 8. That is, O_2 formation was efficient only at pH > 10. These results have been described fully (12) and need not be further explained here. However, it is important to note that, under favourable conditions, reaction between the metalloporphyrin π-radical cation and colloidal $RuO_2.2H_2O$ can result in efficient O_2 generation.

Limited amounts of O_2 were found for the other metal $TSPP^{4-}$ π-radical cations (e.g. Mg^{2+}, Cd^{2+}) although $ZnTSPP^{4-}$ was clearly the best. This raises the important question as to why water oxidation proceeds with $TSPP^{4-}$ but not $TMPyP^{4+}$ complexes. It seems probable that this considerable difference in reactivity is associated with the electronic charge on the porphyrin and it has nothing to do with the respective oxidation potentials. Strong coulombic repulsion between the colloid surface and $TSPP^{4-}$ complexes prevents complexation and keeps the reactants some distance apart. Note, the deprotonated zinc tetrakis (4-hydroxyphenyl)porphine π-radical cation does not react with colloidal $RuO_2.2H_2O$. Thus, the colloid cannot discharge by oxidising any surface-bound porphyrin as might be the case with $TMPyP^{4+}$. In addition, $TSPP^{3-\cdot}$ radicals show less tendency to disproportionate than their $TMPyP^{5+\cdot}$ counterparts and this seems to be a necessary condition for water oxidation. Work is continuing in this area and it is hoped that further experiments with the highly stable $MgTSPP^{3-\cdot}$ radical will help clarify the situation.

Catalysed decay of high valence metalloporphyrins

Oxidation of Pb^{2+} porphyrins with $Br_2^{-\cdot}$ radicals at pH9 results in rapid formation of the π-radical cations which disproportionate(k_D = 5.4×10^7 M^{-1} s^{-1}) to form the porphyrin π-dications. These latter species are unstable with respect to rearrangement to the Pb^{4+} porphyrin.

$$Pb^{II}P \xrightarrow{-e} Pb^{II}P^{+\cdot} \overset{x2}{\rightleftharpoons} Pb^{II}P^{2+} \downarrow Pb^{IV}P$$

As before, the rate of disproportionation was enhanced by the presence of colloidal RuO_2. $2H_2O$.

The resultant Pb^{4+} porphyrins decay slowly via first order kinetics ($k = 1\ s^{-1}$), in some cases partially reforming the original Pb^{2+} porphyrin (27). It was found that colloidal $RuO_2.2H_2O$ had no effect upon the rate of this process and we estimate that $k_{cat} < 10^3 M^{-1} s^{-1}$. No O_2 was formed with or without catalyst.

Mn^{3+} porphyrins are oxidised to the Mn^{4+} compounds, which are stable only at pH > 12. Upon standing in the dark the Mn^{4+} porphyrin reverts back to the original compound without loss. This reduction step involves complex kinetics but the overall rate increases markedly with decreasing pH. Experiments with the Clark electrode have shown that no O_2 is evolved during the reduction process.

Preliminary experiments have shown that the rate of the reduction step is enhanced upon addition of colloidal $RuO_2.2H_2O$. It was found that k_{cat} increased with decreasing pH. This follows the increasing thermodynamic driving force for water oxidation. We are now studying this reaction in detail.

CONCLUSION

Our initial experiments have shown that pulse radiolysis provides a convenient technique by which to obtain kinetic information about any bimolecular interaction between catalyst particles and oxidised radicals in aqueous solution. So far, several types of interaction have been identified. With unstable radicals e.g. $Cl_2^{\cdot-}$, the catalyst can enhance the rate of disproportionation. Such systems seem to have no aptitude to oxidise water to O_2. Stable radicals that do not possess the necessary thermodynamic driving force to liberate O_2 from water can transfer charge to the particle. This charge is stored on the surface of the colloid until equilibrium with the oxidising radical is attained. With stable radicals that are also powerful oxidising agents, discharge of the colloid can lead to oxidation of water to O_2.

To date, we have studied only optical absorption changes associated with the oxidising radicals. We have observed no changes associated with the catalyst itself nor have we been able to monitor conductance changes associated with water oxidation. Such studies should soon become possible. Analysis of the kinetic data has been restricted to simple forms so that we have not resolved any of the individual steps in water oxidation. More detailed analysis might provide useful information regarding the mechanism of O_2 liberation on colloidal oxide surfaces.

ACKNOWLEDGEMENT

We thank the S.E.R.C., G.E. (Schenectady) and the Office of Basic Sciences of the U.S. Department of Energy for financial support of this work.

REFERENCES

1. A. Henglein, J. Phys. Chem., 1981, **85**, 1627.
2. A. Henglein, J. Westerhausen and J. Lilie, Ber. Bunsenges Phys. Chem., 1981, **85**, 182.
3. A. Henglein and J. Lilie, J. Am. Chem. Soc., 1981, **103**, 1059
4. A. Henglein, B. Lindig and J. Westerhausen, J. Phys. Chem., 1981, **85**, 1627.
5. A. Harriman, G. Porter and M.C. Richoux, J. Chem. Soc., Faraday Trans, 2, 1982, **78**, 1985.
6. D. Meisel, W.A. Mulac and M.S. Matheson, J. Phys. Chem., 1981, **85**, 179.
7. M. Venturi, Q.G. Mulazzani and M.Z. Hoffman, J. Phys. Chem., 1984, **88**, 912
8. E. Pramauro and E. Pellizzetti, Abstracts 3rd. Int. Conf. Photochem. Conversion and Storage of Solar Energy, Colorado, U.S.A. August 1980.

9. E. Sutcliffe and M. Neuman-Spallart, Helv. Chim. Acta, 1981, 64, 2148.
10. M. Gratzel, in Energy Resources through Photochemistry and Catalysis, ed. M. Gratzel,(Academic Press, New York, 1983) Chap. 3.
11. K. Kalyanasundaram and M. Gratzel, in Photochemical, Photoelectrochemical and Photobiological Processes, eds. D.O.Hall and W. Palz (D. Reidel, Dordrecht, 1981) Vol. 2, p. 129.
12. P.A. Christensen, A. Harriman, G. Porter and P. Neta, J. Chem.Soc. Faraday Trans. 2, 1984, 80, 1451.
13. A. Henglein, B. Lindig and J. Westerhausen, Radiat. Phys. Chem., 1984, 23, 199.
14. A. Harriman, G. Porter and D. Walters, J. Chem. Soc., Faraday Trans. 2, 1981, 77, 2373.
15. P.A. Christensen, A. Harriman and W. Erbs, J. Chem.Soc., Faraday Trans 2, 1985, 81, 575.
16. A. Mills, A. Harriman and G. Porter, Anal Chem., 1981, 53, 1254.
17. J.F. Endicott, in Concepts in Inorganic Photochemistry, eds. A.W. Adamson and P.D. Fleischauer (Wiley, New York, 1975), p. 81.
18. P. Neta, J. Phys. Chem., 1981, 85, 3678.
19. P. Neta and A. Harriman, J. Chem. Soc., Faraday Trans. 2, 1985, 81, 123.
20. A. Harriman, P.A. Christensen, P. Neta and M.C. Richoux, Sci. Papers Inst. Phys. Chem. Res., 1984, 78, 143.
21. P.A. Christensen, A. Harriman, P. Neta and M.C. Richoux, J. Chem. Soc., Faraday Trans. 2, in press.
22. D.G. Davies, in The Porphyrins, ed. D. Dolphin (Academic Press, New York, 1979) Vol. 3, Chap 4.
23. P. Neta, M.C. Richoux, A. Harriman and L.R. Milgrom, J. Chem. Soc., Faraday Trans. 2, in press.
24. P.A. Christensen and A. Harriman, unpublished results.
25. K. Itaya, N. Shaji and I. Uchida, J. Am. Chem. Soc., 1984, 106, 3423.
26. M.S. Matheson and L.M. Dorfman, in Pulse Radiolysis (M.I.T. Press, Cambridge, Mass.,1969) p. 51.
27. A. Harriman, P. Neta and M.C. Richoux, J. Chem. Soc., Faraday Trans. 2, in press.

PHOTOCATALYSIS AND LIGHT-INDUCED ELECTRON TRANSFER REACTIONS OF TERTIARY AMINES

Lester Y. C. Lee, Kirk S. Schanze, Charles Giannotti,*
David G. Whitten
Department of Chemistry
University of Rochester
Rochester, New York 14627
USA

ABSTRACT. This paper discusses photoinduced electron transfer between various excited acceptors and tertiary alkylamines of various structure. In the first section reactions between indigoid dyes--thioindigo, N,N'diacetylindigo and oxalylindigo--with triethylamine are described. The amine is readily oxidized to a radical cation which, in turn, is easily deprotonated to give a relatively good electron donating free radical. A sequence of electron-proton-electron loss as net hydride donation can take place with acceptors set up to undergo complimentary processes, the latter include both thioindigo and N,N'diacetylindigo. In the second part of the work substituted tertiary amines are photooxidized with various acceptors to give radical cations which can undergo C-C bond cleavage in competition with deprotonation. The significance and possible applications of these reactions is discussed.

1. INTRODUCTION

1.1 Photocatalysis and Photoconversion

The past decade has seen a surge of interest in photochemical reactions which offer the possibility of useful energy conversion and/or storage. Crucial to any potentially useful system are reagents which either absorb light and are themselves converted reversibly into a higher energy form, or mediate (from their excited states or a metastable photoproduct) the conversion of non light-absorbing reagents from an energy poor to an energy rich form. Key to a true photoconversion is the utilization of light as a chemical reagent consumed in the production of energy-storing products. In contrast to photoconversion the term photocatalysis, strictly speaking, refers to the use of light itself as a catalyst or alternatively to the photochemical formation of a catalyst to drive a reaction which, although energetically favorable, does not proceed spontaneously.

*CNRS, Institut de Chimie des Substances Naturelles, F-91190 Gif-sur-Yvette, France.

E. Pelizzetti and N. Serpone (eds.), Homogeneous and Heterogeneous Photocatalysis, 147–159.

Based on the above definitions, which are reasonably clear, it would seem initially easy to differentiate photoconversions and photocatalysis processes. Thus the former would include such well-studied processes as the photoisomerization of norbornadiene to quadricyclane and the photoinduced cleavage of water to hydrogen and oxygen. In contrast photocatalysis could be applied to the use of light to induce cis → trans isomerization of stilbene, the quadricyclane → norbornadiene conversion or the polymerization of olefinic monomers. Again, strictly speaking, a true "photocatalysis" would involve a process in which light is not used as a reagent but in which an excited state undergoes a process or series of reactions which induce net chemical change and ultimately regenerate the original excited state. True examples of this kind of photocatalysis are relatively rare (1,2). A broader, less rigid use of photocatalysis referring to processes in which light or excited states are consumed in forming a catalyst or driving a kinetically slow but spontaneous change has become fairly widespread. This blurring of what should be a fairly precise definition has led to some confusion as to what constitutes a photocatalyst. Further problems have developed due to a mixing of the simple term photocatalyst with the more complex, yet more precise term, light-absorbing catalyst. The latter is indeed what is generally meant in many cases where the term photocatalyst or photocatalysis is used. Clearly in the majority of cases where efficient photoconversion is to be achieved, a critical component is, as mentioned above, a light absorbing reagent which initiates reaction by absorption of light and mediates a series of subsequent dark reactions culminating in net chemical reactions of reagents not directly excited by light simultaneous with regeneration of the absorber. Thus chlorophyll in photosynthesis or the dyes, transition metal complexes or semiconductors employed in various schemes of so-called "artificial photosynthesis" are light absorbing catalysts but not correctly photocatalysts.

Undoubtedly many problems arise in understanding or describing these processes due to the fact that, in general, both homogeneous and heterogeneous photoconversion reactions normally have a single endergonic step (light absorption) followed by a sequence of steps with $\Delta G < 0$ leading to the metastable products in which energy is "stored." In these "subsequent" reactions a number of species may be cycled and thus they too are catalysts in a photoconversion process yet not involved in any way in a photochemical or photophysical step.

Much work in the area of photoconversion in both homogeneous and heterogeneous media has involved either light induced electron transfer from a donor to an acceptor or related forms of charge separation. The photophysics of these processes has been well-studied (3-15) and the present paper will focus not so much on this aspect of light-induced electron transfer but rather on the molecular details of some of the ensuing processes which can culminate either in photoconversion (we refer here to energy storage) or permanent chemical change. By far the most common photochemical redox process is the one-electron quenching of an excited state to form initially a radical ion pair and subsequently, depending upon charge type, microenvironment etc., free radical ions (eq 1). In cases where these products are both stable

$$D^* + A \text{ (or } D + A^*) \rightarrow D\cdot^+, A\cdot^- \rightarrow D\cdot^+ + A\cdot^- \quad (1)$$

the usual fate is "back" electron transfer to regenerate the starting materials in their ground states. Much effort has been spent to devise clever strategies in which either $D\cdot^+$ or $A\cdot^-$ or both can be intercepted to produce secondary redox events and ultimately energy conversion or useful chemical change. Several of the papers in this volume deal with strategies used in both homogeneous and heterogeneous systems.

An interesting variation of the so-called reversible electron transfer process described above is the situation where either $A\cdot^-$ or $D\cdot^+$ or both are unstable and highly reactive such that back electron transfer is avoided or at least averted to some extent without the intercession of added reagents. An early and well-studied example of this type of process was the photoreduction of ketones such as benzophenone by amines (15-17). In most examples studied of this reaction the major factor leading to net reaction is the instability of radical cations formed by the one-electron oxidation of aliphatic amines (18). Studies of photoredox reactions of transition metal complex excited states with amines have been shown to lead to reasonably efficient reaction resulting in substantial energy conversion (19,20); here again the reaction can be attributed to the reactivity of the photogenerated amine cation radical which channels reaction towards products and away from reverse electron transfer (18,21). In the present paper we report studies of the reaction of several tertiary amines with photoexcited electron acceptors--mostly dyes and transition metal complexes. These studies show that, depending on the structure of the amine, a wide variety of permanent and potentially useful photochemical transformations can be carried out which fit into the broad descriptions of photocatalysis as outlined above.

2. MATERIALS AND METHODS

The preparation and purification of the dyes and metal complexes used in this study have been described elsewhere (23). Camphorquinone was provided by Dr. Roy Smith of the L. D. Caulk Co. and several indigo derivatives were gifts of Dr. George M. Wyman. The substituted tertiary amines (1 and 2) were prepared according to methods in the literature (22); the alcohol derivatives (3 and 4) were prepared by aluminum isopropoxide reduction of ketones 1 and 2, respectively. 1-Benzyl-1,4-dihydronicotinamide was prepared by benzylation of nicotinamide followed by reduction of the cationic N-benzylpyridinium salt (34). The reaction studies by conventional photolysis, photochemical esr techniques and laser flash photolysis were as reported in detail in previous papers (20,23).

3. RESULTS AND DISCUSSION

3.1. Photoreduction of Dyes and Transition Metal Complexes by Tertiary Amines

As discussed in the introduction, a number of reagents undergo one-electron redox reactions to yield products which are quite unstable and undergo subsequent reactions in competition with back-electron transfer. Prominent among these reagents are tertiary alkyl amines such as triethylamine. Triethylamine (TEA), for example, reduces excited states of ketones such as benzophenone and related compounds (15-17); a number of studies have established that a rapid sequence of electron-proton transfer quickly converts the excited ketone-amine encounter complex into a neutral radical pair (eq 2) which subsequently

$$\phi_2CO^* + Et_3N: \xrightarrow[\text{transfer}]{e^-} [\phi_2CO\cdot^-, CH_3CH_2\dot{N}^{+}Et_2] \xrightarrow[\text{transfer}]{H^+} [\phi_2\dot{C}OH, CH_3\dot{C}HNEt_2] \quad (2)$$

decays to various products (24-27). Recent work has suggested that the proton transfer step can occur for the benzophenone-TEA system on the scale of picoseconds (25). Studies from our laboratories with metal complexes as the excited acceptors (in initial studies ruthenium (II) complexes with substituted polypyridyl ligands, RuL_3^{2+}) demonstrated that by a sequence of electron, proton, electron transfer the amine could act as a net two-electron donor to reduce two moles of the metal complex (eqs 3-5) (20). The behavior of the amine in this case is an

$$RuL_3^{2+*} + Et_3N \rightarrow RuL_3^{+} + Et_3N\cdot^{+} \quad (3)$$

$$Et_3N\cdot^{+} + Et_3N: \rightarrow CH_3\dot{C}HNEt_2 + Et_3\overset{+}{N}H \quad (4)$$

$$CH_3\dot{C}HNEt_2 + RuL_3^{2+} \rightarrow CH_3CH{=}\overset{+}{N}Et_2 + RuL_3^{+} \quad (5)$$

example of the well-described electrochemical "ece" reaction (28). Since several acceptors could be anticipated to undergo "complimentary" "ece" reactions in which the acceptor is reduced, protonated and then reduced again (28), it became of interest to determine whether suitable combinations of donors and acceptors could be found that would lead to a photochemical two-electron redox conversion.

Evidence that such reactions can occur with some facility has been provided by studies of the photoreduction of indigo dyes such as thioindigo (TI), N,N'diacetylindigo (NDI) and oxalylindigo (OI) with TEA and other amine-donors (23,29). It was found that irradiation of TI in the presence of TEA leads to the two-electron reduction product dihydro- or leucothioindigo TIH_2 in good chemical yield (eq 6).

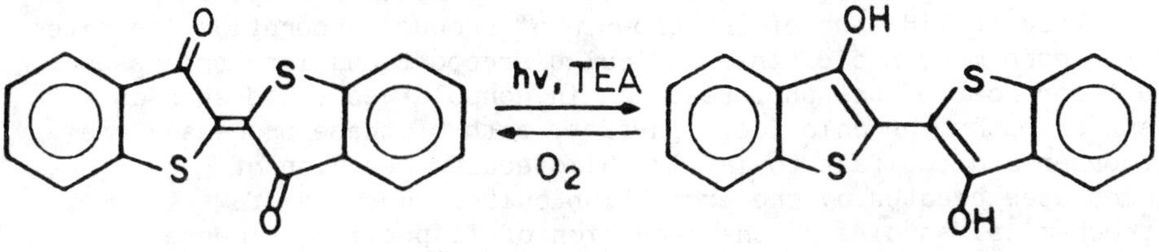

$$\text{TI} \underset{O_2}{\overset{h\nu,\ \text{TEA}}{\rightleftharpoons}} \text{TIH}_2 \qquad (6)$$

This reaction can be reversed to give TI quantitatively in the dark upon treatment with O_2 or air. Quenching studies suggest that TEA and other amines quench the fluorescent singlet of TI by one-electron transfer; one mechanism which would account for the production of TIH_2 would be an electron-proton transfer sequence to give the "semi-reduced" radical TIH•, which could subsequently disproportionate. (Mechanism A, eqs 7-11). Evidence indicating that this mechanism,

$$\text{TI}^* + CH_3CH_2NEt_2 \xrightarrow[\text{transfer}]{e^-} \text{TI}\cdot^-,\ CH_3CH_2\overset{+\bullet}{N}Et_2 \qquad (7)$$

$$\text{TI}\cdot^-,\ CH_3CH_2\overset{+\bullet}{N}Et_2 \xrightarrow[\text{transfer}]{H^+} \text{TIH}\cdot,\ CH_3\dot{C}HNEt_2 \qquad (8)$$

$$\text{TIH}\cdot,\ CH_3\dot{C}HNEt_2 \rightarrow \text{TIH}\cdot + CH_3\dot{C}HNEt_2 \qquad (9)$$

$$\text{TI} + H^+ + CH_3\dot{C}HNEt_2 \rightarrow \text{TIH}\cdot + CH_3CH{=}\overset{+}{N}Et_2 \qquad (10)$$

$$2\text{TIH}\cdot \rightarrow \text{TIH}_2 + \text{TI} \qquad (11)$$

Mechanism A

which has analogies in various quinone photoreductions (30,31), accounts for at least part of reaction 6, comes from esr and laser flash photolysis studies. In the former it is found that photolysis of TI/TEA in an esr spectrometer leads to a sharp five-line spectrum attributable to TIH• that persists only during irradiation. The radical from TEA, $CH_3\dot{C}HNEt_2$, is not detected and hence must not accumulate; that it is formed is indicated by spin trapping experiments in which it reacts with nitrosodurene to form a moderately stable adduct. A relatively long lived transient is detected by laser-flash photolysis which decays by second-order kinetics to form TIH_2 and which is quenched by O_2 to form TI; the transient absorption can be assigned to the semireduced, TIH•. Although both the esr experiments and laser flash photolysis provide support for Mechanism A, the latter provides strong evidence that, for at least several solvents, a faster process accounts for the bulk of the product formation.

Laser flash photolysis studies of the decay of absorption attributable to TIH· and of the grow-in of product absorption indicate that two much more rapid (than radical disproportionation) processes account for most of the photoreaction in nonpolar solvents such as benzene or methylene chloride. In fact, both of these processes are first-order and too fast to involve bimolecular reaction of intermediates created by the excitation pulse. Results of electrochemical studies of the reduction of TI provide evidence suggesting the possibililty of congruent ece reactions of TI and TEA initiated by quenching of $^1TI^*$ by the amine (32,33). When TI is reduced in aprotic solvents it is found to undergo two separate and electrochemically reversible one-electron reductions to generate $TI\cdot^-$ and TI^{2-}, respectively; when it is reduced electrochemically in protic or acidic solvents there is only a single two-electron reduction wave and it is electrochemically irreversible (32,33). In fact electrochemical reduction of TI under these conditions leads to the detection of TIH_2 as a product (33). It is apparent from these studies that $TI\cdot^-$ is readily protonated and that the semireduced species, TIH·, is more readily reduced than TI. This introduces the probability of congruent reaction within the initially formed radical ion pair or successor to the encounter complex, eqs 7, 8, 12-14, Mechanism B.

$$^1TI^* + CH_3CH_2NEt_2 \rightarrow \underset{(a)}{TI\cdot^-,\ CH_3CH_2\overset{+\cdot}{N}Et_2} \qquad (7)$$

$$TI\cdot^-,\ CH_3CH_2\overset{+\cdot}{N}Et_2 \rightarrow \underset{(b)}{TIH\cdot,\ CH_3\dot{C}HNEt_2} \qquad (8)$$

$$TIH\cdot,\ CH_3\dot{C}HNEt_2 \rightarrow \underset{(c)}{TIH:^-,\ CH_3CH{=}\overset{+}{N}Et_2} \qquad (12)$$

$$TIH:^-,\ CH_3CH{=}\overset{+}{N}Et_2 \rightarrow TIH:^- + CH_3CH{=}\overset{+}{N}Et_2 \qquad (13)$$

$$TIH:^- + H^+ \rightarrow TIH_2 \qquad (14)$$

Mechanism B

As mentioned above laser flash photolysis shows two fast decays for absorption in the region where TIH· absorbs and two rapid grow-ins where the product, TIH_2 or $TIH:^-$ absorbs. If it is assumed that reaction 8, (a) → (b), is too rapid for resolution (< 10 ns) under our experimental conditions, the two first order processes could be attributed to reaction 12, (b) → (c). The faster of the two processes (< 10 ns) could be attributed to reaction of a singlet radical pair, for which there is no spin barrier for electron transfer while the second process (~ 200 ns) could be ascribed to reaction of a triplet pair (3(b)) for which the conversion should be slower.

For thioindigo-TEA it is clear that both Mechanisms A and B are operative in several different solvents. Thus for this system it is clear that rates for reactions 9 and 12 are competitive; studies of the reaction by esr suggest that free radical yields are higher in polar solvents such as acetonitrile than in nonpolar solvents such as benzene. The quantum yields for reaction, although generally low in all solvents, are nearly tenfold higher in polar solvents such as acetonitrile (0.03) or dimethylsulfoxide (0.02) than in benzene (0.003) or methylene chloride (0.002). These results suggest that reaction 9 may be relatively faster in polar solvents and that reaction via Mechanism A may be overall more efficient than by Mechanism B. That competition between reactions 9 and 12 should favor reaction 9 in polar solvents is perhaps reasonable; although the individual radicals are neutral they are polar and their diffusion apart may be assisted in more polar media. The differences in overall efficiency via the two mechanisms are probably attributable to the fact that all three pairs (a), (b) and perhaps even (c) can decay back to starting TI and TEA via electron, proton or hydride transfer, respectively, whereas free TIH•, formed on the final surviving intermediate in Mechanism A, must decay to product via the disproportionation (eq 11).

Other indigo dyes also undergo photoreduction in the presence of TEA. Several substituted thioindigo dyes apparently react by the dual mechanisms observed for TI. NDI, which has a similar excited state energy and redox potentials, undergoes photoreduction with TEA to give an analogous product (eq 15); however a mechanistic examination of

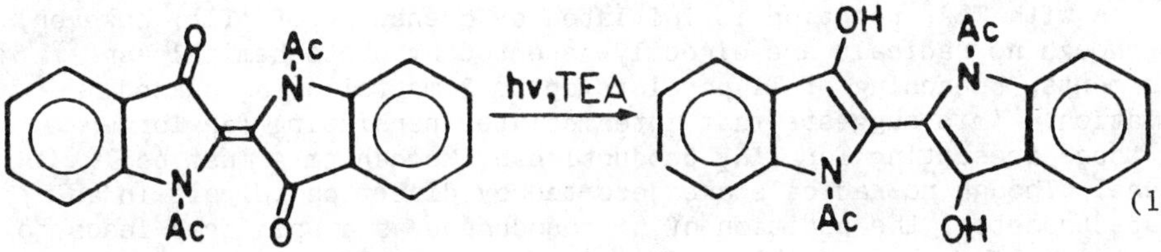

hv,TEA (15)

the reaction indicates significant differences. The most salient observations are that no radicals are detected when the reaction is carried out in a photochemical esr experiment and no long-lived transients are detected in laser-flash spectroscopic studies. Since the reaction is clearly indicated to be initiated by electron-transfer quenching of ^{1}NDI* by fluorescence studies the most reasonable explanation is that the reaction occurs by a variation of Mechanism B; most likely in this case intersystem crossing does not occur and the pairs (a) and (b) are short-lived singlets. The overall quantum yield for NDI reduction is low (0.0003-0.005) and shows little systematic variation with solvent. Interestingly for NDI the photoreduction apparently occurs cleanly by the net two-electron or hydride route; however the factors that favor "vectorial" two-electron transfer also favor energy wasting steps which lead to the starting material.

Oxalylindigo, a <u>cis</u>-fused indigo whose absorption is blue shifted relative to TI and NDI, undergoes photoreduction with TEA but its reaction can be shown to consist of an electron-proton transfer sequence to give the semireduced species, OIH• (eq 16) (23).

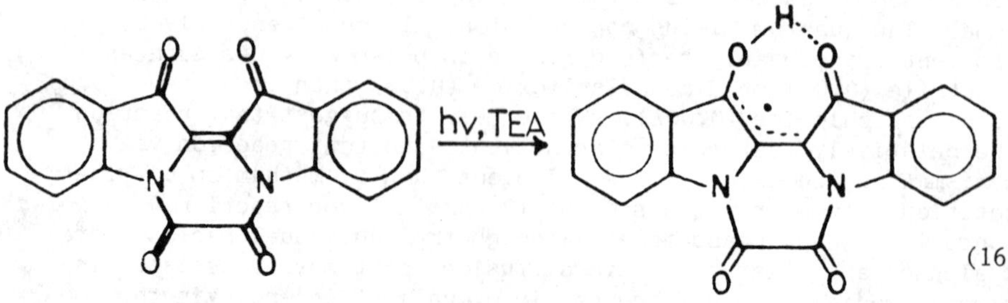

(16)

Although the quantum yield for this reaction is generally higher (ca 0.1) than for TI or NDI, the radical does not disproportionate to give a two-electron product. The termination of the reaction at this stage is easily understood from redox potentials which indicate a wide spacing between the two reduction stages, even in the presence of proton donors (23).

An interesting contrast to the reaction with TEA is offered in the photoreduction of TI by N-benzyl-1,4-dihydronicotinamide (BNAH). Here the same net reduction of TI (eq 6) is observed but the quantum yield is higher (0.13 in benzene) and studies by esr and flash spectroscopy show substantial differences from the reaction with TEA. Here, as with TEA, reaction is initiated by quenching of $^1TI^*$; however, even though no radicals are directly detected by photochemical esr experiments, bleaching of TI persists up to 2 ms following pulsed excitation. This suggests that intermediates persisting far longer than those accounting for TIH_2 production by Mechanism A must be present. Though no radicals are detected by direct photolysis in an esr spectrometer, the addition of nitrosodurene as a spin trap leads to the detection of two radicals. One of these is evidently the adduct of semioxidized BNAH (BNA•) to ND while the other is TIH• as detected previously in TI-TEA photolysis. These results can be most easily interpreted by a radical-chain mechanism (eqs 17-23). This mechanism

$$^1TI^* + BNAH \rightarrow (TI\cdot^-, BNAH\cdot^+)^{1,3} \quad (17)$$

$$(TI\cdot^-, BNAH\cdot^+)^{1,3} \rightarrow (TIH\cdot, BNA\cdot)^{1,3} \quad (18)$$

$$(TIH\cdot, BNA\cdot)^{1,3} \rightarrow (TIH:^-, BNA^+) \quad (19)$$

$$(TIH\cdot, BNA\cdot)^{1,3} \rightarrow TIH\cdot + BNA\cdot \quad (20)$$

$$H^+ + BNA\cdot + TI \rightarrow BNA^+ + TIH\cdot \quad (21)$$

$$TIH\cdot + BNAH \rightarrow BNA\cdot + TIH_2 \quad (22)$$

$$2TIH\cdot \rightarrow TIH_2 + TI \quad (23)$$

incorporates both one-and two-electron transfer paths as outlined in Mechanisms A and B. The moderate quantum yield obtained in this case is probably a consequence of the chain reactions (eqs 21 and 22) which partially offset return to starting material in the radical pairs formed in eqs 17 and 18. The crucial step is reaction 22 which can be attributed to the greater reducing power of ground state BNAH (34,35); such a reaction is unimportant for triethylamine. For TI/BNAH the reduction of dye goes slowly, even in the dark, indicating that no net photoconversion is achieved in the process; photolysis of the product pair formed in eq 18 does not lead efficiently to TI and BNAH although some initial reaction can be observed. When NDI is irradiated in the presence of BNAH reduction of the dye is observed with very low efficiency (ϕ = 1-5 x 10^{-4}); here, as with TEA, the low efficiency may be attributed to reaction solely by the two-electron or net hydride path (Mechanism B) which does not involve free radicals (and hence no chain reaction) but involves rapid and direct formation of products but even more rapid return to starting materials in their ground states.

The overall patterns of photoreduction reported above should be quite general. The coupled sequence of electron-proton-electron-transfer resulting from encounter of an excited acceptor with a donor offers in principle a route for efficient photoconversion. That the overall efficiencies in these reactions studied to date is low can be attributed to the rapid electron, H-atom and possibly hydride transfer paths which lead back to the starting materials. While the electron transfer rates can perhaps be rendered slower for suitable donor-acceptor combinations, the relative importance of the different decay channels has yet to be determined; thus the likelihood of obtaining high overall efficiencies via the two-electron path is still somewhat uncertain.

3.2. C-C Bond Cleavage in Electron Transfer Photoreactions of Tertiary Amines

In the above-discussed examples of photoredox reactions involving tertiary amines as electron donors, the key to obtaining a net two-electron redox process is the rapid deprotonation of the amine radical cation. This process is generally indicated to be quite rapid (24-27) even though the radical cation is a carbon acid which might be expected to deprotonate rather slowly. An alternative reaction of amine cation radicals to give a neutral radical and an iminium ion by C-C bond cleavage might be anticipated as a possibility (eq 24), particularly

$$\text{R-CHR'-}\overset{+\cdot}{\text{N}}\text{R}''_2 \rightarrow \text{R}\cdot + \text{R'CH=}\overset{+}{\text{N}}\text{R}''_2 \qquad (24)$$

in cases where R· is a reasonably stable free radical. Related C-C bond cleavages have been observed for cations generated from ethers in solution (36), for strained hydrocarbons (37,38) and for certain amines in the gas phase (mass spectrometer) (39). Four compounds which possess the features outlined above and which should be reasonable donors in photoinduced electron transfer reactions are the amines containing an adjacent arylketone function (1) and (2) and those with an adjacent benzylic alcohol substituent (3) and (4). Irradiation

Ph-CO-CHPh(N-morpholino) **1**   Ph-CO-CHPh(N-piperidino) **2**

Ph-CH(OH)-CHPh(N-morpholino) **3**   Ph-CH(OH)-CHPh(N-piperidino) **4**

of electron acceptors such as camphorquinone (CQ) or Ru(4,4'-CO_2Et-bpy)$_3^{2+}$ (RuL_3^{2+}) in the presence of compounds 1-4 does lead to electron transfer quenching of the excited acceptor in each case. This reaction may be followed by excited state quenching experiments, or more interestingly, through examination of the consequences of the quenching.

That free radicals are formed in the quenching process is indicated by the finding that in each case the photoprocess can result in efficient polymerization of suitable olefins such as methyl methacrylate or other resins. However, since radicals capable of initiating polymerization can be produced by either deprotonation (23) or C-C bond cleavage (40) the observation of efficient photopolymerization is inconclusive concerning which radicals are formed. Experiments using photochemical-esr and the spin trap, ND, show that for amines 1-4 C-C bond cleavage after electron transfer according to eq 24 is an important path. It is found, for example, that keto-amines 1 and 2 give two trapped radicals on both direct (λ < 320 nm) and acceptor-mediated (λ > 440 nm) photolysis; the trapped radicals exhibit 3- and 6-line hyperfine splittings which can be attributed to the ND-adducts of Ph-CO• and Ph(NR_2)CH•, respectively. The keto-amines (1 and 2 are amino analogs of the widely used benzoin ether photoinitiators and it is to be expected that direct photolysis leads to homolytic cleavage (eq 25) (40). The alcohol derivatives

$$\text{Ph-CO-CHPh(NR}_2) \xrightarrow[\lambda < 320\text{ nm}]{h\nu} \text{PhCO}\cdot + \text{PhCH}\cdot(\text{NR}_2) \qquad (25)$$

3 and 4 are not photoinitiators in the long-wavelength uv and no radicals are detected upon direct irradiation with ND. When an acceptor is irradiated in the presence of 3 or 4 and ND, esr spectra characteristic of the trapped radicals from PhCH(OH)• and $Ph(NR_2)CH$• are detected; in the case of 4 a minor amount of the radical anticipated from deprotonation, $PhCH(OH)C(NR_2)Ph$, is also trapped as its ND-adduct.

Additional evidence that C-C bond cleavage is an important reaction for the photochemically generated radical cations of 1 and 3 is the finding that benzaldehyde and benzyl alcohol, respectively, are major products of the camphorquinone-mediated reactions of these amines in acetonitrile. Although all of the products have not yet been isolated or identified, it is found that the same products are separated and detected by HPLC analysis of samples of 1 irradiated directly in the uv and in the camphorquinone-mediated reaction at $\lambda >$ 440 nm. This strongly implies that the net function of the electron acceptor is as described in eqs 26-29; that is the acceptor mediates

$$A \xrightarrow[vis]{h\nu} A^* \tag{26}$$

$$A^* + R\text{-}CHR'\text{-}NR''_2 \rightarrow A\cdot^-, R\text{-}CHR'\text{-}\overset{+\bullet}{N}R''_2 \tag{27}$$

$$A\cdot^-, R\text{-}CHR'\text{-}\overset{+\bullet}{N}R''_2 \rightarrow A\cdot^-, R'CH{=}\overset{+}{N}R''_2 + R\cdot \tag{28}$$

$$A\cdot^-, R'CH{=}\overset{+}{N}R''_2 \rightarrow A + R'\text{-}CH\text{-}NR''_2 \tag{29}$$

(in the case of compounds 1 and 2 a photoreaction with relatively low energy light which would ordinarily occur only with light absorbed by the substrate at much higher energies. In the case of the alcohols 3 and 4, the acceptor mediated cleavage by eqs 26-29 represent a photoconversion which does not occur efficiently even on direct irradiation of the substrates 3 and 4 at much higher energies.

It is clear that much additional work needs to be done to determine how general this reaction is for substituted tertiary amines. We are currently investigating rates and competition between deprotonation and C-C bond cleavage in several substituted tertiary amines as a function of structure (particularly of substituents in R). However, based on the results obtained thus far, it appears likely that the photoinduced electron-transfer mediated σ-bond cleavage should occur for a variety of donors with appropriate substituents on the σ-bond terminal carbons. The reaction is remarkable in that it involves the use of visible light to induce what appears to be a fairly specific cleavage of a relatively strong σ-bond. It is reasonable to expect that selective reactions induced by visible light could find wide utility for a variety of purposes.

4. CONCLUSION

The results and discussion presented above have focused to a large extent on photoinduced electron transfer reactions in which electron-donor tertiary amines and their radical cations play a dominant role. In each case these reactions have been initiated through activation of a visible or near-uv light absorbing acceptor. The role of the acceptor in the various reactions discussed differs; however in each case the acceptor is, or could be, recycled if the reaction were to be used to produce net chemical conversion. Thus in the case of the indigo dyes, the reduced or dihydro form should be reasonably strong reducing agents which could be recycled in turn by reducing other substrates. Although this reaction has been as yet little explored, the fact that reoxidation by O_2 is very clean, especially for TI, suggests recycling is feasible. In the case of the electron-transfer induced C-C σ-bond cleavage reactions the role of the light-absorbing acceptors camphorquinone and RuL_3^{2+} is apparently already catalytic as indicated by eqs. 26-29; here again it is clear that the homolytic cleavage induced by visible light results in net energy conversion although the immediate products (the free radicals) are so reactive that it is unlikely that they can be harvested in any net energy storing scheme. Thus in the context of the classifications made in the Introduction both of these broad reactions discussed represent photoconversion processes but neither of them involve specifically photocatalysis in the limited sense. On the other hand it is clear, as outlined above, that the electron acceptors used in these studies are light absorbing catalysts and, specifically, that the C-C bond cleavages definitely involve the acceptor to couple a reagent, light, with a reaction (eq 25 for example) which would otherwise not occur due to lack of absorption. If we broaden our definition of photocatalysis to encompass such a process then indeed these latter reactions are examples.

Acknowledgment

We thank the U. S. Department of Energy (contract DE-AC02-84ER13151) and the L. D. Caulk Dentsply Corporation for support of this research.

References

(1) Whitten, D. G.; Wildes, P. D.; DeRosier, C. A. J. Am. Chem. Soc. 1972, **94**, 7811.
(2) Mercer-Smith, J. A.; Whitten, D. G. J. Am. Chem. Soc. 1978, **100**, 2620.
(3) Gafney, H. D.; Adamson, A. W. J. Am. Chem. Soc. 1972, **94**, 8238.
(4) Lawrence, G. S.; Balzani, V. J. Inorg. Chem. 1974, **13**, 2976.
(5) Young, R. C.; Meyer, T. J.; Whitten, D. G. J. Am. Chem. Soc. 1975, **97**, 4781.
(6) Creutz, C.; Sutin, N. J. Inorg. Chem. 1976, **15**, 496.
(7) Creutz, C.; Sutin, N. J. Am. Chem. Soc. 1977, **99**, 241.
(8) Balzani, V.; Moggi, L.; Manfrin, M. F.; Bolletta, F.; Lawrence, G. S. Coord. Chem. Rev. 1975, **15**, 321.
(9) Demas, J. N.; Crosby, G. A. J. Am. Chem. Soc. 1970, **92**, 7262. Ibid. 1971, **93**, 2841.
(10) Foreman, T. K.; Giannotti, C.; Whitten, D. G. J. Am. Chem. Soc. 1980, **102**, 1938.

(11) Monserrat, K.; Gratzel, M.; Tundo, P. J. Am. Chem. Soc. 1980, **102**, 5527.
(12) Navon, G.; Sutin, N. J. Inorg. Chem. 1974, **13**, 2159.
(13) Lin, C. T.; Bottcher, W.; Chou, M.; Creutz, C.; Sutin, N. J. Am. Chem. Soc. 1976, **98**, 6536.
(14) Lin, C. T.; Sutin, N. J. Am. Chem. Soc. 1975, **97**, 3543.
(15) Cohen, S. G.; Baumgarten, R. T. J. Am. Chem. Soc. 1965, **87**, 2996.
(16) Cohen, S. G.; Parola, A.; Parsons, G. H. Chem. Revs. 1973, **73**, 141.
(17) Davidson, R. S.; Wilson, R. J. Chem. Soc. (B) 1970, **71**.
(18) Andrieux, C. P.; Saveant, J.-M. Bull. Soc. Chim. Fr. 1968, 4671.
(19) DeLaire, P. J.; Foreman, T. K.; Giannotti, C.; Whitten, D. G. J. Am. Chem. Soc. 1980, **102**, 5627.
(20) Monserrat, K.; Foreman, T. K.; Gratzel, M.; Whitten, D. G. J. Am. Chem. Soc. 1981, **103**, 6667.
(21) DeLaire, P. J.; Lee, J. T.; Sprintschnik, H. W.; Abruna, H.; Meyer, T. J.; Whitten, D. G. J. Am. Chem. Soc. 1977, **99**, 7094.
(22) Lutz, R. E.; Freck, J. A.; Murphey, R. S. J. Am. Chem. Soc. 1948, **70**, 2015.
(23) Schanze, K. S.; Lee, L. Y. C.; Giannotti, C.; Whitten, D. G. submitted for publication.
(24) Peters, K. S.; Freilich, S. C.; Schaeffer, C. G. J. Am. Chem. Soc. 1980, **102**, 5701.
(25) Peters, K. S.; Schaeffer, C. G. J. Am. Chem. Soc. 1980, **102**, 7566.
(26) Peters, K. S.; Simon, T. D. J. Am. Chem. Soc. 1981, **103**, 6403.
(27) Peters, K. S.; Parry, E.; Rudzki, J. J. Am. Chem. Soc. 1982, **104**, 5535.
(28) Bard, A. J.; Faulkner, L. R. in Electrochemical Methods, Wiley: New York, 1980.
(29) Schanze, K. S.; Giannotti, C.; Whitten, D. G. J. Am. Chem. Soc. 1983, **105**, 6326.
(30) Tanimoto, Y.; Takashima, M.; Itoh, M. J. Phys. Chem. 1984, **88**, 6053.
(31) Roth, H. D. in Chemically Induced Magnetic Polarization; Muus, L. T.; Atkins, P. W.; McLaughlin, K. A.; Pedersen, J. B., Eds.; Reidel: Dordrecht, The Netherlands, 1977; Chapter 4.
(32) Yeh, L. R.; Bard, A. J. J. Electroanal. Chem. 1976, **70**, 157.
(33) Yeh, L. R.; Bard, A. J. Ibid. 1977, **81**, 319.
(34) Martens, F. M.; Verhoeven, J. W. Rec. Trav. Chem. Pays-Bas 1981, **100**, 228.
(35) Fukuzumi, S.; Hironaka, K.; Tanaka, T. J. Am. Chem. Soc. 1983, **105**, 4722.
(36) Arnold, D. R.; Maroulis, A. J. J. Am. Chem. Soc. 1976, **98**, 5931.
(37) Roth, H. D.; Schilling, M. L. M.; Gassman, P. G.; Smith, J. L. J. Am. Chem. Soc. 1984, **106**, 2711.
(38) Roth, H. D.; Schilling, M. L. M.; Wamser, C. C. J. Am. Chem. Soc. 1984, **106**, 5023.
(39) DeJongh, D. C.; Lin, D. C. K.; LeClair-Lanteigne, P.; Gravel, D. Can. J. Chem. 1975, **53**, 3175.
(40) Ledwith, A. Pure Appl. Chem. 1977, **49**, 431.

PHOTOCATALYTIC PRODUCTION OF ASCORBIC ACID. A SECONDARY PHOTOSYNTHESIS IN PLANTS

Fikret Baykut*, Gülçin Benlioğlu* and Gökhan Baykut**
*Department of Chemical Engineering, Faculty of Engineering, University of Istanbul, Laleli, Istanbul, Turkey
**Department of Chemistry, University of Florida, Gainesville, Florida 32611, U.S.A.

ABSTRACT. In this paper a photochemical method to synthesize ascorbic acid is given. The reaction requires oxygen, D-sorbitol (or L-sorbose) and is catalyzed by the flavonoid hesperidin. This synthesis is most probably taking place in plants by solar ultraviolet radiation.

1. INTRODUCTION

Flavonoids are widely distributed in the nature and can be found in many different plants. They are derivatives of flavon.

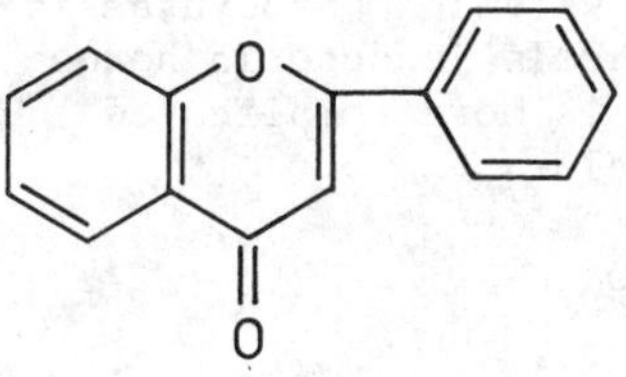

Flavon

The first scientific investigations on flavonoids [1-5] have been performed by Szent-Györgyi, who has shown several different therapeutic effects of them (treatment of capillary resistance anti-inflamatoris factor, etc.) and suggested that flavonoids act synergistically in presence of the vitamin C. Studies[1-4,6] with laboratory animals have led to the conclusion that flavonoids can be used for the treatment of capillary fragility retinal haemorrhage in hypertension, diabetic retinopathy, purpura, rheumatic fever, arthritis, radiation disease, habitual abortion, frostbite, anaphylactic shock and experimentally developed cancer.

Recently some experiments[7,8], again with laboratory animals, have demonstrated that plant extracts containing relatively high amounts of bioflavonoids can be used for cancer treatment. Grapefruit extracts (13% hesperidin in dry substance[9])) prepared in a special way were used for the treatment of mice having fusiformed sarcoma cells[7]) and Ehrlich

E. Pelizzetti and N. Serpone (eds.), Homogeneous and Heterogeneous Photocatalysis, 161–173.

Ascites tumors[8]. Szent-Györgyi and Ruszinak[6] have treated a patient for subcutaneous capillary haemorrhage with "impure" ascorbic acid. The patient was cured while purified vitamin C had no effect in other similar cases. They also have successfully kept guinea pigs on a scorbutagenic diet alive for a longer period of time by using "impure" vitamin C. It has been found that those "impurities" contain hesperidin and other bioflavonoids. Hesperidin is one of the important flavonoids and is found in many plants accompanied with vitamin C.

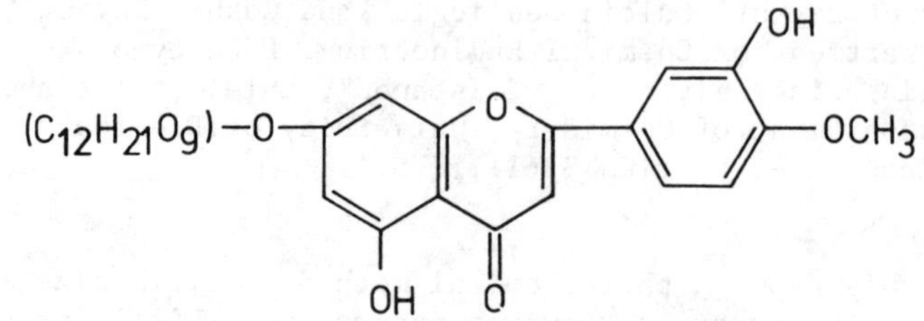

$(C_{12}H_{21}O_9)-O-$ OH OCH$_3$ O OH O

Hesperidin

Vitamin C -ascorbic acid- is produced in the nature in many plants and also in some animal organisms[10], L-ascorbic acid is produced in the livers of some mammals and in kidneys of other vertebrates amphibians and reptiles by glucuronic acid pathway. The suggested biosynthesis pathway starts with D-glucose and leads via D-glucuronic acid, L-gluconic acid, L-gulono-γ-lactone and L-2-oxogulono-γ-lactone to ascorbic acid. The studies on vitamin C biosynthesis pathways in plants have shown that this synthesis is done in hexose phosphate metabolism similar to animals but it is more complicated. The one suggested synthesis pathway is shown below

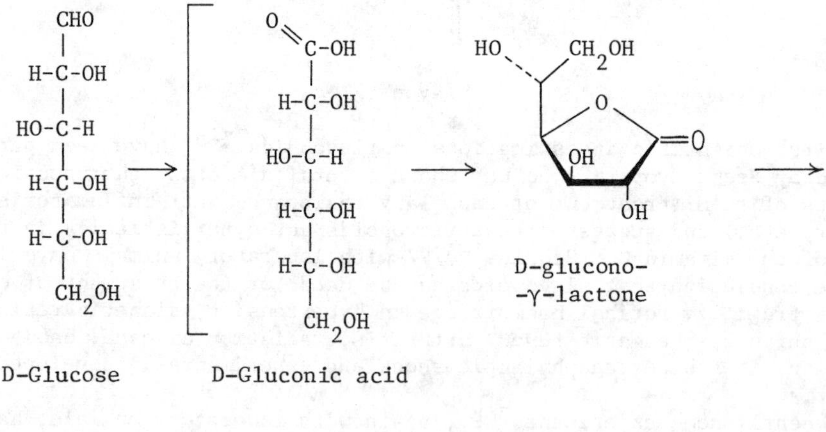

CHO – H-C-OH – HO-C-H – H-C-OH – H-C-OH – CH_2OH → [C(=O)-OH – H-C-OH – HO-C-H – H-C-OH – H-C-OH – CH_2OH] → HO, CH_2OH, O, =O, OH, OH →

D-Glucose D-Gluconic acid D-glucono-γ-lactone

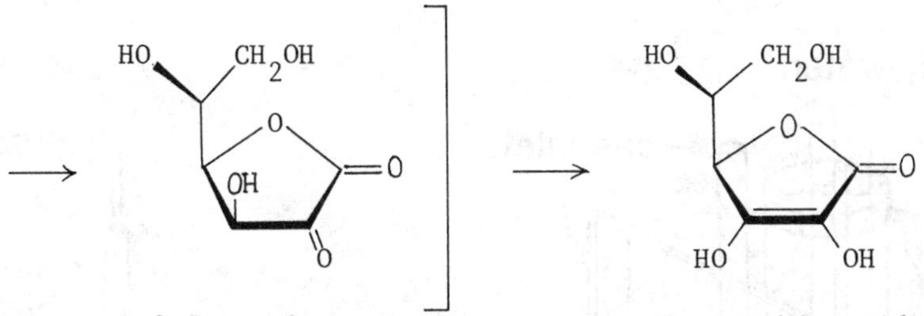

L-2-Oxo-gulono-γ-lactone L-ascorbic acid

This pathway has been studied by ^{14}C and ^{3}H labelling experiments. However, the epimerization mechanism on the C-5 could not be explained completely[10]. Another postulated pathway for the plant biosynthesis of ascorbic acid suggests the formation from D-glucose and D-galactose via D-glucurono-γ-lactone and D-galactronic acid and its methyl ester. The technical production of ascorbic acid starts with D-glucose. After hydrogenation of D-glucose the formed D-sorbitol is aerobically fermented by using acetobacter suboxidans. Acetonation oxidation and hydrolysis rearrangement finally lead to ascorbic acid[10].

The absorption and utilization of UV light in plants is related to the presence and functionalities of various flavonoids[5]. Although the action mechanism is not really known, it is mentioned[5] that the function of bioflavonoids may be important in photosynthetic processes in plants, since the same plants grown at different altitudes show different flavonoid compositions.

In this paper we have investigated one of the possible photoreactions, which can be catalyzed by flavonoids. Using hesperidin as photocatalyst we have shown a new photoreaction leading to ascorbic acid. This photosynthesis may take place in many plants since it requires D-sorbitol, oxygen and solar radiation.

2. MATERIALS AND METHODS

2.1. Chemicals

Hesperidin is a weak acid and therefore soluble in alkaline solutions. Its solubility in neutral and acid media is very poor. However, ascorbic acid decomposes in alkaline solutions by influence of light and oxygen. To be able to work in acidic media instead of hesperidin itself, we used the Na salt of hesperidin phosphoric acid ester, which is soluble in acidic media. The chemicals used in this work, hesperidin phosphoric acid ester-Na-salt, L-ascorbic acid, L(-)-sorbose, D-sorbitol, chloroform, glacial acetic acid, aceton, benzene and methanol were analysis grade chemicals from Merck (Darmstadt, F.R.G.). Kieselgel PF_{254} from Merck was used for thin layer chromatography. Oxygen used in experiments was from Habas (Istanbul, Turkey).

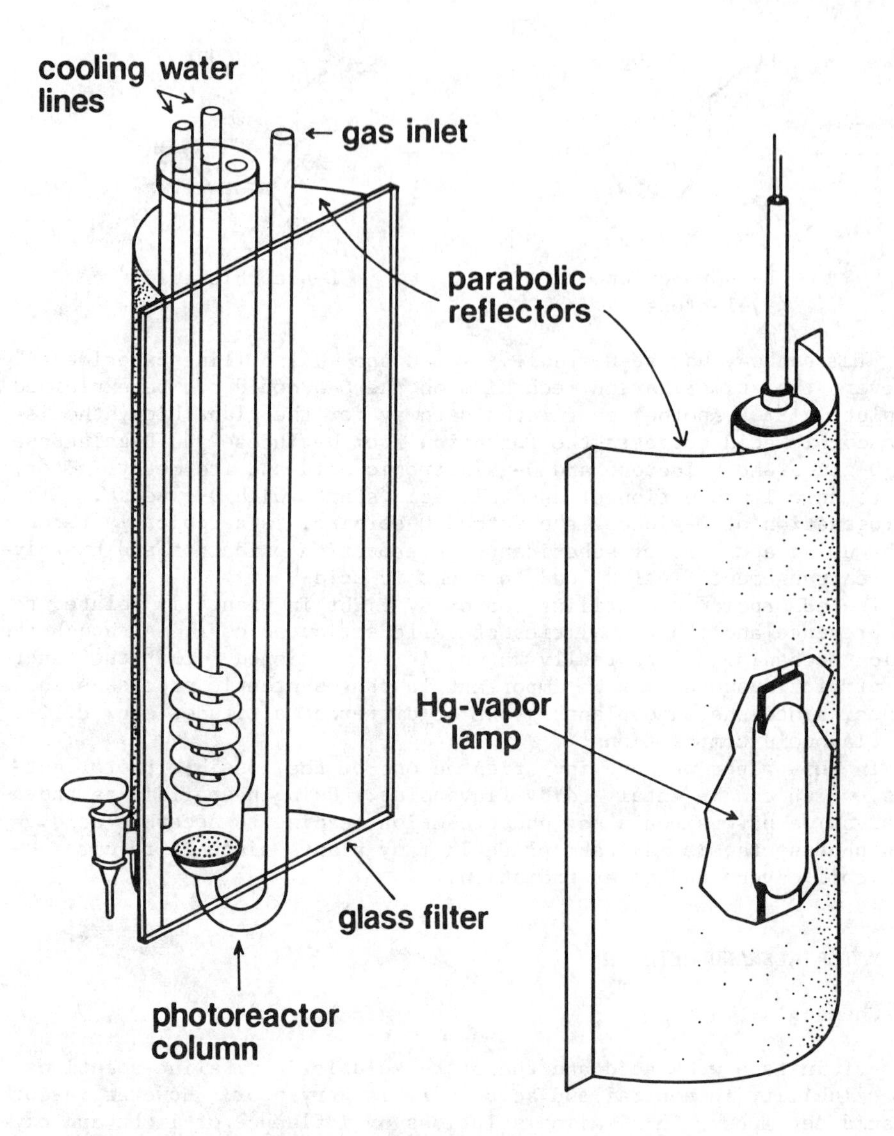

Fig. 1. Photoreactor with the Hg-vapor lamp

2.2. Apparatus

Photochemical experiments were performed by using an Edison 400 HQL high pressure mercury vapor lamp as light source[11]. The lamp was centered on the focal line of a parabolic reflector (after removing the outer glass bulb of the commercially available form of the lamp). The pyrex reactor column was along the focal line of a second parabolic reflector (Fig. 1). Temperature control during the reaction was possible by water flowing through the coils in the reactor. The cooling water was recooled by circulation through a thermostate controlled bath. Pyrex was transmitting the UV radiation above 280 nm. Most of the experiments were carried out by using an additional glass filter between the reactor and the lamp which only transmits wavelengths above 320 nm.

For spectrophotometric determinations a Hitachi (Tokyo, Japan) Model 200A UV/VIS spectrophotometer with a scanning range of 900-185 nm was used. pH measurements were performed using a Metrohm (F.R.G.) E-512 pH-meter. Thin layer chromatographic determinations were made using 2.5 cm x 7.5 cm glass plates and the UV lamp system of Desaga (Heidelberg, F.R.G.).

2.3. Procedures

The pH-dependence of the UV absorption spectra of ascorbic acid is shown in Fig. 2. At pH 2 ascorbic acid shows an absorption maximum at 245 nm, which corresponds to the nondissociated form. At pH=6.4 the mono-

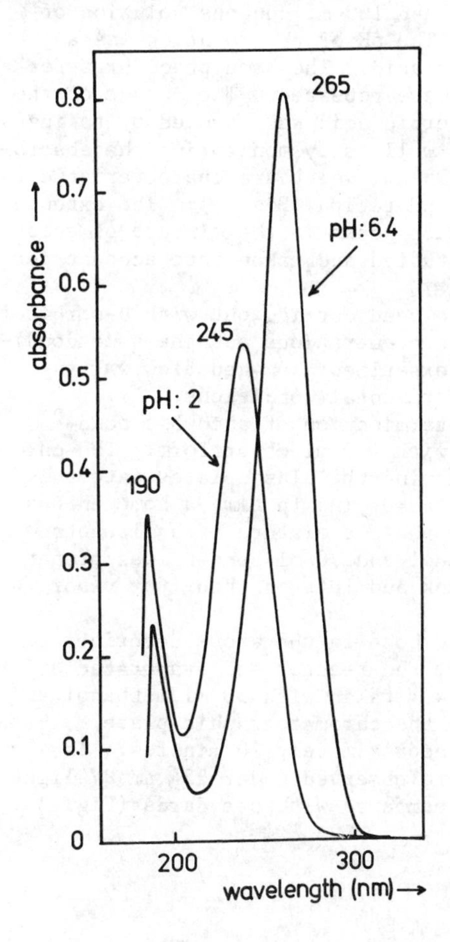

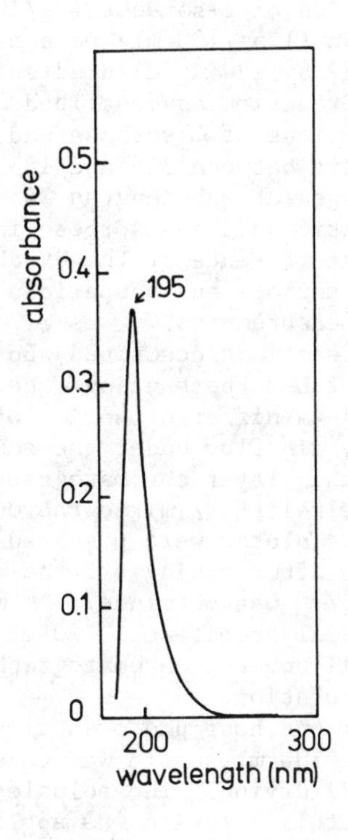

Fig. 2. Samples of absorption spectra of ascorbic acid (left) and sorbose (above) solutions.

dissociated ascorbic acid shows an absorption peak at 265 nm[10]. Therefore our experiments were carried out at pH 6.85 which is very close to pH 6.4 (no difference in absorption spectra observed) and at pH 2. 200 ml aqueous solution of L-sorbose (3g/100 ml) and 100 ml aqueous solution of hesperidin (1.5g/100 ml) were mixed directly before they were put into the photoreactor column. Oxygen was bubbled with a flow rate of 44ml/min while the column was irradiated by the mercury vapor lamp (illuminance:1000 lux). The temperature of the solution was kept constant at 20°C. Twin samples of 2 ml from the solution were taken mostly every 10 minutes. One of the samples was used for the pH determination which was 6.85±0.05 during the entire experiment. The second sample was boiled one minute, diluted with distilled water to a volume of 50 ml and the absorption spectrum in the region between 335 and 185 nm was taken (we obtained the same results also without the heating operation). Because of the absorption peaks of L-sorbose and ascorbic acid the wavelength changes at 265 nm and 195 nm were characteristic for ascorbic acid and sorbose respectively (Fig. 2) at this pH region. 200 ml aqueous solution of L-sorbose (3g/100 ml) and 100 ml aqueous solution of hesperidin (1.5g/100 ml) were mixed. The pH of the solution was adjusted to pH:2 by adding diluted sulphuric acid. The same procedures for the photoreaction as described above were repeated. The change of the concentrations of L-sorbose and L-ascorbic acid was studied by taking the spectra between 335 and 185 nm as well as by monitoring the absorbance changes at wavelengths 245 and 195 nm, which are characteristic for ascorbic acid and sorbose in this pH region (Fig. 2). The extent of the interference of the UV absorption peaks in the mixture of ascorbic acid, sorbose and hesperidin are studied and taken into account during the measurements.

The reactions described above have been carried out with D-sorbitol instead of L-sorbose using the same concentrations and the same conditions. In a different series of the experiments oxygen flow was replaced by air flow under the same experimental conditions.

For thin layer chromatographic determination of ascorbic acid[12] 45g Kieselgel PF_{254} mixed thoroughly with 150 ml chloroform. The chromatographic plates were prepared by dipping the glass plates into this mixture. After taking off the plate and drying in air, a homogeneous "thin layer" was obtained. As mobile phase a mixture of 5 ml aceton 5 ml glacial acetic acid, 20 ml methanol and 70 ml benzene was prepared in a small covered chromatographic tank and left one hour for vapor phase saturation.

After one hour photoreaction of sorbose in the above described conditions a 100 ml sample was taken from the reactor and evaporated at 95°C until drying. The soluble part was taken with 30 ml methanol and 10 µl of this solution was applied on the chromatographic plate. The chromatographic process has required approximately 10 minutes. The plate was dried in the air. Spots were observed under 254 nm UV light of Desaga UV lamp and R_f values were compared with standards (Fig. 3).

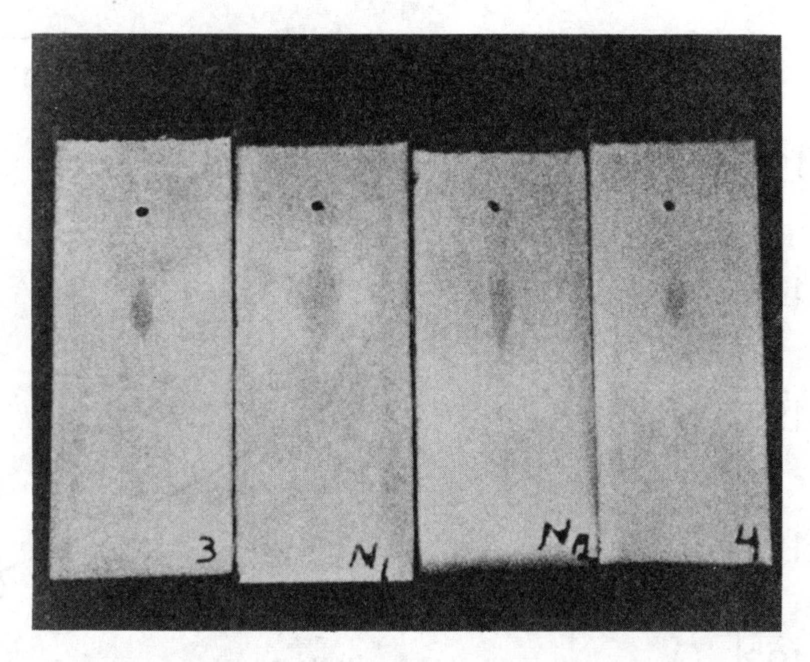

Fig. 3. Thin layer chromatographic spots of ascorbic acid formed by the reaction (called N_1 and N_2) in comparison with standards (called 3 and 4)

3. RESULTS AND DISCUSSION

The absorbance values of L-sorbose and L-ascorbic acid at characteristic wavelengths were monitored during the photochemical process. The change of the absorbance values versus time is displayed in Figs. 4 and 5 for reactions at pH 6.85 and pH 2 respectively. Fig. 4b and 5b clearly show an increase of the ascorbic acid concentration while Figs. 4a and 5a depict a decrease in L-sorbose absorbance values. No change in hesperidin concentration has been observed during the reaction.

The chromatographic spots are shown in Fig. 3 in comparison with ascorbic acid standards. The results of thin layer chromatography also shows that ascorbic acid is produced in this photocatalytic reaction. The replacement of oxygen flow by air still leads to the production of ascorbic acid determined by UV absorption spectra and thin layer chromatography.

All reactions were repeated in the dark and no ascorbic acid formation was observed by either spectrophotometric or chromatographic determination methods.

In our experimental conditions the increase in ascorbic acid concentration in the photoreaction slows down and stops after approximately 70 minutes. A comparative thin layer chromatographic method shows that this limit concentration of ascorbic acid lies at 1% in the reaction

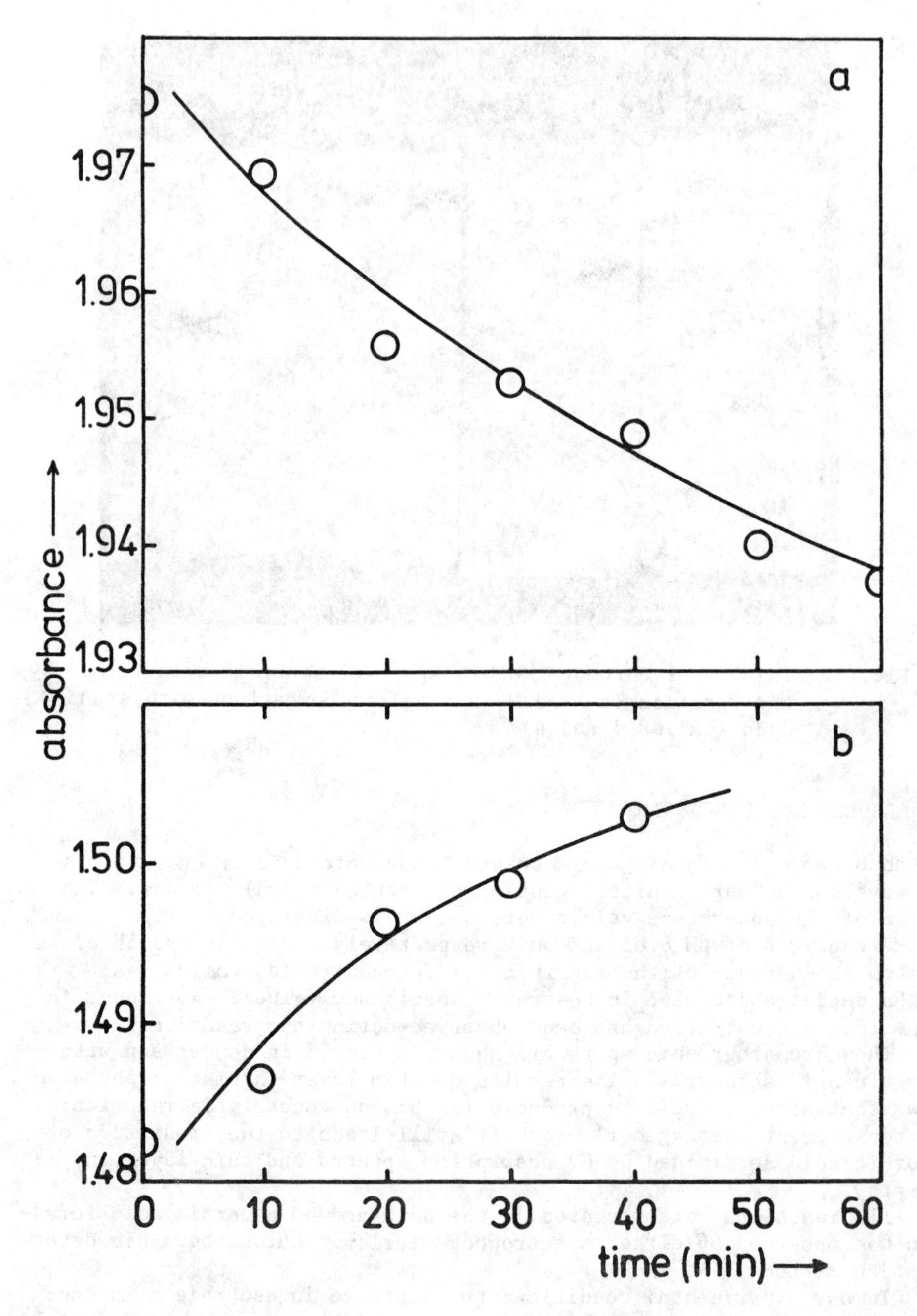

Fig. 4. Decrease of the L-sorbose absorbance (195 nm) vs. time (a) and increase of the ascorbic acid absorbance (265 nm) (b) during the reaction at pH:6.85.

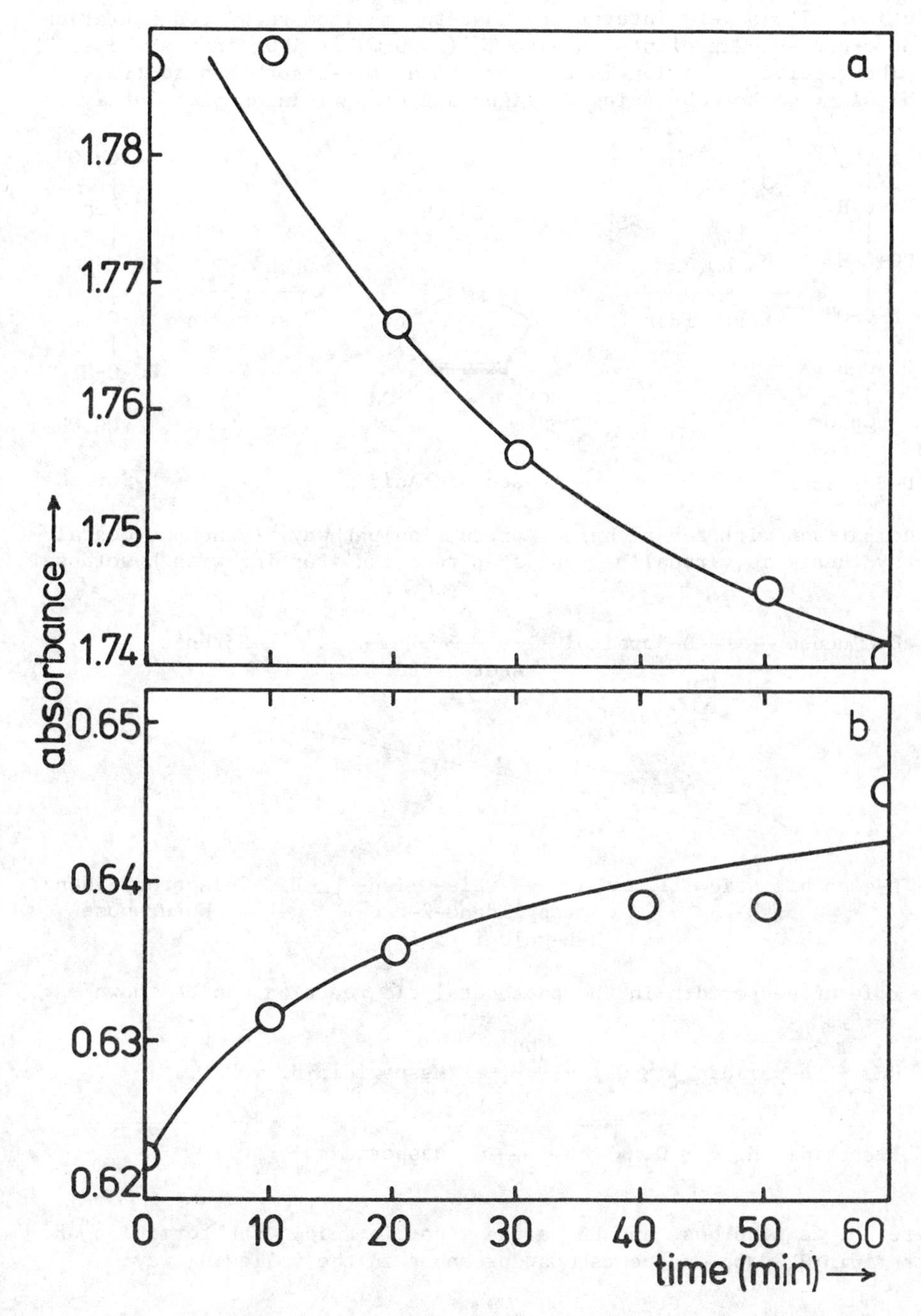

Fig. 5. L-sorbose absorbance (195 nm) decrease vs. time (a) and the increase ascorbic acid absorbance (245 nm) vs. time (b) during the reaction at pH:2.

solution. It is very interesting that the maximum known concentration of ascorbic acid in plants is also 1% (1000 mg in 100g in rose hips)[10].

The observed reaction is the formation of L-ascorbic acid from sorbitol or sorbose by using UV light and hesperidin as photocatalyst:

D-Sorbitol (CH_2OH–HO-C-H–HO-C-H–H-C-OH–HO-C-H–CH_2OH) $\xrightarrow[\text{Hesperidin}]{O_2, h\nu}$ L-Ascorbic Acid $\xleftarrow[\text{Hesperidin}]{O_2, h\nu}$ L-Sorbose (CH_2OH–C=O–HO-C-H–H-C-OH–HO-C-H–CH_2OH)

In comparison with the technical production pathway[10] the photocatalytic synthesis is virtually a one-step reaction starting with L-sorbose:

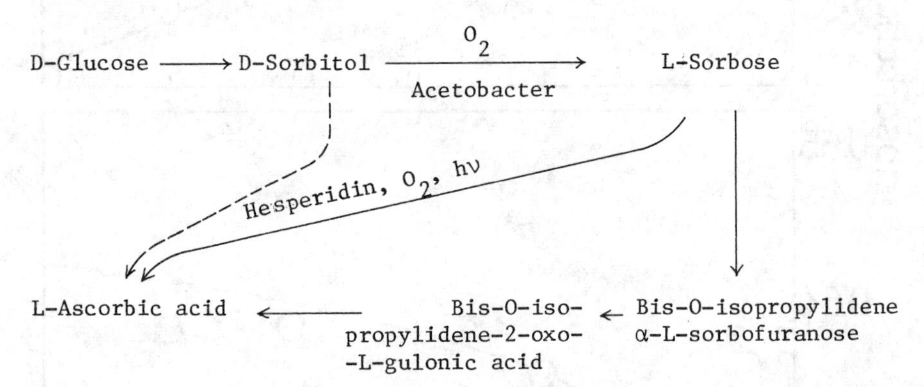

The role of hesperidin in the photocatalytic reaction can be shown as:

$$RH_2 + \text{Hesperidin} + \tfrac{1}{2}\,O_2 \xrightarrow{h\nu} \text{Hesperidin}\cdot H_2 + RO$$

$$\text{Hesperidin}\cdot H_2 + \tfrac{1}{2}\,O_2 \longrightarrow \text{Hesperidin} + H_2O$$

where RH_2 is L-sorbose and RO is the ascorbic acid. The formation of Hesperidin·H_2 complex occurs most probably in the following way:

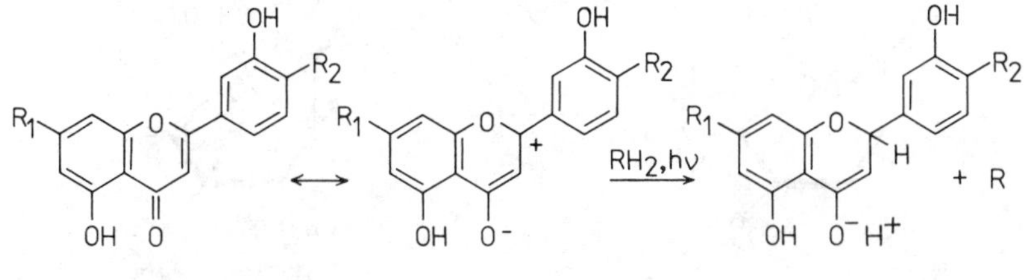

$$\text{Hesperidin} \longleftrightarrow \text{(resonance form)} \xrightarrow{RH_2, h\nu} \text{Hesperidin}\cdot H_2 + R$$

Hesperidin Hesperidin·H_2

$-R_1 = -O(C_{12}H_{21}O_9)$, $-R_2 = -OCH_3$

Hesperidin·H_2 transfers the two hydrogens to the oxygen molecule leading to the formation of a water molecule. The reaction from sorbitol to sorbose can again be catalyzed by Hesperidin.

Another possible mechanism of the photocatalytic ascorbic acid formation may be formulated assuming that excited hesperidin molecules can transfer some energy to O_2 and form singlet oxygen which in turn reacts with sorbose forming ascorbic acid:

$$\text{Hesperidin} \xrightarrow{h\nu} \text{Hesperidin}^* \xrightarrow{O_2} {}^1O_2 + \text{Hesperidin}$$

$${}^1O_2 + RH_2 \longrightarrow RO + H_2O$$

The reaction of sorbose as well as the reaction of sorbitol has led to the same product - ascorbic acid - by using hesperidin and UV irradiation. However, L-sorbose is not found in the nature, while many plants contain high amounts of D-sorbitol. Therefore we assume that this reaction in the nature starts with D-sorbitol and probably proceeds via L-sorbose:

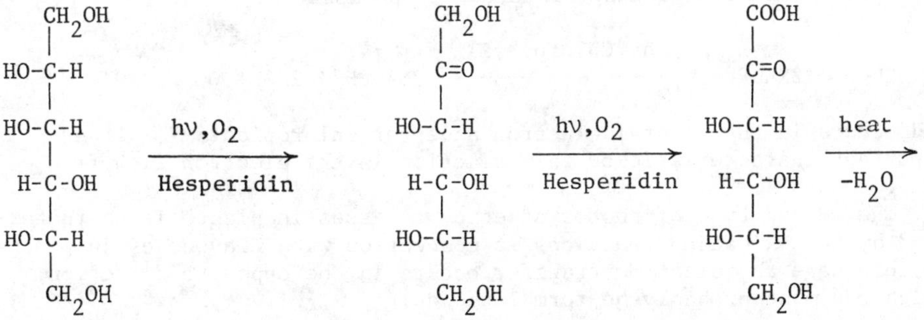

CH_2OH		CH_2OH		COOH	
HO-C-H		C=O		C=O	
HO-C-H	$h\nu, O_2$ →	HO-C-H	$h\nu, O_2$ →	HO-C-H	heat →
H-C-OH	Hesperidin	H-C-OH	Hesperidin	H-C-OH	$-H_2O$
HO-C-H		HO-C-H		HO-C-H	
CH_2OH		CH_2OH		CH_2OH	
D-Sorbitol		L(-) Sorbose		2-Deoxy-2-Oxo-L-gulonic acid	

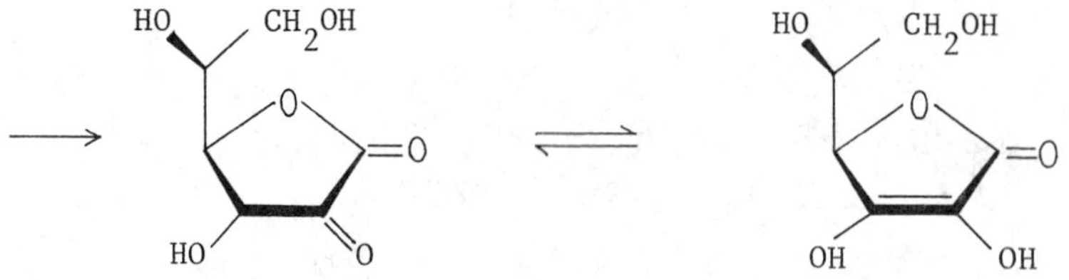

HO CH$_2$OH HO CH$_2$OH

→ ⇌

HO O OH OH

L-ascorbic acid

Since L-sorbose immediately reacts after its formation, it is impossible to find L-sorbose in plants although it forms during this reaction.

Another important fact is that ascorbic acid forms up to a certain concentration and then no increase of the concentration can be determined. We assume that ascorbic acid itself reacts in the presence of UV photons and oxygen and gives secondary products. Therefore this reaction system leads to a steady state concentration of ascorbic acid. This experimental fact can be one of the reasons for the low concentrations of ascorbic acid in plants.

4. CONCLUSION

The first results of our investigation show, that flavonoids, widely distributed in various plants, can perform electron transfer processes under influence of photons. This way they protect many compounds in the plant organism from destructive effects of the solar UV radiation. At the same time they produce other compounds by photocatalytic pathways.

Photochemical processes in plants can be divided into two groups. The first one contains the process based on the photosynthesis. This reaction leads to the formation of sugars by reduction of carbon dioxide with the aid of the natural photosystem. The process involves $NADP^+$ (nicotinamide adenine dinucleotide phosphate):

$$2H_2O + 2NADP^+ \xrightarrow{h\nu,\text{Chlorophyll}} 2NADPH + 2H^+ + O_2$$

and occurs in the natural electron acceptor chloroplast[13]. A very important characteristic of this reaction is the electron flow from water to $NADP^+$.

The second type of photosynthetic processes in plants is characterized by the oxidation reactions in connection with flavonoids in plants. In this case the electron transfer occurs in the opposite direction, which can most probably be formulated as:

$$2NADPH + 2H^+ + O_2 \xrightarrow{h\nu,\text{Flavonoids}} 2NADP^+ + 2H_2O$$

It is possible that plants use some of the O_2, produced by the photosynthesis on chloroplasts, for this second type of photosynthetic reaction, where flavonoids play the key role. Therefore we call these processes secondary photosynthesis, which may be as important as the primary one for production of various natural products. In the secondary photosynthesis there are three main parameters 1) wavelength of the UV radiation, 2) type of flavonoids and 3) types of carbohydrates. Change of these parameters may lead to several different natural photo-products on which the investigations are currently in progress at our laboratories.

5. REFERENCES

1. A. Bensáth, St. Rusznyák and A. Szent-Györgyi, Nature 138, 798 (1936).
2. A. L. Armentano, A. Bensáth, T. Béres, St. Rusznyák and A. Szent-Györgyi, Dtsch. Med. Wschr. 62, 1325 (1936).
3. V. Bruckner and A. Szent-Gyorgyi, Nature 138, 1057 (1936).
4. A. Szent-Györgyi, Z. Physiol. Chem. 255, 126 (1938).
5. G. B. Marini-Bettolo and F. D. Monache, Flavonoids and Bioflavonoids Current research Trends Proceedings of Fifth Hungarian Bioflavonoid Symposium Matrafüred, Hungary, May 25-27, 1977, Eds.:L. Farkas, M. Gábor and F. Kállay; Elsevier, 1977, pp. 1.
6. R. Bognár, M. Rákosi, ibid, pp. 117.
7. A. Şengün, F. Baykut, Ş. Kayahan, B. Berkarda, F. Öz, A. Çevikbaş, U. Çevikbaş, Y. Ceyhan, G. Ulakoğlu, A. Çotuk and A. Kazancigil, Chim. Acta Turc. 10, 69 (1982).
8. A. Şengün, F. Baykut, S. Kayahan, B. Berkarda, Y. Ceyhan, G. Ulakoğlu and A. Çotuk, Chim. Acta Turc. 10, 299 (1982).
9. This value refers to Turkish Grapefruit from South Anatolia (paper in preparation, F. Baykut et al.).
10. Kirk Othmer Encyclopedia of Chemical Technology, 3. Edition, Vol. 24, John Wiley & Sons, New York 1984, pp. 8.
11. G. Baykut, E. Tütem and F. Baykut, Chim. Acta Turc. 11, 2 (1983).
12. a) H. Ganshirst and A. Malzacher, Naturwiss. 47, 279 (1960).
    b) E. Stahl, Thin Layer Chromatography, A Laboratory Handbook, Springer (Berlin), 1965, pp. 235.
13. A. Lehninger, Principles of Biochemistry, Worth Publishers, Inc., New York, 1982, pp. 656.

SPECTROSCOPIC AND ELECTROCHEMICAL STUDIES OF PHOTOCHEMICAL ELECTRON TRANSFER IN LINKED DONOR-ACCEPTOR MOLECULES

James R. Bolton, John A. Schmidt and Aleksander Siemiarczuk
Photochemistry Unit
Department of Chemistry
The University of Western Ontario
London, Ontario, CANADA N6A 5B7

and

Mary D. Archer and Jacquin H. Wilford
Department of Physical Chemistry
University of Cambridge
Lensfield Road, Cambridge CB2 1EP, U.K.

ABSTRACT

Some covalently-linked porphyrin-quinone (PQ) molecules of restricted molecular geometry are described. The excited state P^*Q, where P^* is the porphyrin S_1 state, probably undergoes intramolecular electron transfer to produce the charge-separated state $P^{\dagger}Q^{\cdot}$. The forward rate constant k_{et}^f at 295 K for this electron transfer, measured from fluorescence quenching in PAQ (Scheme 1) compared with its hydroquinone analogue $PAQH_2$, lies in the range $(1 - 230) \times 10^7\ s^{-1}$ and is strongly solvent dependent. The electrochemically measured energy $U_{\pm}$ of $P^{\dagger}Q^{\cdot}$ with respect to the ground state PQ is 1.35 - 1.41 eV and is also somewhat solvent dependent. Rationalization of the solvent dependence of k_{et}^f and of $U_{\pm}$ is attempted both in terms of the Marcus theory of electron transfer and in terms of the Onsager reaction field theory; the latter appears to be more successful.

1. INTRODUCTION

The study of redox catalysts, in which facile electron transfer reactions must occur at the surface of a colloid particle or electrode, is one of the major topics of this workshop. The development of effective catalysts will be greatly aided by knowledge of the structural and environmental factors which control electron transfer processes across interfaces and within defined molecular structures. In this paper, we discuss the effect of solvent on the kinetics and

E. Pelizzetti and N. Serpone (eds.), Homogeneous and Heterogeneous Photocatalysis, 175–188.

thermodynamics of intramolecular photochemical electron transfer reactions between the porphyrin (P) and quinone (Q) moieties in a set of covalently linked PQ molecules of restricted geometry.

The primary energy-conversion step in photosynthesis takes place within reaction-center proteins which span the thylakoid membrane in chloroplasts of green leaves and algae [1,2]. This step consists of a photochemical electron transfer reaction which occurs in ~300 ps with a quantum yield close to unity, transiently generating a potential difference of ~1 V across the membrane. Thus the reaction center acts as a biological solar cell, which converts sunlight to electrical energy at a primary efficiency of ~16% [3]. The detailed structure of the reaction-center protein is undoubtedly the key to this exemplary performance, which allows for a large differential (~10^9) between the forward and back electron transfer rates. This "molecular photodiode" behavior is one of the features we seek to model and hence to understand in our linked PQ molecules.

The recently published crystal structure of the reaction-center protein of a photosynthetic bacterium [4] has greatly enriched our understanding of the *in vivo* donor-acceptor structure. It will be a long time before anyone is able to synthesize anything so sophisticated. However, several groups have made covalently-linked donor-acceptor (usually porphyrin-quinone) molecules which mimic certain aspects of the natural system. The various molecules studied and their photochemical and photophysical properties have been reviewed in recent papers [5-7]. In some of these compounds the rates of both forward and reverse electron transfer have been measured [8]; the results agree with the predictions of Marcus theory [9,10], particularly as regards the observation of the so-called "inverted region" in which electron transfer reactions which are highly exergonic nevertheless proceed more slowly. Our work [5,11-13] has involved the PQ molecules shown in Scheme I, in which a metal-free tritolylporphyrin derivative is linked to p-benzoquinone via one or two amide groups.

In the PAnAQ (n = 2-4) series we have observed fluorescence quenching [11] and electron paramagnetic resonance (EPR) spectra [5] which provide strong evidence for intramolecular electron transfer from the first excited singlet state P^* of the porphyrin to the quinone.

$$PQ \xrightarrow{h\nu} P^*Q \tag{1}$$

$$P^*Q \xrightarrow{k_{et}^{f}} P^{\dot{+}}Q^{\dot{-}} \tag{2}$$

$$P^*Q \xrightarrow{k_{deact}} PQ \tag{3}$$

$$P^{\dot{+}}Q^{\dot{-}} \xrightarrow{k_{et}^{b}} PQ \tag{4}$$

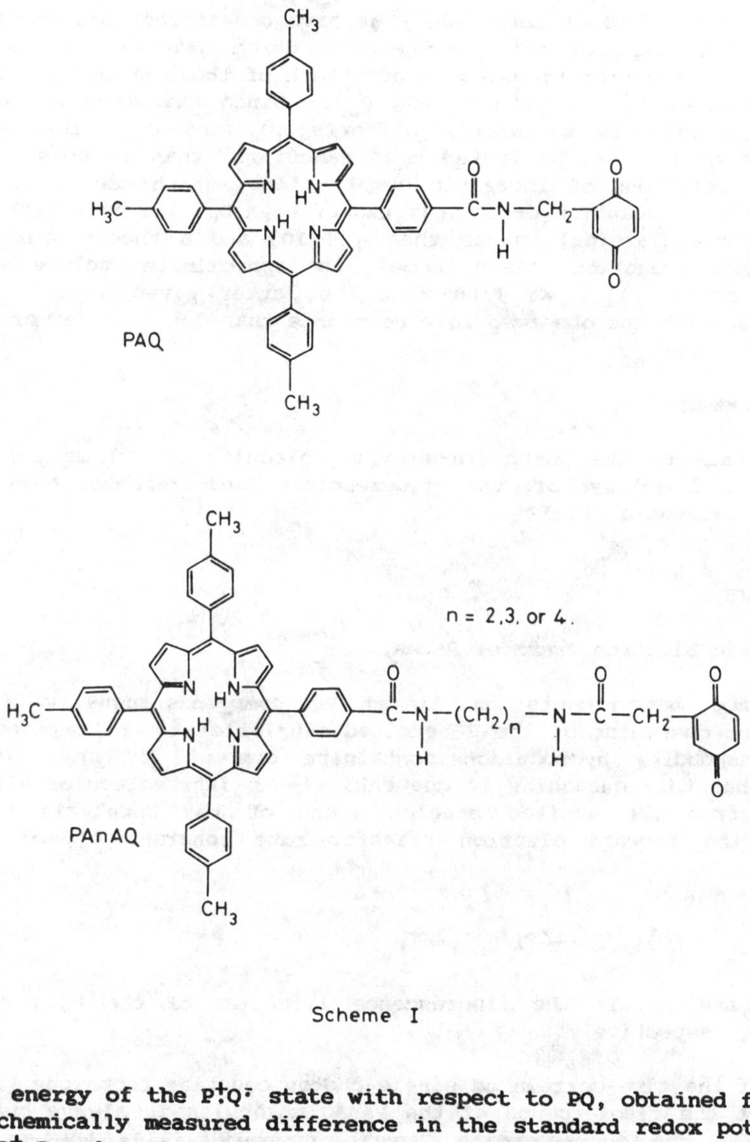

PAQ

n = 2,3, or 4.

PAnAQ

Scheme I

The energy of the $P^{\dagger}Q^{\bar{\cdot}}$ state with respect to PQ, obtained from the electrochemically measured difference in the standard redox potentials of the $P^{\dagger}$,P and Q,$Q^{\bar{\cdot}}$ couples in the linked molecules, is ~1.37 eV [13]. The 0-0 singlet excitation energy of P^* is 1.89 eV, whereas that of the triplet is ~1.45 eV [7a,14]. Thus the formation of $P^{\dagger}Q^{\bar{\cdot}}$ from the triplet state would be only slightly exergonic, whereas its formation from the singlet P^*Q state is substantially exergonic.

Our measured values of k_{et}^f for PAQ in 22 solvents of widely differing static (D_s) and optical (D_{op}) dielectric constants vary by more than two orders of magnitude. We have considered this pronounced variation in the context of theories which predict the solvent dependence of electron transfer rates. Each of these theories predicts a dependence of k_{et}^{f} on both D_s and D_{op}. Since the PQ molecules are soluble in solvents of widely differing D_s and D_{op}, the solvent dependence of k_{et}^{f} can be tested more exactingly than is possible for the redox reactions of inorganic complex ions, which can be examined only in aqueous solution or in solvents of high D_s. We have tested two theories: the classical Marcus theory [9,10] and a theory which uses the Onsager reaction field model to approximate solute-solvent interactions [15,16]. We find that the latter gives a much better correlation with the observed rate constants than does the former.

2. EXPERIMENTAL

The synthesis of the porphyrin-quinone molecules of Scheme I [12,17] and the methodology of the fluorescence measurements have been described elsewhere [11,12].

3. RESULTS

3.1 Forward Electron Transfer Rates

Fluorescence measurements on linked PQ compounds usually show a significant quenching of the P^* excited singlet state as compared with the corresponding hydroquinone-containing compound (PQH_2). If one assumes that this quenching is due entirely to intramolecular electron transfer from the excited singlet state of the porphyrin to the quinone, the forward electron transfer rate constant k_{et}^{f} is given by [10]

$$k_{et}^{f} = 1/\tau_1 - 1/\tau_2 \qquad (5)$$

where τ_1 and τ_2 are the fluorescence lifetimes of the PQ and PQH_2 molecules, respectively.

Use of the time-correlated single-photon counting technique [11,12] shows that the fluorescence of the PAnAQ molecules is always quenched relative to the corresponding $PAnAQH_2$ compounds [11]; however, the decays always contain at least two components, probably as a result of the formation of intramolecular complexes between the porphyrin and quinone entities.

The PAQ molecule exhibits single-exponential fluorescence decays and hence the results are much more easily interpreted in terms of eq. 5. Table 1 shows the fluorescence lifetimes of PAQ and of $PAQH_2$ together with the calculated values of k_{et}^{f} in a variety of solvents.

Table 1. Solvent Dependence of Electron Transfer Rate Constants in PAQ as Derived from Fluorescence Lifetime Measurements

Solvent	D_s[a]	D_{op}[a]	$(\Delta U_{solv} - \Delta U_{ether})$ /eV	τ_1/ns	τ_2/ns	k^f_{et} /10^7 s^{-1}
Ethyl ether	4.33	1.828	0.00	10.77	12.77	1.4
1,2-DME[b]	7.20	1.904	-0.17	10.05	12.41	1.9
MTHF[c]	7.60	1.977	-0.28	9.43	11.69	2.0
Ethyl Acetate	6.02	1.882	-0.12	9.52	11.90	2.1
Acetone	20.7	1.847	-0.17	8.92	12.15	3.0
p-Dioxane	2.21	2.022	-0.17	8.34	11.92	3.6
CH_3CN	37.5	1.806	-0.13	7.39	11.89	5.1
n-PrCN	20.3	1.915	-0.27	7.12	11.31	5.2
Toluene	2.38	2.241	-0.46	4.77	11.7	12.
Benzene	2.27	2.253	-0.47	3.63	11.2	18.
Benzyl ether	4.3	2.375	-0.70	3.45	12.7	20.
n-Butanol	17.5	1.957	-0.32	3.15	12.38	24.
Anisole	4.33	2.301	-0.62	2.45	11.9	32.
Benzonitrile	25.6	2.332	-0.80	2.08	11.7	39.
CH_3CHCl_2	10.0	2.005	-0.34	1.92	10.5	43.
CH_3CHCl_3	7.53	2.068	-0.40	1.69	10.55	50.
CH_2ClCH_2Cl	10.6	2.088	-0.46	1.56	9.61	54.
Cl-Naph [d]	5.04	2.667	-1.03	1.16	10.6	77.
Chlorobenzene	5.62	2.326	-0.68	1.15	11.0	78.
CH_2Cl_2	9.14	2.028	-0.36	1.09	9.14	81.
CH_2BrCH_2Br	4.78	2.369	-0.71	0.52	1.54	130
$CHCl_3$	4.81	2.091	-0.38	0.42	9.05	230

[a] All Values are at 293K, taken from S.L. Murov, *Handbook of Photochemistry*, Marcel Dekker, New York, p. 854 (1973) or W.B. Bunger *Techniques of Organic Chemistry, II, Organic Solvents*, 3rd Ed., John Wiley & Sons, New York (1970).

[b] 1,2-Dimethoxyethane;

[c] 2-Methyltetrahydrofuran;

[d] 1-Chloronaphthalene

3.2 Energy of the Charge-Separated State

As direct spectroscopic measurement of the energy $U_{\pm}$ of the charge-separated $P^{+}Q^{-}$ state is not possible; hence indirect electrochemical assessment is the only method available. The difference between the first porphyrin ring oxidation potential O_1 and the first quinone reduction potential Q_1 gives $U_{\pm}$

$$U_{\pm} = e(O_1 - Q_1) \tag{6}$$

provided that the Gibbs energy and the internal energy can be considered identical, *i.e.*, provided that entropy effects and ion association are negligible and that $P^{+}Q^{-}$ is not significantly stabilized with respect to $P^{+}Q$ and PQ^{-} by coulombic or spin interaction. These are substantial assumptions, requiring scrutiny which we do not here accord to them.

Figure 1 shows a differential voltammogram typical of PAnAQ or PAQ. The usual two porphyrin ring reductions (designated R_1 and R_2) and the two ring oxidations (O_1 and O_2) straddle the Fc ferrocene reference signal. In addition to the reversible porphyrin ring processes (peak fwhm ~96 mV), each of these compounds shows a quasireversible peak Q_1 (peak fwhm ~200 mV) between O_1 and R_1, due to the quinone reduction $Q + e^- \rightarrow Q^{-}$. Cyclic voltammetry or AC voltammetry with scan reversal also shows that the P ring processes are AC-reversible but Q_1 is AC-quasireversible.

Table 2 summarizes our findings for PAQ and PAnAQ. All redox potentials are insensitive to n (for PAnAQ), varying only ~50 mV in CH_2Cl_2. Quinone reduction potentials are, however, strongly dependent on ring substitution and on solvent [13,18].

Table 2 also gives $U_{\pm}$, calculated from eq. 6 for our PQ compounds and for similar molecules studied by Kong *et al.* [19]. In CH_2Cl_2, $U_{\pm}$ is in the range 1.35-1.39 eV. Data on $U_{\pm}$ for a given compound in various solvents are fragmentary but, rather surprisingly, it appears from Table 2 that $U_{\pm}$ increases with D_S, in accord with the trend in coulombic interaction between $P^{+}Q$ or PQ^{-} and a counter ion or in $P^{+}Q^{-}$ itself, but in opposition to the trend in ion-solvent dipole interaction.

Table 2: Redox Potentials[a] of the Compounds of Scheme I and Related Molecules

Compound	Solvent System[b]	R_2	R_1	Q_1	O_1	O_2	$U_{\pm}$/eV[c]
PAQ	TBAH/CH_2Cl_2		-1.66	-0.92	+0.45	?	1.37
PA2AQ	TBAH/CH_2Cl_2	-2.028	-1.672	-0.87	+0.493	?	1.35
PA2AQ	TBAH/CH_3CN		-2.18[d]	-1.41[d]	-	-	1.41
PA3AQ	TBAH/CH_2Cl_2	-2.02	-1.690	-0.90	+0.485	?	1.39
PA4AQ	TBAH/CH_2Cl_2	-2.038	-1.702	-0.90	+0.472	+0.80	1.37
PE3EQ[e]	TBAP/CH_2Cl_2						1.54[f] (est)
PE3EQ[e]	TBAP/DMF						1.70[f] (est)
ZnPE3EQ[g]	TBAP/CH_2Cl_2						1.30[f]
ZnPE3EQ[g]	TBAP/DMF						1.45[f]

[a] Redox potentials are relative to the internal ferrocene/ferricinium couple except where otherwise stated. Error ±0.015 V for PAQ, ±0.025 for PAnAQ.

[b] TBAH = tetrabutylammonium hexafluorophosphate, TBAP = tetrabutylammonium perchlorate, DMF = N,N'-dimethylformamide.

[c] As calculated from eq. 6.

[d] Measured and expressed vs. O_1.

[e] The structure PE3EQ is the same as that of PA3AQ except that the two links are ester (E) groups rather than amide (A) groups.

[f] Taken from Ref 19.

[g] As for PE3EQ except that the tetraphenylporphyrin moiety contains Zn(II).

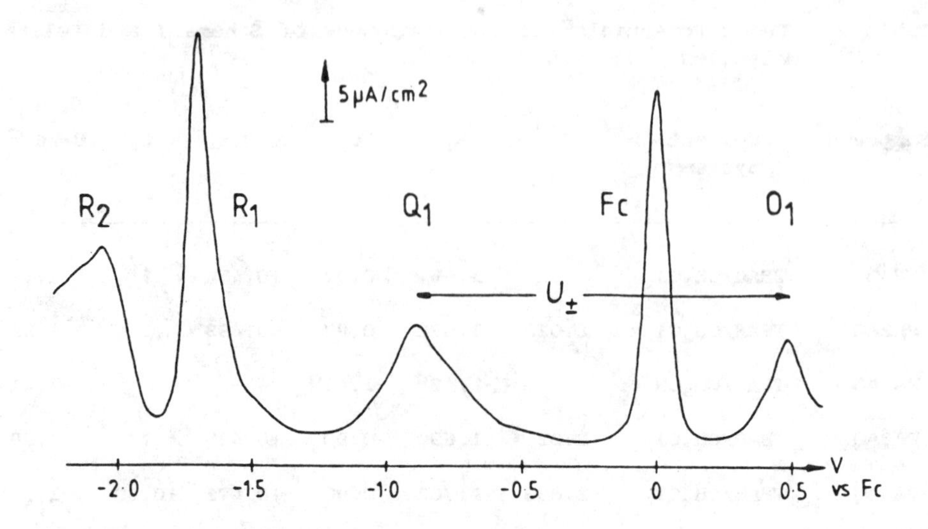

Figure 1: Differential pulse voltammogram of PA3AQ (6×10^{-5} M) in the presence of a small concentration of ferrocene Fc used as a reference. Solvent system 0.1 M tetrabutylammonium hexafluorophosphate in CH_2Cl_2 at 293 K. Scan rate 5 mV s^{-1}, pulse height 50 mV. The energy $U_\pm$ of the charge-separated $P^{+}Q^{-}$ state relative to the ground state PQ, as estimated from eq. 6, is marked. Very similar traces are observed for the other amide-linked PQ compounds listed in Table 2.

4. DISCUSSION

We have attempted to understand our results in terms of the Gibbs energy diagram shown in Fig. 2. We have examined two theories for the rates of electron transfer - the classical Marcus theory and the Onsager reaction field theory. We briefly outline these two theories below.

4.1 Marcus Theory

The Marcus theory [9,10] is a classical theory of electron transfer which assumes that all potential energy surfaces along the reaction coordinate can be considered as parabolae. The Gibbs energy of activation $\Delta G^\dagger$ for a bimolecular redox reaction is given by [10]

$$\Delta G^\dagger = (\lambda/4)(1 + \Delta G^\circ/\lambda)^2 \tag{7a}$$

$$= (1/4)[\lambda + (\Delta G^\circ)^2/\lambda + 2\,\Delta G^\circ] \tag{7b}$$

and the rate constant k_{et} is related to $\Delta G^\dagger$ by

$$\ln k_{et} = A - \Delta G^\dagger/RT \tag{8}$$

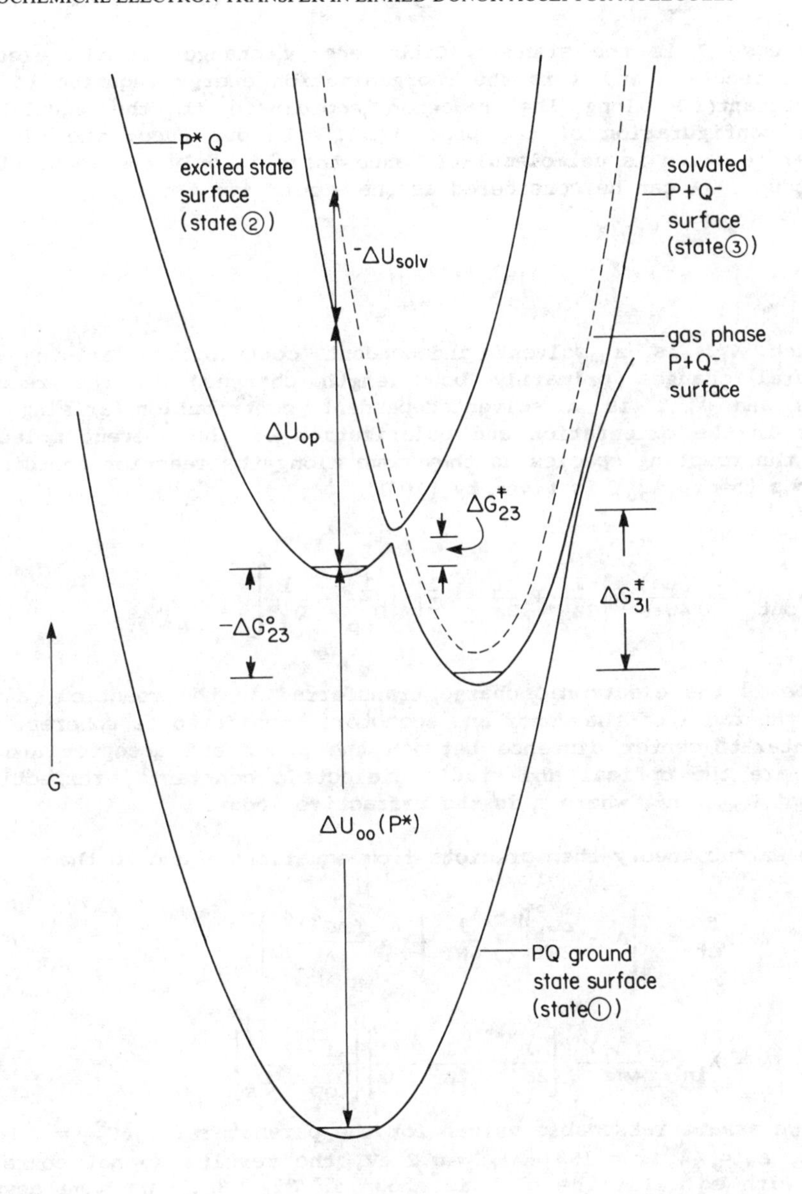

Figure 2. Gibbs energy diagram showing the surfaces for the PQ ground state (1), the $P^{*}Q$ excited state (2) and the charge-separated $P^{\dot{+}}Q^{\dot{-}}$ state (3) as a function of the reaction coordinate. The gas phase $P^{\dot{+}}Q^{\dot{-}}$ curve is not drawn to scale.

ΔG° in eqs. 7 is the standard Gibbs energy change for the electron transfer reaction and λ is the reorganization energy required to move the reactant(s) along the reaction coordinate to the equilibrium nuclear configuration of the product(s). In our case, the electron transfer reaction is unimolecular; hence there is only one reactant and one product. λ can be considered as the sum of two terms

$$\lambda = \lambda_{in} + \lambda_{out} \tag{9}$$

in which λ_{in} is a solvent independent contribution arising from structural changes (primarily bond-length changes) in the reacting species and λ_{out} is a solvent dependent contribution arising from changes in the orientation and polarization of the solvent molecules around the reacting species as they move along the reaction coordinate. In Marcus theory λ_{out} is given by [10]

$$\lambda_{out} = \frac{(\Delta e)^2}{4\pi\epsilon_o}\left[\frac{1}{2a_2} + \frac{1}{2a_3} - \frac{1}{r}\right]\left[\frac{1}{D_{op}} - \frac{1}{D_s}\right] \tag{10}$$

where Δe is the electronic charge transferred in the reaction, a_2 and a_3 are the radii of the donor and acceptor, considered as spheres, r is the center-to-center distance between the donor and acceptor and D_{op} and D_s are the optical and static dielectric constants, respectively Note that $D_{op} = n^2$, where n is the refractive index.

The Marcus theory then predicts from equations 8 and 10 that

$$\ln k_{et}^{f} = \left[A - \frac{\Delta G^\circ}{2RT}\right] - \frac{1}{4RT}\left\{\lambda + \frac{(\Delta G^\circ)^2}{\lambda}\right\} \tag{11a}$$

where

$$\lambda = \lambda_{in} + \frac{(\Delta e)^2}{4\pi\epsilon_o}\left[\frac{1}{2a_2} + \frac{1}{2a_3} - \frac{1}{r}\right]\left[\frac{1}{D_{op}} - \frac{1}{D_s}\right] \tag{11b}$$

If we assume reasonable values for the parameters: $-\Delta G^\circ_{23} = 0.5$ eV, $a_2 = 8$Å, $a_3 = 4$Å, $r = 15$Å; $\lambda_{in} = 0.2$ eV, the results do not correlate at all with eq. 11a; the plot is shown in Fig. 3a. We have assumed that ΔG°_{23} is solvent independent; however, even if we allow for some solvent dependence of ΔG°_{23} and allow a reasonable variation of the other parameters, the fit does not improve significantly.

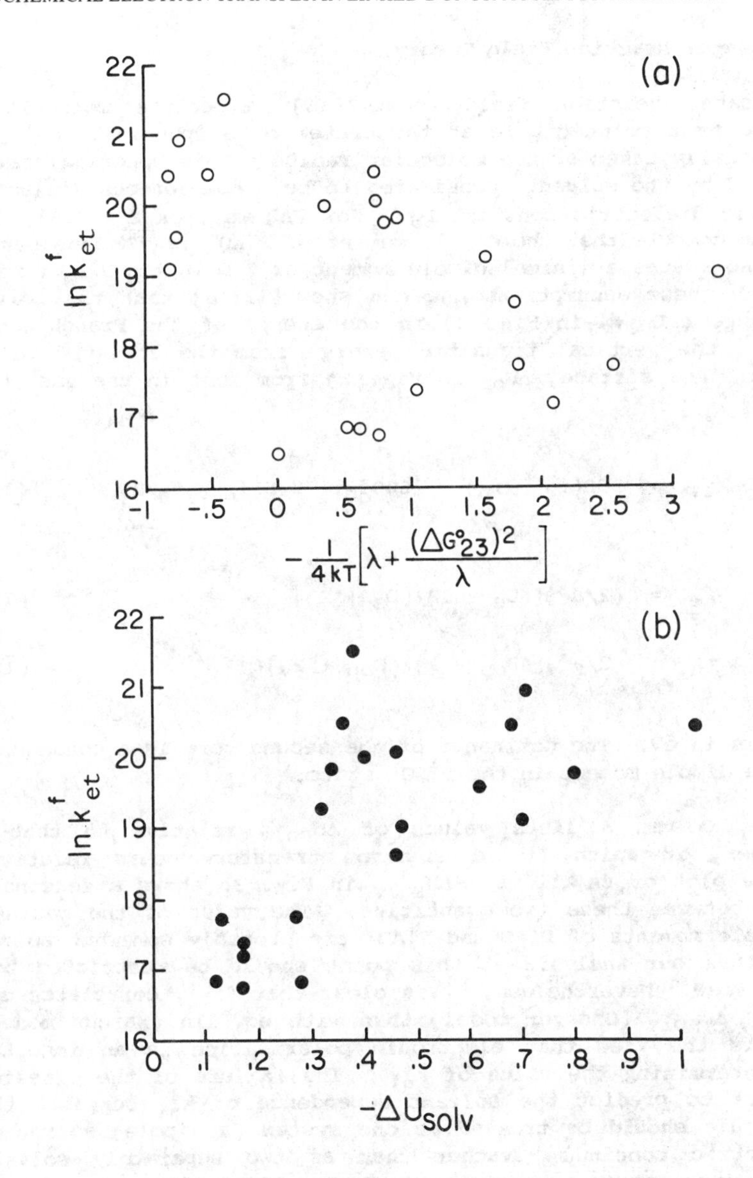

Figure 3. Plot of $\ell n\ k^f_{et}$ according to (a) Marcus theory, eq. 11a and (b) the Onsager reaction-field theory, eq. 12a. In each case the abscissa has been plotted relative to the value for ether taken as zero.

4.2 The Onsager Reaction Field Theory

In the Onsager reaction field theory [16], a solute molecule is approximated by a point dipole at the center of a spherical cavity of radius a, usually taken as the molecular radius. This spherical cavity is surrounded by the solvent, considered to be a homogeneous dielectric with a static dielectric constant D_S. For PAQ we took a ≅ 7.5Å, from which we can compute that the dipole moment of $P^{\dagger}AQ^{\bar{\cdot}}$ is ~72 Debye units (D). We also assume a minimal dipole moment of 2 D in the $P^{*}AQ$ excited state. With these assumptions one can show [12,16] that the solvent induced change (ΔU_{solv} in Fig. 2) in the energy of the Franck-Condon state (*i.e.*, the vertical transition energy from the lowest level of $P^{*}AQ$ to the $P^{\dagger}AQ^{\bar{\cdot}}$ surface, ΔU_{op} in Fig. 2) from that in the gas phase is given by

$$\Delta U_{solv} = -90(f_{\epsilon} - f_n) - 1600 f_n + 2.5(f_{\epsilon} - f_n/2) \quad (12a)$$

where

$$f_{\epsilon} = (2/a^3)[(D_S - 1)/(D_S + 2)] \quad (12b)$$

$$f_n = (2/a^3)[(D_{op} - 1)/(D_{op} + 2)] \quad (12c)$$

and ΔU_{solv} is in eV. The dominance of the second term is a consequence of the large dipole moment in the $P^{\dagger}AQ^{\bar{\cdot}}$ state.

Table 1, Column 4 lists values of ΔU_{solv} relative to that of diethyl ether, in which little electron transfer occurs relatively slowly. The plot of $\ln k^f_{et}$ vs. $-\Delta U_{solv}$ in Fig. 3b shows a reasonable correlation between these two quantities. The value of the radius a and the dipole moments of $P^{*}AQ$ and $P^{\dagger}AQ^{\bar{\cdot}}$ are probably somewhat solvent dependent; thus our analysis at this point should be restricted to a qualitative view. Nevertheless, it is clear that $\ln k^f_{et}$ correlates much better with $-\Delta U_{solv}$ (Onsager model) than with eq. 11a (Marcus model). This supports the view that electronic polarization is an important factor in determining the value of k^f_{et}. The failure of the classical Marcus theory to predict the solvent dependence of k^f_{et} suggests that the PQ molecule should be treated as one system (a dipole) surrounded by a dielectric continuum, rather than as two separately solvated molecules or ions, as is assumed in the Marcus theory [10].

We have not taken into account the solvent dependence of ΔG°_{23} in affecting the rate of electron transfer. An examination of Fig. 2 shows that as $|\Delta G^{\circ}_{23}|$ decreases, the product surface (surface 3) moves to higher energy; this should lead to a larger value of $\Delta G^{\ddagger}_{23}$ and hence a slower rate. At present we only have one test of this prediction. From Table 2 we see that $-\Delta G^{\circ}_{23} = \Delta U_{00}(P^{*}) - \Delta G^{\circ}_{31} = 0.56$ eV for CH_2Cl_2 and 0.50 eV for CH_3CN for PA2AQ. Thus we predict that k^f_{et} should be smaller in CH_3CN as compared with CH_2Cl_2, which indeed is the case.

5. CONCLUSIONS

In this paper we have presented a preliminary analysis of the solvent dependence of electron transfer rate constants in linked donor-acceptor molecules with a restricted geometry. Our initial conclusion from the analysis is that theories which are based on treating the donor-acceptor molecule as a whole (such as the Onsager reaction field theory) will be more successful than those which treat the donor and acceptor separately (such as the classical Marcus theory). We also note that although internal factors, such as the distance and orientation between the donor and acceptor and the nature of the linkage will be important factors, the surrounding solvent environment must be considered in any theory of intramolecular electron transfer.

We plan to conduct further experiments, particularly of electron transfer on the solvent dependence of ΔG°_{23}, to obtain more data against which theories of electron transfer may be tested.

ACKNOWLEDGEMENTS

This work was supported by a Strategic Grant in Energy from the Natural Sciences and Engineering Research Council of Canada (JRB) and by a Grant from the Science and Engineering Research Council of Great Britain (MDA). We also thank Dr. John S. Connolly of the Solar Energy Research Institute for helpful discussions.

REFERENCES

1. Clayton, R.K. *Photosynthesis: Physical Mechanisms and Chemical Patterns*, Cambridge University Press, Cambridge, England (1980).
2. Hall, D.O. and Rao, K.K. *Photosynthesis*, 3rd Ed., Edward Arnold Ltd., London (1980).
3. Bolton, J.R. 'Solar Electricity: Lessons Gained from Photosynthesis', in *Inorganic Chemistry: Towards the 21st Century*, Chisholm, M.H., Ed., ACS Symposium Series No. 211, American Chemical Society, Washington, D.C., pp. 1-19 (1983).
4. Diefenhofer, J., Epp, O., Miki, K., Huber, R. and Michel, H. *J. Mol. Biol.*, 180, 385-398 (1984).
5. McIntosh, A.R., Siemiarczuk, A., Bolton, J.R., Stillman, M.S., Ho, T.-F. and Weedon, A.C. *J. Am. Chem. Soc.*, 105, 7215-7223 (1983).
6. Wasielewski, M.R. and Niemczyk, M.P. *J. Am. Chem. Soc.*, **106**, 5043-5045 (1984).
7. (a) Connolly, J.S. in *Photochemical Conversion and Storage of Solar Energy - 1982*, Part A, Rabani, J., Ed., Weizmann Science Press, Israel pp. 175-204 (1982); (b) Connolly, J.S. and Turner, J.A., 'Status and Prospects for Solar Photochemistry' in *Photochemical Conversions*, Braun, A.M., Ed., Presses polytechniques romandes, Lausanne, Switzerland, pp. 73-130 (1983).
8. Wasielewski, M.R., Niemczyk, M.P., Svec, W.A. and Pewitt, E.B. *J. Am. Chem. Soc.*, 107, 1080-1082 (1985).
9. Marcus, R.A. *J. Chem Phys.* 24, 966-978 (1956); *Annu. Rev. Phys. Chem.* 15, 155-196 (1964).
10. Sutin, N. *Progr. Inorg. Chem.* 30, 441-498 (1983); Sutin, N. and Creutz, C. *J. Chem. Educ.* 60, 809-814 (1983).
11. Siemiarczuk, A., McIntosh, A.R., Ho, T.-F., Stillman, M.J., Roach, K.J., Weedon, A.C., Bolton, J.R. and Connolly, J.S. *J. Am. Chem. Soc.* 105, 7224-7230 (1983).
12. Schmidt, J.A., Siemiarczuk, A., Weedon, A.C. and Bolton, J.R. *J. Am. Chem. Soc.*, 107 (in press).
13. Wilford, J.H., Archer, M.D., Bolton, J.R., Ho, T.-F., Schmidt, J.A. and Weedon, A.C. *J. Phys. Chem.* (submitted).
14. Gouterman, M., and Khalil, G.-E., *Mol. Spectroscopy*, 53, 88 (1974); Harriman, A., *J. Chem. Soc. Faraday Soc. I*, 76, 1978-1985 (1980).
15. Onsager, L. *J. Am. Chem. Soc.* 58, 1486-1493 (1936).
16. (a) Mataga, N. and Ottolenghi, M. in *Molecular Association*, Foster, R., Ed., Academic Press, New York, Vol. 2, p. 40 (1979); (b) Beens, H., PhD. Thesis, Free University of Amsterdam, Amsterdam (1969).
17. Ho, T.-F., McIntosh, A.R. and Weedon, A.C. *Can. J. Chem.* 62, 967-974 (1984).
18. Wilford, J.H. and Archer, M.D. *J. Electroanal. Chem.* (in press).
19. Kong, J.L.Y., Spears, K.G. and Loach, P.A. *Photochem. Photobiol.* 35, 545 (1982).

BIFUNCTIONAL PORPHYRINS: REDOX PHOTOCHEMISTRY OF [MESO(TRITOLYL(PYRIDYL Ru(III) $(NH_3)_4$ L')) PORPHYRIN] (L' = NH_3, PYRIDINE, 5-Cl-PYRIDINE)

Cesar Franco and George McLendon
University of Rochester
Department of Chemistry
River Station
Rochester, NY 14627

Abstract: A series of difunctional porphyrins have been synthesized which contain a redox active Ru(III) moiety attached via a meso pyridyl group to a redox photo active porphyrin ie: [meso(tritolyl)(monopyridylRu(III)$(MH_3)_4$ L') porphyrin] L' = NH_3, pyridine, 5 Cl pyridine. In these complexes, the porphyrin triplet ^{3}excited state is quenched by intramolecular electron transfer, with rates which depend on ΔG as follows: For L = NH_3 ΔE = 0.37V $k = 4 \times 10^4 s^{-1}$ L = pyridine, ΔE = 0.55 V $k = 1.2 \times 10^5 s^{-1}$, L = 5 Cl pyridine ΔE = 0.65 V $k = 2 \times 10^5 s^{-1}$.

Detailed studies of collisional intermolecular quenching of the unsub stituted porphyrin by Ru(III)$(NH_3)_5$ pyridine are reported for comparison with the collisionless, intramolecular systems. Interestingly, the intermolecular rate is significantly faster than the intramolecular process, suggesting an important steric effect in the transition state for electron transfer.

Finally, these results are compared with redox processes for analogous porphyrin quinone systems, and implications for related biological redox mechanisms are discussed.

1. INTRODUCTION

The use of light to promote energy storing oxidation-reduction reactions lies at the heart of most solar conversion schemes, but ignorance of the fundamental parameters which govern such reactions has restricted progress. Within the last few years, much progress has been made in both theory and experiment to pinpoint the molecular parameters which govern rates of electron transfer reactions, but much remains to be learned.[1]

One noteworthy approach has been to study molecules in which two redox active chromophores are confined at a fixed distance and/or orientation.[2-5] With such systems, the effects on rate of exothermicity[6], reactant separation, and other parameters can be measured without the complications inherent in diffusional systems. Although relatively few such systems have been studied in detail, fundamental

E. Pelizzetti and N. Serpone (eds.), Homogeneous and Heterogeneous Photocatalysis, 189–198.

findings have already emerged, including the demonstration of exothermically restricted rates in the previously elusive Marcus "inverted region"[2].

In addition to answering some questions, these new types of bifunctional molecules have raised other questions. For example, pioneering studies of intramolecular electron transfer in Ru(III) substituted cyt c have been reported,[6] in which electron transfer occurs with ΔE ~ 0.1 V over a 12 Å distance, and ket = $50 s^{-1}$. (Here, 12 Å is the measured dis- tance between nearest heme ligand to the imidazole C_2 bound to Ru).

By comparison, Closs and Miller have studied electron transfer in bifunctional steroids.[2] At ~ 12 Å donor-acceptor separation and ΔE ~ 0.1 V, $k_{et} > 10^6 s^{-1}$! Clearly, the nature of the reagents, and/or of the intervening material can dramatically alter the rates of electron transfer reactions.

As part of a more general program to address such questions we recently prepared[5] a new class of bifunctional porphyrin-Ru(III) ammine complexes of general formula I.

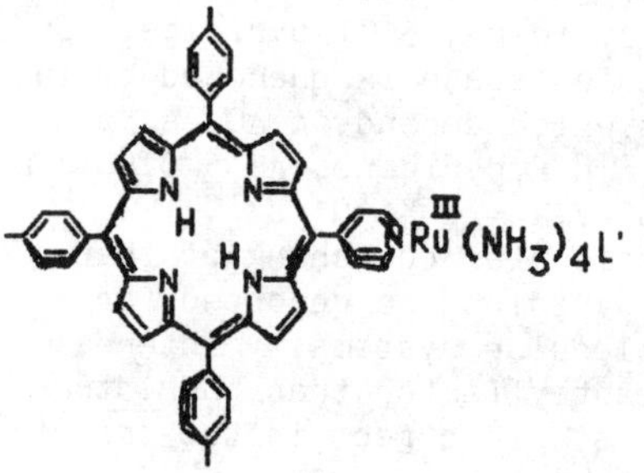

I: L' = NH3, py, 3 Cl pyr

$Ru^{III}(NH_3)_4L'$

In this structure, the pyridine ligand is held essentially perpindicular to the porphyrin π system so that π delocalization is minimized. We here report studies of photoinduced electron transfer in these complexes as a function of ΔG. The rates at fixed reactant orientation are compared with those found in a diffusional encounter, and compared with analogous studies of other bifunctional redox pairs, (eg porphyrin - quinone adducts[8]).

2. MATERIALS AND METHODS

2.1. Materials

The solvents were all reagent grade. In experiments requiring dry solvents, N,N-Dimethylformamide (DMF) was distilled from 4Å molecular sieves (Union Carbide) at reduced pressure. Methylene Chloride were purchased from J. T. Baker as Spectrophotometric Grade Solvent. Solvents were allowed to stand over 4Å molecular sieves.

2.2. Methods

Static emission measurements were made in a Perkin-Elmer MPF-44A spectrofluorimeter. Samples of free ligand porphyrin and modified porphyrin were adjusted for the same concentration, using a Perkin-Elmer Lambda-3 instrument. Excitation wavelengths correspond to the Soret peaks (λ = 480nm), and emission was monitored at the maximum of the mission bend.

Triplet excited state lifetimes were measured using the second harmonic (λ = 532nm) of a Quanta-Ray DCR-2 Nd:YAG laser excitation source. A monitoring tungsten lamp was used. A shutter which was controlled by the computer and synchronized with the laser, minimized the time that the sample was exposed to light. The signal collected from a 1P28 photomultiplier was amplified by a FET probe amplifier and input into a waveform Biomation 6500 transient digitizer. The system was interfaced to a Digital LSI-11/03 computer with controlled the firing of the laser.

The kinetics were monitored by observing the return of the porphyrin signal (Bleaching of the Soret) or by observing the transient triplet-triplet absorption of the 450 nm region, where the triplet-triplet absorption band characteristic of the free base porphyrins predominates[6]. Samples were contained in 1 cm quartz cells and were outgassed by the freeze-pump-thaw method. Measurements were also made in a Ar-purged solution and both methods proved to be efficient in eliminating oxygen from the solution.

The porphyrins under study follow the Beer's law in DMF over a concentration range of 10^{-5} to 10^{-6} M. The fact that the Beer's law holds over such concentration range indicates that oligomer formation and stacks is negligible.

3. RESULTS

The bifunctional porphyrin systems exhibit well defined electrochemistry, as shown in figure 1. The potentials within the adduct are quite similar to those of the isolated chromophores. The porphyrin oxidation potential is slightly raised, no doubt reflecting the coulombic effect of the vicinal Ru(III) center.

Combining this information with the observed onset for the phosphorescence (Eoo = 720 nm = 1.6V) the excited state redox potentials for L = NH_3 can be calculated as 1.6V + 0.45V - 1.6V = 0.37V. These calculations for all the complexes studied in this work are summarized in Table I.

Table I: $E_{1/2}$ Values vs. the NHE in DMF (V).

Compd	$E_{1/2}$ ($Ru^{III/II}$)	$E_{1/2}$ $P^{+/o}$	$E_{1/2}$[b] $^3P^*Ru^{III}/P^+Ru^{II}$
$[P\ Ru(NH_3)_5](PF_6)_3$[a]	0.40	1.47	0.38
$[tr\text{-}P\ Ru(NH_3)_4Py](PF_6)_3$	0.58	1.43	0.55
$[tr\text{-}PRu(NH_3)_4(3\text{-}Cl\text{-}Py)](PF_6)_3$	0.66	1.44	0.64

[a] P = Meso-$H_2TPP(pCH_3)_3$(ePY)
[b] assumes E_{oo}=1.44V

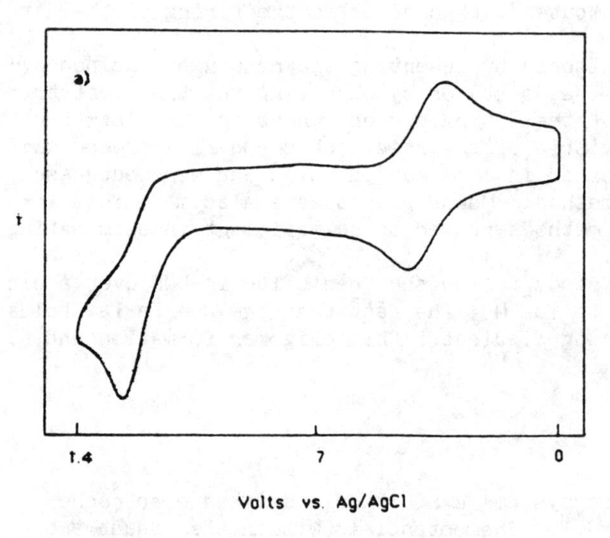

Fig. 1
Cyclic voltammetry is (CH_3CN)0.1M $(Bu)_4NCl$, 25°C, glassy electrode) of complex Ib.

3.1. Excited State Kinetics:

As previously described, excitation of I produces a reactive triplet excited state. The lifetime for triplet decay was measured by monitoring the absorbance at 450 nm (fig 2). When the "trans" ligand on Ru(III) is changed, the rate of excited state decay increases as $^3k_{H2porph} < {^3k_{NH3}} < k_{pyr} < k5_{Clpyr}$. The relationship between triplet decay rate and ΔG is shown graphically in fig. 3. As this figure shows, the data are well described by a Marcus equation: $k = A \exp -(\Delta G-\lambda)^2/4\lambda kT$, assuming a reorganization energy of λ = 0.9V.

Although the observed dependence of rate on ΔG is strongly suggestive that electron transfer is the dominant reaction pathway, energy transfer pathways are not automatically precluded. Dipolar energy transfer can be excluded, since the Ru complex does not absorb appreciably ($E < 20\ m^{-1}\ cm^{-1}$) at the 700 nm wavelength of the porphyrin triplet. A simple heavy atom quenching effect can also be ruled out, since no quenching occurs from the Ru(II) species.

Finally exchange quenching or magnetic dipole quenching is essentially ruled out by the observation that replacement of the Ru N_5 pyr chromophore by a Ru(III) bpy Cl_2 pyr chromophore results in no excited state quenching. This is consistent with electron transfer, since ΔG for electron transfer is endothermic in this case.

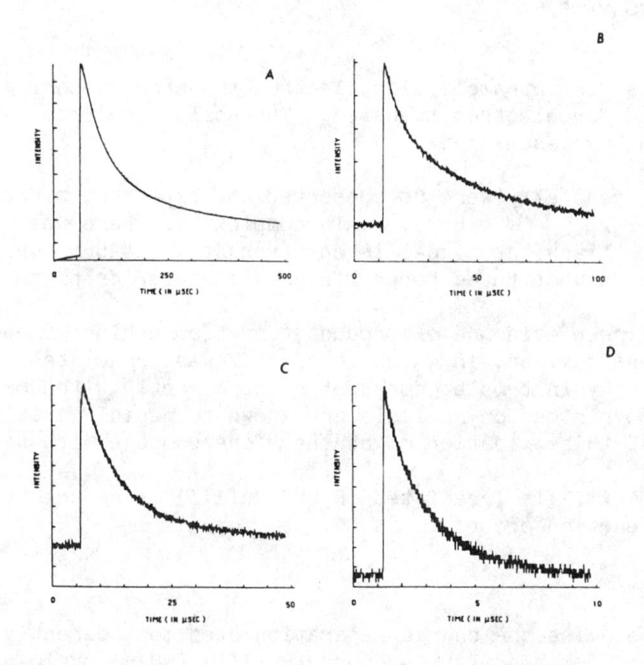

Fig. 2

Triplet decay kinetics, monitored at 460 nm, of H_2porph-pyr-X A. X=H B. X=Ru(III)$(NH_3)_5$ C. X=Ru(III)$(NH_3)_4$(pyr) D. X=Ru(III)$(NH_3)_4$(5Clpyr)

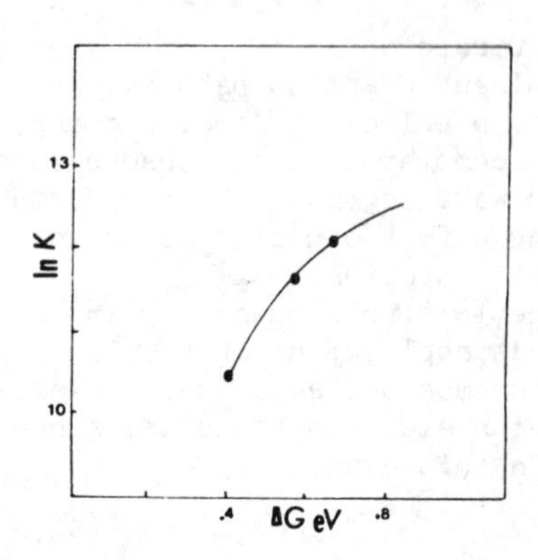

Fig. 3

Plot of rate constants for intramolecular electron transfer quenching (from figure 2) vs. ΔG for electron transfer. The solid line from classical (Marcus) theory assuming λ=0.9V

Reaction products generally were not observed nor expected, since for $\lambda \cong 0.9V$, $k_{back} \geqq k_{forward}$. However, in complex IV, where we expect $kf \approx kb$ a small transient signal is observed at λ = 436nm, which is isosbestic for the unsubstituted porphyrin ground state/ triplet state pair.

To provide more direct evidence of product formation a kinetic competition experiment was devised, in which complex IV was irradiated in DMF. Although the porphyrin triplet does not react directly with DMF ($\tau_{DMF} = \tau_{CH3CN}$), porphyrin cation radicals are known to rapidly oxidize DMF. We reasoned that this oxidation might therefore compete with back recombination.

Indeed, when complex IV is irradiated in DMF Ru(III) is reduced to Ru(II), but no such reaction occurs in CH_3CN.

3.2. Bimolecular Kinetics

Finally, in order to examine the charge separation step most directly, we studied quenching of compound I by <u>unbound</u> (Ru(III) $(NH_3)_5$-pyridine)$^{3+}$.

We expected to observe long lived charge separated products arising by cage escape from the initial ion pair. This was indeed observed. On excitation of I in the prescence of ca 10^{-3}M Ru(III) N_5 pyridine, the soret recovers in two stages. The first decay process is coincident with triplet decay, which includes ca 60% of the total signal with $\tau = 5x10^{-6}$s at [Ru] = $5x10^{-4}$ M. The observed quench rate constant depended linearly on ruthenium concentration, consistent with standard diffusional quenching (Stern Volmer) kinetics.

However, roughly 30% of the signal is observed in a slow, ~ second order decay process with $\tau \sim 10^{-3}$ sec, consistent with cage escape (ϕ ~ 0.3) followed by diffusional recombination. Transient difference spectra of this intermediate at 100 usec show a peak at 450 nm as expected for a Ru (II) N_5 pyridine product.

Two unexpected features of these experiments should be mentioned. First, it is striking that even at low Ru concentrations (Ru ≥ 10^{-4} M the observed intermolecular rate of quenching of I by RuN_5pyr exceeds the corresponding intramolecular quench rate in complex II. We were surprised by this observation, since in complex II, the "quencher" portion of the molecule is obviously already in "collisional" contact with the porphyrin.

A second interesting observation is that at [Ru] > $2x10^{-3}$ M, distinct upward curvature is observed in the Stern Volmer plots consistent with a very efficient static quenching mechanism (figure 4).

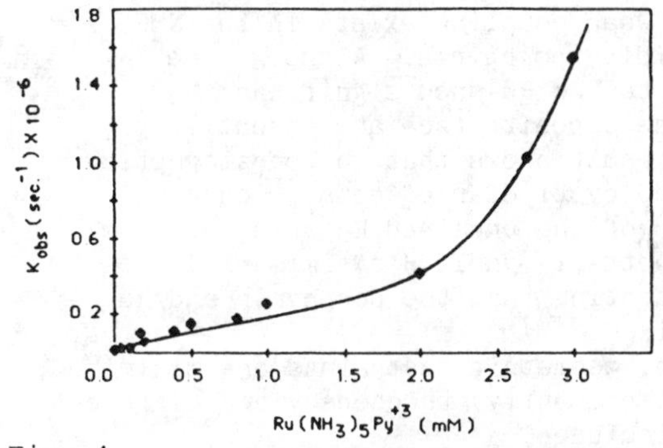

Fig. 4
Quenching of Complex I by free $Ru(NH_3)_5$-pyr
Note the pseudo 1st order bimolecular rate >> intra molecular rate!

4. DISCUSSION

The synthesis of the Ru pendant porphyrins was based on the reasoning that the perpendicular orientation of the porphyrin and pyridine planes would ensure weak electronic coupling, and thus electron transfer would be non adiabatic.

This strategy appears to work surprisingly better than expected. In this regard, it is striking that the bimolecular reaction proceeds much faster than the corresponding unimolecular reaction. The one hundred-fold rate difference observed constitutes a lower limit for this difference, set by the solubility of $(Ru(NH_3)_5 pyr)^{3+}$. At the highest useable Ru concentrations, no rate leveling was observed. Indeed, the rates appeared to increase, suggesting the onset on a static quenching mechanism.

Such results demonstrate that even though the porphyrin and Ru complex are held at vander Waals Contact distance in Compound II the electronic frequency factor for the reaction is not yet maximized, assuming the Franck Condon factors for the intramolecular and intermolecular reactions will be quite similar.

The low intramolecular rate presumably reflects an unfavorable geometry for reaction. However, crystallographic data for meso tetra phenyl porphyrins show that the dihedral angle between the phenyl and porphyrin rings is only ca 65°.[7] Assuming approximately a cosine angular dependence for overlap, substantial wavefunction fixing (and a corresponding rapid rate) is expected, contrary to observation.

Two factors may mitigate against efficient mixing. First, delocalization of the Ru wavefunction into the pyridine may be relatively small. Second, MO calculations show the porphyrin system has a node at the meso carbon thus minimizing efficient wavefunction overlap with the meso substituents.[8]

We suggest that these effects combine to produce unusual nonadiabaticity in a bound difunctional system.

Independent evidence for such weak coupling exists in the NMR spectra of the Ru substituted pyridine porphyrin. Although the porphyrin pyrrole protons closest to the Ru show significant paramagnetic broadening thus on the opposite face are essentially unaffected by the Ru atom. This result shows that the paramagnetic shifts and broadening are dominated by dipolar effects. Contact interactions due to delocalisation of the unpaired Ru spin into the porphyrin system (which would affect all pynole H's similarly) are experimentally negligible, demonstrating that the porphyrin and Ru wavefunctions are indeed weakly mixed.

In the intermolecular reaction, geometric restraints are minimized, so that the diffusional reaction (apparently) proceeds via a different geometry. Several possibilities include:

1. Attack via the ammine face, resulting in the minimum porphyrin-Ru distance.
2. Attack of porphyrin from the top face, resulting in a stacking geometry between the porphyrin and pyridine. We favor the latter explanation for several reasons. First, calculations by Marcus and Siders suggest porphyrin electron transfer is optimized in a stacking geometry (consistent with maximum π-π overlap).

Second, such a path could lead, at high concentration, to ground state stacking (and corresponding static quenching). Such stacking is well established in similar systems eg: tetraphenyl porphyrin - methyl viologen.

In any case, the observation that electron transfer rates can be non adiabatic even at vander Waals contact is surprising, and supports recent suggestions that some diffusional electron transfer reactions may also be nonadiabatic.[1]

Finally, it is interesting to compare the maximum rate observed in a Ru-porphyrin system ($\sim 10^7 s^{-1}$) with that found in a structurally homologous porphyrin quinone system (IV) ($k \sim 10^{11} S^{-1}$). In both cases the acceptor is held at a similar orientation and a nominally similar distance, yet the Ru system reacts at least 10^4 times slower than the quinone system. This reactivity difference cannot arise from Franck Condon effects, since exothermicity is optimised in the Ru system, but not in the quinone system. At optimal ΔG, the data of Waswiewski et al. suggest a rate of ca $10^{12} S^{-1}$ would be appropriate.[3]

Although a number of explanations for this reactivity difference are feasible, we wish to suggest a particularly simple yet powerful explanation.

We wish to suggest that some of the large rate difference between the porph-quinone systems and the porph-pyridyl Ru (III) systems reflects the fact that the latter reaction occurs at the Ru site, which, as noted, already is poorly coupled to the porphyrin ring. In the quin one system, the "effective donor-acceptor distance" is much smaller, and direct electronic coupling is apparently far more stronger, reading to increased rates. This explanation, if correct, has interesting consequen- ces for understanding the relatively slow reactions observed for Ru substituted proteins. It may be that the significant distance in such systems is not the porphyrin-imidazole distance (~11Å) generally quo- ted, but rather the longer porphyrin - Ru distance (>15Å). In this light, much of the apparent difference between the protein reactions and similar model reactions can be accounted for.

4.1. SUMMARY

The surprisingly low rate of electron transfer found in the Porphyrin-Ru(III) system is consistent with calculations of Marcus and Siders.[12]

Such calculations predict that the relative orientation between two π-systems plays an important role in optimizing the rate of electron transfer. The meso pyridine Ru center in the bifunctional system seems weakly coupled to the porphyrin πststen. In the absence of strong aromatic coupling, it appears that transfer occurs directly to the ruthenium atom, probably via a through space pathway. By contrast, in analogous porphyrin quinone systems the through space path is much shorter, consistent with the faster rates observed in such systems.

We concluded that nonadiabaticity can occur, even at the Van der Waal distances between reactants, if the interaction geometry is not optimized.

The observed rates in bimolecular reaction, between unsubstituted porphyrin and the $Ru(III)(NH_3)_5Py$ are 100 times faster than found in unimolecular Porphyrin-Ru(III) systems. This fact might be understood in similar geometric terms. In the bimolecular reaction, π-π overlap of the porphyrin and pyridine may occur, leading to an increase in effective rate.

5. REFERENCES

1) Recent reviews include:
a. M. D. Newton, N. Sutin Ann Rev. Phys. Chem. **35**, 437-480 (1984).
b. T. Guarr, G. McLendon Coord. Chem. Rev. (1985) in press.
2) a) J. R. Miller, L. T. Calcaterra and G. L. Closs, J. Am. Chem. Soc., **106**, 3047-3049 (1984).
b) L. R. Calcaterra, G. L. Closs and J. R. Miller, J. Am. Chem. Soc., **105**, 670-671 (1983).
3) A. D. Joran, B. A. Leland, G. G. Geller, J. J. Hopfield, and P. B. Dervan, J. Am. Chem. Soc., **106**, 6090-6092 (1984).
4) M. R. Wasielewski and M. P. Niemczyk, J. Am. Chem. Soc., **106**, 5043-5045 (1984).

5) C. Franco, G. McLendon Inor. Chem. (1984)
6) J. R. Winkler, D. G. Nocera, K. M. Yocom, E. Bordignon, and H. B. Gray, J. Am. Chem. Soc., **104**, 5798-5800 (1982).
7) J. Hoard in: The Porphyrins (D. Dolphin, editor) Academic N.Y. (1972).
8) M. Gouterman in: The Porphyrins (D. Dolphin, editor) Academic N.Y. (1972)

LINEAR CHAIN PLATINUM COMPLEXES AS PHOTOCATALYSTS

Virginia H. Houlding[‡] and Arthur J. Frank
Solar Energy Research Institute
1617 Cole Boulevard
Golden, CO 80401

ABSTRACT. Linear chain platinum(II) double salts such as Magnus' Green Salt $[Pt(NH_3)_4][PtCl_4]$ and mixed valence complexes with bridging halide ions such as Wolffram's Red Salt $[Pt(etN)_4Cl_2][Pt(etN)_4]Cl_4 \cdot 4H_2O$ were examined as a potentially important new class of low bandgap materials for solar fuel synthesis. Room temperature emission and lifetime measurements on two mixed valence red salts provide evidence for mobile excitonic states at or near the bandgap but rapid photocorrosion in aqueous solution may make them unsuitable as water splitting agents. In the case of the Pt(II) double salts, $[Pt(bipy)_2][Pt(CN)_4]$ and [Pt(bipy)(4-methyl-4'-heptyl-2,2'-bipyridine)]$[Pt(CN)_4]$ (E_{BG} = 2 eV) proved to be highly active photosensitizers for H_2 production in the presence of EDTA and Pt sol. Without Pt sol, very little H_2 is produced, but no evidence for photocorrosion is seen in either system. These results are in marked contrast to Magnus' Green Salt and several other Pt(II) double salts which are ineffective as photosensitizers. From the investigation of these materials, guidelines have been established that should make it possible to design many Pt(II) double salts which are photoactive.

1. INTRODUCTION

Linear chain transition metal complexes may be of special interest to participants of this NATO Workshop as possible photocatalysts for a wide range of applications. Pseudo-one-dimensional materials have been studied extensively for many years[1] stimulated in large measure by the possibility of finding molecular metals and ultimately high temperature superconductors. Our own interest in these materials was that some of them might possess a conduction band or cooperative excited state structure that would facilitate multielectron redox processes such as the light-induced reduction of water at the solid-liquid interface.

Our introduction into this fascinating area of linear chain metal compounds began with mixed valence Pt complexes and Pt(II) double salts in the form of powders. In these one-dimensional materials, the

E. Pelizzetti and N. Serpone (eds.), Homogeneous and Heterogeneous Photocatalysis, 199–211.

respective metal-metal interaction occurs along the chain axis through anion bridges and direct metal ion contact. These materials are intensely colored due to delocalized excited states polarized along the chain. Because the constituent monomers can be modified synthetically, photophysical and photochemical properties can be altered without drastically affecting the overall chain structure.

In this paper, we describe highlights of our results on the photophysical characterization of linear chain mixed valence Pt complexes[2] and Pt(II) Magnus-type double salts.[3] In the latter case, we have found some promising compounds that are active photosensitizers for H_2 production in aqueous suspensions.

2. LINEAR CHAIN MIXED VALENCE PT SALTS

2.1 Background

Linear chain mixed valence Pt salts consist of alternating octahedral d^6 Pt(IV) and square planar d^8 Pt(II) haloamine complexes with axial halide ions from Pt(IV) complexes serving as bridges. In the case of the chloride salts, the corresponding Pt(II) and Pt(IV) monomers are yellow, whereas the linear chain polymers are red due to the electronic interactions of adjacent metal ions via the chloride bridge which give rise to intense $Pt(d^8) \rightarrow Pt(d^6)$ intervalence charge transfer[4] absorption bands in the visible region. Substitution of Cl^- by Br^- or I^- causes a red shift of this transition and gives the complexes a green color. Charge motion along $Pt^{II} \cdots X\text{-}Pt^{IV}$ chain is believed to occur by electron "hopping" assisted by vibrational movement of the halide ion between metal centers as the oxidation state of Pt changes.[5] Little is known about the extent of delocalization of the intervalence charge-transfer excitation and the intrinsic mobility of charge carriers, in part due to the extreme difficulty of obtaining large defect-free single crystals needed for electrical characterization. The problem of defect sites on carrier mobility are expected to become less important in microdispersed systems.[6]

In this section, we explore the photophysical properties of two mixed valence Pt "red salts" powders, $[Pt(etN)_4Cl_2][Pt(etN)_4]Cl_4 \cdot 4H_2O$ (Wolffram's Red; etN = ethylamine) and $[Pt(en)Cl_4][Pt(en)Cl_2]$ ($Pt(en)Cl_3$; en = 1,2-diaminoethane), the structures of which are illustrated in Figure 1.

Important information about the electronic structure of the intervalence charge-transfer region is derived from determinations of diffuse reflectance, emission and excitation spectra, and emission lifetimes. Experimental details on the synthesis of the complexes and spectroscopic measurements are given elsewhere.[2] The excitation spectra were corrected automatically for lamp intensity by using Rhodamine B as a standard quantum counter. The emission spectra were corrected for the wavelength-dependent efficiency of the detection system by using a calibrated light source.[7]

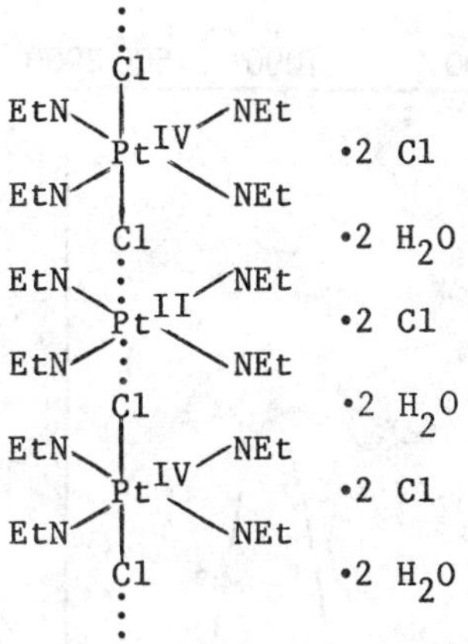

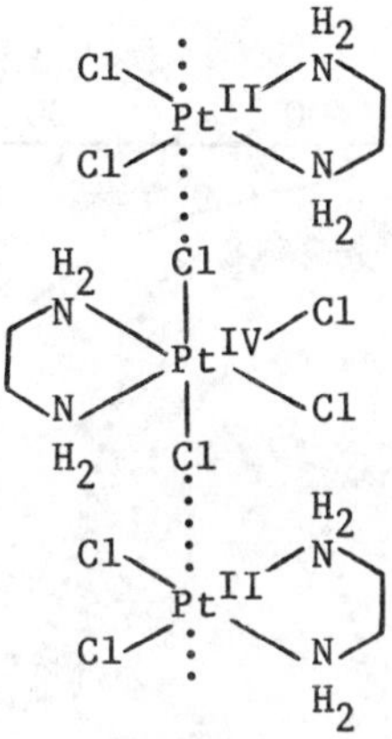

Wolffram's Red (EtN = ethylamine) $Pt(en)Cl_3$

Figure 1. Structures of mixed valence Pt salts

2.2. Results and Discussion

Both salts exhibit strong emission at room temperature,[8] but corrected excitation specta do not follow profiles of diffuse reflectance spectra. In Wolffram's Red, the mismatch is particularly severe (Figure 2). The lack of correspondence is not due to dual emission (e.g., overlapping fluorescence and phosphorescence), since excitation spectra for both salts are found to be independent of emission wavelength. The possibility of the mismatch being an artifact caused by distortion of the diffuse reflectance spectrum can also be dismissed.[2]

The room temperature emission spectra[2] of the two red salts are similar which suggest a common origin for their emissions. Further insight into the nature of the emission may be inferred from the radiative rate constants, k, calculated from quantum yields, ϕ, and lifetime mesurements, τ. The respective ϕ and τ were determined as 0.02 ±0.01 and ≤15 ns for Wolffram's Red and 0.007 ±0.001 and 107 ±2 ns for $Pt(en)Cl_3$. With these numbers, the radiative rate constants of Wolffram's Red and $Pt(en)Cl_3$ are calculated as $\geq 10^6$ s^{-1} and 6.5×10^4 s^{-1}, respectively. Both k values are indicative of a transition that is at least partially allowed. Since the lifetime of $Pt(en)Cl_3$ at room temperature (107 ±2 ns) is more consistent with a spin-forbidden than spin-allowed transition, the two red salt emissions are assigned as spin-forbidden and parity allowed. There are several alternative descriptions of the nature of the emissive states in these crystalline materials. These include emission from isolated monomers, excitons (either localized or delocalized), and defect and/or impurity traps, each of which will be considered in this

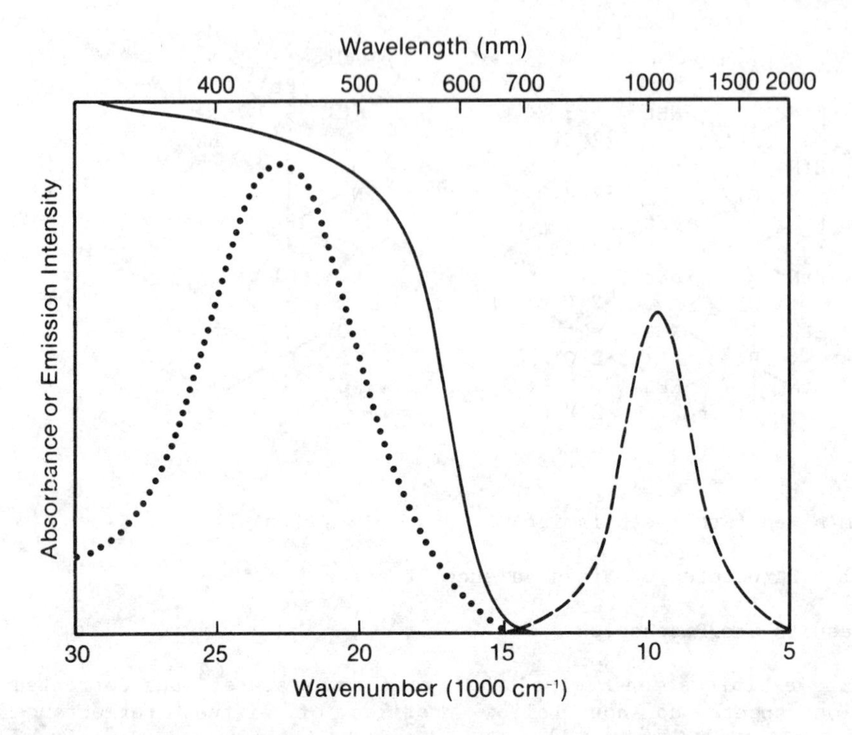

Figure 2. Diffuse reflectance (———), corrected excitation (•••), and corrected emission (---) spectra for $[Pt(etN)_4Cl_2][Pt(etN)_4]Cl_4 \cdot 4\ H_2O$ (Wolffram's Red Salt; etN = ethylamine) at room temperature (Ref. 2).

discussion. Any description of the emissive states of the red salts must be consistent with the observed emission intensity, enormous Stokes shift between absorption and emission, and the mismatch between diffuse reflectance and excitation spectra.

The emissions of the red salts are not attributed to the constituent monomers because the emission and excitation maxima of Wolffram's Red and $Pt(en)Cl_3$ do not correspond to energies expected for transitions in Pt(II) and Pt(IV) monomers.[9-11] Also, emissions in Pt(II) and Pt(IV) haloamine complexes are typically from triplet ligand field states which are both spin- and parity-forbidden. The magnitudes of the radiative rate constants for the two red salts are too large to be consistent with emission from a triplet ligand field state. The emission from $Pt(en)Cl_3$ at 77K[2] is, however, entirely consistent with monomer emission.[11]

Emission from a highly delocalized exciton state is ruled out because of the large Stokes shift. The large distortion of excited state molecular or lattice geometry required to produce such a Stokes shift would erect a significant barrier to exciton migration, thus localizing the exciton. Several points argue against assignment of the emissive state as a localized exciton, however. First, emission from highly distorted excited states would be very weak due to strong radiationless coupling with the ground state, whereas the observed emission intensity is quite strong for the solid state. Second, the low energy side of the absorption band which has been attributed to intervalence charge transfer of the type Pt(II)d_z2 → Pt(IV)d_z2 [12] does not correspond to the excitation spectra, particularly in Wolfram's Red. Thus the emissive state is not the localized intervalence charge transfer state responsible for the intense absorption. Also, the corresponding triplet state is not the emissive state because no transfer of excitation energy occurs between the intervalence charge transfer state and the emissive state.

A more reasonable possibility is that the room temperature emissions arise from highly emissive trap states in the red salt lattices which are populated by mobile excitons produced by light absorption. This description accounts for the lack of correspondence between excitation and diffuse reflectance spectra in Wolffram's Red; the excitonic states produced by absorption in the region around 18,000 cm^{-1} (560 nm) may be far less mobile than the excitonic states created by absorption in the region of ca. 23,000 cm^{-1} (440 nm), and would thus not contribute to trap emissions monitored in the excitation spectrum.

The composition of these salts suggests that dimers formed by interactions of adjacent d^8 square planar Pt(II) units may constitute the emissive traps responsible for their room temperature luminescence behavior. In view of the method of preparation of these salts by partial oxidation of the Pt(II) monomer constituents, it is quite likely that these solid lattices contain defects resulting from adjacent positioning of Pt(II) monomer units. These defect sites would be expected to behave very much like dimeric units due to interactions of the Pt(II) centers.[13] Even preparation of the red salts by mixing equimolar Pt(IV) and Pt(II) constituents is likely to result in dimeric Pt(II) type lattice defects.

Dimeric square planar d^8 Pt(II) complexes typically have a strong d→p absorption in the visible and near-uv region, which overlap with an intense emission band; the d-p transition energies are very dependent on the extent of metal-metal interaction.[14] By analogy, dimeric defect sites in the red salts would be expected to have absorption bands at ca. 13,000 cm^{-1} (770 nm). Weak absorption in this region was therefore carefully sought, and was indeed observed at about 800 nm in diffuse reflectance spectra of Pt(en)Cl_3 preparations in which elemental analysis showed low [Cl]; the intensity of this band diminished as the Cl content was increased. Other evidence also support the occurrence of Pt(II) dimeric defect in the red salt lattices.[2]

The intense structureless absorption bands in the visible, the presence of highly mobile excitonic states, and the cooperative interactions between metal centers makes these materials interesting as light absorbers or possible catalysts in microheterogeneous systems. Unfortunately, Wolffram's Red Salt dissolves in water. Pt(en)Cl_3 is insoluble in water but photodegrades rapidly when irradiated in aqueous suspensions. This instability is absent in nonaqueous noncoordinating solvents such as acetone, which indicates that formation of Pt aquo species is necessary for photodegradation.

2.3. Concluding Remarks

In this investigation, we have shown that emission from Wolffram's Red and Pt(en)Cl_3 originates from emissive trap states which are populated by mobile excitons created by light absorption in the intervalence charge-transfer region. These traps or defect states in the solid lattices are due to the existence of Pt(II) dimers formed by interaction of adjacent d^8 square planar complexes. The presence of the Pt(II) dimers disrupts the migration of excitons along the chain (•••Cl-Pt^{IV}-Cl•••Pt^{II}-Pt^{II}•••Cl-Pt^{IV}-Cl) from the bulk to the surface of the mixed valence Pt particles and account for the disappointingly low photoconductivity of analogous material.[4b] Using current methods of synthesis, it is difficult to avoid incorporation of such defects in the crystal structure. Thus, in order to take advantage of the desirable optical properties of these materials and to harvest the highly mobile excitons in surface redox reactions, new synthesis routes must be developed to reduce the density of defect states in the crystal lattice or else another approach must be sought. One possible approach to circumvent the crystal defect problem is to reduce the size of the linear chain particles to dimensions where the transit time of excitons to travel from the bulk to the surface is much shorter than the time required to trap the excitons. A kinetic analysis[6] of such a mechanism has been described for semiconductors and may be applicable for these materials as well.

A second problem that must be addressed, if the mixed valence salts are to be useful for photosensitization of catalysis in aqueous solutions, concerns their solubility and photoinstability. In aqueous media, Wolffam's Red dissolves and Pt(en)Cl_3 exhibits instability to irradiation. An approach that has been somewhat effective for controlling the latter type of instability has involved the use of hydrophobic polymers to reduce the water activity at the particle interface.[16] In this respect, micelles, vesicles, or microemulsions may also be useful solubilizing media for the material. It may also be possible to suppress photodegradation by surface modifying the material with appropriate redox catalysts as in the case of CdS/RuO_2/Pt.[15] Alternatively, the red salts may be more suitable as sensitizers for nonaqueous processes such as CO_2 reduction and other C_1 chemistry, rather than as agents for photochemical water splitting. Additional research in this area is clearly warranted.

3. Pt(II) LINEAR CHAIN DOUBLE SALTS

3.1. Background

Pt(II) linear chain double salts are composed of square planar d^8 Pt complexes arranged in columnar structures. These double salts exhibit quite different optical properties, photochemical activity, stability, and solubility in aqueous media than Wolffram's Red and $Pt(en)Cl_3$. All the Pt atoms in the linear chain are in the same oxidation state and interact directly with near-neighbor Pt ions rather than through anion bridges. As a consequence of the metal-metal interactions in the solid state, the electronic transitions of the constituent monomers can be greatly perturbed. For example, in the case of Pt(II) haloamines, Pt-Pt interactions can have a pronounced effect upon the d→d transition energies in the linear chain materials.[17] Considerably greater effects occur, however, in the charge-transfer and d-p transitions of many linear chain materials,[18] with red shifts of well over 10,000 cm^{-1} being commonplace for transitions polarized along the chain axis. For many d^8 complexes,[19] increasing red shifts of allowed d^8 transitions occur upon oligomerization to dimers, trimers, tetramers, and extended chain solids, providing direct evidence for delocalization of these excited states along the metal-metal chain.

In this section, we discuss the photophysics and photochemistry of the following Pt(II) double salts and, in some cases, their hydrates: $[Pt(NH_3)_4][PtCl_4]$ (MGS), $[Pt(bipy)_2][PtCl_4]$ (PBCl; bipy = 2,2'-bipyridine), $[Pt(bipy)(MHB)][PtCl_4]$ (PHBCl; MHB = 4-methyl-4'-heptyl-2,2'-bipyridine), $[Pt(bipy)_2][Pt(CN)_4]$ (PBC), and $[Pt(bipy)(MHB)][Pt(CN)_4]$ (PHBC).[3] We have found that the ability of these Pt(II) double salts to photosensitize the reduction of water to H_2 depends on the nature of excited state involved in the redox process. From our study of these materials as particulate photosensitizers for water reduction, we were able to establish guidelines that should make it possible to design many photoactive Magnus-type $[M(II)^{2+}M(II)^{2-}]$ double salts. Experimental details of this investigation are given elsewhere.[3] As in the preceding section, emission and excitation spectra have been corrected. For convenience, the acronyms assigned to the complexes are found in Table I.

3.2. Results and Discussion

3.2.1. MGS. Magnus' Green Salt $[Pt(NH_3)_4][PtCl_4]$ is representative of M(II) linear chain compounds. The structure of MGS, which was determined in 1957, is depicted in Figure 3. It consists of alternating square planar dications $[Pt(NH_3)_4]^{2+}$ and square planar dianions $[PtCl_4]^{2-}$ stacked along the Pt-Pt axis. The Pt-Pt spacings are 3.24 Å and the resulting metal-metal interaction causes a red shift of ca. 4000 cm^{-1} of a d→d transition in $PtCl_4^{2-}$ [20] to 16,700 cm^{-1} (600 nm) in the spectrum of MGS. An even greater red shift is observed for the chain-axis delocalized Pt (d_z2 → p_z) transition.[18] The Pt(d→p) transitions in $PtCl_4^{2-}$ [21] and $Pt(NH_3)_4^{2+}$ [22] are red shifted 11,800 cm^{-1} and 16,500 cm^{-1}, respectively, to 34,500 cm^{-1} (280 nm)[23] in the spectrum of MGS. Metal-metal interaction or spacing has a strong effect on the d-p transition. In K_2PtCl_4 and other Pt chain crystals, the Pt-Pt interaction (4-4.1 Å Pt-Pt spacing) is too weak to justify postulating true chemical bonding[18] and the energy of the d-p transition is essentially that of the monomer. In $[Pt(H_2P_2O_5)_4]^{4-}$ dimers, a Pt-Pt spacing of 2.925 Å [24] results in a

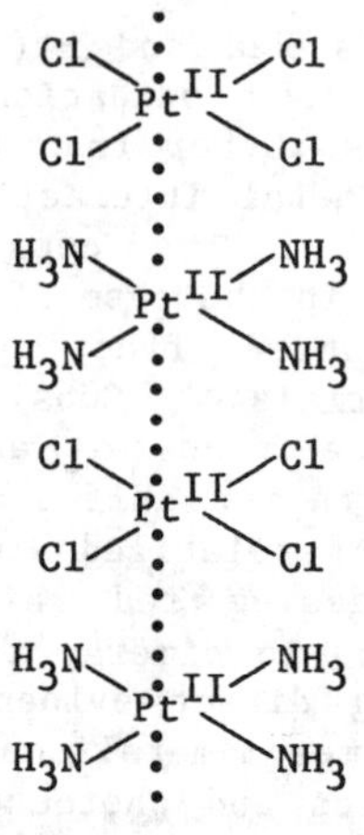

```
Cl\  :  /Cl
    Pt II
Cl/  :  \Cl
     :
H3N\ :  /NH3
    Pt II
H3N/ :  \NH3
     :
Cl\  :  /Cl
    Pt II
Cl/  :  \Cl
     :
H3N\ :  /NH3
    Pt II
H3N/ :  \NH3
     :
```

Figure 3. Structure of Magnus' Green Salt

(d→p) transition energy of 27,200 cm^{-1} (368 nm).[25] The extent of metal-metal interaction in MGS thus falls in between these two cases.

In our studies, we found MGS to be completely nonemissive in the solid state both at room temperature and 77K, for excitation wavelengths between 248 nm and 600 nm. Since K_2PtCl_4 exhibits emission[26] but $Pt(NH_3)_4^{2+}$ does not, vibrational deactivation by the N-H stretching mode probably quenches the emission of $PtCl_4^{2-}$ in MGS. The lowest excited states in MGS are perturbed ligand field states originating from $PtCl_4^{2-}$. Therefore, the absence of emission upon excitation into the (d→p) absorption band shows that coupling between the (d-p) state and the lower energy ligand field states is sufficiently strong to preclude upper excited state emission.

3.2.2. PBCl and PHBCl. In these yellow double salts of Pt bipyridines with $PtCl_4^{2-}$, no appreciable red shift of $PtCl_4^{2-}$ d→d transitions can be seen. Upon excitation in the range of 250-550 nm, PHBCl at room temperature and PBCl at 77K exhibit extremely weak emission at ca. 12,500 cm^{-1} (750 nm) corresponding to the (d→d) emission[27] maximum of K_2PtCl_4 at room temperature. These observations indicate that little or no metal-metal interaction is present in these linear chain materials. The absorption and emission characteristics of PBCl and PHBCl are given in Table 1. We assign the emitting states as ligand field based on the long emission lifetime, large Stokes shift, and very low intensity.

3.2.3. PBC and PHBC. The deep orange double salts of $Pt(bipy)_2^{2+}$ and $Pt(bipy)(MHB)^{2+}$ with $Pt(CN)_4^{2-}$ exhibit an intense asymmetric absorption band in the visible region which overlaps well with a strong emission band (Figure 4). The small Stokes shift indicates little distortion of the excited state structure. Analysis of the emission parameters rule out the possibility that the emission derives from ligand field state. The radiative rate constant of PBC

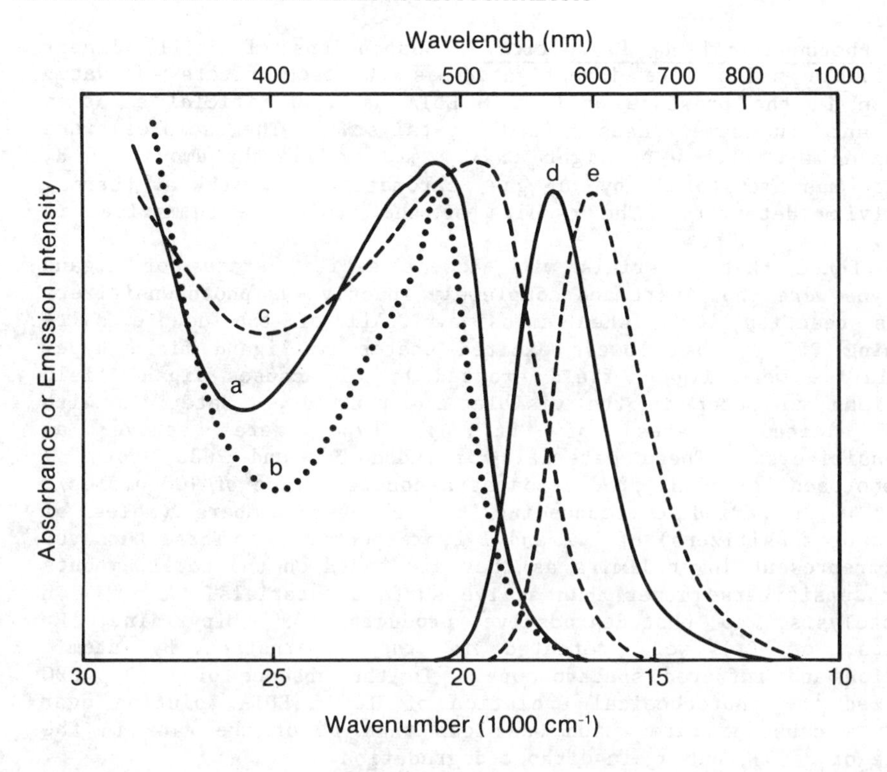

Figure 4. Absorption and emission spectra of PBC (Ref. 3).
a) Diffuse reflectance of PBC•2 H_2O powder
b) Absorption of optically transparent colloidal PBC in water separated from the powder by centrifugation
c) Diffuse reflectance of anhydrous PBC powder
d) Corrected emission of PBC•2 H_2O powder
e) Corrected emission of anhydrous PBC powder

is $k_r = \phi/\tau \simeq 1.2 \times 10^5\ s^{-1}$ based on a room temperature emission quantum yield of 0.002 ± 0.001 and lifetime of 16.8 ± 2.5 ns. A very small emission component with a lifetime of ≤1 ns was also observed. We assign, respectively, the short- and long-lived emission to the spin-allowed and spin-forbidden components of an allowed transition which is most likely the $Pt(p_z \rightarrow d_{z^2})$.[18,25,28] The large magnitude of k_r rules out the possibility that the spin-forbidden emission is ligand field in origin. Upon drying the PBC powder, the absorption and emission peaks corresponding to the d-p transition shift to the red, indicating a decline of the Pt-Pt distance on the removal of water. Such red shifts upon dehydration of the lattice is a typical observation of hydrated linear chain structures.

3.2.4. Photochemical H_2 Production. Suspensions of Pt(II) linear chain double salts were investigated as photosensitizers of water reduction in the presence of 0.05 M EDTA as a sacrificial electron donor, and in some cases, Pt(0) catalyst. The samples were illuminated with 1.1 W of light at $\lambda \geqslant 420$ nm and the amount of H_2 produced was monitored by a gas chromatograph with a thermal conductivity detector. The results of the study are summarized in Table 1.

We found that materials with lowest excited states of ligand field type were photoinert and completely inactive as photosensitizers in this reaction, even when emissive. All of the double salts containing $PtCl_4^{2-}$ had lowest excited states of ligand field type. The relative weak ligand field exerted by Cl^- causes ligand field transitions to occur in the visible and near uv. Materials with lowest excited states of Pt(d-p) type were active as photosensitizers. These materials included PBC and PHBC, both of which photosensitized H_2 production at a constant rate of 100 nmoles/h over a 48 h period corresponding to turnover numbers (moles of H_2/moles of sensitizers) of 1.2 and 1.1, respectively. These turnover numbers represent lower limits as they are based on the total amounts of bulk sensitizers rather than active surface materials. After 48 h of photolysis, no photodegradative products (CN, bipyridine, Pt complexes, or Pt) were detected in the supernatant by atomic absorption and infrared spectroscopy. In the absence of Pt(0), PBC sensitized the photochemical evolution of H_2 in EDTA solution over 48 h at a constant rate which was less than 1% of the rate in the presence of Pt(0), but again without degradation.

3.3. Concluding Remarks

The activity of the Pt(II) Magnus-type double salts to photosensitize the reduction of water depends on the electronic structure of the excited state involved in the redox reaction. The chain-axis delocalized ^{3}Pt(d-p) states of PBC and PHBC were found to be photoactive whereas the ligand-field states of other materials were photoinactive. We showed that strong metal-metal interaction (as in the case of MGS) and relatively long lifetimes (as in the case of PHBCl) were not sufficient conditions for materials to exhibit photosensitization. These results support the conclusion of others[18] that ligand field states are not appreciably delocalized by metal-metal interactions.

The photosensitizing ability of PBC and PHBC upon visible irradiation can be attributed, in part, to two electronic structural effects. First, the materials exhibit sufficiently strong metal-metal interaction that the d-p transitions occur in the visible region. In addition to optically excited surface active chromophores, delocalized (d-p) states may provide a mechanism for transmitting bulk excitation to the surface (as either excitons or even charge carriers) to drive redox reaction at the particle-solution interface. Secondly, strong ligand fields in the constituent monomers serve to raise the ligand field states above the d-p states. Our emission results show that

Table 1. Room temperature photophysical and photochemical data of Pt(II) double salts

Double Salt	Acronym	Hydration Number	Diffuse Reflectance λ_{max}, nm	Emission λ_{max}, nm ($\lambda_{excitation}$, nm)[a]	Emission Lifetime (μs)[a]	Excited state[a] Assignment	nmoles H_2/hr
$[Pt(NH_3)_4][PtCl_4]$	MGS	0	600	none		ligand field	0
$[Pt(bipy)_2][PtCl_4]$	PBCl	2.5	*350, 400sh	750 (400)[b]		ligand field	0
	PBCl[c]	0[d]	red	900 (356)[e]	13[e]	ligand field	--
$[Pt(bipy)(MHB)]PtCl_4]$	PHBCl	0	<350, 400 sh	745 (400)[b]		ligand field	0
$[Pt(bipy)_2][Pt(CN)_4]$	PBC	2	485	570 (485)[b]	0.0168	Pt(d-p)	100 (Pt(0) catalyst)
"	PBC	0[d]	515	610 (515)[b]	0.0185	Pt(d-p)	1 (no catalyst)
$[Pt(bipy)(MHB)][Pt(CN)_4]$	PHBC	3	480	575 (480)[b]		Pt(d-p)	100 (Pt(0) catalyst)
"	PHBC	0[d]	495	620 (515)		Pt(d-p)	--

a) ligand field-$PtCl_4^{2-}$, Pt(d-p) = Pt($5d_{z^2}$-$6p_z$).
b) Emission maximum checked to be independent of excitation wavelength.
c) Produced from violet precipitate.
d) Produced by vacuum dessication of hydrate.
e) Data provided by C. Craig and R. J. Watts.

deactivation of directly excited d-p states by ligand field states of lower energy competes efficiently with redox photosensitization from upper excited states of (d-p) type. Thus the efficiency of photosensitization is greatly improved when the lowest excited state is of the (d-p) type. Using these two criteria, it should be possible to design many Pt(II) Magnus-type materials which will be photoactive.

Acknowledgements

We thank Mr. Carl Craig and Professor Richard J. Watts at the University of California, Santa Barbara, for lifetime and emission data, Dr. John S. Connolly at SERI for access to equipment in his laboratory and assistance with nanosecond lifetime measurements, Mr. Alex Miedaner at SERI for the synthesis of $Pt(bipy)_2^{2+}$ and $Pt(bipy)(MHB)^{2+}$, and Dr. Vincent Miskowski at the Jet Propulsion Laboratory for many helpful discussions. This work was supported by Contract #5083-260-0796 from the Gas Research Institute.

REFERENCES

‡Present address: Corporate Technology Division, Allied Corporation, Morristown, NJ 07960.

1. a. Interrante, L. V., ed.; Extended Interactions between Metal Ions; American Chemical Society:Washington, D.C., 1974.
   b. Keller, H. J., ed.; Low Dimensional Cooperative Phenomena, Plenum Press, New York, 1975.
   c. Keller, H. J., ed.; Chemistry and Physics of One-Dimensional Metals, Plenum Press, New York, 1975.
   d. Hatfield, W. E., ed.; Molecular Metals, Plenum Press, New York, 1979.
   e. Miller, J. S., ed.; Extended Linear Chain Compounds, Plenum Press, New York, 1982.
2. Houlding, H. V.; Craig, C.; Watts, R. J.; Frank, A. J. Inorg. Chem., submitted.
3. Houlding, V. H.; Frank, A. J. Inorg. Chem., October issue, 1985.
4. a. Interrante, L. V.; Browall, K. W.; Bundy, F. P. Inorg. Chem. 1974, **13**, 1158.
   b. Interrante, L. V.; Browall, K. W. Inorg. Chem. 1974, **13**, 1162.
   c. Interrante, L. V.; Bundy, F. P. J. Inorg. Nucl. Chem. 1977, **39**, 1333.
5. Interrante, L. V., reference 16, p. 299.
6. Gratzel, M.; Frank, A. J. J. Phys. Chem. 1982, **86**, 1964.
7. Parker, C. A.; Rees, W. T. Analyst (London) 1960, **85**, 587.
8. Tanino, H.; Kobayashi, K. J. Phys. Soc. Jpn. 1983, **52** 1446.
9. Campbell, J. R.; Clark, R. J. H.; Turtle, P. C. Inorg. Chem. 1978, **17**, 3622.

10. Clark, R. J. H.; Franks, M. L.; Trumble, W. R. Chem. Phys. Lett. 1976, **41**, 287.
11. Fleischauer, P. D.; Fleischauer, P. Chem. Rev. 1970, **70**, 199.
12. Robin, M. B.; Day, P. Adv. Inorg. Chem. Radiochem. 1967, **10**, 247.
13. Isci, H.; Mason, W. R. Inorg. Chem. 1974, **13**, 1175.
14. a. Miskowski, V. M.; Nobinger, G. L.; Kliger, D. S.; Hammond, G. S.; Lewis, N. S.; Mann, K. R.; Gray, H. B. J. Am. Chem. Soc. 1978, **100**, 485.
    b. Fordyce, W. A.; Brummer, J. G.; Crosby, G. A. J. Am. Chem. Soc. 1981, **103**, 7061.
15. Kalyanasundaram, K.; Borgarello, E.; Grätzel, M. Helv. Chim. Acta 1981, **64**, 362.
16. Honda, K.; Frank, A. J. J. Phys. Chem. 1874, **88**, 5577, and references therein.
17. Martin, D. S., Jr., reference 1a, Chapter 18.
18. Anex, B., reference 1a, Chapter 19.
19. a. Balch, A. L.; reference 1e, chapter 1, and references therein.
    b. Schindler, J. W.; Fukuda, R. C.; Adamson, A. W. J. Am. Chem. Soc. 1982, **104**, 3596.
20. Martin, D. S., Jr.; Rush, M. R.; Kroenig, R. F.; Fanwick, P. E. Inorg. Chem. 1973, **12** 301, and references therein.
21. Anex, B. G.; Takeuchi, N. J. Am. Chem. Soc. 1974, **96**, 4411.
22. Isci, H.; Mason, W. R. Inorg. Nucl. Chem. Lett. 1972, **8**, 885.
23. Anex, B. G.; Ross, M. E.; Hedgecock, M. W. J. Chem. Phys. 1967, **46**, 1090.
24. Sperline, R. P.; Dickson, M. K.; Roundhill, D. M. J. Chem. Soc. Chem. Comm. 1977, 62.
25. Fordyce, W. A.; Brummer, J. G.; Crosby, G. A. J. Am. Chem. Soc. 1981, **103**, 7061.
26. This type of lattice structure is often seen in Pt(II) complex crystals. See, for example, Chassot, L.; Müller, E.; von Zelewski, A. Inorg. Chem. 1984, **23**, 4249.
27. Fleischauer, P. D.; Fleischauer, P. Chem. Rev. 1970, **70**, 199.
28. a. Rice, S. F.; Gray, H. B. J. Am. Chem. Soc. 1983, **105**, 4571.
    b. Rice, S. F.; Gray, H. B. J. Am. Chem. Soc. 1981, **103**, 1593.

ELECTRON TRANSFER REACTIONS ON EXTREMELY SMALL SEMICONDUCTOR COLLOIDS STUDIED BY PULSE RADIOLYSIS

O. I. Mićić, M. T. Nenadović, T. Rajh, N. M. Dimitrijević,[1] and A. J. Nozik[2]
[1]Boris Kidric Institute of Nuclear Sciences, Vinca, 11001 Beograd, Yugoslavia
[2]Solar Energy Research Institute, Golden, Colorado 80401, USA

ABSTRACT. Optical effects due to size quantization have been observed for CdS, HgSe and PbSe colloids with particle diameters 20-100 Å in water and acetonitrile. For HgSe and PbSe semiconductors with diameter size less than 50 Å, the optical absorption edge is blue shifted ~2.8 eV. With decreasing particle size, the conduction band edges in bulk colloids are shifted to negative values and cathodic corrosion is the dominant process during illumination of the colloids in the presence of hole scavengers or during injection of electrons by redox couples created by radiation processes. Electron transfer reactions from different electron donors to TiO_2, α-Fe_2O_3, WO_3, p-Cu_2O, HgSe and PbSe colloidal particles were studied by pulse radiolysis. Equilibrium concentration of the reactants can be exploited to derive electron energy levels in the semiconductor colloids. The flat-band potential for TiO_2, α-Fe_2O_3 and WO_3 colloids becomes slightly more negative (0.1 to 0.2 V) than the corresponding single crystal electrodes, which can be explained by the corresponding change in the surface charge on the colloids.

INTRODUCTION

In recent years numerous studies have been made on the use of semiconductor powders and colloids in various photoelectrosynthetic and photocatalytic reactions. If semiconductor particles have small size, their photoelectrochemical properties may not be the same as those of large, single-crystal electrodes. The particle diameters, D_p, can be smaller than the thickness of the space charge layer and, in that case, the details of charge separation may not be the same as in a compact semiconductor electrode.[1-3] The particles also have a large surface-to-bulk ratio, and surface states may therefore be especially important in the interpretation of their photoelectrochemical behavior. Further, in extremely small particles quantization effects can arise due to the confinement of charge carriers in a small space.[4-10] Correlation between the confined electron and hole can result in excitonic effect and/or perturbation of the band structure. These lead to the increase of the effective bandgap.

E. Pelizzetti and N. Serpone (eds.), Homogeneous and Heterogeneous Photocatalysis, 213–226.

The physical nature of the particles prevents investigation of the electron energy levels by impedance measurements which have been used for single crystal semiconductor electrodes. Radiation chemical processes provide the possibility for determining the electron energy levels in the particulate semiconductor since transient strong negative redox potentials can be created in the solutions, resulting in charge transfer between the redox couple and the semiconductor particles:

$$M^{(n-1)+} + coll \rightleftarrows M^{n+} + (e^-)_{coll} \qquad (1)$$

In reaction 1, $(e^-)_{coll}$ represents electrons injected into the semiconductor colloid.

$M^{(n-1)+}$ was produced by radiation chemistry processes following the electron pulse, and electron transfer to the colloids was followed via transient spectrophotometry. In the presence of the $M^{n+}/M^{(n-1)+}$ couple, either the Fermi level or corrosion potential of the semiconductor equilibrates in time with the redox potential of the couple. Under appropriate conditions, when the corrosion potential is above the conduction band the equilibrium potential can be identified with the flat-band potential of the semiconductor colloid. The flat-band potential is a measure of the reducing power of semiconductor materials, and it is related to the electron affinity of the semiconductor and the charge density at the surface.

MATERIALS AND METHODS

All materials used for the preparation of the colloids were commercial products of the highest available purity. Triply distilled water was used throughout. Oxygen was removed by bubbling with argon.

Sols of α-Fe_2O_3, WO_3, and TiO_2 were prepared by hydrolysis of the corresponding salts.[11,12] Cuprous oxide colloids were obtained via the radiolytic reduction of Cu^{2+} ions in alkaline solutions in the presence of 0.05% poly(vinyl alcohol).[13] CdS, PbS, CdSe, PbSe and HgSe colloids were made by controlled precipitation of metal sulfide or selenide in water or acetonitrile in the presence of stabilizer.[9]

The size of the colloid particles was varied by changing the pH of the solution, the salt concentration, and the temperature. Sols of approximately uniform size were obtained by filtration through membrane filters of adequate pore sizes or by centrifugation. The average particle diameter and crystalline structures were estimated by electron transmission microscopy. The crystalline structure of colloids with particle diameter >100 Å was determined by X-ray analysis.

Figure 1 shows the absorption spectra of TiO_2, WO_3, Cu_2O, α-Fe_2O_3, CdS and CdSe colloids which were used in the present study.

Continuous production of the electron adduct of colloids or other products were observed when the colloidal solutions were illuminated with an Osram XBO 150-W xenon lamp. Each irradiation was carried out on a sample of argon-saturated solution of volume 3 cm^3. The light intensity was measured by potassium ferrioxalate actinometer to be 1.2 $\times 10^{-6}$ einstein min^{-1}.

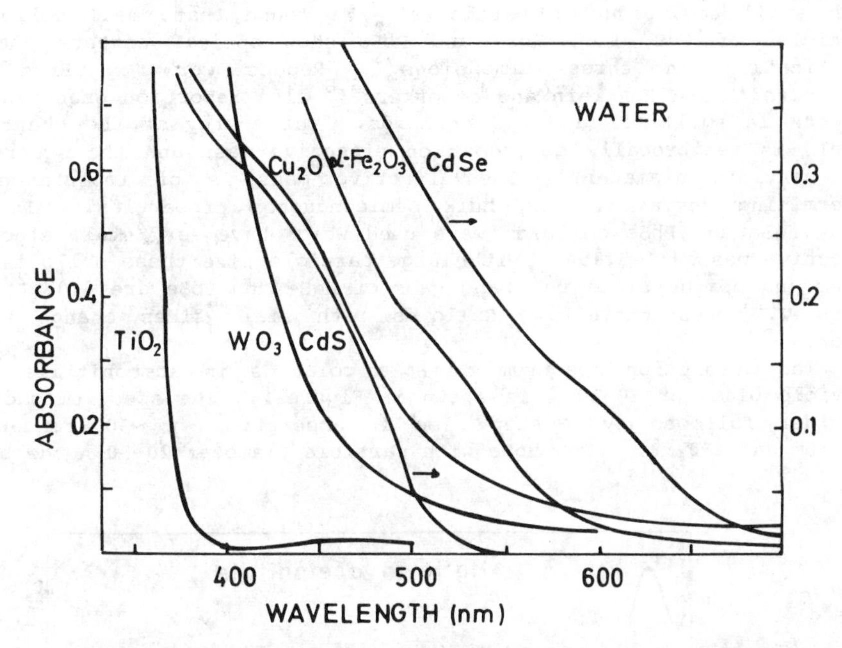

Figure 1. Absorption spectra of colloidal TiO_2 (D_p = 100 Å, 2.2 × 10^{-2} M total concentration), WO_3 (220 Å radius, 1.2 × 10^{-3} M, 0.2% PVA), α-Fe_2O_3 (D_p = 500 Å, 5 × 10^{-4} M, 0.1% PVA) and CdSe (D_p = 300 Å, 2 × 10^{-4} M, 0.05% nafion).

Pulse radiolysis employed a Febetron 707 (Field Emission Corporation) electron accelerator with a pulse duration of 20 ns, the operating conditions being similar to those described previously.[14] The total light path through the cell was 5.1 cm. The absorbed doses were in the range of 5-130 Gy per pulse, measured by using a potassium ferrocyanide dosimeter.[15]

RESULTS AND DISCUSSION

Optical Effects Due to Size Quantization

Quantization effects that arise from confinement of charge carriers in semiconductors with potential wells of small dimensions have been extensively studied in recent years. The vast bulk of the work to date has been done in thin-film, laminar structures.[16,17] These structures are one-dimensional quantum wells; when multiple wells are electronically coupled the resulting layered structures are termed superlattices.[16] Quantum confinement in two dimensions results in quantum well wires, and such systems have recently been reported.[18,19] Quantum confinement in three dimensions should occur with small semiconductor particles. We found that small colloidal particles of CdS, PbS, HgSe and PbSe show optical effects due to quantization in three dimensions. Recent work by Brus and coworkers[4,5] and Henglein and coworkers[6,8] also report on quantization effects in small CdS and ZnS colloids. The energy of the quantized levels is reciprocally dependent on effective mass and the square of the particle diameter. The effective mass is of importance in determining deviation from bulk semiconductor properties. In our study HgSe and PbSe colloids were used which have very small electron effective masses (~0.05). With large particle size these colloids are black and opaque since the band gaps of HgSe and PbSe are 0.07[20] and 0.25 eV,[21] respectively. Colloids with small sizes change their color.

The absorption spectrum of HgSe colloids in acetonitrile with particle diameter 20-30 Å is shown in Figure 2. The spectrum shows a shoulder followed by a step rise in absorption at ~300 nm and a maximum at 250 nm. For HgSe with particle diameter 20-30 Å the band

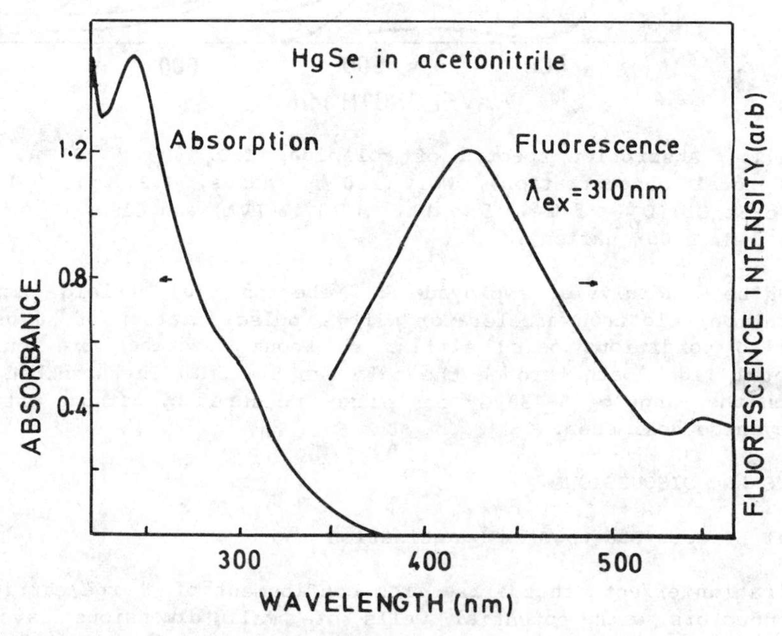

Figure 2. Optical absorption and fluorescence spectra of HgSe (2 × 10^{-4} M)-nafion (0.025%) in acetonitrile prepared at -28°C.

edge is shifted 3 eV (bulk band gap = 0.07 eV). The photoluminescence emission spectrum of HgSe (Figure 2) shows an emission peak at 430 nm with excitation from 270-360 nm. The shape of the emission spectrum is not dependent on the wavelength of the exciting light. These emission spectra also confirm that the absorption shift is simply due to an increased band gap.

The absorption spectra of PbSe colloids with D_p ~ 20 Å and D_p = 50-200 Å are shown in Figure 3. The band gap of PbSe is 0.25 eV, and colloids with large particle sizes (D_p > 200 Å) are black and opaque. Emission spectra of small particle samples show peaks at 440 and 600 nm; the larger particle samples have only one peak at 600 nm. Excitation for these emission peaks were at 340 or 410 nm.

The absorption edge for PbSe sample with D_p = 50-200 Å shows a long tail that arises from the broad distribution of particle sizes; the sample with D_p ~ 20 Å has a narrower size distribution and shows much less of a tail and discrete structure appear. The photoluminescence emission of PbSe is consistent with the absorption spectra. The shift of ~2.8 eV in the effective band gap for ~20 Å is quite dramatic.

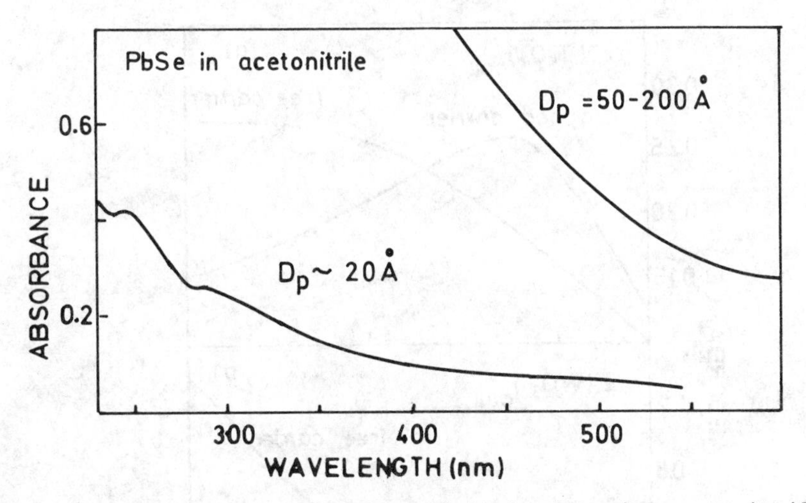

Figure 3. Absorption spectra of colloidal PbSe in acetonitrile. Total concentration 2×10^{-4} M; D_p ~ 20 Å, prepared at -28°C, 0.025% nafion as stabilizer; D_p = 50-200 Å, room temperature, 0.1% nafion.

Nature of Electrons Injected into Semiconductor Particles

The nature of the electrons injected into the semiconductor colloids is of general interest. One question that arises is whether these electrons move freely in the conduction band or are localized at bulk traps or surface traps.

The absorption spectra of the electron adducts of WO_3 and TiO_2 colloids were observed (see Figure 4) by pulse radiolysis techniques. In the presence of 0.1 mol dm^{-3} propanol-2 in acid solution only one type of radical, $(CH_3)_2\dot{C}OH$, is formed which injects electrons into the colloids. The same absorption spectra are obtained in illuminated colloidal solutions. The $(CH_3)_2\dot{C}OH$ radicals have a large negative redox potential for electron donation, $E[(CH_3)_2CO/(CH_3)_2\dot{C}O^- \cdot H^+] = -1.23$ V [22] (vs NHE), and they can react with the sols according to

$$(CH_3)_2\dot{C}OH + TiO_{2coll} \text{ (or } WO_{3coll}) \rightarrow$$

$$(e^-)TiO_{2coll} \text{ (or } (e^-)WO_{3coll}) + (CH_3)_2CO + H^+ \qquad (2)$$

The absorption spectrum of the electron adduct of TiO_2 colloids was observed in earlier work on the γ-radiolysis of aqueous sols of propanol-2 and TiO_2 colloids,[23] and also during photolysis of TiO_2 sols.[3] We find that this absorption spectrum is sensitive to the method of colloid preparation.

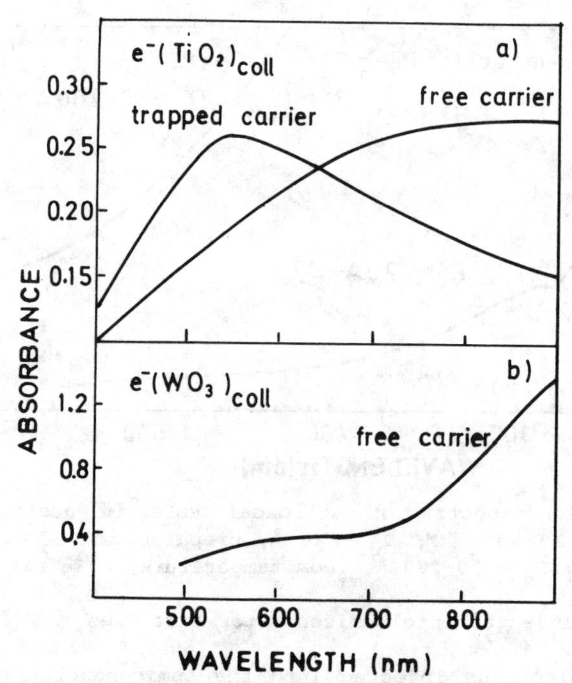

Figure 4. Absorption spectra obtained after irradiation of colloidal solutions in the presence of 0.1 M propanol-2. (a) TiO_2 colloids at pH 2; (b) WO_3-PVA 3.5 mol dm^3 H_2SO_4.

Figure 4a shows the absorption spectra of $(e^-)TiO_2$ intermediates obtained in colloidal sols which were prepared in two different ways. The absorption spectrum with a maximum at 560 nm was obtained from sols produced by dialysis of $TiCl_4$ solution at 0°C (50 Å radius, 0.125 M TiO_2). The broad absorption spectrum was obtained with sols (500 Å radius, 0.025 M TiO_2) which were produced after 2 h of boiling $TiCl_4$ in acid solution.

The curve in Figure 4a with a peak is indicative of localized electrons, while the broad absorption that increases with longer wavelengths above 400 nm is consistent with free carrier absorption.[21] X-ray analysis of the colloids shows that the free carrier absorption is associated with a more crystalline structure, while colloids with localized electrons show mainly an amorphous structure.

A broad absorption was found for $(e^-)WO_{3coll}$ (Figure 4b) that increases rapidly with increasing wavelengths above 700 nm; this absorption also increase with increased electron concentration in the WO_3 colloids. This type of spectrum is consistent with free carrier absorption.

Electrons are long-lived in WO_3 and TiO_2 semiconductor colloids since the Fermi-level potentials of these semiconductors are not negative enough such that reduction of the solvent takes place. Besides the reaction with solvent, electrons can be consumed in the degradation of semiconductor when the conduction band potential is above the cathodic corrosion potential of the semiconductor.

Cathodic Corrosion Process

In order to study the kinetics of semiconductor reduction we chose p-Cu_2O (band gap 1.9-2.0 eV, conduction band potential -1.5 V ns NHE, valence band potential +0.5 V vs NHE) which is unstable as a photocathode since its cathodic corrosion potential is 0.0 V vs NHE:

$$Cu_2O + 2H^+ + 2e^- \rightarrow 2Cu + H_2O \qquad (3)$$

We have used different electron donors, $M^{(n-1)+}$, such as $(CH_3)_2\dot{C}OH$, $\dot{C}O_2^-$, Eu^{2+} and MV^+ to accept a positive hole from Cu_2O since Cu_2O is a p-type semiconductor. The change of the absorption spectrum of Cu_2O sols in Figure 5 after irradiation in the presence of MV^+ can be attributed to the formation of metallic copper with a maximum at 580 nm.

$$\overset{h^+}{\curvearrowright}$$

$$(Cu_2O)_n + mMV^+ \rightleftarrows (Cu_2O)_n^{m-} + mMV^{2+} \quad (pH\ 9\text{-}11) \qquad (4)$$

$$(Cu_2O)_n^{m-} \xrightarrow{mH^+} (Cu^\circ)_m(Cu_2O)_{n-m} + m/2\ H_2O \qquad (5)$$

The absorption spectrum in Figure 5 is similar to the Cu° colloids formed in the absence of Cu_2O by reaction of $(CH_3)_2\dot{C}OH$ with Cu^{2+}.

Pulse radiolysis experiments show that in aqueous solutions of Cu_2O, when Eu^{2+} or MV^+ are formed, the maximum of the copper metal appears after the completion of reactions (4) and (5). However, equilibrium of reaction (4) is established in 100 ms time scale while maximum of $(Cu°)_m(Cu_2O)_{n-m}$ formed in reaction (5) appears 15 seconds after the pulse.

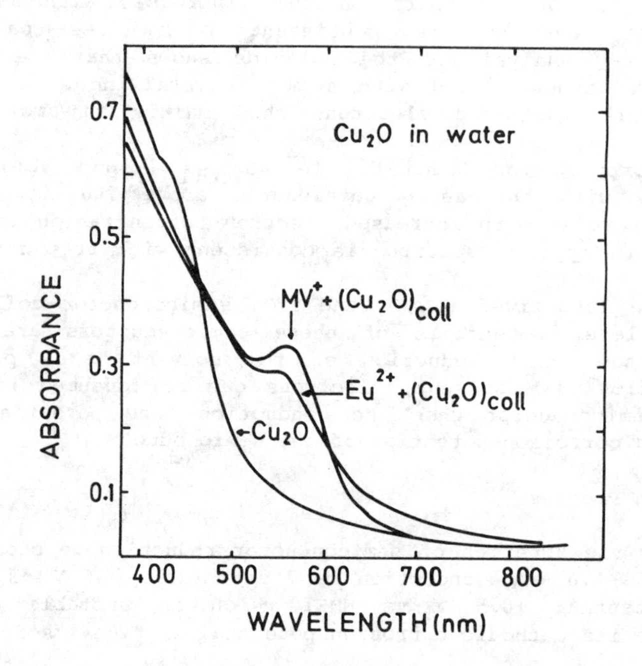

Figure 5. Absorption spectra of colloidal Cu_2O after irradiation (5 kGy) in the presence of propanol-2 and Eu^{3+} or MV^{2+} at pH 6.

It is to be noted that electrons created in the Cu_2O colloids by photochemical processes cannot be transferred to the MV^+ in a reverse reaction since the reaction (5) is fast. However, corrosion is less favorable in an aprotic solvent such as acetonitrile because of the unavailability of protons. Illuminated solution of Cu_2O (monomer concentration 5×10^{-4} mol dm^{-3} containing 0.1% polyethylene glycol and 1.5% of methanol) shows the characteristic absorption spectrum of MV^+ radicals. This means the Cu_2O colloids in acetonitrile can inject electrons into solution.

Determination of Flat-Band Potentials

The equilibrium of the Fermi level (i.e., electrochemical potential) of a semiconductor colloid with the redox potential of a redox couple in solution can be used to determine the flat-band potential of the semiconductor colloids. The concentration of $M^{(n-1)+}$ was always measured (spectrophotometrically) after completion of the electron-transfer reaction (1). This means that the potential for the particles $(E(e^-)_{coll}$ and that of the solution are equilibrated.

$$E(e^-)_{coll} = E°(M^{n+}/M^{(n-1)+}) + 0.059 \log [M^{n+}/M^{(n-1)+}] \qquad (6)$$

For α-Fe_2O_3 and TiO_2 we used the MV^{2+}/MV^+ couple with E° = -0.44 V, and determined the equilibria at different pHs; for WO_3, the Cu^{2+}/Cu^+ couple was used which in the presence of PVA has a standard redox potential +0.29 V (vs NHE) calculated from cyclic voltammograms.

At equilibrium this potential should be equal to the flat-band potential of the particles, E_{fb}, at the pH of the experiment. This identity is only valid if the particles maintain a band structure. It is generally accepted that semiconductor particles with relatively large effective masses and with diameters above about 50-100 Å do exhibit bulk properties and a band structures;[10] hence, the identity in this study is valid since the particles range in diameter from 800 to 70 Å. The variation of $E_{fb}(TiO_2)$ and $E_{fb}(Fe_2O_3)$ with pH and absorbed doses (MV^+ concentration) is shown in Figure 6.

This method for determining flat-band potentials of particles is independent of whether the injected electrons are free or trapped. This is because equilibrium is established between the semiconductor particle and the redox couple in solution. Under these conditions, only the position of the Fermi level in the smeiconductor that is equilibrated with the redox couple is important; all bulk and surface traps that lie below the Fermi level are filled.

Simple calculations show that the magnitude of the injected charge in our experiments (10^{19} to 10^{21} e^-/cm^3) is such that it can only be accommodated through reduction of the semiconductor (i.e., for TiO_2 colloids reaction (7)),

$$x(e^-)_{coll} + TiO_2 + xH^+ \rightarrow TiO_{2-(x/2)} + (x/2)H_2O \qquad (7)$$

leading to a merger of the bulk Fermi level with the conduction band edge. We find that the value of E_{fb} determined at equilibrium from Eq. 6 is not sensitive either to the colloid preparation method or to the particle diameter. However, E_{fb} does depend slightly upon the concentration of the injected electrons and, as expected, upon the pH of the solution. The different electron concentrations were obtained by changing the absorbed radiation dose or the absorbed light intensity and also by changing the particle concentration. Ward *et al.*[27] recently found a light intensity effect on E_{fb} for TiO_2 particulate suspension using the same (MV^{2+}/MV^+) redox couple in solution.

The flat band potentials for WO_3, TiO_2, and α-Fe_2O_3 colloids obtained from radiation chemical and photochemical experiments are about 150-200 mV more negative than those for compact electrodes and

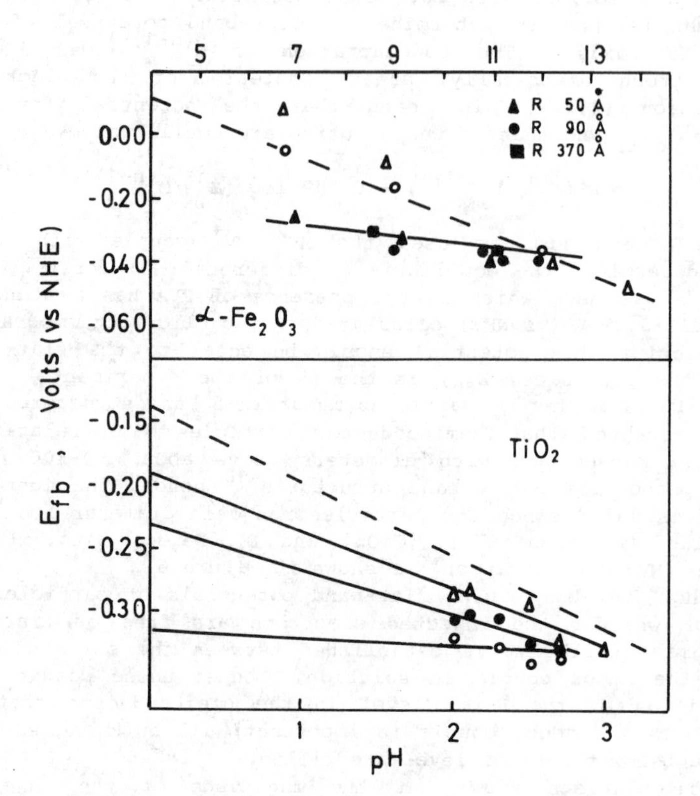

Figure 6. Flat-band potential of α-Fe_2O_3 colloids as a function of pH for colloids with different radius. The dashed line represents literature values for single crystal (ref. 26 (o)) and polycrystalline (ref. 27 (Δ)) electrodes. Flat-band potential of TiO_2 colloids (50-Å radius, 1.3×10^{-2} M TiO_2) as a function of pH for different absorbed doses: (Δ) 10 Gy, (●) 30 Gy, (o) 60 Gy. The dashed line presents the results obtained via flash photolysis experiments (ref. 3).

only show about 10-30 mV change per decade change in electron concentration; this suggests an effect of surface charge. It is well known in photoelectrochemistry that adsorbed negative surface charge shifts E_{fb} to negative values.[29] Higher concentrations of electrons injected into the colloid will change the relative amount of negative surface charge on colloids and shift E_{fb} accordingly. The higher levels of adsorbed surface charge could also interfere with the usual protonic/oxide equilibrium at the surface.

Redox Degradation Potential of Extremely Small Particles

In extremely small particles the conduction band potential is shifted to negative values due to quantization effects. We investigated the redox potential of colloids in the dark via experiments attempting to equilibrate with the redox couple in solution (eq. 1). One question that arises is whether this potential corresponds to the Fermi level of the particle or to the corrosion potential, since for extremely small particles it should be expected that conduction band is above the corrosion potential. HgSe and PbSe colloids were used for which, unfortunately, there is no data available for Fermi levels and corrosion potentials. The cathodic corrosion potential for PbSe and HgSe colloids were determined from differential pulse voltammograms to be -0.716 and -0.172 V vs SCE, respectively.

Various electron donors with redox potential more negative than -0.4 V, such as $NO_2^-C_6H_4CN$, $\dot{C}H_2OH$, $\dot{C}H_3CHOH$, $(CH_3)_2\dot{C}OH$ are able to transfer electrons to HgSe colloids (Figure 7). MV^+ can transfer electrons completely to HgSe and the absorption of MV^+ at 605 nm disappears in 10 ms while an absorption increase at 300-900 nm appears after 20 s. The absorption increase is probably caused by agglomeration of the HgSe colloids when $Hg_m^+(HgSe)_{n-m}$ forms in bulk. MV^+ also transfers electrons to large HgSe particles (500 Å). This means that the colloid potential does not depend upon the particle size. On the other hand O_2^- radicals cannot transfer electrons to HgSe

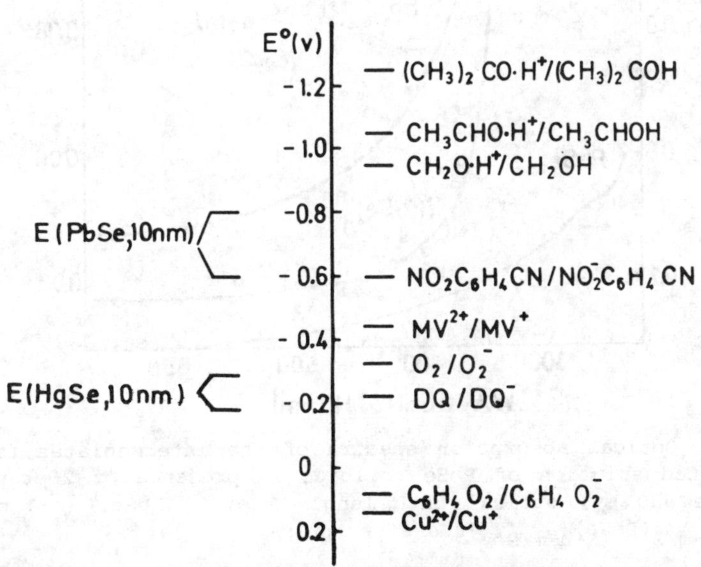

Figure 7. Energy level diagram for equilibration of HgSe and PbSe semiconductors with redox couple potentials in solution.

colloid particles. The potential between 0.3-0.2 V vs NHE is most likely the cathodic corrosion potential of HgSe particles.

For PbSe colloids $NO_2^-C_6H_4\dot{C}N$ and MV^+ radicals do not transfer electrons while $\dot{C}H_2OH$ and $CH_3\dot{C}HOH$ do transfer electrons to the semiconductor (Figure 7). These electron transfer reactions were followed by pulse radiolysis in two steps.

$$mR^{(n-1)+} + (PbSe)_n \rightarrow (PbSe)_n^{m-} + mR^{n+} \tag{8}$$

$$(PbSe)_n^{m-} \rightarrow Pb^{o}_{m/2}(PbSe)_{n-m/2} \tag{9}$$

As can be seen in Figure 8 $(PbSe)_n^{m-}$ is formed after 1 ms while $Pb^{o}_{m/2}(PbSe)_{n-m/2}$ (second intermediate) is formed after 10 s. The $Pb^{o}_{m/2}(PbSe)_{n-m/2}$ absorption spectrum is similar to Pb° colloids formed in radiation chemical process by reaction of $(CH_3)_2\dot{C}OH$ with Pb^{2+}. When irradiated solution of PbSe colloids are exposed to air, Pb° disappears. The rate of oxidation of lead metal formed in the bulk of PbSe is much lower than that of pure lead metal colloids obtained by reaction of $(CH_3)_2\dot{C}OH$ with Pb^{2+}.

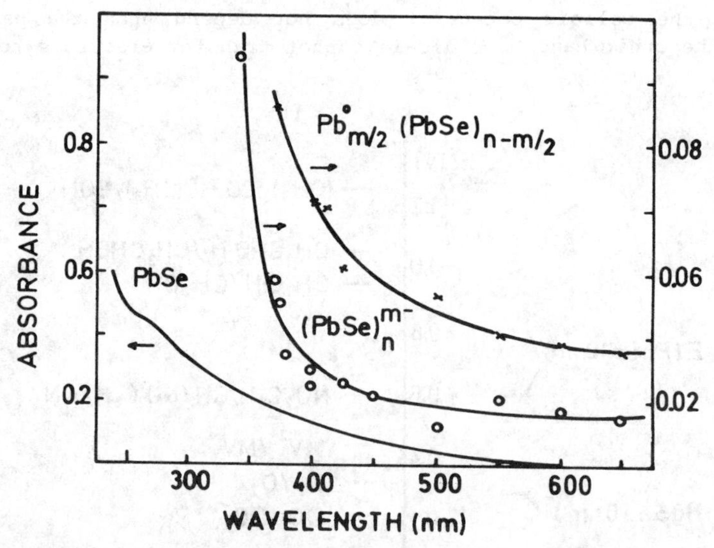

Figure 8. Optical absorption spectra of the intermediates formed in N_2O saturated solution of PbSe colloids in presence of 2-propanol at pH 6, dose 130 Gy, optical path length 1 cm. $(PbSe)_n^{m-}$, 1 ms after pulse; $Pb^{o}_{n-m/2}(PbSe)_{n-m/2}$, 10 s after pulse.

The redox potentials of HgSe and PbSe determined here by observing electron transfer reactions are about 150-300 mV more negative than the corresponding corrosion potentials.

The cathodic corrosion of HgSe and PbSe colloids are dominant processes during the illumination of the colloids in water or acetonitrile solution in the presence of the hole scavenger triethanolamine. Corrosion can be suppressed if an electron acceptor which has potential more positive than the cathodic corrosion potential of semiconductor in the illuminated solution is present. We found that MV^{2+} is reduced to MV^{+} in UV illuminated aqueous solution of $PbSe/SiO_2$ colloids (pH 9) in the presence of TEA as hole scavenger:

$$PbSe \xrightarrow{h\nu} e^- + h^+ \quad (10)$$

$$e^- + MV^{2+} \rightarrow MV^{+} \quad (11)$$

$$h^+ + TEA \rightarrow TEA^{+} \quad (12)$$

These reactions also take place with large PbSe particles. However, it should be pointed out that quantum yield of reaction (11) is much higher for particles with $D_p < 50$ Å.

Size quantization effects with small semiconductor particles should lead to important modifications in the nature of the interfacial electron transfer reactions and attainable power conversion efficiency in the photochemical device.

ACKNOWLEDGEMENT. A.J.N. was supported by the U.S. Department of Energy, Office of Basic Energy Sciences, Division of Chemical Sciences. O.I.M., M.T.N. and T.R. were supported by the U.S.-Yugoslavia Joint Research Fund.

REFERENCES

1. M. Grätzel and A. J. Frank, J. Phys. Chem., **86**, 2964 (1982).
2. D. Duonghong, J. Ramsden, and M. Gratzel, J. Am. Chem. Soc., **104**, 2977 (1982).
3. A. J. Nozik, Appl. Phys. Letts., **30**, 567 (1977).
4. R. Rosetti, S. Nakahara and L. E. Brus, J. Chem. Phys., **79**, 1086 (1983).
5. R. Rosetti, J. L. Ellison, J. M. Bigson and L. E. Brus, J. Chem. Phys. **80**, 4464 (1984).
6. H. Weller, V. Koch, M. Gutierrez and A. Henglein, Ber. Bunsenges. Phys. Chem., **88**, 649 (1984).
7. F. Williams and A. J. Nozik, Nature, **311**, 5989 (1984).
8. A. Fojtik, H. Weller, V. Koch and A. Henglein, Ber. Bunsenges. Phys. Chem., **88**, 969 (1984).
9. A. J. Nozik, Ferd Williams, M. T. Nenadović, T. Rajh and O. I. Micic, J. Phys. Chem., **89**, 397 (1985).
10. L. E. Brus, J. Chem. Phys. **80**, 4403 (1984).
11. N. M. Dimitrijevic, D. Savic, O. I. Micic and A. J. Nozik, J. Phys. Chem., **88**, 4278 (1984).

12. M. T. Nenadović, T. Rajh, O. I. Mićić and A. J. Nozik, J. Phys. Chem., **88**, 5827 (1984).
13. T. Rajh and O. I. Micic, unpublished results.
14. V. Markovic, N. Nikolic and O. I. Mićic, Int. J. Radiat. Phys. Chem., **6**, 227 (1974).
15. J. Rabani and M. S. Matheson, J. Phys. Chem., **70**, 761 (1966).
16. K. Ploog and G. H. Döhler, Adv. Phys., **32**, 285 (1983).
17. R. Dingle, Adv. Solid State Phys., **15**, 21 (1975).
18. P. M. Petroff, A. C. Gossard, R. A. Logan and W. Wiegmann, Appl. Phys. Lett., **41**, 633 (1982).
19. M. Laviron, P. Averbuch, H. Godfrin and R. E. Rapp, J. Phys. Lett. (Orsay, Fr.), **44**, L-1021 (1983).
20. S. L. Lehoczky, J. G. Broerman, D. A. Nelson and C. R. Whitsett, Phys. Rev. B9, 1598 (1974).
21. J. I. Pankove, "Optical Processes in Semiconductors", Dover Pub., Inc., New York, 1971, p. 412.
22. M. Breitenkamp, A. Henglein and J. Lilie, Ber. Bunsenges. Phys. Chem., **80**, 973 (1976).
23. A. Henglein, Ber. Bunsenges. Phys. Chem., **86**, 241 (1984).
24. H. Gerischer, J. Vac. Sci. Technol., **15**, 1422 (1978).
25. R. K. Quinn, R. D. Nasby and B. J. Baugham, Mater. Res. Bull., 1011 (1976).
26. J. H. Kennedy and K. W. Frese, J. Electrochem. Soc., **125**, 723 (1978).
27. M. D. Ward, J. R. White and A. J. Bard, J. Am. Chem. Soc., **105**, 27 (1983).
28. A. J. Bard, F. R. F. Fan, A. S. Gioda, G. Nagasubramanian and H. S. White, Faraday Discuss., **70**, 19 (1980).

NAFION STABILIZED AQUEOUS SOLUTIONS OF COLLOIDAL SEMICONDUCTORS[1]

Olga Micic
Boris Kidric Institute of Nuclear Sciences - Vinca
11001 Beograd, Yugoslavia
and
Dan Meisel
Chemistry Division
9700 S. Cass Avenue
Argonne, IL 60439, U.S.A.

ABSTRACT. Water soluble Nafion polymers have been used to stabilize colloids of CdS, TiO_2 and MoS_2. The perfluorocarbon backbone of the polymer adsorbs on the colloid, and the sulfonate head groups provide electrostatic shielding. The inertness of the stabilizer has been verified in photochemical reactions as well as in fluorescence quenching studies. The electrostatic field has been shown to inhibit charge recombination of anionic photoredox products. Emissive recombination is observed in both CdS and TiO_2 when the latter is prepared at elevated temperatures.

INTRODUCTION

Photochemistry at the electrolyte-solid interface of dispersed semiconductor particles has become an area of intensive research activity following the observation that a variety of redox photochemical processes could be induced at such surfaces.[2,3] The large surface area per unit weight of particles of colloidal sizes, as well as the freedom from problems associated with light scattering, lead to increased interest in photochemistry at colloidal semiconductor interfaces.[4,5] In order to inhibit coagulation of the colloidal particles, it is often necessary to add a stabilizer which adsorbs on the colloid surface and provides the stabilization function through either electrostatic or steric shielding.[6] However, most of the stabilizers that have been utilized in the past (often surfactants or polymers) are quite reactive towards the variety of highly reactive intermediates produced at the particle interface in the course of the photochemical process. Indeed, the effect of the stabilizers on surface properties such as the energy of the recombining electron-hole pair, the amount of adsorbed electron (hole) acceptors, the yield of redox products and rate of their recombination has been recently emphasized for colloidal CdS.[7] Furthermore, the stabilizer itself

E. Pelizzetti and N. Serpone (eds.), Homogeneous and Heterogeneous Photocatalysis, 227–239.

has been shown to interfere in the photochemical[5b,d] or catalytic[8] processes that may occur at the particle interface. The latter process will cause destruction of the stabilizer and eventually coagulation of the sol. To circumvent such complications, sodium hexametaphosphate (HMP) is now routinely used as a stabilizer. However, as a low-molecular weight stabilizer, it may be expected to provide primarily electrostatic stabilization and thus be prone to destabilization caused by high ionic strength. An inert polymeric stabilizer is therefore highly desirable. A stabilizer that will minimize interference in the electron-hole recombination processes or minimize its own chemical reactions at the interface is a perfluorinated polymer stabilizer. In the present report, we describe the preparation and some properties of TiO_2, CdS and MoS_2 colloids stabilized by water soluble Nafion whose general formula is given below:

$$[(CF_2-CF_2)_m - \underset{\substack{| \\ (O - CF_2 - \underset{\substack{| \\ O - (CF_2)_2SO_3H}}{CF} - CF_3)_k}}{CF} - CF_2]_n$$

$m \approx 5$ to 13.5
$n = 1000$
$k = 1, 2, 3...$

The use of Nafion polymers in their film form in a variety of electrochemical processes has been previously described[9] and their utilization in photochemistry has been suggested.[10] A method for dissolving these solids in mixtures of alcohols has been developed,[11] and films cast from the dissolved material on graphite electrodes have been extensively utilized.[12,13] While preparation of colloidal particles, either of metallic[14] or semiconductor[15] materials, inside the water swollen membrane network has been demonstrated, the use of water soluble material as colloid stabilizer has not been reported in the literature.

EXPERIMENTAL

Materials and Colloid Preparation

Nafion solutions, 5% of 1100 EW H-form Nafion in alcohol were obtained from C. G. Processing, Inc. These were diluted and titrated with 0.1 N NaOH solution. Accurate titrations agree well with the specified ion exchange capacity of the polymer (4.6×10^{-3} eq. l^{-1} for 0.5% solutions). All other materials were of highest purity commercially available and were used as received. Water was triply distilled.

Nafion stabilized CdS colloids were prepared by slow dropwise addition of ca. 10^{-2}M Cd^{2+} solutions (unless otherwise stated the perchlorate salt is used) to well-stirred solutions of varying amounts of Na2S and $2.1x10^{-3}$ eq. l^{-1} Nafion (or polyvinylalcohol, PVA). This procedure yields monodispersed CdS particles of 250 Å diameter provided the addition of Cd^{2+} is carried out slowly. Colloids of very similar properties are obtained when the Nafion solutions are originally mixed with the Cd^{2+} solutions as long as the [Cd^{2+}] in this solution is below the maximum capacity of the polymer. As might be expected, this indicates that equilibration of Cd^{2+} among the Nafion sites is much faster than the agglomeration of CdS to form the colloidal particles. The colloidal solution is then twice dialyzed against five liters of water to remove excesses of electrolytes and alcohol. Argon is then bubbled to remove excess of H_2S.

TiO_2 colloids (300 Å radius) were prepared by rapid addition of $TiCl_4$ to 250 ml of ice-cold aqueous 1 M $HClO_4$ solution (~$2.5x10^{-2}$ M Ti (IV)). When temperature effects were checked, the resultant solution was heated at this stage in an open beaker to 95°C without stirring for two hours. The colloidal solutions is then dialyzed against dilute (~10^{-2} M) $HClO_4$ solution to decrease chloride concentration by a factor of 10^4. The resultant stock solution was kept refrigerated at pH 1.3 and diluted with the stabilizer containing solution to the desired concentration prior to the particular experiment. Total titanium concentrations in these solutions were measured using the hydrogen-peroxide method. While no problems are encountered on addition of PVA to the TiO_2 colloidal solutions, rapid coagulation of the sol occurs upon addition of the Nafion solution. Presumably the Nafion polymer strips the particles of their positively charged (H^+ and residual Ti(IV) ions) protective layer. We therefore increase the pH of the colloidal solution to ~9 and then rapidly add the Nafion solution (final 0.1% by weight). The solution is then dialyzed again to remove excess alcohols.

MoS_2 colloids were prepared by dropwise addition of 6 ml of concentrated acetic acid to a well-stirred 100 ml solution of $2x10^{-2}$ M $(NH_6)_2MoS_4$ and $2x10^{-2}$ M Na_2S and 0.1% Nafion. The solution is then warmed for two hours at 40-50°C without stirring. The color of the solution changes during this process from deep red to brown. The resultant MoS_2 colloidal solution is then dialyzed against water to yield a colloid at pH 3.5. No elemental sulfur could be observed on extraction with CS_2. The average particle radius has been determined by TEM to be 175 Å.

Instrumentation

Emission and excitation spectra were measured on a Perkin-Elmer MPF-44B fluorescence spectrometer. Absorption spectra were recorded on a Varian 2300 spectrophotometer. Fluorescence lifetimes were measured using the single photon counting technique in conjunction with a frequency-doubled mode-locked dye laser (λ_{exc} = 315 nm, pulse width ~0.15 ns) pumped by an Ar-ion laser. Details of the system have been previously described.[17] Laser flash photolysis experiments were conducted using a N_2-laser (337 nm, 10 ns width, ~10 mj/pulse). Kinetics of the absorption signals were detected by a IP28 photomultiplier, digitized on a Biomation 8100 transient recorder and processed on an LSI11 computer.

RESULTS AND DISCUSSION

CdS

Absorption spectra of Nafion stabilized CdS colloids at several concentrations are shown in Fig. 1. The onset of the absorption is at 500 nm, slightly blue-shifted from the band gap reported for CdS single crystals (2.38 eV)[24] This may indicate either contributions to the absorption spectra from small particles or small regions of microcrystalinity in the larger particles.[5e,f] At any rate, as can be seen in Fig. 1, Beer's law is satisfactorily obeyed. Emission spectra of these colloids clearly show the red luminescence centered at ~670 nm. This emission has been previously attributed to recombination of photogenerated holes with sulfur vacancies.[16b] We also noticed that the emission develops over long periods of time. After two days of "aging" of the colloid, the emission is fully developed.

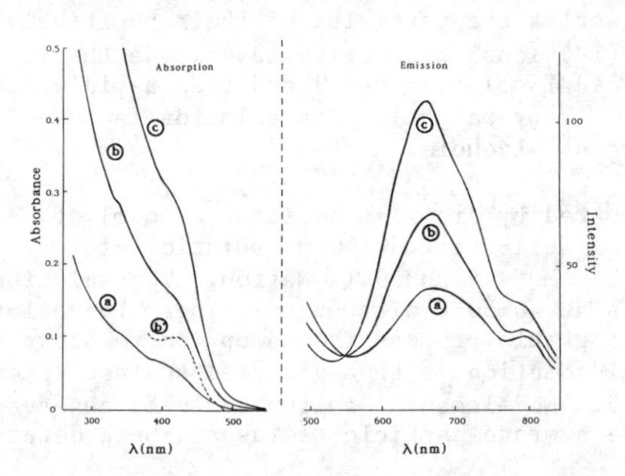

FIG. 1 Absorption and emission spectra of Nafion stabilized CdS colloids. All contain 6.8×10^{-4} M Cd^{2+} and 2×10^{-3} eq./l Nafion. $[S^{2-}] = 1.2\times10^{-4}$ M (a); 3.0×10^{-4} M (b); 5.4×10^{-4} M (c); (b') is excitation spectrum of (b) at λ_{em} = 670 nm; λ_{exc} = 400 nm for emission spectra. Colloids aged for three days.

During this period of time, hardly any change in the absorption spectra is noticeable. This observation indicates then that creation of the sulfur vacancies in the particles is much slower than formation of the particles themselves. This process of formation of imperfection of the CdS particles may result either from reorganization in the crystal lattice or by relatively slow adsorption of the slight excess of Cd^{2+} ions, which were present in the solutions of Fig. 1, onto the surface of the particle. While surface adsorption is usually a diffusion-controlled process, it should be oted that in the Nafion-containing solutions, this process involves an ion-exchange step which may substantially slow the reaction down. However, as the excitation spectrum in Fig. 1 clearly shows the emission originates from band-gap excitation which rapidly degrades to the Fermi level corresponding to the sulfur vacancy.

Observation of direct electron-hole recombination in the emission spectrum is much more difficult. Due to the strong absorption at the same spectral range where this emission is expected, interference from scattering and inner-filter distortions severely limit studies in this range. In the case of CdS low levels of impurities of ZnS further complicate analysis of emission spectra in this range. We therefore avoid studies in this region. Lifetime measurements of the red emission indicate that the decay rate strongly deviates from exponential relaxation (Fig. 2).

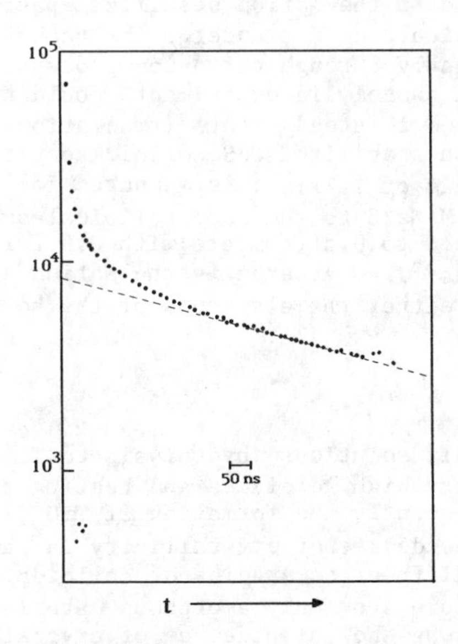

FIG. 2 Decay rate of emission at λ = 650 nm from solution (c) of Fig. 1 measured using the single photon-counting technique.

Neither could it be fit to any of the commonly used kinetic laws for electron transfer reactions in solid matrices. It seems, therefore, that a distribution of decay rate constants may better describe the kinetics of the decay of the red emission. This emission, however, may be contrasted with the long-lived emission observed in Nafion films.[15a]

To check the effect of the stabilizer on the photochemical behavior of the colloid, several flash photolysis and steady photolysis experiments were conducted. When Nafion stabilized colloidal solution, 1.7×10^{-3} M in CdS, containing 5×10^{-4} M of the zwitterionic viologen, 4,4'-bipyridinium-1,'1-bis(trimethylene sulfonate), denoted ZV, was flash photolyzed, formation of the viologen radical anion could easily be detected by following its absorption at 605 nm. This reaction is commonly attributed to scavenging of electrons excited to the conduction band of the CdS colloid. While the formation of $ZV^{\overline{\cdot}}$ followed the laser pulse, no decay in the few μsec range could be observed. The viologen radical, however, decays with a half-life of 700 μsec. At the end of this process, some 20% of the radicals produced seem to remain indefinitely in the solution. The relatively slow decay rate, as compared to the previously reported decay in the μsec range for methylviologen radicals produced by excitation of sodium dodecylsulfate stabilized CdS colloid, indicates higher electrostatic potential field of the negatively charged surface of the Nafion stabilized particles. The fact that some of the radicals do not undergo the back reaction indicates loss of holes, probably through corrosion processes. The observations of the flash photolysis experiments could be extended to the longer periods involved in steady state irradiation. When a similar solution of Nafion stabilized CdS colloid containing ZV is photolyzed, a quantum yield of 1.2×10^{-3} is measured for the stable
Addition of 10^{-2} M Na_2S to the same colloidal solution increases the quantum yield to 0.5 (compare with 0.6 for CdS deposited on SiO_2 colloids[5a]). Apparently the Nafion stabilizer suffers no reaction with either the electrons or the holes produced upon band gap irradiation.

TiO2

In aqueous Ti(IV) salt solutions, hydrolysis to $Ti(OH)_n^{(4-n)+}$ species occurs even at very high acidities and heating accelerates this hydrolysis and consequently the formation of TiO_2.[18] However, heating also increases the degree of crystallinity in the colloidal particles. While x-ray diffraction studies of colloidal TiO_2 prepared at room temperature show only amorphous materials, those prepared at high temperature show high degree of crystalline anatase structure. Nevertheless, even when prepared at nearly water boiling temperatures, amorphous TiO_2 is also present in the colloidal solutions. The higher degree of crystallinity can be correlated with the appearance of emission. In the amorphous material, the abundance of surface imperfections and surface states will shift any recombination emission to energies lower than the band gap energy and thus broaden substantially the emission spectrum. In the following we therefore refer only to TiO_2 colloids prepared at 95°C. Both PVA or Nafion stabilized colloids gave similar results. However, emission was stronger in the latter colloids. It is also interesting to note that the efficiency of hydrogen production from band-gap irradiation of colloidal anatase was found to decrease with the increased amount of amorphous material in these colloids.[19]

Aerated colloidal solutions of TiO_2 show only weak emission. This weak emission can be seen in Fig. 3a relative to a strong Raman band (at 2700 cm^{-1}). Removal of oxygen from the solution strongly enhances the emission intensity (Fig. 3b). The emission peaks at 390 nm, which corresponds closely to the well-established band gap of 3.2 eV for anatase TiO_2.[20] The relatively narrow spectral width of the emission band is a further indication that the recombination process is primarily of free charge carriers with little contribution from surface states recombination. The emission maximum at 390 nm is independent of excitation wavelength in the region 280-350 nm. The excitation spectrum in this region corresponds to the absorption spectrum and shows only structureless increase on going to shorter wavelengths. These two observations lead us to believe that the emissions observed indeed originate in the TiO_2, relatively free of the distortions discussed above. The emission can be seen in Fig. 4 to decay in a monoexponential process with lifetime of 11 nsec.

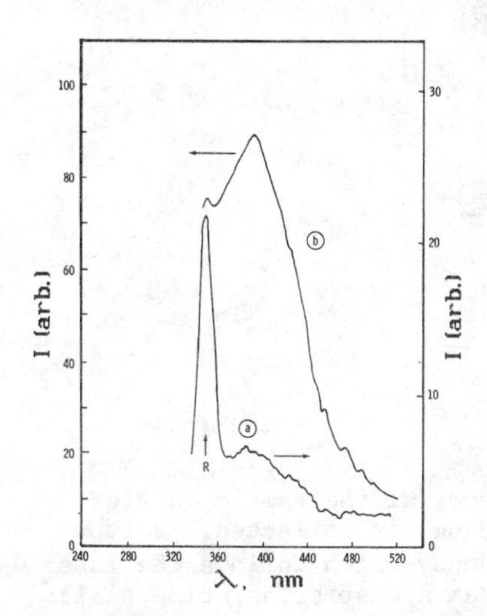

FIG. 3 Emission of TiO_2 colloids: (a) aerated solution; (b) after two hours bubbling N_2O. 0.26 gr/l TiO_2 at pH 2; λ_{exc} = 320 nm. R indicates Raman band.

Emission quantum yields as high as 10^{-2} were obtained; however, even this may be considered a lower limit since the samples were never of complete crystalline structure (the only emitting structure) and since complete removal of oxygen is not certain in these samples.

Further indication to the role of the crystalline structure in the emissive process is obtained from studies of the pH effect on the emission intensity. Increasing the acidity from 10^{-2} M to 1 M of either $HClO_4$ or H_2SO_4 had little effect on the emission intensity for the same amount of titanium. Such an increase in acidity, however, decreased the light absorption at the excitation wavelength of 320 nm. The decrease in absorption can be attributed to dissolution of the amorphous portion of the TiO_2 colloid. Indeed, Ti(IV) ions do not absorb at 320 nm.[21] On the other hand, it has been shown that amorphous TiO_2 is easily dissolved at high acidities, while crystalline TiO_2 is much more resistant to acidic dissolution.[22] At pH 2, the hydrolysis of Ti(IV) ions to form eventually TiO_2 is complete.

As can be inferred from Fig. 3, adsorption of O_2 on the TiO_2 particles severely quenches the emission. The short lifetimes of the emission and the long periods of time required to deaerate the solutions indicate that adsorbed oxygen is the reactive quencher. While bubbling argon through the colloidal solution increases the emission intensity, we find that bubbling N_2O is more efficient in that respect. Even with N_2O, long periods of bubbling are required to obtain substantial elimination of oxygen. As can be seen in Fig. 3,

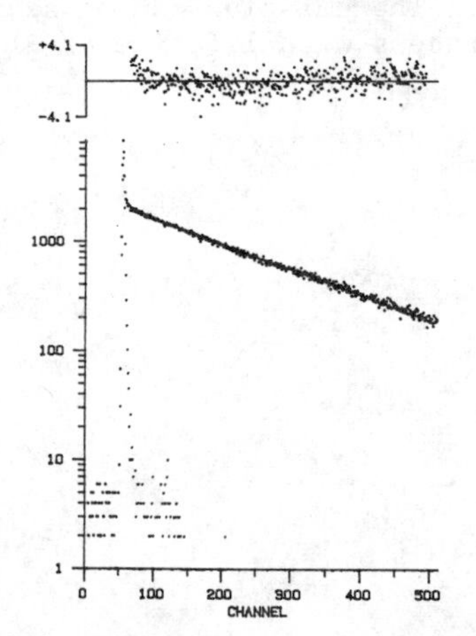

FIG. 4 Time profile of the emission decay of the sample in Fig. 3. λ_{exc} = 315 nm; all emission at $\lambda > 335$ nm is collected. Bottom trace is the laser profile. The fast decay which follows the laser profile is scattered light, primarily Raman scattering; time scale: 58.8 psec per channel.

two hours bubbling of N_2O increases the emission intensity by more than an order of magnitude. However, even after 24 hours of N_2O bubbling, the emission intensity still increases. We also note that merely equilibrating the solution with N_2 after saturation is much less efficient in oxygen removal than bubbling with N_2O. Apparently, constant removal of O_2 into the gas phase is necessary for efficient O_2 desorption. The higher efficiency of N_2O than of Ar in this process is attributed to the higher solubility of N_2O in water and to competitive adsorption/desorption of N_2O/O_2. Adsorbed N_2O evidently is unable to compete with the electron-hole emissive recombination process. Readmission of O_2 to the N_2O saturated solutions slowly reduces the emission intensity. Strong chemisorption of O_2 on the surface of crystalline TiO_2 is well established[23] and the redox level of O_2/O_2^- couple on TiO_2 is located below the conduction band of TiO_2. This strong chemisorption of O_2 is bound to limit the efficiency of this colloid in photochemical water splitting systems through scavenging of the electrons. Warming the TiO_2 colloidal solutions to 50°C for short periods of time while bubbling Ar also increases the emission intensity, presumably by facilitating the desorption of O_2. However, warming for extended periods of time also facilitates agglomeration of the TiO_2 particles. The colloidal solution then turns opaque, and the emission intensity drastically diminishes. The apparent emission spectrum is then shifted slightly to the blue and resembles the one previously observed for TiO_2 suspensions.[5g]

Quenching of the emission in these colloidal systems will depend on a variety of parameters. These will include efficiency of adsorption on the particle surface and the redox potential of the quencher. For the adsorbate to be able thermodynamically to react with the photogenerated electrons or holes, its redox potential has to be between the valence band edge (3.08 V vs. NHE at pH 2 for TiO_2) and the conduction band edge (−0.12 V vs. NHE at pH 2). Since high electrolyte concentrations induce coagulation of the Nafion stabilized colloids, the following experiments were conducted on PVA stabilized TiO_2. Thus, Cd^{2+} ions (up to 0.1 M) have no effect on the emission intensity of the TiO_2 colloids. Adsorption of these ions into the TiO_2 surface is probably inefficient, their reduction potential to Cd^+ is much more negative than the energy level of the conduction band while their oxidation potential to Cd^{3+} is in the appropriate range. On the other hand, the easily reducible Cu^{2+} and methylviologen (MV^{2+}) ions as well as $S_2O_8^{2-}$ ions strongly quench the emission intensity. In none of these quenchers does the quenching efficiency follow a Stern-Volmer dependence on the quencher concentration. Since the observed quenching efficiency is a complex function of adsorption, redox potentials and other surface modifications, simple Stern-Volmer dependence is not expected. In the TiO_2 colloids, further complications may arise due to release of residual absorbed O_2 by the quencher as well as agglomeration of the particles induced by the electrolyte quencher. The halides Cl^-, Br^-, I^- also quench the emission from the colloid. In the concentration range of 10^{-2} - 0.1 M halide (0.26 gr/l TiO_2, pH 2), the efficiency of quenching increases in the expected order of $I^- > Br^- > Cl^-$. However, at lower than 10^{-2} M of halide, the efficiency of quenching by iodide is reduced below that of Br^- and Cl^-. In addition to the above-mentioned complications, it should be noted that the halides strongly complex Ti(IV) ions.

We also noticed that the commonly utilized colloid stabilizer, polyvinylalcohol (PVA) quenches the emission intensity of TiO_2 colloids. The emission intensity of the colloid (0.26 gr/l Ti, pH 2) is reduced ca. 50% upon addition of 0.1% PVA as compared to the Nafion stabilized colloid. This is, in fact, to be expected, since the alcohol could efficiently scavenge the photogenerated holes. The alcohol radical thus produced is, however, a strongly reducing radical. Indeed, flash-photolysis experiments on the PVA stabilized colloids containing MV^{2+} ions show a biphasic formation of MV^+ radicals. Similar observations have been previously reported, and the slower formation of MV^+ has been also attributed to the reaction of MV^{2+} by the alcoholic radicals.[5b,9] Only fast formation (limited by the laser pulse width) could be observed in the Nafion stabilized colloid.

The inertness of the Nafion stabilizer as compared to that of the commonly utilized polyvinylalcohol stabilizer is further exemplified in steady state photolysis of TiO_2 colloids. When 0.5 gr/l of TiO_2 colloid at pH 8.5 stabilized by either 0.1% Nafion or 0.1% polyvinylalcohol (all other parameters identical) is continuously irradiated in the presence of $5x10^{-3}$ M ZV, the yield of the stable $ZV^{\overline{\cdot}}$ radical ion is 3.6 times higher for the PVA stabilized colloid than for the Nafion stabilized one. Scavenging of holes by the alcoholic polymer is responsible for this increased yield.[5b,f]

<u>MoS2</u>

Few experiments have so far been performed on the Nafion stabilized MoS_2 colloids. The absorption spectrum of the colloid shows a broad monotomically increasing absorption covering the entire UV-visible and some of the near IR ranges with an ill-defined onset at app. 700 nm. The band gap of the layered-type structure of MoS_2 semiconductor has been determined to be 1.84 eV (680 nm). The gap for indirect excitation is 1.3 eV (1042 nm).[25] The concentrated $2x10^{-2}$ M MoS_2 colloidal solution is stable for at least one month when stored refrigerated. Since the conduction band edge of MoS_2 is situated at 0.0V vs. NHE, no reduction of viologen is expected upon excitation and none was observed in the flash photolysis experiment.

On the other hand, the highly positive (+1.84 V vs. NHE) position of the valance band and the broad absorption spectrum of this material make it a rather attractive candidate for photo-oxidation of either water or halides. None of these reactions have been observed in our colloidal solutions as was the case in photoelectrolytic cells.

CONCLUSIONS

Dissolved Nafion has been shown in this study to be an efficient colloid stabilizer. The perfluorocarbon backbone adsorbs to the colloid surface, providing primarily electrostatic stabilization through the sulfonate head group. As was expected, this stabilizer is extremely inert and thus free of the undesirable participation in either photochemical or catalytic steps that the colloid might participate in. Additionally, the polyelectrolytic nature of the Nafion polymer provides an electrostatic field that may be utilized in inhibiting back reaction of the photoredox products. It is, however, inferior to uncharged polymeric stabilizers in its sensitivity to high concentrations of electrolytes. The strong binding of cations to this polymer initiates coiling of the polymer which leads to desorption of the colloid surface and eventually to coagulation.

REFERENCES

1. Work at Argonne is performed under the auspices of the Office of Basic Energy Sciences, Division of Chemical Science, US-DOE under contract number W-31-109-ENG-38.
2. a. Kraeutler, B. and Bard, A. J., 'J. Am. Chem. Soc.' **99**, 7729 (1977)
   b. Bard, A. J., 'J. Photochem.' **10**, 59 (1979)
   c. Frank, S. N. and Bard, A. J., 'J. Phys. Chem.' **81**, 1484 (1977)
   d. Reiche, H., Dunn, W. W. and Bard, A. J., 'J. Phys. Chem.' **83**, 2248 (1979)
   e. Izumi, I., Fan, F. and Bard, A. J., 'J. Phys. Chem.' **85**, 218 (1981)
   f. Ward, M. and Bard, A. J., 'J. Phys. Chem.' **86**, 3599 (1982)
   g. Ward, M. White, J. and Bard, A. J., 'J. Am. Chem. Soc.' **105**, 27 (1983).
3. a. Sreva, E. F., Olin, G. R., and Hair, J. R., 'J. Chem. Soc., Chem. Comm.' 401 (1980)
   b. Rao, M. V., Rajeshwar, K., Pal Verneker, V. R. and Dubow, J., 'J. Phys. Chem.' **84**, 1987 (1980)
   c. Reichman, B. and Byvik, C. E., 'J. Phys. Chem.' **85**, 2255 (1981)
   d. Schrauzer, G. N. and Guth, T. D., 'J. Am. Chem. Soc.' **99**, 7189 (1977).

4. a. Borgarello, E., Kiwi, J., Pelizzetti, E., Visca, M. and Gratzel, M., 'J. Am. Chem. Soc.' **103**, 6324 (1981)
b. Duonghong, D., Borgarello, E. and Gratzel, M., 'J. Am. Chem. Soc.' **103**, 4685 (1981)
c. Moser, J. and Gratzel, M., 'Helv. Chim. Acta' **65**, 1436 (1982)
d. Borgarello, E., Kiwi, J. Gratzel, M., Pelizzetti, E. and Visca, M., 'J. Am. Chem. Soc.' **104**, 2996 (1982).
5. a. Henglein, A., 'Ber. Bunsenges. Phys. Chem.' **86**, 241 (1982); ibid, 'J. Phys. Chem.' **86**, 2291 (1982); ibid, 'Ber. Bunsenges. Phys. Chem.' **86**, 301 (1982)
b. Bahnemann, D., Henglein, A., Lilie, J. and Spanhel, L., 'J. Phys. Chem.' **88**, 709 (1984).
c. Rossetti, R., Beck, S. M. and Brus, L. E., 'J. Am. Chem. Soc.' **104**, 7322 (1982); ibid, **106**, 980 (1984)
d. Rossetti, R., and Brus, L., 'J. Phys. Chem.' **86**, 4470 (1982)
e. Rossetti, R., Nakahara, S. and Brus, L., 'J. Chem. Phys.' **79**, 1086 (1983)
f. Brus, L. E., 'J. Chem. Phys.' **79**, 5566 (1983).
g. Chandrasekaran, K. and Thomas, J.K., 'J. Chem. Soc., Faraday Trans. I' **80**, 1163 (1984).
6. Polymeric Stabilization of Colloidal Dispersions, Napper, D. H., Academic Press, Inc., 1983.
7. Kuczynski, J. and Thomas, J. K., 'J. Phys. Chem.' **87**, 5498 (1983).
8. Kopple, K., Meyerstein, D. and Meisel, D., 'J. Phys. Chem.' **84**, 870 (1980).
9. Eisenberg, A. and Yeager, H.L., Eds. "Perfluorinated Ionomer Membranes" ACS Symposium Series No. 180, 1982.
10. Lee, P. C. and Meisel, D., 'J. Am. Chem. Soc.' **102**, 5477 (1980).
11. Martin, C. R., Rhoades, T. A. and Ferguson, J. A., 'Anal. Chem.' **54**, 1641 (1982).
12. a. Rubinstein, I. and Bard, A. J., 'J. Am. Chem. Soc.' **102**, 6642 (1980); ibid, **103**, 5007 (1981).
b. Martin, C., Rubinstein, I. and Bard, A. J., 'J. Am. Chem. Soc.' **104**, 4817 (1982).
c. White, H. S., Leddy, J. and Bard, A. J., 'J. Am. Chem. Soc.' **104**, 4811 (1982).
13. a. Buttry, D. A. and Anson, F. C., 'J. Am. Chem. Soc.' **104**, 4824 (1982).
b. Buttry, D. A., Saveant, J. M. and Anson, F. C., 'J. Am. Chem. Soc.' **88**, 3086 (1984).
c. Gerhardt, G. A., Oke, A. F., Nagy, G., Moghaddam, B. and Adams, R. N., 'Brain Research' **290**, 390 (1984).
14. Lee, P. C. and Meisel, D., 'J. Catal' **70**, 160 (1981).
15. a. Kuczynski, J. P., Milosavljevic, B. H. and Thomas, J. K., 'J. Phys. Chem.' **88**, 980 (1984).

b. Krishnan, M., White, J. R., Fox, M. A. and Bard, A. J., 'J. Am. Chem. Soc.' **105**, 7002 (1983).

16. a. Duonghong, D., Ramsden, J. and Gratzel, M. 'J. Am. Chem. Soc.' **104**, 2977 (1982).
    b. Ramsden, J. and Gratzel, M. 'J. Chem. Soc. Faraday Trans. I' **80**, 919 (1984).
17. Robbins, R. J., Fleming, G. R., Beddard, G. S., Robinson, G. W. Thistlthwaite, P. J. and Woolfe, G. J., 'J. Am. Chem. Soc.' **102**, 6271 (1980).
18. Matijevic, E., Budnik, M. and Meites, L. J. Coll. 'Interf. Sci.' **61**, 302 (1977).
19. Harada, H., Hidaka, H. and Ueda, T., Proc. Xth IUPAC Symp. on Photochem., p. 316 (1984).
20. Adams, M. J., Beadle, B. C., King, A. A. and Kirkbright, C. F., 'Analyst' (London) **101**, 5531 (1976).
21. Nabivantes, B. I. and Kudvitskaya, L. N., 'Russ. J. Inorg. Chem.' (Eng. Trans.) **12**, 616 (1967).
22. a. Lobanov, F. I., Savostina, V. M., Serzhenko, L. V. and Pashkova, V. M., 'Russ. J. Inorg. Chem.' (Eng. Trans.) **14**, 562 (1969).
    b. Wiese, G. R. and Healy, T. W., 'J. Coll. Interf. Sci.' **52**, 452 (1975).
23. Munuera, G., Rives-Arnau, V. and Saucedo, A., 'J. C. S. Faraday I' **75**, 736 (1979).
24. Vuyesteke, A. A. and Sihvonen, Y. T., 'Phys. Rev.' **113**, 40 (1959).
25. a. Kautek, W., Gobrecht, J. and Gerischer, H., 'Ber. Bunsenges. Phys. Chem.' **84**, 1034 (1980).
    b. Tributsch, H., 'Faraday Discus. Chem. Soc.' **70**, 189 (1980).

IN SITU GENERATED CATALYST-COATED COLLOIDAL SEMICONDUCTOR CdS PARTICLES IN SURFACTANT VESICLES AND POLYMERIZED SURFACTANT VESICLES

Yves-M. Tricot[a] and J. H. Fendler[b]
Department of Chemistry and Institute of Colloid and Surface Science, Clarkson University, Potsdam, New York 13676, USA

ABSTRACT. A variety of surfactant vesicles dispersions (anionic, cationic, polymerized or not) were used to in situ generate colloidal semiconductor CdS particles. Catalyst-coating of the CdS colloids produced active systems for visible-light induced electron-transfers and hydrogen generation. The photochemical efficiency was found to depend on both the type of vesicle and sacrificial electron donor. 10% formal quantum yield at 400 nm was achieved with cationic-vesicle-stabilized, Rh-coated CdS in the presence of 1% benzylalcohol. Pronounced quantum size effects were observed by varying concentration and degree of reaction of the Cd^{2+} precursors.

1. INTRODUCTION

The potential of catalyst-coated colloidal semiconductor particles for solar energy conversion devices and more generally photocatalysis or artificial photosynthesis has been recognized by a rapidly growing number of laboratories worldwide.[2-21] Many different strategies have been considered for the generation and the stabilization of these colloids. They include the use of aqueous polymers or polyelectrolytes,[3-5,12,14] inorganic oxide particles,[6-9] macroscopic synthetic membranes,[15-17] water-in-oil microemulsions,[18] non-aqueous solvents[9,12,13] or aqueous surfactant vesicles.[19-21] Because of its optimum characteristics among many available semiconductor materials, CdS has been the most popular and the best described.[3,4,6-13,15-21]

Band gap excitation of CdS by visible light (λ < 520 nm for macroscopic CdS), resulted in promotion of electrons from the valence to the conduction band, and hence in charge separation. CdS conduction band electrons, under favorable conditions, could then reduce water at the catalyst/semiconductor interface.[22] Ideally, dispersed semiconductor particles should be small, uniform and stable. The smaller and the more uniform the semiconductor particles, the greater chance the charge carriers have for escaping to the surface - the site of electron transfer. There is a minimum size, however, before light

E. Pelizzetti and N. Serpone (eds.), Homogeneous and Heterogeneous Photocatalysis, 241–252.

energy absorption occurs at the bulk band gap.[9-12,23] When the size of the semiconductor decreases, the band gap increases, resulting in a blue shift of the absorption edge. This reduces the efficiency of light harvesting, and ultimately leads to the disappearance of semiconductor properties. The smallest diameter for colloidal CdS particles with band gap close to the bulk value is about 50 Å. Such small dispersed semiconductors are difficult to maintain in solution for extended times in the absence of stabilizers. We summarize in this report results obtained by generating and stabilizing catalyst-coated colloidal CdS particles in various types of surfactant vesicles and polymerized surfactant vesicles. Application of these systems to hydrogen generation under visible light irradiation is described, as well as electron-transfer processes and quantum size effects.

2. EXPERIMENTAL SECTION

Preparation of CdS colloids was generally achieved by exposure to H_2S of vesicle-adsorbed Cd^{2+} ions, for negatively charged vesicles, or $Cd/EDTA^{2-}$ complexes for positively charged vesicles. Details of chemicals used and preparation techniques for vesicle and CdS formation have been described for CdS in dihexadecylphosphate (DHP) vesicles,[19] in dioctadecyldimethylammonium chloride (DODAC) vesicles,[20] and in vesicles made of $[n\text{-}C_{15}H_{31}CO_2(CH_2)_2]_2N^+[CH_3][CH_2C_6H_4CH=CH_2], Cl^-$ (1).[21] Characterization by optical absorbance, fluorescence, atomic absorption, electron-microscopy, light-scattering, laser flash photolysis and adsorption techniques have also been described,[19-21] as well as the irradiation and hydrogen detection techniques.[19-21]

3. RESULTS AND DISCUSSION

3.1 H_2 generation and electron-transfer with CdS in DHP vesicles

The activity of vesicle-stabilized, catalyst-coated colloidal CdS was first demonstrated in DHP vesicles, using Rh as catalyst and thiophenol (PhSH) as sacrificial electron donor.[19] Figure 1 illustrates the mechanism of this photosensitized H_2 generation. The proposed position of the CdS particle (partially buried in the vesicle bilayer) was supported by the following observations:

(a) CdS particles generated from externally adsorbed Cd^{2+} ions did not precipitate, even after months; therefore, they had to remain bound to the vesicle interface. (b) CdS fluorescence was efficiently quenched by PhSH, which was located in the hydrophobic membrane; therefore, the colloidal CdS particles had a direct contact with the inner part of the membrane. (c) The CdS particle retained access to the surface where it originated, since entrapped polar electron acceptors such as methylviologen (MV^{2+}), while unable to penetrate the DHP membrane, could also quench the fluorescence of inner-surface-generated CdS particles. However, this quenching decreased with time, showing a gradual penetration of the CdS toward

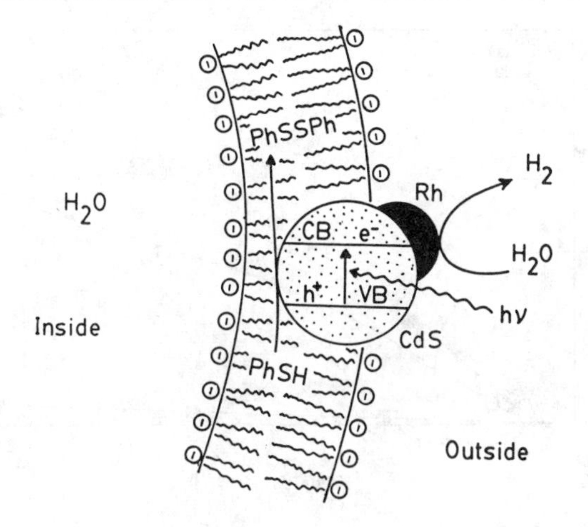

Figure 1. An idealized model for CdS-sensitized water photoreduction by PhSH in aqueous DHP vesicles. The position of the colloid in the vesicle (represented here as generated on the outside surface) is based on fluorescence quenching experiments.

the middle of the bilayer. (d) CdS particles at the vesicle interiors remained at the inner surface of the membrane, since externally added quencher such as MV^{2+} and Rh^{3+}, while adsorbed at the outer surface of DHP vesicles, did not quench inner-surface CdS fluorescence, even after several weeks.

Although CdS could be located selectively at the inner or outer surface of the vesicles, symmetrically organized samples were found easiest to prepare and most reproducible. No significant effect of CdS location upon the photochemical activity for H_2 generation was observed. Therefore, symmetrically organized samples were used to investigate the effects of temperature, catalyst and electron-donor concentrations and mode of Rh^{3+} reduction on H_2 generation. 25°C was found optimal and Figure 2 shows the effect of catalyst concentration together with several blank experiments. The maximum H_2 generation rate in this system was approximately 5 μmol/h, and the total H_2 amount that could be produced was limited by the tolerable concentration of PhSH by the DHP vesicles, which was 1:1 PhSH:DHP molar ratio. Characterization of this system by dynamic light-scattering determined 800-1000 Å diameter vesicles, little affected by the presence of Rh-coated CdS particles. The size of the CdS particles could be estimated by electron microscopy to be about 40 Å in diameter. Further characterization was obtained by probing this system's

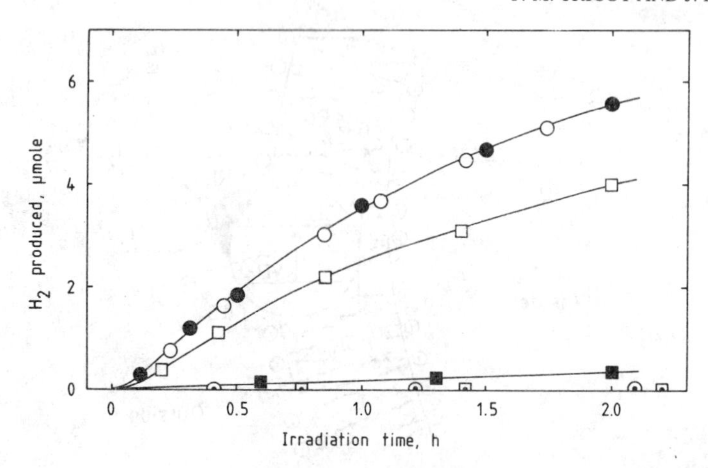

Figure 2. Hydrogen production at 25°C in deaerated solutions as a function of catalyst (rhodium) concentration (UV reduced) during the first 2 h or irradiation using 350-nm cutoff and water filters, in the presence of 10^{-3} M of PhSH as electron donor with $2 \cdot 10^{-4}$ M CdS and $2 \cdot 10^{-3}$ M DHP. (●) Rh 1.2 x 10^{-4} M, (○) Rh 0.8 x 10^{-4} M, (□) Rh 0.4 x 10^{-4} M, (■) no Rh, (⊡) no Rh and no CdS (only PhSH), (⊙) Rh^{3+} (unreduced) and no CdS (with PhSH).

photochemical activity with laser flash photolysis. Electron-transfer from the conduction band of laser-flash-irradiated CdS to methylviologen (MV^{2+}) produced the reduced species $MV^{+\cdot}$, which could be detected by its absorption in the visible, at 400 or 605 nm,[24] as depicted in Figure 3.

Varying of $MV^{+\cdot}$ concentration resulted in proportional amounts of MV^{+} until a plateau was reached (Figure 4). This suggested a saturation by MV^{2+} of either the CdS or the vesicle surface, implying in the latter case fast diffusion of MV^{2+} between DHP headgroups and the CdS surface. However, no significant change was found whether CdS and MV^{2+} were located at both surfaces or only at the outer surface of the vesicles. This different organization changed the density of MV^{2+} on the DHP-water interface, and would, therefore, result in a different saturation concentration of MV^{2+}. It should be pointed out that with $3 \cdot 10^{-3}$ M DHP concentration, up to 10^{-3} M of MV^{2+} or Cd^{2+} could be adsorbed when distributed at both surfaces (up to $5.5 \cdot 10^{-4}$ M when only at the outer surface). Because of the unknown effect of CdS on MV^{2+} adsorption, and the possible difference in electron-transfer quantum yield, it is difficult to compare directly the results from outer-surface and symmetrically distributed CdS. In the latter case, the electron-transfer process was also probed by competition with H_2 generation, in the presence of Rh catalyst, previously reduced by visible light at the CdS surface in the presence of an electron donor.

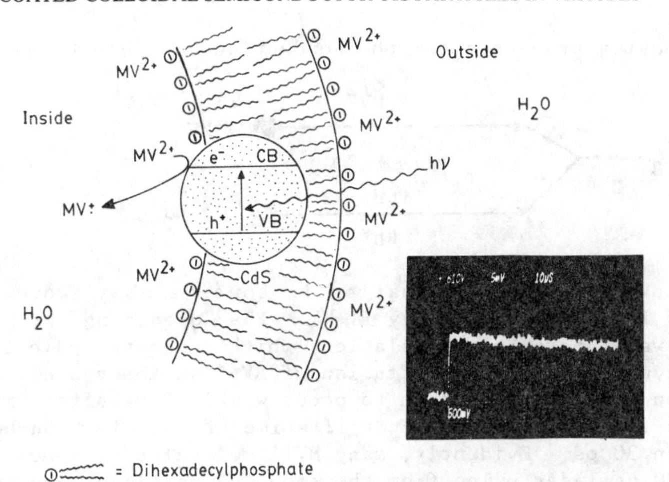

Figure 3. An idealized mechanism of photoinduced electron transfer from CdS conduction band to methylviologen (MV^{2+}), resulting in the formation of methylviologen radical cation ($MV^{+\cdot}$). The colloidal CdS, as represented, was generated at the inner surface of the vesicles. Its exact location was based on fluorescence quenching experiments. Insert: oscilloscope trace showing the formation of $MV^{+\cdot}$, by the absorbance change at 396 nm, after a 355 nm, 20 ns laser pulse.

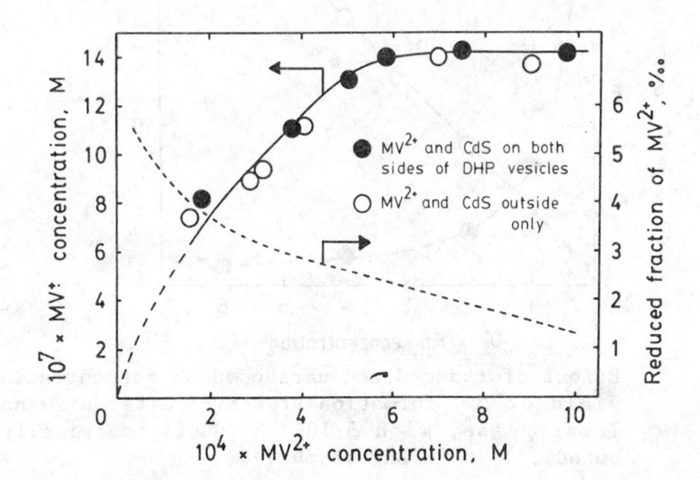

Figure 4. Effect of MV^{2+} concentration on the formation of $MV^{+\cdot}$ radical cation after a 20 ns, 355 nm laser pulse, with $4 \cdot 10^{-4}$ M CdS (at both sides) or $2.5 \cdot 10^{-4}$ M CdS (outside only) and $3 \cdot 10^{-3}$ M DHP at pH 7-8.

The two competing processes for the conduction band electrons are:

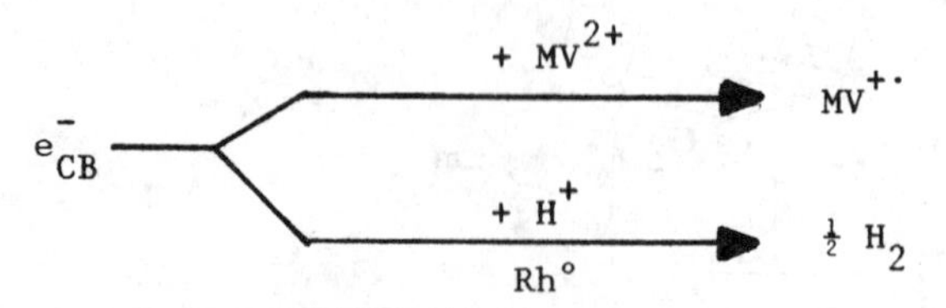

Figure 5 shows that $MV^{+\cdot}$ formation is considerably reduced by the presence of Rh^o, albeit not by Rh^{3+}. The "quenching" of $MV^{+\cdot}$ formation follows a Stern-Volmer relation, which suggests again that this process is controlled by the diffusion of MV^{2+} at the vesicle surface. The formation of $MV^{+\cdot}$ was shown to occur within 1 ns after excitation by an 30 ps laser pulse,[4] and the lifetime of the electron-hole pair was less than 30 ps. Evidently, many $MV^{+\cdot}$ molecules were being formed during our 20 ns laser pulse from the same CdS particle, and surface-diffusion of MV^{2+} was maintaining its presence at the semiconductor interface.

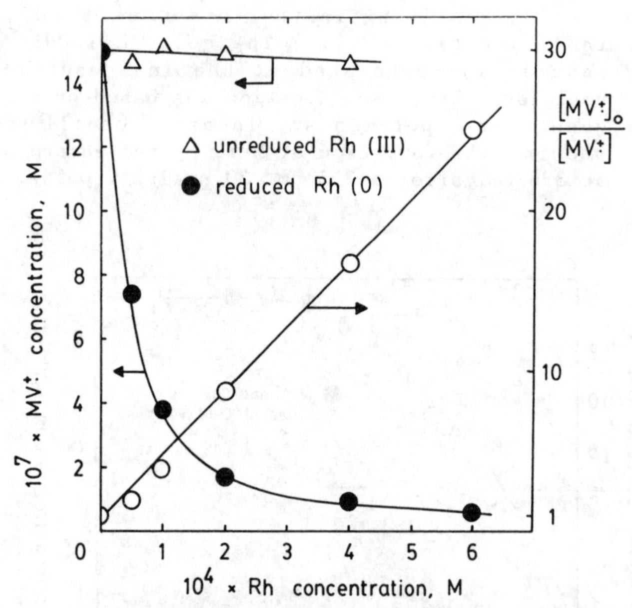

Figure 5. Effect of reduced and unreduced Rh concentration on the yield of $MV^{+\cdot}$ formation from MV^{2+} after a 20 ns, 355 nm laser pulse, with $5 \cdot 10^{-4}$ M CdS (symmetrically distributed), $3 \cdot 10^{-3}$M DHP at pH 7-8.

The slope of the Stern-Volmer plot in Figure 5 is the ratio of the rate constants for H_2 generation and $MV^{+\cdot}$ formation, divided by the concentration of CdS-absorbed MV^{2+}. Unfortunately, none of these parameters could be derived directly from the present data.

3.2 Quantum size effects and H_2 generation with CdS in anionic and cationic vesicles

The relatively low H_2 generation rate obtained from Rh-coated CdS in DHP vesicles (5 µmol/h) and the limitation due to the destabilizing effect of the electron donor (PhSH) prompted us to develop better systems. Apart from anionic DHP vesicles, colloidal CdS particles were in situ generated in cationic DODAC vesicles, as well as in vesicles prepared from unpolymerized and polymerized 1 (see experimental section). To ensure effective adsorption of Cd^{2+} and Rh^{3+} at both surfaces of cationic vesicles, EDTA was added at concentrations equal to the sum of Cd^{2+} and Rh^{3+} concentrations. Adsorption of the complex Cd/EDTA on DODAC vesicles was shown to reach nearly the same adsorption density than Cd^{2+} on DHP vesicles.[20] Polymerization of 1 was achieved by uv irradiation for 60 minutes and was followed by the decrease of the styrene absorbance around 250 nm. After 60 minutes, complete disappearance of the styrene moiety was observed, which was taken to correspond to complete polymerization. This process has been studied in detail elsewhere and was shown to produce polymers with chain length of approximately 20 monomers and led to cleft formation.[25] The optical properties of the colloidal CdS particles formed in different vesicle systems vary significantly, as revealed in Figure 6. For instance, colloidal CdS formed in unpolymerized 1 had an absorption edge at 463 nm compared to 498 nm when formed in polymerized 1. These absorption edges, or wavelengths corresponding to

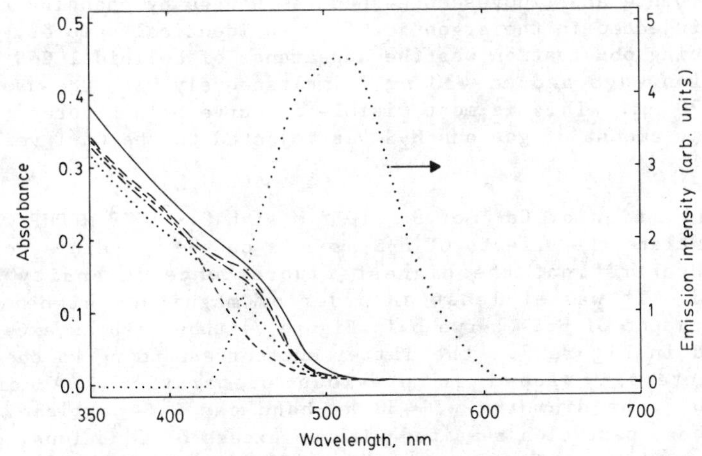

Figure 6. Absorbance spectra of colloidal CdS in various vesicular media or protected by hexametaphosphate (HMP), and fluorescence emission of CdS in dihexadecylphosphate (DHP) vesicles. (——) in polymerized 1 vesicles; (---) in unpolymerized 1 vesicles; (—) in HMP aqueous solution; (-·-·) in dioctadecyldimethylammonium chloride (DODAC) vesicles; (···) in DHP vesicles. $2\cdot10^{-4}$ M CdS, $2\cdot10^{-3}$ M surfactant or $1.3\cdot10^{-3}$ M HMP.

the band gap energy[26] were obtained by intersecting the slope of the absorbance curve at its inflexion point with the baseline in the spectra of Figure 6. The spectra were measured using the vesicle dispersions before CdS formation and, therefore did not contain the scattering contribution of the vesicles. In DHP or DODAC vesicles, as well as in hexametaphosphate-stabilized aqueous dispersions, colloidal CdS had intermediate band gaps and shapes of absorbance spectra. The differences presumably arose from different CdS sizes or size distributions in the various media. Clearly, it appeared that CdS formed in unpolymerized 1 had a diameter equal or smaller than 50 Å.[11] Particles with absorption edges close to 500 nm, as in vesicles prepared from polymerized 1 were probably around 100 Å in diameter. The slope of the absorbance curves has been taken as an indication of the degree of crystallinity of the particles.[8-12] The lowest one appeared then to correspond to CdS in DHP vesicles. However, only these particles showed a significant, albeit very weak, fluorescence. As the onset of fluorescence was at higher energy that the band gap of the CdS particles, it might be inferred that there was a distribution of particle sizes. The small ones having a higher band gap energy,[9,12] fluorescence at wavelengths shorter than the absorption edge of the big particles, which are responsible for most of the absorbance, was possible.

A clear confirmation of this hypothesis was obtained by generating CdS in DHP vesicles by slow injection of argon-diluted H_2S, in carefully controlled volumes.[23] Figure 7 shows the large variations in absorbance and fluorescence spectra induced by changing the amount of H_2S injected in the argon gas flow to identical samples. The most interesting observation was the appearance of colloidal CdS having an absorption edge around 430 nm, simultaneously with the usual one at about 490 nm. This is most visible in curve b in Figure 7, where an equimolar amount of gaseous H_2S was injected to the Cd^{2+} vesicle dispersion.

A concentration of Cd^{2+} of $8 \cdot 10^{-4}$ M with $2 \cdot 10^{-3}$ M DHP was chosen to illustrate the effects of H_2S amounts on the properties of CdS. At this concentration, the highest fluorescence intensity could be obtained. It was at least an order of magnitude higher with the optimum amount of H_2S (curve b in Figure 7) than with an excess of H_2S (curve d in Figure 7), the latter one corresponding to the fluorescence intensity found in previous preparations, According to Henglein,[9] the diameter of 430 nm-band gap CdS particles is about 25Å. These particles remained with an excess of Cd^{2+} ions, which was reported to enhance fluorescence.[6-9] 430 nm band gap, fluorescing CdS could be stabilized only at the inner surface of DHP vesicles. This selectivity was understood in terms of protection against aggregation with CdS from other vesicles, restricted diffusion at the inner water/DHP interface, and reduced effect of aging by dissolution-precipitation equilibrium.[23]

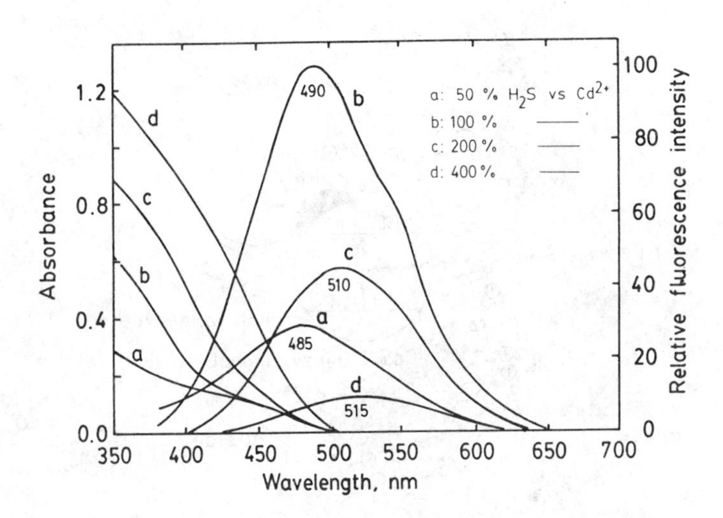

Figure 7. Absorbance and fluorescence emission spectra (under 350 nm excitation) from Cd^{2+}/CdS 8 · 10^{-3} M symmetrically distributed in 2 · 10^{-3} M DHP vesicles at pH 8, after slow injection of argon-diluted, controlled volumes of gaseous H_2S.

The different vesicle systems shown in Figure 6 were tried for visible-light-induced H_2 generation. Polymerization of 1 vesicles affected the capability of EDTA to act as an electron donor for Rh-coated CdS particles. The role of EDTA was first to reverse the charge of Cd^{2+} and Rh^{3+} ions in order to control the growth of the CdS particles at the surfaces of the vesicles. It was hoped that EDTA could subsequently be used as electron donor in the CdS-photosensitized water reduction. However, EDTA was found inefficient in DODAC and unpolymerized 1 vesicles.[21] On the other hand, EDTA was efficient in polymerized 1 vesicles. This effect was understood as an increased access to CdS of the aqueous phase and EDTA, resulting from the cleft formation upon polymerization of 1.[25] Unfortunately, the EDTA amounts were still severely limited by their destabilizing effect on the cationic vesicles. To overcome this problem, several alcohols were tried as electron donors, and among them, benzylalcohol was found by far the most efficient. It could be added up to saturation in water (ca. 4% vol.) without destroying the vesicles. Only 1% benzylalcohol gave a maximum rate, at 40%, of over 200 µmol H_2/h (4.5 ml H_2/h) from 25 ml samples containing only 0.82 mg of CdS/Rh catalyst (2 · 10^{-4} M CdS and 4 · 10^{-5} M Rh). Figure 8 illustrates the mechanism and proposed structure of the polymerized 1 vesicle system under visible light irradiation.

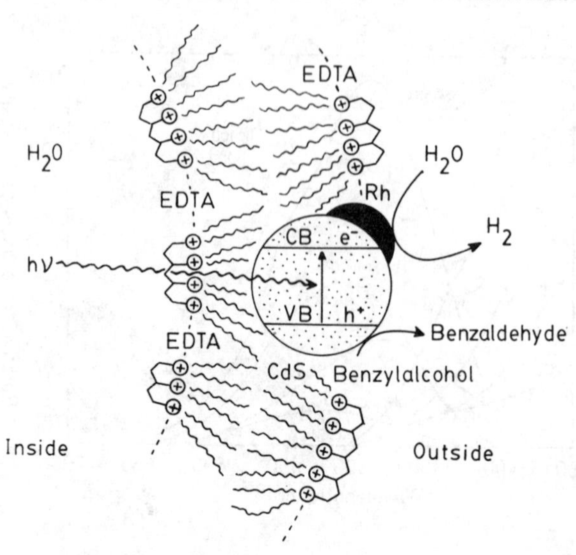

Figure 8. An ideal scheme for hydrogen generation with benzylalcohol as sacrificial electron donor, in polymerized 1-vesicle-stabilized, Rh-coated, colloidal CdS.

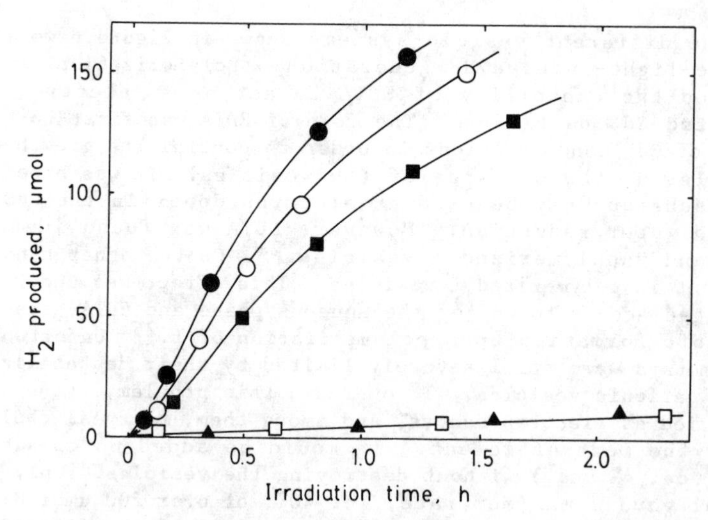

Figure 9. Effect of stabilizing medium on hydrogen production from 2 · 10^{-4} M CdS, 4 · 10^{-5} M Rh, at 40°C in the presence of 1% benzylalcohol. (●): polymerized 1 vesicles; (○): DODAC vesicles; (■): unpolymerized 1 vesicles; (□): DHP vesicles; (▲): HMP. $2 \cdot 10^{-3}$ M surfactant or $1.3 \cdot 10^{-3}$ M HMP.

Under optimal conditions, various vesicle-stabilized CdS systems were compared, together with aqueous HMP-stabilized CdS, as shown in Figure 9. It appeared immediately that all three cationic vesicle systems produced hydrogen highly efficiently, whereas both negatively charge media (DHP vesicles and HMP) were comparatively inefficient.

The nature of this charge effect has not yet been elucidated. It probably related to the surface charge of CdS and its interaction with the charges of the vesicles, and/or to the presence of charged intermediate species. Further work is in progress to develop new combinations of semiconductors and vesicles, improve existing systems and get deeperunderstanding of the mechanism of interfacial processes in colloidal semiconductor particles.

ACKNOWLEDGEMENT. The authors wish to thank the U.S. Department of Energy for financial support and acknowledge the fruitful participation of the collaborators whose names are mentioned in the references.

REFERENCES AND NOTES

1. a) Present address: Department of Material Research, The Weizmann Institute of Science, Rehovot 76100, Israel. b) Present address: Department of Chèmistry, Syracuse University, Syracuse, New York 13210.

2. Fox, M. A. Ed. "Organic Phototransformations in Nonhomogeneous Media;" Am. Chem. Soc. Symp. Ser. 278; Am. Chem. Soc: Washington, D.C. 1985,

3. Ramsden, J.J.; Grätzel, M. J. Chem. Soc. Far. Trans. I 1984, 80, 919-933.

4. Serpone, N.; Sharma, D.K.; Jamieson, M.A.; Grätzel, M.; Ramsden, J. J. Chem. Phys. Lett. 1985, 115, 473-476.

5. Moser, J.; Grätzel, M. J. Am. Chem. Soc. 1984, 106, 6557-6564.

6. Henglein, A. Ber. Buns. Phys. Chem. 1982, 86, 302-305.

7. Henglein, A.; Gutiérrez, M. Ber. Buns. Phys. Chem. 1983, 87, 852-857.

8. Weller, H.; Koch, U.; Gutiérrez, M.; Henglein, A. Ber. Buns. Phys. Chem. 1984, 88, 649-656.

9. Fojtik, A.; Weller, M.; Koch, U.; Henglein, A. Ber. Buns. Phys. Chem. 1984, 88, 969-977.

10. Rossetti, R.; Nakahara, S.; Brus, L.E. J. Chem. Phys. 1983, 79, 1086-1088.

11. Brus, L.E. J. Chem. Phys. 1984, 80, 4403-4409.

12. Rossetti, R.; Ellison, J.L.; Gibson, J.M.; Brus, L.E. J. Chem. Phys. 1984, 80, 4464-4469.

13. Ramsden, J.J.; Webber, S.E.; Grätzel; M. J. Phys. Chem. 1985, 89, 2740-2743.

14. Dimitrijevic, N.M.; Savic, D.; Micic, O.I.; Nozik, A.J. J. Phys. Chem. 1984, 88, 4278-4283.

15. Mau, A.W.-H.; Huang, C.-B.; Kakuta, N.; Bard, A.J.; Campion, A.; Fox, M.A.; White, J.M.; Webber, S.E. J. Am. Chem. Soc. 1984, 106, 6537-6542.

16. Meissner, D.; Memming, R.; Kastening, B. Chem. Phys. Lett. 1983, 96, 34-37.

17. Kuczynski, J.P.; Milosajevic, B.H.; Thomas, J.K. J. Phys. Chem. 1984, 88, 890 - 896.

18. Meyer, M.; Wallberg, C.; Kurihara, K.; Fendler, J.H. J. Chem. Soc. Chem. Comm. 1984, 90-91.

19. Tricot, Y.-M.; Fendler, J.H. J. Am. Chem. Soc. 1984, 106, 7359-7366.

20. Rafaeloff, R.; Tricot, Y.-M.; Nome, F.; Fendler, J.H. J. Phys. Chem. 1985, 89, 533-537.

21. Tricot, Y.-M.; Emeren, Å; Fendler, J.H. J. Phys. Chem. in press.

22. Grätzel, M. "Energy Resources through Photochemistry and Catalysis", Academic Press, New York, 1983.

23. Tricot, Y.-M.; Fendler, J.H. J. Phys. Chem., submitted.

24. Watanabe, T.; Honda, K. J. Phys. Chem. 1982, 86, 2617-2619.

25. Reed, W.; Guterman, L.; Fendler, J.H. J. Am. Chem. Soc., 1984, 106, 1987-1907.

26. More precisely, the absorption edge was proposed to be the wavelength corresponding to the energy of formation of the "lowest exciton", or the lowest energy electron/hole pair confined in the volume of the colloidal particle.[9,12] Due to space restriction, this exciton has less energy than a "free" electron/hole pair, which defines the band gap energy.

DEVELOPMENT OF CATALYSTS FOR WATER PHOTOREDUCTION : IMPROVEMENT, POISONING AND CATALYTIC MECHANISM.

Edmond Amouyal
Laboratoire de Physico-Chimie des Rayonnements
Université Paris-Sud
91405 Orsay Cedex
France

ABSTRACT. In model photosystems for hydrogen production from water, the catalyst plays a dominant role. In the present paper, various types of heterogeneous catalysts - colloidal metals, metal deposited on semiconductor powders, metal and metal oxide powders - are reviewed and compared. When added to the aqueous solutions, metallic ions, sulphur and nitrogen compounds serve as poisons of the catalytic hydrogenation of the electron relay, and thus improve the H_2 production efficiencies. Taking into account the two competitive reactions, a catalytic mechanism on the particle surface is proposed and discussed.

1. INTRODUCTION

The renewed interest for redox catalysts has been stimulated by studies on the direct conversion of solar energy into chemical energy. Among the different approaches, the photochemical splitting of water into molecular H_2 and O_2 has received much attention. Indeed, several sacrificial model systems capable of producing hydrogen (1-9) and oxygen (10, 11) separately from water have been proposed and described. These systems often include four components : a photosensitizer, an electron-transfer relay, a sacrificial electron donor and an electron-transfer catalyst. The first half systems of H_2 generation from visible light used acridine yellow (1), proflavine (5), $Ru(bipy)_3^{2+}$ (2 - 4) as photosensitizers, and PtO_2 (1, 4), K_2PtCl_6 (1, 2) or colloidal Pt (3) as catalysts. On the other hand, oxygen was photoproduced using a similar half system with RuO_2 as a catalyst (10, 11).

One of the most intensively studied systems (12) involves $Ru(bipy)_3^{2+}$ as a photosensitizer, methyl-viologen MV^{2+} as an electron relay, EDTA as an electron donor and colloidal Pt as a catalyst (3, 13). The main reactions leading from visible-light excitation of the sensitizer to H_2 production are the following :

$$Ru(bipy)_3^{2+} \xrightarrow{h\nu} Ru(bipy)_3^{2+*} \quad (1)$$

$$Ru(bipy)_3^{2+*} \longrightarrow Ru(bipy)_3^{2+} \quad (2)$$

E. Pelizzetti and N. Serpone (eds.), Homogeneous and Heterogeneous Photocatalysis, 253–265.

$$Ru(bipy)_3^{2+*} + MV^{2+} \longrightarrow Ru(bipy)_3^{3+} + MV^{+\cdot} \quad (3)$$

$$Ru(bipy)_3^{3+} + MV^{+\cdot} \longrightarrow Ru(bipy)_3^{2+} + MV^{2+} \quad (4)$$

$$Ru(bipy)_3^{3+} + EDTA \longrightarrow Ru(bipy)_3^{2+} + EDTA^{+} \quad (5)$$

$$MV^{+\cdot} + H^{+}(H_2O) \overset{Pt}{\rightleftharpoons} MV^{2+} + 1/2\ H_2 \quad (6)$$

Difficulties arise from undesirable reactions such as $MV^{+\cdot}$ dimerization (reaction 7)

$$2\ MV^{+\cdot} \rightleftharpoons (MV^{+\cdot})_2 \quad (7)$$

and the irreversible hydrogenation of methyl viologen (reaction 8)

$$MV^{+\cdot}\ (\text{or } MV^{2+}) \xrightarrow{H_2,\ Pt} MV^{+}H \xrightarrow{H_2,\ Pt} \text{reduction products} \quad (8)$$

Besides the EDTA consumption (reaction 5), the latter reaction catalyzed by colloidal Pt is the main factor which limits the longevity of the system (14 - 16). Of major importance is thus the inhibition of the catalytic hydrogenation. Such reaction may be prevented (i) by finding electron relays whose structure would be less sensitive to hydrogenation (13, 15, 17 - 19), (ii) by using more specific catalyst such as RuO_2 (20, 21) (iii) and/or by adding to the solution catalyst poisons such as sulphur compounds, glutathione in particular (22, 23).

The present paper intends to review the H_2 production efficiencies of new catalysts and to compare them with those obtained from Pt compounds (23), then to examine the effect on these efficiencies of glutathione and various types of catalyst poisons (24) i.e. compounds containing sulphur or nitrogen and metallic ions. The aim of the review is finally to throw some light on the reaction mechanism on the catalyst surface.

2. HYDROGEN PRODUCTION EFFICIENCIES OF HETEROGENEOUS CATALYSTS

The development of very efficient catalysts for H_2 generation is of utmost importance in order to achieve a complete photochemical water-splitting system. The $Ru(bipy)_3^{2+}/MV^{2+}$/EDTA/colloidal Pt model system has been recently characterized in a more precise way (23, 25) as regards to the optimization of hydrogen quantum yield Ø (1/2 H_2). An optimum yield Ø (1/2 H_2) = 0.171 ± 0.02 was found for the following parameter values : pH 5, $Ru(bipy)_3^{2+}$ 5.65×10^{-5} M, MV^{2+} 3×10^{-3}M, EDTA 0.1 M and colloidal Pt 1.92×10^{-5} M. This system can then be used to test and compare H_2 production catalysts of various types : colloidal metals, metal deposited on semiconductor powders, metal and metal oxide powders, as long as one optimizes the catalyst concentration for each of them.

2.1. Colloidal metals

The study of such catalysts has developed considerably since we have first shown (3) that colloidal Pt and Au can mediate the water photoreduction in a model system. These metal hydrosols can be prepared either chemically or radiolytically. In a recent study (23), we have used colloidal Pt which had been chemically prepared (26) and have compared the latter with different colloidal metals obtained in our laboratory from radiolysis means (27, 28).

Table I : Quantum yields for H_2 formation (23) from the irradiation (λ_{exc} = 453 nm) of pH 5 aqueous solutions of $Ru(bipy)_3^{2+}$ 5.65 x 10^{-5} M, MV^{2+} 3 x 10^{-3} M, EDTA 0.1 M and various colloidal metals prepared radiolytically (27, 28).

N°	Catalyst	Particle diameter (Å)	Metal concentration (M) [a]	Ø (1/2 H_2) [b]
1	Pt [c]	16 - 1000	1.92 x 10^{-5}	0.171
2	Pt [c][d]	"	"	0.105
3	Pt	15	2 x 10^{-5}	0.170
4	Pt [e]	—	2 x 10^{-5}	0.100
5	Ir	12	2 x 10^{-5}	0.173
6	Ir	40	"	0.173
7	Ir	< 8	"	0
8	Os	< 50	5 x 10^{-5}	0.160
9	Ru	"	6 x 10^{-5}	0.139
10	Rh	"	4 x 10^{-5}	0.080
11	Co	"	2 x 10^{-5}	0.066
12	Ni	"	4 x 10^{-5}	0.060
13	Pd	"	5 x 10^{-4}	0.056
14	Ag	"	2 x 10^{-5}	0.050
15	Au	"	10^{-4}	0.042
16	Cu	"	10^{-5} - 10^{-3}	0
17	Cd	"	"	0
18	Pb	"	"	0

(a) Since it is difficult to obtain a precise determination of the concentration of the reduced metal, the given optimum concentration may be considered as an indication. (b) Corrected for light-scattering effects by the colloidal particles. (c) Chemically prepared. (d) Aerated solution. (e) Without PVA.

The corresponding results are gathered in Table I. The catalysts of group VIII, in particular the platinides (Pt, Ir, Os) are highly efficient for H_2 production with the $Ru(bipy)_3^{2+}/MV^{2+}/EDTA/$ catalyst model system. It is well known indeed that these metals have the lowest overpotential for H_2 production in water electrolysis cells. As to the other metals, their catalytic efficiency cannot be readily correlated with the overpotential values.

The relative order of metal catalytic activity for H_2 evolution turns out to be as follows : Ir, Pt, Os > Ru > Rh > Co, Ni, Pd, Ag, Au > Cu, Cd, Pb. The highest Ø (1/2 H_2) is thus obtained for Ir hydrosols Ø (1/2 H_2) = 0.173 ± 0.020. Unlike Rafaeloff et al (29), we have noticed for Os and Ru a catalytic efficiency similar to that of Pt, but a poor activity for Pd. Such discrepancy could be a result of the sample preparation conditions.

As in the case of chemically prepared Pt, Ø (1/2 H_2) depends on the catalyst concentration, the optimum yields for Pt and Ir radiolytically prepared being obtained for the same metal concentration of 2×10^{-5} M (Table I).

An optimum concentration, different from 2×10^{-5} M, may mean that the metal salt reduction upon γ-irradiation is not complete - the precise concentration of the reduced metal is indeed difficult to determine -, but it could also indicate a specific effect of the metal nature.

Table I shows that radiolytically prepared Pt is just as efficient as the chemically prepared Pt. Moreover, it presents the same efficiency even though the sizes are different (Compare N° 1 and 3). This result confirms (16, 30) the absence of any particle size effect over a wide range of size on the Pt catalyzed H_2 generation. Ir exhibits the same behaviour except for very small particles (diameter < 8 Å, experiment N° 7) in which case H_2 yield drops dramatically. This would suggest that there is a **minimum size** for the catalytic generation of hydrogen. As expected, colloidal Pt is less efficient without PVA and also when used in non deaerated solutions but the yield remains high. From a practical point of view it is important to note that non noble metals such as colloidal Co and Ni catalyze the reaction fairly efficiently but, in the same experimental conditions, the H_2 evolution stops within one hour for Ni. However, it seems that the preparation conditions of the colloid have to be better defined.

2.2. Metal and metal supported powders

Pt black powders are less efficient than colloidal Pt (Table II) but if one uses Pt supported on TiO_2, the catalytic activity is nearly the same for a similar particle size (Table II N° 28 and Table I N° 3) and for an optimum Pt percentage of 0.5 % corresponding to a Pt concentration of 6×10^{-5} M ; it does not appear better contrary to literature results (31) from which Pt/TiO_2 was found to be more efficient than Pt/PVA. It should be recalled however that in the latter case the reaction parameters had not been optimized which renders a comparison difficult.

Though the Pt percentage is different, Table II seems to indicate that the nature of the support plays a role, the semiconductor TiO_2 giving higher yields than Al_2O_3 which is in turn more efficient than

SiO_2. Pt/Fe_2O_3 decomposes in aqueous solutions and H_2 production stops very quickly. Even though in our experiments where λ_{exc} = 453 nm the catalyst does not absorb light, this support effect can be compared with Pichat et al (32) results in the case of the UV photocatalytic H_2 generation from liquid methanol using exactly the same batches of catalysts. This effect may be also due in our experiments to the distinct particle sizes and to the metal contents. The use of non noble metals much cheaper than precious ones, is economically worthwhile if one can improve the efficiency and the longevity of this kind of catalyst. Table II shows that Ni powder is astonishingly more efficient than colloidal Ni. Furthermore the H_2 volume obtained and the H_2 evolution duration are much higher. We would like to point out that Raney Ni though more finely divided than Ni powder is less efficient and behaves like colloidal Ni i.e. the H_2 evolution stops within one hour. If we use TiO_2 as a support - which for Pt leads to the best results - the efficiency referred to a unit mass is superior. Table II shows that this efficiency depends on the Ni percentage and is maximum for a 4.83 content, $\varnothing'(1/2\ H_2)$ = 0.108, and for a Ni particle size of about 150 Å for the three samples investigated.

Table II : $\varnothing'$ $(1/2\ H_2)^{(a)}$ for metal and metal-supported powders (same experimental conditions as in table I) (23).

N°	Catalyst	Metal loading (weight %)	Particle diameter (Å)	Catalyst amount added (mg)	$\varnothing'(1/2\ H_2)^{(a)}$
20	Co	-	-	14	0.060
21	Ni	-	-	16	0.110
22	Raney Ni	-	-	2	0.036
23	Pt black	-	-	14	0.058
24	Pt/Fe_2O_3	0.5	-	12	(b)
25	Pt/SiO_2	6.4	20	12	0.074
26	Pt/Al_2O_3	0.6	30-35	12	0.120
27	Pt/TiO_2	0.05	20-25	12	0.073
28	Pt/TiO_2	0.5	20	12	0.160
29	Pt/TiO_2	5	20-25	12	0.084
30	Ni/TiO_2	0.5	183	8	0.056
31	Ni/TiO_2	4.83	135	8	0.108
32	Ni/TiO_2	13.8	150	8	0.051

(a) $\varnothing'$ $(1/2\ H_2)$ is a value uncorrected for light scattering effects. It corresponds to a lower limit of $\varnothing$ $(1/2\ H_2)$. (b) Decomposes.

2.3 Metal oxide powders

Ruthenium oxides, known to be good catalysts of water oxidation in O_2 have been found (20,21) to efficiently mediate the H_2 formation from water without catalyzing the hydrogenation of the electron relay (Table III). RuO_2 and IrO_2 codeposited on zeolite lead to the highest yield (23) Ø' (1/2 H_2) = 0.102 ± 0.020. This yield (non-corrected for light scattering) is comparable to the one obtained with the colloidal Pt but the necessary catalyst concentration is higher by about two orders of magnitude.

PtO_2 which was the catalyst used in two of the first model systems of water photoreduction (1, 4) leads to much lower yields (approximately 55 % less than colloidal Pt). It is worth noting that TiO_2 has given a better yield than RuO_2, but it is less stable. It should be remarked from an economical point of view that several oxides, particularly TiO_2, Fe_2O_3, Sm_2O_3, CeO_2, MnO_2 and ZnO are of very low cost.

Table III : Ø' (1/2 H_2)[(a)] for metal oxide powders (same experimental conditions as in Table I)(23).

N°	Catalyst	Catalyst amount added (mg)	Ø'(1/2 H_2)[(a)]
40	RuO_2, x H_2O	3	0.084
41	RuO_2/TiO_2	5	0.087
42	RuO_2 + IrO_2/zeolite	12	0.102
43	PtO_2, x H_2O	22	0.072
44	TiO_2	12	0.086
45	Fe_2O_3	4	0.056
46	MnO_2	3	0.042
47	WO_3	6	0.039
48	Sm_2O_3	9	0.026
49	Nb_2O_5	6	0.026
50	ZnO	2	0.024
51	CeO_2	4	0.018

(a) Same as note (a) of Table II

3. IMPROVEMENT OF HYDROGEN PRODUCTION BY CATALYST POISONING

During the visible-light irradiation of the classical model system $Ru(bipy)_3^{2+}/MV^{2+}$/EDTA/colloidal Pt, MV^{2+} hydrogenation (reaction 8) catalyzed by colloidal Pt was shown to occur (14 - 16) and a completely reduced bipiperidine derivative has been isolated (16). Besides the EDTA consumption, this irreversible reaction represents the other main limitation of the system. As for H_2 production, MV^{2+} hydrogenation depends on the concentration of colloidal Pt and a competition between both reactions was put forward (14, 30). This suggests that H_2 production can be improved by elimination of the hydrogenation reaction. Recently, it was shown (22) that sulphur compounds (cysteine and glutathione) used as catalyst poisons, can efficiently retard the MV^{2+} degradation. Therefore, in the course of an attempt to optimize H_2 quantum yields $\emptyset$ (1/2 H_2) (23, 25), we have examined the effect of Pt concentration on $\emptyset$ (1/2 H_2) in the presence and in the absence of glutathione (Fig. 1).

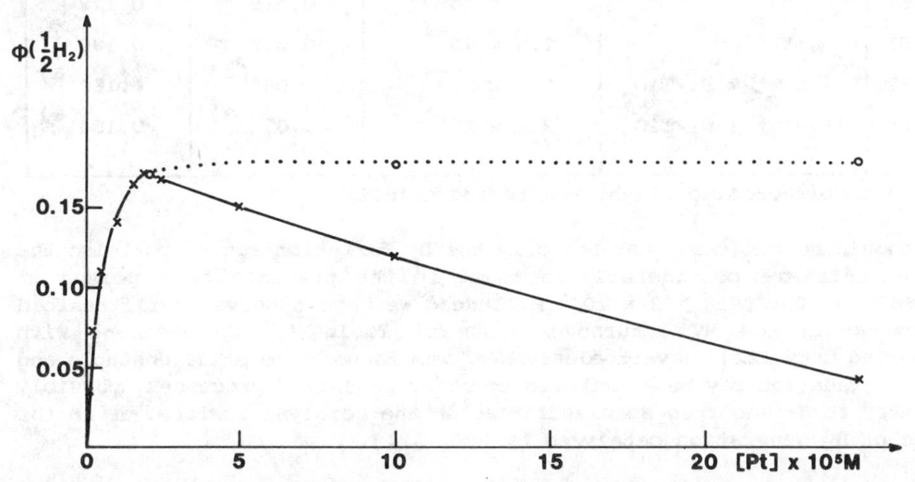

Figure 1 : H_2 formation quantum yields vs concentration of colloidal Pt (23, 24). a) Experimental conditions : Ru (bipy)$_3^{2+}$ 5.65 x 10^{-5} M, MV^{2+} 3 x 10^{-3} M, EDTA 0.1 M and pH 5(—x—). b) Same as (a) but in the presence of glutathione 2 x 10^{-3} M.(...o...).

This study is now extended (24) to other poisons, nitrogen compounds and metallic ions, which represent main types of poisons besides sulphur compounds (33, 34).

Fig. 1a illustrates that H_2 quantum yields increase with increasing Pt amounts up to a maximum value reached at [Pt] = 2 x 10^{-5}M and then decrease for higher concentrations. We have attributed (14, 30) such decrease to the MV^{2+} ($MV^{+\cdot}$) hydrogenation reaction 8 catalyzed by colloidal Pt. Table IV and Fig. 1b demonstrate that if hydrogenation catalyst poison, such as glutathione (22), is used $\emptyset$ (1/2 H_2) yields increase and the values obtained are identical to the optimum H_2 yield, $\emptyset$ (1/2 H_2) = 0.180, for all the Pt concentrations investigated. The resulting plateau thus reflects the existence of a $\emptyset$ ($MV^{+\cdot}$) optimum, $\emptyset$ ($MV^{+\cdot}$) = 0.181, determined under the same optimized conditions but in the absence of catalyst (23, 25). In other words, the Pt catalysis of the MV^{2+} hydrogenation is totally inhibited by glutathione and the expected maximum of $\emptyset$ (1/2 H_2) is observed in all the Pt concentration range (Table IV).

Table IV : Ø (1/2 H_2) in the presence and in the absence of glutathione 2 x 10^{-3} M acting as a poison (same experimental conditions as in Table I).

N°	Catalyst	Metal concentration (M)	Ø (1/2 H_2)	
			without poison	with poison
60	Pt	2 x 10^{-5}	0.171	0.171
61	"	10^{-4}	0.119	0.178
62	"	2.5 x 10^{-4}	0.042	0.180
63	5 wt % Pt/TiO_2	6.2 x 10^{-4}	0.084 [a]	0.160 [a]
64	13.8 wt % Ni/TiO_2	3.8 x 10^{-3}	0.051 [a]	0.104 [a]

(a) uncorrected for light scattering effects.

It should be remarked that not only the H_2 formation rates, but also the H_2 amounts are considerably improved in the presence of a poison in particular for [Pt] > 2 x 10^{-5} M. Indeed we have observed a fifteenfold increase in the MV^{2+} turnover numbers (Table V) in agreement with reported data (22). Nevertheless, MV^{2+} was found to be still unstable and its degradation may be attributed to other undesired processes, possibly related to H· adsorbed species formed on the catalyst surface, as in the case of H_2 generation catalyzed by RuO_2 (21).

Table V : Catalytic turnover numbers in the presence and in the absence of poison.

	without poison	with 2,2'-bipyridine
MV^{2+}	10	150
Ru $(bipy)_3^{2+}$	> 400	> 1300
Pt [a]	> 200	> 370

(a) Colloidal Pt concentration of 2.5 X 10^{-4} M

As for colloidal Pt, a metal content optimum is observed for Ni/TiO_2 and Pt/TiO_2 (Table II). Such an optimum has been observed also by Prahov et al (35) in the case of H_2 production from aliphatic alcohols by UV illumination of Ni/TiO_2 powders (same batch) as in the case of Pt/TiO_2 powder (optimum for 0.5 % Pt percentage) and was attributed to an electron transfer from the titania to the metal. In our case where the supported metals do not absorb light, it is more likely that this optimum is related to the metal catalyzed MV^{2+} hydrogenation (reaction 8). Indeed we observe that H_2 yields increase up to the optimum in the presence of glutathione for a Ni content of 13.8 % and for a Pt amount of 5 % (Table IV). Though the mechanism is different, we cannot exclude however the participation of TiO_2 not only as a support but as a co-catalyst. It should be remarked that Pt/TiO_2 is a less efficient hydrogenation catalyst than colloidal Pt (Table IV, compare experiments N° 62 and N° 63).

Various poisons have been tested. The results obtained so far (24) show that another sulphur compound, i.e. thiophene, behaves similarly to glutathione, with a similar poisoning efficiency. Nitrogen compounds, pyridine and 2,2'-bipyridine, lead to an efficiency of about 80 % whereas metallic ions, K^+ and Sn^{2+}, have very little influence on the MV^{2+} hydrogenation. It is well established (33, 34) that poison molecules inhibit catalysis as a result of a strong adsorption at the catalytic surface and blocking of the active sites. All the results suggest that (i) the poison, $MV^{2+}(MV^{+}\cdot)$ and H^+ are adsorbed on the metal particle, (ii) nitrogen compounds having a free electron pair in the valence shell behave as poisons (33) ; this is the case not only of pyridine and 2,2'-bipyridine, but also of EDTA which is always present in the solution, (iii) two types of active sites are implied, one for H_2 production and the other one for MV^{2+} hydrogenation. Indeed, H_2 formation requires two metal atoms per site whereas hydrogenation involves about five atoms per site (34). Thus the inhibition of MV^{2+} $(MV^{+}\cdot)$ hydrogenation depends mainly on a competition between methyl-viologen and poison molecules adsorption for the same sites. Finally, a similar poisoning efficiency is observed for sulphur compounds having an electron acceptor character, and for nitrogen compounds with an electron donor character. This seems to indicate that the poison does not modify the electronic properties of the metallic atoms located near the hydrogenation sites.

4. CATALYTIC MECHANISM ON METALLIC PARTICLES

As a result of the competition between H_2 formation and MV^{2+} hydrogenation, the H_2 catalytic production (equilibrium 6) masks a more complex mechanism (27, 36-40). Two $MV^{+}\cdot$ decay processes have been observed by several authors (27, 38 - 40) in pulse radiolysis experiments of MV^{2+}/colloidal metal systems, namely a fast process due to the diffusion controlled electron transfer reaction of $MV^{+}\cdot$ to Pt and a slower process with a kinetic order varying from one to two, the latter process being attributed to a desorption step leading to H_2 evolution. Because of a possible ionizing radiation effect on metal colloids (41) and of the absence of buffer and EDTA (acting as an electron donor and a catalyst poison), it seems more likely (12) that the kinetic order is one in continuous photolysis experiments in the presence of EDTA. In other words only one catalyst particle would be involved. The poisoning effects (24) as well as the pH dependance of the catalyzed MV^{2+} $(MV^{+}\cdot)$ hydrogenation reaction (14, 21) show that this undesired reaction is tightly linked to the useful H_2 reaction and that MV^{2+} $(MV^{+}\cdot)$, H^+, EDTA and to a lesser extent Ru $(bipy)_3^{2+}$ are adsorbed on the catalyst surface. We are thus led to propose the following mechanism (12) :

4.1 An electron transfer from $MV^{+\cdot}$ to a colloidal particle

$$Pt + MV^{+\cdot} \rightleftharpoons Pt\ MV^{+} \rightleftharpoons Pt^{-}\ MV^{2+} \rightleftharpoons Pt^{-} + MV^{2+} \quad (9)$$

leading to a multicharged particle through successive reaction sequence 9 according to the electron-accumulating model of Henglein (36, 37). For the nth charge (which implies a nth $MV^{+\cdot}$) :

$$Pt^{n-} + MV^{+\cdot} \rightleftharpoons Pt^{n-}\ MV^{+} \rightleftharpoons Pt^{(n+1)-}\ MV^{2+} \rightleftharpoons Pt^{(n+1)-} + MV^{2+} \quad (10)$$

It should be remarked that all the steps of these reactions are reversible, due to the thermal stability of MV^{2+} in the presence of H_2 and colloidal Pt (21).

4.2 The above charging reactions are immediately followed by the protonation of the species formed (38) in particular :

$$Pt^{n-} + H^{+} \rightleftharpoons Pt^{(n-1)-}H \quad (11)$$

and

$$Pt^{n-}MV^{+} + H^{+} \rightleftharpoons Pt^{(n-1)-}H\ MV^{+} \quad (12)$$

The protonation rates depend on the nature of metal (overpotential). They are fast for Pt but slower for Au (39). One may consider that H, $MV^{+\cdot}$ and MV^{2+} are adsorbed on the colloidal particle surface and the amount of each adsorbed species depends on the metal nature.

4.3 The rate limiting desorption step leading to H_2 formation may involve either a H atom transfer between two particles as for example

$$Pt^{(n-1)-}H + Pt^{(n-1)-}H \rightleftharpoons 2\ Pt^{(n-1)-} + H_2 \quad (13)$$

$$Pt^{(n-1)-}H\ MV^{+} + Pt^{(n-1)-}H\ MV^{+} \rightleftharpoons 2\ Pt^{(n-1)-}\ MV^{+} + H_2 \quad (14)$$

or more likely an electrochemical desorption on the same particle :

$$Pt^{(n-1)-}H + H^{+} \rightleftharpoons Pt^{(n-2)-} + H_2 \quad (15)$$

$$Pt^{(n-1)-}H\ MV^{+} + H^{+} \rightleftharpoons Pt^{(n-2)-}MV^{+} + H_2 \quad (16)$$

$$Pt^{n-}H\ MV^{2+} + H^{+} \rightleftharpoons Pt^{(n-1)-}MV^{2+} + H_2 \quad (17)$$

4.4 Methyl viologen hydrogenation (and dimerization, reaction 7) competes with the desorption step and becomes more efficient as the pH increases (14, 21) and for [Pt] > 2 x 10^{-5} M (23, 30) :

$$Pt^{(n-1)-}H\ MV^{+} \rightleftharpoons Pt^{(n-1)-} + MV^{+}H \qquad (18)$$

$$Pt^{n-}H\ MV^{2+} \rightleftharpoons Pt^{n-} + MV^{2+}H \qquad (19)$$

Such MV^{+} H and MV^{2+} H species (42) lead to stable hydrogenation products.

5. CONCLUSION

In summary, a large selection of catalysts is now available (Tables I – III) (23) which produce H_2 very efficiently and, for some of them (Ni/TiO_2, metal oxides), are of very low cost. However an important limitation arises from the catalytic destruction of the electron relay during the overall processes. To minimize the undesired hydrogenation of the electron relay, and therefore to increase the longevity of model systems, it would appear preferable to use ruthenium oxides which are not hydrogenation catalysts, or to add to the solution a catalyst poison such as sulphur and nitrogen compounds which present free electron pairs in the valency shell (33) ; it is in particular the case of EDTA. Among these poisons, glutathione totally inhibits the hydrogenation reaction. This constitutes a rare example of a catalytic process which can be selectively directed toward one of the competitive reaction path. Such poisoning effects have brought some new informations concerning these surface reactions and a catalytic mechanism is proposed which implies H^{+}, MV^{2+} ($MV^{+\cdot}$) and poison adsorption on the catalyst surface. Such a mechanism will be tested on the basis of pulse radiolysis experiments in the presence of catalyst poison. The results will be reported elsewhere (43).

Acknowledgements

It is a pleasure to thank P. Koffi for his cooperation and Drs J. Barbier, A. Bernas and P. Pichat for helpful discussions.

REFERENCES

1 B.V. KORIAKIN, T.S. DZHABIEV and A.E. SHILOV, Dokl.Akad.Nauk., S.S.S.R., 233, 620, (1977)
2 J.M.LEHN and J.P. SAUVAGE, Nouv.J.Chim., 1, 449, (1977)
3 A. MORADPOUR, E. AMOUYAL, P. KELLER, and H. KAGAN, Nouv.J.Chim., 2, 547, (1978).
4 K. KALYANASUNDARAM, J. KIWI, and M.GRÄTZEL, Helv.Chim.Acta, 61, 2720, (1978).
5 A.I. KRASNA, Photochem.Photobiol., 29, 267, (1979).
6 G.M. BROWN, B.S. BRUNSCHWIG, C. CREUTZ, J.F. ENDICOTT, and N. SUTIN, J.Amer.Chem.Soc., 101, 1298, (1979).
7 I. OKURA and N. KIM THUAN, (a) J.Mol.Catal., 5, 311, (1979)., (b) J.Mol.Catal., 6, 227, (1979).

8 P.J. DELAIVE, B.P. SULLIVAN, T.J. MEYER and D.G. WHITTEN., J.Amer. Chem.Soc., 101, 4007, (1979).
9 M. GOHN and N. GETOFF, Z. Natürforsch., 34a, 1135, (1979).
10 J.M. LEHN, J.P. SAUVAGE, and R. ZIESSEL, Nouv.J.Chim., 3, 423, (1979).
11 J.KIWI and M. GRÄTZEL, Ang.Chem.Int.Ed.Engl., 18, 624, (1979).
12 E. AMOUYAL, Riken's Quarterly, 78, 220, (1984) and references therein.
13 E. AMOUYAL and B. ZIDLER, Isr.J.Chem., 22, 117, (1982).
14 P. KELLER, A. MORADPOUR, E. AMOUYAL, and H. KAGAN. Nouv.J.Chim., 4, 377, (1980).
15 P. KELLER, A. MORADPOUR, E. AMOUYAL and H. KAGAN, J.Mol.Catal., 7, 539 (1980).
16 P. KELLER and A. MORADPOUR, J.Amer.Chem.Soc., 102, 7193, (1980)
17 E. AMOUYAL, B. ZIDLER and P. KELLER, Nouv.J.Chim., 7, 725, (1983).
18 V. HOULDING, T. GEIGER, U. KÖLLE and M. GRÄTZEL, J.Chem.Soc. Chem.Comm., 681, (1982).
19 A. LAUNIKONIS, J.W. LODER, A.W.H. MAU, W.H.F. SASSE and D. WELLS, Isr.J.Chem., 22, 158, (1982).
20 E. AMOUYAL, P. KELLER and A. MORADPOUR, J.Chem.Soc.Chem.Comm., 1019, (1980).
21 P. KELLER, A. MORADPOUR and E. AMOUYAL, J.Chem.Soc. Far.Trans. I, 78, 3331, (1982).
22 O. JOHANSEN, A. LAUNIKONIS, J.W. LODER, A.W.H. MAU, W.H.F. SASSE, J.D. SWIFT and D. WELLS, Aust.J.Chem., 34, 2347, (1981).
23 E. AMOUYAL and P. KOFFI, (a) 5th International Conference on Photochemical Conversion and Storage of Solar Energy, Osaka, Japan, August 1984, Book of Abstracts, p. 29. (b) J.Photochem., (1985), in press.
24 E. AMOUYAL and P. KOFFI, to be published.
25 E. AMOUYAL and P. KOFFI, C.R.Acad.Sci., Série II, 300, 199, (1985).
26 L.D. RAMPINO and F.F. NORD, J.Amer.Chem.Soc., 63, 2745, (1941).
27 M.O. DELCOURT, N. KEGHOUCHE and J. BELLONI, Nouv.J.Chim., 7, 131, (1983)
28 J. BELLONI, J-L. MARIGNIER, M.O. DELCOURT and M. MINANA, Fr. Patent n° 84.09196 (1984).
29 R. RAFAELOFF, Y. HARUVY, J. BINENBOYM, G. BARUCH and L.A. RAJBENBACH, J.Mol.Catal., 22, 219, (1983).
30 E. AMOUYAL, D. GRAND, A. MORADPOUR and P. KELLER, Nouv.J.Chim., 6, 241, (1982).
31 J.M. LEHN, J.P. SAUVAGE and R. ZIESSEL, Nouv.J.Chim., 5, 291, (1981).
32 P. PICHAT, J.M. HERRMANN, J. DISDIER, H. COURBON, and M.N. MOZZANEGA. Nouv.J.Chim., 5, 627, (1981).
33 E.B. MAXTED, Advances in Catalysis, Academic Press, 3, 129, (1951).
34 J. BARBIER, "Catalyse par les métaux", Editions du CNRS, 305, (1984).
35 L.T. PRAHOV, J. DISDIER, J.M. HERRMANN, and P. PICHAT, Int.J. Hydrogen Energy, 9, 397, (1984).
36 A. HENGLEIN, J.Phys.Chem., 83, 2209, (1979).
37 A. HENGLEIN and J. LILLIE, J.Amer.Chem.Soc., 103, 1059, (1981).

38 D. MEISEL, W.A. MULAC, and M.S. MATHESON, J.Phys.Chem., 85, 179, (1981).
39 M.S. MATHESON, P.C. LEE, D. MEISEL, and E. PELIZZETTI, J.Phys.Chem., 87, 394, (1983).
40 M. BRANDEIS, G.S. NAHOR, and J. RABANI, J.Phys.Chem., 88, 1615, (1984).
41 M. HAISSINKI, "Actions Chimiques et Biologiques des Radiations", Masson, 15, 185, (1971).
T.W. EBBESEN, J.Phys.Chem., 88, 4131, (1984).
42 S. SOLAR, W. SOLAR, N. GETOFF, J. HOLCMAN and K. SEHESTED, J.Chem.Soc., Far.Trans. I, 78, 2467, (1982).
43 E. AMOUYAL, M.O. DELCOURT and M. GEORGOPOULOS, to be published.

RADIOLYTIC METHODS OF PREPARATION OF COLLOIDAL AND HETEROGENEOUS REDOX CATALYSTS AND THEIR APPLICATION IN LIGHT-INDUCED H_2 GENERATION FROM WATER

Y. Haruvy, R. Rafaeloff and L. A. Rajbenbach
Chemistry Department
Soreq Nuclear Research Center
Yavne 70600, Israel

ABSTRACT. Radiolytic methods have been employed to prepare highly stable Pt hydrosols of low polydispersity, supported on poly-N-methylolacrylamide (PNMAM) as well to form Pt clusters embedded within nylon-grafted-acrylamide (Ny-g-AM) membranes. The metal hydrosols and the Pt bearing polymeric foils were used to catalyze visible-light induced H_2 generation from water. The mechanistic features of the radiolytically-induced reduction of Pt(IV) to Pt(O) and of the concurrent polymerization processes occurring in aqueous solutions of H_2PtCl_6 in the presence of NMAM are discussed. The relatively high catalytic effectiveness of the Pt bearing Ny-g-Am recoverable membranes is ascribed to the accumulation of ions within the aqueous domain of the metal bearing polymeric grids. The catalytic performance of Pt sols and Pt bearing membranes is compared with that of analogous radiolytically prepared Pd, Os and Ru systems. Some physical characteristics of the Pt catalytic systems are presented.

1. INTRODUCTION

The study of visible-light-induced electron transfer reactions received widespread attention during the last few years with most of the effort directed towards photo-induced cleavage of water. Redox catalysis plays an important role in the water cleavage reaction cycle and considerable effort has been invested in the development of stable, reproducible high-activity noble-metals suspensions. The catalytic properties of such systems depend upon the nature of the stabilizing polymeric supports and, to a significant extent, upon the preparative methods employed which govern the dispersity and size of the metal clusters.

The customary methods of preparation of noble-metals hydrosols involve the reduction of appropriate metallic ions with reducing agents such as: H_2, $NaBH_4$, citrates, hydrazine sulfate, etc. Polyvinyl alcohol, polyvinylpyrrolidone polyacrylates and carbowax 20-M are examples of macromolecular supports commonly employed as protective agents.

E. Pelizzetti and N. Serpone (eds.), Homogeneous and Heterogeneous Photocatalysis, 267–273.

It has been known for some time [1-4] that radiolysis of aqueous solutions of silver and gold salts in the presence of hydroxyl radical scavengers, such as isopropanol, leads to the formation of zero valent metals. Recently the radiolytic method has been employed for the preparation of Pt sols by subjecting aqueous solutions of H_2PtCl_6 and isopropanol in the presence of polyvinyl alcohols to ionizing irradiation [5]. At the same time an effort was initiated in our laboratory [6] to employ ionizing radiation to induce a train of simultaneously occurring reduction and polymerization processes in aqueous solutions of noble-metals salts in the presence of N-methylolacrylamide. N-methylolacrylamide was chosen to serve both as an effective OH radical scavenger and a building block of a polymeric support to stabilize the metal hydrosols. Subsequently the radiolytic method of reduction of noble metal salts has been employed by us to induce formation of metallic clusters embedded within nylon-grafted-acrylamide membranes. The rationale for the use of Ny-g-Am foils as a depository of metal clusters is presented in section 3.

2. COLLOIDAL REDOX CATALYSTS

2.1. Preparation and Characterization of Radiolytically Formed Pt Hydrosols

Deaerated aqueous solutions containing $4x10^{-3}$ mol dm^{-3} H_2PtCl_6 (0.08 wt% Pt) or other noble-metals salts and $8x10^{-2}$ mol dm^{-3} NMAM were irradiated with a Co-60 source at a dose rate of 0.12 Mrad/h to a total dose of 5.7 Mrads. ESCA analysis of the dark-irradiated solutions confirmed the total reduction of Pt(IV) ions to the metallic state. The size of the Pt aggregates dispersed within the solution was determined by transmission electron microscopy. The measured diameters of the Pt clusters were found to range from 8 to 24 Å with an average value of 17 Å and a mean deviation of 4 Å. Assuming a spherical shape for the colloidal Pt particles it can be calculated that the average particle contains about 170 atoms. The size of the Pt-PNMAM particles, determined by light-scattering technique, indicates a rather large hydrodynamic radius of 1200 Å. The latter value is significantly higher than that observed in the highly effective Pt hydrosols stabilized by Carbowax 20-M, where a hydrodynamic radius of 180 Å was determined [7].

The radiolytically formed low polydispersity sols are characterized by very high stability as evidenced by the fact that storage for a period of 30 months did not result in flocculation and only a minor increase of 30% in the average Pt particle size was observed. The Pt-PNMAM sols were also found to be remarkably stable in the presence of normally effective flocculating agents. Neither $MgSO_4$ nor NaCl, when present in the dispersion medium at saturation concentrations, induced flocculation. Variation of solution pH in the range 1.6 to 9 was also found ineffectual in inducing metal precipitation.

2.2. Mechanistic Features of γ-Radiation-Induced Formation of Pt Hydrosols

The gamma radiation induced reduction of Pt(IV) to Pt(O) and the concurrent polymerization of NMAM is initiated by radiolytically formed primary species. In the relatively dilute aqueous solutions used in the preparation of Pt sols, ionizing radiation is absorbed predominantly by the solvent. The primary radical species formed in the radiolysis of water are hydrated electrons e^-_{aq}, H atoms and OH. The yield of these species in terms of G values (molecules per 100 eV) is 2.65, 0.6 and 2.65, respectively [8]. OH radicals will disappear from the system by reaction with NMAM while the other two primary species can react with both solutes. The rate constants of H and e^-_{aq} with $[PtCl_6]^{2-}$ and NMAM are very high, approaching the diffusion-controlled limit [9]. Under the experimental conditions employed it can be estimated that practically all H and OH species as well as at least 2/3 of e^-_{aq} react with NMAM. The involvement of NMAM derived radicals, formed in the latter reactions, in the reduction of Pt(IV) is supported by the finding that at relatively low dose irradiation (200 krads) no monomer depletion was observed. This finding is to be contrasted with the observation that similar irradiation of $8x10^{-2}$ mol dm^{-3} NMAM solution yielded a highly viscous solution in which practically all the NMAM ($<10^{-5}$ mol dm^{-3}) was consumed.

A detailed discussion of the complex reaction sequence leading to reduction of Pt(IV) to Pt(O) and the concurrent polymerization of NMAM is beyond the scope of this presentation. The notable features of the complex reduction-polymerization processes, most likely involved in the gamma radiation-induced formation of Pt hydrosols, can be summarized schematically as follows: (a) In the early stages of radiolysis electrons which escaped reaction with NMAM and monomer derived radicals, reduce Pt(IV) to Pt(III). The reaction of NMAM radicals species with Pt(IV) proceeds via the formation of Pt(IV)/ NMAM radical complex. The latter species reverts to Pt(III) state by internal electron transfer and quite likely may also engage in a reaction with the monomer, initiating a polymerization process. Upon depletion of Pt(IV) ions the polymerization process is initiated by NMAM radicals reaction with the monomer. (b) Pt(II) formed by disproportionation of Pt(III) species is stabilized by NMAM and accumulates in the system until the concentration of NMAM is depleted by polymerization below that of Pt(II). (c) The formation of metallic Pt results mainly from a dark reaction between Pt(II) and the polymeric alcohol-(PNMAM). The latter reaction is catalyzed by Pt(O). The thermal reduction of Pt(II) to Pt(O) which is inhibited by the presence of olefins has been also observed by us to occur in aqueous solutions of MeOH and of isopropanol. However, the time scale of these reduction processes was higher by several orders of magnitude than that occurring in the presence of PNMAM. The occurrence of a dark reaction between Pt(II) and PNMAM which is suppressed by the presence of free NMAM accounts for the observation that metallic Pt formation becomes noticeable only after a critical dose of radiation has been delivered to the system to deplete NMAM concentration below that of Pt(II). (d) The remarkable stability of the radiolytically formed

Pt/PNMAM sols most likely reflects the occurrence of coordination between Pt(IV) and NMAM or NMAM-derived radicals in the early stage of radiolysis, leading to a specific metal-polymer morphology. The latter assumption is supported by the finding that radiolysis of H_2PtCl_6-PNMAM solutions in the presence of OH radical scavengers such as isopropanol or free NMAM yielded sols of inferior stability.

2.3. Mediation of Visible-Light Induced Reduction of Water

The catalytic effectiveness of the radiolytically formed Pt sols was assayed in the visible light-induced hydrogen generation from water. The illuminated system was made up of $Ru(bipy)_3^{2+}$ serving as a photosensitizer, MV^{2+} functioning as an electron relay agent, EDTA as a sacrificial electron donor and the radiolytically prepared Pt sol as a catalyst. A detailed description of the experimental conditions has been given in an earlier communication [9].

The salient features of the catalytic performance of the Pt-PNMAM sols can be summarized as follows: (a) the evolution rates of H_2 mediated by Pt-PNMAM sols compare favorably with those in which the water reduction cycle was catalyzed by Pt sols formed by reduction of Pt(IV) by H_2 or citrates in the presence of stabilizing agents such as polyvinyl alcohol or carbowax 20-M. (b) Prolonged storage (30 months) does not impair the catalytic effectiveness of the radiolytically prepared hydrosol. (c) The H_2 generation rate is significantly less reduced by increasing the pH of the photolytic system from 5 to 6 than reported by other groups. The latter observation most likely reflects the reduced tendency of the Pt-PNMAM system to catalyze the pH dependent parasitic hydrogenation reaction of methylviologen. (d) The catalytic effectiveness of Pt surpasses that of Pd, Os and Ru sols prepared by analogous procedure (see Fig. 7 of Ref. 6).

3. HETEROGENEOUS Pt REDOX CATALYSTS

3.1. Macromolecular Support

The effort to form Pt clusters anchored to an insoluble polymeric support was undertaken in order to try to develop a catalytic species which combines a performance comparable to that of its microheterogeneous analogue, together with an ease of recovery and the possibility of repeated use. In order to minimize the leaching of the metal species from its support, the host material has to bear a large number of pendant ligand groups which can bind transition metal ions. Another critical requirement of the host support material is the ease of accessibility of reactants to its internal volume. Swellable gel-type membranes bearing a large number of functional groups capable of forming complexes with transitional metal ions appear to present a suitable type of matrix to anchor catalytic species.

It was thus of interest to evaluate the catalytic effectiveness of Pt and other noble-metals clusters embedded within nylon-grafted-acrylamide gel type membranes. We have recently shown [10-12] that radiolytic techniques are highly suited to prepare gel-type membranes based on nylon-6 films. Electron-beam preirradiation of nylon foils followed by immersion at 50°C in aqueous solutions of acrylamide and its derivatives leads to the formation of grafted copolymers containing up to 1300% of AM [11]. In such highly grafted films the nylon backbone constitutes only a minor component which imparts mechanical support and prevents fluidization of the water soluble PAM. The strategy employed in anchoring of metallic clusters within the membranes involved absorption of the metal ions from aqueous solution followed by their reduction by means of exposure to Co-60 irradiation in the presence of an OH radical scavenger.

3.2. Preparation and Characterization of Pt Clusters Embedded within Membranes

One g of 120 m thick dry Ny-g-AM foil, containing approximately 90 wt% AM, was immersed in an ampoule containing 50 ml aqueous solution of 0.1 wt% Pt in the form of H_2PtCl_6 and 0.1 wt% AM. The contents of the ampoule were left at room temperature under occasional stirring for 24 hours. Subsequently, the solution was deaerated and irradiated with a Co-60 source to a total dose of 5.7 Mrads. The irradiated foil was vigorously rinsed with water and dried in a vacuum oven. The Pt content of the dry foils, determined by flame spectroscopy, was found to amount to 1.5 wt%. The Pt content of the films was found to match the uptake of H_2PtCl_6 during the 24 hours immersion period, representing approximately 30% of the platinic acid originally present in the solution. ESCA analysis of the irradiated films confirmed the reduction of Pt(IV) to the metallic state. The diameter of the embedded Pt clusters determined by transmission electron microscopy was found to range between 23-48 Å with a mean value of 34 ± 1.5 Å.

The strong anchoring of the Pt clusters to the host material was confirmed by flame spectroscopy analysis of the metal bearing polymeric foils recovered from the photochemical reaction system following 24 hours illumination. The elution of Pt into the aqueous medium was minimal, the loss of metal amounting to 1-3%. The high uptake of water and inorganic ions by the Pt/Ny-g-AM foils is evidenced by the drastic changes in its resistivity upon immersion in electrolyte solutions. The dry Pt/Ny-g-AM film exhibits a resistivity of about 5×10^{12} ohm cm. This value is incidentally comparable to that of pure nylon-6 film, the resistivity of which amounts to 2.2×10^{13} ohm cm [13]. However, following a 2-minute immersion in saturated KCl solution, the resistivity of the films drops dramatically and becomes practically equal to that of the KCl solution.

Osmium, palladium and ruthenium metallic clusters were anchored to Ny-g-AM foils by a procedure analogous to that employed in the case of Pt. However, the oxidation state was not characterized and the cluster size was not determined. Deposition of Pt clusters on commercial polyvinyl alcohol and cellophane films by the radiolytic method resulted in a lower metal uptake than that observed for Ny-g-AM films. The weight percentage of the embedded Pt in polyvinyl alcohol and cellophane was found to amount to 0.86 and 0.46, respectively, as compared with 1.5 for Ny-g-AM.

3.3. Comparison of Catalytic Effectiveness of Microdispersed and Immobilized Pt Clusters on H_2 Evolution from Water

The relative effectiveness of Pt bearing Ny-g-AM films was established by following H_2 evolution rates over a prolonged period of time (55 hours) in the $Ru(bipy)_3^{2+}/MV^{2+}$/EDTA system catalyzed by Pt/Ny-g-AM. These data were compared with those in a colloidal system under identical conditions of illumination, reactants and Pt concentrations. The rate of H_2 evolution in the heterogeneous system was found to be lower by about 20% than in the colloidal one. However, the total H_2 yield (plateau value) was lower by only about 6% (see Fig. 4 of Ref. 14).

A priori one could expect the catalytic effectiveness of the embedded metallic particles at equal Pt concentration to be significantly lower than in colloidal systems. The reduction in the catalytic efficiency of the heterogeneous system could be ascribed to diffusional encounter limitations, the relatively larger size of the immobilized Pt clusters and the somewhat reduced accessibility of reactants to the catalytic sites. The higher than expected catalytic effectiveness of the Pt/Ny-g-AM system could result from microenvironmental effects within the aqueous domains of the swollen membrane. Spectroscopic analysis has shown that the local concentration of $Ru(bipy)_3^{2+}$, MV^{2+} and EDTA within the aqueous fraction of the polymeric gel was higher by a factor of 5 to 10 than that in the bulk of the solution. The existence, within the Ny-g-AM membrane structure of domains of enhanced reactants concentration in proximity to catalytic sites establishes favorable conditions to impair the thermal back electron transfer reaction. The occurrence of the latter reaction in homogeneous solutions constituting the main factor degrading the quantum efficiency of the H_2 generation process.

3.4. The Catalytic Efficiency of Recovered Pt/Ny-g-AM

Multiple usage of the immobilized catalyst system was probed following 40-hour illumination cycles. After recovery and rinsing the foils were reintroduced into fresh reactants solutions. After 3 reaction cycles about 85% of the original activity was still preserved.

4. CONCLUSIONS

Radiolysis of aqueous solutions of noble-metals salts and N-methylol-acrylamide provides a convenient method for the formation of highly stable metallic sols of low polydispersity. In addition, gamma radiolysis of systems made up of nylon-acrylamide copolymer films immersed in aqueous solutions of metal salts and acrylamide leads to the production of immobilized metallic clusters within gel-type polymeric grids. The use fo recoverable, swellable membranes bearing catalytic sites, which can form domains of local relatively high reactants concentration, appears to present an attractive way of enhancing photo-products yields in solar energy conversion processes.

5. REFERENCES

1. G. Czapski and A. O. Allen, J. Phys. Chem. 70, 1859 (1966).
2. A. Henglein, J. Phys. Chem. 83, 2209 (1979).
3. G. Westerhausen, A. Henglein and J. Lillie, Ber. Bunsenges. Phys. Chem. 85, 182 (1981).
4. K. Kopple, D. Meyerstein and D. Meisel, J. Phys. Chem. 84, 870 (1980).
5. M. O. Delcourt, N. Keghouche and J. Belloni, Nouv. J. Chim. 6, 131 (1982).
6. R. Rafaeloff, Y. Haruvy, J. Binenboym, G. Baruch and L. A. Rajbenbach, J. Mol. Catal. 22, 219 (1983).
7. J. Kiwi, in: _Energy Resources through Photochemistry and Catalysis_ (M. Gratzel, ed.), Academic Press, 1983, p. 297.
8. M. S. Matheson and L. M. Dorfman,, _Pulse Radiolysis_, MIT Press, Cambridge MA, 1969.
9. L. A. Rajbenbach and D. Meisel, unpublished results.
10. Y. Haruvy and L. A. Rajbenbach. J. Appl. Polym. Sci. 26, 3065 (1981).
11. Y. Haruvy, L. A. Rajbenbach and J. Jagur-Grodzinski, J. Appl. Polym. Sci. 27, 2711 (1982).
12. Y. Haruvy, L. A. Rajbenbach and J. Jagur-Grodzinski, Polymer 25, 1431 (1984).
13. D. B. Freeston, C. H. Nicholls and M. T. Pailthorpe, Polym. Photochem. 1, 85 (1981).
14. R. Rafaeloff, Y. Haruvy, G. Baruch, I. Schoenfeld and L. A. Rajbenbach, J. Mol. Catal. 24, 345 (1984).

THE EFFECT OF PROMOTERS ON THE PHOTOCHEMICAL WATER CLEAVAGE IN SUSPENSIONS OF Pt-LOADED TiO_2 WITH INCREASED LIGHT TO CHEMICAL CONVERSION EFFICIENCY

J. Kiwi
Institut de Chimie Physique
Ecole Polytechnique Fédérale
CH-1015 Lausanne
Switzerland

ABSTRACT. The addition of Nb and Mg cation to TiO_2 shows that promoter incorporation has a beneficial effect on H_2 production in TiO_2 dispersions under irradiation. Concentrations of 1% Nb, 7.5% Li and 1-2% Mg have been shown to be optimal in promoting interfacial charge transfer. The point defect model is used as a basis for the discussion of the electronic character of the solids and as such provides an indicator of the reactivities as a function of the increasing concentration of added dopant. Dopants added to TiO_2have been shown to affect the ratio Ti^{+4}/ Ti^{+3} and also the amount of O^{-2} in the TiO_2 lattice. In some cases the promoting ions introduce lattice defects of the Schottky type in the bulk TiO_2. It was found that not all of the doping-ions incorporated into TiO_2 promote catalytic activity. A few monolayers of the modified interface (e.g; TiO_2-Ba^{+2}) also effectively increase the surface efficiency when enhancing H_2 production in TiO_2 slurries under u.v. irradiation.

1. INTRODUCTION

During the last few years there has been a surge of interest in finding chemical processes which are capable of quantum storage of light energy. Work from different laboratories has been reported in recent reviews (1-5). In recent years, much importance has been given to the study of foreign-ion doping of non-metallic semiconductors (6). In the case of variable stoichiometry in non-metallic compounds such as TiO_2, the introduction of acceptor ions such as Li^+ and Mg^{+2} or Nb^{+5} as donor ions leads to changes in stoichiometry and electrical conductivity in the composite material (7). It is our aim to assess the catalytic efficiency due to ion-doping and to understand the origin of the ion-doping effect on the hydrogen generation capacity of TiO_2-promoted systems. The strategy followed in the course of this work was to isolate promoter and support effects by restricting the initial studies to unpromoted Pt-TiO_2.

It has been known for many years (8) that TiO_2 can accomodate a

E. Pelizzetti and N. Serpone (eds.), Homogeneous and Heterogeneous Photocatalysis, 275–302.

considerable amount of defects in the bulk and at the surface. The alteration of catalytic properties in semiconductors by introducing ions of different valency was first observed by Verwey (9). In previous studies, the action of alkaline-dopant ions has been shown to have beneficial catalytic effects in general (10) and on TiO_2 in particular (11). More recently, electrode materials doped with Li and Mg (12) have shown a better performance in oxygen reduction processes. The cations used as promoters Nb^{+5}, Li^+, Mg^{+2} enter the TiO_2 lattice during the doping of TiO_2 (as described in the experimental section) due to their similarities in ionic radii to Ti^{+4} (0.63 Å) (13). Their ionic radii are: Nb^{+5} (0.66 Å), Li^+ (0.68 Å) and Mg^{+2} (0.78 Å). The need to carry out the present studies arises from the fact that until now detailed studies have not been made reporting the effect of Nb, Li and Mg-ions doping on TiO_2 and its effect on the photocatalytic properties of the resulting materials.

We also wish to report on the effect of Ba^{+2}-ions added in the electrolyte when Pt-TiO_2 has been photolysed under U.V. light. Ba^{+2} apparently acts as a peroxide scavenger forming an insoluble peroxide which removes peroxo compounds from the surface of the catalyst. The water photocleavage is enhanced 3 times under these conditions, the overall conversion yield increasing to 0.7%. This observation shows that a few monolayers of the modified surface, due to added Ba^{+2} in the electrolyte, suffice to confer to the surface a different capacity for H_2 generation.

2. EXPERIMENTAL SECTION

U.V. irradiations were carried out with a Rofin lamp. Volume of irradiation flasks was always 25 cc. Samples were degassed with Ar and the H_2 produced or O_2 photoadsorbed were analyzed by means of a Gow-Mac conductivity detector. Details on photolytic experiments have been reported previously (21). Catalyst loading by Pt via ion-exchange from $Pt(NH_3)_4(OH)_2$ was used. The same technique was employed as described in reference (21). Samples of Li-doped TiO_2 were prepared by Dr. Panek (Bayer AG, Krefeld, Uerdingen, West Germany). By a slurry technique, Li-doped TiO_2 and Mg-doped TiO_2 was prepared, similar to the one employed by Teichner et al. (22) for the preparation of lithiated NiO. In order to produce Li or Mg doping within a reasonable time (three hours), an intimate mixture of LiOH and TiO_2-Bayer was heated at temperatures between 400-700° C. The actual Li and Mg content of the catalysts was analyzed by atomic absorption by Dr. Panek. BET areas were measured in a Micromeritics 2205 instrument. Peroxotitanates were determined by titration with $KMnO_4$ 10^{-3} M at pH 0. Details of this determination in which the liquid as well as the catalyst precipitate were agitated 4 hours with permanganate have been reported before (21). Further confirmation for the formation of peroxide was obtained by using o-tolidine as a redox indicator (23). In acid media the formation of a yellow color with an adsorption maximum at 433 nm is observed. This peak is characteristic for the two-electron oxidation product of o-tolidine formed by the reaction with peroxide (24). Under acidic conditions the extinction coefficient for the two-electron oxidation product of o-tolidine was assessed at 1.1 10^{-3} $M^{-1}cm^{-1}$.

Mobilities were measured on the Li-doped anatase particles situated in the stationary planes of a Mark II microelectrophoresis apparatus (Rank Brothers, Cambridge, England). Determination of the conduction band position of Li-doped anatase powders was carried out following a technique recently reported for these kind of measurements (25). EPR measurements were performed on a Varian type EC-365 X band spectro meter. The relative intensities of the EPR signals were measured at 4.2° K.

TiO_2 doped with $Nb^{+5}$was prepared by digesting an appropiate amount of Nb_2O_5 together with Ti O SO_4 in H_2SO_4. Thermal hydrolysis was continued according to the Blumenfeld procedure (8) to obtain anatase. TiO_2 rutile was obtained from the anatase by dissolving the latter in NaOH, precipitation as Na_2TiO_3, and subsequent peptizing in HCl. Upon heating the HCl solution to 70° C hydrolysis occurs, yielding a precipitate. X-ray diffraction analysis performed with the powder reveals rutile structure with >95% cristallinity. Nb^{+5} doping of rutile was carried out by adding the appropriate amount of Nb_2O_5 to the HCl solution prior to hydrolysis.

3. IMPLICATIONS FOR WATER PHOTOCLEAVAGE THROUGH Nb-DOPED TiO_2

The effect of TiO_2 structure, RuO_2 loading and Nb_2O_5-doping on photoinduced water photosplitting is shown in Figure 1 (14). As seen from this Figure, anatase with 0.1% RuO_2 and 0.4% Nb_2O_5 produced the highest yields for H_2 evolution.

To explain the effect of Nb_2O_5 while increasing the yields of H_2, it is necessary to invoke the solid-state reactions of Nb_2O_5 and TiO_2 at high temperature. As a product of such reaction inhomogeneous doping of TiO_2 with Nb-species is attained. To identify the species present in the composite catalyst EPR-spectra of Nb-doped TiO_2 was taken. This spectrum (15) indicated that Nb occupies substitutional Ti^{+4} sites in the form of Nb^{+4} ($4d^1$). As shown in Figure 2, spectrum A represents Nb^{+4}, in agreement with previous EPR studies of Nb^{+4} in single crystals of TiO_2 (16). The relative intensities of the Nb^{+4} spectral lines for 0.1, 0.5 and 1.0% samples were found to be 0.18, 0.44 and 1.0 showing in experimental error a good correlation with the added initial Nb^{+5}. Therefore it is possible to establish that no saturation appears in the incorporation of Nb^{+4} into TiO_2 up to this level. The spectrum of the species B in Figure 2 agrees well with those of a spectrum found for Ti^{+3} intersticial ions (17).

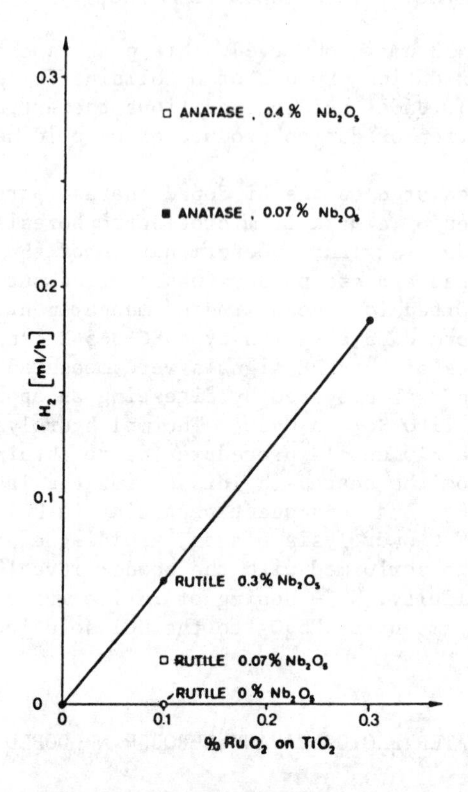

Figure 1. Influence of catalyst structure and composition on the visible light-induced (λ > 400 nm) decomposition of water. Conditions : 500 mg of TiO_2/L loaded with 40 mg of Pt, Nb_2O_5 doping, and RuO_2 loading variable, pH 4.7, $Ru(bpy)_3^{2+} = 10^{-4}$ M, $MV^{2+} = 5 \times 10^{-3}$ M, T = 25° C.

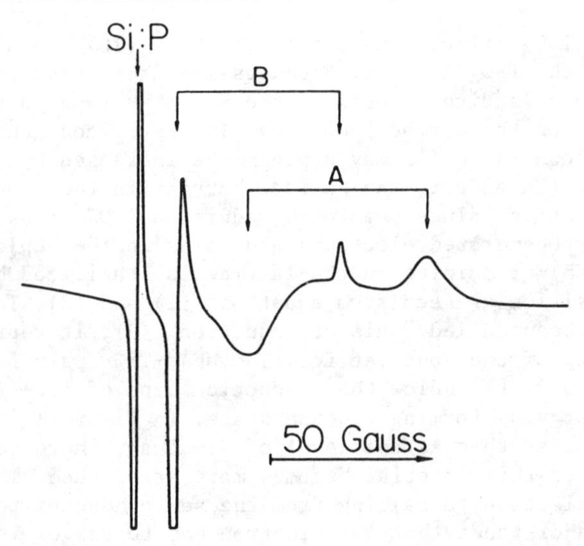

Figure 2. EPR spectrum of 0.5% Nb-doped TiO_2 recorded at 4.2° K and 9.4 GHz. Trace A (Nb^{+4}); trace B (Ti^{+3}).

Samples were checked against TiO_2 powders after H_2 absorption, indicating that Nb^{+5} added also exists in the crystal lattice as penta valent ion. Therefore, only a certain fraction of the initially added Nb^{+5} substitutes Ti^{+4} reducing an equivalent amount of Ti^{+4}-ions to the trivalent state Ti^{+3} entering substitutionally Nb^{+4} on Ti^{+4} sites.
Equation (1) below using Hauffe's (10) notation shows such a process:

$$Nb_2O_5 \rightarrow 2Nb^{+4}/Ti/ + \tfrac{1}{2}O_2 \tag{1}$$

where $Nb^{+4}/Ti/$ denotes Nb ions on TiO_2 lattice positions. Such sites act as donor impurities, as readily seen in

$$2Nb^{+4}/Ti/ \rightarrow 2Nb^{+5}/Ti/ + 2e^-_{CB} \tag{2}$$

The formation of Ti^{+3}, as experimentally observed by the EPR technique, can be accounted for by the equation

$$2e^-_{CB} + 2Ti^{+4} \rightarrow 2Ti^{+3} \tag{3}$$

due to conduction band electrons (e^-_{CB}). Moreover, the formation of O^{-2} lattice defects is a possible process taking place due to e^-_{CB} as shown by eq. (1) and

$$2e^-_{CB} + \tfrac{1}{2}O_2 \rightarrow O_2^{-2} \tag{4}$$

Nb added to TiO_2 affects the ratio Ti^{+4}/Ti^{+3} (18) and also the amount of O^{-2} in the TiO_2 lattice. From eqs 1-4 it is seen that Nb doping would introduce lattice defects of the Schottky type in the bulk (and consequently on the surface) of TiO_2. The net excess charge due to O^{-2} present, as shown in eq. 4, may explain the increased H_2 yields reported in Fig. 1. (In effect, the surface barrier in the doped material moves to more negative values due to the additional O^{-2} present and the separation of photogenerated electrons and holes in the depletion layer increases.) The Nb^{+5} remaining in the lattice is stabilized by a nearby $Ti^{+4} \rightarrow Ti^{+3}$ transition as seen from equations (2) and (3). The induced Ti^{+3} states have been called "valence induction" (9); it represents an extrinsic property of the host lattice. The Nb^{+5}-Ti^{+3} pair formed is a donor level at $\sim$ 0.04 [V] below the conduction band of TiO_2 (19) and minimizes its energy by forming a bound state. In Figure 1 it is also important to consider that as the level of Nb-dopant increases (in TiO_2) the band bending of this material becomes more pronounced. This would in turn facilitate electron tunnelling from the semiconductor to the electrolyte since the distance that the electron has to travel is reduced. This observation may also explain the cathodic shift of the flat band potential reported for rutile (20) when doped with Nb_2O_5.

4. Li-DOPED ANATASE BASED CATALYST POWDERS ACTIVE IN WATER PHOTOCLEAVAGE UNDER U.V. LIGHT

4.1. Photocatalytic H_2 Evolution on Li-Doped Catalysts

Photocatalytic H_2 Evolution on Li-Doped TiO_2. Figure 3 presents the results of the H_2 evolution rate for UV irradiations of a platinized catalyst of TiO_2 (Bayer) containing different amounts of Li. In all cases a flux of 170 mW/cm^2 of a 150-W Xe lamp was used; 0.05% Pt loading on TiO_2 was employed in all cases. In the present experiments it is readily seen that Li doping affects the rate of hydrogen generation and that the sample 0.05% Pt/TiO_2-7% Li represents the most favorable situation for H_2 evolution. Different loadings of Li have been prepared by heating LiOH and TiO_2 for 15 h at 200° C and for 2-3 h at 500° C. The 200° C preparation temperature is relative low, below $\sim$ 0.3 of the Tammann temperature for TiO_2. From this observation it follows that the mobility of the Li ions at 200° C inside the bulk of TiO_2 is small. It is difficult then to imagine that substitutional incorporation of Li could occur at 200° C. Therefore, it is unlikely that bulk Li doping has taken place. Li ion doping will take place on the surface of TiO_2. In work related to Li doping of NiO (26) it has been observed that Li^+ ions begin to enter the TiO_2 lattice at $\sim$ 400° C. The fact that Li^+ does not have the same valence state as Ti^{+4} means that the samples prepared at 500° C (with consequent remarkable increase in the observed H_2 yields) involve an extra charge introduced by these ions.

TiO_2 is an n-type excess semiconductor. When Li^+ is added to TiO_2,

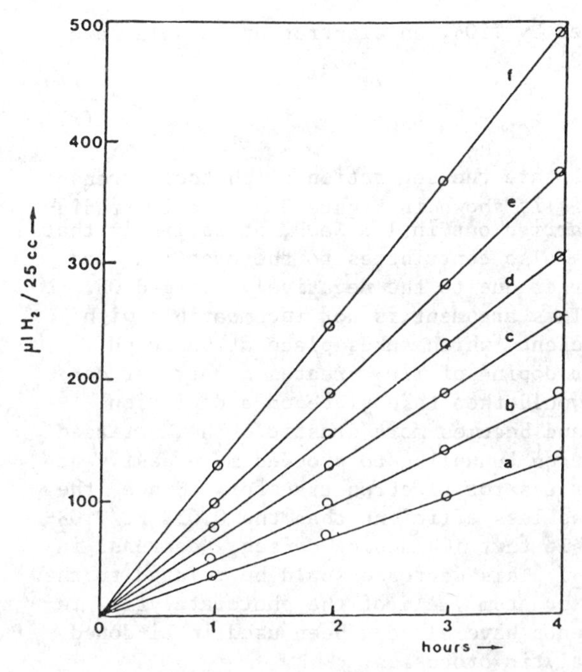

Figure 3. Hydrogen evolution induced by UV light in 1 N NaOH solutions using 40 mg of catalyst 0.05% Pt/TiO_2 (Bayer) with an added amount of atomic Li at 500°C: (a) 0%, (b) 0.5%, (c) 2.7%, (d) 5%, (e) 10%, (f) 7%.

an extra charge is introduced in the system due to this acceptor. Using the Kröger-Vink notation, one may express the defects introduced into TiO_2 in the following way:

$$O_o \;=\; \tfrac{1}{2}O_2 + V_o^{\bullet\bullet} + 2n \tag{5}$$

or

$$2Ti_{Ti} + O_o \;=\; \tfrac{1}{2}O_2 + V_o^{\bullet\bullet} + 2Ti_{Ti}' \tag{6}$$

where O_o = oxygen ions on normal lattice sites, $V_o^{\bullet\bullet}$ = oxygen anion vacancy doubling positive charged, n = electrons in lattice, Ti_{Ti} = Ti metal on a normal lattice position, Ti_{Ti}' = Ti metal on a normal lattice position with a negative charge. Since Li-TiO_2 will equilibrate with the air atmosphere. This means that the incorporation of Li as Li_2O proceeds without adding extra oxygen from the gas phase in the equilibrium state. If Li_2O is incorporated into TiO_2 as acceptor, this acceptor will be compensated at low oxygen activities by oxygen vacancies. Balancing the equation for Li incorporation with respect to charge mass and lattice sites

$$Li_2O \;=\; 2Li_{Ti}''' + O_o + 3V_o^{\bullet\bullet} \tag{7}$$

From equation (5)-(7) it follows that the acceptor doping through Li has resulted in the introduction of a sizable concentration of oxygen vacancies. These oxygen vacancies would be in direct proportion to the Li concentration used. This increase in oxygen vacancy concentration is consistent with the experimental findings reported in Figure 3 up to a level of 7% in Li doping.

When UV light is absorbed by TiO_2, an electron and a hole are formed as shown by equation (8):

$$h\nu\ (>E_g) + TiO_2 \rightarrow e_{CB} + h^+ \qquad (8)$$

Li doping seems to facilitate the separation of photoelectrons and holes at the barrier. This is shown in Figure 3. As all irradiations in Figure 3 have been carried out in 1 N NaOH, it is likely that the excess of OH^- basic groups also contributes to the creation and maintenance of a Schottky barrier due to the negatively charged OH^- groups on the surface (27). This argument is not incompatible with an increase in tunneling efficiency which takes place at basic pH values. It is well known that doping of TiO_2 creates a narrower depletion layer. Band bending would then take place on a depletion layer with a narrower width and becomes more drastic. The increased band bending would allow electron tunneling to proceed more easily at the interface, enhancing the rates for electron transfer. Since the 0.05% Pt/TiO_2-10% Li sample was less efficient than the 0.05% Pt/TiO_2-7% Li (Figure 3), it is possible that 10% doping causes a decrease in the negative potential of TiO_2. This decrease would be related to the corresponding decrease in the quantum yield of the photocatalytic reduction reaction. Such arguments have already been used in Li-doped surfaces involved in photocatalytic processes.

Figure 4 presents results for H_2 evolution when 40 mg of 0.05% Pt/TiO_2-7% Li is irradiated in 1 N NaOH solution. Samples of the catalyst have been prepared at the temperatures shown in this figure. Since doping was carried out between 0.3 and 0.5 of the fusion temperature of TiO_2, Li ions have enough mobility to penetrate quite homogeneously in the bulk and at the surface. But heating at these relatively high temperatures produces a decrease in the BET area, sintering the samples. Since the bulk diffusion of Ti ions controls the surface area of these catalysts, the sintering process of the host species is reflected in reduced BET areas. Undoped TiO_2 (Bayer) was determined to be 180-200 m^2/g. Determining the BET area for samples doped at 400-700° C, and dividing the H_2 yields at each temperature (Figure 4) by the area available in 40 mg of catalyst, one arrives at the plot of H_2 yields vs. temperature shown in Figure 4b. The maximum activity per milligram of catalyst in Figure 4a is found at 500° C. But since the BET areas of the catalyst doped at 500 and 600° C are 33 and 20 m^2/g, respectively, the activity per square meter of catalyst at 600° C is higher. This is shown in Figure 4b.

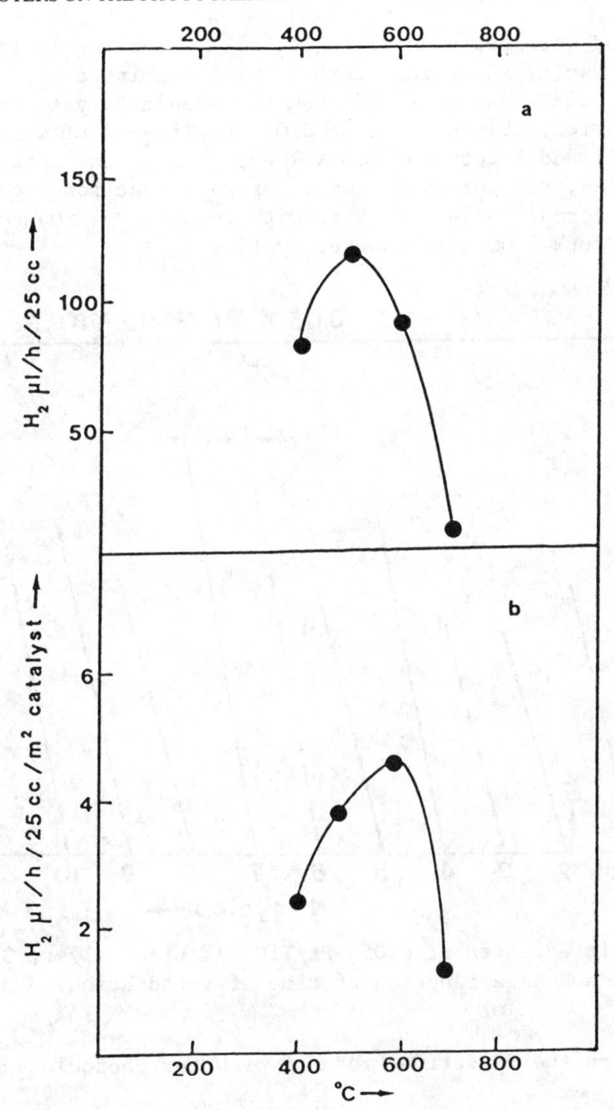

Figure 4. (a) Media rate per hour of hydrogen produced under UV irradiation when 40 mg of 0.05% Pt/TiO_2-7% Li catalyst is irradiated in 1 N NaOH. Temperature of diffusion of lithium into TiO_2 is shown on the abscissa. (b) Same as in part a - hydrogen yields as a function of m^2 BET area of the $Li-TiO_2$ catalyst prepared at different temperatures.

Figure 5 shows repetitive cycles for H_2 formation when 0.05% Pt/TiO_2-7% Li is irradiated several times. The rates of H_2 production in a 4-h period were reproducible up to 15%. Bubbling in the first three recyclings (as shown in Figure 5) will eliminate H_2. In 10 recyclings, 5 mL of H_2 was obtained from the same catalyst over a 40-h irradiation period. Since 35 mg of 0.05% Pt/TiO_2-7% Li was added (4.5×10^{-4}mol) and a total of 6.3×10^{-4}mol of H_2 was obtained during these cycles, the turnover number for H_2 production exceeds 1. Therefore, H_2 formation is catalytic with respect to TiO_2 as reported previously in work from this laboratory (21).

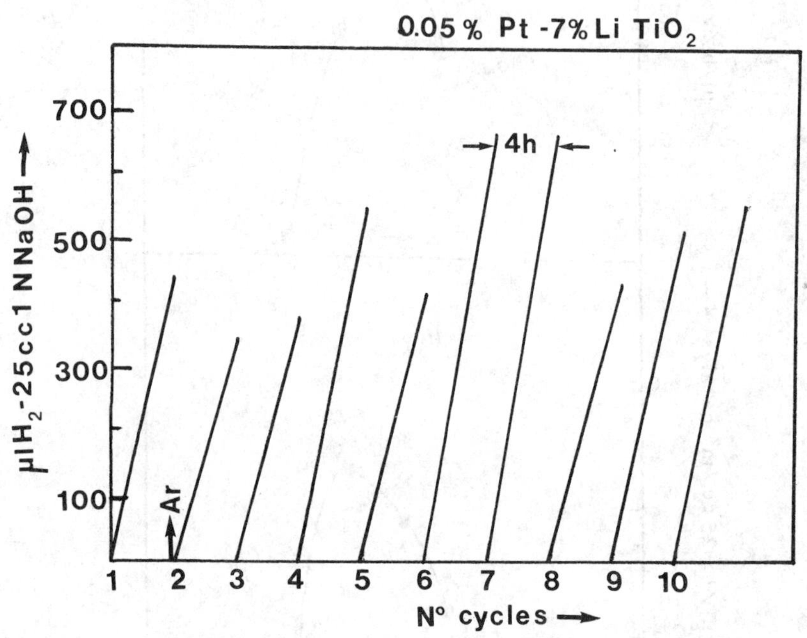

Figure 5. UV irradiation of 0.05% Pt/TiO_2-7% Li catalyst; 35 mg in 25 cm^3 of 1 N NaOH as a function of time of irradiation.

4.2. Studies on the Oxidative Products of Water Photocleavage

Figure 6 (28) presents the results of irradiations carried out for 8 hours on 25 cm^3 solutions when 40 mg of catalyst 0.05% Pt-7% Li is added in solution. As seen from this figure in column a, the pH plays an important role in the amount of hydrogen obtained. The oxidative product formation is shown in columns b and c. When solutions are centrifuged after irradiation and the supernatant is titrated with 10^{-3}M $KMnO_4$ at pH 0, the optical absorption decreases at 525 nm. In a second step, the catalyst precipitate separated from the solution is also reacted with 10^{-3}M $KMnO_4$ (at pH 0) and shows that oxidative products are formed in the bulk. This is shown in column c for the different pHs under study.

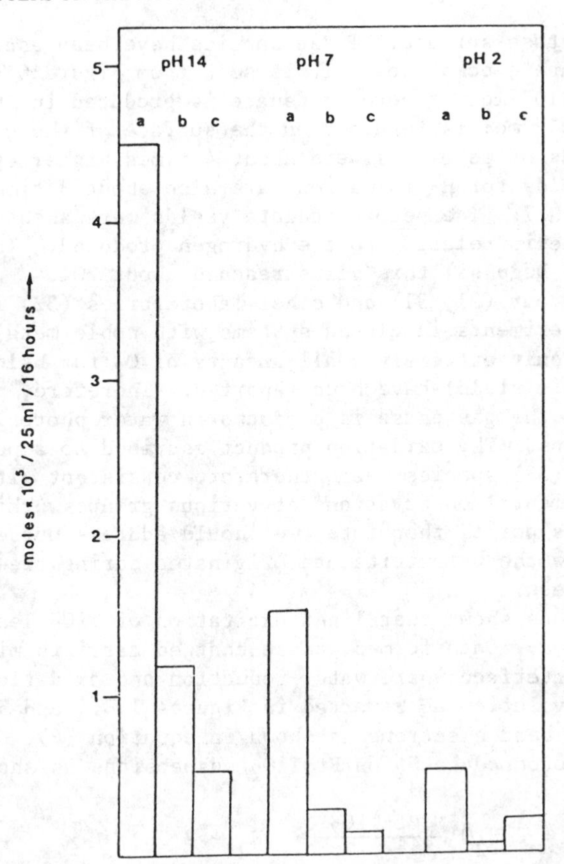

Figure 6. Amount of hydrogen produced (column a) when 25 cm^3 of solution is irradiated with 40 mg of 0.05 Pt/TiO_2-7% Li catalyst at pH 14,7, and 2 for 8 h. In column b, the amount of peroxytitanate in the liquid is shown in each case. In column c, the amount of peroxytitanate particles associated with the catalyst surface is shown in each case.

From the difference between the blank and the photolyzed sample, one derives the quantity of permanganate consumed by reaction with peroxide. Data reported in Figure 6 assumes that 2.5 moles of peroxide are oxidized by one mole of $KMnO_4$. Recent work on TiO_2 systems irradiated in gas and liquid media (29,30) have reported peroxide formation (O_2^{2-}) by EPR and by infrared (IR) spectroscopy when TiO_2 is irradiated in the UV. These species were reported to have been retained on the TiO_2 surface in all cases. IR spectral bands between 800 and 900 cm^{-1} appear on UV irradiation due to the various modes of vibrations of Ti-O in the anatase surface. These species have been ascribed to surface peroxytitanate complexes. It is seen from Figure 6 that at pH 14 about 1.2 x 10^{-5}mol of peroxytitanate is produced in the liquid phase and 0.6 x 10^{-5} mol is retained on the surface of the catalyst powder. The yields of peroxides were about 4 times higher at pH 14 than at pH 7. Yields for H_2 formation are also about 3 times higher at pH 14 than at pH 7. The peroxytitanate yields were about 50% of the amount possible in relation to the hydrogen produced. In the 0.05% Pt/TiO_2 (P25 Degussa) this value reached about 20%.

From work in our (21, 31) and other laboratories (32) involving water cleavage experiments in closed systems with noble metal loaded TiO_2 suspensions, only extremely small amounts of O_2 far below 10% of their stoichiometric yield) have been reported. Therefore, practically no oxygen in the gas phase is produced in water photocleavage of these suspensions. The oxidation product ascribed to a peroxide or peroxititanate (33) species seems therefore consistent with the most recent experimental observations of various groups working in this area. At this point, therefore, we should address ourselves to the question of how the peroxititanate originates during the water photocleavage process.

Equation (8) has shown that light excitation of TiO_2 leads to e^-_{CB} and h^+ formation. Once formed, these charged carriers migrate to the solid-liquid interface where water reduction and oxidation take place. Hydrogen evolution as reported in Figures 1,3,4 and 5 would involve conduction band electrons as shown in equation (8). These electrons would photoproduce H_2 on Pt-TiO_2 suspensions as shown by equation (9):

$$2e^-_{CB} + 2\,H_2O \xrightarrow{Pt-TiO_2} H_2 + 2\,OH^- \qquad (9)$$

while valence band holes (h^+) lead to peroxide formation:

$$2h^+ + 2H_2O \longrightarrow H_2O_2\ (TiO_2) \qquad (10)$$

where H_2O_2 (TiO_2) is an unidentified peroxide or peroxititanate free or associated with TiO_2. Since no O_2 has been found in closed systems, if present, is unlikely to escape into the gas phase since photo-uptake by the TiO_2 particulate has been reported to be a rapid process (34). This is shown in equation (11):

$$O_2 + 2e^-_{CB} + 2H^+ \longrightarrow H_2O_2\ (TiO_2) \qquad (11)$$

Rapid chemisorption of O_2 onto TiO_2 has been previously reported by the same laboratory (35) via reduction with conduction band electrons producing $HO_2\cdot$ radicals as detected by EPR methods. Since O_2 is reduced to $O_2\cdot^-$ ($HO_2\cdot$) by conduction band electrons, one should not overlook the fact that part of the available electrons for proton reduction is consumed in this way, lowering the quantum yield for H_2 formation (21,23,28.31).

To rationalize the results obtained in Figure 6, the results seem consistent with the following experimental observations: (a) OH basic groups on the surface are more abundant at basic pH. More intermediate peroxide and peroxytitanate are detected in basic media. Since this intermediate is the precursor of the H_2, it explains the decrease observed in H_2 yields as the pH becomes more acidic. (b) Adsorption of H_2O_2 causes an intense yellow color (36). In our experiments long-time irradiations (50 h) produced an intense yellow color but only in alkaline media. (c) Since peroxytitantes originate from intermediates that are formed from mobile OH groups at the TiO_2 surface, the loss of available OH basic groups will lead to a more difficult surface diffusion step for the remaining OH groups (37). The activity for oxidative product formation at more acidic pHs will be therefore lowered consistent with the results shown in Figure 6.

In order to determine kinetically the validity of the proposed mechanism as stated in equations (8) through (11), photoadsorption experiments were carried out to show the time scale of this phenomena. In this way the events taking place in reactions (9) and (11) could be correlated, lending further support to the mechanism leading to water photocleavage. Figure 7 presents the kinetics of oxygen photoadsorption for (a) 0.05% Pt/TiO_2, (b) 0.05% Pt/TiO_2-0.5% Li, and (c) 0.05% Pt/TiO_2-7% Li. Runs were carried out in alkaline solutions at pH 14; 375 μL of O_2 was injected in the three runs. Results are shown on the left-hand side in Figure 7. On the right-hand side of this figure the concomitant hydrogen production is shown as a function of time. The observed rate for H_2 production (samples a-c) was the same as the reported rates in Figures 3 and 5, once the initial oxygen added was consumed. From these oxygen adsorption experiments, it is readily seen that the O_2 adsorption capacity of TiO_2 is changed due to Li doping. This observation has been reported before on Li-NiO and Li-ZnO samples adsorbing oxygen from the gas phase (38). The adsorbed fraction of oxygen increases (as shown in Figure 7) when the Li content in the sample is higher. The decreasing rate observed for the oxygen adsorption in the three samples under study is due to the boundary layer produced on O_2 adsorption which makes subsequent absorption slow. A Schottky barrier is formed at the surface due to oxygen chemisorption as shown by equation (12) and will move to more

$$e^-_{CB} + O_2 \longrightarrow O_2^- \qquad (12)$$

negative values during the O_2 photoadsorption process. Such a process would take place at high pH since O_2^- at lower pH is converted to HO_2·. Figure 7 shows then that doping TiO_2 with Li at 500° C develops the capacity for oxygen chemisorption of these catalysts. This capacity is not negligible since up to one monolayer of O_2 on TiO_2 can be adsorbed in liquids. When the light is turned off, the O_2 stays irreversibly adsorbed, which is a characteristic of photoadsorption processes.

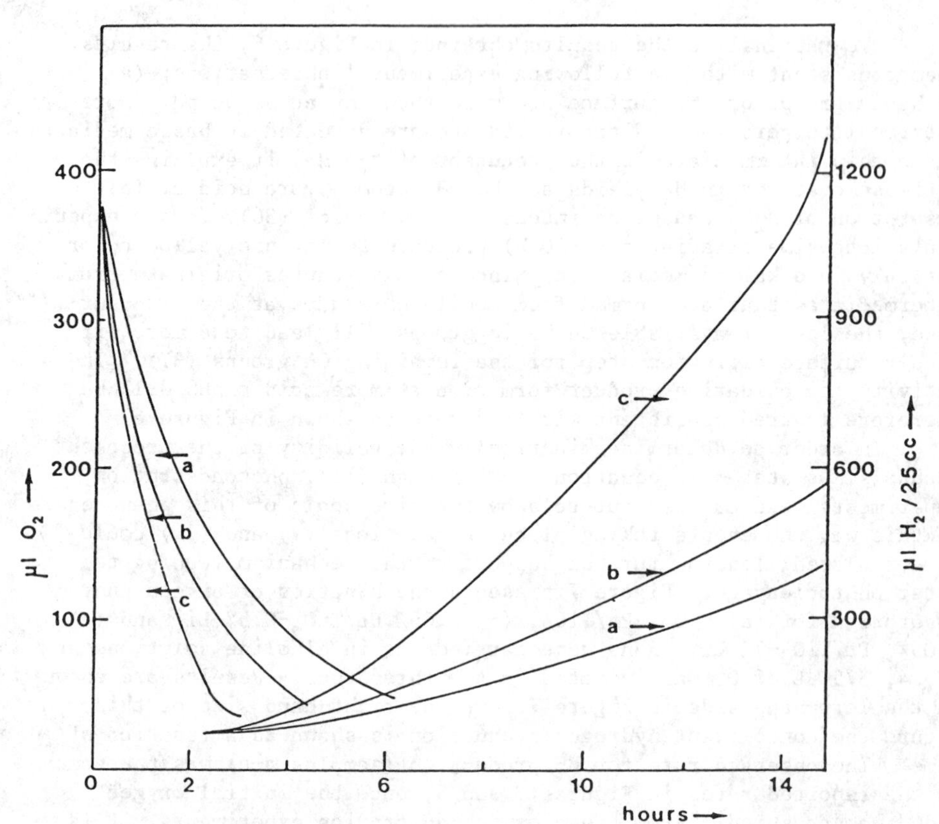

Figure 7. O_2 photoadsorption (left ordinate) and photoproduced H_2 (right ordinate) when solutions containing 375 μL of O_2 are irradiated in 25 cm^3 of degassed 1 N NaOH containing 40 mg of the following catalysts: (a) 0.05% Pt/TiO_2, (b) 0.05% Pt/TiO_2-0.5% Li, (c) 0.05% Pt/TiO_2 7% Li. The lines plotted in the figure are drawn through points taken at 30 min. intervals in the O_2 photoadsorption experiments.

From Figure 7, it is readily seen that the nature of active sites on the catalyst is similar for O_2 adsorption or H_2 evolution. H_2 evolution is only possible when more than 90% of the O_2 added initially has been chemisorbed. The question as to whether the Li doping increases the intrinsic activity of the active sites but not their number remains to be considered. By using equations (5) through (7) elaborated upon previously in this paper, it is readily seen that Li^+-ion compensates for electron deficiencies in the existing oxygen vacancies. Active catalytic sites could, therefore, be associated with oxygen vacancies created in the doped system. Since in Figure 7, photoadsorption occurs preferentially in the 7% Li^+_TiO_2 system (over lower

Li-doped systems) then we are led to believe that Li does affect the number of active catalytic sites up to this doping level. This observation would lend further support to the H_2 evolution results shown in Figure 7 and, more important, previously reported in Figure 3.

Formation of surface peroxides has been reported on TiO_2 under various conditions either by absorption of H_2O_2 (45,33) or involving light irradiation processes (30,46,21). There are 4 forms of superficial peroxides possible on TiO_2 that, if present in relatively high amounts, (irradiations of more than 1 day in an active catalyst) are detrimental to further H_2 production by the catalyst. Figure 5 has shown recycling of 4 hours for H_2 evolution for a considerable number of cycles. This short recycling time seems to prevent accumulation of peroxititanates on the TiO_2 surface, hindering further H_2 evolution. Structure a and b shown below will be possible for μ-peroxodimers, in which (a) is more likely than (b) since it coordinates one Ti only. The possibility of structures (a), (b), (c) to co-exist will

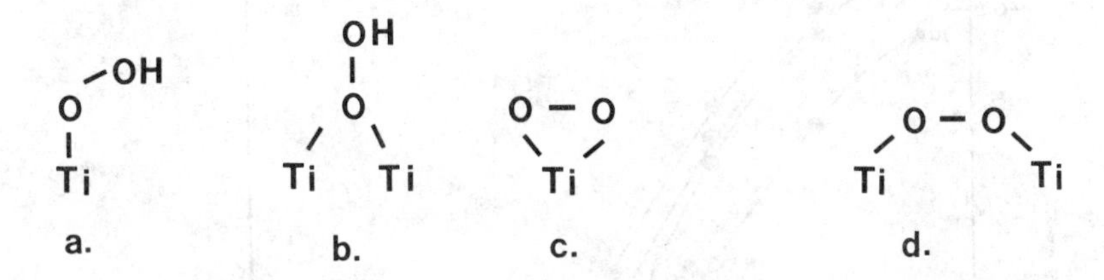

depend on the kinetics, stability constant and reversibility of the postulated structures. The most favored structure would be (d), having stereochemistry and bonding distances that favor protonation to peroxititanate (47). In this case, equation (10) could be rewritten:

$$\mathrm{Ti\,OH} + 2h^{+} \longrightarrow \mathrm{Ti}\diagup\overset{\mathrm{H}}{\mathrm{O}} — \overset{\mathrm{H}}{\mathrm{O}}\diagdown\mathrm{Ti} + 2H^{+} \qquad (10a)$$

4.3. Electrochemical Characterization of Li-Doped Catalyst Powders.

Figure 8 presents the results obtained for electrophoretic mobilities of the following samples: (a) TiO_2 (Bayer), (b) TiO_2-2.7% Li, (c) TiO_2-7% Li and, (d) TiO_2-10% Li. A solution of 10^{-2}M ionic strength made of appropriate concentrations of NaOH and HCl was used in all electrophoretic measurements. This allows a constant width for the diffuse layer for all the particles under observation. From Figure 8 it is readily seen that a shift takes place in the isoelectric point (IEP) of anatase as Li doping increases from 5.4 (TiO_2 (Bayer) to 6.3 (TiO_2-7% Li). Since the anatase used in this work was obtained by low-temperature hydrolysis from titanyl sulfate, sulfate ions cause the hydrolyzed TiO_2 to show an IEP at a more acid pH than TiO_2 (P25 Degussa) (IEP) 6.6). Thus, some adsorption of sulfate takes place. Higher values for the IEP in the Li-doped samples indicate that positive charges carried by the Li are determining in the observed mobilities around the IEP. Electrophoretic mobilities are therefore sensitive in the region where the observed mobilities for species tend to zero. The species under consideration are $TiOH_2^+$ and TiO^-. Electron microscopy studies (28) show that particles of Li-doped TiO_2 catalyst aggregate in solution. Care was taken to follow the mobility of the smaller aggregates during electrophoresis. They consist of smaller particulate aggregates per kinetic unit and better reflect the electric field interaction on the particles under observation.

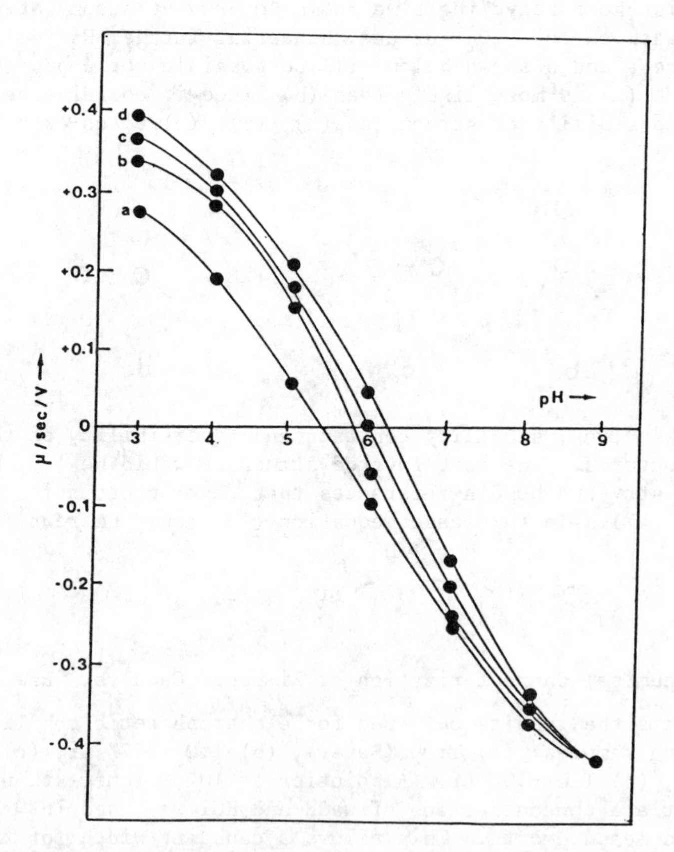

Figure 8. Electrophoretic mobility as a function of pH for 10^{-2}N (NaOH + HCl) solutions of (a) TiO_2 (Bayer), (b) TiO_2-2.7% Li, (c) TiO_2-7% Li, (d) TiO_2-10% Li.

Conduction band determination as shown in Figure 9 was carried out according to an electrochemical method (39) of collecting the photogenerated charge on a Pt flag electrode immersed in the irradiated suspensions. The solution used had a composition as described in the legend to Figure 9. Our reaction vessel was of the same design and volume as employed in ref. 8. The voltage variation induced under irradiation was followed in a recorder which was fit to register current variation by means of an appropriate resistance. Typical data from such experiments are shown in Figure 9. Since the redox potential of the MV^{2+}/MV^{+} couple is pH independent, in traces a-c, the observation of the pH effect is related to changes within and/or on the surface of the Li-doped semiconductor powders used. The source of this effect is the shift of the Fermi energy level E_f with pH as shown by equation (13). Extrapolation of the two lines $\Delta i/\Delta t$ when large and small $\Delta i/\Delta t$ changes were observed afforded a

$$E_f = E_f(\text{pH } 0) - 0.059\text{pH (at } 25^{\circ}\text{C)} \tag{13}$$

precision of $\pm$ 0.1 pH unit in the determination of pH_o. The intersection point (pH_o) represents the pH value in equation (13) when the Fermi level (E_f) equals the redox potential of the couple MV^{2+}/MV^{+}. For traces a (TiO_2) (Bayer), b (TiO_2) (Bayer)-2.7% Li) and c (TiO_2) (Bayer)-7% Li), the pH_o values found were 6.5, 7.5 and 9.0, respectively. From the pH_o values obtained in equation (13) and with the knowledge that the MV^{2+}/MV^{+} redox potential is -0.69 V vs. SCE, E_f values of -0.23, -0.19, and -0.13 V are obtained for the compounds shown in traces a-c. Li doping shifts pH_o in equation (13) to more basic pH values.

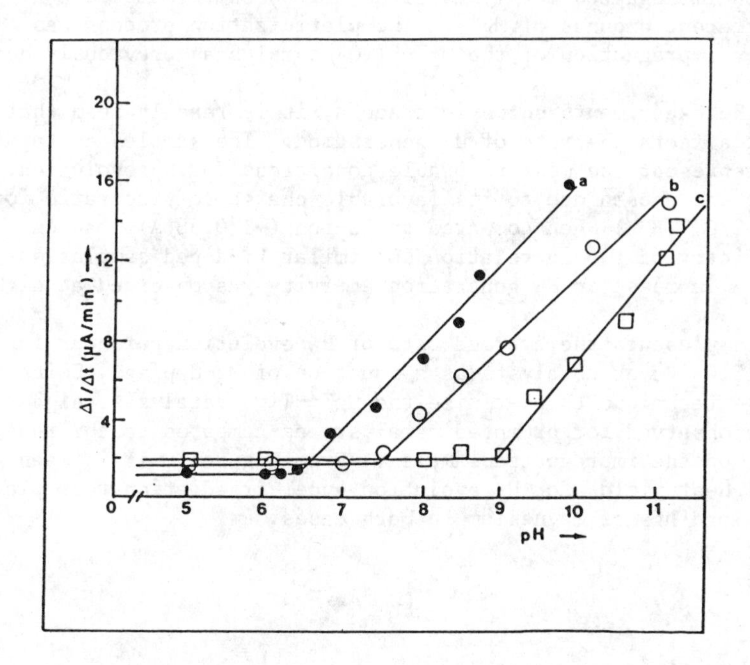

Figure 9. Dependence of rate of the change of photocurrent with time ($\Delta i/\Delta t$) on pH for stirred and N_2- purged TiO_2 suspension photocell. Cell conditions: TiO_2 powder (250 mg); H_2O (100 mL); [NaOAc] = 1.0 M; [KNO_3] = 0.1 M; [MV^{2+}] = 1 mM. Platinum collector electrode at -0.20 V vs. SCE: (a) TiO_2 (Bayer), (b) TiO_2 (Bayer)-2.7% Li, (c) TiO_2 (Bayer) -7% Li.

5. THE EFFECT OF PROMOTER AT THE SEMICONDUCTOR INTERFACE

In order to further test alkaline-doped TiO_2 catalysts active in water photolysis, we extended our investigations to Mg^{+2}-TiO_2 doped systems (40). The Mg^{+2}-TiO_2 doped samples were prepared by Dr. Panek (Bayer AG, Krefeld, Uerdingen, West Germany) using the slurry technique as outlined by Teichner in reference (22). In order to produce bulk magnesium-doping within a reasonable time, a dried slurry of each Mg^{+2}-TiO_2 sample was heated for two hours at 500°C. X-ray diffraction analysis was performed on all samples up to 12.5% Mg^{+2}-content. No evidence for titanate formation ($MgTiO_3$ or Mg_2TiO_4) was found. This experimental observation indicates that lattice substitution by Mg^{+2} has taken place in the doping process without formation of a new phase. About 1000°C has been shown necessary to form $MgTiO_3$ under the present experimental conditions (8).

Figure 10 presents the average rate per hour for H_2 evolution measured during the first four hours of photolysis. Trace a shows 0.5% Pt-TiO_2 impregnated catalysts of Mg^{+2}-TiO_2 Bayer samples containing different amounts of Mg^{+2}. The platinization process was performed by impregnation of the Mg^{+2}-TiO_2 samples as previously described (21).

From the experiments shown in trace a, it is readily seen that Mg^{+2}-doping affects the rate of H_2 generation. The samples containing 1% and 2% represent the most favorable conditions for H_2 evolution. Mg^{+2}-doping was chosen due to its favorable charge to size ratio for this ion (+2/0.7.8 Å) when compared to Li-ion (+1/0.68 Å). An enhancement factor of two in relation to similar Li-doped samples was observed. A decrease in H_2 generation activity was observed at higher Mg-doping.

Trace b presents the average rate of H_2 evolution per hour for 0.05% Mg^{+2}-TiO_2 Bayer catalysts as a function of Mg-doping. Using the ion-exchange technique to platinize the Mg^{+2}-TiO_2 catalysts, higher yields were observed for promoted catalysts as compared to the runs carried out on the impregnated samples. From Figure 10 it is seen that the highest yields for H_2 evolution under irradiation took place at similar loadings of magnesium in both cases.

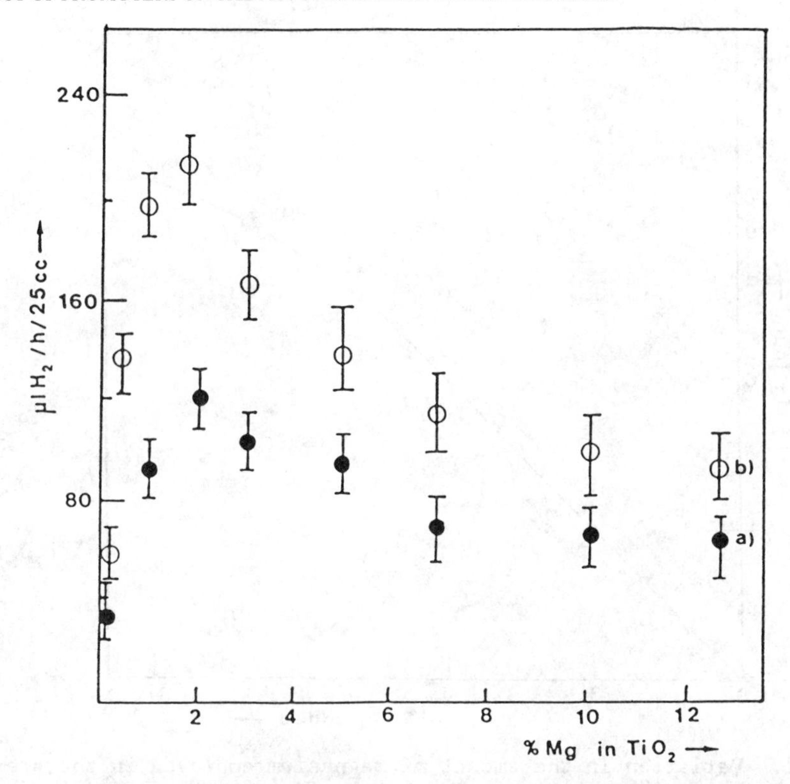

Figure 10 a) Media rate per hour of hydrogen induced by u.v. in 25 cc 1 N NaOH solutions using 40 mg impregnated 0.5% Pt-Mg^{+2}-TiO_2 catalyst. All irradiations were carried out with an Xe lamp 90-100 mW/cm^2.
b) Media rate per hour of hydrogen under u.v. irradiation when 40 mg exchanged 0.05% Pt-Mg^{+2}-TiO_2 was irradiated in 1N NaOH.

Figure 11 presents the results for the atomic percentage of Mg^{+2} found vs. the amount of Mg^{+2} added during the preparation. A 1:1 linear correlation is formed; we ascribe this to the substitutional doping of Ti^{+4} with Mg^{+2}. At high Mg^{+2}-dopant levels not all the Mg^{+2} initially added is incorporated in the TiO_2 solid.

In order to explain the results obtained in Figure 10 we use again the point defect model as a simple qualitative model throughout this study only as a basis for the discussion of the electronic character of the solids. As such, it should provide an indicator of the reactivities as a function of the increasing concentration of added magnesium (6,7,41).

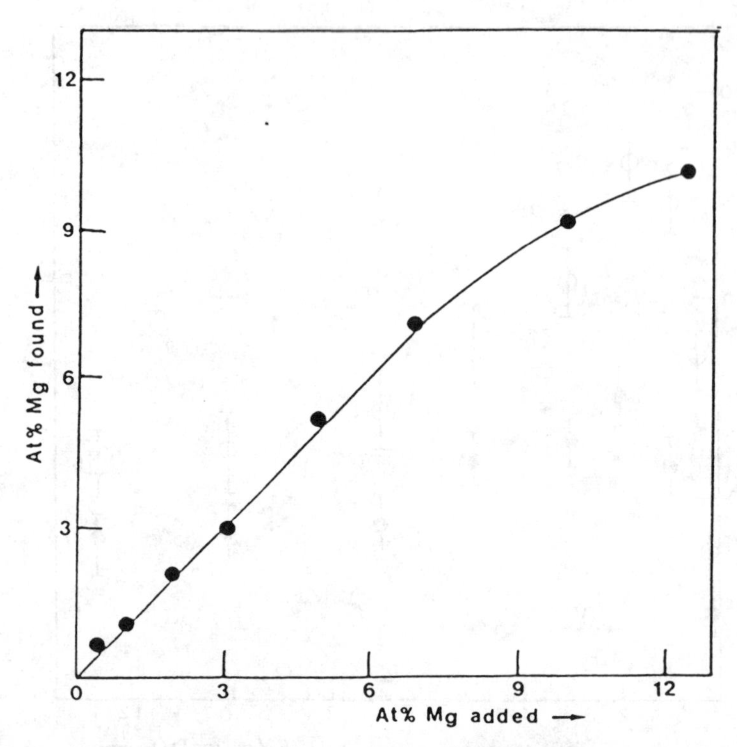

Figure 11. Variation in the amount of magnesium employed in the preparation of the Mg^{+2}-TiO_2 samples and the amount of magnesium found by atomic absorption in the actual samples.

TiO_2 is an n-type metal excess semi-conductor. Compensation of n-type character is apparently achieved by magnesium-doping. Using the Kroger-Vink (42) notations, the defects introduced into TiO_2 may be expressed in the following way:

$$O_o = \frac{1}{2}O_2 + V_o^{\cdot\cdot} + 2e' \quad (14)$$

$$\text{or } 2Ti_{Ti} + O_o = \frac{1}{2}O_2 + V_o^{\cdot\cdot} + 2Ti'_{Ti} \quad (15)$$

$$K_1 = [V_o^{\cdot\cdot}]\,[e']^2\, P_{O_2}^{\frac{1}{2}} \quad (16)$$

where: O_o : oxygen ions on normal lattice sites

$V_o^{\cdot\cdot}$: oxygen anion vacancy doubly positive charged

e' : electrons in lattice

Ti_{Ti}: Ti on a normal lattice position

Ti'_{Ti}: Ti on a normal lattice position with a negative charge

P_{O_2} : partial pressure of the oxygen atmosphere

In addition to equations (14)-(16), lattice incorporation of MgO could be stated by equations (17)-(20), using Hauffe's notation (10, 37) h^+ being holes in the valence band:

$$O_o = \tfrac{1}{2}O_2 + V_o \quad (17)$$

$$MgO + \tfrac{1}{2}O_2 = Mg''_{Ti} + TiO_2 + 2h^+ \quad (18)$$

Adding equations (4) and (5):

$$MgO + O_o = Mg''_{Ti} + V_o + TiO_2 + 2h^+ \quad (19)$$

Since $V_o + 2h^+ = V_o^{\cdot\cdot}$

$$MgO + O_o = Mg''_{Ti} + V_o^{\cdot\cdot} + TiO_2 \quad (20)$$

Magnesium has been introduced in the catalysts under oxidizing conditions at 500°C, i.e. as MgO, and under these conditions the titanium will exist almost entirely as Ti^{+4}-ion (6). Equations (14) through (20) then indicate that Mg^{+2} added increases the concentration of ionized vacancies. This would, in turn, compensate n-type conductance.

Since a decrease in the observed H_2 evolution rate seems to take place above 2% Mg^{+2}-doping, it is possible that the added Mg^{+2}-ion would neutralize the strongest acidic surface OH-group (43) of TiO_2 above this level. Residual sites (above this level of magnesium doping) then show also residual activity contributing to the hydrogen generation process (44).

It has been observed in Figure 10 that the formation of H_2 during photolysis is not accompanied by the appearance of O_2 in the gas phase. Instead, peroxotitanium complexes are formed which were analyzed by $KMnO_4$ titration as previously reported (21). It is readily seen from Figure 12 that peroxotitanate detected by this method corresponds to about 40% of the possible stoichiometric amount in relation to the H_2 evolved during the reaction.

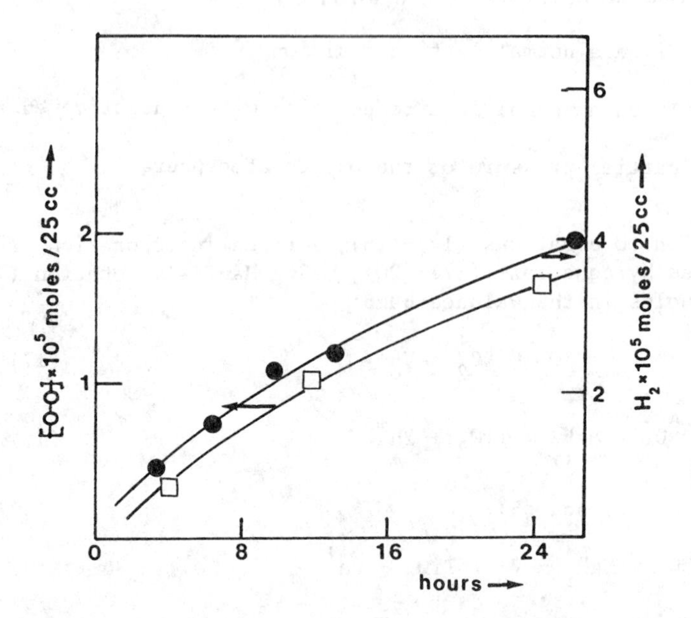

Figure 12. Amount of peroxotitanate as a function of irradiation time for 1N NaOH solution of 40 mg 0.05% Pt-7% Mg^{+2}-TiO_2 catalyst. H_2 production as a function of time is shown on the right hand side.

6. THE EFFECT OF Ba^{+2}-IONS ON SUSPENSIONS LEADING TO H_2 EVOLUTION UNDER U.V. LIGHT

The effect of Ba^{+2} ions on hydrogen production from irradiation of TiO_2/Pt dispersions in 0.1 M NaOH is illustrated in Figure 13. The volume of H_2, detected in the head space above the solution, is plotted as a function of irradiation time. In the absence of Ba^{+2}, the hydrogen generation rate ($r(H_2)$) is initially 100 μl/h which corresponds to an overall light to chemical energy conversion efficiency (η) of 0.12%. The rate decreases with irradiation time until after ca. 10 h a plateau is obtained where $V(H_2)$ = 700 μl. Addition of 0.05 M $Ba(OH)_2$ induces an increase in $r(H_2)$ which is initially 600 μl/h. This corresponds to an η-value of 0.7%. The rate of hydrogen generation decreases also in this case with time. However, even after accumulation of 1.2 ml of hydrogen, the photostationary state is not yet reached since there is still a net H_2 generation. Note that in Figure 13, by introducing $Ba(OH)_2$, the OH^- concentration is doubled, which could affect the hydrogen yields. It is important to avoid $BaCO_3$ contamination of the catalyst. In the preparation of $Ba(OH)_2$/NaOH mixtures, we first dissolved the appropriate amount of $Ba(OH)_2$ in water. After filtering, the NaOH pellets were added to the $Ba(OH)_2$ solution and, if necessary, the filtration was repeated.

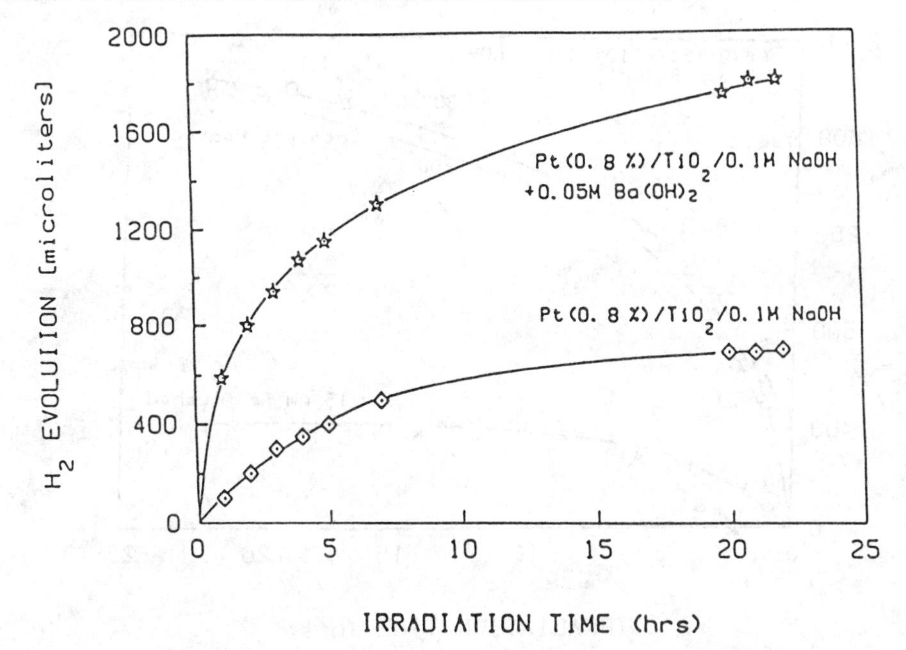

Figure 13. Hydrogen generation during irradiation of aqueous dispersions of TiO_2/Pt (0.8%). 50 mg catalyst in 25 ml 0.1 M NaOH were irradiated with the $\lambda > 320$ output of a 150 W Xe lamp. The figure illustrates the drastic improvement in the hydrogen yields induced by Ba^{2+} ions.

Thus, the augmentation of the efficiency for hydrogen generation observed in Figure 13 is primarily due to Ba^{+2} and not to the increase in OH^- concentration. In fact, the deviation in the $r(H_2)$ values between runs with different lots of catalyst was smaller than 5% if during the loading of TiO_2 (P-25) with Pt the exchange procedure is employed (21).

This improvement in H_2 output could not be explained as being due to increased absorption of Ba^{+2}-ions nor should these ions intervene in the interfacial electron transfer leading to water photocleavage. Based on these observations, we reasoned (23) that peroxiscavengers, e.g. a simple inorganic ion like Ba^{+2} which forms an insoluble Ba-peroxide in solution reduces the surface concentration of peroxo-species formed during the photolysis. It has previously been mentioned (21,28,46,48) that peroxo-species produced during this reaction are partially adsorbed at the TiO_2 interface hindering further H_2 evolution by the catalyst. To test the involvement of Ba^{+2}-ions as a scavenger for photogenerated peroxides forming insoluble barium peroxide, we have carried out regeneration of the catalyst at 300°C, destroying the peroxides produced during irradiation. This is shown in Figure 14.

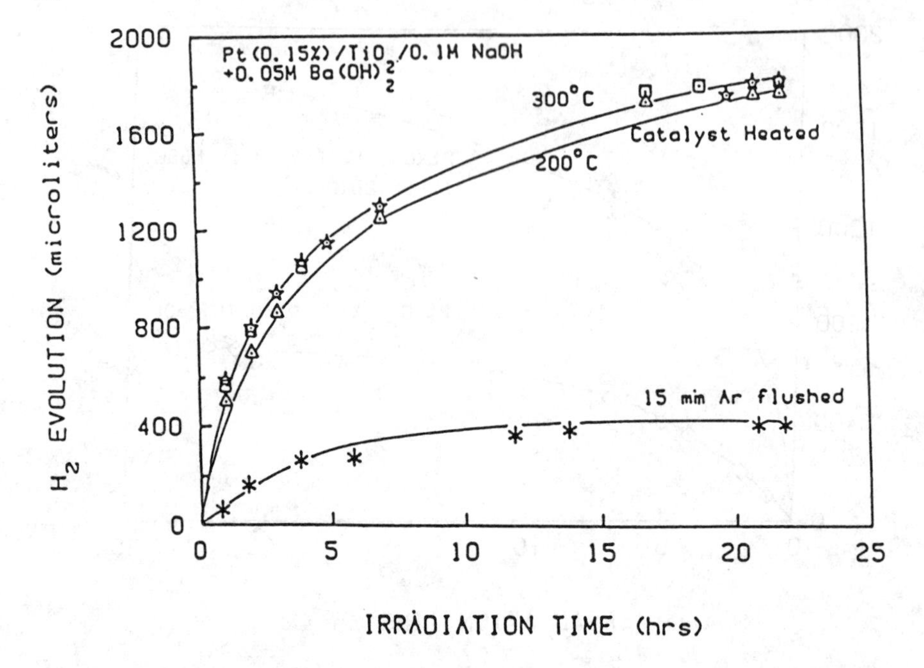

Figure 14. Recycling of the water cleavage catalyst by inert gas flushing and thermal treatment.

- ⊡ First run of the water cleavage experiment in the presence of 0.05 M $Ba(OH)_2$, under conditions as in Figure 1.
- ✱ After reaching the photostationary state dispersion was flushed with Ar for 15 min. and irradiation repeated.
- ☆ △ Particles were filtered and treated under Ar for 2 hours at 200°C and 300°C, respectively. The catalyst was redispersed in 25 ml NaOH (0.1 M)/$Ba(OH)_2$ ((0.05 M) and irradiation repeated.

In this figure, the curve denoted by represents a repetition of water photolysis with Ba^{+2}-ions containing TiO_2/Pt dispersions under the same conditions as in Figure 13. After 20 h of illumination, when ca. 1.8 ml of H_2 had accumulated above the solution, the photostationary state was attained and there was practically no further H_2 generation. Flushing the dispersion with Ar for 15 min. restored only partially its photocatalytic activity since the initial value of $r(H_2)$ was 75 µl/h and a plateau was reached already after accumulation of 400 µl H_2. In contrast, heating the catalyst for 2 hours under Ar at 300°C reactivated the catalyst entirely as shown by the fact that the hydrogen generation curve observed after such a treatment is indistinguishable from that obtained in the first run. Lowering the temperature for the thermal treatment to 200°C appears to give a slightly lower hydrogen yield. Therefore, thermal treatment decomposes the peroxides and as shown in Figure 14, it has been established that when a TiO_2/Pt suspension containing 0.05M $Ba(OH)_2$ and 0.1 M NaOH was irradiated under conditions as in Figure 13, ca. 2 ml of hydrogen were produced. At this time, the catalyst was centrifuged and divided into two parts. The first was immediately examined for peroxide by the permanganate method (21) while the second was first heated under Ar at 300°C and then subjected to analysis. This test showed that the peroxide formed during irradiation was completely eliminated by the heat treatment. The accumulation of these species at the surface therefore decreases the water photolysis efficiency leading to the leveling-off observed in H_2 evolution as observed in Figures 13 and 14. The accumulation of peroxide during the water photolysis, therefore, provides a key to the understanding of the effect of Ba^{+2} ions on the quantum yield of hydrogen formation. There is evidence from the work by Boonstra and Mutsaers (45) that H_2O_2 is specifically adsorbed at the TiO_2 surface, producing a peroxo-complex with one of the possible structures outlined at the end of section 4.2. The insoluble Ba^{+2}-peroxide reducing the surface concentration of peroxo-species can be stated:

$$Ba^{+2} + HO_2^- \rightarrow BaO_2 \times aq + H^+ \tag{21}$$

The solubility product of $BaO_2 \times 8H_2O$:

$$K = [Ba^{+2}] \times [H_2O_2] \; [OH^-]^2 \tag{22}$$

is o.85 x 10^{-11} at 0°C (18) which, at the Ba^{+2} and OH^- concentrations employed in Figures 13 and 14 would correspond to an upper limit of 3.7 x 10^{-8} M in peroxide concentration.

Further confirmation for the formation of peroxide was obtained by using o-tolidine as a redox indicator (23). A yellow color peak with an absorption maximum at 433 nm is observed. This is characteristic for the two-electron oxidation product of o-tolidine formed by the reaction with peroxide (49). The permanganate titration gives a higher ration, i.e. 2:1, which could be the consequence of partial decomposition of peroxide during the neutralization process (45).

Another effect of Ba^{+2}-ions was observed when photouptake of O_2 was carried out as shown in Figure 15. In these experiments, 1 ml of O_2 was injected in the dispersion and its consumption under band-gap irradiation monitored by gas chromatography. The results shown in Figure 15 indicate that the photo-uptake of O_2 is markedly retarted in the presence of $Ba(OH)_2$. The presence of Ba^{+2} ions at the TiO_2-water interface appears to also retard the reduction of oxygen by conduction band electrons and the H_2 recombination with O_2, as is apparent from the results in Figure 15. The effective action of Ba^{+2} ions could be rationalized knowing that alkaline-earth metal ions are strongly adsorbed onto TiO_2 (43), which should favor their reaction with peroxides associated at the surface.

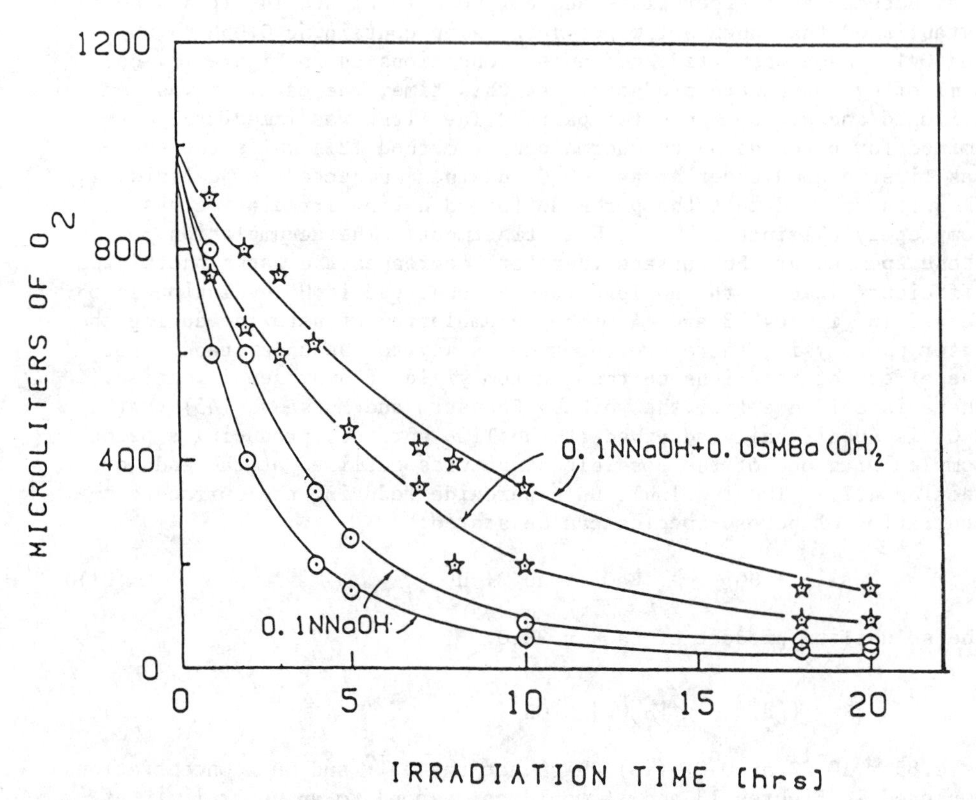

Figure 15. Effect of Ba^{+2} ions on the photo-uptake of oxygen in 25 ml TiO_2/Pt suspensions. Conditions as in Figures 13 and 14. Successive injection of 2 x 1 ml of O_2 after deaeration of the dispersion with Ar. The upper curve corresponds to a second injection in each case.

○ 0.1 M NaOH.
☆ 0.1 M NaOH + 0.05 M $Ba(OH)_2$.

We conclude, therefore, that Ba^{+2} ions enhance water photocleavage. The overall efficiency of 0.7% obtained is close to the theoretical limit of ca. 2% expected for an absorber with 3.2 eV bandgap and a light source such as our Xe lamp whose spectral intensity distribution resembles the emission spectrum of the sun and, with an output of 90 mW/cm^2, is close to the solar radiance under AM1 conditions.

7. REFERENCES

1. Energy Resources Through Photochemistry and Catalysis, M. Grätzel, ed., Academic Press, New York, 1983.
2. Photogeneration of Hydrogen, A. Harriman and M. West, eds., Academic Press, London, 1982.
3. Photochem. Convers. Storage Sol. Energy (Proc. IV Int. Conf.), J. Rabani, ed., Weizmann Science Press of Israel, Jerusalem, 1982.
4. Photochem. Convers. Storage Sol. Energy (Proc. III Int. Conf.), J.S. Connolly, ed., Academic Press, New York, 1981.
5. J. Kiwi, K. Kalyanasundaram and M. Grätzel, Struc. & Bonding, 49, 37 (1981).
6. P. Kofstadt, Non-stoichiometry, diffusion and electrical conductivity in binary metal oxides, Wiley, New York (1972). R.A. Smith, Semiconductors, Cambridge Univ. Press, London (1979).
7. Semiconductors, N. Hannay, Reinhold, New York, (1959).
8. Titanium,J. Barksdale, Ronald Press, New York (1966)-
9. E. Verwey, P. Haaijmann, F. Romeign and G. Osterhout, Philips Res. Rept. 5, 173 (1949). E. Verwey, Philips Research Reports, 5, 173 (1950).
10. K. Hauffe, S. Schlosser, DECHEMA-MONOGRAPHIE, 26, 222 (1956). The Structure of Metallic Catalysts, J. Anderson, Academic Press, New York, (1975).
11. Spillover of Adsorbed Species. Studies in Surface Science and Catalysis, 11. Elsevier Scientific Pub. Co., Amsterdam (1982).
12. A. Tseung and H. Bevan, J. Mater. Sci. 5, 604 (1970).
13. R. Shanon, Acta. Cryst. A 32, 751 (1976).
14. E. Borgarello, J. Kiwi, E. Pelizzetti, M. Visca and M. Grätzel, J. Am. Chem. Soc., 103, 6324 (1981).
15. J. Kiwi, J. Süss and S. Szapiro, Chem. Phys. Letts., 106, 135 (1984).
16. P. Zimmerman, Phys. Rev. B8, 3917 (1973).
17. J. Conessa, G. Munuera, A. Munoz, V. Rives, J. Sanz and J. Soria, Inter. Symp. Spillover Adsorbed Species, Lyon, France (Sept. 1983).
18. N. Eror, J. Solid State Chem., 38, 281 (1981).
19. W. Harris and R. Wilson, Ann. Rev. Mater. Sci., 8, 99 (1978).
20. P. Salvador, Solar En. Maters. 2, 413 (1980).
21. J. Kiwi and M. Grätzel, J. Phys. Chem., 88, 1302 (1984).
22. S. Teichner, Adv. Catal., 20, 107 (1969).
23. B. Gu, J. Kiwi and M. Grätzel, Nouv. J. Chimie (1985).
24. W. Hansen, T. Kuwana and R. Osteryoung, Anal. Chem., 38, 1810 (1966).
25. M. Ward and J. White, J. Am. Chem. Soc., 105, 27 (1983).
26. Contact Catalysis, Z. Szabo, D. Kallo, eds., Elsevier, New York (1976).
27. F. Wagner and G. Somorjai, J. Am. Chem. Soc., 102, 5494 (1980).

S. Ferrer and G. Somorjai, Surf. Sci., 94, 41 (1980).
S. Ferrer and G. Somorjai, J. Phys. Chem., 85, 1464 (1981).
28. J. Kiwi and C. Morrison, J. Phys. Chem., 88, 6146 (1984).
29. A. Gonzales-Elipe, G. Munuera and J. Soria, J. Chem. Soc. Faraday Trans. 1, 75, 748 (1979).
30. G. Munuera, J. Navio, "Proceedings of the 4th Nat. Meeting on Adsorption", Sevilla, September 1979, p. 25.
31. E. Yesodharan, S. Yesodharan and M. Grätzel, Sol. Energ. Mat., 10, 287 (1984).
32. A. Mills and G. Porter, J. Chem. Soc. Faraday Trans. 1, 78, 3659 (1982).
G. Blondeel, A. Harrimann and D. Williams, Solar Ener. Mat., 9, 217 (1983).
33. J. Muhlebach, K. Muller, G. Schwarzenbach, Inorg. Chem., 9, 2381 (1970).
M. Mori, M. Skibata, E. Kyuno and S. Ito, Bull. Chem. Soc. Japan, 29, 904 (1956).
34. C. Jaeger and A. Bard, J. Phys. Chem., 24, 3146 (1979).
35. A. Bard, Science, 207, 138 (1980).
36. A. Boonstra and C. Mutsaers, J. Phys. Chem., 80, 1694 (1976).
37. K. Hauffe, H. Raveling and D. Rein, J. Phys. Chem., 104, 89 (1977).
38. G. Shobaky, P. Gravelle and S. Teichner, Bull. Soc. Chim. Fr. 3251 (1968).
F. Stone, Adv. Catal., 13, 1 (1962).
39. M. Ward, J. White and A. Bard, J. Am. Chem. Soc., 105, 27 (1983).
40. J. Kiwi and M. Grätzel, J. Phys. Chem., submitted (1985).
41. D. Dowden, Surface Science, 2, 215 (1979) Int. Atomic Energy Vienna.
The Physical Basis for Heterogeneous Catalysis, E. Drauglis and R. Jaffe, Plenum Press, New York (1975).
42. Solid State Physics, F. Scitz and D. Turnbell, eds., Academic Press, New York, 3, 307 (1956).
43. G. Parfitt, Prog. Surf. Membr. Sci., 11, 181 (1976).
H. Boehm, Disc. Faraday Soc., 52, 264 (1971).
44. Catalysis, J. Anderson and M. Boudart, eds., vol. 3, Springer Verlag, Berlin (1983).
45. A. Boonstra and C. Mutsaers, J. Phys. Chem., 79, 1694 (1975).
46. E. Yesodharan and M. Grätzel, Helv. Chim. Acta, 66, 2145 (1983).
47. The Chemistry of Peroxides, P. Patai, J. Wiley, New York (1983).
48. G. Brown and J. Darwent, J. Phys. Chem., 88, 4596 (1984).
49. W. Hansen, T. Kuwana and R. Osteryoung, Anal. Chem., 38, 1810 (1966).

METALLIC CATALYSTS ON SEMICONDUCTORS: TRANSPARENCY AND ELECTRICAL CONTACT PROPERTIES

Adam Heller
AT&T Bell Laboratories
600 Mountain Avenue
Murray Hill, New Jersey 07974
U.S.A.

ABSTRACT. Incorporation of Group VIII metal catalysts in photoelectrodes changes the fill factor, the photocurrent and the photovoltage of photoelectrochemical cells. A loss in photocurrent can be avoided by using a transparent metallic catalyst, i.e. by controlling the microstructure of the metal. Metal films are transparent if made of poorly interconnected subwavelength size metal particles. Similarly, suspensions of metal particles are transparent if the particles are much smaller than the wavelength of the transmitted light. The barrier heights and the photovoltages depend on the interfacial chemistry of the semiconductors and the catalysts. They can be modified by hydrogenation/dehydrogenation of the interface. Hydrogenation increases the barrier heights and the photovoltages of *p*-type semiconductor/catalyst contacts, and decreases these in *n*-type semiconductor/catalyst contacts.

1. INTRODUCTION

Semiconductor based photoelectrochemical systems consist of macroscopic or microscopic solar cells, to which the physics of these cells applies.[1,2] In wire-connected photoelectrodes we deal with electrical contacts to large single crystalline or polycrystalline semiconductor surfaces. In suspensions of semiconductor crystallites we also deal with electrical contacts, which now constitute microscopic anode and cathode zones on the semiconductor surface.

Incorporation of catalytic metals like Pt, Rh, Ru, Pd, RuO_2, etc. in the surface of a semiconductor crystal, or in that of a dispersed semiconductor crystallite, has three major effects: It accelerates a desired chemical reaction, and thereby improves the fill factor; it decreases the flux of photons penetrating the semiconductor; and it introduces a new electrical contact with a different barrier height, i.e. with different band bending. In addition, submonolayer amounts of catalysts may also increase or decrease the recombination velocity of the photogenerated minority charge carriers.

We shall see that light absorption and reflection by metallic catalysts depends on the size of the metal particles and on their electrical connectedness; and that the properties of metallic catalyst/semiconductor contacts depend on the partial pressure of hydrogen and on the presence or absence of a chemical reaction between the semiconductor and the metallic catalyst.

E. Pelizzetti and N. Serpone (eds.), Homogeneous and Heterogeneous Photocatalysis, 303–315.

2. EFFECTS OF METALLIC CATALYSTS ON QUANTUM EFFICIENCIES

The quantum or current efficiencies of photoelectrochemical reactions depend on the fraction of photons penetrating the semiconductor in the absence of a catalyst. The loss is determined by the index of refraction of the semiconductor, n_{sem}, and by its extinction coefficient, k_{sem}, which are higher than the corresponding values for solutions, n_{sol} and k_{sol}. For most solutions $k_{sol} = 0$. At normal incidence the fraction of reflected photons, l_{ref}, is

$$l_{ref} = \frac{(n_{sol} - n_{sem})^2 + k_{sem}^2}{(n_{sol} + n_{sem})^2 + k_{sem}^2} . \tag{1}$$

For example, l_{ref} is 0.206 for the InP/1M $HClO_4$ interface at $\lambda = 600$ nm.

The fraction of photons lost increases when the semiconductor is metallized, mostly because extinction coefficients of metals, k_{met}, are substantial.[3] Thus, approximately half of the 500-800 nm photons are lost to absorption in a 7 nm thick, continuous and dense platinum film. The absorption and reflection losses depend, however, on the microstructure of the metals and can be reduced.[4,5]

Because electron-hole recombination rates and electrochemical reaction rates are potential dependent, the quantum or photocurrent efficiencies vary with potential. Nevertheless, quantum efficiencies are essentially independent of potential at sufficiently negative potentials in photocathodes, and at sufficiently oxidizing potentials in photoanodes, if the minority carrier diffusion lengths exceed the absorption lengths. The latter condition usually holds for high quality semiconductors. In the potential independent region, photoelectrodes are precise actinometers. Each photon penetrating the semiconductor creates an electron-hole pair, and each photogenerated charge carrier contributes to the Faradaic current. Thus, by measuring the photocurrent in the potential independent region it is possible to measure the fraction of photons that traverses a metallic catalyst film. This fraction can be independently measured also by ellipsometry, which yields the average thickness of the film d, its index of refraction n_{met}, and its extinction coefficient k_{met}. Note that the fraction of photons absorbed, l_{abs} is given by

$$l_{abs} = 1 - e^{-4\pi k_{met} d/\lambda} \tag{2}$$

λ being the wavelength of light *in vacuo*. The parameters n and k are related to the complex dielectric function $\tilde{\epsilon} = \epsilon_1 + i\epsilon_2$ according to

$$\tilde{\epsilon} = \tilde{n}^2 = (n + ik)^2 \tag{3}$$

where $\tilde{n}$ is the complex refractive index. For normal incidence, the relative reflection loss at a two-phase boundary between an optically thick metal film and a transparent ambient is given by Eq. 1, except that n_{sem} is replaced by n_{met} and k_{sem} by k_{met}.

Recent work in our laboratory revealed that absorption and reflection losses introduced by metal films on surfaces of semiconductors can be drastically reduced by controlling the microstructure of the metals:[4–6] When the dimensions of the metal particles comprising the films are much smaller than the wavelength of the incident light, and when the particles are electrically not well interconnected, the metal films become transparent.

The optical properties of transparent metals are fully described by Bruggeman's effective medium theory.[7,8] According to this theory, the effective dielectric function of a system, $\tilde{\epsilon}_{\text{eff}}$, made of sufficiently small, isolated particles with a dielectric function $\tilde{\epsilon}_{\text{met}}$, occupying a volume fraction f_{met}, dispersed in a fluid with a dielectric function $\tilde{\epsilon}_{\text{sol}}$, which occupies a volume fraction $f_{\text{sol}} = 1 - f_{\text{met}}$, is obtained by solving the equation

$$f_{\text{met}} \left(\frac{\tilde{\epsilon}_{\text{met}} - \tilde{\epsilon}_{\text{eff}}}{\tilde{\epsilon}_{\text{met}} + \kappa \tilde{\epsilon}_{\text{eff}}} \right) + f_{\text{sol}} \left(\frac{\tilde{\epsilon}_{\text{sol}} - \tilde{\epsilon}_{\text{eff}}}{\tilde{\epsilon}_{\text{sol}} + \kappa \tilde{\epsilon}_{\text{eff}}} \right) = 0 \tag{4}$$

where κ is the screening coefficient. For an isotropic 3-dimensional system of small particles in a fluid $\kappa = 2$. For an isotropic 2-dimensional system $\kappa = 1$, and $\kappa = 0$ for thin vertical rods on a normally illuminated surface.

In a typical experiment we characterized a platinum film on a p-InP crystal by ellipsometrical, photoelectrochemical and electron microscopic means. The ellipsometric measurements yielded an average film thickness of 33.7 nm, a platinum volume fraction $f_{\text{met}} = 0.5$, a screening coefficient $\kappa = 0.8$, and the $\langle\tilde{\epsilon}\rangle$ spectrum shown in Fig. 1 (solid line). The theoretical curve, calculated from Eq. 4 by using the independently measured dielectric functions of dense platinum and of 1M $HClO_4$, is represented by the dashed line. The sample was prepared by photoelectroplating platinum on the (100) p-InP surface from a 1M $HClO_4$ solution containing $1.2{\times}10^{-5}$ mol L^{-1} Pt(IV) over a 15h period. The low concentration of Pt(IV) assured mass transport limited plating, an essential condition for obtaining porous films made of subwavelength size particles with low connectedness between the platinum particles.

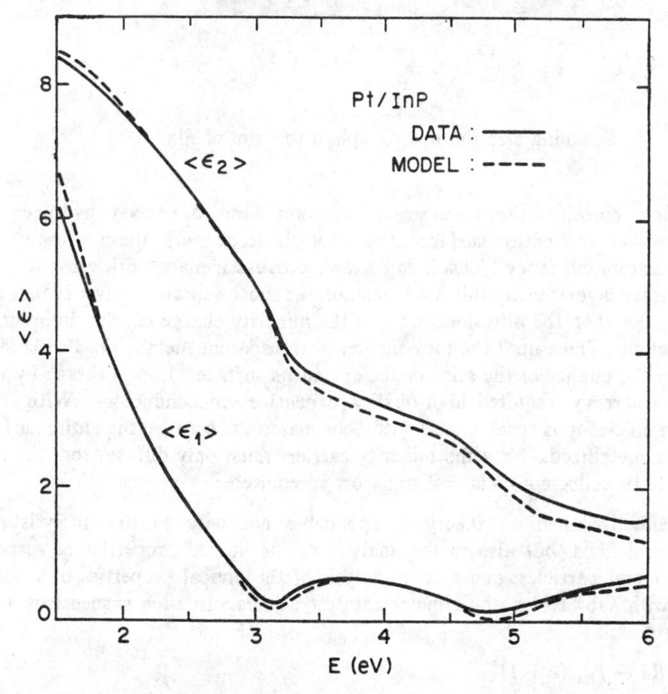

Figure 1. Comparison of the measured dielectric spectrum (solid line) with that calculated from the Bruggeman effective medium theory for a 32.6 ± 0.8 nm thick Pt film on p-InP. The volume fraction of the metal in the film is 0.50 ± 0.02.

The naked eye sees no change in the appearance of the p-InP crystal surface after the electrodeposition of the transparent platinum film, nor does one see a change by optical microscopy. Scanning electron microscopy reveals, however, that the film consists of porous, spherical particles with an average diameter of 80 nm (Fig. 2). These particles consists, in turn, of still smaller particles, with an average diameter of 5 nm, which aggregate in grape-like structures seen in the transmission electron micrograph of Fig. 3. Because of the low connectedness between both the 80 nm and the 5 nm particles, the sheet resistance of the film is at least 10^6 times greater than that of a dense Pt film having an equal metal mass per unit area. While a dense Pt film of equal mass per unit area transmits only 3% of visible or near infrared photons, the film of Figs. 1-3 transmits 92% of these photons.

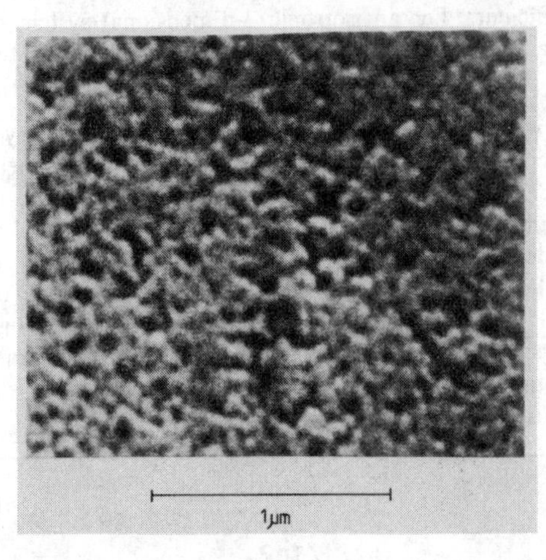

Figure 2. Scanning electron micrograph of the film of Fig. 1.

Because of their porosity, the transparent platinum films are good hydrogen evolution catalysts. One can cover the entire surface of a photoelectrode with these without causing a substantial loss in quantum efficiency. This is important, because if made with conventional Pt, the catalyst films must either be extremely thin (<1 nm), or one must deposit catalyst islands and space these at distances smaller than the diffusion length of the minority charge carriers in order to assure efficient carrier collection. Thus, until the introduction of transparent metals, the Pt island coverage required depended on the quality of the semiconductor and its surface. Low coverage by Pt islands, resulting in high transparency, required high quality, expensive semiconductors. With transparent metals high quantum efficiency is reached also with poor materials, because the entire surface of the semiconductor can be metallized. Now the minority carriers must only diffuse from the absorption depth to the surface to be collected; no lateral transport is required.

Bruggeman's effective medium theory is applicable not only to the analysis of optical properties of thin metal films, but also to the analysis of the optical properties of suspensions of subwavelength-sized metal particles, and to the analysis of the optical properties of semiconductor crystallites covered with subwavelength-diameter catalyst islands. In such suspensions $f_{met} \ll 1$, while $|\epsilon_{met}| \gg \epsilon_{sol}$ and

$$\tilde{\epsilon}_{eff} \simeq \tilde{\epsilon}_{sol}/[1 - f_{met}(\kappa + 1)] \tag{5}$$

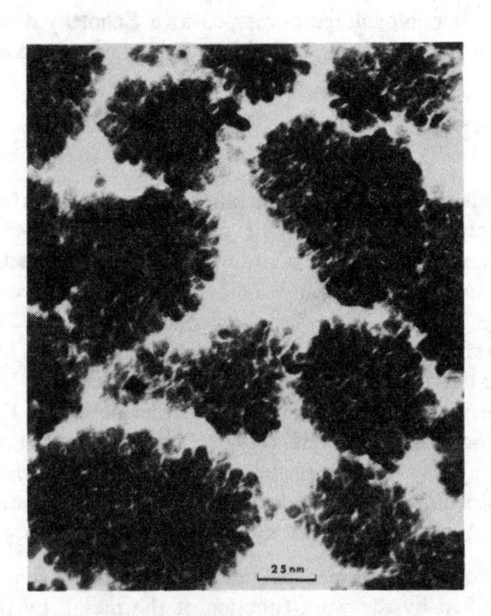

Figure 3. Transmission electron micrograph of the top layer of the platinum particles constituting the film of Figs. 1 and 2.

Note that the dielectric function of the metal does not appear in Eq. 5 and that the dielectric function of the suspension is increased over that of the transparent solution by a constant factor proportional to the volume fraction occupied by the metal. Because $\kappa \approx 2$ and $f_{met} << 10^{-3}$, the dielectric properties of subwavelength size particle suspensions, and thus their optical properties, are essentially identical with those of the electrolytic solution in which the particles are suspended, i.e. $k_{eff} \approx k_{sol}$ and $n_{eff} \approx n_{sol}$. (One should, nevertheless, keep in mind that small particles of Cu and Au, with band-to-band or plasma transitions in the visible region do absorb light, and that transparency can be achieved only in metals having short wavelength plasma transitions.)

In both films and suspensions of metal particles transparency results from screening. The photon field is excluded from the metal particles so that their actual composition becomes irrelevant. The photons are now "squeezed" into the ambient fluid surrounding the particles, thereby increasing the effective field strength within the voids and the apparent refractive index of the filling fluid. As a result, losses typically associated with light transmission through metallic media vanish: the absorption loss (Eqs. 2 and 3) is reduced to its ideal limit of nil. Moreover, by keeping f_{met} small one can minimize reflection, as evident from Eqs. 1 and 3.

3. ELECTRICAL CONTACTS BETWEEN METALLIC CATALYSTS AND SEMICONDUCTORS, BARRIER HEIGHTS, OHMIC CONTACTS AND PHOTOVOLTAGES

In metallic catalyst activated photoelectrodes most of the photogenerated minority charge carriers pass through the metal-semiconductor or metal-insulator-semiconductor (MIS) contacts. This is the case also when only a small part of the surface of the semiconductor is coated with the metal, because the rate of the electrochemical reaction (and thus of the photocurrent density) is much greater on the catalyst coated zone than at the bare semiconductor surface. For this reason, metallic catalyst activated semiconductor electrodes are best described as metal-semiconductor (Schottky) contacts, or as metal-insulator-semiconductor (MIS) contacts.

The maximum or open circuit photovoltage developed at a Schottky diode, V_{oc}, which is also the maximum bias voltage reduction that one can attain in a photoelectrolytic cell, is given by eq. 6

$$V_{oc} \simeq \phi_B + \frac{kT}{q} \ln \frac{J_{sc}}{A^{**}T^2} \tag{6}$$

where ϕ_B is the barrier height, k is the Boltzmann constant, $T$is the temperature, q is the charge of the electron, J_{sc} is the light limited or short circuit photocurrent and A^{**} is the Richardson constant, modified by optical phonon scattering, quantum mechanical reflection and tunneling of carriers at the metal semiconductor interface.[9] Eq. 6 holds when the dark current is dominated by thermionic emission of majority carriers from the semiconductor to the metal, i.e. when the semiconductor and its metal interface are of adequate quality for both the bulk and the interface recombination currents to be smaller than the thermionic emission current. V_{oc} will, of course, be smaller if either bulk or interface recombination is rapid. Furthermore, the V_{oc} values of Eq. 6 are reached only if the semiconductor doping level is below 10^{17} cm^{-3}. At higher doping levels the barrier narrows and leakage of majority carriers to the metal increases because of tunneling. Thus, in order to reach either high photovoltage or high bias reduction in photoelectrolysis, one must have a high barrier; a low-doped, good quality semiconductor; and slow interfacial recombination.

Ideally, when an electrical contact is formed between a metal and a semiconductor, the height of the resulting barrier is determined by the work function of the metal, by the electron affinity of the semiconductor, by the type and concentration of charge carriers and by the magnitude and direction of the interfacial dipole. Ideal or nearly ideal contacts are formed only in the absence of an interfacial chemical reaction between the metal and the semiconductor. Interfacial chemical reactions are best avoided by forming MIS contacts, i.e. by forming a stable, non-reactive tunnelable insulator film on the semiconductor prior to its metallization. Upon subsequent metallization one must not damage the insulating film. Thus, for example, silicon can react with platinum in the absence of an interfacial SiO_2 layer to form a Pt-Si metallic alloy interface. Obviously, the contact is now between Si and the metallic alloy and not between Pt and Si. In binary semiconductors the consequences of interfacial reactions are more complicated. In the absence of a reaction-blocking layer, not only is the chemistry of the metal layer in immediate contact with the semiconductor changed, but the interfacial reaction also depletes preferentially one of the constituents of the semiconductor. For example, gold reacts with GaAs to form an Au-Ga alloy[10] and the interface contains an excess of As. Thus, the contact is not between Au and GaAs, nor simply between an Au-Ga alloy and GaAs, but between GaAs and severely Ga depleted (As enriched) GaAs. Such a layer is heavily self doped and may produce a tunnelable junction. Similarly, electrical contacts between e-beam evaporated Ru, Rh, Os, Ir or Pt films and air exposed (100) faces of n-GaAs do not show the characteristics expected for contacts between the metals and n-GaAs: Although the work functions of these catalysts vary between 4.71 eV for Ru to 5.65 eV for Pt, there is less than 0.05 eV variation in the barrier heights of their n-GaAs contacts.[11] Again the probable reason for the similar barrier heights is As enrichment of the semiconductor surface, either by alloying of Ga with the catalyst or by air exposure. Oxidation in air results in the net reaction

$$4GaAs + 3O_2 \rightarrow 2Ga_2O_3 + 4As$$

Note that Ga_2O_3 is thermodynamically more stable than As_2O_3, and that the reaction

$$2GaAs + As_2O_3 \rightarrow Ga_2O_3 + 2As$$

is spontaneous. Interfacial reaction with metals is not unique to GaAs. When a carefully prepared *n*-CdS (A plate) is platinized by *e*-beam evaporation of the metal, the contact formed does not have the characteristics expected for Pt on *n*-CdS. The observed characteristics suggest that an interfacial layer of platinum sulfide is formed, along with a Cd enriched, heavily self doped n^+-CdS surface layer.[12]

As mentioned earlier, barrier heights at metal semiconductor contacts are lowered when the rate of electron-hole recombination in the bulk or on the surface of a semiconductor are high. These rates are drastically increased if the surface of the semiconductor is mechanically damaged by grinding, polishing, sawing or ion bombardment. All introduce lattice defects and through these a sufficiently high density of states in the forbidden gap to greatly accelerate recombination.

Semiconducting oxides like TiO_2 and $SrTiO_3$, as well as semiconductors with stable surfaces like InP, do not appear to react at ambient temperature with Pt or Rh. Also CdS does not appear to react at ambient temperature with Ru or with Rh. In all of these systems the contacts are well behaved. The barrier heights vary when metallic catalysts with different work functions, forming different interfacial dipoles, are employed. The difference between a semiconductor that reacts at its interface and one that does not is seen in Table 1, where the barrier heights formed upon Pt, Rh or Ru metallization of (100) faces of *n*-GaAs are compared with those formed with (100) faces of *p*-InP.[11,12] The greatest difference in *n*-GaAs (between Pt and Rh) is 0.1 eV. That in *p*-InP is 0.4 eV.

TABLE I
Barrier Heights (eV) for *n*-GaAs and for *p*-InP

METAL	*n*-GaAs	*p*-InP
Ru	+0.98	−0.57
Rh	+0.90	−0.31
Pt	+1.00	−0.72

Chemical reactions can produce leaky junctions not only because of tunneling but also because of chemical inhomogeneity of the interface. Whenever the interfacial chemistry is not unique, i.e. whenever microscopic regions of the metal-semiconductor interface have different chemistries, carriers recombine by thermionic emission at the sites having the lowest barriers.[13]

In contacts between platinum group metals and semiconductors that are well behaved, i.e. in which there is no chemical reaction at the interface and where the semiconductor surface is not damaged, the barrier heights are drastically changed in a hydrogen atmosphere.[12] Under hydrogen, the absolute magnitude of the barrier height increases in contacts of *p*-type semiconductors like *p*-InP and decreases in *n*-type semiconductors like *n*-TiO_2, *n*-$SrTiO_3$ and *n*-CdS. The changes are completely reversed in air or in oxygen. Repeated cycling neither increases nor decreases the change. The magnitude of the change depends on the semiconductor and the metal. For any single semiconductor and for different catalysts hydrogen exposure produces barriers of similar height, though the barrier heights differ in air. Examples of these changes are shown in Table II. The values for *n*-CdS/Pt and *n*-$SrTiO_3$/Ru are in parentheses because of apparent interfacial reaction between the metal and the semiconductor.

TABLE II

Reversible Changes in Barrier Heights in Hydrogen and in Air (in eV).

Semiconductor	n-TiO_2 (Rutile)		n-$SrTiO_3$		n-CdS		p-InP	
Atmosphere	H_2	Air	H_2	Air	H_2	Air	H_2	Air
Ru	0.0	0.1	(0.2)	(0.7)	0.75	1.28	−0.85	−0.57
Rh	0.0	0.6	0.1	1.0	0.75	1.00	−0.82	−0.31
Pt	0.0	0.5	0.1	0.6	(1.63)	(1.84)	−0.91	−0.72

Hydrogen exposure also improves the quality factors of those well-behaved contacts that are particularly poor in air. The quality factor of a solar cell, A_0, is defined by Eq. 7

$$V_{oc} = A_0 \frac{kT}{q} \left[\ln \frac{J_{sc}}{J_0} + 1 \right] \tag{7}$$

where J_0 is the dark current density and the other terms have their earlier defined meaning. In an ideal diode $A_0 = 1.04$. An A_0 value close to ideal implies little recombination or leakage at the interface. While hydrogen brings the poorer, chemically well-behaved diodes closer to ideal, it has little effect on diodes with apparent chemical reaction at the interface (n-CdS/Pt, n-$SrTiO_3$/Ru).

TABLE III

Hydrogen Effects on Diode Quality Factors

Semiconductor	n-$SrTiO_3$		n-CdS		p-InP	
Atmosphere	Air	H_2	air	H_2	Air	H_2
Ru	(3.3)	(2.9)	2.3	1.0	1.2	1.4
Rh	2.9	1.0	1.5	1.4	1.1	1.2
Pt	7.2	1.3	(1.7)	($\sim$2)	1.2	1.2

The observed effects are consistent with reversible hydrogenation of the catalyst-semiconductor interface. After hydrogenation the contacts are essentially between metal-like hydrogen at the interface and the semiconductor, and the interfacial dipole changes because the interfacial hydrogen can be ionized to protons. When the metal and the semiconductor react at their interface, as is the case in contacts between n-GaAs and metallic Pt, Ir, Os, Pd or Rh and between n-CdS and Pt, the new interfaces are either not hydrogenated or only partially hydrogenated in a hydrogen atmosphere, possibly because the interface is now thermodynamically more stable and not subject to spontaneous hydrogenation.

Changes in barrier height upon hydrogenation have a profound impact on the performance of hydrogen evolving photoelectrodes and suspended microelectrodes. In photocathodes, such as those made with p-InP and Pt, Rh or Ru, the increase in barrier height translates to a higher V_{oc}, i.e. to a reduction in the bias needed for photoelectrolysis. In n-TiO_2 the contacts of the same metals become truly ohmic upon hydrogenation. In n-$SrTiO_3$ they become nearly ohmic, and the barrier heights of n-CdS are substantially reduced (Table II).[12] This reduction in barrier height can greatly increase the quantum efficiency of photoelectrochemical processes on suspended semiconductor crystallites of >0.1 μm diameter.[4]

In a suspended photolyzing semiconductor crystallite having a microcathode and a microanode, the photovoltage can not exceed the difference between the barrier heights at the two microelectrodes. If the cell is perfectly symmetrical (Fig. 4, left) $V_{oc} = 0$; If there is some asymmetry, V_{oc} can not exceed the difference between the height of the microanode and microcathode barriers (Fig. 4 center); When one of the contacts is ohmic, the photovoltage limit is the barrier height of the non-ohmic contact (Eq. 6). For photolysis to take place V_{oc} must exceed $\Delta G/nF$ where ΔG is the Gibbs free energy change of the reaction, n is the number of the electrons transferred and F is Faraday's constant. Hydrogenation of a catalyzed microcathode of a suspended n-type semiconductor crystallite can increase the asymmetry, lower the barrier to electron transport to the cathode catalyst, and increase V_{oc}.

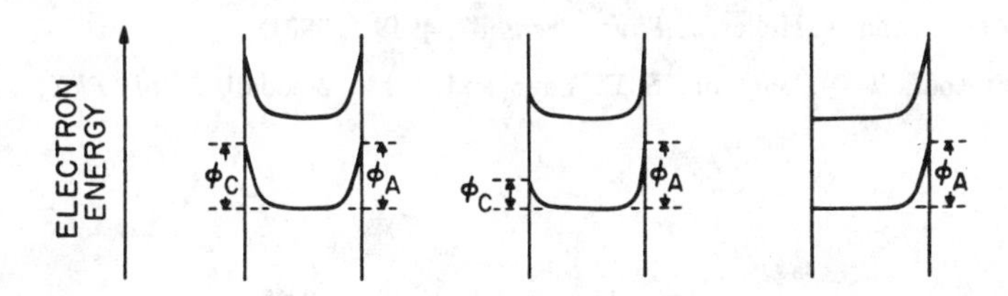

Figure 4. The photovoltage of a microcell, consisting of a suspended semiconductor crystallite with a microanode and a microcathode, can not exceed the difference between the barrier heights of the two microelectrodes, ϕ_A and ϕ_C. In a symmetrical cell (left) the photovoltage is zero. In an asymmetrical cell (center) the photovoltage is less than $|\phi_A - \phi_C|$. If the microcathode is an ohmic contact (right), the photovoltage limit is ϕ_A.

4. CONCLUDING REMARKS

In the absence of a metallic catalyst certain photoelectrode reactions, including hydrogen evolution, are slow and much of the photovoltage is lost at photocurrents typical of solar cells. Such a loss implies a low fill factor. The catalyst increases the fill factor. Its presence on the surface impacts, however, also on the limiting or short circuit photocurrent and on the open circuit photovoltage. The photocurrent changes because a metallic catalyst absorbs and reflects light. The photovoltage changes because incorporation of the catalyst creates a new electrical contact, with a new barrier height. In this lecture we have seen that by controlling the microstructure of platinum metal it is possible to create essentially transparent platinum films. We have also seen that the barrier heights of contacts (or microcontacts) between semiconductors and catalysts and thus also the photovoltages of photoelectrochemical cells (or of microcells consisting of a suspended metal-catalyzed crystallites) depend on the interfacial chemistry. This chemistry can be substantially changed by reversible hydrogenation/dehydrogenation.

REFERENCES

(1) S. J. Fonash, *Solar Cell Device Physics*, Academic Press, New York, 1981.

(2) H. J. Hovel, 'Semiconductors and Semimetals,' Vol. **11**, *Solar Cells*, Academic Press, NY, 1975.

(3) G. Hass and L. Hadley, *Optical Properties of Metals* in *Am. Inst. Phys. Handbook*, D. E. Gray, Coordinating Editor, McGraw Hill, NY, 1972, Chapter 6.

(4) A. Heller, *Science,* **223**, 1141 (1984).

(5) J. D. Porter, A. Heller and D. E. Aspnes, *Nature* (London), **313**, 664 (1985).

(6) A. Heller, D. E. Aspnes, J. D. Porter, T. T. Sheng and R. G. Vadimsky, *J. Phys. Chem* 00, 0000 (1985).

(7) D. E. Aspnes, *Thin Solid Films*, **89**, 249 (1982).

(8) D. A. G. Bruggeman, *Ann. Phys.* (Leipzig) **24**, 636 (1935); ibid., 679.

(9) Ref. 2, p. 122.

(10) M. B. Panish, *J. Electrochem. Soc.* **114**, 516 (1967).

(11) D. E. Aspnes and A. Heller, *J. Vac. Sci. Technol.,* **B1**, 602 (1983).

(12) D. E. Aspnes and A. Heller, *J. Phys. Chem.* **87**, 4919 (1983).

(13) J. L. Freeouf, T. N. Jackson, S. E. Laur and J. M. Woodall, *Appl. Phys. Lett.* **40**, 634 (1982).

FENDLER - As always, I found your lecture highly stimulating and educational. It was pleasing to hear that Pt on CdS is not as efficient as Rh and Ru. Indeed, we observed that efficient photosensitized H_2 evolution on CdS needed Rh rather than Pt. However, these experiments were carried out on dispersed particles incorporated into vesicles. To what extent can your data be transferred into the much more complex colloidal systems?

NOZIK - The work of O. Micic and ourselves on the flat band potential of such colloids suggests that there the relationships persist also in colloidal semiconductors.

SERPONE - On our work on H_2S cleavage, we reported quantum efficiencies of about 50% for CdS/Rh photocatalysts. We can explain this now from what you just said about the decrease in barrier height upon hydrogenation. We have prepared our CdS/Rh in the presence and absence of air, and yet the quantum efficiencies are identical. I fail to understand why CdS/Rh prepared in air should be just as efficient as the one prepared under argon.

HELLER - The n-CdS/Rh contacts prepared in air may not differ from that prepared in the basence of air after all the oxygen has been stripped by reaction with photogenerated hydrogen and the metal has reached equilibrium with hydrogen, i.e., is saturated with the gas. The "incubation" or "delay" time will however be larger for the sample prepared in air.

BRUS - Could you comment on the changes expected, both optically and catalytically, if you had a square, periodic array of Pt particles on the surface, instead of a random distribution, assuming that the particle number density is the same in both cases?

HELLER - Yes. The case of periodic subwavelength wide metal wires with subwavelength spacing between the wires has been analyzed in far-infrared and microwave studies. If the electric vector of the radiation is in the direction of the wires, the absorption is strong. If the vector is perpendicular to the wires, the radiation passes - (as it does in the parent metal films that I discussed). This is the basis for the IR and microwave wire polarizers. Thus, with a set of parallel wires we pass polarized light. With a double set of wires running perpendicular to each other, the loss will be substantial. If, however, we chop up the wires into segments that are much smaller than the wavelength, the radiation will pass with little loss.

LEWIS - Can you give us your thought and any experiments which relate to a decrease in the work function upon hydrogenation? Is this due to an interfacial dipole introduced by H_2, or due to a deoxygenation of grain boundaries in the polycrystalline metal films?

HELLER - Repeated increase/decrease of the partial pressure of hydrogen leads to corresponding increase/decrease of barriers of p-InP/Group VIII metal contacts and to decrease/increase of barriers of n-TiO_2/Group VIII metal contacts when the hydrogen is diluted by an inert gas such as N_2 or Ar. This and experiments on bare (non-metallized) p-InP hydrogen evolving photocathode contacts suggest that it is the presence of hydrogen at the interface that is important.

MANASSEN - Gary Hodes has deposited two Pt electrodes on a CdS crystal, one giving an ohmic contact and the second a barrier. When illuminated in a sulfide solution, hydrogen is only evolved at the ohmic contact. Your talk has addressed the solid state aspects of the micro-electrochemical cell formed on the semiconductor particle. I would like to stress again the importance of the solution composition. We have to consider the kinetics of dissolved redox species with the two electrodes which may be influenced as well by the reacting ions as the chemical nature of the metal electrode. These kinetic influences, as it were, 'bias' the micro-electrochemical cell.

HELLER - I fully agree with you on the importance of the kinetic parameters; mass transport electron/hole recombination kinetics and adsorption/desorption kinetics can all be very important.

FRANK - Regarding the nature of the metal and semiconductor and the treatment of the semiconductor surface, what has been your experience in preventing or reducing the coalescence of metal islands on the semiconductor surface?

HELLER - Grains of the high-melting group VIII refractory metals (Pt, Rh, Re, Ru, Ir) do not appear to grow in 1 M $HClO_4$ solutions in our experiments which lasted 1-3 days and were run at ambient temperatures. In strongly complexing solutions and also in lower-melting, soft metals like Au, there is growth. Transparent metal films are readily obtained in the above group VIII refractory metals, but not with Au.

TSUBOMURA - I would like to comment on the reason why the work functions of metals deposited on semiconductor surfaces are reduced by hydrogen: We believe that the surface of metal (having granular structure) is covered by adsorbed oxygen when exposed in air, thus forming a M^+-O^- type surface double layer and increasing the work function of the metal. Hydrogen reacts with O_{ads} on the metal surface and removes it as H_2O, thus recovering the Fermi energy of the metal to the original level, which is as high as that of the n-TiO_2. The clearest experimental evidence for this is that when Pt (or Pd) is deposited on TiO_2 in vacuum and the i-v curve directly measured without breaking the vacuum, the TiO_2/Pt contact is ohmic. In other words, when air is introduced, the contact shows considerable amount of Schottky barrier. Cf. H. Tsubomura et al., Surface Science, 92, 400 (1980); J.Appl.Phys., 52, 5705, 6227 (1981); J.Electrochem.Soc., 129, 444 (1982).

HELLER - Our results disagree with this suggestion. The hydrogen effects on the barrier heights are reversed not only by oxygen, but also by replacing the hydrogen with nitrogen or argon.

TSUBOMURA - I should like to make a further comment on the mechanism of ohmic contact formation between metal and semiconductor for the case where metal loaded semiconductor (for example, TiO_2/Pt) is acting as effective photocatalyst for hydrogen production. Dr. Heller says that this is an auto-catalytic process, i.e., when H_2 is produced by some means, it decreases the work function of the metal and makes the contact ohmic. Nakato and myself proposed a theory recently stating that in a metal-semiconductor contact where the metal is in the form of a small particle placed here and there on the semiconductor surface, as is the case for the powder photocatalysts, the effective Schottky barrier can change by illumination if the rate of reaction of the electrons with the solute is different from that of the holes. So, the Fermi level of the metal can automatically go up till the metal forms ohmic contact with the semiconductor. Thus, a smooth electron transfer from excited semiconductor to metal can be explained without assuming the auto-catalytic effect of hydrogen produced. Cf. Y. Nakato, H. Tsubomura, J.Photochem., 29, 257 (1985), a Special Issue for IPS-5, Osaka (1984).

HELLER - Our results disagree with this suggestion. We see reversible hydrogen effects also in electrical contacts made with thick, dense and continuous films of group VIII metals, such as platinum, rhodium, and ruthenium and semiconductors like n-TiO_2 (rutile) and n-$SrTiO_3$ (perovskite).

MUNUERA - In connection with the interpretation of the effect of gases in modifying the nature of the metal-semiconductor contact, I would like to draw your attention to the fact that H_2 is adsorbed on TiO_2/Rh and other similar systems in two forms (see for instance: G. Munuera et al. in 'Studies in Surface Chemistry and Catalysis', vol. 17, 149 (1983), Elsevier). A strong irreversible adsorption and a weak reversible form which is able to generate Ti^{3+} species (detected by ESR) as well as H^+ spillover from the metal to the support. In my view, this weak hydrogen should be involved in the origin of the phenomenon. A second point is in connection with $SrTiO_3$. Since J.M. Lehn published his well known work on $SrTiO_3/Rh$, it has been assumed that $SrTiO_3$ is much better than TiO_2 to produce H_2 from water when loaded with noble metals. From your results on the changes in surface barrier with H_2 it seems to me that TiO_2/Rh should be better than $SrTiO_3/Rh$. Can you comment on this?

HELLER - A measure of performance is the difference between the barrier heights of the two microelectrodes on the semiconductor particle, when the microanode is in equilibrium with the oxygen it evolves and the microcathode is in equilibrium with the hydrogen it evolves. In n-TiO_2 (rutile) this difference is smaller than in n-$SrTiO_3$ (perovskite) when the microelectrodes are made with Rh. We made no measurements on barriers of n-TiO_2 (anatase).

ON THE NATURE OF THE INHIBITION OF ELECTRON TRANSFER AT ILLUMINATED P-TYPE SEMICONDUCTOR ELECTRODES

D. Meissner , Ch. Sinn and R. Memming
Institut f. Physikalische Chemie,
Universität Hamburg,
Laufgraben 24,
D-2000 Hamburg 13, FRG

P.H.L. Notten and J.J. Kelly
Philips Research Laboratories,
5600 JA Eindhoven, The Netherlands

ABSTRACT. A rather strong inhibition of the cathodic photocurrent is reported for various p-type semiconductor electrodes which would lead to a severe loss of energy when such electrodes are applied in solar energy conversion systems. In this paper results are presented for various redox systems at GaAs and InP which exhibit quite different behavior with respect to the inhibition of the cathodic photocurrents. The nature of the inhibition is discussed in terms of two different models, in one case an efficient back reaction via the valence band (recombination shunt in the electrolyte), in the other strong recombination via surface states is assumed.

1. INTRODUCTION

In connection with solar energy conversion, p-type semiconductor electrodes are more favorable than n-type materials because they are more stable. This experience is based on the observation that in most cases hydrogen is formed in a photocathodic process at p-type electrodes whereas the cathodic reduction of the semiconductor does not play a dominant role [1]-[3]. On the other hand a large overvoltage was found with respect to the flatband potential of the electrodes [3]-[6]. This result shows that the reduction of water or protons is inhibited at semiconductors. In principle there are two possibilities of overcoming this problem: (i) by depositing a suitable catalyst for hydrogen evolution, such as e.g. platinium [3],[4], (ii) by adding a suitable electron acceptor to the electrolyte, i.e. a reversible redox couple, the reduced species of which is capable of reducing H_2O in the electrolyte in the presence of colloidal Pt-catalyst. It is clear that only redox couples of a sufficiently negative standard potential are of interest here. In several cases it was found that also the reduction of various redox couples at photocathodes, e.g. p-GaAs, is inhibited [4],[5]. In order to obtain some more information on this inhibition effect we

E. Pelizzetti and N. Serpone (eds.), Homogeneous and Heterogeneous Photocatalysis, 317–333.

investigated the charge transfer processes between semiconductor electrodes and redox systems in more detail by comparing properties of GaAs and InP electrodes.

2. EXPERIMENTAL

The GaAs-crystals with a carrier density of about $10^{17} cm^{-3}$ were perchused from MCP, England. The InP-crystals with a carrier density of about $10^{18} cm^{-3}$ were grown in the Philips Research Laboratory, Eindhoven. In the experiments the [100]-faces were in contact with the electrolyte. The GaAs-electrodes were etched in a mixture of H_2SO_4(98%)/H_2O_2(30%)/H_2O(3:1:1) and the InP-electrodes in HCL.

The electrochemical measurements were performed in a conventional cell using a saturated calomel electrode (SCE) as a reference electrode using standard electrochemical equipment. The impedance measurements were made by using a dynamic method at 20 KHz described elsewhere [8], [9]. For illumination a 150 W Xe-lamp was used.

3. RESULTS AND DISCUSSION

3.1. Model for H_2-evolution at p-GaAs and p-InP

In Fig. 1 photocurrent-potential curves are given for p-GaAs [5] and p-InP electrodes [10]. It is quite evident from this figure that the onset potential of photocurrent (U_{on}) differs from the flatband potential U_{fb} by several hundred millivolts in solutions free from any redox system, i.e. the hydrogen evolution is strongly inhibited. Impedance measurements have further shown that the potential distribution changes upon illumination as illustrated by the Mott-Schottky curves in Figs. 2 and 3 [5],[10]. The shift of these curves depends on the light intensity and saturates at higher intensities as shown in the insert of Fig. 2. The maximum shift of the flatband potential was about 0.2 - 0.3 V.

The negative shift of the flatband potential and the corresponding change of the potential distribution must be due to an accumulation of electrons at or near the surface of the semiconductor. It should be emphasized that the shift of the capacity curves was measured in the potential range where the photocurrent is constant. Accordingly the accumulation of charges occurs even during electron transfer.

We interpreted this effect by trapping of electrons, excited by light, in surface states. The cathodic shift of the flatband potential corresponds to an upward shift of the energy bands at the electrode surface as illustrated in Fig. 4. Since the energy of surface states also moves upward with respect to the empty states of the electron acceptor in the solution an electron transfer via surface states becomes to be more effective (Fig. 4c). A detailed kinetic study with p-GaAs [5] has shown that the potential dependence of the normalized photocurrent is nearly independent of light intensity. A quantitative evaluation of the model presented in Fig. 4 [11] agrees with this experimen-

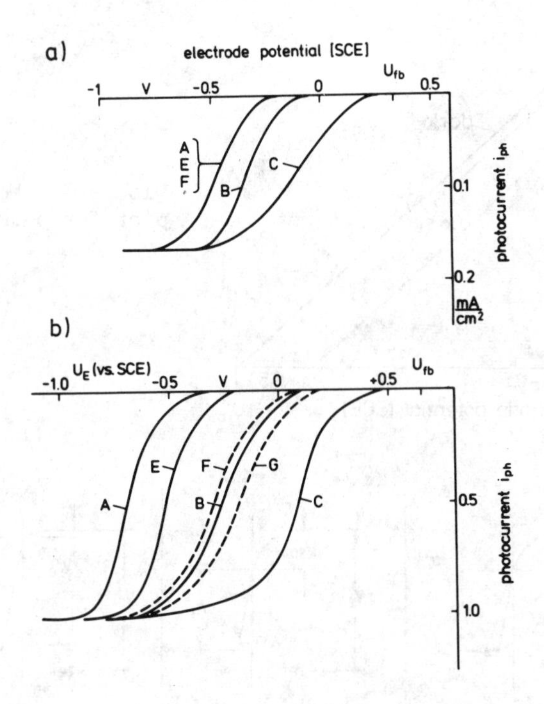

Figure 1. Photocurrent vs. electrode potential in 1M H_2SO_4
a) p-GaAs, b) p-InP. curve A: without redox system; B: [Cosep]$^{3+}$;
C: Fe^{3+}; E: Cr^{3+}; F: Eu^{3+}; G: MV^{2+}
measurements at GaAs by lock-in technique

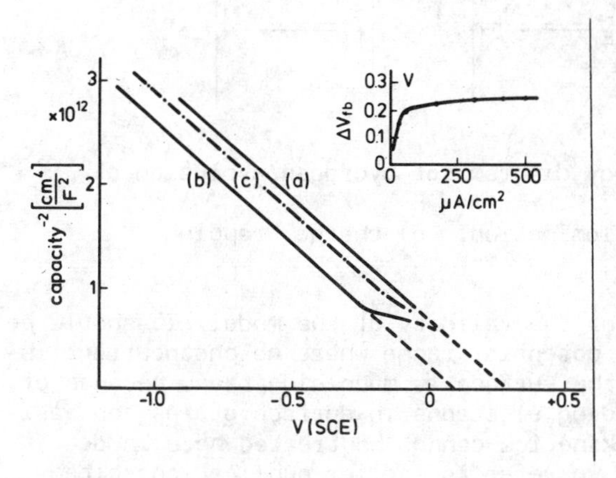

Figure 2. Mott-Schottky plot for p-GaAs in 1M H_2SO_4. a) in darkness, b) and c) during illumination;
b) with Eu^{3+},
c) with [Cosep]$^{3+}$

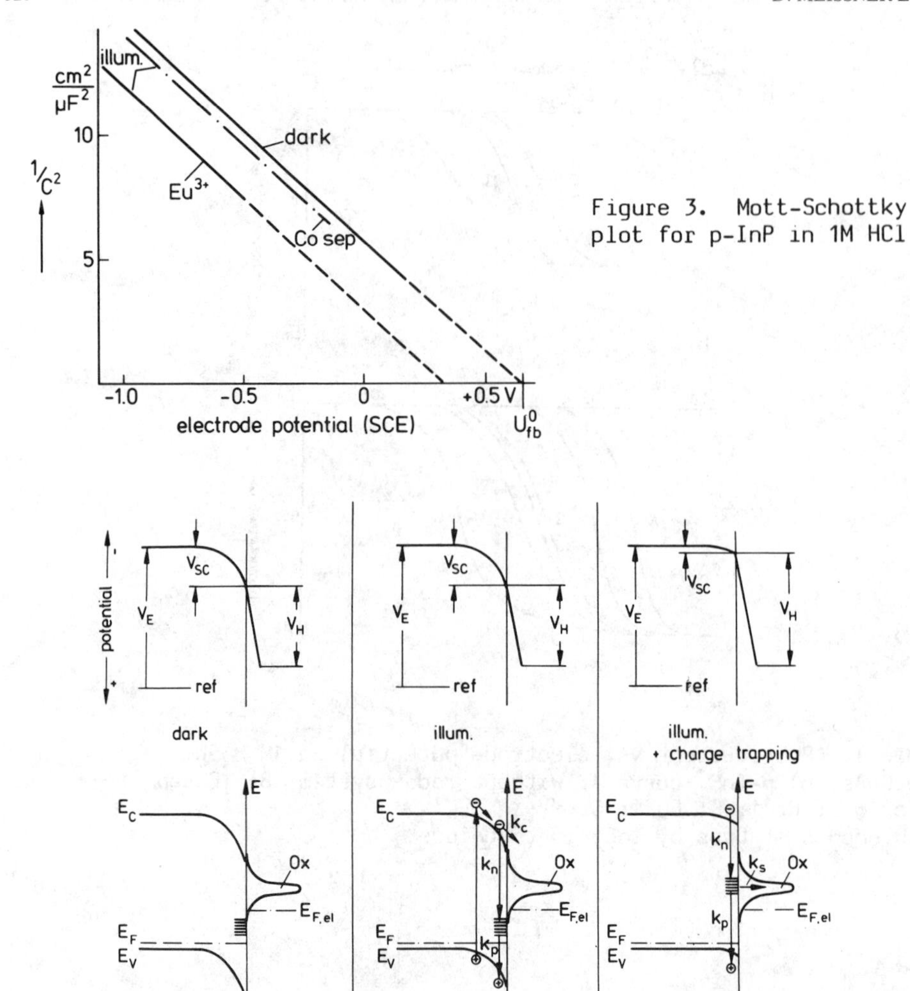

Figure 3. Mott-Schottky plot for p-InP in 1M HCl

Figure 4. Potential and energy diagram for hydrogen evolution at p-type electrode
a) in darkness, b) during illumination, c) charge trapping

tal result which is a proof for the validity of the model. It should be further mentioned that in the potential range where no photocurrent is observed the hole density at the surface is much higher and because of the relatively small band bending electrons in surface states can easily recombine with holes. The kinetics cannot be treated here in detail. For further information we refer to earlier publications cited above.

3.2. Charge transfer processes between semiconductor electrodes and redox systems

3.2.1. Light induced reduction reactions at p-type electrodes. According to Fig. 1a and b the onset potential of photocurrent can occur at less cathodic potentials in the presence of suitable redox systems. There seems to be a general trend that the onset potential moves closer to the flatband potential if the standard potential of the redox couple becomes more positive which is reasonable from the thermodynamic point of view. For illustration of the energy parameters the positions of the energy bands at the surface of GaAs and InP and the standard potential of various redox couples are given in Fig.5. These couples may be clas-

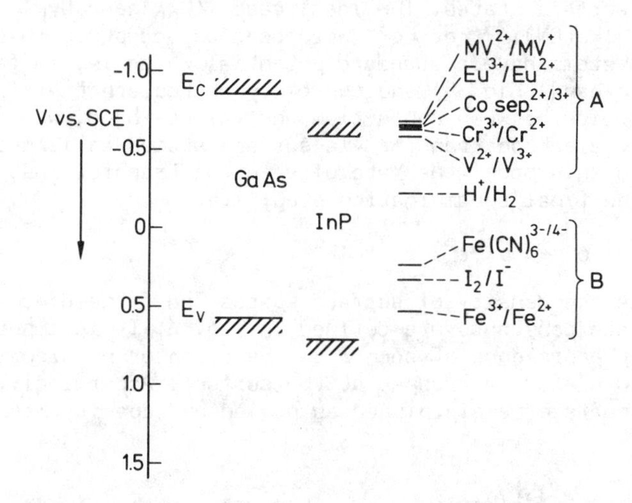

Figure 5. Energy levels at the surface of GaAs and InP and standard potentials of redox couples

sified by two types of systems: In type A we have couples with a standard potential negative with respect to the standard hydrogen potential, and in type B those with a positive standard potential. All A-type couples listed in Fig. 5 are rather close in their potential. It is evident from the photocurrent-potential curves in Fig. 1 that at p-InP the onset of photocurrent occurs at less cathodic potentials with respect to the H_2-evolution for all redox systems (type A and B). In the case of p-GaAs a similar effect was only observed with B-type redox systems and with Co(III)sepulchrate (A-type). With other A-type acceptor molecules such as Eu^{3+}, V^{3+} and Cr^{3+} the photocurent-potential dependence at GaAs is identical to that found without any redox system. It is important to note, however, that also in the case of the latter electron acceptors the photocurrent is due to the reduction of the redox system and not due to H_2-formation as proved by rotating ring-disc experiments [12]. These results are rather surprising from the thermodynamic point of view because the reduction of type A acceptors would occur more easily at GaAs than at InP, i.e. the conduction band occurs at higher energies (see Fig. 5). These observations indicate that the light induced electron transfer is not only determined by thermodynamic but also by kinetic parameters.

As mentioned before it is theoretically expected that the photocurrent onset occurs close to the flatband potential U_{fb}. At potentials being negative with respect to U_{fb}, the energy bands are bent downward so that at p-type electrodes the electrons created by light excitation are driven towards the surface by the corresponding electric field. They are easily transferred across the interface if an electron acceptor molecule of suitable energy is present. Results which show a considerable difference between U_{on} and U_{fb} are usually interpreted by strong recombination of electron-hole pairs within the space charge layer or via surface states. The ideal case, (U_{on} near U_{fb}) was found if e.g. $[Fe(CN)_6]^{3-}$ or Fe^{3+} were used as acceptors (B-type). Since these systems have a standard potential which is, in fact, close to the valence band (Fig. 5) and the empty states are therefore located mainly at energies between conduction and valence band, we interpreted this result by electron transfer via surface states as illustrated in Fig. 6 [5]. In this model the rate of electron transfer must be large compared to the final recombination step, i.e.

$$k_s f N_t c_{ox} \gg k_p f N_t P_s \tag{1}$$

in which N_t is the density of surface states and f the occupation factor. The rate constants are defined in Fig. 6. It is important to note that in the presence of such a B-type acceptor no minority carriers are accumulated or trapped at the surface. Accordingly, the band edges at the surface remain pinned as proved by capacity measurements.

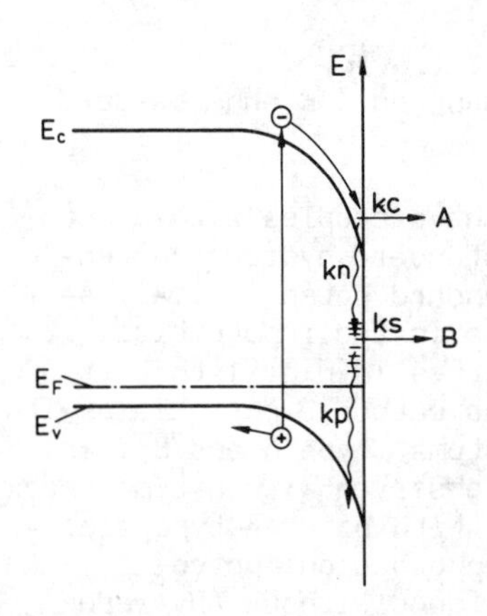

Figure 6. Energy diagram for electron transfer to A-type and B-type acceptor

In the case of the A-type couples one should expect an electron transfer directly from the conduction band to the acceptor molecules because the standard potentials are negative (see Fig. 5). Here the transfer rate (rate constant k_c) has to compete with the first recombination step (Fig. 6), i.e. only with systems where the condition

$$k_c n_s c_{ox} > k_n n_s (1-f) N_t \tag{2}$$

is fulfilled a photocurrent can be expected. Since a less negative onset potential U_{on} was found with most redox systems at p-InP but not at p-GaAs although the energy conditions of the first are less favorable, one might suppose a higher density of surface states (higher N_t) to be present on the GaAs surface. An alternative model is discussed below (Fig. 11).

3.2.2. Oxidation reactions at n- and p-type electrodes. Before discussing the kinetics in more detail it is necessary to analyze also the oxidation process of the redox systems (type A) at GaAs and InP. Corresponding results obtained are presented in Fig. 7.

Considering first the curves obtained in the dark (Fig. 7a-d) one can recognize that for instance Eu^{2+}-ions are easily oxidized at p-GaAs and not at p-InP. With n-type electrodes just the opposite result was found, i.e. an additional dark current occurs at n-InP but not at n-GaAs. It must be concluded from these results that in the case of GaAs the oxidation of Eu^{2+} proceeds via the valence band and for InP via the conduction band. Similar observations were made with Co(II) sepulchrate (see Fig. 7a) and also with V^{2+} (not shown).

These results are rather puzzling because of two reasons: First the reduction and oxidation of several redox systems occur via different energy bands of GaAs, i.e.

$$\left.\begin{array}{l} Ox + e^- \rightarrow Red \\ Red + h^+ \rightarrow Ox \end{array}\right. \quad \text{at GaAs} \tag{3}$$

Secondly, it is surprising that a rather small difference in the position of bands of InP with respect to GaAs leads to a change of the mechanism in that at InP both, reduction and oxidation, occur via the conduction band, i.e.

$$Ox + e^- \rightleftarrows Red \qquad \text{at InP} \tag{4}$$

It is further remarkable that the anodic dark current at p-GaAs in the presence of Eu^{2+} or $[Co(II)sep]^{2+}$ occurs even at rather negative potentials (around -0.35 V) at which the band bending is relatively large (~ -0.65 eV). The anodic current reaches a plateau which is dependent on concentration and rotation speed of the electrode, indicating that the oxidation reaction is limited by diffusion of ions towards the electrode. In the case of $[Co(II)sep]^{2+}$ the currents are not very reproducable because the reduced species of this redox couple is not very stable [13].

The fact that reduction and oxidation of A-type redox couples occur via different energy bands of GaAs can only be understood in terms of a large reorientation energy λ of these redox couples. Assuming for Eu^{2+}/Eu^{3+} a λ-value being similar to that of Fe^{2+}/Fe^{3+} for which λ = 1 - 1.2 eV was determined by electrochemical methods [14], then the energy distribution of filled and empty states can be

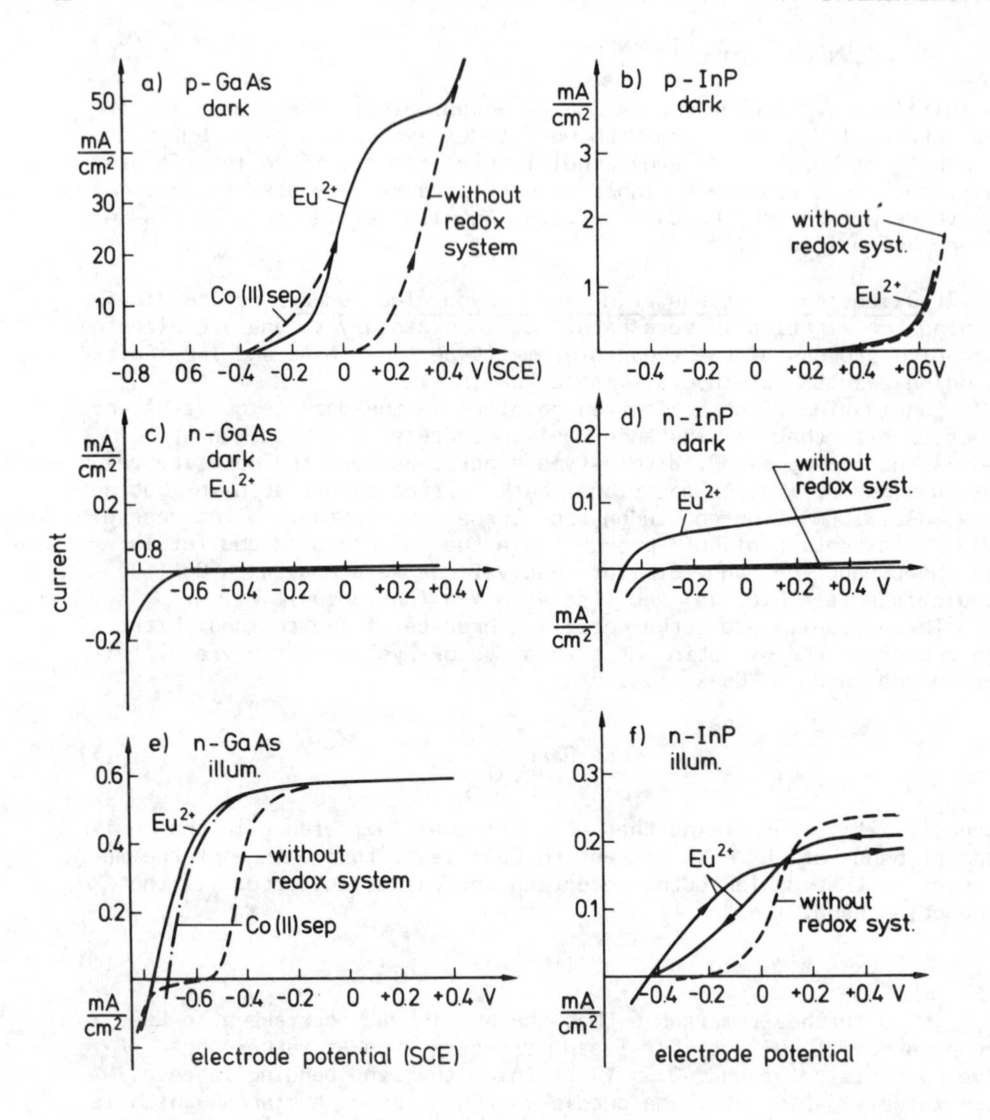

Figure 7. Anodic current vs. electrode potential.
Left: GaAs, right: InP. a) and b) p-type, c) - f) n-type
measurements at GaAs with Eu^{2+} at 5000 rpm. All others without rotation

calculated [15] according to

$$D_{ox} = D_{ox}^{o} \quad \exp\left[-\frac{(E-E_{F,el}+\lambda)^2}{4\,kT\lambda}\right] \tag{5}$$

$$D_{red} = D_{red}^{o} \exp\left[-\frac{(E-E_{F,el}-\lambda)^2}{4\,kT\lambda}\right] \tag{6}$$

The corresponding energy distributions for λ = 1 eV are shown in Fig. 8a, in which D_{ox} and D_{red} are plotted on a logarithmic scale and $D_{ox}^{o} = D_{red}^{o} = 1$. For comparison we have also plotted the corresponding distributions for $[Co\ sep]^{2+/3+}$ which has the same standard redox potential [13] (Fig. 8b). In the latter case the reorientation energy is expected to be much smaller because of the rigid sepulchrate cage around the cobalt ion. We assumed a value of 0.4 eV, a value found for instance for $[Fe(CN)_6]^{3-/4-}$ [14].

Although we have assumed a Gaussian type of distribution according to eqs. (5) and (6) which may be too simple [16], Fig. 8 illustrates

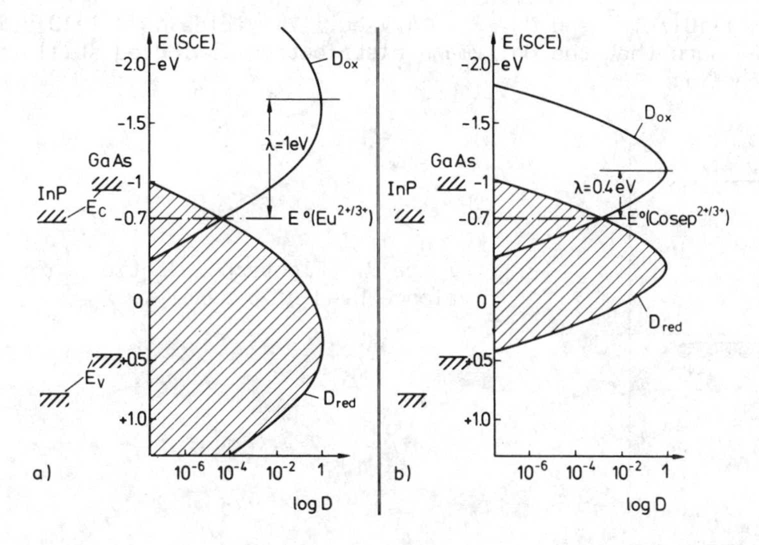

Figure 8. Distributions of electron states in the a) Eu^{3+}/Eu^{2+} and b) $[Co\ sep]^{2+/3+}$ redox systems

quite clearly that the energy levels of the redox systems are distributed over a large range compared to the bandgaps of the semiconductors used here. One can also see immediately that the electrons consumed in the reduction of Eu^{3+} at GaAs originate from the conduction band whereas holes are required for the oxidation process. This energy diagram does not give much information concerning the role of surface states within the gap because of the wide energy distribution of states in the redox system. The situation is similar for $[Co\ sep]^{2+}$, although a direct electron transfer from filled states of this redox system into the valence band should be difficult because of the low density of states for $\lambda \leq 0.5$ eV. Since we found a large oxidation current at p-GaAs (see Fig. 7a) we have to conclude that either surface states are involved or that the energy states tail off differently.

Before applying this model to reactions at InP the oxidation processes at p-GaAs must be analyzed in more detail. As already mentioned above the oxidation of Eu^{2+}, V^{2+} and $[Co(II)sep]^{2+}$ occurs even at very negative potentials. Taking for instance an electrode potential of -0.1 V vs. SCE the downward band bending amounts to $-U_{sc} = 0.4$ V as shown in Fig. 9 and an anodic current of about 20 mA cm^{-2} was measured (Fig. 7a). The question arises whether such a high current density can be carried by the holes if one assumes Boltzmann distribution for the majority carriers according to which the surface hole density p_s is given by

$$p_s = p_o \exp\left(\frac{eU_{sc}}{kT}\right) \tag{7}$$

(p_o hole density in the bulk).
With $p_o = 1 \cdot 10^{17} cm^{-3}$ and $U_{sc} = -0.4$ V eq. (7) yields $p_s = 1 \cdot 10^{10} cm^{-3}$
It can be shown that the Boltzmann distribution is indeed still valid (see appendix).

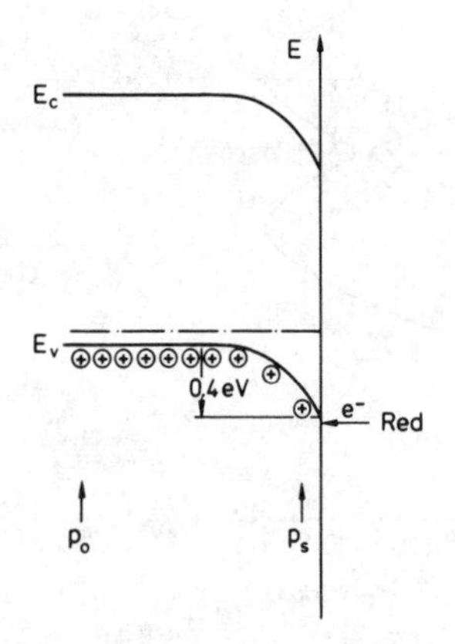

Figure 9. Electron injection into the valence band of p-GaAs

Returning now to the energy diagram in Fig. 8 and applying it to InP one can understand that an oxidation of Eu^{2+} occurs as an electron injection into the conduction band - as is clear from measurements with n-InP in the dark (Fig. 7d) - since the energy bands are lower than for GaAs. Such an injection of electrons into the conduction band is not detectable with p-InP because the electrons are driven back by the electric field. On the other hand there are many more occupied energy states of this redox system around the upper edge of the valence band than around the conduction band so that a valence band process should be even more favorable. This conclusion seems to be in contradiction to the experimental result because no additional anodic current could be detectable with any redox couple. This cannot be due to other bulk properties of InP or its doping. One can also show for InP that an anodic current should be visible already at rather negative potentials. Because of uncertainties in the distribution of energy states we also investigated the oxidation of Eu^{2+} at p-InP in electrolytes of higher pH in which the energy bands of InP are located at higher energies. However, also in this case no additional oxidation current was found.

These considerations show quite clearly that the electron transfer from Eu^{2+} or V^{2+} into the valence band of InP must be inhibited by another process. Now it is well known that an In_2O_3-layer is easily formed on InP-electrodes. Since In_2O_3 is also a semiconductor with a large bandgap of the order of 3.5 eV, one must assume that in a thin layer of the oxide the energy levels have a large gap. A corresponding energy diagram is presented in Fig. 10. We assume the upper level of

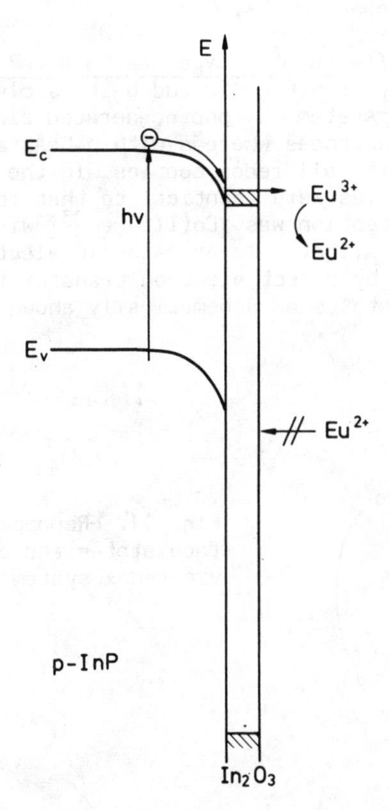

Figure 10. Energy levels of InP and thin layer of In_2O_3

In_2O_3 to be located near the conduction band of InP because the redox systems can be easily reduced by electron transfer via the conduction band. The reverse process via the conduction band would also operate.

An electron transfer from the redox system into the valence band, however, would be inhibited because of the large energy barrier for holes. If the oxide layer is very thin then electrons could only reach the valence band by tunneling through the oxide layer. The tunneling probability T can be estimated from the usual tunneling equation. One obtains $T \sim 10^{-3}$ for a layer thickness d = 20 Å and a barrier height of 2.2 eV and $T \sim 10^{-7}$ for d = 50 Å. According to this model the layer thickness should range in the order of some tens of Ångströms.

Finally, it should be mentioned that the anodic photocurrent at the n-type electrodes also occurs much closer to the flatband potential in the presence of A-type redox systems (Fig. 7e and f). This indicates again the efficiency of the valence band process. It is a little surprising that it occurs with n-InP because of the oxide layer. However, in this case the holes are driven towards the surface because of the upward band bending which increases the surface hole concentration. Possibly the oxide is more easily reduced at n-InP. This point must be investigated by further measurements.

3.2.3. Analysis of photoeffects at p-type GaAs and InP. From section 3.2.1. and especially from Fig. 1a and b it is clear that the reduction of A-type redox systems by photogenerated electrons occur rather easily at p-InP electrodes whereas with p-GaAs a relatively high inhibition was observed with all redox couples. In the latter case all photocurrent potential curves were identical to that found without a redox system. The only exception was $[Co(III)sep]^{3+}$ with which the photocurrent occured at a slightly lower cathodic electrode potential. The reduction may proceed by direct electron transfer from the conduction band or via surface states as schematically shown in Fig. 11.

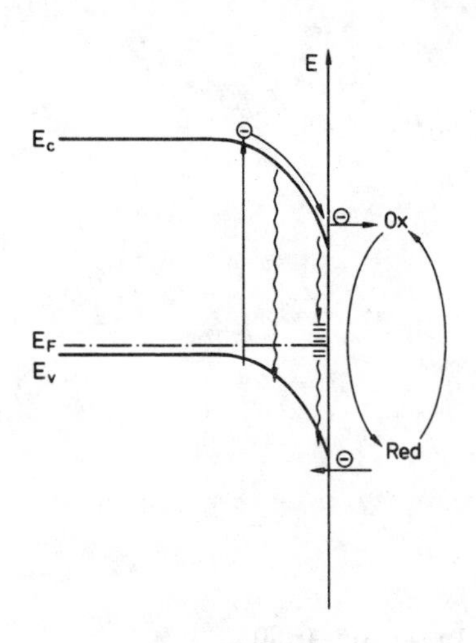

Fig. 11. Recombination via surface states and current shunt via redox system

It is interesting to note that the anodic dark current (oxidation of redox couple) starts to rise around the same potential at which the onset of photocurrent was also observed (compare Fig. 7a with Fig. 1a), and reaches a high value in that potential range where no cathodic photocurrent is detectable. This leads to a model of inhibition as follows:

In a solution containing only the oxidized species of a redox system, Ox, the latter can be reduced by electron transfer from the conduction or via surface states as schematically shown in Fig. 11. The reduced species, Red, formed in this process can be oxidized again by electron injection into the valence band if the Red-ion remains at the electrode for a sufficient time as also shown in Fig. 11. In consequence we have a cathodic current initiated by light and an equal anodic dark current. These two partial currents compensate each other to zero if the Red-ion cannot diffuse away from the surface. Accordingly, we would have a recombination shunt for electrons and holes in the electrolyte just in that range where no net photocurrent is detected. The results obtained with InP are in qualitative agreement with this model. Since in this case no electron transfer from Eu^{2+} into the valence band seems to be possible a net photocurrent is visible.

As already mentioned above the model is only valid if the reduced species (Red) remains at the surface for a sufficient time and if it does not diffuse away from the surface before it is reoxidized. The influence of diffusion can be estimated as follows:

Using a rotating disc electrode the cathodic photocurrent is determined by the light intensity and is related to the surface concentration of the oxidized species by:

$$i_{ph} = F \frac{D}{\delta} \left(c^{s}_{ox} - c^{\infty}_{ox}\right) \tag{8}$$

in which F is the Faraday constant, D the diffusion constant, δ the thickness of the diffusion layer and c^{s}_{ox} and c^{∞}_{ox} the concentrations of the oxidized species at the surface and in the bulk of the solution, respectively. Assuming for a moment that the reduced species is not reoxidized then the concentration of the reduced species c^{s}_{red} at the surface is given by

$$c^{s}_{red} = c^{s}_{ox} - c^{\infty}_{ox} \tag{9}$$

$(c^{\infty}_{red} = 0)$

This concentration (c^{s}_{red}) represents then the maximum surface concentration under stationary conditions, which can be calculated from eqs. (8) and (9). Taking typical values such as $i_{ph} = 10^{-3} Acm^{-2}$, $c^{\infty}_{ox} = 10^{-2}M$, $D = 10^{-5} cm^{2} s^{-1}$ and $\delta = 6 \cdot 10^{-6}$ m (calculated for 5000 rpm) one obtains:

$$c^{s}_{red} = 7 \cdot 10^{-4} \text{ Mol/l}$$

which is a relatively small value compared to c^{∞}_{ox}. Including now the reoxidation of the reduced species the corresponding oxidation current is given by:

$$i^{+}_{v} = F\, k\, c^{s}_{red} \tag{10}$$

The rate constant k can be determined from the dark experiments performed with electrolytes to which the reduced species was added (Fig. 7a). Taking a current value of $2 \cdot 10^{-2}$ Acm^{-2} ($U_E \sim -0.1$ V) and $c_{red} = 5 \cdot 10^{-2}$M one obtains $k = 4 \cdot 10^{-3}$ cm s^{-1}. Inserting this value and the surface concentration c^s_{red} derived above into eq. (10) one obtains

$$i_v^+ = 3 \cdot 10^{-4} \text{ A cm}^{-2} .$$

This value amounts to about 30% of the cathodic photocurrent ($i_{pu} = 1 \cdot 10^{-3}$ Acm^{-2}). Since the rate constant k for the oxidation of the Red-species varies exponentially with band bending one could conclude that i_v^+ and i_{pu} are of the same order of magnitude and could compensate to zero. This result would be consistent with our model. On the other hand i_v^+ should depend on the rotation speed ω because c^s_{red} varies with the thickness of the diffusion layer δ ($\delta \sim 10^{-1/2}$) which was not found yet. It should be emphasized, however, that the above estimate on diffusion does not give any information on the time during which the Red-ion formed during the reduction process, sticks to the surface. If this time would be sufficiently long it would still be reoxidized before it diffuses away from the surface.

Another possibility for explaining the large overvoltage for the photocurrent at p-GaAs would be that strong electron hole recombination via surface states takes place as also indicated in Fig. 11. Then the photocurrent is expected to occur at a band bending at which the surface hole density p_s becomes small enough to decrease the recombination rate to a very low level. The rather low inhibition observed with p-InP must then be interpreted on the basis of a much lower density of surface states as was suggested by Heller et al. [7] for the InP/In_2O_3 interface.

In principle it is very difficult to distinguish between the two models because both, the anodic dark current and the recombination, are essentially determined by the surface hole concentration. At the moment there is only one piece of experimental evidence against the model of a recombination shunt in the electrolyte: In the latter model one expects that the rise of cathodic photocurrent and anodic dark current would match exactly. This condition seems not to be fulfilled for $[Co\ sep]^{2+/3+}$ because the photocurrent occurs at less cathodic potentials than that found with the other systems whereas the onset of the anodic dark current was identical for all redox systems at GaAs. This result is still uncertain because it is rather difficult to obtain reproducible anodic dark currents due to the instability of the $[Co(II)sep]^{2+}$. It is further interesting to note that $[Co\ sep]^{2+/3+}$ is the only A-type redox couple with which a photocurrent onset at less cathodic potentials at p-GaAs was observed. This could be due to a very fast transfer rate as also found in photochemical electron transfer reactions where $[Co(III)sep]^{3+}$ was used as an electron acceptor [17].

4. CONCLUSIONS

The nature of the inhibition of the cathodic photocurrent has been analyzed in terms of two models. In one case the inhibition, which is especially pronounced for p-GaAs, can be interpreted by recombination of electron hole pairs via surface states within the semiconductor. In the other model a transfer of photogenerated electrons was postulated compensated by a reverse transfer of electrons into the valence band. The latter model is based on the observation that reduction and oxidation of various redox couples occur across different energy bands of GaAs which is due to the very large reorientation energy λ of most redox couples in aqueous solutions. This effect could lead to severe losses of energy (small photovoltage) if such a redox couple is applied in a regenerative semiconductor-liquid photovoltaic cell [18],[16]. Consequently, any further research should concentrate on redox couples exhibiting relatively small reorientation energies.

Acknowledgements

The authors are indebted to Prof. B. Kastening, R. Reinicke, J. Lauermann (all Hamburg) and to J.E.A.M. v.d. Meerakker (Eindhoven) for many valuable discussions. We thank Prof. F. Scandola for supplying the cobalt sepulchrate. The Hamburg group also acknowledges the financial support for part of this work by the Deutsche Forschungsgemeinschaft.

APPENDIX

In Fig. 7a-b it was shown that the oxidation of Eu^{2+} is a valence band process. The corresponding anodic current at p-GaAs starts to rise around $U_E \sim -0.35$ V (Fig. 7a). For instance at $U_E = -0.1$ V, i.e. at a downward band bending of $U_{sc} = 0.4$ V, a current of 20 $mAcm^{-2}$ was measured. Since at this potential the surface hole density amounts only to $p_s = 10^{10} cm^{-3}$ the question arises whether the Boltzmann distribution is still valid. This can be proved as follows:

If the current would be entirely determined by the transport of holes and not by the kinetics at the interface itself, it is determined by the conductivity σ in the space charge layer and the electric field E_s is given by

$$j = \sigma E_s \tag{11}$$

The electric field across the space charge layer can be quantitatively obtained by integrating the Poisson equation. An excellent approximation is also given by

$$E_s = \frac{U_{sc}}{d_{sc}} \tag{12}$$

in which d_{sc} is the thickness of the space charge region given by (see e.g. [19])

$$d_{sc} = 2\, L_{D,eff} \left[\frac{e\, U_{sc}}{kT} - 1\right]^{1/2} \tag{13}$$

The effective Debye length is determined by the hole density according to

$$L_{D,eff} = \left(\frac{\varepsilon\, \varepsilon_o\, kT}{2\, p_o\, e^2}\right)^{1/2} \tag{14}$$

With $p_o = 1 \cdot 10^{17} cm^{-3}$ we have $L_{D,eff} \approx 1 \cdot 10^{-6} cm$ and $d_{sc} = 8 \cdot 10^{-6} cm$ for $U_{sc} = 0.4$ V. Inserting this value into eq. (12) the field strength amounts to $E_s = 5 \cdot 10^4 Vcm^{-1}$. In order to obtain a current of 20 $mAcm^{-2}$ a conductivcity of $\sigma = 4 \cdot 10^{-7} (\Omega\, cm)^{-1}$ is required at this field strength (eq.(11)). Assuming a mobility of holes of $\mu_p = 4 \cdot 10^2 cm^2 V^{-1} s^{-1}$ one can estimate the hole density required for the above conductivity by the equation

$$p = \frac{\sigma}{e \mu_p} \tag{15}$$

One obtains $p = 6 \cdot 10^9 cm^{-3}$, a value which is of the same order of magnitude as the surface hole density given by the Boltzmann distribution. This rough estimate shows that the current can just be carried by holes without disturbing the Boltzmann distribution.

References:

[1] H. Gerischer in 'Topics in Appl. Physics' (B.O. Geraphin ed.) Vol. **31**, p. 115, Springer-Verlag, Berlin 1979.

[2] R. Memming, Electrochim. Acta, **25**, 77 (1980).

[3] A. Heller in 'Photochemical Conversion and Storage of Solar Energy' (J. Rabani ed.), The Weizmann Science Press, Israel, p. 63 (1982).

[4] R. Memming and J.J. Kelly in 'Photochemical Conversion and Storage of Solar Energy' (J.S. Connally ed.) Academic Press, New York, p. 243 (1980).

[5] J.J. Kelly and R. Memming, J. Electrochem. Soc., **129**, 730 (1982).

[6] M.P. Dare-Edwards, A. Hammelt and J.B. Goodenough, J. Electroanal. Chem., **119**, 109 (1981).

[7] D.E. Aspnes and A. Heller, J. Phys. Chem., **87**, 4919 (1983).

[8] J.E.A.M. v.d. Meerakker, J.J. Kelly and P.H.L. Notten, J. Electrochem. Soc., **132**, 638 (1985).

[9] K. Schröder and R. Memming, Ber. Bunsenges. Phys. Chem., **89**, 385 (1985).

[10] K. Tubbesing, R. Memming and B. Kastening, J. Electroanal. Chem., submitted.

[11] R. Memming in 'Photoelectrochemistry, Photocatalysis and Photoreactors' (M. Schiavello ed.), Nato Ser. C. Vol. **146,** D. Reidel Publ. Comp. Dordrecht, p. 107 (1984).

[12] R. Memming, J. Electrochem. Soc., **125,** 117 (1978).

[13] J.J. Creaser, R. Geue, J. Mac B. Harrowfield, A.J. Herlt, A.M. Sargeson, M.R. Snow and J. Springborg, J. Am. Chem. Soc., **104,** 6016 (1982).

[14] R. Memming and F. Möllers, Ber. Bunsenges. Phys. Chem., **76,** 475 (1972).

[15] H. Gerischer in 'Phys. Chemistry' (H. Eyring, D. Henderson, W. Jost eds.), Vol. **IX A,** Academic Press, New York, p. 463 (1979).

[16] H. Gerischer in 'Photoelectrochemistry, Photocatalysis and Photoreactors', (M. Schiavello, ed.), D. Reidel Publ. Comp., Dordrecht, p. 39 (1984).

[17] M.A. Rampi Scandola, F. Scandola, A. Indelli and V. Balzani, Inorg. Chim. Acta, **76,** L69 (1983).

[18] W. Kautek and H. Gerischer, Electrochim. Acta, **27,** 355 (1982).

[19] R. Memming in 'Compresensive Treatise of Electrochemistry' (B.E. Conway et al., eds.),Vol. **7,** Plenum Press, New York, p. 529 (1983).

THE IMPORTANCE OF SOLUTION KINETICS IN PHOTOELECTROCHEMICAL PHENOMENA

J. Manassen, D. Cahen, G. Hodes, R. Tenne and S. Licht
The Weizmann Institute of Science
Rehovot 76100
Israel

ABSTRACT. The importance of solution effects for photoelectrochemical performance is discussed at the hand of experiments done at photoelectrodes. The solution composition is shown to be of crucial importance for photoelectrochemical behavior and the significance of this for photocatalytic phenomena is discussed.

INTRODUCTION

There is much similarity between the physics and chemistry of phenomena occurring at photoelectrodes and at semiconductor photocatalytic particulate systems. The great advantage of studying phenomena at photoelectrodes is that the registered photocurrent can be easily measured with great accuracy. We have studied, in detail, the influence of solution composition, especially in polysulfide solutions, and have come to conclusions which are also of importance for photocatalysis at semiconductor particulate systems. We shall describe some relevant older work together with more recent results and discuss their importance.

RESULTS

If a CdSe photoelectrode is illuminated in a polysulfide solution, it shows an initial maximum photocurrent which generally decreases to a steady state value in a matter of seconds to minutes. Several of these photocurrent/time curves are illustrated in Fig. 1. The explanation given for this behaviour is that the peak indicates the initial hole flow through the interface, and the decrease of current towards the steady state is due to losses caused by the non-ideal solution kinetics. This can be expressed by defining the normalized ratio $(I_p-I_s)/I_p$, where I_p is the peak current and I_s is the steady current. The higher this ratio, the larger are the losses.

This phenomenon of peak and steady current is actually a form of concentration polarization. Ions from the solution are oxidized at the

E. Pelizzetti and N. Serpone (eds.), Homogeneous and Heterogeneous Photocatalysis, 335–341.

photoanode, therefore the rate of transport of the material to be oxidized to the electrode and of the oxidized material from the electrode has to be rapid. If it is rapid enough compared to the rate of oxidation, no decrease in current will occur after onset of illumination (curve A in Fig. 1). This phenomenon is intimately connected with the stability of the photoelectrode. If holes arrive at the surface of the semiconductor and are not scavenged by solution species, they either recombine with electrons or oxidize the surface itself, which leads to photocorrosion.

If this hypothesis is correct, chopping of the light source ought to lead to reduction in the losses, because after each light pulse there is a dark period in which the ionic transport in the solution continues. In Fig. 2, for example, the difference between peak and steady current is plotted as a function of light intensity and it can be seen that this difference is appreciably smaller for the chopped light over the whole range of light intensity, which is in agreement with the above described picture.

Fig. 3 shows very clearly that degradation of the photoelectrode (decrease in output) begins, when the light chopper is turned off, illustrating the connection between solution kinetics and photocorrosion.

It is likewise expected that other factors will increase the rate of transport in the solution will improve the photoelectrochemical behaviour and in particular, stability. It is known that the dissolution of neutral salts generally will decrease the Debye radius of the ions, because of which they are less sensitive to drag and move faster. Fig. 4 shows that the stability of the photoelectrode improves appreciably with the addition of NaCl to the solution, which again is in accordance with the picture given.

The actual composition of polysulfide solutions has been more recently considered by us in some detail. In a polysulfide solution we dissolve together S, S^{--} (and usually OH^-). Ions of the form S_x^{--} as well as HS^- are formed, x being 2,3,4 or 5. These species have each their own physical properties.

Although solutions of sodium sulfide are rather basic it is customary to add additional base, when preparing polysulfide solutions for photoelectrochemical purposes. One of the expectations being that the equilibrium:

$$OH^- + HS^- = H_2O + S^{--}$$

will be moved to the right, which will increase the concentration of the oxidizable sulfide ion. It appears however that in spite of different published values in the literature, the equilibrium constant of the reaction is smaller than 10^{-17}, which means that even in concentrated alkali the concentration of free sulfide will be negligible. Sor for that reason no added alkali is necessary. On the other hand, it appears that the addition of OH^- is detrimental to photoelectrochemical activity. Its addition to the solution increases the concentration of S_3^{--}, while S_4^{--} decreases. The optical absorbtion of S_3^{--} reaches by 50 nm more into the visible than that of S_4^{--} and therefore less light reaches the electrode after the addition of OH^-, which is expressed in a decrease in the current output. Addition of OH^- also causes the redox potential of the

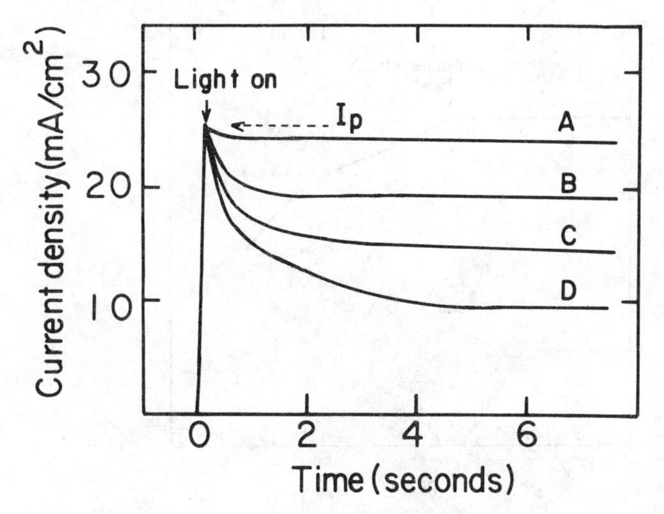

Figure 1. Photocurrent density curves vs. time for a CdSe electrode in polysulfide. A. 0.9M S^{--}, $6x10^{-3}$M Se^{--}; B. 0.9M S^{--}, 1M S; C. 0.25M S^{--}, 0.2M S, T=44°C; D. As in C at 24°C.

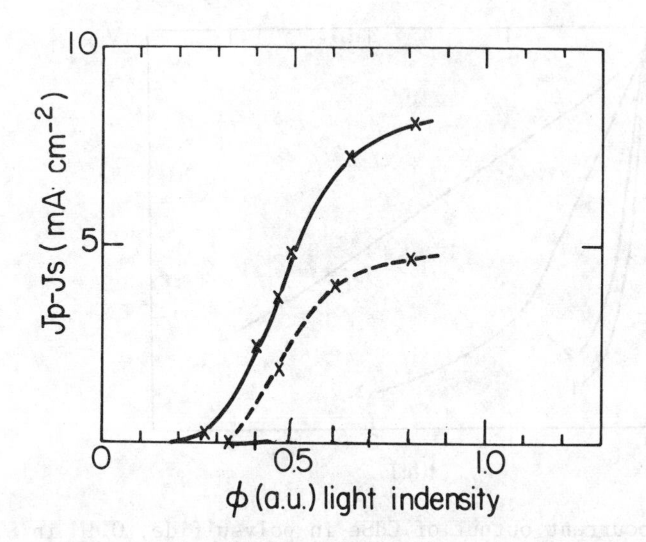

Figure 2. The difference between peak and steady photocurrents vs. light intensity. (——) continuous light. (----) chopped light (50 hz).

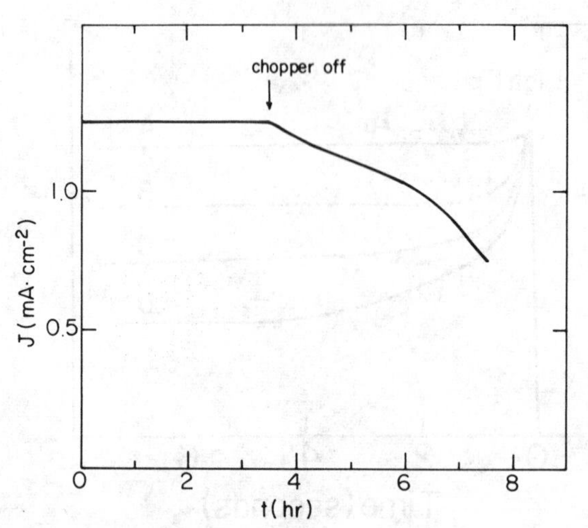

Figure 3. Photocurrent output of a CdSe electrode in polysulfide 0.5M in S, S^{--} and OH^{-}. After 4 hours the chopper (45 hz) was switched off.

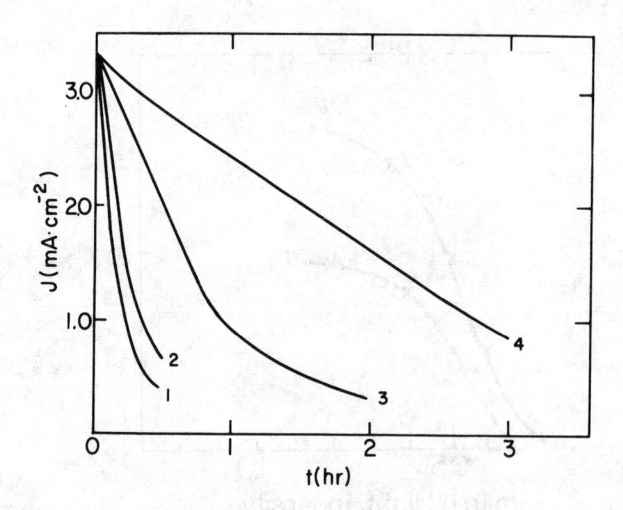

Figure 4. Photocurrent output of CdSe in polysulfide, 0.1M in S̀, S^{--} and OH^{-} as a function of time. 1. Without NaCl added; 2. 1.5M in NaCl; 3. 3M in NaCl; 4. 5M in NaCl.

solution to move towards a more negative value which is only partially compensated by shift in the flat band potential, by which the open circuit voltage decreases. So here we see a rather complicated interdependence of solution composition and photoelectrochemical performance.

The addition of certain ions to the polysulfide solution, like Se^{--} or copper ions has a positive effect on photoelectrode performance. This is a clear case of catalysis of the charge transfer, which may be of great importance for photocatalysis. Possible mechanisms have been suggested for these effects.

It is a well known fact that the rate of ionic transport in solution is dependent on the cationic species and therefore it may be expected that also photoelectrochemical activity will be somewhat cation dependent. Fig. 5 shows this effect to be considerable, when the power curves of the same electrode are measured in Li, Na, K and Cs polysulfide solution. In order to verify that this is because of different solution kinetics, the normalized ratio of peak to steady was measured as a function of light intensity and Fig. 6 shows the losses (larger ratio) to be of the expected order.

This accumulated experience led to the results given in Fig. 7, where the conversion efficiency of 7.7% of a CdSeTe monocrystal in the standard 1M Na_2S, 1M S, 1M OH^- polysulfide solution, could be boosted to 12.7% by proper manipulation of the solution composition. These values were obtained in natural sunlight.

We may reasonably ask ourselves in how far is this experience applicable to photocatalysis by particulate semiconductor species ?

Particulate systems have a much greater surface to bulk ratio than electrodes, and therefore the current density, when electron transfer occurs, will be much lower. On the other hand in photoelectrochemistry the redox couples are chosen for their rapid kinetics and still show the polarization phenomena described. In photocatalysis the reacting system is chosen because of the reaction one wants to perform without necessarily taking into account its electrochemical behaviour; the reacting molecules may be kinetically much more sluggish. Therefore in spite of the large surface area of the photocatalyst, polarization phenomena cannot be exluded. One way to check this would be to study the influence of chopping of the light source on the rate of the reaction and the stability of the semiconductor. Also, the addition of neutral salts to the reacting system in different concentration could give useful information.

An additional difference is that in particulate systems both minority and majority carriers undergo chemical reactions with species dissolved in the liquid, which increases the importance of solution composition compared to photoelectrochemistry, where, (at least ideally), only the minority carriers react at the photoelectrode. In general it can be stated in view of the above, that for a complete description of a photocatalytic system, the influence of solution composition should be taken into account.

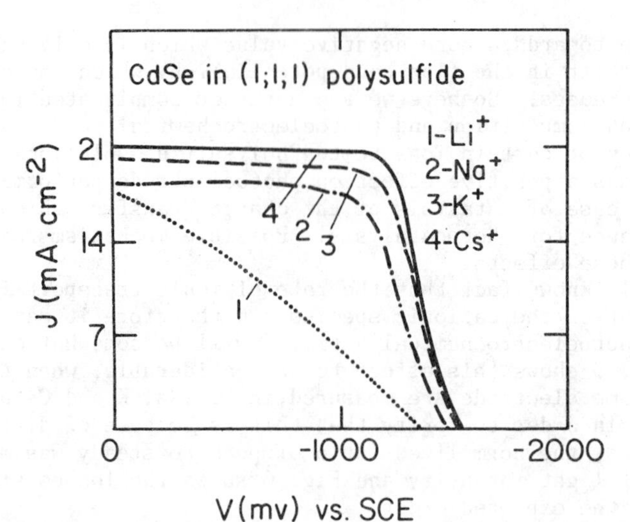

Figure 5. Power curves of CdSe in polysulfide solutions containing different cations.

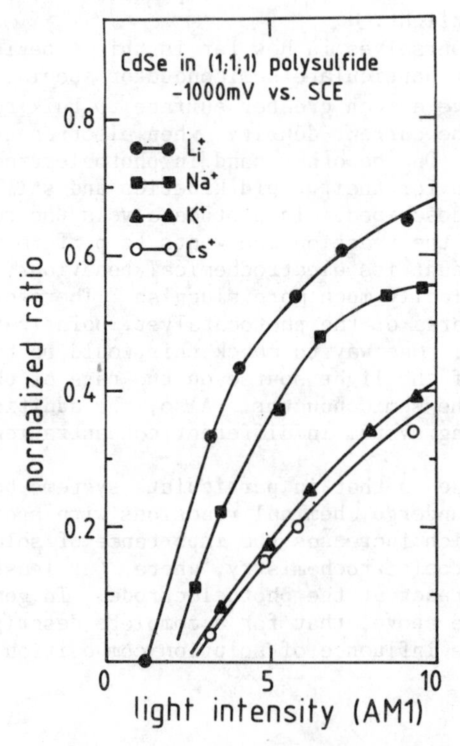

Figure 6. Normalized ratio of peak to steady photocurrent of CdSe in polysulfide solution as a function of light intensity.

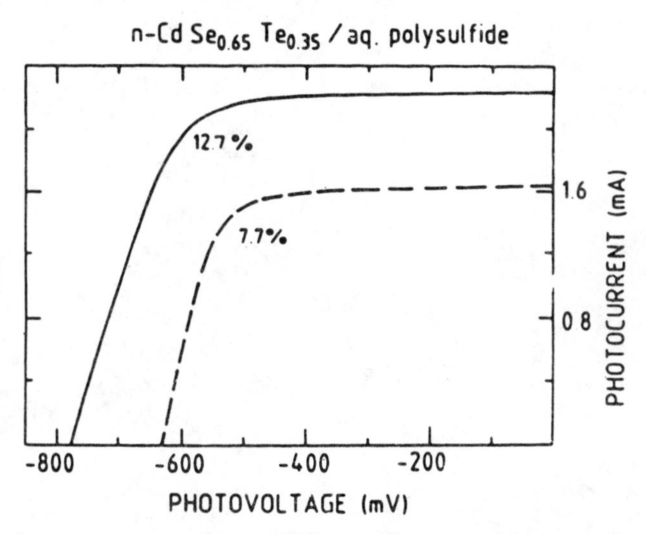

Figure 7. Power curves of a n-$CdSe_{0.65}Te_{0.35}$ monocrystal in natural sunlight.
(- - -) 1M in S, S^{--} and OH^-. (——) 1.8M Cs_2S, 3M S, 0.5 mM $CuSO_4$.

REFERENCES

1. D. Lando, J. Manassen, G. Hodes and D. Cahen. J. Amer. Chem. Soc. 101, 3969 (1979).
2. R. Tenne. J. Electroanal. Chem. 143, 113 (1983).
3. R. Tenne. J. Electrochem. Soc. 129, 143 (1982).
4. S. Licht, R. Tenne, G. Dagan, G. Hodes, J. Manassen, D. Cahen, R. Triboulet, J. Rioux and C. Levy-Clement. Appl. Phys. Lett. 46, 608 (1985).
5. S. Licht, R. Tenne, H. Flaisher and J. Manassen. J. Electrochem. Soc., in press.

MECHANISTIC ASPECTS OF SEMICONDUCTOR PHOTOELECTROCHEMISTRY: THE BEHAVIOR OF SI AND GAAS IN NONAQUEOUS SOLVENTS

Nathan S. Lewis*, Mary L. Rosenbluth, Louis G. Casagrande,
and Bruce J. Tufts
Department of Chemistry
Stanford University
Stanford, CA USA 94305

ABSTRACT. Kinetic methods for identification of the dominant recombination mechanism of photogenerated carriers can yield important information concerning the current-voltage behavior of semiconductor/liquid junctions. We describe here the application of these techniques to systems based on n-Si and n-GaAs photoelectrodes. The use of nonaqueous solvents and outer sphere redox couples allows separation of the overpotential losses associated with fuel formation from losses inherent to the solid/solvent junction itself. Control over surface chemical interactions in both liquid and Schottky junctions can be attained by photoelectrochemical treatments to reduce interface recombination processes. Calculations of the theoretical upper limit on the photovoltage for a given semiconductor/liquid interface can be useful in assessing the prospects for improved current-voltage performance in response to surface modification procedures.

*Address correspondance to this author.

1. INTRODUCTION

There are three major problems involved in the harvesting of solar energy: 1) design of an efficient charge separation process; 2) conversion of these charges into useful work, either through an electrical load or through the production of fuels; and 3) stability of the photosensitizer/collection system under solar irradiation. Much effort has been devoted to finding a system which simultaneously meets all of these requirements (1-5). To date, only large band gap semiconductors, with band gaps (E_g) >2.5 eV, have provided stable photoanode materials in aqueous solution which are capable of oxidizing water to oxygen. Extensive work has thus been devoted to fuel production from single crystals and particulate systems based on these semiconductors. However, it is unlikely that materials with such large band gaps could result in efficient conversion of solar energy under terrestrial conditions (6).

The investigation of this photovoltaic process requires a clear

E. Pelizzetti and N. Serpone (eds.), Homogeneous and Heterogeneous Photocatalysis, 343–362.

distinction between recombination losses which are inherent to charge separation and storage action, as opposed to inefficiencies which result from overpotentials unique to the particular fuel system under study. A detailed understanding of these processes has rarely been addressed in the literature of particulate or single crystal semiconductor/liquid junctions. Some of the key questions which need to be resolved are: 1) How can we predict and design semiconductor/liquid interfaces to produce a given voltage, current, and efficiency? 2) What are the mechanistic features which control charge transfer at these interfaces? 3) What are the unique chemical properties of the semiconductor interface, and how does one manipulate these properties to beneficially affect the system performance? 4) How can one minimize the rapid surface recombination processes which are likely to occur in particulate systems? Only after gaining further insight into these questions is it likely that rational progress can be made in the area of integrating the production of fuels with the photosensitization and charge separation processes at semiconductor/liquid interfaces.

The goal of our work is to elucidate the mechanistic features which are involved in the charge separation and storage events at the semiconductor/liquid interface. In order to minimize kinetic limitations at the solid/liquid interface, we typically limit our studies to electrochemistry of outer sphere redox couples. Additionally, unwanted decomposition or passivation processes are minimized by using nonaqueous solvent systems. Finally, for our semiconductors, we utilize single crystals of small band gap, potentially practical, semiconductors. A key feature of these materials is that their bulk properties can be controlled by proven solid state techniques. This allows variation of the parameters of the solid, and rational interpretion of the effects of such variation on the properties of our semiconductor/liquid systems.

In this work, we shall summarize and highlight our studies of n-Si (7-10) and n-GaAs (11,12) semiconductor electrodes in contact with nonaqueous solvents. These systems have been chosen for study because a wealth of scientific and technological information is available regarding the solid state and interfacial properties of these solids (13). Additionally, the insight gained into the fundamental behavior of these liquid junctions is expected to impact the solid state interface theories which describe the behavior of technologically important Si and III-V based semiconductor devices.

1.1. Background and "Ground Rules"

Because we are primarily interested in the fundamental behavior of the semiconductor/liquid interface, we do not emphasize the optimization of cell losses merely to improve efficiency. All of our liquid junction solar cell results are "unoptimized", and are not corrected for optical reflection or absorption losses, concentration polarization losses, or uncompensated resistance losses. We also point out that our typical experimental apparatus consists of three-electrode cells which are operated under potentiostatic control of the working electrode. However, with proper cell design, it has been shown that such three-electrode measurements do correctly predict the behavior of two-electrode photovoltaic devices (14,15). Also, it is important to emphasize that our cell parameters do not result merely from photocorrosion processes; we have verified photoelectrode stability (at solar photocurrent densities) in all of our systems for many more coulombs of charge passed through the cell than would be required to completely corrode or passivate the semiconductor crystal. However, the achievement of long term cell stability is not a major focus of our studies, and application of our results to practical solar energy storage systems may require further research into cell stability.

1.2. Comparisons of Semiconductor/Liquid Junctions with Schottky Barriers

A recurrent theme of these studies is the correlation of liquid junction behavior with the behavior of Schottky barrier systems. For most covalent semiconductors, in particular, Si and GaAs, it is well known that there is a non-ideal dependence of the barrier height of the junction, ϕ_b, on the work function of the contacting metal overlayer (13,16-18). This lack of change in barrier height is thought to arise in some cases from a stabilization by change in surface charge (from emptying or filling of surface states), and is referred to as Fermi level pinning. For silicon-based interfaces, the junction chemistry has been related to the formation of metal-silicides (13), while for GaAs, several suggestions have been proposed to explain the electrical properties of the junction, including the presence of elemental As at the metal/GaAs interface (19-22), metal-induced gap states (23), or junction-induced defect states (24,25). One of the features of liquid junctions is the ability to prepare a semiconductor surface that undergoes a substantially different chemical interaction than metal barriers. Furthermore, in principle one can control the electrochemical potential of the solution (by variation of the redox couple) without changing the dominant chemical interactions between the solid and the solution/electrolyte. We feel that such correlations with the solid state theory and technology, and possible modifications of these solid state models, may well be the most valuable scientific result of our work.

2. THE BEHAVIOR OF N-SI/ALCOHOL JUNCTIONS

An investigation of n-Si/methanol interface properties is summarized in Figure 1. For a variety of outer-sphere redox couples, we observe an initial open circuit voltage, V_{oc}, which behaves "ideally" over a range of solution redox potential, $E(A^+/A)$. For very negative $E(A^+/A)$, we observe poor rectification properties and low V_{oc} values. Variation in the redox potential increases the calculated barrier height, and yields an increased V_{oc} as expected. Additionally, V_{oc} reaches an upper limit at very positive $E(A^+/A)$ values.

We are currently attempting to complete mechanistic studies of the region where V_{oc} changes in response to $E(A^+/A)$, and we will not concentrate on the behavior of this region at present. We do observe improvements in the behavior of all redox systems after photoelectrochemical anodization of the n-Si in methanol solvent (vide infra), and this leads to steady state V_{oc} values which are typically larger than are reported in Figure 1. This behavior is described in detail in a separate publication. However, we do wish to point out here some assumptions which are implicit to previous treatments correlating a linear V_{oc} vs $E(A^+/A)$ dependence with "ideal" behavior, because the theory is somewhat more complicated than for Schottky barrier systems. Previous theories have adopted the tenet that the rate determining step involves a "bimolecular process", which is first order in the concentration of majority carriers at the surface and is first order in the concentration of solution acceptor. Implicit assumptions regarding this region are: 1) the recombination mechanism does not change as $E(A^+/A)$ is varied; 2) majority carrier thermal emission is the dominant recombination mechanism; and 3) the reorganization energies, Franck-Condon factors, and specific adsorption properties of the redox couple-electrolyte-solvent system are constant throughout the series of redox reagents used in the study. Even with these assumptions, and with the additional assumption of "reversible" electron transfers with relatively low reorganization energies, the Marcus-Gerischer model for electron transfer (26-28) does not predict a strictly linear dependence of V_{oc} on $E(A^+/A)$. This can be seen explicitly from equation (1), where ν is the transmission coefficient, λ is the reorganization energy, N_c is the effective density of states in the conduction band, and E_{cb} is the energy of the bottom of the conduction band.

$$J_o = \tfrac{1}{2}q\nu N_c \exp[E_{cb}-E(A^+/A)]\left(\frac{kT}{\pi\lambda}\right)^{\frac{1}{2}}[A^+]\exp[-(E_{cb}-E_{ox})^2/4\lambda kT] \qquad (1)$$

Assuming a constant frequency factor and fixed band edges for all redox systems, the dependence on $E(A^+/A)$ will be similar to Schottky barriers except for terms which contain the solution reorganization energy, λ. We have calculated the predicted dependence of V_{oc} on $E(A^+/A)$ for the ferrocene series, assuming a reorganization energy of 0.7 eV. 0.7 eV is the approximate λ value which can be obtained from the Marcus theory expression for self exchange rates, combined with measured electron exchange rates of various substituted ferrocene/ferricenium systems (29). We find that within the error of typical V_{oc} measurements for liquid junctions, one should obtain linear behavior of V_{oc} vs. $E(A^+/A)$. Thus, within the limitations of the assumptions stated above, we calculate that the barrier to electron transfer is primarily in the semiconductor itself, and that the redox system should contribute little to the overall activation energy for the interfacial electron transfer. These predictions are currently being tested in our laboratory. However, these assumptions might not be appropriate for all semiconductor/liquid junctions; thus, "ideal" junction behavior need not necessarily result in a linear V_{oc} vs. $E(A^+/A)$ plot over this potential range. A detailed analysis of the mechanistic features in this interesting region will be reported separately.

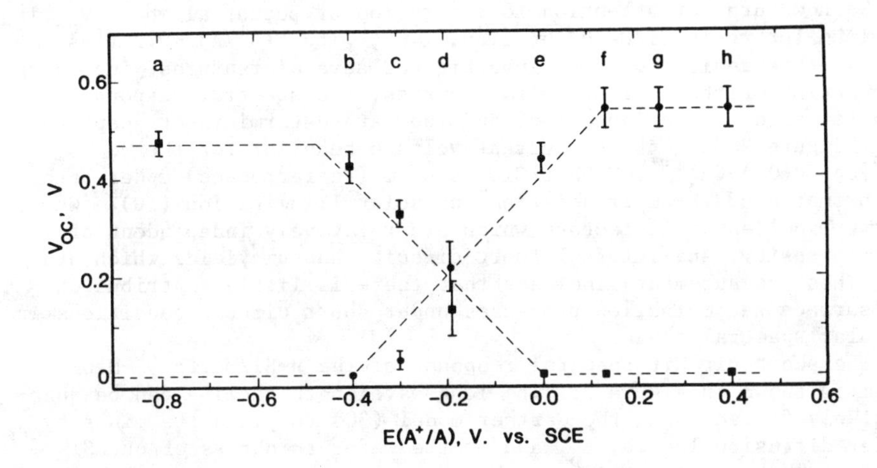

Fig. 1. Open circuit photovoltage vs. solution redox potential for n-Si and p-Si photoelectrodes in 1.0M KCl/CH_3OH solution. The redox couples used were: (a) cobaltocene$^{+/0}$, (b) N,N'-dimethyl-4,4-bipyridinium dichloride$^{2+/+}$, (c) N,N-dibenzyl-4,4'-bipyridium dibromide$^{2+/+}$, (d) decamethylferrocene$^{+/0}$, (e) N,N,N',N'-tetramethylphenylenediamine$^{+/0}$, (f) dimethylferrocene$^{+/0}$, (g) ferrocene$^{+/0}$ and (h) acetylferrocene$^{+/0}$. A tungsten-halogen bulb was used to provide light intensities which yielded short-circuit photocurrent densities of 25-30 mA/cm^2

We now turn our attention to the region of potential where V_{oc} is a maximum for the n-Si/MeOH junction, i.e., with $E(A^+/A)>+0.2$ V vs. SCE. In this region, we have investigated several features, including the behavior of the current-voltage curves, the spectral response characteristics of the junction, and the rate-determining transport step. Figure 2 depicts the current-voltage behavior for the n-Si/$[Fc\text{-}CH(OH)\text{-}CH_3]^{+/0}$-MeOH-$LiClO_4$ system (Fc=ferrocene) under both monochromatic 632.8 nm irradiation and solar illumination (10). We observe excellent fill factors which are relatively independent of light intensity, and internal short circuit quantum yields which are 1.0. These measurements indicate that there is little contribution from surface recombination processes under short circuit conditions in the solar spectral range.

The short circuit spectral response of the n-Si/$Me_2Fc^{+/0}$-MeOH system is shown in Figure 3. The long wavelength decline can be quantitatively fitted using the Gartner model (30) to yield the minority carrier diffusion length, L_p, and/or the wafer thickness of our Si samples. The modest decline at short wavelengths contrasts with the spectral response of typical p-n Si homojunctions, which show pronounced declines in external quantum yield at the shortest wavelengths due to front surface recombination losses. Other features of the current-voltage curves in Figure 2 include the lack of large cathodic currents in forward bias as would be displayed in conventional diodes, and the shape of the anodic portion of the current-voltage characteristic. The cathodic current in this cell is limited by mass transport of the Fc^+ to the electrode surface. Independent measurements at higher $[Fc^+]$ (but identical short circuit photocurrent density) yield identical V_{oc} values to those displayed in Figure 2 (31).

The fill factor depicted in Figure 2 is reduced by concentration overpotential losses and uncompensated series resistance losses. Obviously, minimization of these losses would result in improved cell performance. An important distinction must be made between fundamental losses due to interface states and poor electrode kinetics, as opposed to concentration overpotential losses and series resistance losses, which are factors of the cell design. In our system, we can assign the major losses to the latter factors (10). This prediction has been verified by construction of a thin-layer two electrode solar cell from this system (14), as well as from independent measurements of these losses at Pt electrodes in our three-electrode electrochemical cells (10). Our assignment of the cell-oriented losses as dominant features of these current-voltage curves can also explain the poor fill factors previously observed by other workers for a variety of semiconductor electrodes in both aqueous and nonaqueous solvents (32-34), and does not require the postulation of fundamental losses due to interface states to explain this aspect of the current-voltage behavior.

It is also important to understand the quantitative features which control the voltage of this n-Si/liquid junction system. Attempts to understand the origin of photovoltages which are independent of $E(A^+/A)$ and are substantially less than E_g have resulted in

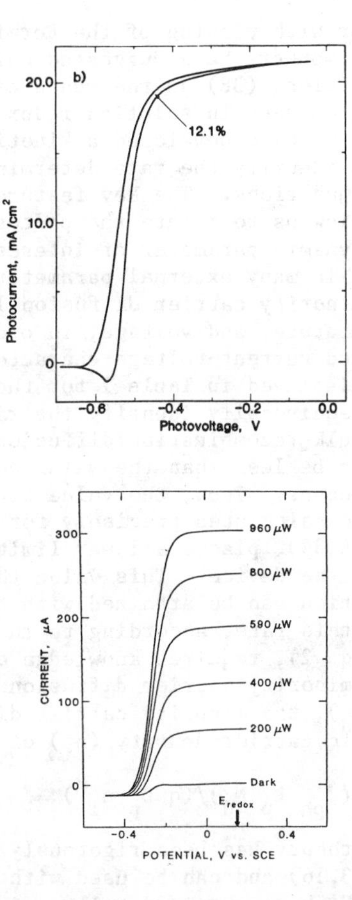

Fig. 2. Current-voltage characteristics (sweep rate 50 mV/s) of n-Si/1.5 M $LiClO_4/CH_3OH$ junctions under 70 mW/cm^2 of ELH-type irradiation. The curves are photoelectrode efficiencies measured under potentiostatic control of the working Si photoanode. (a) Improvement in illuminated J-V characteristics of a 0.2-Ω cm resistivity, (100) oriented, n-Si anode. The mirror-finished electrode surface has been etched for 10 s with 48% aqueous HF, and rinsed with CH_3OH; the redox couple is 0.2 M $(CH_3)_2Fc$/0.5 mM $(CH_3)_2Fc^+$. Numbers on dashed curves refer to J-V scan sequence from the initial scan in the positive voltage direction. After initial oxide growth (3-10 scans), a stable current-voltage characteristic is obtained (solid lines). The substantial increase in voltage and improvement in fill factor indicate the passivation of surface recombination by chemical modification (oxide growth) of the n-Si/CH_3OH interface. (b) Light J-V characteristics at 296 K of 0.2-Ω cm resistivity, (100) oriented, matte-textured (see ref. 10) n-Si anodes in a stirred 0.2 M (1-hydroxyethyl)Fc/0.5 mM (1-hydroxyethyl)Fc^+/1.5 M $LiClO_4/CH_3OH$ solution.

association of such behavior with pinning of the Fermi level by surface states (35). Other workers have suggested carrier inversion processes (36,37) or hot carriers (38) as the means of generating constant photovoltages with changes in solution redox potential. In order to test these models, we have developed a kinetic approach (7,8,31) which allows us to identify the rate determining step for carrier transport at these junctions. The key feature of these experiments is that they allow us to relate the photostationary state variable V_{oc} to the thermodynamic parameter of interest, ϕ_b. The approach involves variation in many external parameters, such as majority carrier density, minority carrier diffusion length, photocurrent density, cell temperature, and voltage, in order to examine their effects on the observed current-voltage characteristics.

Data of this type is displayed in Table 1 for the n-Si/$Me_2Fc^{+/0}$-MeOH system. These data unequivocally identify the rate determining step of this interface as bulk recombination/diffusion. The total recombination rate can never be less than the value due to any one individual recombination process. Thus, the value for bulk recombination/diffusion, which can be calculated precisely for transport to any semiconductor barrier (13,16,39), places a lower limit on the sum of the recombination rates for the device. This value therefore places an upper limit on the V_{oc} which can be attained with this system. A quantitative evaluation of this rate, according to the well-known Shockley diode equation, (eqn 2), requires knowledge of the photocurrent density (J_{ph}), the minority carrier diffusion length (L_p), the majority carrier density (N_d), the minority carrier diffusion coefficient (D_p), and the intrinsic carrier density (n_i) of the material.

$$V_{oc} = (kT/q) \ln [(J_{ph} L_p N_d)/(q D_p n_i^2)] \qquad (2)$$

For Si homojunctions, this theory has been rigorously tested in a variety of circumstances (13,16) and can be used with confidence for liquid junctions as well. We have obtained values of the necessary bulk constants for Si from the recent literature, including the intrinsic carrier density (16), minority and majority carrier mobilities and lifetimes, and resistivity vs. carrier density relationships (40). L_p can be determined either using the near-infrared spectral response data (Figure 3) or from the ASTM surface photovoltage method (41). The model utilized in equation (2) also requires knowledge of the photocurrent density at V_{oc}. The value of J_{sc} has been determined by the short circuit current at a given incident light intensity, and this minority carrier current density is assumed to flow at V_{oc} as the current which bucks the recombination current.

The data in Table 1 also allow for <u>predictive variation</u> in the V_{oc} of these junctions. Maximization of L_p*N_d will lead to the highest voltage performance of these systems, and lack of simultaneous control over L_p and N_d can account for the lack of change of current-voltage properties observed in previous studies of Si/liquid junctions. For the longest lifetime material, the wafer thickness will be a limiting factor on the "effective" measured L_p value. Quantitative treatments of this correction, including the effects of a

Table 1

Semiconductor	Solvent	J_{ph} mA/cm^2	N_{maj} cm^{-3}	D_{min} cm^2/sec	n_i cm^{-3}	L_{min} μm	V_{oc} (theory)[a] V	V_{oc} (observed) V
n-Si	CH_3OH	17.0	7.7×10^{15}	11.65	1.45×10^{10}	300	0.60	0.57–0.58
n-Si	CH_3OH	17.0	9.3×10^{14}	11.65	1.45×10^{10}	700	0.56	0.51–0.52
p-Si	CH_3CN	21.0	2.1×10^{16}	34.97	1.45×10^{10}	400	0.61	0.48–0.50
n-GaAs	CH_3CN	22.0	8.0×10^{15}	10.9	1.8×10^{6}	2	0.94	0.72–0.75
n-GaAs	H_2O	22	10^{16}	10.9	1.8×10^{6}	2	0.94	0.72–0.75
p-InP	H_2O	25	2.4×10^{17}	119	1.5×10^{7}	2	0.86	0.66–0.68
n-CdTe	CH_3CN	15	10^{16}	1.7	6.9×10^{5}	1	1.01	0.70–0.72
n-$MoSe_2$[b]	CH_3CN	20	7×10^{16}	5	1.3×10^{10}	5	0.6	0.6
n-$MoSe_2$[b]	H_2O	25	10^{17}	5	1.3×10^{10}	5	0.6	0.6
n-WSe_2[b]	H_2O	15	10^{17}	5	1.9×10^{10}	5	0.6	0.7

a) Calculated based on equation 2. The theoretical value is calculated for an infinite crystal tickness, and thus ignores any further lowering of V_{oc} due to the large surface recombination velocity of the back ohmic contact in typical experimental systems. The correction is on the order of 5%, and is not significant to the accuracy of the present values available for semiconductor/liquid junctions.

b) The accuracy of these values are limited both by the dispersion in experimental V_{oc} values and some uncertainty concerning the value of n_i for these crystals. We have used $E_g(MoSe_2) = 1.06$ eV, $E_g(WSe_2) = 1.16$ eV and approximate $N_c = N_v = 10^{19}$ for these materials.

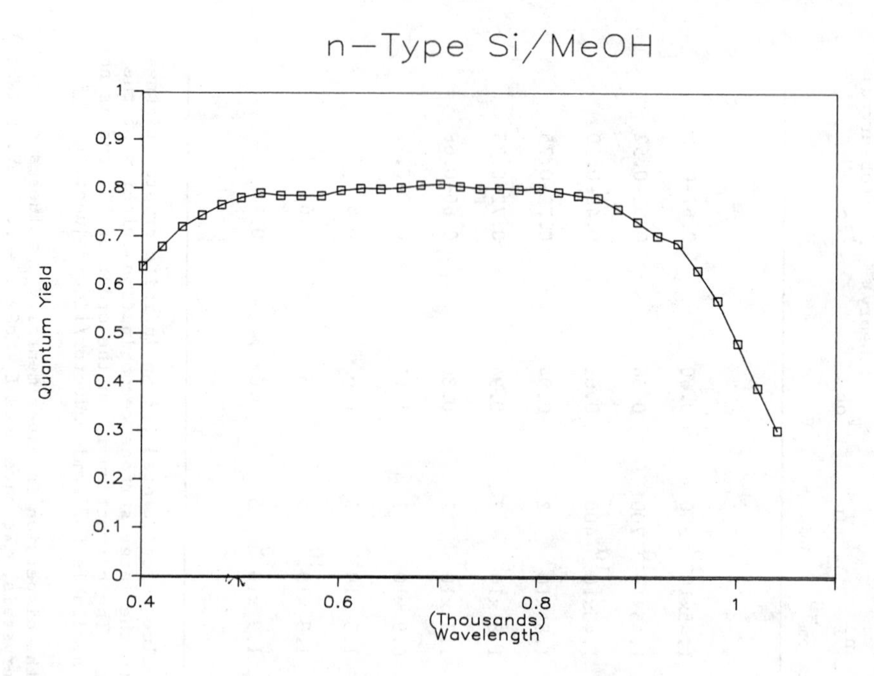

Fig. 3. Short circuit spectral response of a 1.5-2Ωcm resistivity, (100) oriented, N-Si sample in $Me_2Fc^{+/0}$-$LiClO_4$-MeOH.

finite back surface recombination velocity, are obtained by modification of the boundary conditions for the diffusion equation, and can be found in standard solar photovoltaic texts (16,42). Importantly, our n-Si/liquid junctions do not deviate from the bulk recombination/diffusion limit even at very large V_{oc} values. For instance, the 635 mV open circuit voltage for the 0.2 ohm-cm resistivity, float zone n-Si is one of the highest voltages reported to date for any n-Si photovoltaic system under these conditions, even including values reported for highly sophisticated back surface field and concentrator n-p homojunction systems (43). This implies that the n-Si/MeOH junction is of such high quality that all other recombination processes have small rates compared to the bulk recombination rate. Higher V_{oc} values at this photocurrent density and temperature might be obtained with the availability of more highly doped, long lifetime Si substrates (31); however with this unprocessed 0.2 ohm-cm resistivity material, our n-Si/liquid junctions have reached the theoretical limit on voltage performance for any semiconductor junction fabricated with this material (and used under our conditions; i.e., room temperature, 20 mA/cm^2 photocurrent density, no back surface fields, etc.).

At extremely high dopant densities, the minority carrier lifetime is limited by Auger recombination processes (44); thus, in practice for Si samples, there is an upper limit on the product of L_p*N_d which can be attained by variation of N_d. Doping the semiconductor to degeneracy will increase other rate constants, such as tunnelling through the barrier, and will yield lower measured photovoltages. We also observe that when such rates are greater than the bulk recombination/diffusion rate, the temperature dependence of the system indicates a different rate determining step for carrier transport (31). Within the kinetic framework developed for solid state junctions which we have applied to semiconductor/liquid interfaces, kinetic studies as a function of L_p, N_d, temperature, photocurrent density, and cell voltage are sufficient in most cases to uniquely identify the rate determining step for transport in a particular system.

We must then ask the important questions: What interfacial chemistry is responsible for the excellent behavior of the n-Si/MeOH system, and can this chemistry be utilized in other applications? We have made substantial progress in answering both of these problems. Photoelectrochemical growth of an oxide overlayer which is thin enough to allow tunnelling, but which reduces surface recombination losses in the operating cell, has been identified by a number of experimental techniques. Oxide growth has been directly identified by XPS experiments on samples which have been exposed to irradiation under AM2 conditions (31). Correlation of oxide growth with the current-voltage properties of the liquid junction indicates that the V_{oc} of the n-Si/-MeOH-$Me_2Fc^{+/0}$ cell attains the bulk recombination/diffusion limited value at extremely low coverages of this oxide overlayer. Thicker oxides can still support solar photocurrent densities without appreciable series resistance drops; however, if the oxide is allowed to grow (in sufficiently wet solutions), it eventually becomes thick enough to degrade the fill factor.

We have verified that this oxide remains intact under conditions necessary to obtain improved performance of metal-Si Schottky barriers by formation of metal-insulator-semiconductor type structures. Photoelectrochemical treatment prior to junction formation with transparent overlayers of Au, Pt, or Pd yields striking improvements in junction performance (10,31). For these MIS type systems, we observe much larger V_{oc} values than are observed for direct Schottky barriers which have been fabricated without the intervening chemical oxide layer. An example is the n-Si (1.5 ohm-cm resistivity)/Pd junction, which displays a V_{oc} of 520 mV at J_{sc}=20 mA/cm^2. Evidence that pinning of the Fermi level has been substantially reduced has been obtained by exposure of these silicon/oxide/metal contacts to hydrogen gas. For Pt group metals, hydrogen gas is thought to induce an interfacial dipole which results in a lower effective barrier height for the n-Si/metal barrier (45-47). Consistently, we observe much lower voltages in these solid state systems upon exposure to hydrogen-containing ambients. This change is completely reversible upon exposure to air (31), and no effect of hydrogen or air exposure is observed for direct n-Si/Pd barriers. Thus, control over surface recombination processes which can be achieved at semiconductor/liquid junctions can be transferred to solid state systems.

3. EXTENSION OF KINETIC RESULTS FOR SI/MEOH TO OTHER SEMICONDUCTOR/LIQUID SYSTEMS: TRANSPORT KINETICS OF N-GAAS-BASED JUNCTIONS

We can now attempt to extend this kinetic treatment to other semiconductor/liquid junctions, in order to understand the rate determining processes in these systems. Calculation of the upper limit on V_{oc} requires knowledge of the parameters in equation (2), but many of these important variables have not been accurately determined for crystals utilized in typical photoelectrochemical studies. However, good approximations can be made for the intrinsic carrier density from the relation $n_i^2=N_cN_v\exp(-E_g/kT)$, with the use of optical values for E_g and approximating the effective densities of states in the valence and conduction bands as 10^{19} cm^{-3}. The comparison between theory and experimental results for several semiconductor/liquid junctions is depicted in Table 2. For the electrodes in which experiment agrees with these calculations, no change in surface chemistry could produce larger V_{oc} values under the stated conditions; thus, attempts at surface modification to improve V_{oc} values of such systems would not seem worthwhile. Furthermore, these calculations apply in principle to particulate systems and fuel forming photoelectrodes as well as to our single crystal, regenerative cells. These kinetic calculations are also valuable in defining the maximum output voltage which can be attained from any particular semiconductor sample to drive a photoelectrolysis reaction.

It can be seen from Table 2 that improved V_{oc} values can be obtained for many systems. In the junctions with verifiable discrepancies between bulk recombination/diffusion theory and experimental

Table 2

Resistivity, ρ ohm-cm	Waferthickness, W microns	Hole Diffusion Length, L_p microns	V_{oc} (theory)[a] Volts	V_{oc} (exptl)[b] Volts
0.20	315	195	0.630	0.630
0.60	370	165	0.589	0.593
1.50	390	190	0.566	0.568
1.95	330	85	0.540	0.550
1.70	240	45	0.528	0.523
1.70	240	12	0.492	0.495
1.70	240	8.0	0.482	0.470
1.70	240	5.0	0.470	0.462

a) Values are calculated from Eqn 2, taking $n_i = 1.45 \times 10^{10}$ cm^{-3}. We have assumed a large back surface recombination velocity. When $L_p \geq W/2$, the true L_p might exceed W; if this is the case, assumption of a large back surface recombination velocity leads to theoretical V_{oc} values 15-20 mV greater than calculated.

b) Open circuit photovoltages relative to a Pt foil counterelectrode. Variation in V_{oc} is typically ± 5mV from sample to sample. For samples with $\rho \geq 7$ ohm-cm, we observe higher V_{oc} than predicted by Eqn 2, due to the onset of high level injection. All V_{oc} values were measured with ELH irradiation sufficient to provide short circuit photocurrent densities of 20 mA/cm^2 on each sample.

data, it is clear that some other recombination process must provide the rate-determining step. A particularly prominent example of such a system is n-GaAs; thus, we have concentrated recently on elucidating the transport kinetics of n-GaAs (100-oriented)/CH_3CN junctions.

Application of kinetic techniques to the n-GaAs/Fc-CH_3CN junction yields several important conclusions regarding the fundamental chemistry at III-V surface barrier interfaces (12). The temperature dependence of V_{oc} (constant illumination intensity, 100 mW/cm^2 of ELH-type tungsten-halogen illumination) indicates a transport activation barrier of 1.4±0.1 eV, which is in excellent accord with the value of E_g for GaAs. Variation of illumination intensity leads to pronounced changes in the slope of the V_{oc} vs. T plots, but produces only small changes in the extrapolated intercept. We also find similar activation energies in experiments where the photocurrent density is held constant, and V_{oc} is monitored as a function of T.

These data rule out thermionic emission over a pinned surface barrier as the rate determining transport step for this system. Artificial suppression of the electron injection current, due to low concentrations of solution acceptor, is ruled out because the photovoltage is independent of the absolute concentration of Fc^+ ($E(A^+/A)$ is held fixed). Additionally, a full analysis of the $\ln(J_{sc})$-V_{oc} behavior as a function of temperature indicates that the diode quality factor is relatively independent of temperature (1.3±0.1 from 210-300 K), indicating that a recombination-tunnelling mechanism is unlikely to dominate the junction transport (48). The mechanism which is most consistent with our transport data on this system is recombination/-generation by surface trapping levels. Analysis of the Shockley-Read-Hall model for recombination indicates that this mechanism could produce intercepts of V_{oc} vs. T plots which equal E_g, which is consistent with the results presented in Figure 4.

Conventional transport theories contain many definitions of the "activation energy" for a junction, depending on the particular transport mechanism involved. These energies typically are determined from activation plots of the saturation current (J_o) extracted from forward bias I-V measurements in the dark. If the diode quality factor is not unity, these plots yield fundamentally different information than a V_{oc} vs. T plot. The V_{oc} vs. T plots yield information for all the temperature dependent variables, while the $\ln J_o$ vs. 1/T plots only contain information concerning the portion of the kinetics which is voltage dependent as well. For instance, if the mechanism is thermionic emission over the barrier, then a plot of $\ln J_o/T^2$ vs. 1/T will likely correlate with the barrier height as measured by a capacitance vs. voltage technique, whereas the V_{oc} vs T plot may also reflect the effects of variations in interface charge with changes with the cell voltage. Similarly, for recombination/generation in the depletion region, the simple Sah-Noyce-Shockley model (uniformly distributed trap sites throughout the depletion region of a single trap energy, diode quality factor = 2) predicts that V_{oc} vs. T plots will yield an intercept of E_g, but $\ln J_o$ vs. 1/T plots will yield the trap energy.

When the diode quality factors are not close to 1.0 or 2.0, the interpretation of forward bias plots becomes complicated, and a simple

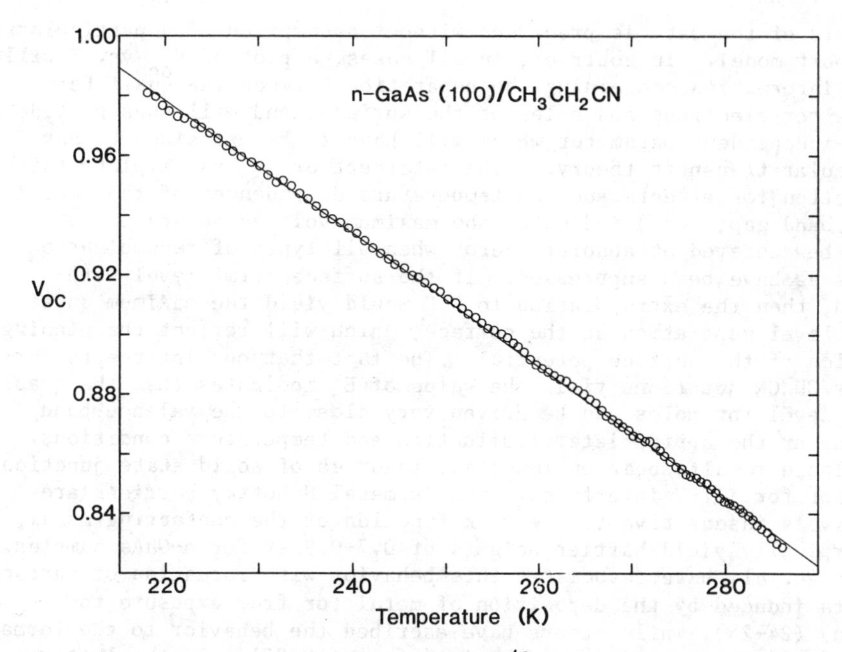

Fig. 4. V_{oc} vs. T for the n-GaAs/$FeCp_2^{+/0}$-$LiClO_4$-CH_3CH_2CN system. Voltages were measured between an n-GaAs working electrode (0.05-0.2 cm^2) and a Pt wire poised at the solution potential. The cell is immersed in an ethanol solution in a strip-silvered dewar flask, and the cooling rate is typically 5 K/min.

analysis of the data is precluded without assumption of a particular transport model. In contrast, in all cases, a plot of V_{oc} vs. T will yield information concerning the separation between the quasi Fermi levels for electrons and holes at the surface, and will thus provide a model-independent parameter which will have to be explained by any particular transport theory. The intercept of V_{oc} vs. T plots (after correction for effects such as temperature dependences of the mobilities, band gap, etc.) indicates the maximum voltage separation which would be achieved at absolute zero, when all types of recombination processes have been suppressed. If the surface Fermi level were pinned, then the extrapolation to 0 K would yield the maximum quasi Fermi level separation at the surface, which will reflect the pinning position of the surface potential. The fact that our intercepts for n-GaAs/CH_3CN junctions yield the value of E_g indicates that the quasi Fermi level for holes can be driven very close to the valence band edge under the appropriate illumination and temperature conditions.

These results bear on important theories of solid state junction behavior for III-V interfaces. n-GaAs/metal Schottky barriers are relatively insensitive to the work function of the contacting metal, and typically yield barrier heights of 0.7-0.9 eV for n-GaAs samples. Spicer et. al. have associated this behavior with formation of surface defects induced by the deposition of metal (or from exposure to oxygen) (24-25), while others have ascribed the behavior to the formation of elemental arsenic at the interface (19-22). In the latter theory, elemental arsenic establishes an effective work function which determines the junction barrier height, while in the former, the defects are intrinsic to the material and are present after essentially any junction formation process (the Unified Defect Model). Thus, these two models differ in predicting a rational strategy to effect changes in junction transport. Our liquid junction results are apparently inconsistent with the predictions of the Unified Defect Model concerning the activation barriers for n-GaAs samples which have been exposed to oxygen.

Reference to equation (1) indicates that for an outer sphere redox system such as ferrocene, with λ=0.7-1.0 eV, that the measured activation energy will be dominated by the activation parameters of the semiconductor itself. Thus, the activation energy observed for our liquid junction systems indicates that the transport barrier is much higher for liquid junctions than would be predicted by the "2/3 rule" (49,50) or other empirical theories based on solid state Schottky junction results. These results strikingly indicate that liquid junctions of GaAs do not possess the same inherent limitations on carrier transport which are found in Schottky barriers. If the dominant recombination mechanism is surface trapping with an activation barrier of the band gap, then in principle, chemical manipulation of these trap levels should lead to improved junction behavior and could ultimately produce bulk recombination/diffusion limited systems, as is found for n-Si/MeOH junctions. In contrast, if the Fermi level is pinned by defects which are formed with essentially all types of contacts, then the upper limit on the attainable V_{oc} will reflect these inherent limitations on the band bending in the semiconductor.

Our transport results for n-GaAs do not directly locate the surface Fermi level, which still may reside substantially above E_{vb}; however, the V_{oc} of 1.05 V at 200 K and 10 Suns establishes a lower limit on the band bending in the n-GaAs/CH_3CH_2CN-$Fc^{+/0}$ junction. This value is in excess of predictions based on pinning by intrinsic defect levels, but would be consistent with the minimization of As at the interface via chemical reactions at the CH_3CH_2CN-$Fc^{+/0}$ junction. It is also interesting to compare these junction properties to the current-voltage behavior of n-GaAs anodes in basic Se^{2-} solutions (51). The poorer fill factors typically obtained from etched n-GaAs surfaces in the aqueous KOH-Se^{2-} electrolyte likely reflect the thermodynamic properties of As and Ga oxides at this pH, resulting in excessive interfacial elemental As, which leads to high interface recombination rates (52). Thus, the passivation of surface recombination by Ru ions, which is necessary in aqueous basic Se^{2-} media, is not required to obtain excellent fill factors and high V_{oc} values for n-GaAs surfaces in nonaqueous solvents. Further spectroscopic experiments are underway in order to understand the detailed surface chemistry occurring in these liquid junction systems.

Finally, insight into the chemical source of recombination has allowed us to rationally improve the current-voltage properties of the n-GaAs system. Our previous results on etched surfaces in the n-GaAs/CH_3CN-$Fc^{+/0}$ system yielded V_{oc} values of 0.72-0.75 V at J_{sc}=20 mA/cm^2 (11). Methods which minimize the amount of excess elemental As at the interface are expected to yield improved junction behavior; thus, anodic oxidation at low current densities could yield a surface with reduced elemental As. Aspnes has performed spectroscopic ellipsometric studies of anodic oxides on n-GaAs and has found essentially no contamination from elemental As if the oxidation is performed at low current density (53). Consistently, we observe that photoelectrochemical cycling of n-GaAs in CH_3CN-$Fc^{+/0}$ electrolyte yields improved cell behavior, and this cycling results in V_{oc} values of 810-830 mV (T=300 K) at J_{sc}=20 mA/cm^2. The insolubility of the Ga and As oxides in CH_3CN will allow us to investigate directly the influence of such surface overlayers on n-GaAs junction properties, and such studies, including spectroscopic ellipsometric probes of overlayer composition, will be reported separately. It is certainly encouraging to be able to rationally modify and improve the interface properties of the n-GaAs system based on an understanding of the junction chemistry of recombination sites.

4. CONCLUSIONS AND SUMMARY

Kinetic methods for identification of the dominant recombination mechanism of photogenerated carriers can yield important information concerning the current-voltage behavior of the semiconductor/liquid junctions. The use of nonaqueous solvents and outer sphere redox couples allows separation of the overpotential losses associated with fuel formation from losses inherent to the semiconductor/liquid junction itself. Control over surface chemical interactions in both liquid and Schottky junctions can be attained by photoelectrochemical treatments to reduce interface recombination processes. The development of these kinetic methods can, in principle, allow identification of the dominant recombination step for most semiconducting photosensitizers, including single crystal, thin film, and particulate semiconductors. Finally, calculations of the theoretical upper limit on the photovoltage for a given semiconductor/liquid interface can be useful in assessing the prospects for improved current-voltage performance in response to surface modification procedures.

5. ACKNOWLEDGEMENT

We acknowledge support from the Department of Energy, Office of Basic Energy Sciences, Grant DE-FG03-84ER13222 for the III-V work, and the National Science Foundation, Grant CHE-8312692 for studies of Si surfaces. NSL also acknowledges support under the Presidential Young Investigator Program from the National Science Foundation, and PYI matching funds which were generously provided by Monsanto Co., the Exxon Educational Foundation, Mobil Corp. and the IBM Young Faculty Development Award Program.

6. REFERENCES

1. Nozik, A. Ann. Rev. Phys. Chem., 1978, 299, 89.
2. Wrighton, M.S. Acc. Chem. Res., 1979, 12, 303.
3. Heller, A. Acc. Chem. Res., 1981, 14, 154.
4. Kiwi, J.; Kayanasundaram, K.; Gratzel, M., in "Structure and Bonding", Vol. 49, K. Jorgensen, Ed., Springer-Verlag, Berlin, 1982.
5. Rajeshwar, K., J. Appl. Electrochem., 1985, 15, 1.
6. Shockley, W.; Queisser, H.J. J. Appl. Phys., 1961, 32, 510.
7. Rosenbluth, M.L.; Lieber, C.M.; Lewis, N.S. Appl. Phys. Lett., 1984, 45, 423.
8. Lewis, N.S., J. Electrochem. Soc., 1984, 131, 2496.
9. Cogan, G.W.; Gronet, C.M.; Gibbons, J.F.; Lewis, N.S. Appl. Phys. Lett., 1984, 44, 539.
10. Gronet, C.M.; Lewis, N.S.; Cogan, G.; Gibbons, J. Proc. Natl. Acad. Sci., U.S.A., 1983, 80, 1152.
11. Gronet, C.M.; Lewis, N.S. Appl. Phys. Lett., 1983, 43, 115.
12. Casagrande, L.G.; Lewis, N.S., J. Am. Chem. Soc., in press.
13. Sze, S.M. "Physics of Semiconductor Devices", 2nd Ed., John Wiley and Sons, New York, NY, 1981.
14. Gibbons, J.F.; Cogan, G.W.; Gronet, C.M., Lewis, N.S. Appl. Phys. Lett., 1984, 45, 1085.
15. Orazem, M.E.; Newman, J. J. Electrochem. Soc., 1984, 131, 2582.
16. Fahrenbruch. A.L.; Bube, R.H., "Fundamentals of Solar Cells", Academic Press, New York, NY, 1983.
17. Mead, C.A.; Spitzer, W.G. Phys. Rev., 1964, 134A, 713.
18. Williams, R.H. J. Vac. Sci. Technol., 1981, 18, 929.
19. Freeouf, J.L.; Woodall, J.M. Appl. Phys. Lett., 1981, 39, 727.

20. Thurmond, C.D.; Schwartz, G.P., Kammlott, G.W., Schwartz, B. J. Electrochem. Soc., 1980, 127, 1366.
21. Heller, A.; Miller, B.; Lewerenz, H.J.; Bachmann, K.J. J. Am. Chem. Soc., 1980, 102, 6555.
22. Chang, C.C.; Citrin, P.H.; Schwartz, B. J. Vac. Sci. Technol., 1977, 14, 943.
23. Tersoff, J. Phys. Rev. Lett., 1984, 52, 465.
24. Spicer, W.E.; Lindau, I.; Skeath, P., Su, C.Y. J. Vac. Sci. Technol., 1980, 17, 1019.
25. Lindau, I.; Chye, P.W.; Garner, C.M.; Pianetta, P.; Su, C.Y.; Spicer, W.E. J. Vac. Sci. Technol., 1978, 15, 1332.
26. Marcus, R.A.., J. Chem. Phys., 1965, 43, 679.
27. Gerischer, H., in "Physical Chemistry", Vol. 9A, Eyring, M.; Henderson, D.; Yost, W. Eds., Academic Press, New York, 1970.
28. Morrison, S.R. "Electrochemistry at Semiconductor and Oxidized Metal Electrodes", Plenum Press, New York, 1980.
29. Pladziewicz, J.R.; Espenson, J.H. J. Am. Chem. Soc., 1973, 95, 56.
30. Gartner, W.W. Phys. Rev., 1959, 116, 84.
31. Rosenbluth, M.L.; Lewis, N.S., submitted for publication.
32. a) Baglio, J.A.; Calabrese, G.S., et. al. J. Electrochem. Soc., 1982, 129, 1461.
b) Calabrese, G.S.; Lin, M.S.; Dresner, J.; Wrighton, M.S. J. Am. Chem. Soc., 1982, 104, 2412.
33. Legg, K.D.; Ellis, A.B.; Bolts, J.M.; Wrighton, M.S. Proc. Natl. Acad. Sci., U.S.A., 1977, 74, 4116.
34. Kohl, P.A.; Bard, A.J. J. Electrochem. Soc., 1979, 126, 603.
35. a) Bard, A.J.; Bocarsly, A.B.; Fan, F.-R.F., Walton, E.G., Wrighton, M.S., J. Am. Chem. Soc., 1980, 102, 3671.
b) Bard, A.J.; Fan, F.-R.F.; Gioda, A.S., Nagasubramanian, G.; White, H.S. Discuss. Farad. Soc., 1980, 70, 19.
36. Kautek, W., Gerischer, H., Ber. Bunsenges, Phys. Chem., 1980, 84, 645.
37. Turner, J.A.; Manassen, J.; Nozik, A.J. Appl. Phys. Lett., 1980, 37, 488.
38. Boudreaux, D.S.; Williams, F.; Nozik, A. J. Appl. Phys., 1981, 51, 2158.
39. Shockley, W., Bell. Syst. Tech. J., 1949, 28, 435.
40. a) Thurber, W.R.; Mattis, R.L.; Liu, Y.M.; Filliben, J.J. J. Electrochem. Soc., 1980, 127, 2291.
b)Thurber, W.R.; Mattis, R.L.; Liu, Y.M.; Filliben, J.J., J. Electrochem. Soc., 1980, 127, 1807.
c) Dziewior, J.; Silber, D. Appl. Phys. Lett., 1979, 35, 170.
41. Annual Book of ASTM Standards, 1978, F391-78, 795.
42. Fonash, S.J., "Solar Cell Device Physics", Academic Press, New York, NY, 1981.
43. Blakers, A.W.; Green, M.A.; Jiqun, S.; Keller, E.M.; Wenham, S.R.; Godfrey, R.B., Szpitalak, T.; Willison, M.R. IEEE Elect. Device Lett., 1984, EDL-5, 12.
44. Dziewior, J.; Schmid, W., Appl. Phys. Lett., 1977, 31, 346.
45. Ruths, P.F.; Ashok, S.; Fonash, S.J.; Ruths, J.M. IEEE Trans. Elect. Dev., 1981, ED-28, 1003.

46. Armgarth, M.; Soderberg, D.; Lundstrom, I. Appl. Phys. Lett., 1982, 41, 654.
47. Aspnes, D.E.: Heller, A., J. Vac. Sci. Technol., 1983, B1, 602.
48. Riben, A.R., Feucht, D. L. Sol. State Electron., 1966, 9, 1055.
49. Mead, C.A. Sol. State Electron., 1966, 9, 1023.
50. Fan, F.-R.F.; Bard, A.J. J. Am. Chem. Soc., 1980, 102, 3677.
51. a) Parkinson, B.A.; Heller, A.; Miller, B. J. Electrochem. Soc., 1979, 126, 954.
b) Heller, A.; Parkinson, B.A.; Miller, B., Appl. Phys. Lett., 1978, 33, 512.
52. Heller, A., ACS Symp. Ser., 1981, 146, 57.
53. Aspnes, D.E.; Schwartz, G.P.; Gualtieri, G.J.; Studna, A.A.; Schwartz, B. J. Electrochem. Soc., 1981, 128, 590.

CHARGE INJECTION INTO SEMICONDUCTOR PARTICLES - IMPORTANCE IN PHOTOCATALYSIS

Marye Anne Fox
Department of Chemistry
University of Texas
Austin, TX 78712-1167

ABSTRACT. Charge injection into irradiated semiconductors represents a new method for effecting photocatalytic oxidations. The advantages afforded by this method include: (1) high chemoselectivity from preferential adsorption effects and from the relative positions of the substrate oxidation potential and the semiconductor valence band, (2) controlled secondary reactions of primary reactive intermediates, and (3) the possibility of restricting the number of electrons exchanged photocatalytically. Several physical probes (transient laser flash spectroscopy, laser-induced coulostatic measurements, and time-resolved sensitization) have proved useful in characterizing the requisite electron exchange. The photooxidation of organic alcohols adsorbed on tungsten oxides varying in composition from discrete chemical species ($WO_2(OR)_2$) through heteropolytungstates ($M_3PW_{12}O_{40}.10\ H_2O$) to extended semiconductor structures (WO_3 powders) exemplify the power of these new routes.

1. INTRODUCTION

1.1. Charge Separation

Irradiation of semiconductors as either colloids, powders, or single crystals with photons of greater energy than the band gap causes an electron-hole pair separation.[1,2] Through efficient interfacial electron transfer, the photogenerated hole is filled by oxidation of an adsorbed donor. The highly energetic conduction band electron can effect, in similar fashion, the reduction of an adsorbed acceptor. These electron exchanges thus generate proximate, surface-bound reactive intermediates whose subsequent chemical reactivity can be significantly influenced by the semiconductor surface on which they are produced.

To a first approximation, the thermodynamic permissibility of such photoinduced electron exchanges can be predicted from the relative positions of the valence and conduction bands of the semiconductor and from the oxidation and reduction potentials of the donor and acceptor

E. Pelizzetti and N. Serpone (eds.), Homogeneous and Heterogeneous Photocatalysis, 363–383.

respectively, Figure 1. Such diagrams also place a predictive limit on the maximum obtainable photovoltage employing such systems as photoelectrochemical cells.

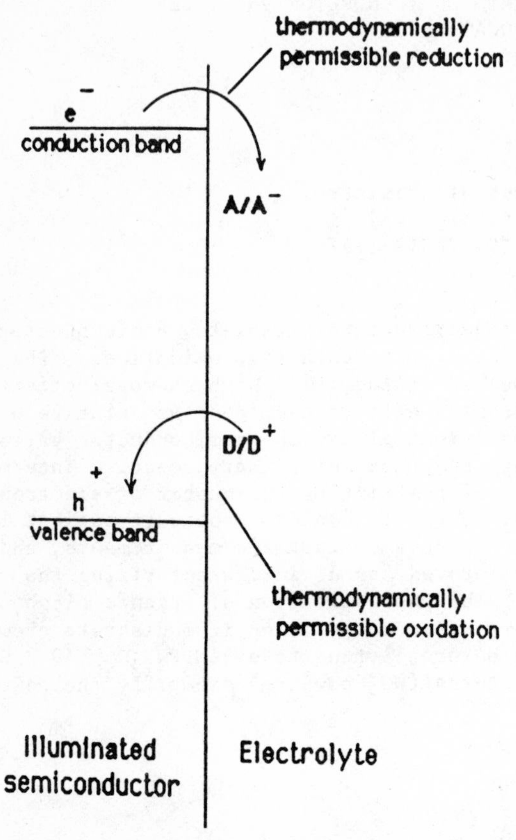

Figure 1. Thermodynamic Constraints for Electron Exchange at Illuminated Semiconductor-Electrolyte Interfaces

The semiconductor band positions can be estimated from theory based on the electronegativities of the constituent atoms and the zero ζ potential[3] or can be determined by experiment. For example, flat-band positions can be obtained by the characterization of the change in the space-charge capacitance produced by the production of a depletion layer at the surface of a semiconductor,[4] by the determination of the potentials for the onset of photocurrent on irradiated semiconductor electrodes,[5] by differential stress measurements with attached piezoelectric detectors,[6] or by the observation of the dark voltammetric response of redox couples spanning a potential range above and below the flatband position.[5,7] The redox potentials for reversible couples can be obtained by a variety of standard electrochemical methods, perhaps most conveniently by cyclic voltammetry. Although the peak potentials of irreversible redox couples lack thermodynamic significance, these values nonetheless can provide a reasonable basis for the calculation.

The ability to observe demonstrable chemical conversions on such photoelectrochemical devices depends critically on the inhibition of the reverse electron exchange, which is often highly exothermic. A significant advantage afforded by these irradiated semiconductor surfaces is found in the space-charge region encountered at the solid-liquid interface when the semiconductor is immersed in an electrolyte containing the redox couple of interest. The induced band bending provides a mechanism for the physical separation of electrons and holes. In n-doped materials, the photogenerated electron moves from the site of excitation near the interface toward the bulk of the solid, as the photogenerated hole migrates toward the surface. As the particle size decreases from macroscopic crystals to powders to colloids to even smaller aggregates, the character of the space-charge region is altered and the possibility of irreversible electron exchange may ultimately depend on the kinetics of secondary thermal reactions or on the presence of a contacting metal catalyst. The construction of integrated, multicomponent photoelectrochemical systems is thus sometimes necessary to observe chemical conversions, particularly for redox couples for which a substantial overpotential is commonly encountered.[8]

The most thoroughly studied photocatalytic system involves water spitting, the concurrent oxidation of water to hydrogen and reduction to oxygen. Because both steps are multiple electron events with substantial kinetic barriers, the most efficient systems must rely on redox mediators, intermediate electron carriers. Although the oxidation of water to hydroxy radical occurs relatively efficiently, the subsequent conversion to oxygen requires additional catalysis.

Because of their high reactivity and low chemical selectivity, hydroxy radicals generated photoelectrochemically in aqueous solutions are capable of initiating the oxidation of a variety of organic materials. Control of these hydroxy-mediated reactions, however, is nearly impossible. For example, upon irradiation of benzene-saturated aqueous suspensions of TiO_2, phenol formation can be observed, although the major isolable product is carbon dioxide derived from further indiscriminate oxidation.[9]

1.2. Selective Organic Reactions

If chemical selectivity is to be retained for organic redox reactions, it is necessary to employ a redox inactive solvent. The primary electron exchange can then be forced to involve the substrate of interest, and improved chemical control can be achieved. Among nonaqueous solvents, acetonitrile has a dielectric constant appropriate for charge separation. It can also function as a medium in which semiconductors can be suspended and in which substrates of interest can be dissolved. Furthermore, it has been shown to be effective as a solvent for many photoelectrochemical cells.[10]

An example of an efficient organic oxidation induced by

irradiation of a semiconductor powder suspended in a nonaqueous solvent is shown in eqn 1.

$$Ph_2C{=}CH_2 \xrightarrow{TiO_2^*} Ph_2C{=}O \qquad (1)$$

Here, the carbon-carbon double bond of 1,1-diphenylethylene suffers oxidative cleavage, producing benzophenone in nearly quantitative chemical yield.[11] Although the quantum efficiency of this process is only modest (Φ = 0.05)[12] because of electron-hole recombination within the semiconductor particle or because of back electron transfer between adsorbed radical ion pairs, the reaction is quite clean and easy to work up.

Subsequent mechanistic study with standard physical organic techniques has shown that the critical organic intermediate in this transformation is the olefin radical cation. This conclusion is based on linear free energy plots of rate data for substituted aryl olefins, Figure 2,[13] by comparison with authentic homogeneous photochemical

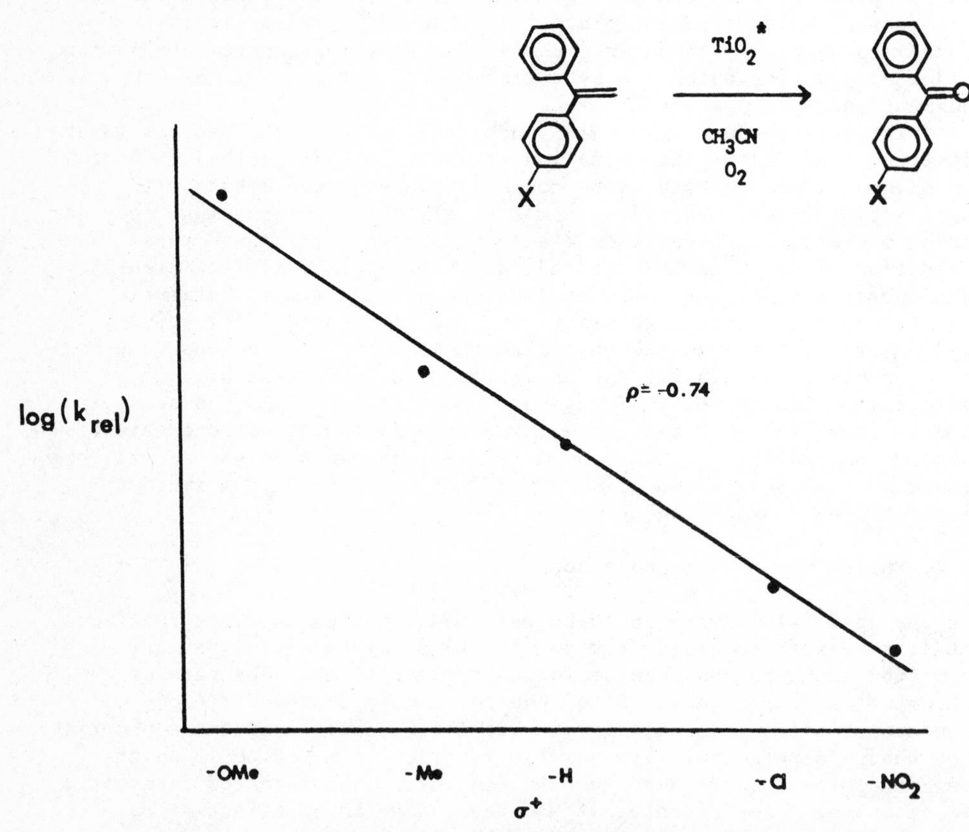

Figure 2. Hammett Plot for Photochemically Induced Oxygenation of Substituted Diphenylethylenes

routes for preparation of the same radical ion pair,[14] and by direct observation of transient formation of absorptive radical ion analogues on colloidal anatase suspensions.[12] The possibility that the reaction occurs via secondary reaction of photogenerated singlet oxygen or superoxide is unlikely, for control experiments demonstrate that different reactivity is to be expected from these reagents[11] than is observed under our photocatalytic oxygenation conditions.

The mechanism for the reaction can therefore be represented most simply as shown in Scheme 1.

$$TiO_2 \longrightarrow e^- + h^+$$

$$h^+ + D \longrightarrow D^{\cdot +}$$

$$e^- + O_2 \longrightarrow O_2^{\cdot -}$$

$$D^{\cdot +} + O_2^{\cdot -} \longrightarrow DO_2 \qquad \text{product}$$

$$D^{\cdot +} + O_2 \longrightarrow DO_2^{\cdot +} \xrightarrow{+e^-} DO_2 \rightarrow \text{product}$$

Scheme 1. Mechanism for Semiconductor-mediated Photooxygenation

Here, photoexcitation of the semiconductor creates an electron-hole pair. The hole transforms an adsorbed olefin to the radical cation which is either trapped by adsorbed oxygen or by photogenerated superoxide. In the former case, reduction of the surface-bound intermediate produces an oxygenated species identical to that accessible from the latter route. Under Lewis acid catalysis, this intermediate cleaves, producing oxygenated product. The oxygenated intermediate, depending on its chemical structure, may also initiate chain oxygenation amplifying the chemical conversion observed from a single photoevent.

A wide range of other organic functional group transformations can be initiated by charge injection into irradiated semiconductor surfaces. For example, condensation products formed from amine oxidation are known, eqn 2,[15]

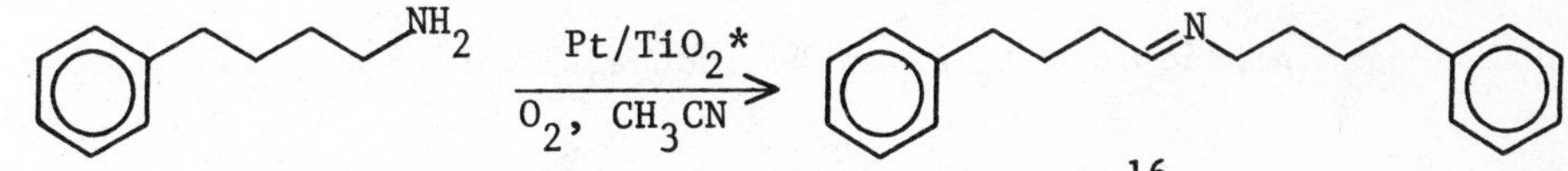

$$Ph(CH_2)_4NH_2 \xrightarrow[O_2,\ CH_3CN]{Pt/TiO_2^*} Ph(CH_2)_3CH{=}N(CH_2)_4Ph \qquad (2)$$

as are oxidation products from thioethers, eqn 3,[16]

$$(n\text{-}Bu)_2S \xrightarrow[H_2O,\ O_2]{TiO_2^*} (n\text{-}Bu)_2SO \qquad (3)$$

benzyl ethers, eqn 4,[17]

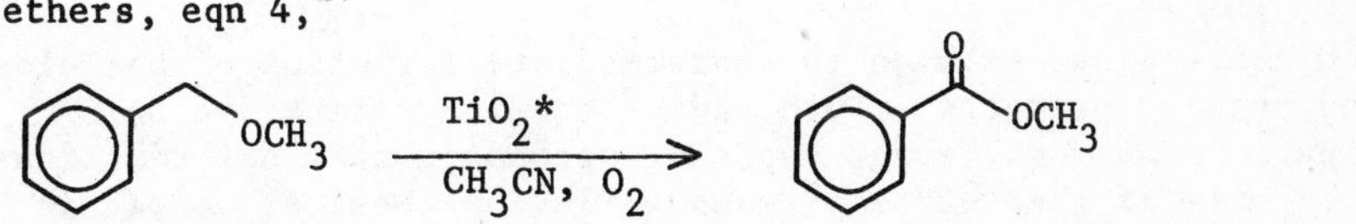

$$PhCH_2OCH_3 \xrightarrow[CH_3CN,\ O_2]{TiO_2^*} PhC(=O)OCH_3 \qquad (4)$$

alcohols, eqn 5,[18]

$$n\text{-}C_7H_{15}CH_2OH \xrightarrow[\text{PhH, } O_2]{TiO_2^*} n\text{-}C_2H_{15}CHO \tag{5}$$

amides, eqn 6,[19]

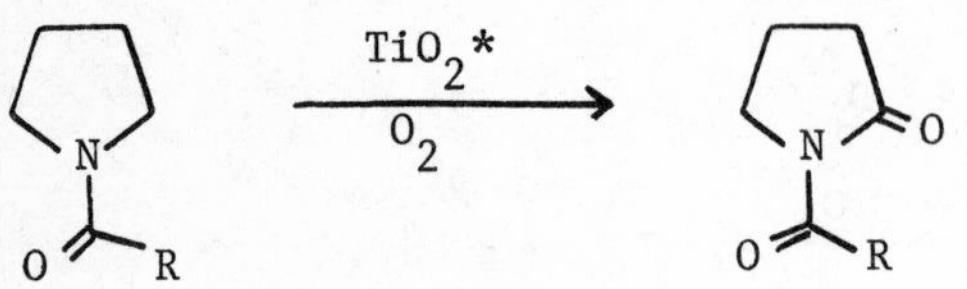

$$\xrightarrow[O_2]{TiO_2^*} \tag{6}$$

and carboxylic acids, eqn 7.[20]

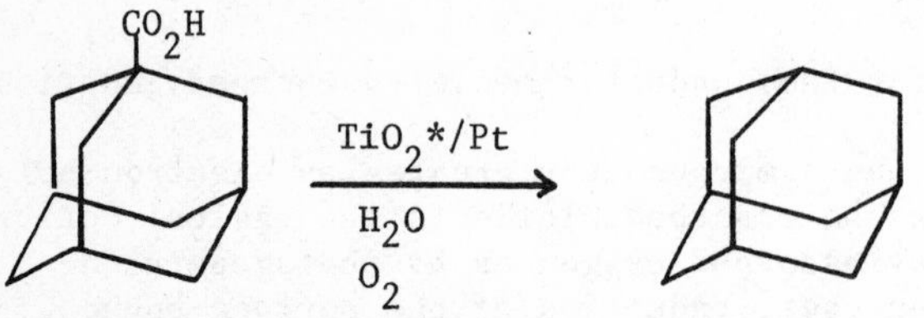

$$\xrightarrow[H_2O,\ O_2]{TiO_2^*/Pt} \tag{7}$$

Since protons can be reduced on these irradiated surfaces, these materials can also be used as efficient hydrogenation catalysts, eqn 8.[21]

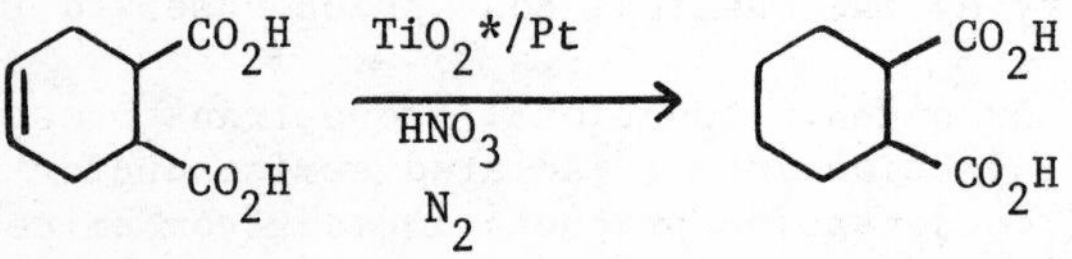

$$\xrightarrow[HNO_3,\ N_2]{TiO_2^*/Pt} \tag{8}$$

Even carbon-carbon bond formation can be induced upon irradiation, eqn 9.[22]

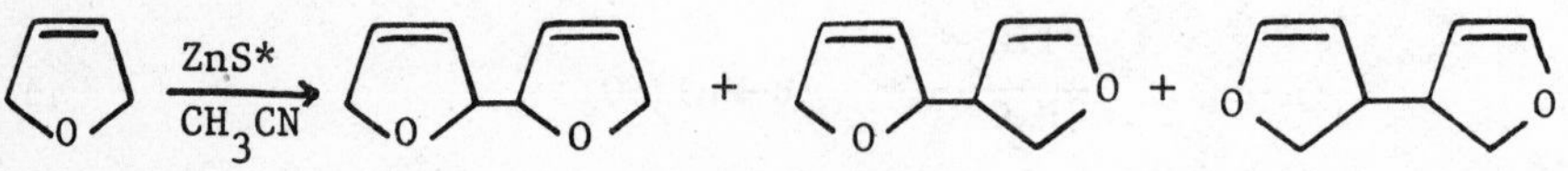

$$\xrightarrow[CH_3CN]{ZnS^*} \quad + \quad + \tag{9}$$

All such conversions rely on the intermediate formation of one electron oxidized organic products. Such radical cations are known to participate in many different types of secondary chemical reactions, and it is only if the course of such follow-up chemical steps can be controlled that a unique chemistry will be attainable on these surfaces.

2. ADVANTAGES OF PHOTOCATALYSIS

2.1. Controlled Reactivity

The ability of an irradiated semiconductor surface to specifically activate a desired organic substrate to redox chemistry relies on thermodynamic and physical access to the donor's electron density. As discussed earlier, a given molecule can function as an electron donor only if its oxidation potential lies at a potential less positive than the valence band of the semiconductor catalyst. By judicious choice of the semiconductor, the position of the valence band, and hence of its excited state oxidizing power, can be varied rationally.

Table 1 lists the band positions for several common semiconductors.[2,23]

Table I. Band Positions[a] for Some Common Semiconductor Photocatalysts

semiconductor	valence band (V vs. SCE ± 0.1 V)	conduction band (V vs. SCE ± 0.1 V)
TiO_2	+3.1	-0.1
SnO_2	+4.1	+0.3
ZnO	+3.0	-0.2
WO_3	+3.0	+0.2
CdS	+2.1	-0.4
CdSe	+1.6	-0.1
GaAs	+1.0	-0.4
GaP	+2.2	-1.0
SiC	+1.6	-1.4

[a]Band positions in water at pH 1.

Since organic functional group oxidation potentials can be altered by structural modification and by substitution pattern, it is easy to identify many different types of groups which lie either above or below a given valence band. Furthermore, since the valence band positions of many semiconductors, especially of metal oxides, are pH dependent, this relative positioning can also sometimes be controlled by choice of pH and/or electrolyte. An example of the influence of the specific semiconductor on the course of a photocatalyzed oxidation can be seen in the contrasting course of oxidative reactivity of tartrate: on TiO_2, this substrate suffers decarboxylation, while on CdS oxidation of the alcohol group occurs.[24]

A second significant factor governing selective reactions on irradiated semiconductors derives from the requirement for preadsorption for efficient electron exchange. If differential adsorption can select for one substrate over an alternate reactant, the former will be selectively oxidized. The sterically unencumbered environment of a primary alcohol makes its adsorption onto metal oxide surfaces quite easy, compared with a more congested secondary alcohol. Consistent with this ordering, Pattenden and coworkers have shown that excellent chemical yields of aldehydes can be attained on irradiated TiO_2 powders, whereas much poorer yields of ketones result from secondary alcohols.[18] A parallel inorganic example of this selectivity can be found in the strong kinetic preference for the oxidation of arsenous in the presence of bromide on TiO_2,[25] where thermodynamic considerations dictate that both materials should be readily oxidized.

2.2. Surface Direction of Secondary Reactions

The chemistry ultimately observed after charge injection also depends on the course of secondary thermal reactions. Since the reactive intermediates formed by electron exchange on the irradiated semiconductor surface are themselves strongly absorbed, access to potential reaction partners will be influenced dramatically by the presence of the surface.

In some cases, the course of chemical reaction can be completely diverted by such surfaces from usual reaction pathways. In the oxidation of benzyl alkyl ethers (eqn 4), for example, the presumed intermediate is the α-alkoxybenzylhydroperoxide 1 derived from oxygenation of the highly stabilized radical formed upon deprotonation of the initial radical cation, eqn 10.

$$PhCH_2OR^{+\cdot} \xrightarrow{-H^+} Ph\dot{C}HOR \xrightarrow{O_2,\ e^-,\ H^+} PhCH(OOH)(OR)\ (\mathbf{1}) \xrightarrow[-H_2O]{TiO_2} PhC(=O)OR \quad (10)$$

In solution this species is stable, even with respect to normal pressure distillation. On TiO_2, however, this intermediate fragments, giving rise to the observed benzoate esters.

Similarly, completely divergent behavior is observed for the radical cation of 1-methylnaphthalene when generated in homogeneous solution and when adsorbed on the surface of an irradiated TiO_2 photocatalyst, eqn 11.[14]

$$\text{1-methylnaphthalene } (CH_3) \xrightarrow[CH_3CN,\ O_2]{TiO_2{}^*} \text{2-}(COCH_3)C_6H_4CO_2H; \qquad \text{1-methylnaphthalene} \xrightarrow[CH_3CN,\ O_2]{C_6H_4(CN)_2{}^*} \text{1-naphthaldehyde } (CHO) \quad (11)$$

On the semiconductor, the surface directs the rapid bimolecular combination of the radical cation with adsorbed oxygen or superoxide to oxygenate the electron-deficient ring, whereas in solution the radical cation can be more slowly deprotonated, generating a radical which leads to ultimate side chain oxygenation.

2.3. Control of the Number of Electrons Exchanged

On metal surfaces, oxidation or reduction of an adsorbate will ensue as the Fermi level is adjusted to a position near the redox potential. If the species formed after electron exchange is itself redox active at the applied potential, a second (or more) electron will also be passed. Thus, conventional electrochemical experiments often produce materials formed by multiple electron transfer.

In principle, excited semiconductors offer an alternate method for the control of subsequent reactivity. In the ground state, the usual semiconductors are poor oxidation catalysts by virtue of the high lying conduction band position. They become good photocatalysts only upon excitation, where a hole is created in the highly oxidizing valence band. After an adsorbed species fills the hole, its oxidizing power is turned off. At low light flux, therefore, where holes do not accumulate in competition with interfacial electron transfer, it should be possible to restrict redox events to single electron exchanges.

An example which demonstrates that this concept can be realized is found in the photocatalytic oxidation of vicinal diacids. These species suffer two electron oxidations under preparative electrolysis at a metal electrode, eqn 12.[26,27]

CO_2H / CO_2H $\xrightarrow[-2CO_2]{-2e^-}$ (12)

In contrast, monodecarboxylation characteristic of single electron charge injection is observed as a major route on irradiated TiO_2 powders, eqn 13.[21]

CO_2H / CO_2H $\xrightarrow[CH_3CN]{TiO_2{}^*}$ CO_2H (13)

Photoelectrocatalysis thus offers several unique ways to control chemical reactivity: selectivty in the initial charge injection event, direction of reaction of the redox intermediates, or restriction of the redox reaction to one electron events.

3. PHYSICAL PROBES OF CHARGE INJECTION

Mechanistic studies of the mode of charge injection can be significantly aided by the availability of methods for time-resolved characterization of intermediates generated at the semiconductor-electrolyte interface. In our research program, we have employed three such methods: direct detection of transient absorption or emission of intermediates formed by laser spectroscopy, laser coulostatic flash studies, or time-resolved photosensitization experiments.

3.1. Laser Flash Spectroscopy: Detection of Transient Intermediates

Although the direct optical detection of transient intermediates formed on single crystalline electrodes, on amorphous powders, or on supported semicondcutors is difficult, the problem can be greatly simplified if optically transparent colloidal dispersions of the catalyst are employed. If we assume that most colloidal dispersions consist of particles of sufficient dimension to retain their macroscopic semiconducting properties, we can infer that analogous intermediates would be generated in the other systems as well.

Many studies have been reported in which inorganic or organic ions participate as redox substrates or relays on aqueous colloidal suspensions of metal oxide or other semiconductors.[28] Analogous experiments can also be conducted in nonaqueous media for purely organic redox reactants. For example, a colloidal suspension of TiO_2 can be prepared by controlled hydrolysis of titanium tetra-isopropoxide in acetonitrile,[12] a route parallel to that reported by Bard and by Graetzel in aqueous solution.[29,30] The resulting material, which resists significant precipitation for at least several days, has high optical transparency, high absorbance in the ultraviolet, and the ability to photocatalyze organic reactions which occur readily on powder suspensions.

In a typical experiment, such a suspension is made approximately 0.01 M in redox-active substrate. The resulting mixture is excited with a laser pulse (third harmonic of a Q-switched Nd:YAG laser (355 nm)) and the formation of a transient spectrum can be monitored.[31] By collecting data at various time intervals after the flash, time-resolved spectra are obtained.

In this way, both oxidative and reductive transients can be detected. For example, in the presence of trans-stilbene a transient spectrum is obtained, Figure 3, which corresponds both in spectral features and in lifetime to that of an authentic sample of the trans-stilbene cation radical generated in the same medium by pulse radiolysis. Similarly, a spectrum characteristic of the one-electron reduction product of methyl viologen can be detected, Figure 4, and its kinetic growth and decay monitored, when the suspension contains the methyl viologen dication. Thus, either oxidation or reduction can be monitored by this optical technique. Standard quenching experiments can then provide information regarding contrasting reactivity on surfaces and in homogeneous solvent cages.

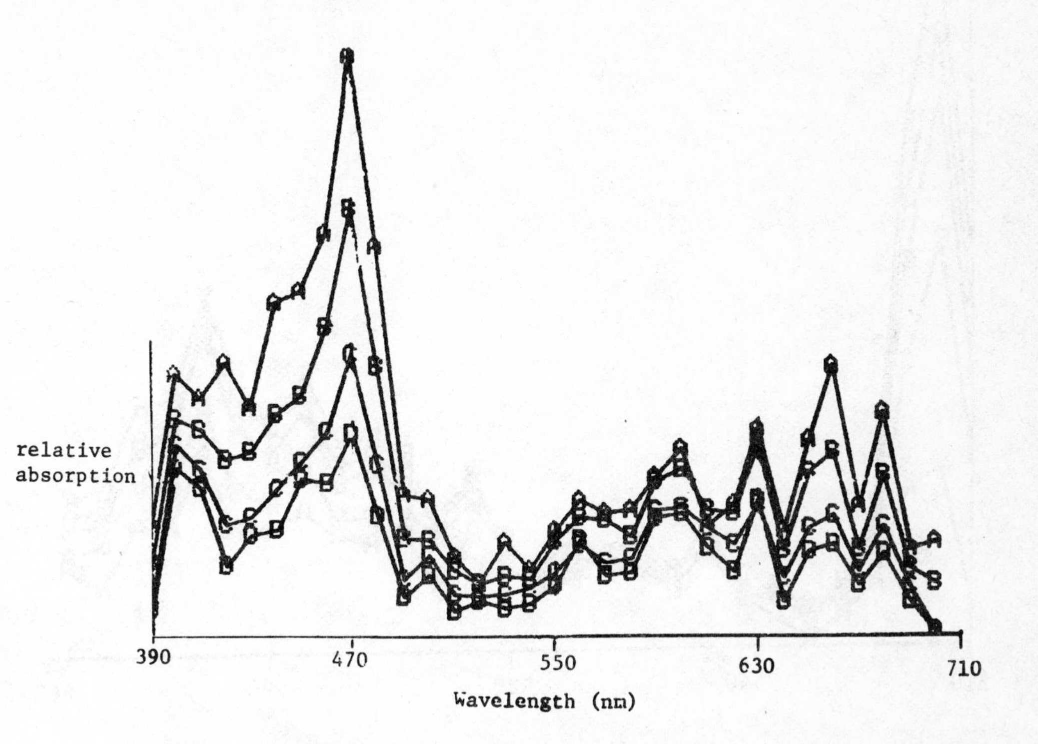

Figure 3. Transient production of the trans-stilbene radical cation sensitized by colloidal TiO_2 (excitation wavelength = 355 nm, 0.01 M trans-stilbene in colloidal TiO_2 suspended in 0.1 M perchloric acid in acetonitrile). Times after flash, in microsec: A, 13; B, 52; C, 95; D, 126.

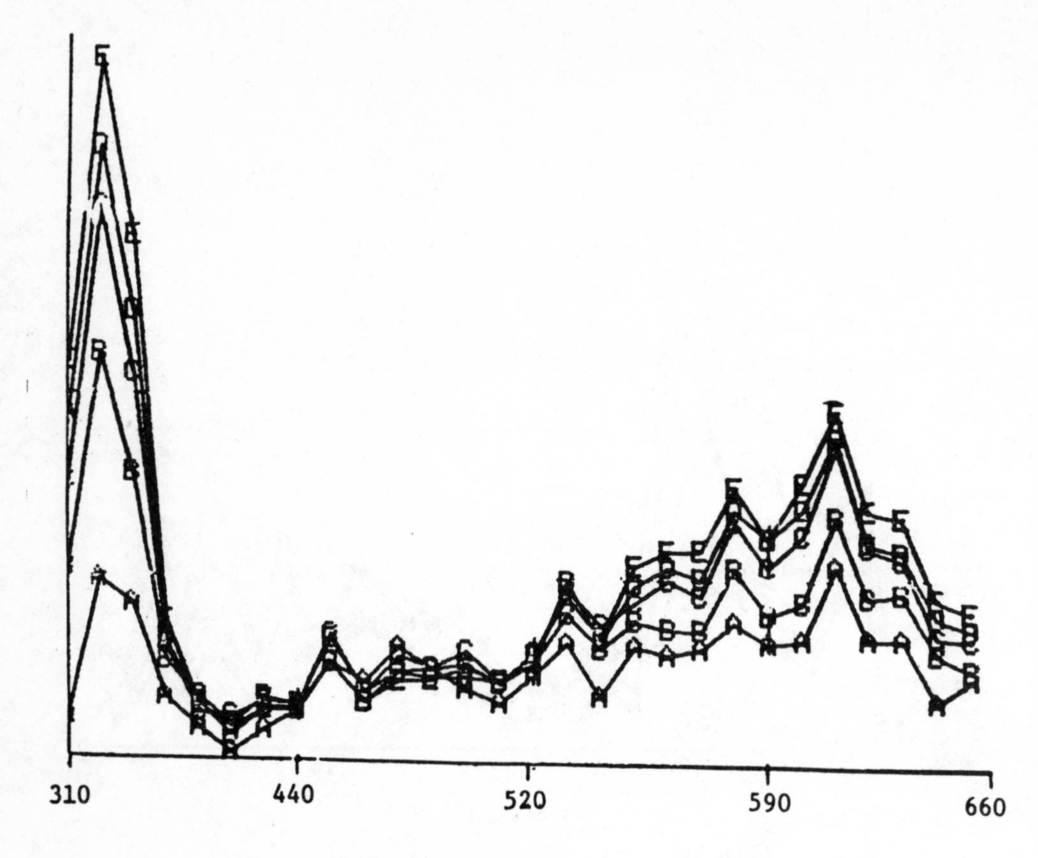

Figure 4. Growth of the absorption spectrum of methyl viologen monocation sensitized by colloidal TiO_2 (excitation wavelength = 355 nm, 1 X 10^{-5} M methyl viologen dication in colloidal TiO_2 suspended in 0.1 M perchloric acid in acetonitrile). Times after flash, in microsec: A, 0.4; B, 1; C, 2.2; D, 4.4; E, 10.

3.2. Laser-induced Coulostatic Flash Studies

When a semiconductor is dipped into a redox-active electrolyte, charge exchange, accompanied by band bending, occurs across the interface until equilibrium is established. Laser coulostatic measurements involve descriptions of the perturbation of the electrode-electrolyte equilibrium by an instantaneous pulse of charge induced by band-gap photons absorbed in the space-charge region. The subsequent time-resolved return of the electrode potential to equilibrium is then observed at open circuit.[32]

The overall photoelectrochemical process is thus a sequence of several events: 1) the photoinduced formation of an electron-hole pair within the semiconductor and its migration to the surface; 2) electron-hole recombination; 3) heterogeneous electron transfer at the electrode-electrolyte interface; and 4) redox equilibration of redox participants within the electrolyte.

These processes can be resolved temporally. When single crystalline TiO_2 immersed in acetonitrile containing 0.2 M $LiClO_4$, a rapid photopotential rise is observed immediately after the flash (0 - 100 nsec), followed by a slower second rise component (0.1 to 3 microsec) in the presence of a redox couple, followed by a two-component photopotential decay (lifetimes 3-5 microsec and approximately 20 microsec).[33]

The time domains of these potential excursions can be ascribed to different events. The initial potential excursion is the result of electron-hole separation in the space charge layer. As the pulsed electrode attempts to attain its new equilibrium potential, electron-hole recombination and dark interfacial charge transfer ensue. The latter process will depend on the identity of the electrolyte and will usually dominate the initial part of the potential decay. The final open circuit dark equilibrium is attained as the solution and semiconductor Fermi levels adjust independently.

Such studies are crucial for a reasonable physical depiction of the electron-hole recombination process and can be effectively used to differentiate rates of charge injection at native and chemically modified surfaces. The differences in turn will be very important for evaluation of recyclable heterogeneous electrocatalysts activated by light signals.

3.3. Photosensitization of Redox Reactions at Semiconductor Surfaces

Usually electron-hole pair separation is accomplished at semiconductor-liquid interfaces by using band-gap excitation. The utility of this route is marred by the unavailability of stable low band gap semiconductors which are responsive to large fractions of the visible and infra red spectral regions.

An alternate method by which charge injection can be induced would be to have an adsorbed dye molecule, converted to its excited state by photoexcitation, transfer either energy or electrons to activate the semiconductor. If the dye's ground state reduction potential (and, hence, its lowest unoccupied molecular orbital) lies above the conduction band, charge injection can occur, producing an equivalent situation to that obtained by oxidative hole capture by an adsorbed electron donor. The electron transfer event can itself be monitored by standard quenching techniques.

Xanthene dyes, for example, have been used to initiate photoelectrochemical responses in illuminated semiconductors. The introduction of a colloidal dispersion of TiO_2 to an acetonitrile solution of erythrosin B causes spectral shifts in the ultraviolet spectrum characteristic of adsorption and dye aggregation. A dye molecule adsorbed on the surface had a significantly shorter singlet excited state lifetime (250 psec) compared to one in homogeneous solution (1.6 nsec).[34] The photosensitization occurred more efficiently from the singlet than from the triplet state of the dye, an observation consistent with previous inferences derived from photocurrent measurements.

These photosensitization studies are important for the design of polymer-coated multicomponent semiconductor systems in which the dye molecules act as photoresponsive antennae, funnelling energy or electrons to the desired solid-liquid junction.

4. A RECENT EXAMPLE: DISCRETE SPECIES TO EXTENDED STRUCTURES

An example of how these concepts of charge injection from organic donors into illuminated semiconductors can provide useful mechanistic information can be found in our recent study of alcohol oxidation of tungsten oxides of varying composition.

4.1. Heteropolyoxoanions as Oxidation Catalysts

A large family of light responsive anions are known which contain varying numbers of tungsten atoms surrounded by oxide, sometimes surrounding a single central heteroatom.[35] These species are often referred to as heteropolyoxoanions or as "tungsten blues" in description of the intense blue color of the mono- or di-reduced forms of the catalyst. Several studies have also shown that photoexcitation promotes the catalyst reduction in the presence of organic donors.[36,37]

Our interest in such photocatalysts lies in the central role that they can play in developing a continuum between discrete intermolecular interactions and the semiconductor-adsorbate system we wish to understand.

4.2. Preliminary Results[38]

Although the emission of these complexes is very weak, single photon counting techniques allowed for fluorescence lifetime measurements and for fluorescence quenching studies in the presence of alcohols. With either the sodium and tetraalkylammonium salts of $PW_{12}O_{40} \cdot 10\ H_2O$, pre-complexation with the alcohol can be clearly observed both in water and in acetonitrile. This conclusion is also bolstered by chemical shift and proton exchange studies by nmr spectroscopy. We can therefore profitably compare the photochemistry of the continuum 2 - 4

$$\mathrm{RO{-}W(=O)_2{-}OR} \qquad PW_{12}O_{40}^{3-} \cdot ROH \qquad WO_3 \mid ROH$$

2   3   4

in which the first member 2 is a discrete light responsive organometallic molecule, the second 3 an associated donor-acceptor complex, and the third 4 an authentic example of a powdered semiconductor-adsorbate system. In all three cases, long wavelength ultraviolet irradiation results in high chemical yields of carbonyl oxidation products (aldehydes from primary alcohols and ketones from secondary alcohols).

Mechanistic investigation of the photocatalyzed benzhydrol oxidation, eqn 14,

$$Ph_2CH(OH) \xrightarrow[CH_3CN,\ O_2]{(NPr_4)_3PW_{12}O_{40}^*} Ph_2C{=}O \quad (14)$$

revealed that electron transfer ensued either upon excitation of benzyhdrol in the presence of 3 or 4, upon excitation of $(Ph_2CHO)_2MoO_2$, or upon excitation of 3 or 4 in the presence of benzyhydrol. Although fluorescence quenching was accompanied by the rise of singly and doubly reduced molybdate, no transient formation of radical ions or radicals derived from benzophenone (all of which have known spectral features and well-defined lifetimes[39]) could be detected. We infer therefore that photoexcitation produces an oxidized organic intermediate which itself is oxidized further in a rapid net two electron transfer. Slow disproportionation of the starting molybdate completes the cycle, giving monoreduced catalyst. This route is also consistent with preliminary photocurrent measurements and with laser-induced coulostatic profiles.

We expect that detailed kinetic descriptions of such systems will lead to a better characterization of the charge injection reactions occurring at semiconductor interfaces.

ACKNOWLEDGMENT. Financial support for this research by the National Science Foundation and by the Robert A. Welch Foundation is gratefully acknowledged.

5. REFERENCES

1. Bard, A.J. Science, 1980, **207**, 139.

2. Nozik, A.J. Ann. Rev. Phys. Chem. 1978, **29**, 189.

3. a) Butler, M.A.; Ginley, D.S. in Semiconductor Liquid-Junction Solar Cells; Heller, A., ed.; Electrochemical Society, Princeton, N.J., 1977, p. 290; b) Gomes, W.P.; Cardon, W.F. Ibid., p. 120; c) McCaldin, J.O.; McGill, T.C. in Thin Film Semiconductors and Interfacial Reactions; Poate, J., ed.; Wiley-Interscience, New York, 1977.

4. Gerischer, H. in Physical Chemistry: An Advanced Treatise; Vol. **IXA**, Eyring, H.; Henderson, D.; Jost, W., eds.; Academic Press, New York, 1970.

5. Nakatani, K.; Tsubomura, H. Bull. Chem. Soc. Japan 1977, 50, 783.

6. Handley, L.J.; Bard, A.J. J. Electrochem. Soc. 1980, 127, 338.

7. Kohl, P.A.; Bard, A.J. J. Am. Chem. Soc. 1977, 99, 7531.

8. Mau, A.W.H.; Huang, C.-B.; Kakuta, N.; Bard, A.J.; Campion, A.; Fox, M.A.; White, J.M.; Webber, S.E. J. Am. Chem. Soc. 1984, 106, 6537.

9. a) Izumi, I.; Dunn, W.W.; Wilbourn, K.O.; Fan, F.R.F.; Bard, A.J. J. Phys. Chem. 1980, 84, 3207; b) Izumi, I.; Fan, F.R.F.; Bard, A.J. J. Phys. Chem. 1981, 85, 218.

10. Frank, S.N.; Bard, A.J. J. Am. Chem. Soc. 1975, 97, 7427.

11. Fox, M.A.; Chen, C.C. J. Am. Chem. Soc. 1982, 103, 6757.

12. Fox, M.A.; Lindig, B.; Chen, C.C. J. Am. Chem. Soc. 1982, 104, 5828.

13. Fox, M.A.; Chen, C.C. Tetrahedron Lett. 1983, 24, 547.

14. Fox, M.A.; Chen, C.C.; Younathan, J.N. J. Org. Chem. 1984, 49, 1969.

15. Fox, M.A.; Chen, M.J. J. Am. Chem. Soc. 1983, 105, 4497.

16. Davidson, R.S.; Pratt, J.E. Tetrahedron Lett. 1983, 24, 5903.

17. Fox, M.A.; Pincock, J.L.; Pincock, A. Tetrahedron 1985, 0000.

18. Hussein, F.H.; Pattenden, G.; Rudham, R.; Russell, J.J. Tetrahedron Lett. 1984, 25, 3363.

19. Pavlik, J.W.; Tantayanon, S. J. Am. Chem. Soc. 1981, 103, 6755.

20. Kraeutler, B.; Bard, A.J. J. Am. Chem. Soc. 1978, 100, 2239; 1978, 100, 5985.

21. Fox, M.A.; Park, K.H. J. Am. Chem. Soc., submitted for publication.

22. Yanagida, S.; Azuma, T.; Kawakami, H.; Kizumoto, H.; Sakurai, H. Chem. Comm. 1984, 21.

23. Gerischer, H.; Willig, F. Top. Curr. Chem. 1976, 61, 33.

24. White, J.; Bard, A.J. J. Phys. Chem. 1985, 89, 1947.

25. Dabestani, R.; unpublished result.

26. Radlick, P.; Klem, R.; Spurlock, S.; Sims, J.J.; van Tamelen, E.E.; Whitesides, T. Tetrahedron Lett. 1968, 5117.

27. Pleininger, H.; Lehnert, W. Chem. Ber. 1967, 100, 2427.

28. Graetzel, M. Energy Resources through Photochemistry and Catalysis, Academic Press, New York, 1983.

29. Hardee, K.L. Bard, A.J. J. Electrochem. Soc. 1975, 122, 739.

30. Duonghong, D.; Borgarello, E.; Graetzel, M. J. Am. Chem. Soc. 1981, 103, 4685.

31. Rodgers, M.A.J.; Foyt, D.C.; Zimek, Z.A. Radiat. Res. 1978, 75, 296.

32. a) Perone, S.P.; Richardson, J.H.; Deutscher, S.B.; Rosenthal, J.; Ziemer, J.W. J. Electrochem. Soc. 1980, 127, 2580; b) Richardson, J.H.; Perone, S.P.; Deutscher, S.B. J. Phys. Chem. 1981, 85, 341.

33. Kamat, P.V.; Fox, M.A. J. Phys. Chem. 1983, 87, 59.

34. Kamat, P.V.; Fox, M.A. Chem. Phys. Lett. 1983, 104, 379.

35. Pope, M.T.; Barrows, J.N.; Jameson, G.B. J. Am. Chem. Soc. 1985, 107, 1771.

36. a) Papaconstantinou, E. Chem. Comm. 1982, 12; b) Papaconstantinou, E.; Dimotikaliv, D.; Politou, A. Inorg. Chim. Acta 1980, 46, 155.

37. a) Ward, M.D.; Brazdil, J.F.; Grasselli, R.K. J. Phys. Chem. 1984, 88, 4210; b) Yamase, T.; Kurozumi, T. J. Chem. Soc., Dalton Trans. 1983, 2205.

38. Fox, M.A.; Cardona, R.; Gaillard, E. J. Am. Chem. Soc. 1985, 107, 0000.

39. Simon, J.D.; Peters, K.S. J. Am. Chem. Soc. 1982, 104, 6142 and references cited therein.

KIWI - What is the role of ZnS on CdS when deposited jointly on SiO_2, to increase due to 'synergic effects' the H_2 photoproduced in your system?

FOX - ZnS passivates surface states and electron-hole recombination sites on the photoactive CdS particle. The SiO_2 is simply one of several possible inert supports on which the CdS semiconductor films can be mounted.

HELLER - Concerning the point raised by Dr. Kiwi, Zn^{2+} exchange at II-VI semiconductor surfaces leading to passivation has been described by several groups, including one at the Weizmann Institute and another at Simon Fraser University. The interpretation of the effect has been the shifting of surface states away from the midgap region. This reduces recombination.

FOX - Our results agree and simply show that the effects from single crystalline electrodes also have analogy for colloids, thin films and supported thin films of the same semiconductor. We did find that the number of surface states is greater in these alternate dispersions than on single crystals.

FRANK - How do you account for the 0.6 V (SCE) difference between the flat band potential of the single crystal and the colloidal CdS?

FOX - Passivation of surface states by surface-bound Zn^{2+} (or ZnS) which allows reduction to occur only from the more energetic conduction band. This allows for H_2 generation, which is not thermodynamically permissible if the electron is mediated by the lower energy surface states.

MANASSEN - Hammett plots may be confusing. One might expect a different behaviour of electron transfer is rate determining.

FOX - We believe the Hammett plots do not test the initial electron transfer which is virtually instantaneous with the original excitation. Rather we believe they test relative rates of subsequent reactions and indicate that substituents which stabilize the radical cation (perhaps by inhibiting back transfer) can improve the overall oxygenation rate.

McLENDON - In your Hammett plot, how do you preclude the possibility that the relative rates actually reflect the equilibration $R^+ + R' \longrightarrow R^o + R'^+$ which might produce rather strong linearity but for a mechanistically different reason?

FOX - The Hammett plot probably does not test the initial electron exchange (which is essentially without a barrier) but instead relates to substituent effects on the relative rates of back electron transfer (or recombination) to that with oxygen or superoxide which ultimately leads to oxidative cleavage.

OLLIS - For the high yield organic conversions shown early in your talk, what were typical quantum yields?

FOX - Typically 0.01 - 0.05 for 4-6 hr irradiation with conventional photoreactors.

MEMMING - (1) Comment: Our membranes are different because each CdS particle contacts both solutions. Not really comparable with Nafion membranes. (2) Why do you try to improve H_2 evolution? The anodic process at CdS and similar compounds is much more difficult and irreproducible. If the anodic process does not work in a reliable way, then also the other reaction does not go. (3) Unfortunately, many literature data concerning band positions for CdS are wrong. A detailed study in our laboratory has shown that all measurements (photocurrent, impedance, etc...) depend very much on scan rate, prepolarization and cleaning of the surface. What kind of pretreatment did you use?

FOX - (1) Your membranes are certainly different than ours. I refer to them simply because they represent a multicomponent supported semiconductor system which predates ours. (2) We have aimed at particle and surface morphological optimization before addressing the more difficult (and more important) question of oxidation. (3) The pretreatment of our CdS particles depended on the method of their generation. Details are presented in the original literature. We agree with your observation of critical dependence on pretreatment method. In fact, the ZnS blocking of surface states illustrates on such sensitive effect.

LEWIS - (1) We know that free O_2^-, especially in benzene as a solvent, must be very difficult to form energetically. Also, in water, we know that Pt evolution of H_2 should occur before O_2^- formation. Do you have any comments on why the Pt chooses to make O_2^- as opposed to form H_2, and can you offer an explanation regarding how and whether free O_2^- is formed in benzene? (2) Can you rule out peroxy-TiO_2 species as oxidants for the organic reactions as opposed to superoxide as a reductive intermediate to couple with the organic intermediates?

FOX - (1) We are not positive that free O_2^- is the active oxygen source. Certainly oxygen itself can trap radical cations. Most of our work has been done in CH_3CN, where superoxide is stable and easily formed and where it may indeed be an intermediate. Kinetic characteristics of these oxygenations seem not to be affected by solution phase superoxide additives. So whatever the oxygenation reagent is, it must be surface-bound. To my knowledge, detailed mechanistic studies in Pattenden's reactions in benzene have not been conducted. (2) The involvement of surface peroxo-titania may certainly be postulated, so long as a mechanistic corollary allows for their replenishment by chemisorbed O_2.

BOLTON - You have emphasized electron transfer between the cathodic and anodic sites. However, in a complete cell involving H_2 evolution on one side and O_2 evolution on another separated by a membrane, protons must also be transported across the membrane. Have you considered how to achieve this efficiently?

FOX - We have not yet coupled our H_2 generation reactions (with sacrificial S^{2-}) to a counterelectrode O_2 evolution. In principle, however, the proton transport problem could be handled by adding a salt bridge in which electrolyte ions (rather than protons or hydroxide) can move to either side of the membrane and hence to neutralize charge.

SAKATA - You succeeded in mild oxidation of a dicarboxylic acid by using TiO_2. That reaction is different from that of Pt electrode. However, if you use a porous metal electrode with small applied bias, could you not drive the same reaction?

FOX - We have not tried porous electrodes, but this effect may not only be attributed to increased surface area, but also to the fact that electrogenerated intermediates formed in the primary electron transfer on metal surfaces are themselves more readily oxidized than the original substrate. With low current density on irradiated semiconductors, the ratio of holes to adsorbed reactant is low so that one electron events apparently prevail. At high current densities on metal electrodes, multiple electron redox events predominate.

KISCH - What is the other product formed in the photooxidation of diphenylethylene to benzophenone? Is singlet oxygen involved?

FOX - Formaldehyde is the primary product which is further oxidized to carbon dioxide. Singlet oxygen seems not to be involved since different products are obtained from tetramethylethylene with $^1O_2$ or with $Pt/TiO_2/O_2$, and because trans-stilbene (which is cleaved efficiently by photoelectrochemical methods) is not active with $^1O_2$ (according to Chris Foote). Singlet oxygen traps (e.g., glyoxylic acid) do not inhibit the reaction.

PICHAT - For oxidations you mentioned high chemical yields. However, you indicated quantum yields in only one or two cases. I am afraid that quantum yields which can be calculated from published papers are most often low. In your opinion, does that mean that the practical interest of oxidations over illuminated semiconductors would be limited to transformations leading to products with high added value?

FOX - Quantum yields depend sensitively upon the preparation and doping of the semiconductor, as well as upon surface morphology. Typically, quantum yields (at 390 nm) of 1 to 5% are obtained. Practical utility would seem to be limited to chemoselective transformations.

NOZIK - Could you summarize in a simple way, for the non-organic chemist what is the essential uniqueness of the organic chemistry possible at the illuminated semiconductor-liquid interface compared to either dark organic electrochemistry or homogeneous organic chemistry.

FOX - The heterogeneous surface allows for control (probably kinetics of competing reactions) of chemical reactions which follow the primary electron transfer. We show that surface recombination of intermediates can be more rapid than radical ion deprotonation. We also find lower local concentrations of radical ions (and hence less radical ion dimerization) than generated at the double layer of metal electrodes. Finally photoactivation of surface at low light flux can limit the number of oxidizing equivalents available per adsorbed molecule.

HYDROGEN EVOLUTION AND SELECTIVE ORGANIC PHOTOSYNTHESIS CATALYZED BY ZINC SULFIDE

Horst Kisch
Institut für Anorganische Chemie der Universität
Erlangen-Nürnberg
Egerlandstr. 1
D-8520 Erlangen

ABSTRACT. UV-irradiation of bis-(cis-1,2-dicyano-1,2-ethylene-dithiolate)zincate(II) in aqueous tetrahydrofuran (THF) or 2,5-dihydrofuran (2,5-DHF) leads to cubic n-ZnS which photocatalyzes an efficient hydrogen evolution coupled to a preparative useful dehydrodimerization of the cyclic ether. C-C bond formation occurs regioselectively in the case of THF leading to 2,2'-bitetrahydrofuryl exclusively. No reaction is observed with 2,3-dihydrofuran, six-membered and open chain ethers. This demonstrates that the semiconductor catalyst is able to differentiate between minor structural variations in the substrate. Mechanistic investigations reveal that the reductive and oxidative primary processes of the excited zinc sulfide are strongly coupled. C-C bond formation occurs only if the excited state reduces water but not dinitrogen oxide or zinc ions.

Introduction

The use of semiconductor powders as photocatalysts for chemical transformations mainly concentrated on the cleavage of water (1). Only a few organic photoreactions have been reported recently (2), like the photo-Kolbe reaction (3) and the oxidation of amides (4) and olefins (5). In all these cases the products are known compounds and the preparative utility seems uncertain. Herein we report on a chemoselective photosynthesis of previously unknown dehydrodimers of a cyclic ether (6). This synthetic useful C-C bond formation is catalyzed by n-ZnS, a highly selective photocatalyst capable to differentiate between minor structural changes in the substrate. Recently it was found that zinc sulfide may also catalyze the dehydrodimerization of amines and alcohols (7).

E. Pelizzetti and N. Serpone (eds.), Homogeneous and Heterogeneous Photocatalysis, 385–395.

Experimental Part

Polychromatic irradiations on a preparative scale were carried out with a 120 mL quartz immersion lamp apparatus fitted with a high-pressure mercury lamp (Philips HPK 125W, $\lambda \geqslant 248$ nm). Hydrogen was measured with a continously recording gas burette (8). Monochromatic irradiations were conducted with a low-pressure mercury lamp ($\lambda \geqslant 254$ nm) and the total amount of H_2 in solution and in the gas phase was determined by gas-chromatography.

The zinc sulfide photocatalyst [ZnS] was prepared by irradiation ($\lambda \geqslant 248$ nm) of $[ZnL_2][NBu_4]$ (1), L = cis-1,2-dicyano-1,2-ethylenedithiolate, in 2,5-dihydrofuran (2,5-DHF)/water = 14/1 (v/v). The white precipitate was washed with water and dried in a desiccator. x-ray fluorescence microanalyses indicated a ratio of Zn/S = 1/0.86. No zinc sulfide was formed when the irradiation was conducted under an atmosphere of dinitrogen oxide.

Dehydrodimers 2 - 4: 50 mg (0,06 mmol) of 1 were irradiated ($\lambda \geqslant 248$ nm) in 2,5-DHF/water = 14/1 for 42 h. This resulted in the formation of 2.7 L (o.12 mol) of hydrogen. Vacuum distillation yielded 13 g of a fraction containing 2, 3, 4 in the ratio of 1:2:1, respectively.

Experiments on the chemoselectivity of the reaction were performed by irradiating ($\lambda \geqslant 248$ nm) a suspension of 5 mg (0,05 mmol) of ZnS (9) in 10 mL of water in the presence of 10 mL of the appropriate ether.

In the competition experiments there were used 8.3 mg (0,01 mmol of 1 as catalyst precursor; only 2 - 4 but no mixed dehydrodimers were obtained in the systems 1,4-dioxane/2,5-DHF/water = 10/1/9 or 35/1/2 and THF/2,5-DHF/water = 10/1/9 (v/v).

Further experimental details can be found in ref.6e.

Results and Discussion

UV-irradiation of metal dithiolenes in homogeneous solution of aqueous tetrahydrofuran (THF) gives rise to a catalytic hydrogen formation (6). With zinc dithiolenes the heterogeneous photocatalyst [ZnS] is formed during the initial induction period (6c). If THF is replaced by 2,5-dihydrofuran (2,5-DHF), a tenfold rate increase is observed, resulting in turnover numbers of about 4000 (mol of hydrogen per mol of zinc dithiolene). The mechanism of this unusual photoproduction of a semiconductor from a soluble coordination compound is not understood at present. In the case of the maleonitriledithiolate complex 1 (M = Zn, tetrabutylammonium as gegenion, see Fig. 2) zinc sulfide is formed but the corresponding cadmium and mercury complexes do not photodecompose to their metal sulfides (6d). Formation of [ZnS] strongly depends upon the excitation wavelength and is only efficient when $\underline{\underline{1}}$ is irradiated within the high-energy band at λ = 270 nm (Fig. 1). This indicates

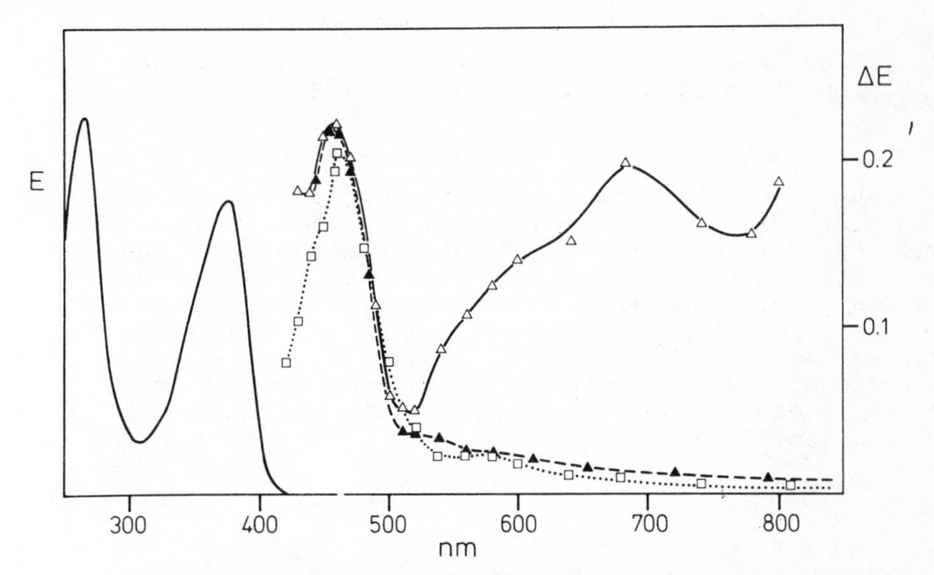

Fig. 1: Transient spectra obtained upon laser flash photolysis of $Na_2[ZnL_2]$, L = cis-1,2-dicyano-1,2-ethylenedithiolate, at λ = 308 nm. In water, measured 20 (△) and 200 ns (▲) after flashing. In EtOH/water (1/1, v/v), measured 200 ns (□) after the flash. —— absorption spectrum of $\underline{\underline{1}}$.

that the reaction may occur from a higher excited state of $\underline{\underline{1}}$. While the fluorescence and phosphorescence of $\underline{\underline{1}}$ at 77 K do not show a wavelength-effect (10), the same dependence is observed in laser flash experiments at room temperature. The observed transients at λ = 480 nm ($t_{1/2} \sim$ 50 µs) and λ = 700 nm ($\tau \leq$ 0.1 µs) are assigned to the complex radical anion and the solvated electron on the basis of quenching and pulse radiolysis experiments (11). Thus, the process corresponds to a photooxidation of $\underline{\underline{1}}$ and an electron injection into the solvent (Fig. 2). Since these transients are observed only when excitation is performed at 248 or 308 nm but not at 353 nm, it seems likely that the fragmentation of $\underline{\underline{1}}$ into [ZnS] is induced by electron transfer to the solvent. This is corroborated by complete inhibition of the photodecomposition of $\underline{\underline{1}}$ in the presence of dinitrogen oxide as electron scavenger (12).

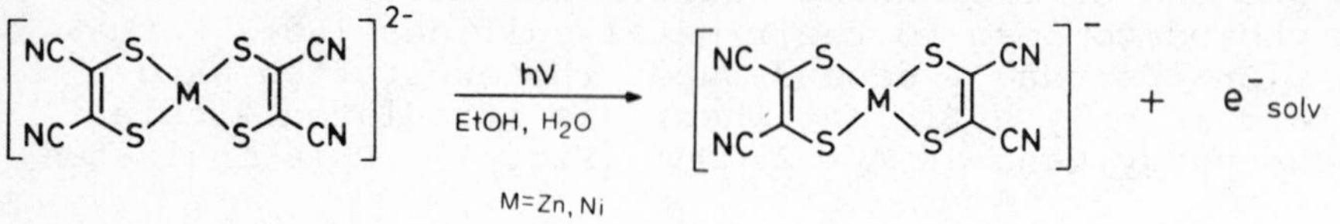

Fig. 2: Photoinduced electron transfer to solvent upon excitation of metal dithiolenes at 248 and 308 nm.

Hydrogen production is contingent on the presence of both 2,5-DHF (or THF) and water. Cleavage of water is indicated by the appearance of more than 90% D_2 in the initial gas phase if D_2O is used. However, this is due to an exchange process since a complete material balance shows that no water is consumed. 2,5-DHF is dehydrodimerized to the novel compounds 2, 3, 4 (Fig. 3) obtained in the ratio of 1:2:1, respectively. This mixture of regioisomers is easily isolated in high yield by simple distillation.

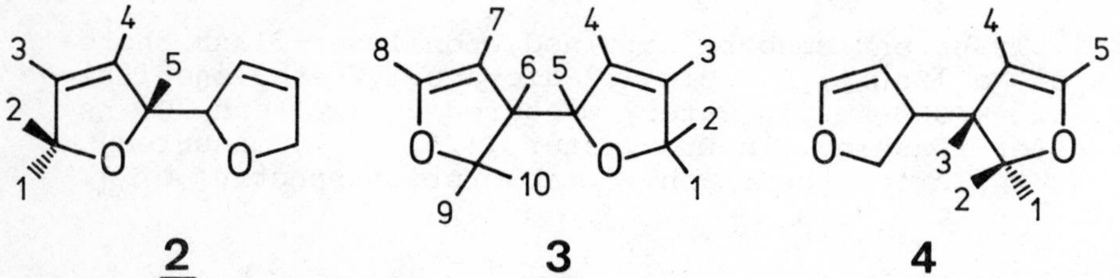

Fig. 3: Dehydrodimers 2 - 4 obtained from 2,5-dihydrofuran (2,5-DHF)

Further separation and structural characterization is discussed elsewhere (6e). The reaction is very clean and produces small amounts of furan and 2,3-dihydrofuran as by-products. Product development as a function of irradiation time indicates that hydrogen evolution is coupled to dehydrodimerization (Fig. 4). This is further substantiated by the finding that both processes are inhibited by electron scavengers like dinitrogen oxide, SF_6 or Zn(II).

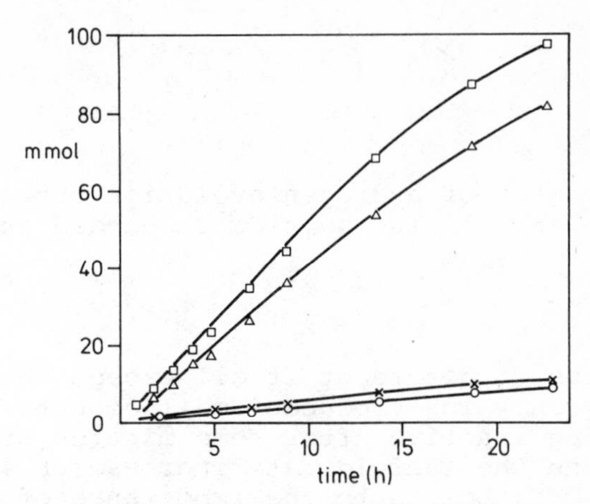

Fig. 4: Formation of hydrogen (□), dehydrodimers $\underline{\underline{2}}$ - $\underline{\underline{4}}$ (△), furan (x) and 2,3-dihydrofuran (○) as a function of irradiation time. [1] = 5×10^{-4} M in 120 mL of 2,5-DHF/water= 14/1 (v/v), λ_{irr} = 248 nm.

Neglecting the minor amounts of by-products, the observed reaction can be described by equation 1. It is endothermic by about 13 kcal/mol. In the case of THF the 2,2'-bitetra-

$$2\,RH \xrightarrow[H_2O]{h\nu,\ ZnS} R-R + H_2 \qquad (1)$$

RH:

hydrofuryl is formed exclusively.

The nature of the products $\underline{\underline{2}}$,$\underline{\underline{3}}$,$\underline{\underline{4}}$ and their statistical ratio of 1:2:1, respectively, point to intermediate ether radicals formed by hydrogen abstraction from the α-position of the cyclic ether. Subsequent dimerization leads to the observed dehydrodimers. Accordingly, the same products are obtained if these radicals are generated in homogeneous solution via H-abstraction by OH-radicals. However, in this case the reaction is less selective, producing in addition about 30% of 3-hydroxytetrahydrofuran from 2,5-DHF and 2,3'-bitetrahydrofuryl from THF (6e).

The high selectivity of the [ZnS] photocatalyst is demonstrated by the strong influence of minor structural changes in the substrate (Fig. 5).

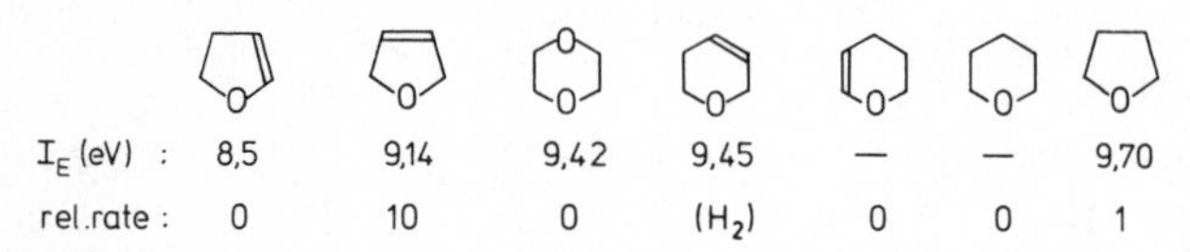

Fig. 5: Relative rates of hydrogen evolution obtained by irradiating ($\lambda \geqslant 248$ nm) zinc sulfide suspended in the aqueous ether.

Six-membered ethers do not react at all except the allylic 5,6-dihydro-2H-pyran which produces small amounts of hydrogen before becoming inactive after some minutes of irradiation time. This and the tenfold rate increase on substituting 2,5-DHF for THF, points to the importance of allylic stabilization of the postulated radicals. However, this is not the major determining factor since THF is a much better substrate than 5,6-dihydro-2H-pyran. Noteworthy is the inertness of the enolic 2,3-dihydrofuran as compared to the highest reactivity obtained with the allylic 2,5-DHF.

The rate of hydrogen evolution increases linearly with light intensity and does not change when the temperature is varied from 20 to 50°C. Changing the apparent pH-value from 4 to 10 does not significantly influence the reaction rate. Above pH 10 the conventionally prepared zinc sulfide (9) dissolves and the rate strongly decreases. This is in sharp contrast to [ZnS] prepared from $\underline{1}$ which induces a rate increase (13).

[ZnS] as prepared from $\underline{1}$ has sphalerite structure (6e) and n-type semiconductor character. The latter follows from the observation of an anodic photocurrent in THF/water (14) exhibiting an onset potential of -0.6 V (vs.NHE). This is different to the flat-band potential (-1.64 V in water of pH 7) of an Al-doped single crystal electrode (15) and points to charge recombination via a surface state in the case of the [ZnS] powder (16). The band-gap energy of the latter is obtained from diffuse reflectance spectra as 3.67 eV (12).

Irradiation of zinc sulfide in aqueous suspension immediately results in photocorrosion (17) yielding zinc(O) and sulfur. No corrosion and no chemical reaction occurs in a dry 2,5-DHF suspension. However, hydrogen evolution and dehydrodimerization are observed in aqueous 2,5-DHF. In this case photocorrosion is almost completely suppressed and zinc sulfide becomes a very stable photocatalyst. When ZnS is dehydrated, the catalytic activity strongly decreases. 2,5-DHF has to contain at least 6% (v/v) of water to induce hydrogen evolution. From this value and the surface area of 17 m^2/g it may be concluded that only fully hydrated ZnS catalyzes reaction 1 (12).

Scheme 1 summarizes the basic reaction steps, neglecting other important details like adsorption and desorption processes. The common electron - hole model

$$[ZnS(H_2O)_x(RH)_y] \underset{}{\overset{h\nu}{\rightleftarrows}} {}^*[ZnS(H_2O)_x(RH)_y] \qquad (a)$$

$$^*[ZnS(H_2O)_x(RH)_y] \xrightarrow[H_2O]{} [ZnS(H_2O)_x(RH)_y] + H_{ad} + OH_{ad} \qquad (b)$$

$$^*[ZnS(H_2O)_x(RH)_y] \xrightarrow[H_2O]{RH} [ZnS(H_2O)_x(RH)_y] + H_{ad} + OH^- + RH^{+\bullet}_{ad} \qquad (c)$$

$$OH_{ad} + RH_{ad} \longrightarrow H_2O + R^{\bullet}_{ad} \qquad (d)$$

$$RH^{+\bullet}_{ad} \longrightarrow R^{\bullet}_{ad} + H^+ \qquad (e)$$

$$R^{\bullet}_{ad} \longrightarrow 1/2\ R_2 \qquad (f)$$

$$H_{ad} \longrightarrow 1/2\ H_2 \qquad (g)$$

Scheme 1

is not sufficient to explain the observed inhibition of both hydrogen and dehydrodimer formation by dinitrogen oxide or zinc(II) ions. That contrasts the colloidal zinc sulfide system where only hydrogen evolution but not aceton and pinacol formation from iso-propanol are suppressed by dinitrogen oxide (18b). Thus, different to previous experience, the reductive and oxidative primary processes seem to be strongly coupled and 2 - 4 are formed only if water is reduced but not dinitrogen oxide or zinc(II) ions. Similar to problems encountered in ZnO catalyzed photochemistry (19) it may be more appropriate to assume that light absorption effects the entire ZnS/aqueous ether interface. The dominating role of water becomes understandable if the photoreactive surface complex composed of an active site with adsorbed water and ether (RH) molecules, dissociates water into adsorbed hydrogen and OH radicals (step b). The latter abstract hydrogen from adsorbed RH to produce water and the ether radical R• (step d). Product formation occurs by dimerization of R and H to 2 - 4 and hydrogen. Recombination of R• with H is not observed since no incorporation of deuterium into non-reacted ether or into the products is found when D_2O is used. This indicates that the oxidative and reductive processes occur at different sites (20).

Although step b would explain the unique role of water, it seems rather unlikely on the basis of thermodynamic considerations (6e). However, it cannot be completely ruled out since the adsorbed water may be coordinated to a surface zinc(II) centre resulting in a weakening of the H-OH bond. More likely seems a direct oxidation of the ether (step c) followed by deprotonation of the radical cation to the ether radical R• (step e). Here again, water would be reformed, however, the unique role of water and the effects of dinitrogen oxide or zinc(II) are more difficult to explain. One may assume that these inhibitors either prevent the formation of the surface complex photoreactive in reaction 1 or that they just very efficiently compete with step c (21).

Diffuse reflectance spectra of the heterogeneous catalyst as isolated from the reacting mixture indicates the presence of small amounts of Zn(O). This may suggest that hydrogen evolution is catalyzed by metallic zinc as observed in the case of colloidal zinc sulfide (18a). Contrary to that, complete inhibition of hydrogen and concomitant formation of Zn(O) occurs upon addition of 3% (mol % relative to ZnS) of Zn(II) to the reacting suspension. This suggests that Zn(O) catalyzes only when present in very low concentration or that it exerts no catalytic effect at all (22).

The failure of 2,3-dihydrofuran, six-membered and acyclic ethers to undergo reaction 1 cannot be rationalized on the basis of their reducing properties (Fig. 5). It rather points to selective adsorption processes at the zinc sulfide surface. This is corroborated by the fact that addition of only 10% of 2,5-DHF to THF/water/ZnS completely inhibits dehydrodimerization of THF in favor of 2,5-DHF. In the case of homogeneous competition one would expect formation of 2,2'-bitetrahydrofuryl and mixed dehydrodimers since the rates of steps c should be the same at this concentration ratio as indicated by the relative rates of hydrogen evolution (Fig. 5).

Conclusion

Cubic n-ZnS is photochemically produced from a soluble zinc dithiolene. This heterogeneous photocatalyst enables a chemo- and regioselective dehydrodimerization of cyclic ethers. Conventionally prepared ZnS may be used also but shows less long-term stability. The reaction is coupled to reduction of water to hydrogen. However, no water is consumed due to its reformation in subsequent reaction steps. Oxidative and reductive primary processes of the excited semiconductor are strongly coupled since both are inhibited by electron scavengers.

Acknowledgment: The author is highly indebted to his co-workers mentioned in the references and to Dr. R. Millini. Most of the experimental work was performed at the Max-Planck-Institut für Strahlenchemie, Mülheim a.d. Ruhr. Part of the work was supported by A. Krupp von Bohlen und Halbach-Stiftung.

Notes and References

1) For a recent summary see e.g. "Energy Resources through Photochemistry and Catalysis"; Grätzel, M. Ed.; Academic Press: 1983.
2) Summary: Fox, M. A. Acc. Chem. Res. 1983, 16, 314.
3) **Kraeutler, B; Bard, A.J. J.Am.Chem. Soc.** 1977, 99, 7729; 1978, 100, 5985.
4) Pavlik, J.W.; Tantayanon, S.J. J. Am. Chem. Soc. 1981, 103, 6755.
5) a) Formenti, M.; Julliet, F.; Teichner S.J.; C.R. Acad. Sci. Paris 1970, 270, 138. b) Kanno, T.; Oguchi, T.; Sakuragi, H.; Tokumaru, K. Tetrahedron Lett. 1980, 21, 467. c) Fox, M.A.; Chen,C.-C. J. Am. Chem. Soc. 1981, 103, 6757.
6) a) Henning, R.; Schlamann, W.; Kisch, H. Angew. Chem. 1980, 92, 664. b) Battaglia, R.; Henning, R.; Kisch, H. Z. Naturforsch. 1981, 36b, 396. c) Bücheler, J.; Zeug, N.; Kisch, H. Angew. Chem. 1982, 94, 792. d) Battaglia R.; Henning, R.; Dinh-Ngoc, B.; Schlamann, W.; Kisch, H. J. Mol. Catal. 1983, 21, 239. e) Zeug, N.; Bücheler, J.; Kisch, H. J. Am. Chem. Soc. 1985, 107, 1459.
7) Yanagida, S.; Azuma, T.; Kawakami, H.; Kizumoto, H.; Sakurai, H. J. Chem. Soc., Chem. Commun. 1984, 21.
8) Bogdanovic, B.; Spliethoff B. Chem. Ing. Techn. 1983, 55, 156.
9) Kurian, A.; Suryanarayana, C. V. J. Appl. Electrochem. 1972, 2, 223.
10) Fernandez, A.; Kisch, H. Chem. Ber. 1984, 117, 3102.
11) Fernandez, A.; Görner, H.; Kisch, H. ibid. in the press.
12) Bücheler, J.; Kisch, H. unpublished.
13) [ZnS] contains small amounts of an THF-polymer which may prevent dissolution in the alkaline medium.
14) Sprünken, R.; Bücheler, J.; Kisch, H. unpublished results.
15) Fan, F.-R.; Leempoel, P.; Bard, A.J. J. Electrochem. Soc. 1983, 130, 1866.
16) ESCA-spectra of [ZnS] point to the presence of Zn(I) due to a Zn(3p) binding energy of 1.6 eV less than in conventionally prepared zinc sulfide (12).
17) a) Schleede, A. Chem. Ber. 1923, 56, 386. b) Shionoya, S. J. Chem. Soc. Jap. 1950, 71, 461. c) Platz, H.; Schenk, P.W. Angew. Chem. 1936, 49, 822. d) Schleede, A.; Herter, M.; Kordatzki, W.; Z. Phys. Chem. 1923, 106, 386.
18) a) Henglein, A.; Gutierrez, M. Ber. Bunsenges. Phys. Chem. 1983, 87, 852. b) Henglein, A.; Gutierrez, M.; Fischer, Ch.-H. ibid. 1984, 88, 170.
19) Steinbach, F. Top. Current Chem. 1972, 25, 117.

20) Similarly it was proposed that electron injection from adsorbed dyes into the conduction band of CdS and subsequent reduction occur at different sites, Takizawa, T.; Watanabe, T.; Honda, K. J. Phys. Chem. 1978, 82, 1391.
21) In the language of the band model this suggests that a strongly coupled electron-hole pair (exciton) may be the photoreactive excited state for reaction 1. The role of excitons in photoelectrochemistry is described by Gerischer, H.; Ber. Bunsenges. Phys. Chem. 1973, 77, 771.
22) Platinization of ZnS-powder does not enhance hydrogen evolution in the presence of sodium sulfide. Reber, J.-F.; Meier, K. J. Phys. Chem. 1984, 88, 5903.

PHOTOCATALYTIC ORGANIC SYNTHESIS BY USE OF SEMICONDUCTORS OR DYES

T. Sakata

Institute for Molecular Science
Myodaiji Okazaki 444
Japan

ABSTRACT.
The results of electrochemical and photocatalytic measurements indicate the important role of RuO_2 as a reduction catalyst on irradiated n-type semiconductors. Quantitative analyses of photocatalytic hydrogen production from water and various organic compounds demonstrate that water is always involved as oxidizing agent in these reactions. The product analysis and pH dependence of the reaction of organic acids show the existence of new reaction paths besides photo-Kolbe reaction. In the case of lactic acid the reaction depends on the kind of semiconductor. Photocatalyses of semiconductors and dyes are applied to amino acid synthesis from keto- or hyroxy- carboxylic acids in ammonia water. The quantum yields of photocatalytic amino acid synthesis using visible light are about 20%- 40%. Moreover the reactions are highly selective and ,in the case of hydroxy-carboxylic acids,depend strongly on the kind of semiconductor.

1. INTRODUCTION

Rapid progress is mow being made in research in the field of photocatalysis. During past several years surface modification of particulate semiconductors such as TiO_2,$SrTiO_3$ and CdS by the addition of Pt,Pd or RuO_2 has been found to increase the photocatalytic activity of the semiconductor by a factor of $10-10^3$(1-7).Catalytic roles of such metal and oxide catalysts on semiconductors have been investigated(8-14). Besides water splitting,various photocatalytic reactions such as the oxidation of cyanide(15,16),sulphite(16,17),sulfide(18,19),acetate(20), hydrocarbons(21-26),amines(27,28)and other substances(29-36) have been reported. We have demonstrated taht hydrogen production from the photocatalytic reactioins of water with various organic compounds such as alcohols(38,39),carbohydrates(39),hydrocarbons(41,42),artificial high polymers(41,43) and biomasses(43,44) using powdered semiconductor photocatalysts. In these reactions, organic molecules are oxidized and water is reducted to produce hydrogen. In several cases hydrogen production is very efficient. The quantum yields of hydrogen evolution amount to more than 50%(45). These photocatalytic reactions can also be applied to

E. Pelizzetti and N. Serpone (eds.), Homogeneous and Heterogeneous Photocatalysis, 397–413.

organic redox reactions(29): Electrons in the conduction band can be used for reduction of unsaturated bonds such as C≡C,C=C,C=N,and C=O. And holes in the valence band can be used for the oxidation reactions. In some cases, organic dye molecules show a high activity as a photocatalyst. In this paper we report some results of photocatalytic reactions of organic acids and amino acid synthesis by use of powdered semiconductors and dyes.

2. CATALYTIC PROPERTIES OF RuO_2 ON N-TYPE SEMICONDUCTOR UNDER ILLUMINATION (46,47)

For the lastfew years RuO_2 has often been used in photocatalytic water splitting systems. In these systems RuO_2 has been always considered to play a role as an oxidation catalyst. In order to elucidate the catalytic role on n-type semiconductors, the photoelectrochemical behaviors of single crystal TiO_2 and CdS by depositing a small amount of RuO_2 and Pt were investigated. A remarkable decrease of photoanodic current due to the hydrogen evolution were observed by depositing RuO_2 on TiO_2 surface. In the photovoltage measurement of a TiO_2 single crystal on which RuO_2 and Pt are deposited separately in two regions, the potentials of RuO_2 and Pt shifted into the negative direction under illumination. Similar behavior was observed for CdS single crystal on which RuO_2 and Pt are deposited. A good photocatalytic activity of particulate RuO_2/TiO_2 was observed for hydrogen evolution from ethanol-water(1:1)mixture. These results suggest that RuO_2 functions as a reduction catalyst for any reaction in which hydrogen is produced.

Recently Nakabayashi et al. measured the separation factor of hydrogen evolved from an equmolar mixture of H_2O and D_2O in the photocatalytic decomposition of water on particulate Pt/TiO_2(48). The Separation factor S is defined by the following equation:

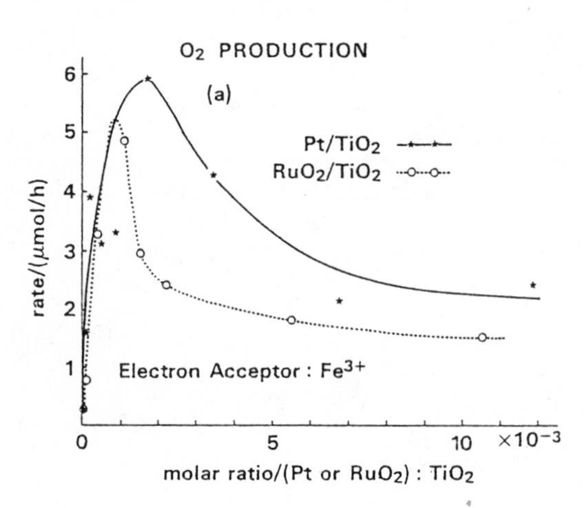

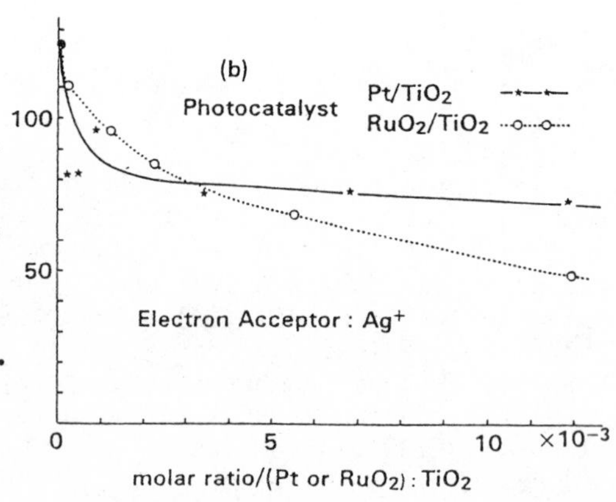

Figure 1. Dependence of the oxygen production rate on the molar ratio of RuO_2(or Pt)to TiO_2 (47).

$S=(C_H/C_D)gas/(C_H/C_D)soln$, where C_H and C_D are the amount of hydrogen and deuterium atom,respectively. The photocatalytic separation factor, obtained as 5.3 for Pt/TiO_2 catalyst,agreed well with the electrochemical separation,5.0,on Pt electrode. Their result indicates clearly that hydrogen is evolved on Pt loaded on powdered TiO_2 in the photocatalytic decomposition of water. We applied the same method to the present system. The separation factor for hydrogen production from an ethanol-water mixture was 7.2±0.7 on particulate RuO_2/TiO_2 photocatalyst. This value agreed well with the electrochemical separation factor,7.0±0.5, obtained on a thin-film Ru electrode. This result indicates that hydrogen is evolved on the reduced RuO_2 on powdered TiO_2. Interestingly,the separation factro ,5.4±0.5,obtained on powdered Pt/TiO_2 photocatalyst for the ethanol-water mixture is quite close to the value for the reduction of water,5.3,obtained by Nakabayashi et al. and also to the electrochemical one,5.0,obtained on Pt electrode in water(48). This supports our previous idea that hydrogen is produced not from ethanol through a dehydrogenation process but from the reduction of water(38,39).

In order to clarify whether RuO_2 on TiO_2 catalyzes oxygen evolution under irradiation,we measured the oxygen evolution rate under irradiation using Fe^{3+} or Ag^+ as an electro acceptor. Their reactions are written as

$$H_2O + 2Fe^{3+} \longrightarrow 2Fe^{2+} + 2H^+ + 1/2\ O_2 \qquad (1)$$

$$H_2O + 2Ag^+ \longrightarrow 2Ag + 2H^+ + 1/2\ O_2 \qquad (2)$$

Figure 1,a and b, shows the dependence of oxygen evolution rate on the molar ratio of RuO_2 or Pt to TiO_2,using Fe^{3+} and Ag^+ as electron acceptors,respectively. The dependence of the oxygen evolution of the Fe^{3+} system behaves quite differently from that of the Ag^+ system as shown in this figure. The oxygen evolution rate of the Fe^{3+}system shows a maximum agaist the concentration of RuO_2 or Pt on TiO_2 as shown in Figure 1a, whereas the rate decreases monotonically with increasing concentration of RuO_2 or Pt on TiO_2 as shown for the Ag^+ system in Figure 1b. Moreover, the oxygen evolution rate is much larger for the Ag^+ system than for the Fe^{3+} system,even though the other conditions like light intensity and experimental setup are the same. In the Ag^+system, bubbles of oxygen are evolved vigorously. In accordance with the progress of reaction 2, Ag metal is deposited on the TiO_2 surface. The quantum yield of oxygen evolution is about 25%. This high efficiency would be explained by the irreversibility of the reaction(49). As seen in Figure 1b, TiO_2 shows the best activity,while the loading of RuO_2or Pt on TiO_2 decreases the efficiency of oxygen evolution. This cannot be explained by the catalytic effect of RuO_2 on the oxygen evolution. The best activity of oxygen evolution by use of TiO_2 suggests that oxygen evolution is easy on the TiO_2 surface under irradiationn. This is understandable because of the highly positive potential(2.79V vs.NHE at pH 0) of holes produced in the valence band(50). In this case, a good catalytic property would not be indispensable to the oxygen evolution. The potential of RuO_2 and Pt deposited on TiO_2 single crytal electrode was shifted in the negative direction in the presence of Fe^{3+} or Ag^+ under irradiation. From these results,RuO_2 was concluded to be a reduction site even for oxygen evolution reaction in the presence of a strong electron acceptor such as Fe^{3+} and Ag^{+}. The maximum of the oxygen evolution rate,which was observed in the presence of Fe^{3+},was interpreted as being caused by a rectifying action of the metal/semiconductor junction(47).

3. HYDROGEN PRODUCTION AND COMPLETE DECOMPOSITION OF ORGANIC MOLECULES IN WATER

Table 1. Examples of photocatalytic hydrogen production from organic compounds and water

Reactant	Photocatalytic reation	ΔG^0 (kJ/mol)	$\Delta\varepsilon^0$ (free energy change per elect. transferred
Carbon	1) $C + 2H_2O \longrightarrow 2H_2 + CO_2$	63 kJ/mol	+0.16 eV
	2) $C + H_2O \longrightarrow H_2 + CO$	92	+0.48
Alcohols	3) $CH_3OH + H_2O \longrightarrow 3H_2 + CO_2$	9.0	+0.02
	4) $C_2H_5OH + H_2O \longrightarrow 2H_2 + CH_4 + CO_2$	−34	−0.07
	5) $C_2H_5OH - 3H_2O \longrightarrow 2CO_2 + 6H_2$	97	+0.08
Carbohydrate (glucose)	6) $C_6H_{12}O_6 + 6H_2O \longrightarrow 6CO_2 + 12H_2$	−32	−0.01
Hydrocarbon	7) $C_6H_6 + 12H_2O \longrightarrow 6CO_2 + 15H_2$	356	+0.12
	8) $C_{16}H_{34} + 32H_2O \longrightarrow 16CO_2 + 49H_2$	1232	+0.14
Amino acid	9) $H_2NCH_2COOH + 2H_2O \longrightarrow 3H_2 + NH_3 + 2CO_2$	47	+0.06
Polymer	10) $(CH_2CHC1) + 4H_2O \longrightarrow 5H_2 + 2CO_2 + HC1$		
	11) $(CH_2NHCONH) + 3H_2O \longrightarrow 2H_2 + 2NH_3 + 2CO_2$		

Several examples of photocatalytic reactions which have been investigated in this laboratory are shown in Table 1. The evidence taht H_2O is involved in these reactions and is reduced photoelectrochemically was obtained by using D_2O instead of H_2O. For the decomposition of ethanol in D_2O 88% of the evolved gas was D_2,10% was DH and 2% was H_2 (39). In order to confirm whether the reactions proceed quantitatively,the complete decomposition was carried out for methanol,ethanol,sugar(40) and n-hexadecane(42). The amount of H_2 and CO_2 produced agreed well with the theoretical value calculated from the quantity of the starting material and the equations shown in Table 1. These results indicate clearly that water is involved in the reactions as an oxidizing agent. Usually oxygen is not evolved,but is captured by the carbon in the organic compounds, and water is split with the aid of the reducing power of the organic compounds. In many cases light energy is stored through the reactions, although the increase in free energy per transferred electron is rather small as shown in Table 1. Interestingly, methane is formed in addition to hydrogen in the case of ethanol; this can be explained by the photocatalytic decomposition of acetic acid accumulated as a reaction intermediate(39). This reaction is decribed in Table 1,eqn.(4). The ratio of hydrogen to methane expected from this equation is 2.0. However,the experimental value is about 14, which is much larger than the expected value. This can be explained well by assuming that reactions(5) and(6) in Table 1 both take place. The result of the complete decomposition of ethanol indicates that the photogenerated electrons and holes are consumed four times as fast for reaction(5) in table 1 as for reaction(4).

4. PHOTOCATALYTIC REACTION OF ORGANIC ACID AND WATER —— New Reaction Paths Besides Photo-Kolbe Reaction(51).

The above result for the photocatlytic reaction of ethanol shows that the flow of photogenerated electrons and holes is controlled on the photocatalyst. In order to clarify this type of phenomenon the photocatalytic reaction of organic acids was investigated(51). Krauetler and Bard(20,52) found that in the presence of Pt/TiO_2 the hydrocarbon was formed by the following reaction which they denoted the photo-Kolbe reaction:

$$RCOOH \longrightarrow RH + CO_2 \qquad (3)$$

Sato(53) has recently reported photocatalytic synthesis of ethane from acetic acid in the vapour phase. Yoneyama et al.(54) investigated the factors influencing product distribution in the photocatalytic decomposition if aqueous acetic acid in Pt/TiO_2. We measured the rates of production of hydrogen,RH and CO_2 during the decomposition of several organic acids on Pt/TiO_2 and found that the rate of production of hydrogen is large even compared with that of the hydrocarbon. This result suggests that reactions other than reaction(3) take place. Further evidence supporting this proposal is the pH dependence of the rates of production of hydrogen and methane from acetic acid as shown in Fig.2. The ratio of hydrogen to methane depends strongly on the pH, and increases with increasing pH. Above pH 8.8 for Pt/TiO_2(anatase) (pH 7.1 for Pt/TiO_2(rutile)) only hydrogen is evolved and almost no methane is produced. The formation of methanol(CH_3OH) and glycol acid ($HOCH_2COOH$) in the aqueous medium suggests that a new reaction path in which water is involved as an oxidizing agent exists since they react with water to produce hydrogen and CO_2(38,39,51). This new reaction is written as

$$CH_3COOH + 2H_2O \longrightarrow 2CO_2 + 4H_2$$

$$\Delta G^\circ = 78.5 \text{ kJ mol}^{-1} \qquad (4)$$

Reaction(4) has two characteristic features. The first is that water is involved in the reaction as an oxidizing agent, and the second is that a substantial amount of free energy is stored (ΔG°= -52.3 kJ mol^{-1} for the photo-Kolbe reaction of acetic acid(reaction(3)). The non-Kolbe type electrode reaction of organic acids in which olefins and alcohols are produced is known

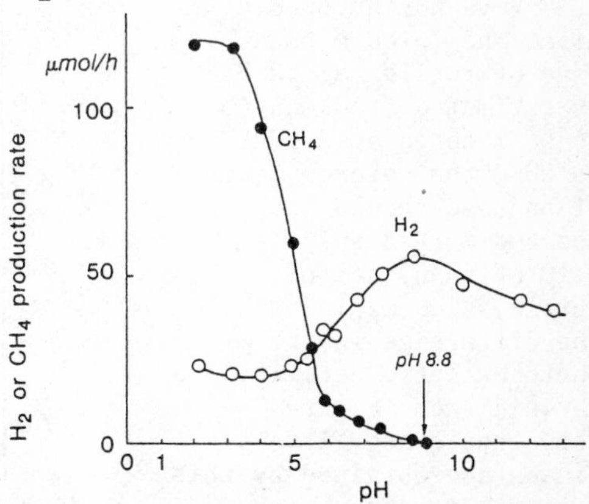

Figure 2. Dependence of the rates of production of H_2 and CH_4 on pH for the reaction of acetic acid with anatase

as the Hofer-Moest reaction (55). Since methanol and glycol acid are produced as reaction intermediates, reaction(4) has a close similarity to the Hofer-Moest reaction.

The pH dependence of the reaction was explained by the pH dependence of the valence band position of TiO_2 (i.e., the pH dependence of oxidation power) and the concentration of OH^-. As shown in the following equation, the valence band of rutile ($TiO_2(r)$) is located deep and the photogenerated holes in the valence band have an enough power to oxidize CH_3COO^-.

$$U_{vb} = +2.79 - kT/e \cdot [pH] \quad \text{V vs NHE} \qquad (5)$$

The oxidation power of the photogenerated holes decreases with increasing pH, since the valence band is shifted in the negative direction according to eqn.(5). Therfore, at high pH, the oxidation of CH_3COO^- becomes difficult and the oxidation of OH^- ($OH^- + p \longrightarrow \cdot OH$) begins to overwhelm that of CH_3COO^-, since the valence band of TiO_2 is located deeply enough to oxidize OH^- even under alkaline conditions. As seen in Fig.2, CH_4 is not produced above pH 8.8 for anatase. A similar dependence of CH_4 production rate on pH was observed for rutile. In the case of rutile, CH_4 is not produced above pH 7.1. The pH value above which CH_4 formation is provented, relates presumably to the critical potential of the oxidation of CH_3COO^-. When the pH value of 7.1 for rutile is used in eqs.(5), 2.37 V vs NHE is obtained as the potential of the oxidation of CH_3COO^- at this pH. If the direct oxidation by the photogenerated hole in the valence band is assumed, this potential is thought to be equal to the critical potential of the oxidation of CH_3COO^- on the rutile surface in aqueous medium. If the critical potential of the oxidation of CH_3COO^- on the anatase surfae is the same with that on the rutile surface, the valence band must be located at 2.37 V vs NHE at pH 8.8. Since the valence band edge of rutile at pH 8.8 is estimated from eqn.(5) to be located at 2.27 V vs NHE, the valence band of anatase should be located more deeply by 0.10 eV than that of rutile, which explains well the difference in the photocatalytic activity between them. Figure 3 shows the energy levels of anatase obtained by this analysis, together with those of rutile.

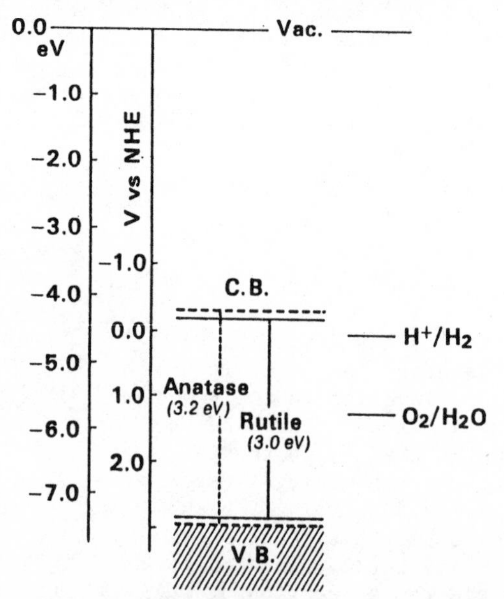

Figure 3. Energy diagram of anatase and rutile(pH 0.0)

5. EFFECT OF SEMICONDUCTORS ON PHOTOCATALYTIC REACTION OF LACTIC ACID (56)

The specificity of the photocatlyst to a given reaction is important to control the reaction. In the photocatalytic reaction of lactic acid in water, a clear difference was observed in the reaction products, dependeng on the kind of semiconductor(56).

When the glass bulb containing the photocatalyst suspended in the lactic acid solution was irradiated with white light from the Xe lamp, gas bubbles evolved vigorously. The gaseous products were H_2 and CO_2 for Pt/TiO_2 and only H_2 for Pt/CdS. The quantum yields of H_2 production are quite high as shown in Table 2. The wavelength dependence of the quantum yield indicated that the band-gap excitation of the semiconductor is essential to the reaction. Since a clear difference was observed in the gaseous products for Pt/TiO_2 and Pt/CdS , the reaction products in aqueous medium were also analyzed. Table 3 shows the results after 4 h of irradiation. Since a large excess of lactic acid,about 30 mmol,was used as the reactant, the result in this table should be considered as for an initial atage of the reaction. As shown in this table, for Pt/TiO_2 ,the amounts pf H_2,CO_2, and CH_3CHO are nearly equal. This result suggests

Table 2 Quantum Yields for H_2 Production from Lactic Acid–Water (1:1 vol) Solution[a] (56)

cat.	wavelength/nm	quantum yield
Pt/TiO_2	420	0.03
	400	0.43
	380	0.64
	360	0.71
Pt/CdS	520	0.08
	500	0.12
	480	0.21
	460	0.26
	440	0.38

[a] Quantum yields were based on incident photon flux, which was measured by a thermopile (Eppley Lab. Inc.).

Table 3 Photocatalytic Reaction Products from Lactic Acid–Water (1:10 vol) Solution[a] (56)

	prod/mmol					
cat.[b]	H_2	CO_2	CH_3CHO	C_2H_5OH	CH_3-COOH	CH_3-COCOOH
Pt/TiO_2	1.21	1.43	1.08	0.047	0.151	0.02
Pt/CdS	1.20	0.015				0.80

[a] Irradiated with 1-kW Xe lamp (under 500-W operation) for 4 h.
[b] 300 mg of catalyst was used. In the case of Pt/CdS, the photocatalyst prepared by photochemical deposition of Pt showed a poorer activity than the photocatalyst by mechanical mixing. Therefore, in the present experiment the photocatalyst prepared by the latter method was used. There was no dependence of the distribution of the reaction products on the preparation method.

that the reaction(1) in Figure 4 takes place. Ethanol,acetic acid and pyruvic acid were produce as minor products. The production of ethanol and acetic acid can be explained by decarboxylation of lactic acid and the oxidation of acetaldehyde,respectively(39). On the other hand,for Pt/CdS,the main products are hydrogen and pyruvic acid. Only a trace amount of CO_2 was produced and no acetalydehyde was detected,which is in strong contrast with case of Pt/TiO_2 . Since the products were quite different from that for Pt/TiO_2, we can not apply reaction(1) in Fig.4 to this system. The reaction(2) in Fig.4 is proposed for Pt/CdS.

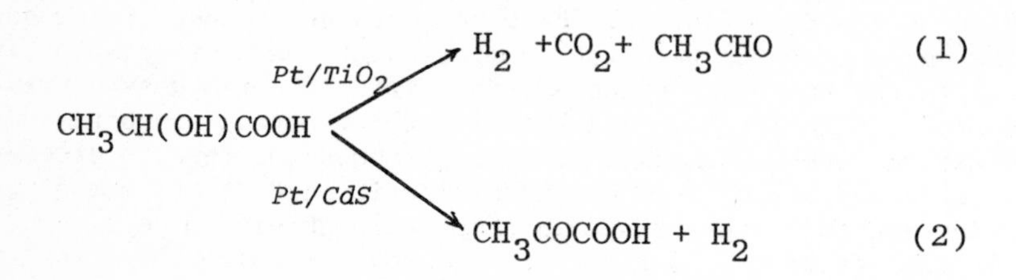

Figure 4. Depencence of photocatalytic reaction of lactic acid on the kind of semiconductor.

The ratio of the yield of H_2 to that of pyruvic acid is a little larger than the ratio(1.0) expected from the reaction (2) in Fig.4. Liquid chromatography indicates taht some unidentified compounds are produced as minor products together with pyruvic acid, which might explain the above discrepancy. The decomposition of pyruvic acid produced from reaction (2) in Fig.4 can be discounted because of the small yield of CO_2.

The anodic current due to the oxidation of lactic acid begins to rise at about 1.1 V vs.SCE at a glassy carbon electrode in 0.5M K_2SO_4 solution. Since the valence band edge of CdS is located at 1.6 V vs. SCE(57), lactic acid can be oxidized with CdS as well TiO_2. A clear difference in oxidation power of the hole in the valence band. It is known that ethanol can be decomposed efficiently with both of these two semiconductor photocatalysts. However,acetic acid is decomposed with TiO_2 but not with CdS (41), because the oxidation potential of acetic acid located more positively than the valence band edge of CdS. The situation in the case of lactic acid looks similar to the case of acetic acid. Because the valence band of TiO_2 is located deeply enough to oxidize carboxylate,decarboxylation could occur for TiO_2. On the other hand, the valence band of CdS is located less deeply than that of TiO_2. Therefore CdS photocatalyst could oxidize only the OH group of lactic acid. In order to see the effect of electrode potential on the reaction products,a glassy carbon electrode was used changing the electrode potential and electrolysis of lactic acid was performed. Interestingly, the products do not depend on the electrode potential for this electrode. Acetaldehyde path(reaction(1) in Fig.4) is dominant in this case. This result seems to suggest the importance of the difference in the adsorption properties of lactic acid and in catalysis in the oxide surface of TiO_2 and sulfide surface of CdS. Anyway, the above results indicate the possibility of controlling photocatalytic reactions by selecting a semiconductor with a suitable oxidizding power and catalytic property.

It is known that CdS suffer a corrosion under irradiation(58). In order to determine the amount of CdS dissolved during the photocatalytic reaction,cadmium was analyzed by atomic absorption spectroscopy. In the sample before irradiation,0.1 mmol of Cd was detected. This quantity was not changed even after 130 h irradiation. Since 28.3 mmol of hydrogen was produced during the reaction, more than 99.6% of the photogenerated holes were consumed for the photocatalytic reaction even if all of 0.1 mmol of Cd was assumed to have been produced from the photocorrosion. In the analysis by polarography, the amount of Cd^{2+} was smaller than 3.6 μmol. These results indicate that the photocorrosion of CdS is negligibly small in the present reaction,suggesting a very rapid and efficient oxidation of lactic acid.

6. HIGHLY EFFICIENT PHOTOCATALYTIC PRODUCTION OF AMINO ACIDS FROM ORGANIC ACIDS-AMMONIA-WATER BY USE OF DYE OR SEMICONDUCTOR(64-66)

Amino acid production by irradiating a mixture of various organic compounds and ammonia with UV light has been investigated mainly from the view point of chemical evolution(59,60). In 1979, Reiche and Bard demonstrated that various amino acids were produced from methane-ammonia-water with Pt/TiO_2 photocatalyst under near UV irradiation(61,62). Very recently Kawai et al. reported not only the formation of various amino acids but also that of polypeptide from glucose-ammonia-water with Pt/TiO_2(63). The efficiency of the above reactions is rather low and the reaction mechanism seems to be very complicated because of the simultaneous production of various amino acids. We report here highly efficient photocatalytic synthesis of amino acids from organic acids in ammonia water by use of various dyes and semiconductors as photocatalysts(64-66). The quantum yields of these reactions amounts to 10-40% with visible light without metal catalyst such as Pt. Moreover the reactions are selective and depends strongly on the kind of semiconductors.

Among various organic acids,keto-,hydroxy- and unsaturated carboxylic acids were found to produce efficiently the corresponding amino acids. The reactions are classified into three types(65). The typical examples are shown in Figure 5. In the first type of reactions, keto-carboxylic acid ,ammonia and a sacrificial reagent such as triethanol amine are used. The sacrificial agent is used to reduce imino acid which is produced from keto-carboxylic acid and ammonia. The assumed reaction scheme of this type is expressed as,

$$RCOCOOH + NH_3 \rightleftharpoons RC(=NH)COOH + H_2O \quad (6)$$

$$RC(=NH)COOH + 2H\cdot \longrightarrow RCH(NH_2)COOH \quad (7)$$

Reaction(6) proceeds thermally to produce imino acid(67). Reaction (7) is driven photochemically in the present case. These reactions resemble the Knoop reaction(67). In the Knoop reaction, imino acid is reduced thermally by using hydrogen Pt catalyst or by ethanol and alkali **metal.** On the other hand, in the present photocatalytic reaction,excited dye or semiconductor oxidizes a sacrificial agent like triethanolamine. In the case of reductive quenching of the excited dye, the dye itself is reduced. For fluorescein derivatives and $Ru(bpy)_3^{+2}$, they are known to undergo one electron reduction under alkaine condition(68,69,70). These reduced dyes are thought to reduce the iminoacid,since they have a strong reducing power. The other possibility is the oxidative quenching of the excited dye, presumably at the lowest triplet state. In this case, the imino acid is reduced firstly by the excited dye and the oxidzed dye extracts an electron from the sacrificial agent. The reaction scheme based in the reduction quenching of the excited dye is shown in Figure 6.

In the case of semiconductor photocatalyst,hydrogen atoms adsorbed on the semiconductor surface, which are produced by the photocatalytic reaction of water(or H^+) with electrons in the conduction band of the semiconductor ,are thought to be responsible for the reduction of the

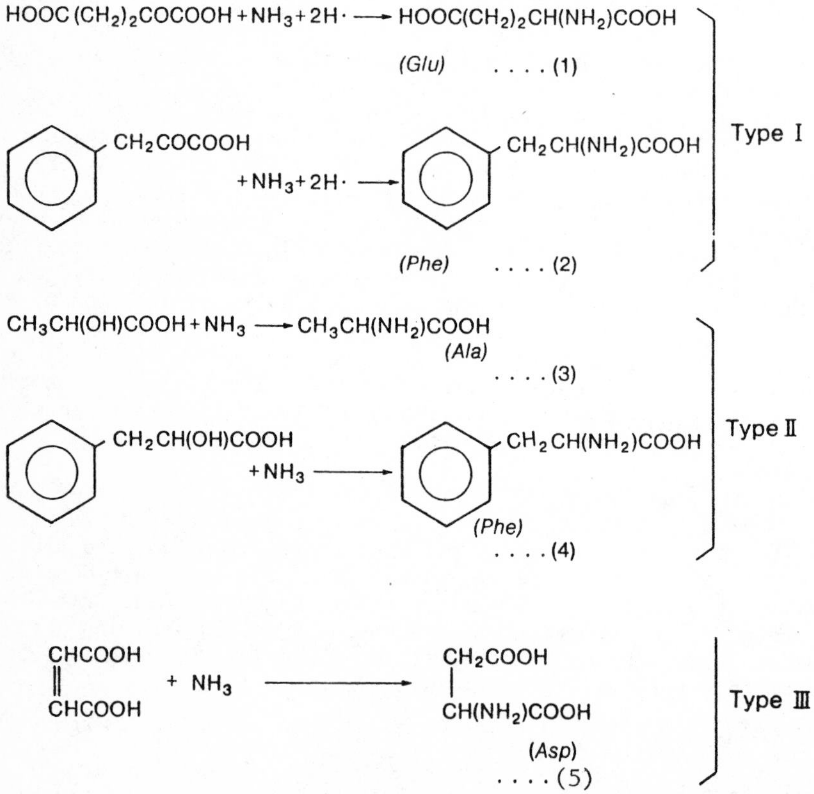

Type I

$$HOOC(CH_2)_2COCOOH + NH_3 + 2H\cdot \longrightarrow HOOC(CH_2)_2CH(NH_2)COOH \quad (Glu) \quad \ldots (1)$$

$$C_6H_5CH_2COCOOH + NH_3 + 2H\cdot \longrightarrow C_6H_5CH_2CH(NH_2)COOH \quad (Phe) \quad \ldots (2)$$

Type II

$$CH_3CH(OH)COOH + NH_3 \longrightarrow CH_3CH(NH_2)COOH \quad (Ala) \quad \ldots (3)$$

$$C_6H_5CH_2CH(OH)COOH + NH_3 \longrightarrow C_6H_5CH_2CH(NH_2)COOH \quad (Phe) \quad \ldots (4)$$

Type III

$$CHCOOH{=}CHCOOH + NH_3 \longrightarrow CH_2COOH{-}CH(NH_2)COOH \quad (Asp) \quad \ldots (5)$$

Figure 5. Typical examples of the photocatalyticsynthesis of amino acids from organic acid-ammonia-water mixtures using dyes or powdered semiconductors as photocatalysts.

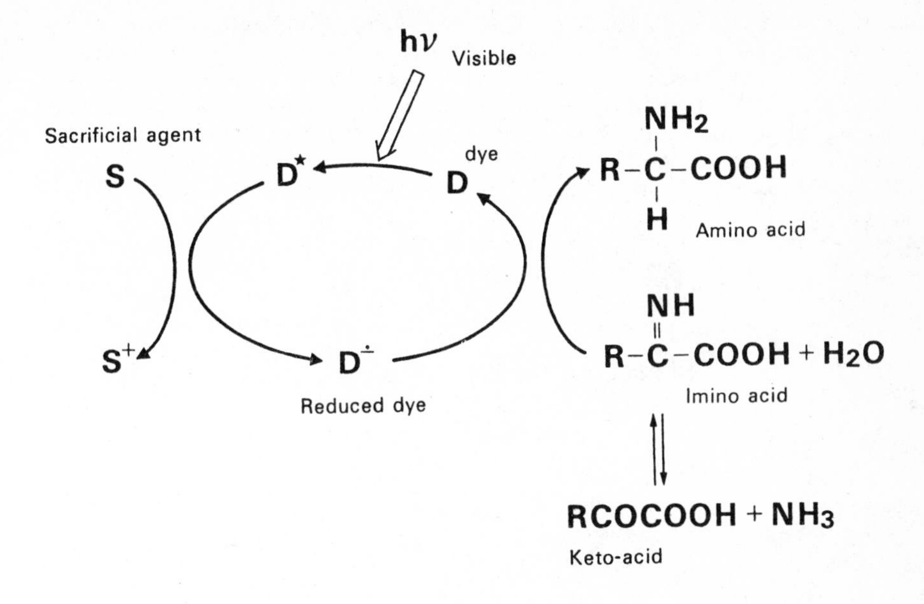

Figure 6. Reaction scheme of photo-Knoop reaction based on the reductive quenching of the excited dye.

imino acid.

Unlike the Knoop reaction Pt is not necessary in the present reaction, although the addition of Pt on the semiconductor surface increase the photocatalytic activity. Various kinds of dyes and semiconductors were found to function as good photocatalyst. For instance, 2-keto-glutaric acid in ammonia water(7% NH_3) produced glutamic acid (Glu) of 1.4 mmol after 10 hours irradiation(λ >430 nm) with ZnTPPS as photocatalyst. In the similar reactions,pyruvic,glyoxylic,phenylpyruvic, p-hydroxyphenylpyruvic,3-mthyl-2-oxo-butanoic,4-methyl-2-oxo-pentanoic and 3-methyl-2-oxo-pentanoic acids were found to produce efficiently the corresponding amino acids,i.e., alanine(Ala),glycine(Gly),phenylalanine(Phe),tyrosine(Tyr),valine(Val),leucine(Leu) and isoleucine(Ile), respectively. Various dyes such as ZnTPPS,riboflavine,dibromofluorescein, $Ru(bpy)_3^{+2}$ and Cu-chlorophyll worked as good photocatalysts. By analogy with the Knoop reaction,this type of reaction may be called a photo-Knoop In this reaction not only dyes but also various semiconductors such as TiO_2,CdS and WSe_2 work as photocatalyst.

Interestingly, amino acids were produced without a sacrificial agent such as triethanolamine, although the production rate was decreased, depending on the kind of keto-carboxylic acid. In the absence of a sacrificial agent,NH_3 is thought to work as an electron donor. Recently Tabushi et al. reported efficient reduction of keto-carboxylic acid derivatives to dehydroderivatives on a functionalized micelle by using ZnTPPS as photocatalyst(71). Their reactions seem to have a close similarity with the present reactions.

In the second type of reactions, hydroxy-carboxylic acids are used instead of keto-carboxylic acids. Ala,Gly,Phe,Val and Leu were produced efficiently from lactic ,glycol, phenyllactic ,2-hydroxy-3-methyl butyric and 2-hydroxy-isocaproic acids, respectively. In this type of reaction no sacrificial agent is necessary, because hydroxy-carboxylic acids works as the reducing agents.(see reaction(8)).

The addition of a sacrificial agent such as triethanolamine(TEOA) suppresses the production rate of amino acid as is shown in Figure 7. In the case of alanine synthesis from lactic acid and ammonia,

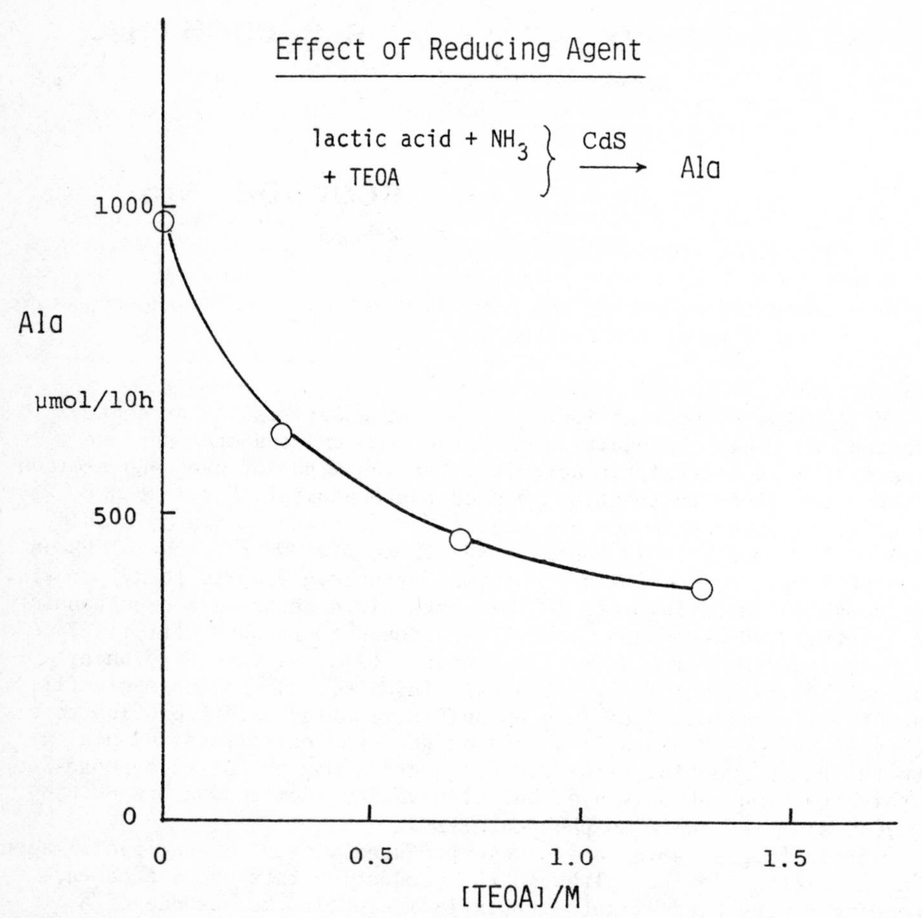

Figure 7. The effect of triethanolamine(TEOA) on the photocatalytic production rate of Ala from lactic acid and ammonia.
Experimental condition: water ;40ml, ammonia water(28%);10ml, lactic acid; 5ml; CdS; 200 mg.
light source : a 500W Xe lamp.

pyruvic acid was formed and accumulated in the aqueous medium. From these results the reaction scheme is thought to be as follows.

$$\mathrm{RCH(OH)COOH} \xrightarrow[\text{cat.}]{h\nu} \mathrm{RCOCOOH} + 2\mathrm{H}\cdot$$
$$\mathrm{RCOCOOH} \xrightarrow{\mathrm{NH_3}} \mathrm{RC(=NH)COOH} + \mathrm{H_2O}$$
$$\mathrm{RC(=NH)COOH} \xrightarrow{2\mathrm{H}\cdot} \mathrm{RCH(NH_2)COOH} \qquad (8)$$

This type of reaction can be regarded as a modification of the first type of reaction and can also be called a photo-Knoop reaction. In the second type of reaction, almost all dyes,which has been examined untill now, did not show any photocatalytic activity. Moreover,the activity depends strongly on the kind of semiconductor as shown in Table 4. This fact is in strong contrast to the case of the first type of reaction. This is explained well by the fact that the main product in the aqueous medium in the photocatalytic decomposition of lactic acid is pyruvic acid for Pt/CdS and acetaldehyde for Pt/CdS, as was described in the previous section. Therefore, in the case of TiO_2,amino acids are not formed efficiently. A similar result was obtained for the formation of glycine from glycol acid. In the presence oxygen, the formation of amino acids was suppressed strongly. A similar effect was observed in the case of the first type of reactions. This is expected from reactions (6)-(8),since in the presence of oxygen

Table 4. Dependence of alanine production from lactic acid and ammonia on the kind of semiconductor photocatalyst.

Reaction	Photocatalyst	Rate(μmol/10h)
$CH_3CH(OH)COOH + NH_3 \longrightarrow CH_3CH(NH_2)COOH$	CdS	2420
	ZnS	34
	TiO_2	12
	MoS_2	0
	Fe_2O_3	0
	GaP	0
	WSe_2	0

light: 500W Xe lamp

Table 5. Quantum yields of photocatalytic amino acid production by use of dyes or semiconductors.

	Reactant	Photocatalyst	Product	Quantum Yield (%)
Keto-carboxylic acid	2-ketoglutaric acid	ZnTPPS	Glu	32
		CdS		35
	pyruvic acid	Br_2Fluo	Ala	16
	phenylpyruvic acid	ZnTPPS	Phe	12
	p-hydroxyphenyl pyruvic acid	ZnTPPS	Tyr	16
	3-methyl-2-oxo-butanoic acid	Acridine Yellow	Val	6
Hydroxy-carboxylic acid	lactic acid	CdS	Ala	20
	2-hydroxy-3-methyl butyric acid	CdS	Val	19
	phenyllactic acid	CdS	Phe	6
	2-hydroxy-iso-caproic acid	CdS	Leu	23

λ=440nm

the excited electrons are consumed to reduce oxygen instead of imino acid. Instead of ammonia , NH_2OH was found to be effective as a nitrogen source for the photocatalytic amino acid production. In this case, oxy-imino acid is thought to be reduced to produced amino acid.

In addition to the photo-Knoop reactions, we tried additive reactions of NH_3 to the unsaturated carboxylic acids such as maleic and cinnamic acids. Several reactions were successful. For instance,aspartic acid was produced by the addition of NH_3 to maleic acid by use of TiO_2 as a photocatalyst as shown in Figure 5.

Letokhov et al. reported an efficient production of aspartic acid (Q.Y.∼0.35) by making use of two photon excitation of maleic acid with a picosecond UV laser at a very high laser power of $3x10^8$ W/cm^2(72). In our method ,however, the reactions proceed with visible light under a mild conditions. Table 5 shows some examples of these reactions and the quantum yields(Q.Y.'s) of the reaction. As is shown in this table, the quantum yields of these reactions are very high with visible light. Because of the simplicity and high efficiency,this method might be one of the promising means for amino acid synthesis. High efficiency also seems to suggest an important role of photocatalytic processes in the chemical evolution not only by semiconductors (61,62) but also by dyes,since various dyes such as porphyrines have been postulated to have exsisted before the evolution of photosynthesis by microorganisms(73,74).

ACKNOWLEDGEMENTS

The auther wishes to acknowledge the valuable assistance of his collaborators whose work is cited in the references.

REFERENCES

1. J.M.Lehn,J.P.Sauvage and R.Ziessel,Nouv.J.Chim.,4,623(1980)
2. T.Kawai and T.Sakata,Chem.Phys.Lett.,72,87(1980).
3. S.Sato and J.M.White,Chem.Phys.Lett.,72,85(1980).
4. K.Domen,S. Naito,H. Soma,T. Ohnishi and K. Tamaru, J.Chem.Soc.Chem. Commun.,543(1980).
5. F. T. Wagner and G. A. Somorjai, J.Am.Chem.Soc.,102,5494(1980).
6. D.Dounghong,E. Borgarello and M. Grätzel, J.Am.Chem.Soc.,103,4685 (1981).
7. K. Kalyanasudaram, E. Borgarello and M. Grätzel,Helv.Chim.Acta,64,362 (1981).
8. J. M. Lehn, J. P. Sauvage,R. Ziessel and L. Hilaire, Israel J. Chem.,22,168(1982).
9. Y. Nakato and H. Tsubomura, Israel J. Chem.,22,180(1982).
10. J. R. Darwent and A. J. Mills, J.Chem.Soc.,Faraday Trans.2,78,359 (1982).
11. G. Blondeel, A. Harriman, G. Porter, D. Urwin and J. Kiwi, J.Phys.Chem.,87,2629(1983).
12. E. Amouyal,P. Keller and A. Moradpour, J.Chem.Soc.,Chem.Commun.,1019 (1980).
13. D. Meissner, R. Memming and B. Kastening, Chem.Phys.Lett.,96,34 (1983).
14. A. Heller, E. Aharon-Shalom, W. A. Bonner and B. Miller, J.Am.Chem.Soc.,104,6942(1982).
15. S.N. Frank and A. J. Bard, J.Am.Chem.Soc.,99,303(1977).
16. S.N. Frank and A. J. Bard, J.Phys.Chem.,81,1484(1977).
17. M. Matsumura, Y.Saho and H. Tsubomura, J.Phys.Chem.,87,3807(1983).
18. D. H. M. Thewissen, A. H. A. Tinnemans, M. Eeuwhorst-Reinten, K. Timmer and A. Mackor, Nouv.J.Chim.,7,191(1983).
19. E. Borgarello, W. Erbs and M. Gratzel, Nouv.J.Chim.,7,195(1983).
20. B. krauetler and A. J. Bard, J.Am.Chem.Soc.,100,5985(1978).
21. A. H. Boonstra and C. A. H. A. Mutsaers, J.Phys.Chem.,79,2025(1975)
22. C. Yun, M. Anpo, S. Kodama and Y. Kubokawa,J.Chem.Soc.,Chem.Commun. 609(1980).
23. I. Izumi,W. W. Dunn, K. O. Wilbourn,Fu-Ren F.Fan and A. J. Bard, J.Phys.Chem.,84,3207(1980).
24. M. Fujihira, Y. Sato and T. Osa, Nature,293,206(1981).
25. M. Fijihira, Y. Sato and T. Osa,Chem Lett.,1053(1981).
26. S. Sato and J. M. White,Chem.Phys.Lett.,70,131(1980).
27. S. Nishimoto, B. Ohtani, T. Yoshikawa and T. Kagiya, J.Am.Chem.Soc. 105,7180(1983).
28. S.Yanagida, T. Azuma, H. Kawakami, H. Kizumoto and H. Sakurai, J. Chem.Soc.,Chem.Commun.,21(1984).
29. M. A. Fox, Acc.Chem.Res.,16,314(1983).
30. T. Watanabe, T. Takizawa and K. Honda, J.Phys.Chem.,81,1845(1977).
31. M. Miyake, H. Yoneyama and H. Tamura, Bull.Chem.Soc.Jpn.,50,411 (1977).
32. T. Inoue, A. Fujishima, S. Konishi and K. Honda, Nature,277,637 (1979).

33. T. Kanno, T. Oguchi, H. Sakuragi and K. Tokumaru, Tetrahedron Lett.,21,4673(1980).
34. M. A. Fox and C. Chen, J.Am.Chem.Soc.,103,6757(1981).
35. S. Yanagida, T. Azuma and H. Sakurai, Chem.Lett.,1069(1982).
36. J. Bucheler, N. Zeug and H. Kisch, Ang.Chem.,94,792(1982).
37. M. Matsumura, M. Hiramoto, T. Iehara and H. Tsubomura, J.Phys.Chem. 88,248(1984).
38. T. Kawai and T. Sakata,J. Chem.Soc.,Chem.Commun.,694(1980).
39. T. Sakata and T. Kawai, Chem.Phys.Lett.,80,341(1981).
40. T. Kawai and T. Sakata, Nature,286,474(1980).
41. T. Sakata and T. Kawai, J.Synth.Org.Chem.Jpn.,39,589(1981).
42. K. Hashimoto, T. Kawai and T. Sakata, J.Phys.Chem.,88,4083(1984).
43. T. Kawai and T. Sakata, Chem.Lett.,81(1981).
44. T. Sakata and T. Kawai, Nouv.J.Chim.,5,279(1981).
45. T. Sakata and T. Kawai, *Energy Resources through Photochemistry and Photocatalysis*, M. Grätzel Ed., Academic Press,1983,pp.331-358.
46. T. Sakata, T. Kawai and K. Hashimoto, Denki Kagaku,51,79(1983).
47. T. Sakata, K. Hashimoto and T. Kawai, J.Phys.Chem.,88,5214(1984).
48. S. Nakabayashi,A. Fujishima, K. Honda, Chem.Phys.Lett.,102,464 (1983).
49. T. Sakata, T. Kawai and K. Hashimoto, Chem.Phys.Lett.,88,50(1982).
50. M. Tomkewicz, J.Electrochem.Soc.,126,1505(1979).
51. T. Sakata, T. Kawai and K. Hashimoto, J.Phys.Chem.,88,2344(1984).
52. B. Krauetler and A. J. Bard, J.Am.Chem.,100,2239(1978).
53. S. Sato, J.Chem.Soc.,Chem.Commun.,26(1982).
54. H.Yoneyama, Y.Takao, H. Tamura and A. J. Bard, J.Phys.Chem.,87,1417 (1983).
55. H.Hofer and M. Moest, Ann.Chem.(Warsaw),323,284(1902).
56. H. Harada, T. Sakata and T. Ueda, J.Am.Chem.Soc.,107,1773(1985).
57. R. Williams, J.Vac.Sci.Technol.,13,12(1976).
58. O. M. Kolb and H. Gerischer, Electrochim. Acta,18,987(1973).
59. C. Sagan and B. N. Khare, Science,173,417(1971).
60. K.Hong, J. H. Kong and R. S. Becker, Science,184,984(1974).
61. H. Reiche, A. J. Bard, J.Am.Chem.Soc.,101,3127(1979).
62. W. W. Dunn, Y. Aikawa and A. J. Bard, J.Am.Chem.Soc.,103,6893(1981)
63. T. Kawai, M. Fujii, S. Kambe and S. Kawai, a) Abstr. Fall Meet. Electrochemical Society of Japan,1983,Abstract E214,p.211, b) Spring Meet. Electrochemical Society of Japan,1984, Abstract, p.106, c) The fifth international conference on photochemical conversion and storage of solar energy, Osaka,1984, book of abstract,p.143 and p.145.
64. T. Sakata, Denki Kagaku,53,15(1985).
65. T. Sakata, J. Photochem.,29,205(1985).
66. T. Sakata and K. Hashimoto, Nouv.J.Chim.,in press.
67. F. Knoop and H. Oesterlin, Z.Physiol.Chem.,148,294(1925);170, 186(1927).
68. K. Hashimoto, T. Kawai and T. Sakata, Chem.Letters,709(1983),
69. K. Hashimoto, T. Kawai and T. Sakata,Nouv.J.Chim.,8,693(1984).
70. C. Creutz, N. Sutin, B. Brunschwig, J.Am.Chem.Soc.,101,1297(1979).
71. I. Tabushi, S. Kugimiya and T. Mizutani, J.Am.Chem.Soc.,105,1658 (1983).

72. V. S. Letokhov, Yu. A. Matveetz, V. A. Semchishen and E. V. Khoroshilova, Appl.Phys.,B26,243(1981).
73. H. Gaffron : *The Origin of pribiological Systems and of their Molecular Matrices*, S. W. Fox ed., Academic Press, p.437(1965).
74. M. Calvin, Science,130,1170(1959).

PHOTOCATALYTIC OXIDATION OF ORGANIC COMPOUNDS WITH HETEROPOLY ELECTROLYTES. ASPECTS ON PHOTOCHEMICAL UTILIZATION OF SOLAR ENERGY

E. Papaconstantinou, P. Argitis, D. Dimoticali, A. Hiskia
and A. Ioannidis
Chemistry Department, N.R.C. Demokritos, Athens, Greece

ABSTRACT. Heteropoly electrolytes (HPC) of molybdenum and tungsten are photosensitive in near visible and ultra violet light in presence of a great variety of organic compounds. Photoexcitation results in "reversible" stepwise reduction of HPC with concomitant oxidation of organic compounds. Photoreduction proceeds to the extent that the rate of photoreduction is matched by the back reoxidation by H^+ or oxygen. At this stage a steady state is obtained at which H_2 is produced with a quantum yield of 1%, whereas, maximum quantum yield for alcohol oxidation in presence of oxygen is 10-15%. Coupling the photoreduced and oxidized species into a photogalvanic (photofuel) cell a current is produced with an efficiency of 1-2 electron per 100 photons. Incorporation of vanadium into HPC shifts the absorption toward the visible, but there seems to be no improvement in photochemistry.

INTRODUCTION

The chemistry of Heteropoly compounds (HPC) is of diversified interests[1]. Apart from the academic interest, they are also of practical importance. Molybdenum and tungsten oxides and sulfides as well as HPC are extensively used as catalysts in a variety of commercially important chemical processes, such as hydrocracking, hydrogenation, isomerization, polymerization etc.[2] The formation of the blue reduction products of HPC, the so called heteropoly blues (HPB), finds application in analytical chemistry in the colorimetric determination of several elements such as P, Si, As, Ge and in biochemistry for uric acid, sugar etc.[3] Their ability to accept and release electrons has found use in the efforts to elucidate the various steps in photosynthesis[4]. The motion of the electrons within the crystal lattice can be affected by applying an external potential. Several oxides in an externally applied electric field acquire a blue color which disappears by reversing the polarity. The electrochromism of these compounds is of interest in its possible use in digital display devices[5]. Their potentiality in photochemical work has been recognized in the past[6] and has been patented for possible use in photography[7]. Various Russian workers have investigated the photochemistry of molybdenum and tungsten for analytical purposes[8].

E. Pelizzetti and N. Serpone (eds.), Homogeneous and Heterogeneous Photocatalysis, 415–431.

It is well documented that HPC are capable of multielectron reductions in distinct reduction steps without decomposition. Apparently, this is of fundamental importance in their catalytic activity. On the other hand compounds that are multielectron reducing reagents are generally in demand as potential reductants for nitrogen, carbon and for splitting water.

This paper deals, generally, with the photochemistry of HPC of the general formula $A_aX_xO_m^{n-}$, where $A = P, Si, Fe, H_2$, $X = Mo$ or W and $\frac{a}{x} = \frac{1}{12}$ and $\frac{2}{12}$, as well as their potentiality in photocatalysis and solar energy research. It presents an outline of the research pursued in our laboratory the last few years plus some new developments in the photocatalytic oxidation of organic compounds, and sensitization by vanadium.

A. Photochemistry, basic studies

It is now well documented that HPC are photosensitive in presence of a great variety of organic species, upon exposure to visible and UV light producing HPB products. The organic compounds involve alcohols, glycols, hydroxy acids, carboxylic, dicarboxylic acids. Quantum yields vary with HPC, organic reagents and with reduction step, dropping to ~10% in going from first to second reduction step at least for $PW_{12}O_{40}^{3-}$, (PW_{12}^{3-}), Tables I and II.

TABLE I Comparative photoreduction of various HPC 5×10^{-4} M, in the presence of various concentrations of Isopropyl alcohol at 254 nm.[12b]

Isopropyl alcohol M	Φ of 1-electron reduction products			
	PW_{12}^{4-}	SiW_{12}^{5-}	FeW_{12}^{6-}	$H_2W_{12}^{7-}$
0.1	0.12	a	a	a
1.0	0.13	0.05	a	a
5.0	0.12	0.09	0.02	a
10.0		0.10	0.03	a

[a]Values less than 0.01

The formation of the blues and the extent of reduction could be easily followed by the characteristic spectra of HPB , for instance Fig. I. The photoproduction of HPB is accompanied by oxidation of organic species. Thus for instance, in presence of HPC, isopropyl alcohol forms acetone and ethyl alcohol forms acetaldehyde.

The extent of reduction depends on the organic reagent and on the HPC used and this relates to the ease with which they accept electrons. Thus, for instance $[P_2Mo_{18}O_{62}]^{6-}$, ($P_2Mo_{18}^{6-}$) in presence of isopropyl alcohol accepts photochemically up to 6 electrons, whereas PW_{12}^{3-} accepts ~2 electrons. More on this will be discussed below in connection with hydrogen production. Table III shows the maximum number of electrons

TABLE II Quantum yield of formation of the 1-electron reduction product of PW_{12}^{3-} at 254 nm, in the presence of various organic compounds in $HClO_4$, 0.1 M[12b].

Organic additive	M	PW_{12}^{3-},M	$\Phi(PW_{12}^{4-})$
CH_3OH	1	5×10^{-4}	0.11
C_2H_5OH	1	5×10^{-4}	0.15
$CH_3CH_2CH_2OH$	1	5×10^{-4}	0.15
$(CH_3)CHOH$	1	5×10^{-4}	0.13
$(CH_3)COH$	1	5×10^{-4}	0.08
CH_2OHCH_2OH	0.5	5×10^{-4}	0.02
$CH_2OHCOOH$	1	1×10^{-3}	0.10
$CH_3CHOHCOOH$	1	5×10^{-4}	0.07
CH_3COOH	0.01	5×10^{-4}	0.03
$(COOH)_2$	0.5	1×10^{-3}	0.02
$CH_2(COOH)_2$	0.01	5×10^{-4}	0.02
$(CH_2)_3(COOH)_2$	1	1×10^{-3}	>0.01
$(CH_2OHCH_2)_3N$[a]	0.1	1.5×10^{-4}	0.05
$CH_2(NH_2)COOH$	1	1×10^{-3}	<0.01
$CH_3CH(NH_2)COOH$	0.5	5×10^{-4}	0.01

[a] $HClO_4$, 0.26 M. A white precipitate is formed at PW_{12}^{3} concentrations larger than 2×10^{-4} M.

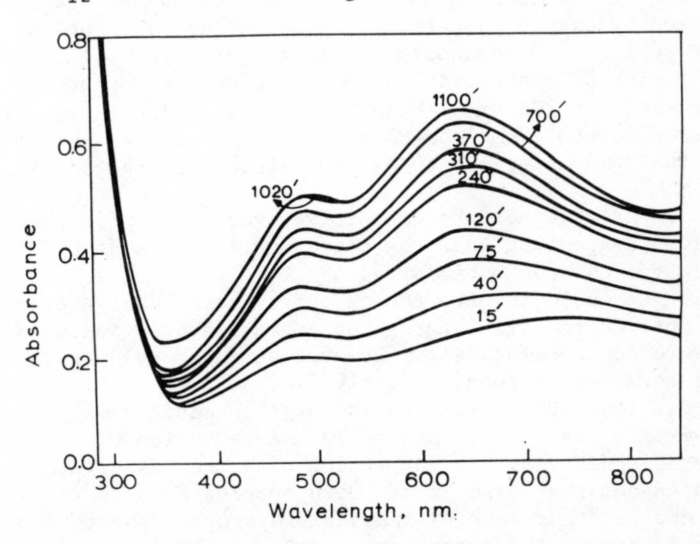

Fig. I Spectra of photoreduced PW_{12}^{3-} in which the successive formation of 1-electron (751 nm) and 2-electron reduction products is shown. Photolysis with 254 nm. PW_{12}^{3-} 1×10^{-4}M, propan-2-ol 2 M., in 0.1 M $HClO_4$; Solutions deaerated with Ar. Time is indicated on spectra.

added photochemically on a series of 12-tungstates in presence and absence of Pt catalyst.

TABLE III Maximum number of electrons added photochemically in the presence of isopropyl alcohol (1.0 M) in 0.1 M $HClO_4$[a] [21b]

Compd	max.no. of added electrons		Compd	max.no. of added electrons	
	without Pt^b	with Pt^b		without Pt^b	with Pt^b
PW_{12}^{3-}	2.2	1.3	FeW_{12}^{5-}	1.4	0.3
SiW_{12}^{4}	1.7	0.9	$H_2W_{12}^{6-}$	1.2	0.2

[a]Conditions: [HPC] = 5×10^{-4} M; photolysis with Xe 150-W lamp without filter (deaeration with Ar).

[b]Values within 20%; concentration of Pt catalyst $\sim5\times10^{-6}$M.

The relative rates of HPB formation are a function of HPC concentration and organic additive increasing with concentration of organic additive up to at least 5-10 M for $P_2Mo_{18}^{6-}$, whereas PW_{12}^{3-} needs considerably less alcohol, Fig. II, III.

The spectra of the nonreduced HPC of both molybdates and tungstates are characterized by oxygen to metal charge transfer bands at the near visible and UV regions and no absorption in the visible, (all metals are Mo(VI) or W(VI) with d^o configuration). The reduced forms present broad absorption bands around 700 nm. These bands are attributed to metal to metal charge transfer ($M^{5+} \rightarrow M^{6+}$, M=Mo or W) and are responsible for the blue color of the compounds, and to d-d transitions of the d^1 metal ions[9,10,11].

The reduced HPC are not photosensitive over ∿500 nm in presence of organic additives suggesting that the above mentioned bands of HPB are not responsible for the photochemistry.

Excitation seems to involve the oxygen to metal charge transfer of the HPC. Attempts to identify aggregates or any association between HPC and alcohols were not conclusive[12]. However stronger evidence that aggregates are formed were found recently[13].

The photoexcited HPC react with the organic additives by inter-electron or hydrogen transfer resulting in the reduction of HPC and oxidation of organic additives. In connection with the last statement, it should be pointed out that experiments with species such as tertiary butyl alcohol and Fe^{2+}that lack of removable hydrogens showed that, at least to a great extent, the redox reactions take place by hydrogen abstraction. With tertiary butyl alcohol the photoredox reaction was considerably slower, whereas, no photoreaction took place with PW_{12}^{3-} and Fe^{2+}. Designating 12-tunstates as W_{12}^{n-}, and using Me_2CHOH,the photochemical reaction mechanism may be presented as follows:

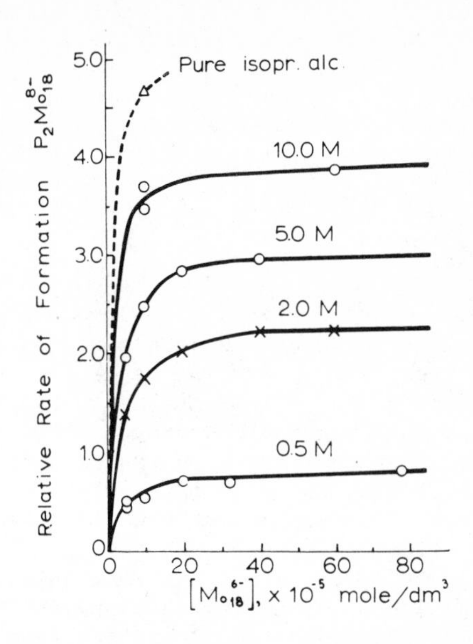

Fig. II Variation of the relative rates of formation of $P_2Mo_{18}^{8-}$ with concentrations of $P_2Mo_{18}^{6-}$ for various concentrations of isopropyl alcohol, in 0.1M $HClO_4$. High pressure Hg-lamp with pyrex filter (solutions deaerated with Ar).

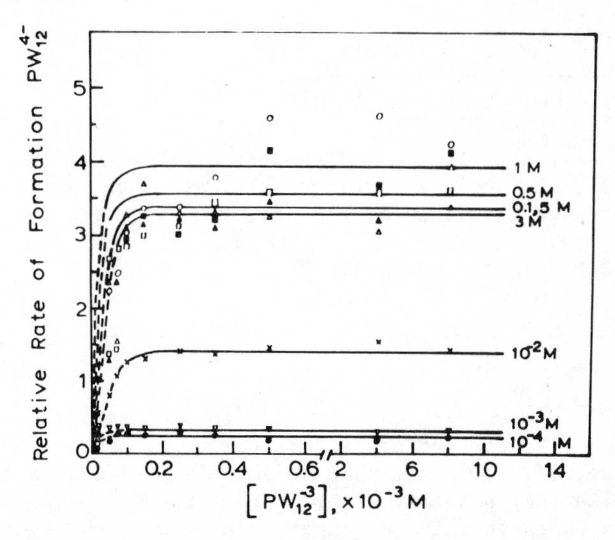

Fig. III Variation of the relative rates of formation of PW_{12}^{4-} with concentration of PW_{12}^{3-} for various concentrations of isopropyl alcohol in 0.1M $HClO^4$. High pressure Hg-lamp with 252 nm filter, (solutions deaerated with Ar).

$$W_{12}^{n-} \xrightarrow{h\nu} {}^*W_{12}^{n-} \quad [1]$$

$$^*W_{12}^{n-} \xrightarrow{k_{-1}} W_{12}^{n-} \quad [2]$$

$$^*W_{12}^{n-} + Me_2CHOH \xrightarrow{k_Q} W_{12}^{(n+1)} + Me_2\dot{C}OH + H^+ \quad [3]$$

$$W_{12}^{n-} + Me_2\dot{C}OH \xrightarrow{k_R} W_{12}^{(n+1)-} + Me_2CO + H^+ \quad [4a]$$

$$W_{12}^{(n+1)} + Me_2\dot{C}OH \rightarrow W_{12}^{(n+2)-} + Me_2CO + H^+ \quad [4b]$$

$$W_{12}^{(n+2)-} + W_{12}^{n-} \rightarrow 2W_{12}^{(n+1)-} \quad [4c]$$

Reactions 4a,b,c are known to take place. They proceed until all W^{n-} has been converted to $W^{(n+1)-}$. Reactions 4a,b,c are essentially reaction 4a. Photoreduction proceeds, with a similar mechanism, to higher reduction steps depending on conditions. This mechanism was originally proposed mainly on the basis that removable hydrogen atoms are required for photoreduction of HPC and that the resulting radicals are powerful reducing reagents known to react with HPC with rates close to diffusion controlled[14,15]. The mechanism subsequently was adopted by Darwent[16] and verified by isotope effect by Ward[17]

The rate determining step for $W_{12}^{(n+1)-}$ formation and consequently acetone formation is reaction 3. This has been proved by using deuterated Me_2CDOH. In this case, it was found that the rate of acetone formation dropped by about four times[17]

From reactions 1,2 and 3 (reactions 4 are too fast to interfere), applying the steady state approximation for $^*W_{12}^{n-}$ we obtain, from reaction 3

$$R = \frac{d\left[W_{12}^{(n+1)-}\right]}{dt} = \frac{k_Q\left[Me_2CHOH\right]I_o}{k_{-1}+k_Q\left[Me_2CHOH\right]} \quad (7)$$

When $k_a\left[Me_2CHOH\right] >> k_{-1}$, $R \simeq I_o$, i.e. independent of $\left[Me_2CHOH\right]$ as shown in Figures II and III. This happens, for $P_2Mo_{18}^{6-}$, at concentrations of isopropyl alcohol greater than 5-10 M, whereas, for PW_{12}^{3-} greater than 0.1 M.

Taking the reciprocal of expression 7

$$\frac{1}{R} = \frac{1}{I_o} + \frac{k-1}{k_Q I_o\left[Me_2CHOH\right]} \;.$$

Fig. IV shows a plot of the reciprocal of the rate of formation of $P_2Mo_{18}^{8-}$ vs $1/\left[Me_2CHOH\right]$ for various concentrations of $P_2Mo_{18}^{6-}$. A good straight line is observed. However, this is not always true, at least with other HPC, indicating, possibly, that another type of mechanism, i.e., formation of a photosensitive aggregate precursor might operate. This is under investigation.

B. Utilization of HPC in solar energy research

It is generally accepted that for a system to be suitable for solar energy

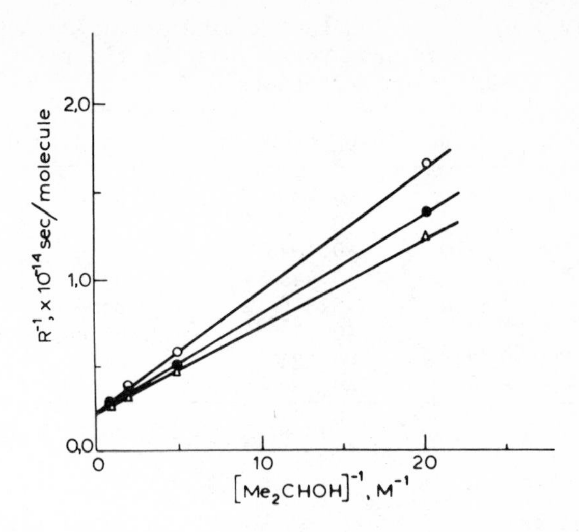

Fig. IV Reciprocal of the rate of formation of $P_2Mo_{18}^{8}$ vs the reciprocal of isopropyl alcohol concentration. $[P_2Mo_{18}^{6-}]$: ΔΔ, 4×10^{-4}M; ●●, 2×10^{-4}M; oo, 1×10^{-4}M. Photolysis with high pressure Hg-lamp with pyrex filter.

conversion and storage, it has to meet the following general specifications: (a) to absorb considerably in visible, (b) to have a long life excited state, (c) to participate in redox reactions that are able to store the absorbed energy, (d) to have a photoredox cycle of about one million times.

These are difficult requirements to match and the efforts are along various systems that can fulfill one or the other requirement. We will present below our efforts in utilizing HPC in solar energy research and this is done along three directions: (a) hydrogen production, (b) photocatalytic oxidations of organic compounds, (c) photogalvanic cells. Although there is interrelation between these three cases, we will present them, for clarity reasons, rather separately.

1. H_2 production

It has been pointed out previously that HPC are photosensitive at near visible and UV light, in presence of a great variety of organic additives[12]. Excitation is accompanied by reduction of HPC and oxidation of the organic species, see reaction 1,2,3,4. The addition of electrons drives the redox potential towards more negative values which, as is known are oxidized back depending on their redox potential to the original nonreduced species, by atmospheric oxygen. This is used in the photocatalytic oxidation of organic compounds; see below[15]. If, however, oxygen is excluded, the redox potentials are negative enough to cause from thermodynamic point of view, reduction of H^+ [18]. Table IV shows UV spectral data and half-wave potentials for the first two 1-electron reductions for oxidized 1:12 tungstates[25]. This may be

TABLE IV UV spectral data for oxidized 1:12 Tungstates and half-wave potentials for the first two 1-electron reductions[25]

anion	λ^a, nm	$E_{1/2}$^{b} V	pH dependence
PW_{12}^{3-}	265.0	-0.023	none
		-0.266	none
SiW_{12}^{4-}	262.0	-0.187	none
		-0.445	none
FeW_{12}^{6-}	264.0	-0.349	none
		-0.577	below pH 4.0
$H_2W_{12}^{6-}$	257.5	-0.581	below pH 4.9
		-0.730	

^{a}In 1 M sulfuric acid. ^{b}vs see, in 1 M sulfate at 25°C

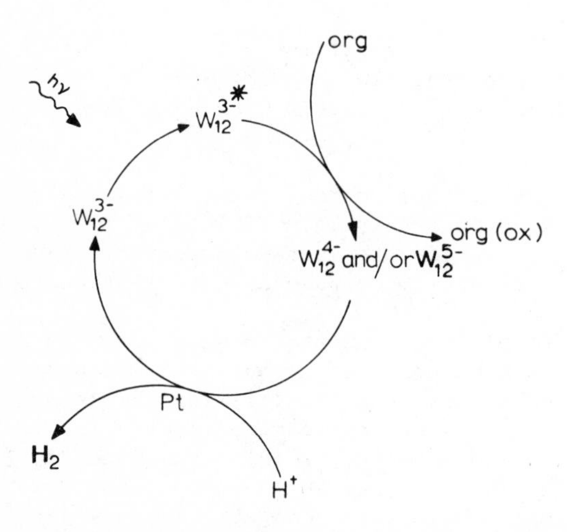

Fig. V Schematic diagram of H_2-evolution, where "Org" is a reductant that undergoes irreversible oxidation and Pt is a suitable platinum catalyst[21b].

accomplished in presence of Pt catalyst to overcome the high reduction potential necessary for the reduction of atomic hydrogen, $H^+ + e^- \rightarrow H$, $E^o = -2,2V$, and produce instead molecular hydrogen, $2H^+ + 2e \rightarrow H_2$, $E_7^o = -0,41V$, or $E_o = 0,0V$, where E_7, E_o are potentials at pH 7 and 0 respectively[19]. The feasibility of H_2-evolution by reduced SiW_{12}^{4-} was established by Russian workers, by reducing SiW_{12}^{4-} with amalgamated zin zinc[20]. H_2-evolution has now been obtained photochemically by various laboratories[21].

It turns out that reduced HPC are capable of H_2-evolution with and without Pt catalyst[21b]. Whereas they are not useful as sensitizers, they absorb very little in the visible, they seem to be useful as electron carriers (relays). Photoreduction proceeds to the extent that the rate of photoreduction matches the back reoxidation by H^+. The reduction step at which this takes place depends on the HPC used. Thus, for instance, with PW_{12}^{3-} this happens at a higher reduction step than $H_2W_{12}^{6-}$ which is reduced at more negative reduction potentials. When Pt catalyst is present the reduction potential at which the steady state H_2-evolution is obtained is even lower (Table III). Apparently HPC also serve as heterogeneous catalysts, in the absence of metallic Pt but not as effective. The cycle for H_2-evolution is depicted in Fig. V. The rate of H_2-production is limited by the low quantum yield of HPB production. Reoxidation of HPBs' by H^+ presented generally as

$$2W_{12}^{(n+1)} + 2H^+ \xrightarrow{k_H} 2W_{12}^{n-} + H_2 \tag{5}$$

seems to be 100% efficient. One can observe, Table III, that for PW_{12}^{3-}, reaction 5 involves roughly the $W_{12}^{(n+2)-}$ species, whereas for $[FeW_{12}O_{40}]^{5-}$, (FeW_{12}^{5-}) and $[H_2W_{12}O_{40}]^{6-}$, $(H_2W_{12}^{6-})$ it involves mainly the first reduction step, i.e., $W_{12}^{(n+1)-}$.

The rate of H_2-evolution, reaction 5, was followed by the change in absorbance of the HPB bands at ∿750 nm. This was done by carefully excluding oxygen, first by deaeration and secondly by repeating cycles of photoreduction and back reoxidation in which the left over oxygen was consumed. Typical rate constants of H_2-evolution at the steady state for 1:12 tungstateswere $(5.6\pm3.1)\times10^{-4}s^{-1}$ and $(2.6\pm1.8)\times10^{-4}s^{-1}$ with and without Pt respectively. The total production of hydrogen, measured by gas chromatography, verified reaction 5. About 1 ml of hydrogen was generally produced from 10 ml solution composed of HPC, 5×10^{-4}M, isopropyl alcohol 1M, in 0.1M $HClO_4$, after 24 hours of photolysis with a 150W Xe lamp. This amounts to an efficiency of the order of 1 molecule of hydrogen per 100 photons absorbed. The low yield represents the low efficiency of photoproduction of higher reduction products as mentioned before.

With hydrogen production a second steady state is established (reactions 3,4,5) that involves the reduction step from which H_2 is produced. Reaction 4 is too fast so that it can be safely ignored. The H_2 produced matches, stoichiometrically, the number of electrons added photochemically. Assuming for simplicity that $W_{12}^{(n+1)-}$ is the reduction step from which H_2 is produced, and using steady states for $^*W_{12}^{n-}$ and $W_{12}^{(n+1)-}$, we obtain from reaction 5

$$R(H_2) = \frac{d[W_{12}^{(n+1)-}]}{dt} = \frac{k_Q[Me_2CHOH]I_o}{k_{-1}+k_Q[Me_2CHOH]}$$

This expression is identical with expression 7. They both represent the rate of formation of some reduced tungstate, in this case $W^{(n+1)-}$. At sufficiently high concentrations of isopropyl alcohol $k_Q[Me_2CHOH] \gg k_{-1}$ and the rate depends only in I_o.

This expression, for the rate of H_2-evolution, has been recently presented by Darwent pursuing also research along these lines[21c]. The last equation is rearranged to

$$\frac{1}{R(H_2)} = \frac{1}{I_o} + \frac{k_{-1}}{k_Q I_o [\text{Me CHOH}]}$$

A plot of the reciprocal of the initial rate of H_2-evolution vs $1/[Me_2CHOH]$ gave a good straight line[21c].

As has been pointed out, the overall reaction is

$$Me_2CHOH \rightarrow Me_2CO + H_2 \quad (8)$$

i.e., the production of H_2 is matched by concomitant production of acetone. The quantum yield for acetone production is $\sim$0.01, which represents the low q.y. of that particular reduction step. In the case of PW_{12}^{3-}, for instance it represents the second reduction step, i.e., $PW_{12}^{4-} + e^- \rightarrow PW_{12}^{5-}$.

2. Photocatalytic oxidation of organic compounds in presence of oxygen

It has been stated earlier that a variety of organic compounds undergo photocatalytic oxidation in presence of HPC, Table II. In this case reaction 5 is replaced by

$$W_{12}^{(n+1)-} + 1/4\ O_2 + H^+ \rightarrow W_{12}^{n-} + 1/2\ H_2O \quad (6)$$

The second order rate constant for reoxidation of the blues by H^+ at the steady state, reaction 5, as has been discussed earlier, is $\sim 5 \times 10^{-3} M^{-1}s^{-1}$, whereas, reoxidation of the first reduction product by oxygen, reaction 6, is considerably faster. This reaction is pH dependent, but it will not concern us at present. Using laser flash photolysis to generate the reduced HPC, the second order rate constant for the reaction

$$PW_{12}^{4-} + 1/4\ O_2 + H^+ \rightarrow PW_{12}^{3-} + 1/2\ H_2O$$

was $\sim 88\ M^{-1}s^{-1}$ [15]. Recently this work was expanded to other HPC for which k was found to vary from 75, for the above reaction, to 6.5×10^3, 2.8×10^5 and $1.75 \times 10^5 M^{-1}s^{-1}$, for SiW_{12}^{5-}, FeW_{12}^{6-} and $H_2W_{12}^{7-}$ respectively

The upper limit in the q.y. for the photocatalytic oxidation of organic compounds will be the q.y. of formation of the first reduction step, Table I. This is indeed the case where the steady state is established at the earliest possible stage of photoreduction, i.e. before the photoreduction of the second step becomes significant. Using, for simplicity, only the first reduction step, and applying the steady state approximation for W_{12}^{n-} (reactions 1 and 6) we get

$$\frac{d[W_{12}^{n-}]}{dt} = k[H^+][O_2][W_{12}^{(n+1)-}] - \Phi_1 I_o = 0$$

and
$$[W_{12}^{(n+1)-}] = \frac{\Phi_1 I_o}{k[H^+][O_2]} = \frac{\Phi_1 I_o}{k'}$$

where Φ_1 is the q.y. of the first reduction step, I_o is the intensity of radiation, assuming 100% absorption, and $k' = k[H^+][O_2]$
From reaction 6

$$\frac{d[W_{12}^{(n+1)-}]}{dt} = k[H^+][O_2][W_2^{(n+1)-}] = \Phi_1 I_o$$

Since, as we have discussed earlier, $W_{12}^{(n+1)-}$ oxidizes stoichiometrically organic compounds, the q.y. for the oxidation products will be the same. Table I shows that the best candidates for maximum yield of photooxidized organic compounds are PW_{12}^{3-} and $S_1W_{12}^{4-}$. Using $S_1W_{12}^{4-}$ and Me_2CHOH in presence of oxygen we were able to obtain, with a pyrex filter, q.y. for Me_2CO $\sim$0.1 .

3. Vanadium sensitized photochemistry of HPC

There is an advantage using HPC in solar energy conversion and storage as hydrogen in that one compound works as photosensitizer, relay and catalyst. In reality, apart from sacrificial reagent it is one component system. However, the light of excitation is limited to near visible and UV light essentially below 400 nm, at least for tungstates. Hence these compounds while have been shown to be good relays they are not effective photosensitizers.

Our efforts have been to incorporate chromophores to move the absorption toward the visible light. The most obvious and least troublesome was to incorporate vanadium in HPC using the known mixed heteropoly vanadates known to absorb considerably in the visible.

We have selected $[PV_2^VMo_{10}O_{40}]^{5-}$ and $[PV_2^VW_{10}O_{40}]^{5-}$ designated for simplicity as $V_2^VMo_{10}$ and $V_2^VW_{10}$. These compounds were selected because of their stability vs other mixed HPC. Fig. VI show the spectra of the nonreduced mixed HPC relative to the corresponding molybdates and tungstates for comparison.

Without getting into details, to be discussed in a forthcoming publication, the following conclusions were drawn.

Incorporation of vanadium into HPC of molybdenum and tungsten has shifted the absorption to higher wave lengths thus allowing photochemistry to be performed well into the area of high intensity solar light; Fig. VI. This however was not without cost: a. Generally incorporation of vanadium into HPC, mainly tungstates, makes the reduced HPC less stable than the corresponding "classical" heteropoly blues of the Keggin (1:12) and Dawson (2:18) structures[1] . Similar instability has been observed during photolysis. However this may not be a serious

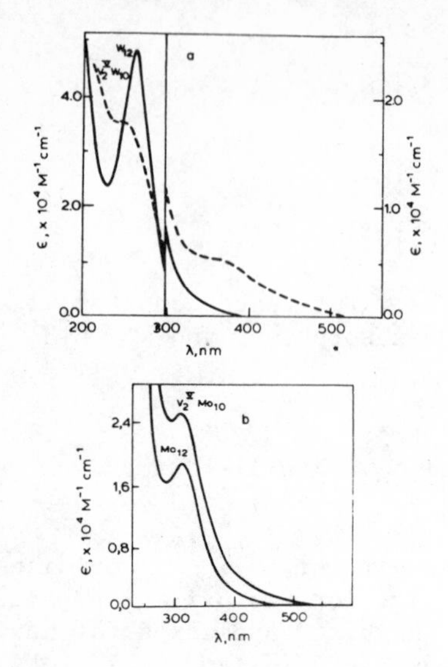

Fig. VI Spectra of nonreduced HPC showing the shift in absorbance toward the visible with incorporation of V^{5+} ions.

$W_{12} = PW_{12}O_{40}^{3-}$, $V_2^V W_{10} = PV_2W_{10}O_{40}^{5-}$

$Mo_{12} = PMo_{12}O_{40}^{3-}$, $V_2^V Mo_{10} = PV_2Mo_{10}O_{40}^{5-}$

problem, from a practical point of view, since catalysts of this kind operate effectively in thermal reactions[22]. An equilibrium is established between the mixed HPC and the decomposition products thus allowing the system to operate despite partial decomposition. b. Quantum yields of the one electron photoreduced product obtained at the $O\text{-}V^V$ CT bands are about four times less for tungstovanadates and about ten times less for molybdovanadates relative to quantum yields at the $O\text{-}W^{VI}$ and $O\text{-}Mo^{VI}$ CT hands respectively. c. Irradiation with a 400 nm cut off filter, i.e., in the area of $O\text{-}V^V$ CT bands would place a maximum of only two electrons on vanadium atoms. No further reduction that would place electrons on molybdenum or tungsten, as has been shown electrochemically, was observed on prolonged irradiation. Under these conditions no hydrogen was produced. As far as $V_2^V W_{10}$ was concerned, reduction did not proceed beyond the addition of two electrons on vanadium atoms even with UV irradiation and no hydrogen was produced under any circumstances. On the other hand $V_2^V Mo_{10}$ would be reduced past two electrons when photolyzed with $\lambda < 400$ nm ($O \rightarrow Mo^{VI}$ bands). Under these circumstances hydrogen was evolved with q.y. <0.01. It has been stated earlier that photoreduction of HPC in presence of at least alcohols proceeds mainly by hydrogen transfer through formation of free radicals which are known to reduce thermally HPC[14]. Similar stepwise reduction mechanism is suggested for the mixed HPC.

It has been reported that 1:12 tungstates produce hydrogen with an efficiency of about one molecule of hydrogen per 100 photons[21b]. No reasonable direct comparison with the hydrogen produced by those tungstates can be made because of the specificities that are presented with molybdovanadates: a. The part of the absorption in the visible area due to incorporation of vanadium is useless as far as hydrogen production is concerned. b. Instability involved with higher reduction products makes the picture not so clear.

4. Photogalvanic cells

Photogalvanic cells are based on the potential difference created between the photoreduced and oxidized forms of HPC. Whereas, photogalvanic cells are easy to construct with many chemicals that undergo photoredox reactions, regenerative photogalvanic cells limit the use of chemicals[23].

The ease of reoxidation of some HPC by oxygen regenerates the photogalvanic cells (in this case, strictly speaking, fuel cells) composed of HPC with various organic reagents which are consumed during the process. Other photogalvanic cells using molybdenum compounds but not HPC have been reported[24].

The electrolyte in both cells was composed of an HPC and organic reagent. The pH was kept low with $HClO_4$ to avoid base hydrolysis of HPC. Photolysis of the electrolyte in the "light" half cell turned the solution blue, whereas, the electrolyte in the "dark" half-cell remained colorless. Fig. VII shows the photopotential produced in a solution of PW_{12}^{3} in presence of isopropyl alcohol, whereas, the corresponding spectra of the photoreduce PW_{12}^{3-} could be easily followed in the visible; Fig. I.

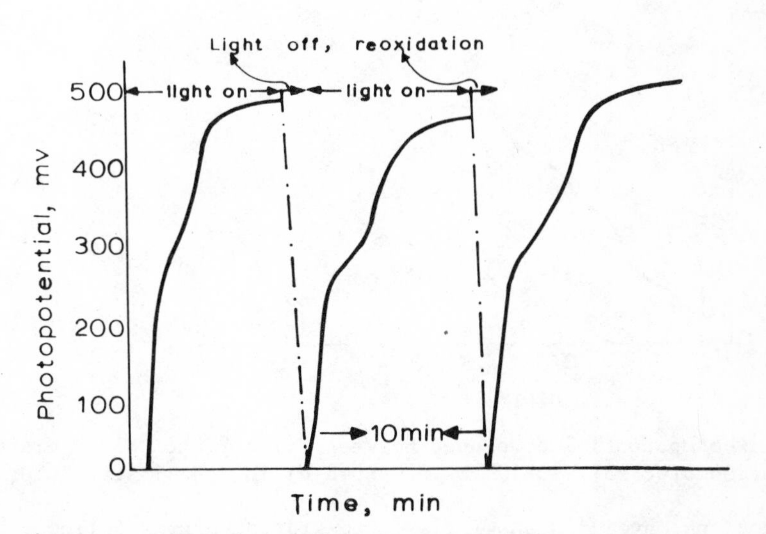

Fig. VII Photopotential cycles between "light" and "dark" electrolytes composed of PW_{12}^{3-}, 1×10^{-4}M; propan-2-ol, 0.8M, in 0.1M $HClO_4$, deaerated with Ar. Light source 150W Xe lamp; reoxidation by atmospheric oxygen[26]

The photopotential developed between the "light" and "dark" half-cells produced a current of ∿150 mA which dropped asymptotically to zero zero[26]. At the same time the dark electrolyte turned blue thus, indicating flow of electrons (reduction) to the electrolyte in the dark half-cell. Leaving the electrolyte open in the air brings about reoxidation to the colorless PW_{12}^{3-} form, thereby establishing the "dark" electrolyte to its original form. Further illumination restores the photopotential and the cycle is repeated. The cycles were repeated 20-30 times with no loss in activity of PW_{12}^{3-} except alcohol had to be replenished.

The photopotential develops in two steps corresponding to the stepwise addition of two electrons as shown in Fig. VII. This is to be expected for it is in reality a photoredox titration similar to the one obtained with reducing reagents, for instance, Cr^{2+} or polarography[25]. The redox steps in the two half-cells were followed by the characteristics spectra of the reduced forms.

The photochemical reactions are 1.2,3,4a and 4b whereas the electrochemical reaction is 4c. The electrolyte of the "dark" half-cell is reoxidized by oxygen, reaction 6, whereas, the electrolyte of the "light" cell is further photoreduced, reactions 1,2,3,4a and 4b.

Similar results have been obtained with $P_2Mo_{18}^{6-}$. Fig. VIII shows the potential developed with $P_2Mo_{18}^{6-}$ and isopropyl alcohol. The corresponding spectra are shown in Fig. IX.

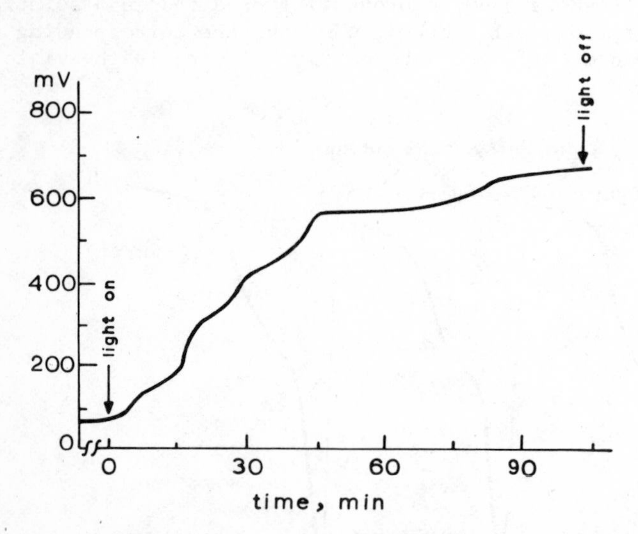

Fig. VIII. Photopotential developed between "light" and "dark" electrodes in solution of $P_2Mo_{18}^{6-}$ 5×10^{-5}M, propan-2-ol 1M, pH ∿1 with $HClO_4$.

Typical reactions, avoiding showing excited states, are as follows:

$$P_2Mo_{18}^{6-} + Me_2CHOH \xrightarrow{h\nu} P_2Mo_{18}^{8-} + Me_2CO + 2H^+ \quad (9)$$

$$P_2Mo_{18}^{8-} + Me_2CHOH \xrightarrow{h\nu} P_2Mo_{18}^{10-} + Me_2CO + 2H^+ \quad (10)$$

$$P_2Mo_{18}^{10-} + Me_2CHOH \xrightarrow{h\nu} P_2Mo_{18}^{12-} + Me_2CO + 2H \quad (11)$$

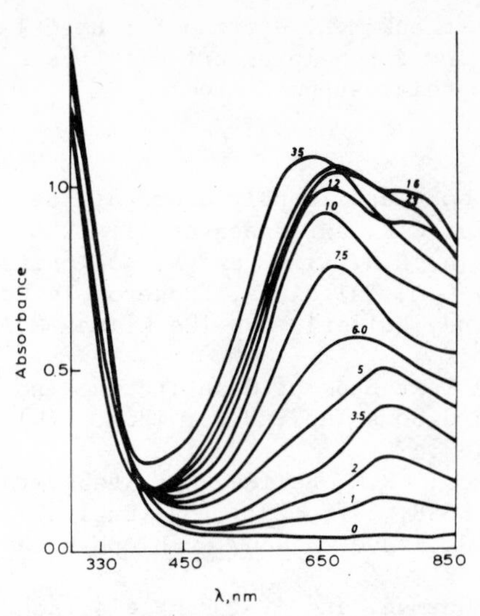

Fig. IX The corresponding spectra of photoreduced $P_2Mo_{18}^{6-}$ of Fig. VIII. Time is indicated on spectra.

Overall electrochemical reaction

$$P_2Mo_{18}^{12-} + P_2Mo_{18}^{6-} \rightarrow P_2Mo_{18}^{10-} + P_2Mo_{18}^{8-} \quad (12)$$

In the dark half-cell

$$P_2Mo_{18}^{10-} + 1/2\ O_2 + 2H^+ \rightarrow P_2Mo_{18}^{8-} + 2H_2O \quad (13)$$

(Protonation of molybdates has been omitted for clarity).
Reaction 13 brings back the electrolyte in the $P_2Mo_{18}^{8-}$ oxidation state; the oxidation state $P_2Mo_{18}^{6-}$ is not obtained with atmospheric oxygen. Photolysis regenerates the $P_2Mo_{18}^{12-}$ oxidation state, reaction 11, so that after the first cycle the electrochemical reaction is

$$P_2Mo_{18}^{12-} + P_2Mo_{18}^{8-} \rightarrow 2P_2Mo_{18}^{10-}$$

This cell, though, seems to be unstable when photoreduction proceeds to addition of 6-electrons.

In principle, the standard free energy involved with PW_{12}^{3}, calculated from $E_{1/2}$ values, is ∿24 KJ (reaction 4c), whereas, the corresponding value for $P_2Mo_{18}^{6-}$ is 33 KJ (reaction 12).

The results then, so far, may be summarized as follows: (a) HPC seem to have potential as electron carriers (relays) in hydrogen production; (b) the photocatalytic oxidation of organic compounds might very well be of practical use, considering the great variety of organic species that are oxidized, and the great number of HPC that exist. (c) Photogalvanic cells might be able to efficiently combine the photocatalytic oxidation of organic compounds with the concomitant production of H_2 and electricity.

Acknowledgements. We thank M.Z. Hoffman for useful discussions and comments and D. Arapoglou for helping out with the diagrams. EP and M.Z.H. acknowledge financial support from NATO.

References

1. M.T. Pope, "Heteropoly and Isopoly Oxometalates" in Inorganic Chemistry Concepts 8, Ed. C.K. Jørgensen et al. 1983.
2. See for instance: (a) C.H. Kline and V. Kollonitsch, Ind.Eng.Chem., 1965, 57, 53. (b) G.A. Tsigdinos, "Heteropoly Compounds of Molybdenum and Tungsten", Bulletin Cdb-12a Climax Molybdenum Co. of Michigan, USA.
3. (a) A.I. Vogel, "A Text Book of Quantitative Inorganic Analysis", Third Ed., Wiley and Sons Inc., N.Y., 1966. (b) H. Wu, J. Biol. Chem., 1920, 43, 189.
4. (a) See for instance: R.M. Bekina, A.F. Lebedeva, and V.A. Shuvalov, Dokl. Akad. Nauk. SSSR, 1976, 231, 739 (English translation). (b) S.P. Berg and S. Izava, Bioch. et Biophysica Acta, 1977, 460, 206.
5. R.J. Cotton, A.M. Guzman, and J.W. Rabalais, Acc.Chem.Res., 1978, 11, 170.
6. M. Rindl, S. African J. Sci., 1916, 11, 362.
7. L. Chalkley, J. Phys. Chem., 1952, 56, 1084.
8. See for instance: (a) A.A. Nemodruk and E.V. Bezrogova, Zhur. Anal. Khim., 1969, 24, 292 (English translation); (b) S.A. Morosanova, N.Ya. Kolli, and T.G. Kushnirenka, ibid. 1977, 32, 96; (c) M.N. Ptushkina, L.I. Lebedeva, ibid. 1979, 49, 1433.
9. (a) G.M. Varga, Jr., E. Papaconstantinou and M.T. Pope, Inorg. Chem. 1970, 9, 662; (b) E. Papaconstantinou and M.T. Pope, Inorg. Chem., 1970, 9, 667.
10. J.M. Fruchart, G. Herve, J.P. Launay, and K. Massart, J. Inorg. Nucl. Chem., 1976, 38, 1627.
11. H. Soo and M.T. Pope, Inorg. Chem., 1972, 11, 1441.
12. E. Papaconstantinou, D. Dimotikali and A. Politou, Inorg. Chim. Acta, 1980, 46, 155; b. D. Dimotikali and E. Papaconstantinou, ibid., 1984, 87, 177.
13. C.R. Hill, private communication.
14. E. Papaconstantinou, J. Chem. Soc., Faraday 1, 1982, 78, 2769.
15. E. Papaconstantinou, J. Chem. Soc., Chem. Commun., 1982, 12.
16. J.R. Darwent, J. Chem. Soc., Chem. Commun., 1982, 798.
17. M.D. Ward, J.F. Brajdil, and R.K. Grasselli, J. Phys. Chem., 1984, 88, 4210.
18. E. Papaconstantinou, D. Dimoticali, A. Ioannidis, and P. Argitis, J. Photochem., 1981, 17, 171.
19. See for instance: a. "Photogeneration of Hydrogen", Royal Institution Symposium, Eds A. Harriman and M.A. West, Academic Press, N.Y. 1982. b. "Energy Resources through Photochemistry and Catalysis", M. Gratzel Ed., Academic Press, N.Y., 1983.
20. E.N. Savinov, S.S. Saidkhanov, V.N. Parmon and K.I. Zamaraev, React. Kinet. Catal. Lett., 1981, 17, 407.
21. a. S.S. Saidkhanov, A.I. Kokorin, E.N. Savinov, A.I. Vokov, V.N. Parmon, J.Mol.Catal. 1983, 31, 365. b. A. Ioannidis and E. Papa-

constantinou, Inorg. Chem., 1985, 24, 439. c. R. Akid and J.R. Darwent, J. Chem. Soc. Dalton Trans., 1985, 395.
22. I.V. Koznevnikov and K.I. Matveev, Russ. Chem. Rev., 1982, 51, 1075.
23. See for instance: W.J. Albery and A.W. Foulds, J. Photochem., 1981, 15, 321 and references therein.
24. a. M.H. Navidi, H.G. Brittain and A. Hetter, Science, 1970, 169, 980. (b) T. Yamase and T. Ikawa, Inorg. Chim. Acta, 1979, 37L, 529.
25. M.T. Pope and G.M. Varga Jr., Inorg. Chem., 1966, 5, 1249.
26. E. Papaconstantinou and A. Ioannidis, Inorg. Chim. Acta, 1983, 75, 235.

PHOTOFORMATION OF HYDROGEN IN LIQUID WATER IN THE PRESENCE OF Pt/TiO_2 CATALYST AND ORGANIC IMPURITIES

I. Ait Ichou, D. Bianchi, M. Formenti and S.J. Teichner
U.A. 231 (CNRS), University Claude Bernard (Lyon I)
69622 Villeurbanne Cedex
France.

SUMMARY : The photoformation of hydrogen from liquid water in the presence of Pt/TiO_2 catalyst is not a photocatalytic reaction of water splitting. The gazeous products formed are composed of hydrogen and CO_2. Oxygen is not detected. The formation of CO_2 stems out from the presence in water of an organic "sacrificial" substance. In the absence of an organic impurity the buildup of the peroxydic species on the surface of TiO_2 limits the reaction of photoformation of hydrogen. It is suggested that this hydrogen is evolved through the photolysis of surface OH groups of TiO_2.

INTRODUCTION

The catalytic photodecomposition of water in the presence of TiO_2 and U.V. light is questioned for many reasons. The amount of hydrogen collected is often correlated with the amount of surface OH groups on TiO_2 (1, 2) and the reaction proceeds only to a very limited extent, oxygen not being detected. On prereduced TiO_2 only hydrogen gas is collected and its amount matches the amount of Ti^{3+} ions which are photooxidized by H_2O, the reaction not being catalytic (3). No reaction is observed on preoxidized dehydroxylated TiO_2 (4). A metal, like Pt, is often associated with TiO_2 as it is supposed to collect hydrogen adsorbed species and release them as a dihydrogen gas (reverse spill-over)(5). A prereduced Pt/TiO_2 system produces in the presence of water vapour and U.V. light a limited amount of hydrogen which is again correlated with the reoxidation of titania by H_2O, as for the system without Pt (3). In the absence of water vapour no hydrogen is photo-released which seems to show that OH groups (if they remain on a sample reduced at 700°C) are not photolyzed. In the presence of liquid water hydrogen is released in a higher amount but the reaction is still limited by the thermal back reaction with oxygen on Pt(3). The use of organic scavengers of oxygen, eventually formed by photosplitting of H_2O, was proposed ("sacrificial" system)(6), where the products of the photoreaction are H_2 and mainly CO_2 (with some other products of the destruction of the organic substance). It has been shown in the previous

E. Pelizzetti and N. Serpone (eds.), Homogeneous and Heterogeneous Photocatalysis, 433–443.

work (7) that an alcohol like isopropanol is photodehydrogenated catalytically in the vapour phase in the presence of Pt/TiO_2. As the amount of gazeous hydrogen released (together with acetone) is not increased in the presence of water vapour it was concluded that, at least, in this gas-solid photocatalytic system the alcohol is not a "sacrificial" compound. In the liquid phase isopropanol is photodecomposed even without the Pt/TiO_2 catalyst (7). In the presence of Pt/TiO_2 the rate of reaction is increased by a factor 2. For the same Pt/TiO_2 catalyst a limited photodecomposition of pure liquid water was observed through the build-up of a pressure of gas collected in a batch reactor, equipped with a manometer filled with dibutylphtalate, in order to increase the sensitivity. However the composition of the gas phase was not known at that time. In particular, for various mixtures of isopropanol with water it was not established if the alcohol was a "sacrificial" agent or if it was merely dehydrogenated, because of the lack of the analysis of the gas phase (CO_2 content). In the present work this problem is carefully examined. It is shown that in all cases envisaged (even for "pure" water) a photoproduction of CO_2 is always recorded. Oxygen is never detected. The photorelease of hydrogen may be also correlated with the scavenging of oxygen by TiO_2 under the form of a peroxo species if the amount of the organic "sacrificial" substance is very small.

EXPERIMENTAL

A batch, constant volume, reactor with immersed mercury lamp (Hanovia, 100 W), screened by water jacket, is connected to a manometer filled either with mercury or with dibutylphtalate or water, in order to increase the sensitivity in the measurement of gas pressure. The catalyst (50 mg) is dispersed in the liquid medium (60 cm^3) by a continuous magnetic stiring. The detailed method of the preparation, the properties and the labelling of Pt/TiO_2 catalysts are fully described elsewhere (8). After the impregnation of TiO_2 (P25, Degussa) by a solution of H_2PtCl_6 the system is dried, calcined in O_2 at 400°C and reduced by H_2 at 400°C. The amount of Pt in the catalyst is 1 % (by weight) and its dispersion is of the order of 50 %. Before the run the liquid in the photoreactor is degassed at room temperature (boiled at reduced pressure) during 30 min. in order to evacuate the dissolved oxygen and carbon dioxide. Bi-distilled water is used throughout. The composition of the gaz phase (completed with He to the atmospheric pressure in a sampling device) is obtained by mass spectroscopy.

RESULTS AND DISCUSSION

1. Experiments with dibutylphtalate

It has been shown previously (7) that "pure" liquid water at 25°C is photodecomposed into a gas of unknown composition in the presence of Pt/TiO_2 catalyst, the manometer being filled with dibutylphtalate. No gaz is evolved in the presence of pure TiO_2 (without Pt) or of Pt/Al_2O_3 catalyst. The unknown photoreaction requires therefore a simultaneous presence of photoactive TiO_2 and of the precious metal. The results obtained during the present investigation, in the photodecomposition of water, with the dibutylphtalate in the manometer, are shown on fig. 1, giving the pressure (in mm of dibutylphtalate, 1 mm pressure of dibutylphtalate in 200 cm^3 volume represents 0.02 cm^3 of gas S.T.P. or 9 x 10^{-4} mmole) on the ordinate and the time on stream on abscissa. The total amount of gas produced after 1680 min is therefore of the order of 10 cm^3 (500 mm pressure of dibutylphtalate). The reaction does not seem to be terminated. The composition of the gas phase, analyzed by mass spectroscopy, after sampling with He, and the ratio R = H_2/CO_2 are given in table I.

TABLE I

Experiment	Manometric liquid	Liquid in the reactor	H_2 (%)	CO_2 (%)	R = H_2/CO_2
Fig. 1	dibutyl-phtalate	60 cm^3 H_2O	5.62	1.52	3.7
A	Hg + 1 drop of dibutyl-phtalate	60 cm^3 H_2O	3.56	0.72	4.9
Fig. 2a	Hg	60 cm^3 H_2O + 0.3 g of o-phtalic acid	1.86	3.13	0.6
Fig. 2b	Hg	60 cm^3 H_2O + 0.5 cm^3 of acetic acid	2.44	3.52	0.7
Fig. 3b	Hg	60 cm^3 H_2O + 0.02 cm^3 of acetone	4.2	1.55	2.7
Fig. 3c	Hg	60 cm^3 H_2O "pure"	2.7	0.4	6.7
Fig. 4b	n-heptane	60 cm^3 n-heptane + 0.5 cm^3 water	1.9	0.2	9.5
Fig. 4c	H_2O	60 cm^3 H_2O "pure	2.0	0.15	13.0

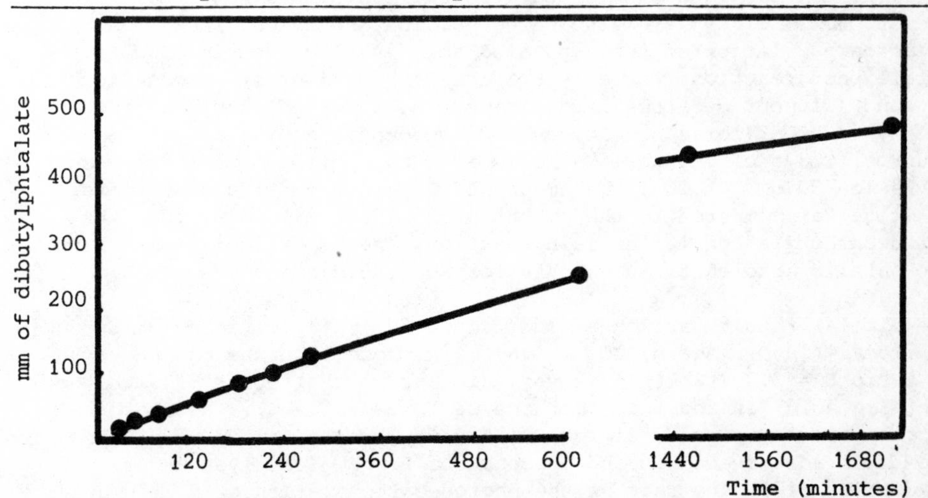

Figure 1. Photodecomposition of water in the presence of Pt/TiO_2 catalyst (50 mg). The pressure is measured with a dibutylphtalate manometer.

No oxygen was detected in the gas phase. The high proportion of CO_2 (about 20 % or 2 cm^3 of generated gas) seems to indicate that some "sacrificial" substance is present in the system. Besides organic impurities and CO_2, eventually dissolved in water, the only organic substance present in the photoreactor is the vapour pressure of dibutylphtalate. The solubility in water of this compound is (at 25°C) of the order of 0.04 g in 100 cm^3 of H_2O (9). Assuming a saturated solution, 0.024 g of dibutylphtalate in 60 cm^3 of H_2O, completely oxidized, would release 30 cm^3 of CO_2. It appears therefore that in these conditions the photosplitting of water is mainly due to the presence of a "sacrificial" organic compound.

In a second experiment (A, table I) the manometer was cleaned, filled with mercury and only a small drop of dibutylphatalate was deposited on the surface of mercury. After the photoreaction in the same conditions, during 15 hours, the ratio R was found equal to 4.9 (table I). This result shows that the vapour pressure of a small amount of the organic liquid still allows the build-up of conditions for the presence of the "sacrificial" organic substance in liquid water.

2. Experiments with other organic compounds

The dibutylphtalate could be hydrolyzed in the conditions of the photoreaction. Also it is of interest to study the behaviour of the o-phtalic acid as a "sacrificial" compound dissolved directly in water, with mercury as a manometric liquid. Figure 2a shows the build-up of the pressure (in 135 cm^3 volume) when 0.3 g of o-phtalic acid are dissolved in 60 cm^3 of liquid water in the photoreactor. After 2 hours of reaction 0.8 cm^3 of gas are released which is an amount comparable to that observed after 2 h in the experiment of Fig. 1 with dibutylphalate. The ratio R = 0.6 is now inversed (table I) as CO_2 is now a main reaction product. Traces of CH_4 are also detected. This behaviour shows that the photodecarboxylation of the o-phtalic acid is also contributing to the overall photoreaction. For a "sacrificial" behaviour of an organic compound (without decarboxylation) every H_2 produced should be accompanied by 1/2 CO_2 (for a non-oxygenated compound), giving R = 2. The amount of 0.3 g of o-phtalic acid is able to provide by a total decarboxylation 80 cm^3 of CO_2. The amount of CO_2 (0.5 cm^3) released after 2 h (fig. 2a) compared to the amount of H_2 (0.3 cm^3) shows that the photodecarboxylation is the main reaction. The "sacrificial" behaviour of o-phtalic acid contributes only for about 0.3 cm^3 of H_2.

Similar results are observed (figure 2b) with acetic acid (0.5 cm^3 in 60 cm^3 of H_2O). Again, CO_2 is the main product in the gas phase and the ratio R = 0.7 (table I) is very similar to that observed with the o-phtalic acid. In addition, methane is formed (2.7 %). The photoreaction of acetic acid, giving CO_2 and CH_4, was observed even with pure liquid acetic acid in the presence of Pt/TiO_2 catalyst under irradiation (10). The rate of the photodecarboxylation of acetic acid (fig. 2b) is however much higher than for o-phtalic acid (fig. 2a) and is practically proportional to the concentration of the acid (fig. 2c).

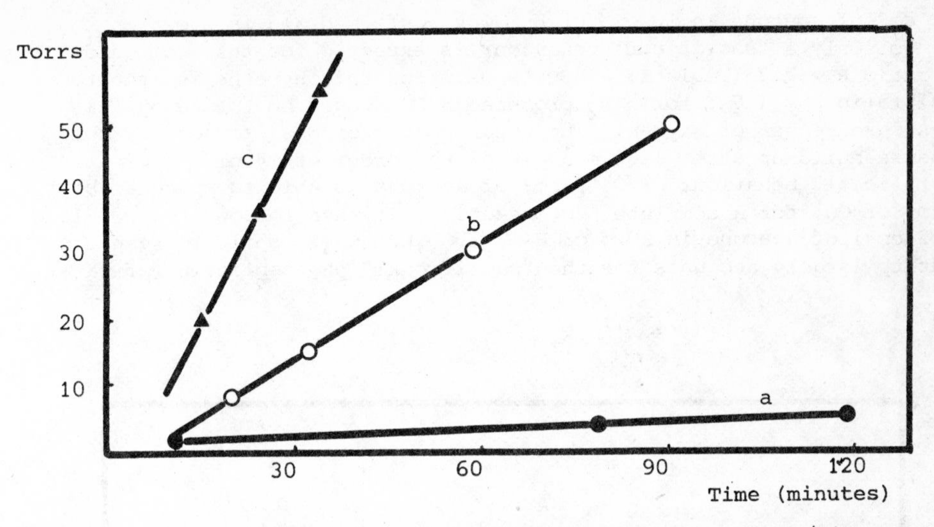

Figure 2. Photodecomposition of water in the presence of Pt/TiO_2 catalyst (50 mg). The pressure is measured with a mercury manometer. a) Water (60 cm^3) containing 0.3 g of o-phtalic acid, b) Water (60 cm^3) containing 0.5 cm^3 of acetic acid, c) Water (60 cm^3) containing 1 cm^3 of acetic acid.

For 1 cm^3 of the acid in 60 cm^3 of H_2O the slope is practically twice that found for 0.5 cm^3 of acid. The formation of H_2 shows however that a fraction of acetic acid is consumed in a "sacrificial" reaction, otherwise only CH_4 and CO_2 would be formed.

Coming back to the experiment with dibutylphtalate it may be observed that in the case of hydrolysis of this compound the o-phtalic acid formed would contribute to the production of CO_2 by decarboxylation whereas butylalcohol would be eventually photodehydrogenated (8) and then it would follow a "sacrificial" behaviour. Also, in connection with the previous work concerning the photocatalytic dehydrogenation of isopropanol in the gaz phase (8) experiments with a small amount (0.02 cm^3) of isopropanol in 60 cm^3 of H_2O were performed. Pure liquid isopropanol is easily photodehydrogenated in the presence of Pt/TiO_2 but may be also photodecomposed without this catalyst (7). Figure 3a shows the results. The pressure build-up (70 torrs) after 1200 min. corresponds to the formation of 11.4 cm^3 of gas (in 135 cm^3 volume). A mere photodehydrogenation of 0.02 cm^3 of isopropanol would produce 6 cm^3 of H_2. The excess of the gaz collected accounts for the "sacrificial" behaviour of isopropanol. Indeed, a ratio R close to 3 is found. But if CO_2 is well observed at the end of the reaction (1200 min.) no traces of this gas (only H_2) are detected after 60 min. This behaviour shows that the catalytic photo-dehydrogenation of isopropanol proceeds easier than its "sacrificial" photooxidation. As the product of the dehydrogenation of isopropanol is acetone, a similar experiment with

0.02 cm^3 of acetone in 60 cm^3 of H_2O was performed and is shown on fig. 3b. Only a "sacrificial" behaviour is expected for this compounds. The ratio R = 2.7 (Table I) accounts well for this hypothesis (theoretical ratio R = 1.7 ; for 5 H_2 produced 3 CO_2 would be formed by this organic scavenger of oxygen). The total amount of gas produced by the pressure build-up after 1200 min. is of the order of 14 cm^3. The "sacrificial" behaviour of 0.02 cm^3 of acetone is able to produce about 30 cm^3 of CO_2 for a complete photoreaction. It then follows that 0.015 g (0.02 cm^3) of acetone in 60 g of H_2O, or 0.025 % (25 ppm) of organic impurity largely accounts for the "sacrificial" photosplitting of water.

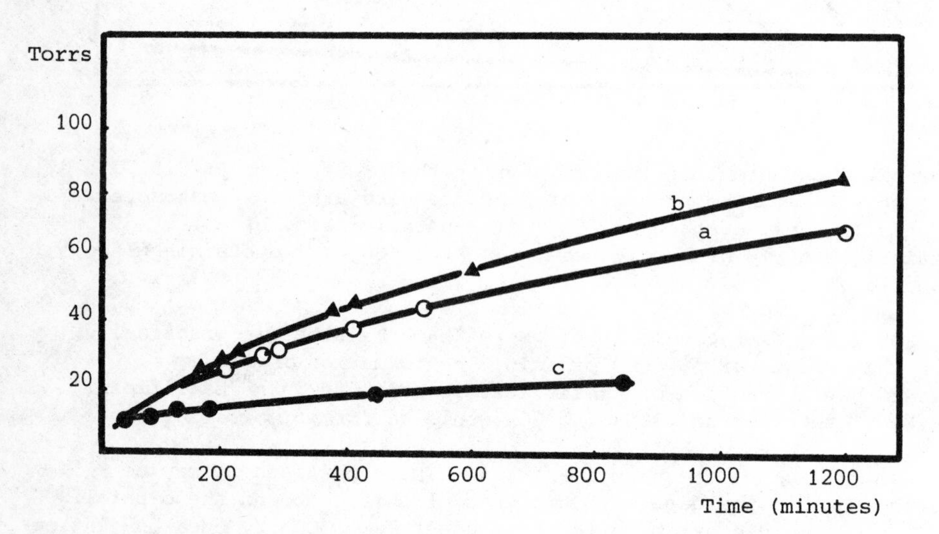

Figure 3. Photodecomposition of water in the presence of Pt/TiO_2 catalyst (50 mg). The pressure is measured with a mercury manometer a) Water (60 cm^3) containing 0.02 cm^3 of isopropanol, b) Water (60 cm^3) containing 0.02 cm^3 of acetone, c) Water "pure".

3. Experiments with "pure" water

Pure water (deoxygenated and decarbonated by boiling under reduced pressure) was submitted to the photoreaction in the presence of 50 mg of Pt/TiO_2 catalyst, the manometer being filled with mercury. The results are given in fig. 3c. The pressure build-up during 600 min. is very similar to that observed for water with dibutylphtalate in the manometer (fig. 1). The amount of the gas collected after 850 min. (in 135 cm^3 volume) is of the order of 4 cm^3. The ratio R = 6.7 (table I) is the highest now observed. It has been pointed out previously that for a purely "sacrificial" behaviour (no decarboxylation, no oxygen in the organic scavenger), this ratio should be of the order of 2. Clearly, a net hydrogen excess is here produced with respect to any "sacrificial" impurity eventually present in the system. The amount of CO_2 formed (in 4 cm^3 of total gas) is of the order of 0.5 cm^3. This represents an organic impurity level in 60 g of H_2O of the order of 0.45 ppm of carbon. It is obvious that, unless a very severe experimental conditions of purity are established, any photosplitting of water is suspected to occur, in part, in "sacrificial" conditions. In the present case hydrogen gaz excess is observed. As oxygen is not detected in the gas phase (it should exceed the proportion of CO_2) it is assumed that this oxygen is build-up as peroxo species on the surface of TiO_2 (11). In the case of TiO_2 (P25 Degussa) the density of titanium ions is 2/nm^2 (12). With 50 mg of TiO_2 present in the reactor (specific surface area 50 m^2/g) 5 x 10^{18} titanium surface ions are available to build a peroxo species of the type Ti-O-O-Ti (11). Assuming that a pair of Ti ions is able to fix two oxygen atoms as a peroxo species 5 x 10^{18} oxygen atoms would be fixed and would correspond to the escape of 5 x 10^{18} molecules of H_2 or 2 x 10^{-1} cm^3 of H_2. Grätzel (11) found that this saturation limit for the fixation of oxygen as peroxo species on TiO_2 surface is experimentally well exceeded. Also this author prefers a limit for oxygen uptake corresponding to 22 peroxide molecules per nm^2 (13). This figure sets a limit for hydrogen gas evolution of 2.2 cm^3. Out of 4 cm^3 of total gas collected after 850 min. (fig. 3c) CO_2 accounts for 0.5 cm^3 (see above) and H_2 for 3.5 cm^3. This amount of H_2 is close to the previous limit (2.2 cm^3) based on the saturation of the surface of TiO_2 by peroxo species.

Summing up, during the photoreaction carried out on water (as pure as possible) gaseous hydrogen is evolved i) in part as a result of the "sacrificial" reaction of some organic impurity with oxygen, ii) in remaining part as a result of the scavenging of oxygen by TiO_2, under the form of a peroxo species. It does not seem therefore that the photosplitting of water is of a catalytic nature. The limit for H_2 evolution is set by the level of organic impurity in water and by the saturation limit of peroxo species on the surface of TiO_2. These species are probably not formed (or they are destroyed by the organic scavenger) as long as the organic impurity level exceeds this saturation limit. This is the case of all experiments reported above with various organic substances in the system.

4. Experiments with n-heptane

If some organic compound is unable to exhibit a "sacrificial" behaviour, the gas formed by the irradiation of Pt/TiO_2 catalyst immersed in this organic compound, in the absence of water, would be essentially H_2 and would come from the photolysis of the surface OH groups of TiO_2. Such a behaviour seems to be encountered with n-heptane. Figure 4a shows the pressure build-up in the photoreactor (135 cm^3 volume) containing 60 cm^3 of n-heptane with 50 mg of Pt/TiO_2 catalyst, and is measured with the manometer filled with n-heptane. In about 400 minutes 0.4 cm^3 of gas are collected. This small amount does not allow the analysis of the gas phase with enough confidence. However, assuming that only H_2 is evolved (see below) by photolysis of OH groups of TiO_2, the following calculations can be made. Assuming that the surface of TiO_2 is fully hydroxylated (9 OH/nm^2) and that the photolysis of 2 OH gives out 1 molecule of H_2, 50 mg of TiO_2 (50 m^2/g) represent 2 x 10^{19} OH which by photolysis would give out 1.8 x 10^{-5} mole of H_2 or 0.4 cm^3 of H_2. In fact, after about 400 minutes, this amount is collected (fig. 4a). The reaction is however not terminated. By assuming that the limit of the pressure build-up would be practically achieved after 1400 minutes and would be of the order of 5 torr (2.5 torr in 400 minutes), this would set up a limit for the volume (S.T.P.) of the gas collected of the order of 0.8 cm^3 and would correspond to the photolysis of 18 OH/nm^2 . It appears therefore that the amount of the gas evolved in this experiment matches quite well a reasonable density of OH groups on the surface of a TiO_2 support.

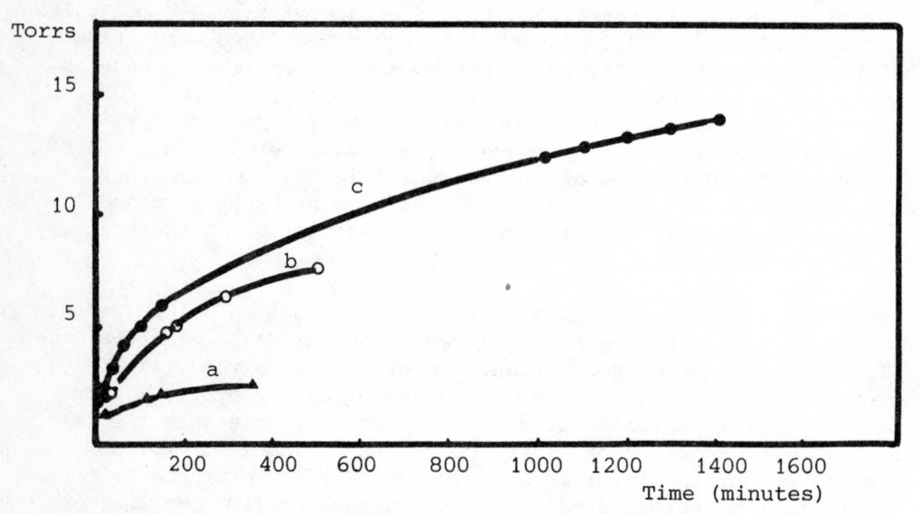

Figure 4. Photoformation of the gas phase in the presence of Pt/TiO_2 catalyst (50 mg). a) The photoreactor contains 60 cm^3 of n-heptane. The pressure is measured with a n-heptane manometer. b) The photoreactor contains 60 cm^3 of n-heptane and 0.5 cm^3 of water. The pressure is measured with a n-heptane manometer. c) The photoreactor contains 60 cm^3 of "pure" water. The pressure is measured with a water manometer.

In the same conditions as for the experiment of figure 4a a small amount of water (0.5 cm^3) was dispersed in n-heptane. The pressure build-up, shown on figure 4b, is increased only by a factor 3. The ratio R found is 9.5 (table I), whereas the amount of the gas collected after about 425 minutes is 1.2 cm^3. The amount of CO_2 is therefore only of the order of 0.1 cm^3, the essential part of the gas (1.1 cm^3) being formed by H_2. Also it is possible to assume that in this experiment, as in the previous one (with pure n-heptane), the paraffin is not a "sacrificial" compound but may contain some "sacrificial" impurity. The reaction of figure 4b does not seem to be terminated after 425 minutes. A complete dissociation of 0.5 cm^3 of H_2O would produce 622 cm^3 of H_2

and about 310 cm^3 of CO_2 from n-heptane. This is not the trend shown by curve 4b. The amount of H_2 evolved after 425 minutes (1.1 cm^3) and the non "sacrificial" behaviour of n-heptane are again easily correlated with the photolysis of about 20 OH/nm^2, but not with the photolysis of the first fraction (after 425 minutes) of 0.5 cm^3 of water present.

A decisive argument in favour of this mechanism (photolysis of OH's) is given by the experiment of figure 4c. The corresponding curve exhibits almost the same trend as the curve of figure 4b. It represents however the pressure build-up observed when the photoreactor is filled with 60 cm^3 of pure water, containing 50 mg of Pt/TiO_2, the manometer being also filled with water. (For conveniance, the pressure on the ordinate of figure 4 is given in torrs). The ratio R found is 13 (table I) which is the higher one recorded. The volume (S.T.P.) of the gas collected after 1400 minutes is of the order of 2.2 cm^3, value which is close to that observed (3.5 cm^3) in the experiment of figure 3c, also performed with "pure" water. The conditions of the purity of water are improved in the experiment of figure 4c as R = 13 instead of R = 6.7 for the experiment of figure 3c. Because of the smaller level of the organic "sacrificial" impurity for the experiment of figure 4c the total amount of gas (2.2 cm^3) collected after 1400 minutes, may be smaller than that (4 cm^3) collected after 850 minutes for the experiment of figure 3c, due to the difference in the kinetics of the "sacrificial" and not "sacrificial" photolysis of OH's. The similar trend of curve 4c to that of curve 4b, despite the fact that the conditions for the eventual photolysis of water are tremendously improved going from 0.5 cm^3 of water in n-heptane (figure 4b) to 60 cm^3 of pure water (figure 4c), tends to confirm that the evolution of H_2, even in the presence of pure water excess (almost in the absence of the "sacrificial" impurity) is correlated with the photolysis of OH's on the surface of the rehydroxylated (in water) TiO_2. The organic scavenger of oxygen contributes only to a minor fraction (0.1 cm^3) of hydrogen collected, the main part of this gaz (2.1 cm^3) being evolved by the scavenging action towards oxygen of TiO_2 itself which forms a peroxo species (11).

5. The formation of the peroxydic species on the surface of TiO_2

The experimental conditions for the detection of the peroxydic species through the formation of a blue colour with o-tolidine are described in detail by M. Grätzel (11, 14). As the Pt/TiO_2 catalyst used by this author is obtained by the photoreduction of H_2PtCl_6 impregnated TiO_2 whereas in the present case the catalyst is obtained by the thermal reduction of the precursor in H_2 at 400°C (5), the formation of the peroxy species was checked for the present catalyst. With a freshly reduced (H_2 at 400°C) Pt/TiO_2 catalyst (50 mg) immersed in water (60 cm^3) a blue colour is developped in the presence of o-tolidine even without any irradiation. No such colour was observed by M. Grätzel before the irradiation of the H_2PtCl_6 impregnated TiO_2. In the present case the development of the blue colour without irradiation may stem out from the following interactions. At 400°C, in H_2, Pt containing

TiO_2 may be partially reduced (15). When this reduced catalyst is handled in air, O_2^- ions are fixed on its surface (16). These ions exhibit peroxydic character and therefore the blue colour observed with o-tolidine in the absence of irradiation may be attributed to these ions. Also the blank with o-tolidine, in the absence of irradiation, was performed on Pt/TiO_2 catalyst dispersed in 60 cm^3 of water, after bubbling hydrogen, at room temperature, during 30 minutes, through the dispersion. No blue colour was then observed with o-tolidine. The same blank is achieved if the dispersion of the catalyst in water is evacuated under reduced pressure. A new batch of the dispersion of Pt/TiO_2 catalyst in 60 cm^3 of pure water was treated by one of the previous methods of removing O_2^- ions from the surface of TiO_2, and was irradiated during 12 h. The blue intense colour was then observed through the addition of o-tolidine. The conditions for the observation (and the build-up) of the peroxydic species are therefore the same for Pt/TiO_2 catalyst prereduced in H_2 as for the photoreduced Pt/TiO_2 catalyst.

No blue colour with o-tolidine was observed after irradiation in the conditions of the experiments of figure 4a and b (in the presence of 60 cm^3 of n-heptane, pure or with 0.5 cm^3 of H_2O). This negative behaviour of o-tolidine may be due to the requirement of a water solvent for the development of the blue colour.

Conclusions

The photoformation of H_2 in pure liquid water in the presence of Pt/TiO_2 catalyst is a very limited reaction and it does not stem out from the photodecomposition of water. Hydrogen gas is evolved by the photolysis of TiO_2 surface or sub-surface OH groups. This reaction requires the presence of the photonic activator (TiO_2, because Pt/Al_2O_3 is inactive) and of the precious metal (pure TiO_2 is inactive). Oxygen remains on the surface of TiO_2 as it is never detected in the gas phase and is under the form of peroxydic species detected by the blue colour reaction with o-tolidine. These peroxo species are able to oxidize into CO_2 an organic "sacrificial" compound. They are therefore destroyed, the surface of TiO_2 is rehydroxylated in water and the formation of H_2 and CO_2 can continue until the "sacrificial" compound is exhausted (7). In the absence of the "sacrificial" organic substance the limit for the evolution of H_2 is set by the saturation conditions of the surface of TiO_2 towards the build-up of the peroxo species.

References

(1) H. van Damme and W.K. Hall, J. Amer. Chem. Soc., 1979, 101, 4373.
(2) R.S. Magliozzo and A.I. Krasna, Photochem. Photobiol., 1983, 38, 15.
(3) S. Sato and J.M. White, Chem. Phys. Lett., 1980, 72, 83.
(4) T. Sakata, T. Kawai, T. Koiso and M. Okuyama, Adv. Hydrog. Ener., 1982, 2, 773.
(5) I. Ait-Ichou, M. Formenti and S.J. Teichner in "Spillover of Adsorbed Species", G.M. Pajonk, S.J. Teichner and J.E. Germain Eds., Elsevier Sci. Pub. Co., Amsterdam, 1983, p. 63

(6) T. Sakata and T. Kawai, Nouv. J. Chimie, 1981, 5, 279.
(7) I. Ait-Ichou, M. Formenti and S.J. Teichner in "Catalysis on the Energy Scene", S. Kaliaguine and A. Mahay Eds., Elsevier Sci. Pub. Co., Amsterdam 1984, p. 297.
(8) I. Ait-Ichou, M. Formenti, B. Pommier and S.J. Teichner, J. Catal., 1985, 91, 293.
(9) Handbook of Chemistry and Physics, 43d Edition.
(10) B. Kreutler and A.J. Bard, J. Amer. Chem. Soc., 1978, 100, 2239, 5983.
(11) M. Grätzel, Personal communication
J. Kiwi and C. Morrison, J. Phys. Chem., 1984,88, 6146.
(12) H.P. Bochum, Disc. Farad. Soc., 1971, 52, 264.
(13) J.R. Harbour, J. Tromp and M.L. Hair, Canad. J. Chem., 1985, 63, 204.
(14) Boe Gu, J. Kiwi and M. Grätzel, Nouv. J. Chimie, in press.
(15) T. Huizinga and R. Prins, J. Phys. Chem., 1981, 85, 2156.
(16) M. Formenti, H. Courbon, F. Juillet, A. Lissatchenko, J.R. Martin, P. Mériaudeau and S.J. Teichner, J. Vac. Sci. Techn., 1972, 9, 947.

SURFACE PROPERTIES OF CATALYSTS. IRON AND ITS OXIDES; SURFACE CHEMISTRY, PHOTOCHEMISTRY AND CATALYSIS

G. A. Somorjai and M. Salmeron
Materials and Molecular Research Division
Lawrence Berkeley Laboratory and Department of Chemistry
University of California
Berkeley, California 94720 USA

ABSTRACT. The results of selected surface science studies of catalytic processes are described. Our conclusions, drawn from many such studies, are that reactions which occur over metal based catalysts can be divided into three classes: 1) those occurring on the bare metal surface, 2) those occurring on active overlayers bound to the metal, and 3) those occurring on coadsorbate modified metal surfaces. Case histories are given describing catalytic processes fitting each category. The surface properties of iron and its oxides are then described. We review the photochemical studies using iron oxides that are aimed at the photocatalytic dissociation of water.

1. INTRODUCTION

As originally defined by Berzelius, a catalyst is a material that increases the rate of a chemical reaction without itself being either consumed or changed (1). While the increase of reaction rates is often desiderable, control of reaction selectivity or product distribution is also a key ingredient in a successful catalytic process. The catalyst selects among available reaction channels (thermodynamically allows) to yield desirable products while blocking the formation of undesirable species. In many cases involving solid catalysts these pathways consist of an adsorption step, followed by a surface reaction and finally desorption of the product. The understanding of such processes rests upon our molecular level understanding of the adsorption/desorption reactions, the structure of catalytic surfaces and the coordination of reaction intermediates on these surfaces.

The methodology used in the study of catalysis has changed drastically over the past 15 years. Classical methods required high surface area samples ($>10m^2$) to generate measurable amounts of product. Presently, catalyst samples with surface areas as low as 1 cm^2 are sufficient for study by a number of techniques under ultra-high vacuum (UHV) conditions. Such small samples can have a high degree of homogeneity and can be obtained in single crystal form. The use of such single crystal as model catalysts has greatly accelerated the rate of development of catalysis science.

E. Pelizzetti and N. Serpone (eds.), Homogeneous and Heterogeneous Photocatalysis, 445–477.

Many techniques have been developed for the direct study of surfaces and have been applied to catalysis. A partial list of such techniques is presented in Table 1. Catalysts studied include platinum for hydrocarbon conversion (2), iron and rhenium for ammonia synthesis (3, 4), and silver for the partial oxidation of ethylene (5). The hydrogenation of carbon monoxide has been studied over a large number of metal surfaces including iron (6), nickel (7), ruthenium (8) and molybdenum (9). The oxidation of carbon monoxide, shown to have oscillatory behavior, has been studied over platinum catalysts (10, 11). Finally, platinum and rhodium have been used to catalyze the hydrogenation of ethylene (12) and molybdenum to hydrodesulfurize thiophene to butenes (13). These investigations have lead to an increased understanding of catalytic mechanisms and the role of surface structure in catalysis.

The atomic scale characterization of catalytic surfaces has often been performed under UHV conditions (10^{-10}–10^{-9}) torr. These conditions are necessitated by the need to keep sample surfaces clean of contamination and the need to use electron or ion spectroscopies. At pressures of 10^{-6} torr a clean surface would be completely covered by an adsorbed monolayer of the background gas in a period of about 1 second. In order to keep surfaces clean for periods of hours the UHV conditions are necessary. Ions and electrons interact strongly with condensed matter and thus are ideal for the study of surfaces because of their short penetration depths. The use of these probes also requires working in vacuum.

Much of the criticism of a UHV technique as viable tools for the study of catalysis has been based on the argument that chemisorbed species formed under these conditions may bear no relationship to those formed at industrial catalytic pressures. In order to overcome this difficulty, we have designed an apparatus for combined UHV surface analysis and high pressure catalytic reaction studies over small surface area samples (14). In addition to being equipped with a number of surface analytical techniques, this UHV chamber has an internal environmental cell that can enclose the sample, isolating it from vacuum and forming part of a closed loop batch reaction that can be pressurized to about 30 atm (Fig. 1). In this manner we can characterize the catalytic surface before and after exposure to reaction conditions and attempt to correlate surface characteristics with the kinetic parameters (rate, product distribution, activation energy) of the catalytic reaction.

The surface science studies of catalytic reactions, that have been performed to date indicate that there are three types of metal based catalytic systems:

1. The first encompasses reactions that occur directly on a bare metal surface. These catalytic processes are characterized by their sensitivity to the atomic structure of the surface.
2. The second class includes reactions occurring on an overlayer that is irreversibly bound to the metal surface. Such reactions involve adsorbates that are weakly bound to this overlayer and thus, are insensitive to the structure of the underlying metal surface.
3. Finally, reactions on coadsorbate modified surfaces fall into the third category. These catalysts contain additives that alter the bonding of reactants to the surface, or block specific reactions sites, **altering reaction pathways.**

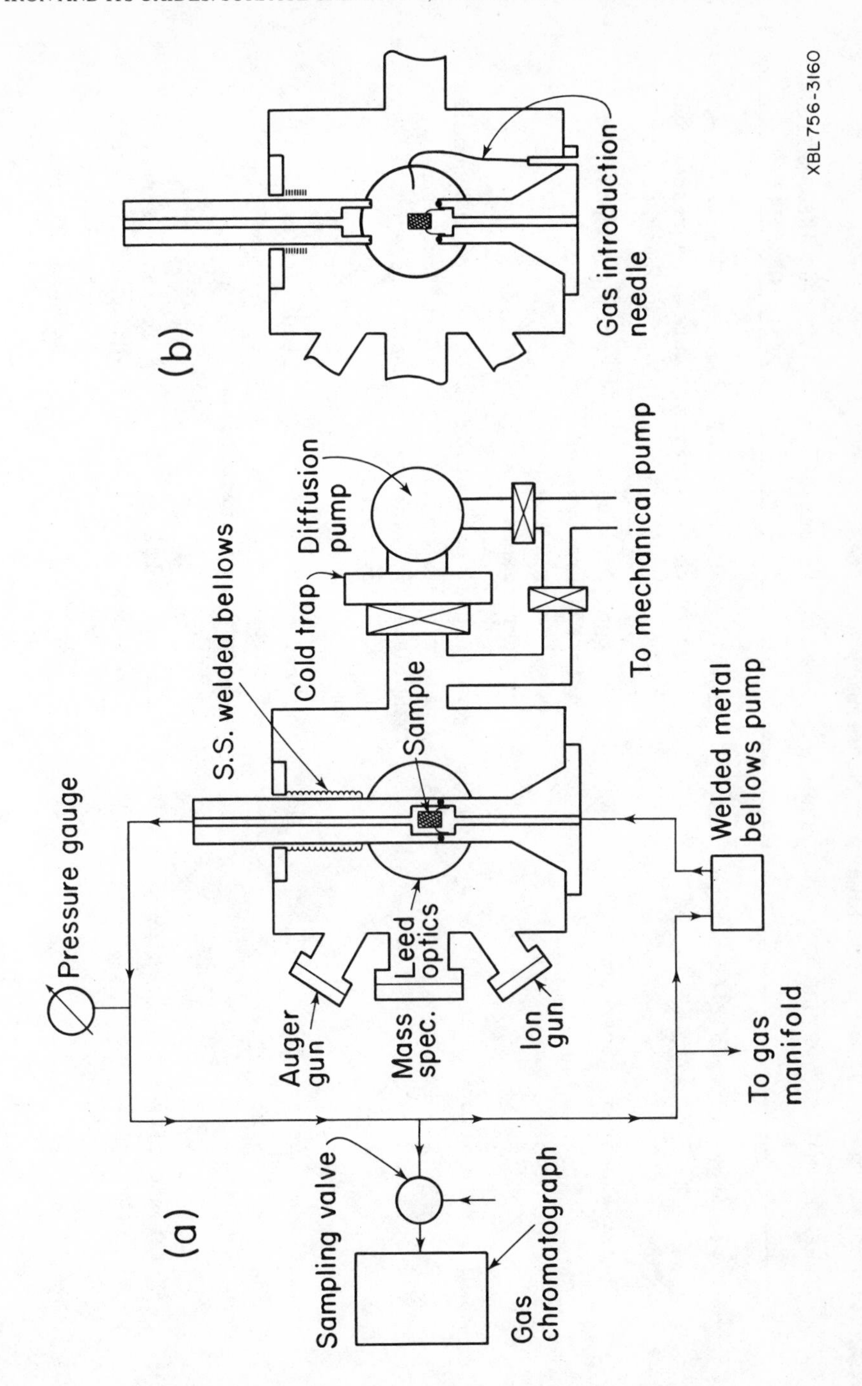

Figure 1.- a) Schematic diagram of the high pressure/UHV apparatus with sample enclosed in the high pressure cell. b) HP/UHV apparatus with cell open.

TABLE 1. - Table of some of the frequently utilized surface characterization techniques to determine the structure and composition of solid surfaces. Adsorbed species present at concentrations of 1% of a monolayer can be readily detected.

SURFACE ANALYSIS METHOD	ACRONYM	PHYSICAL BASIS	TYPE OF INFORMATION OBTAINED
Low energy electron diffraction	LEED	Elastic backscattering of low energy electrons	atomic surface structure of surfaces and of adsorbed gases
Auger electron spectroscopy	AES	Electron emission from surface atoms excited by electron x-ray or ion bombardment	surface composition
High resolution electron energy loss spectroscopy	HREELS	Vibrational excitation of surface atoms by inelastic reflection of low energy electrons	Structure and bonding of surface atoms and adsorbed species
Infrared spectroscopy	IRS	Vibrational excitation of surface atoms by absorption of infrared radiation	Structure and bonding of adsorbed gases
X-ray and ultraviolet photoelectron spectroscopy	XPS UPS	Electron emission from atoms	Electronic structure and oxidation state of surface atoms and adsorbed species

Ion scattering spectroscopy	ISS	Inelastic reflection of inert gas ions	Atomic structure and composition of solid surfaces
Secondary ion mass spectroscopy	SIMS	Ion beam induced ejection of surface atoms as positive e negative ions	Surface composition
Extended X-ray absorption fine structure analysis	EXAFS	Interference effects during x-ray emission	Atomic structure of surfaces and adsorbed gases
Thermal desorption spectroscopy	TDS	Thermally induced desorption or decomposition of adsorbed species	Adsorption energetics composition of adsorbed species

Below, we shall present case histories of surface science studies of catalytic processes representing each of the three categories. Following these we shall focus on the surface chemistry of iron and its oxides and review the photochemistry of iron oxides, and its possible catalytic role in photosynthesis in the pre-chlorophyl era of the planet.

2. CATALYSIS ON METAL SURFACES: THE AMMONIA SYNTHESIS REACTION

The synthesis of ammonia from stoichiometric mixtures of H_2 and N_2, has been studied over clean iron and rhenium single crystal surfaces having a variety of crystallographic orientations. The reactions were performed at temperatures between 500K and 800K and at total pressures of 20 atm. The surfaces were fully characterized using LEED and AES to ensure cleanliness and to determine atomic surface structure.

Ammonia synthesis has also been studied extensively over high surface area catalysts. The conclusions drawn from these studies are that the rate determining step of this reaction is the dissociative adsorption of N_2, and that the activity of the catalyst is strongly dependent on its atomic structure. The proposal has been made that iron surface atoms having seven-fold coordination are an important component of the active site (15).

Studies of the chemisorption of nitrogen on the low Miller index planes of iron single crystals show dramatic variations with crystallographic orientation (16). Initial sticking coefficients vary in the ratio 60:3:1 on the Fe(111), Fe(100) and Fe(110) surfaces respectively. We have studied the synthesis of ammonia over these surfaces (Fig. 2), and found that the activity follows the same trend as the sticking coefficient, the relative rates being 418:25:1. The similarity between the dependences of the sticking coefficients and reaction rates on surface structure is a good indication that N_2 chemisorption is the rate limiting step for this reaction. Examination of the sketches of the surface structures in Fig. 2 shows that activity depends upon the atomic scale roughness of the surface. The Fe(111) surface, being the most open and corrugated, is also the most active, while the Fe(110) surface, the most closely packed, is much less active. Roughening of the two least active surfaces by argon ion bombardment increased their catalytic activity by a factor of six (17). It should also be noted that the most active surface, Fe(111), contains atoms that have C_7 coordination.

The ammonia synthesis reaction was also studied over rhenium single crystal surfaces (4). Once again the reactivity varies drastically with surface structure, ranging over more than three orders of magnitude among the surfaces studied (Fig. 3). Higher activity was associated with surfaces having open topography. It is apparent that surface roughness is requisite for the dissociative chemisorption of N_2, however, the detailed geometry of the active sites cannot be uniquely determined from our results. Although both the Fe(111) and Re(1120) surfaces contain atoms having seven-fold coordination, the most active Re surface, Re(1121), does not. Atomically rough surfaces, having high coordination hollows, seems to be a general requirement for the dissociative adsorption of N_2.

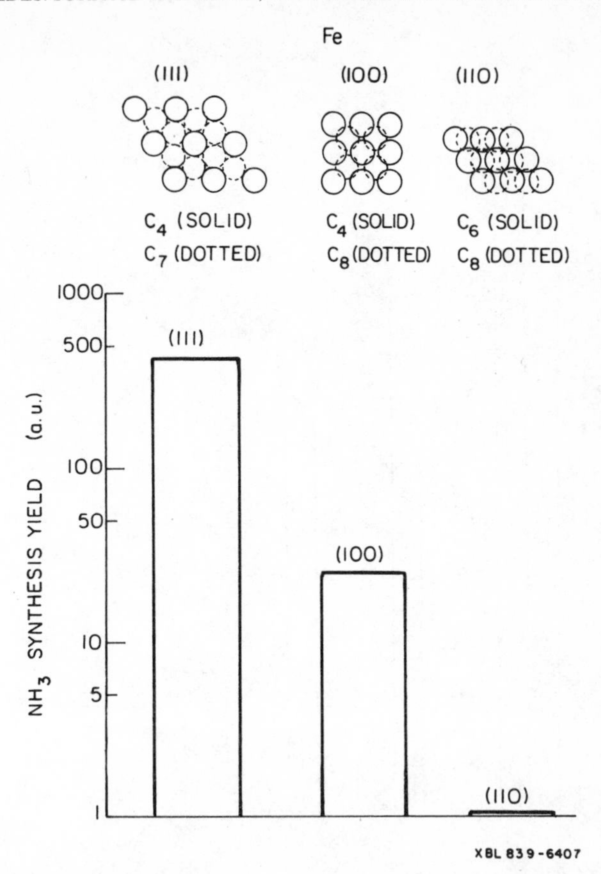

Figure 2.- Rates of NH_3 synthesis over three low Miller index Fe surfa̲ces.

It appears that reactions occurring directly over metal surfaces are sensitive to the local surface structure. Open or rough surfaces that expose sites of high coordination in either the first or second atomic layer of the surface show especially high catalytic activity. Surface roughening can be introduced by using high Miller index surfaces with high densities of steps and kinks. The bottoms of the steps expose sites of high coordination to the reactant molecules. In addition to the ammonia synthesis reaction, H_2/D_2 exchange at low pressures (37) and the hydrogenolysis of n-hexane and lighter alkanes (38,43) have shown similar structure sensitivity. Recent theoretical studies have related the electronic structure of high coordination sites with observed catalytic activity. These show an interesting correlation, however, the detailed theoretical understanding of the surface structure dependence of such catalytic reactions is still in its initial stages (18).

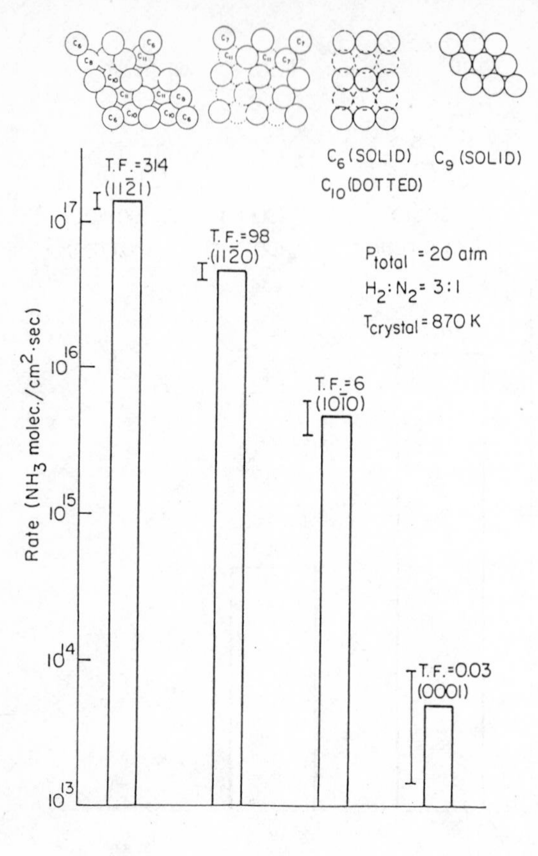

Figure 3.- Rate of NH_3 synthesis over four Re surfaces.

3. CATALYSIS ON AN ACTIVE OVERLAYER: THE HYDROGENATION OF ETHYLENE

Platinum and rhodium single crystals are good catalysts for the hydrogenation of ethylene at temperatures of 300-400 K and atmospheric pressures. The surfaces were analyzed before and after reaction using AES, LEED, TDS, HREELS, and a ^{14}C radiotracer technique that monitors the rate of removal of labeled carbon from the surface. Kinetic parameters for ethylene hydrogenation over a Pt(111) surface compare well with those found using other forms of platinum, (films, small particles) as can be seen from Table 2. Activation energies for the reaction are all between 9 and 11 kcal/mole, and the rates all have close to -0.5 and 1.0 order dependence on ethylene and hydrogen pressures, respectively. Furthermore, the absolute values of the reaction rates, expressed as turnover frequencies (ethane molecules formed per surface Pt atom per second), are all within the same order of magnitude, a clear indication that the reaction is not strongly affected by the structural details of the catalyst surface.

TABLE 2.- Comparison of the Kinetic Parameters for Ethylene Hydrogenation over Different Platinum Catalysts

Catalyst	log rate[a]	a[b]	b[b]	Ea, kcal/mol	ref
Platinized foil	1.9	-0.8	1.3	10	30
Platinum evaporated film	2.7	0	1.0	10.7	31
1% Pt/Al_2O_3	--	-0.5	1.2	9.9	32
Platinum wire	0.6	-0.5	1.2	10	33
3% Pt/SiO_2	1.0	--	--	10.5	34
0.05% Pt/SiO_2	1.0	0	--	9.1	35
Pt(111)	1.4	-0.6	1.3	10.8	12

a Rate in molecule/(Pt atom s), corrected for the following conditions: T=323 K, $P_{C_2H_4}$=20 torr, P_{H_2}= 100 torr.

b Orders in ethylene (a) and hydrogen (b) partial pressures.

Extensive studies of the chemisorption of ethylene on single crystal metal surfaces have been performed (19,20). Over Pt(111) and Rh(111) surfaces, ethylene undergoes a series of structural and chemical transformation as the temperature is increased. At low temperatures, the molecule adsorbs undistorted parallel to the surface. At about 300 K a complete structural rearrangement occurs and a new surface moiety, ethylidyne (CCH_3), is formed. During the conversion of ethylene to ethylidyne a hydrogen atom from ethylene bonds to the surface and subsequently recombines with another H atom to desorb as H_2. A hydrogen atom shift occurs within the remaining C_2H_3 species, to produce a methyl group bonded to the second carbon in such a way that the carbon-carbon bond is perpendicular to the surface and the whole moiety sits on a three-fold hollow site (Fig. 4). This structure has been characterized by a wide variety of techniques (21-24). Ethylidyne is stable up to temperatures of 400 K at which point further decomposition leads to the formation of CH and C_2H fragments, and ultimately to the creation of a graphitic overlayer.

One of the most important results of our studies is the observation that ethylidyne fragments are present on the metal surface even after the ethylene hydrogenation reaction. The results of TDS, LEED and HREELS experiments on a Pt(111) surface after ethylene hydrogenation are compared with similar data from ethylidyne in Fig. 5. The LEED pattern after the reaction reveals a p(2x2) ordered overlayer, similar to that obtained for ethylidyne, although the extra spots are more diffuse (an indication of partial disorder). The TD and HREEL spectra after reactions show peaks that are primarily attributable to ethylidyne. The extra features are most probably due to minority CH_x fragments (25).

The stability of ethylidyne in the presence of high pressures of hydrogen and ethylene has been studied using HREELS and radiation tech-

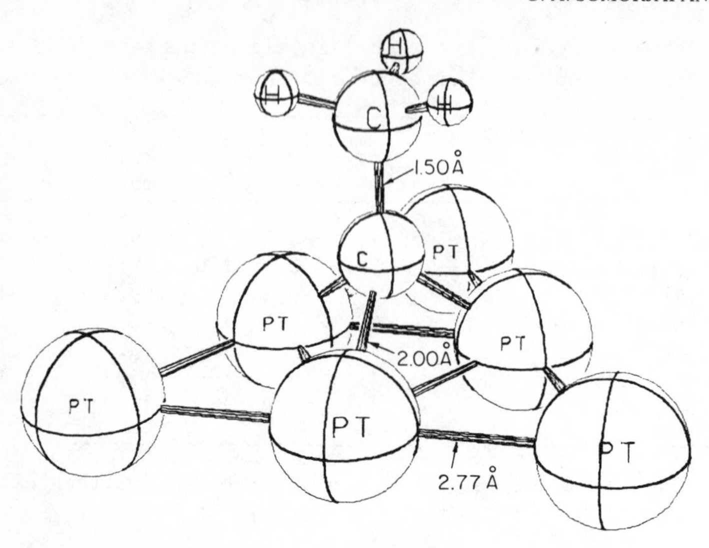

Figure 4.- The ethylidyne moiety adsorbed on the Pt(111) surface.

niques with deuterated and ^{14}C labeled ethylidyne respectively (24-27). Fig. 6 compares the rates of ethylene hydrogenation over Pt and Rh surfaces with rates for ethylidyne hydrogenation and the rate of incorporation of surface hydrogen into the methyl group of ethylidyne. It is clear that both processes are orders of magnitude slower than the overall rate of ethylene hydrogenation and thus, that neither can be an intermediate step in the reaction pathway.

From our results it is apparent that ethylidyne forms and that it is permanently present on the metal surface during ethylene hydrogenation reactions. The stability of this species under reaction conditions suggests that it is not, in fact, a direct intermediate for the formation of ethane. Rather, we propose that it facilitates the transfer of hydrogen, dissociated by the metal surface, to ethylene that is weakly adsorbed to the ethylidyne overlayer. Such a mechanism is illustrated in Fig. 7. It is clear that olefin hydrogenation reactions occur on metal surfaces that are covered with a carbonaceous overlayer. With this in mind, it should be noted that alkylidyne moieties, analogous to ethylidyne, have been observed after propylene and butylene adsorption (19,20) suggesting a similar mechanism for the hydrogenation of other olefins. Further studies are necessary to prove this proposition.

The ethylene hydrogenation reaction is probably the most distinctive example of a catalytic reaction occurring on an active overlayer rather than a metal surface. The role of the metal in this case is to

Evidence for the Presence of Ethylidyne after Atmospheric Hydrogenation of Ethylene over Pt (111)

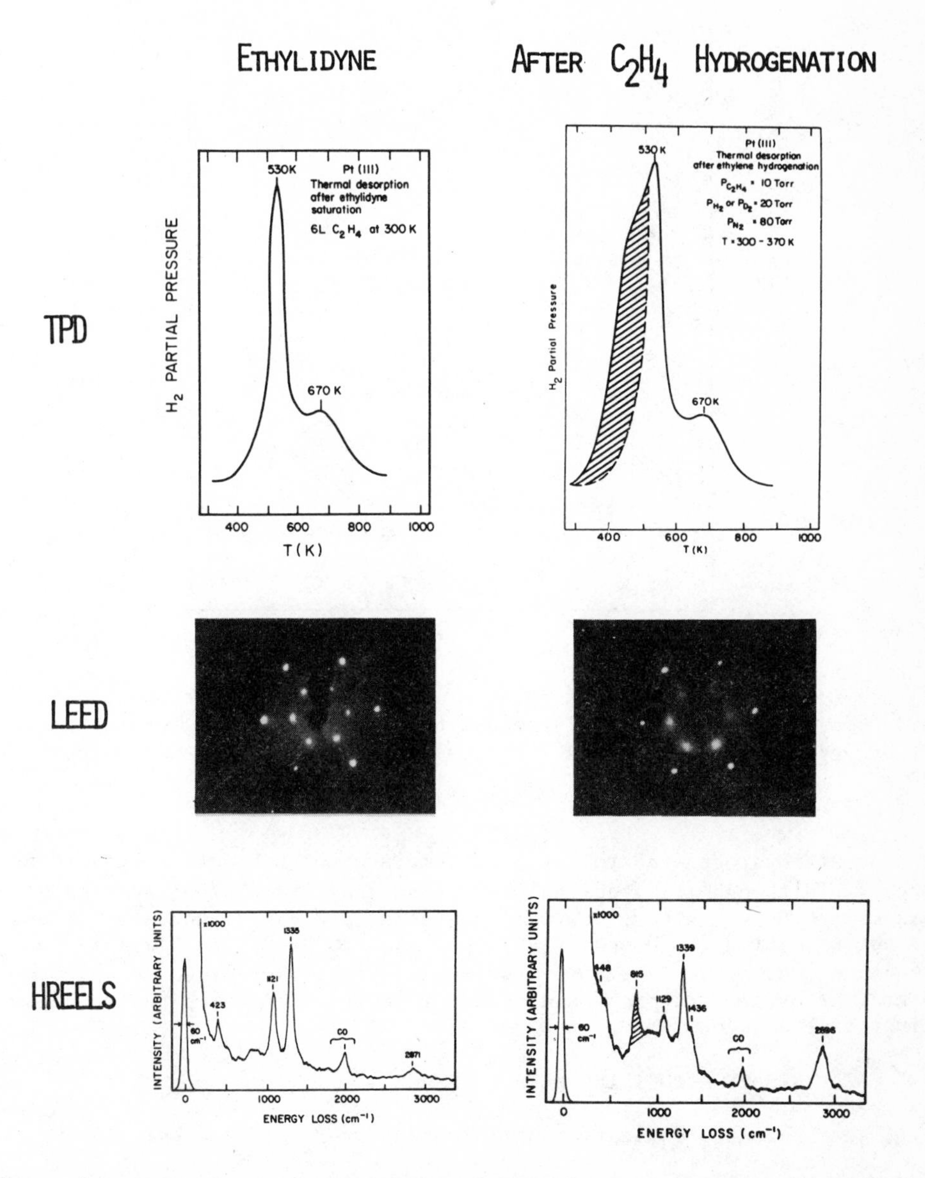

Figure 5.- Comparison of TPD, LEED and HREELS measurements on a Pt(111) surface after C_2H_4 hydrogenation with those on an ethylidyne covered surface.

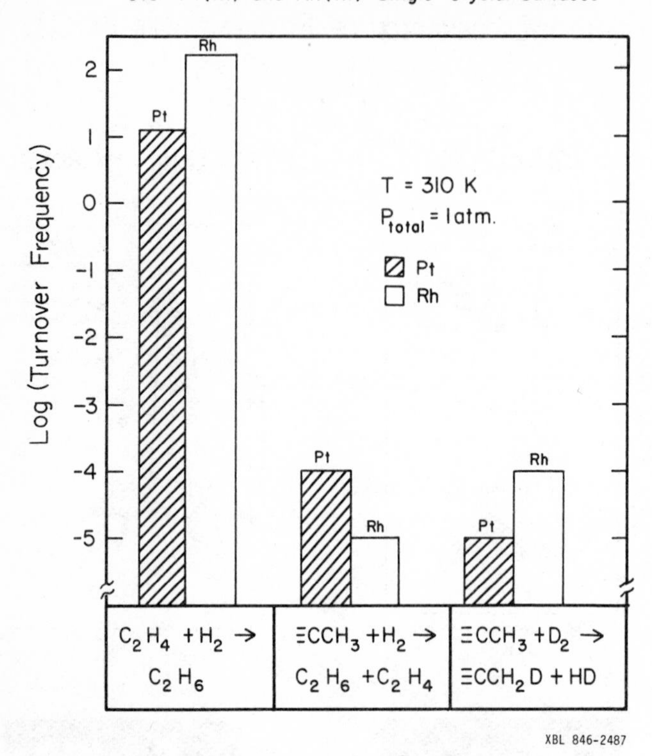

Figure 6.- Comparison of the rates of ethylene hydrogenation over Pt and Rh with those for ethylidyne hydrogenation and those for exchange of hydrogen between the metal surface and the methyl group of ethylidyne.

dissociate hydrogen and to act as a template for the active ethylidyne species. Although it is quite likely that many examples of such catalysis exist there are very few well characterized cases. One similar case is the oxidation of CO and Pt surfaces at high pressures (10,11). The reaction occurs over a platinum oxide layer that serves as an active source of oxygen and plays a dominant role in determining the reaction kinetics and mechanism.

4. CATALYSIS BY A MODIFIED METAL SURFACE

4.1. The Hydrodesulfurization of Thiophene over Sulfided Molybdenum Surfaces

The extraction of sulfur from sulfur containing organic molecules is

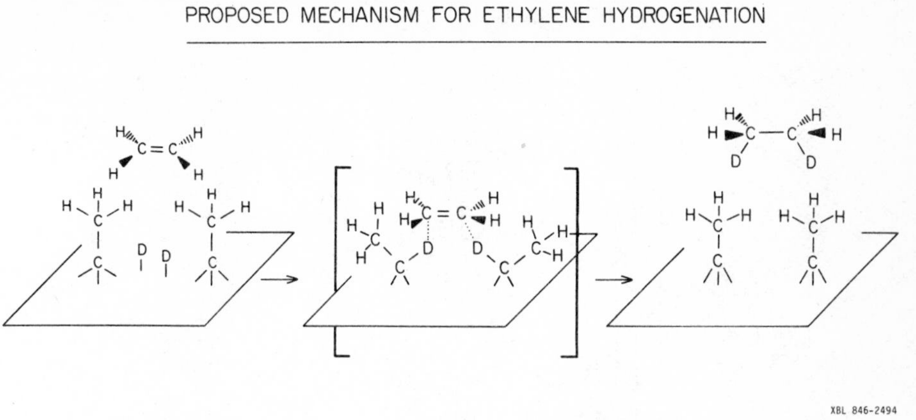

PROPOSED MECHANISM FOR ETHYLENE HYDROGENATION

XBL 846-2494

Figure 7.- Proposed model for ethylene hydrogenation mechanism over an ethylidyne covered surface.

achieved by the hydrodesulfurization (HDS) process. We have recently studied the hydrodesulfurization of thiophene (C_4H_4S) over Mo(100) single crystal surfaces. The reactions are performed at temperatures of 550-700 K using mixtures of 2.5 torr of thiophene in 1 atm. of hydrogen. A typical product distribution is shown in Fig. 8 compared with that from a high surface area MoS_2 catalyst (28). The similarity between the two distributions suggest that the reaction mechanisms over the two catalysts are identical.

The chemisorption of sulfur on the Mo(100) surface has been studied in detail under UHV conditions (29). At monolayer coverages sulfur is adsorbed atomically in a lattice with a p(2x1) structure with respect to the substrate. Heating the surface to temperatures between 1400K and 1800K results in the desorption of sulfur and the formation of additional overlayer structures as the coverage decreases. The LEED patterns, proposed real space structures, and coverages at which they occur are presented in Fig. 9.

At low coverages ($\leq$ 0.5 monolayers) sulfur atoms adsorb on four-fold hollow sites (30). The heat of desorption from this site, estimated from thermal desorption data, is about 110 kcal/mole. The appearance of a second peak in the TD spectra at higher coverages indicates the creation of a second, more weakly bound, sulfur adsorption state. The formation of a p(2x1) lattice at monolayer coverages suggests that the additional sulfur atom could in fact be bound to a different site, like in a bridged configuration (Fig. 9). Structures composed of both hollow site and bridge-bonded sulfur can be alternatively proposed to account for the LEED patterns observed for coverages between 0.5 and 1.0 monolayers.

Studies of the rate of HDS of thiophene over clean and sulfided surfaces show that low sulfur coverages result in a decrease in reaction

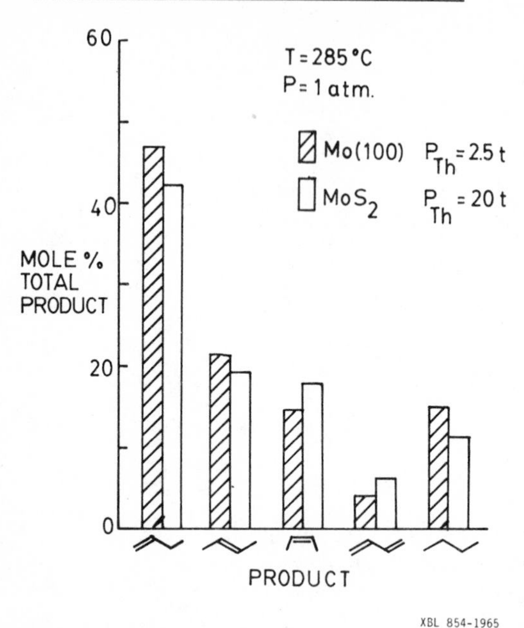

Figure 8.- Comparison of thiophene HDS product distributions over Mo(100) and MoS_2 catalysts.

rates (Fig. 10). The molybdenum surface is always covered with sulfur after reactions, and thus, the fact that the rate is affected by initial sulfiding of the surface indicates that the sulfur present on the surface during reactions is chemically different from that deposited beforehand. Isotope labelling studies using ^{35}S indicate that the sulfur deposited under UHV conditions remains on the surface during the reaction, retarding the reaction rate, while sulfur from thiophene is removed from the surface as H_2S, without forming a strong surface bond. Sulfiding of the Mo surface to coverages greater than ∿0.6 monolayers does not result in significant further reduction in the HDS rate. This observation is directly related to the formation of a weakly bound sulfur species at high coverages. This sulfur species is removed from the surface under reaction conditions desorbing as H_2S and thus, does not have any effect on the reaction rate. Although the effect of sulfiding the catalyst is primarily one of site blocking, reducing overall activity, it also induces a change in the product selectivities. The adsorbed sulfur selectively blocks sites for hydrocarbon hydrogenation and thus reduces the rates of butane and butene production without affecting the rate of butadiene production. As a result the presence of sulfur shifts the selectivity of the reaction towards the production of butadiene.

Finally, it appears that the HDS of thiophene occurs either via direct, concerted hydrogenation of C-S bonds producing H_2S, or through

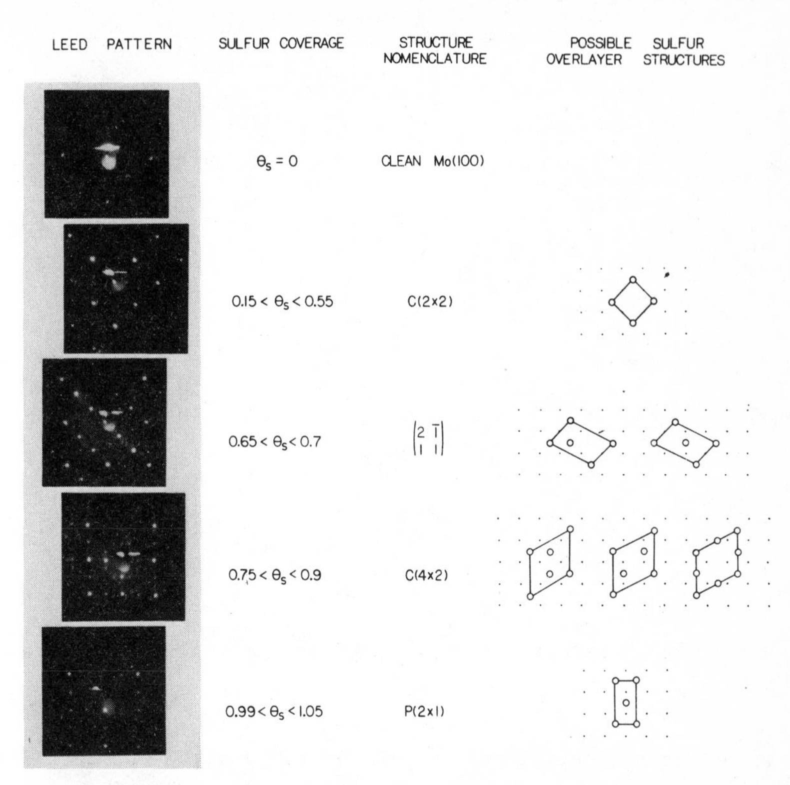

Figure 9.- LEED patterns and proposed real space structures for sulfur on the Mo(100) surface at various coverages.

some weakly bound, easily reduced Mo-S intermediate. In either case the process does not proceed via deposition of sulfur into the fourfold hollow site, from which it cannot be reduced under our reaction conditions. This sulfur species, bound in fourfold hollow sites, blocks hydrocarbon hydrogenation, modifying the HDS product distribution.

4.2. The Conversion of n-Hexane over Gold-Modified Platinum Surfaces

The reforming of n-hexane over platinum catalysts occurs via a number of reactions including isomerization, cyclization, dehydrocyclization, and hydrogenolysis. Most of these are structure sensitive and have been studied in detail because of their importance as models for the processes used to produce high octane gasoline. Gold itself is inactive as a catalyst for these molecular rearrangements. However, when added to a Pt single crystal surface it will alter the product distribution markedly. This is shown for the Pt(111) face as a function of gold coverage in Fig. 11. The rate of isomerizatoin increases while the rates of dehydrocyclization and hydrogenolysis drop quickly with gold coverage.

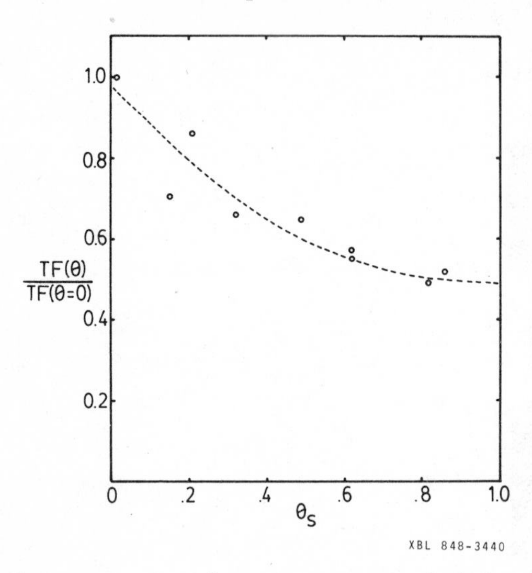

Figure 10.- Initial rate of thiophene HDS vs. initial sulfur coverage on the Mo(100) surface.

Thus, low concentrations of gold drastically change the product distribution (41).

This observation is usually explained by pointing out the important role that the high symmetry three-fold platinum sites play in the dehydrocyclization and hydrogenolysis reactions. The addition of gold to the surface decreases the concentration of these sites more rapidly than the low symmetry bridging and top sites. Thus, the rates of reactions requiring high symmetry sites (dehydrocyclization and hydrogenolysis) are reduced more rapidly than those needing the low symmetry sites. This phenomenon is often called the ensemble effect.

In order to explain the large increase in the rate of isomerization, one must invoke the formation of mixed Pt-Au sites that are more active than the pure Pt. Such sites may be ensembles having Pt_2Au or $PtAu_2$ stoichiometry. Thus, alloying of gold with Pt can change the reaction selectivity both by site blocking and by producing new catalytic sites.

4.3. The Hydrogenation of Carbon Monoxide over Potassium Doped Transition Metal Surfaces

The hydrogenation of carbon monoxide produces alkanes, alkenes and alcohols selectively or in a mixture over many transition metal surfaces. Potassium is used as an additive in many catalysts for this reaction to increase activity and modify product distributions. These promoters increase the molecular weights of the products and increase the selecti-

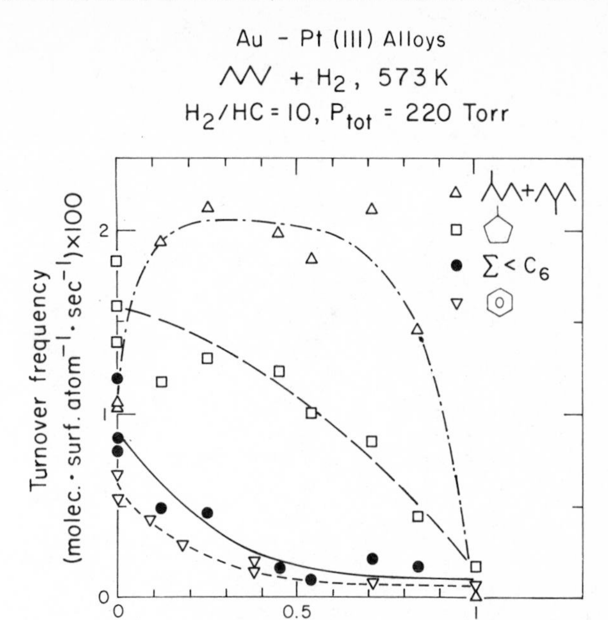

Figure 11.- The dependence of the initial rates of product formation from n-hexane conversion on gold coverage over Au-Pt alloys.

vity towards alkanes and unsaturates. The first explanation of this phenomenon in terms of an electronic effect was given by Blyholder who suggested that potassium donates electrons to the metal which in turn increases the amount of back donation to the CO molecules. Increased donation into the $2\pi^*$ orbital increases the metal-adsorbate bond strength while simultaneously weakening the C-O bond.

Surface science studies have shown that such effects are, in fact, observed. The heat of adsorption of CO on the Rh(111) surface is increased by 12 kcal/mole (from 25 kcal/mole to 37 kcal/mole) when coadsorbed with potassium. Vibrational studies using HREELS revealed a continuous decrease of the CO stretching frequency from 1850 cm^{-1} to 1550 cm^{-1} with increasing potassium coverage (42) (Fig. 12). Finally, isotope scrambling experiments using $^{13}C^{16}O$ and $^{12}C^{18}O$ indicate that potassium will induce molecular dissociation, unobserved in the absence of the alkali metal (43). The enhanced ability to break CO bonds, induced by the presence of potassium, results in a higher surface carbon coverage and, as a result, a decrease in the hydrogen coverage. The decrease in the surface H:C ratio in turn favours carbon-carbon bond formation to produce higher molecular weight products. The reduced rate of C-H bond formation favours the production of unsaturated hydrocarbons.

The modification of the catalytic properties of a metal surface by

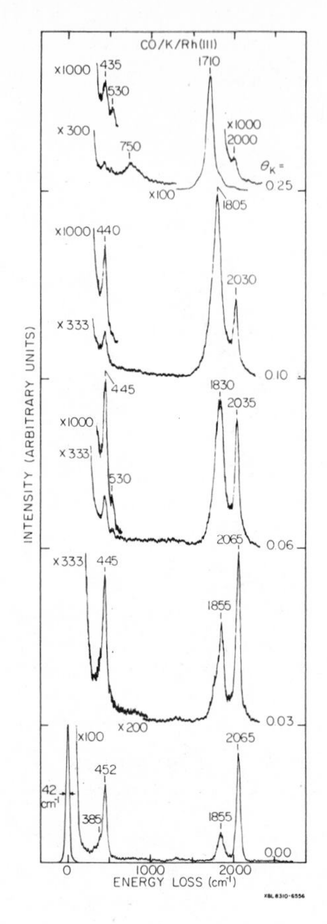

Figure 12.- HREELS spectra of CO coadsorbed with K on the Rh(111) surfa ce at various Θ_K.

potassium differs from the previous cases in that it is an electronic effect. In the case of sulfur on the Mo(100) surface the modification is achieved by selective site blocking while the addition of gold to the platinum surface both blocks sites and creates new sites. An electronic effect that occurs through the metal substrate is expected to be more long range in nature. This has been observed to be the case using HREELS (Fig. 12). The addition of potassium at low coverages induces a uniform and continuous shift in the CO stretching frequency. For such an effect to be observed at $\Theta_K \lesssim 0.05$ the influence of potassium cannot be highly localized.

5. SELECTED PHYSICAL AND CHEMICAL PROPERTIES OF IRON OXIDES

Iron is the most abundant transition metal that is present on our planet. The metal and its oxides have desirable structural, magnetic, electrical and chemical properties that are employed in a multitude of applications. It has a body-centered cubic crystal structure at room temperature, but undergoes a phase-transformation of face-centered cubic structure at higher temperatures that involve volume change, as well. Iron oxides have a complex phase diagram that is shown in Figure 13, consisting of at least three distinct iron-oxygen compounds (44). Stoichiometric hematite, α-Fe_2O_3 is an intrinsic n-type semiconductor with a band gap of 2.2 eV, which is in visible light, but is an insulator at room temperatures (45,46). This iron oxide is inexpensive, stable under acidic and basic aqueous conditions, and has useful optical properties, that is, good matching between the bandgap and the solar spectrum, and a large absorption coefficient. It is possible to produce a less resistive semiconducting oxide material by reducing some of the $iron^{3+}$ to $iron^{2+}$ state. The hematite is then a mixed valence compound with enhanced conductivity at room temperature which is due to a hopping process for electrons between Fe^{2+} and Fe^{3+} ions (47). The Fe^{2+} ions can be introduced by producing oxygen deficiencies or by adding a dopant, which induces a charge compensation process. However, the corundum α-Fe_2O_3 phase has a low solubility for divalent ions, since the Fe^{3+}_{2-x}-Fe^{2+}_{x}-O_3 stoichiometry induces the formation of the Fe_3O_4 spinel phase. Thus, it is difficult to prepare homogeneous doped semiconducting samples of Fe_2O_3 without Fe_3O_4 phase inclusions (48). The divalent oxide FeO is a p-type semiconductor with a bandgap also near 2.3 eV. The mixed oxide Fe_3O_4 is a low resistivity compound, almost metallic.

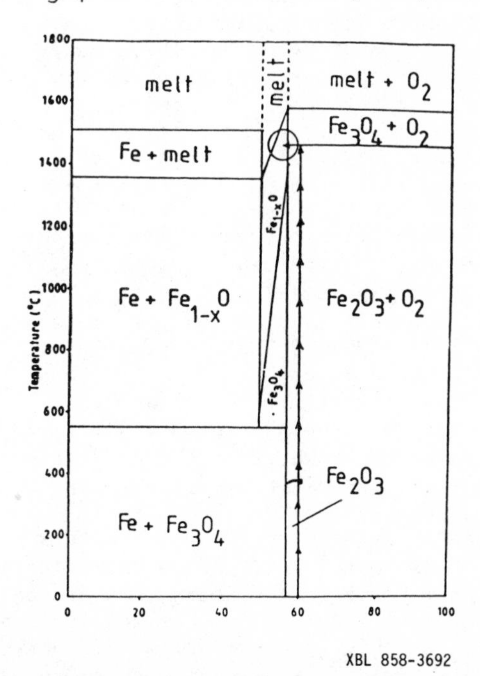

Figure 13.- Phase diagram of the Fe-O system.

This discussion of iron oxide chemistry is divided into three parts: first, we discuss the surface chemistry of iron and iron oxide as revealed by modern surface science studies. Then we discuss the photochemistry of iron oxide, as this material appears promising for the photocatalyzed reactions of water and nitrogen.

6. THE SURFACE CHEMISTRY OF IRON AND IRON OXIDES

Many of the crystal faces of metallic iron have been subjected to low energy electron diffraction studies. The surface structure of the (110), (111) and (100) faces show no reconstruction. However, there is strong relaxation of the more open (111) and (100) surfaces. That is, the interlayer spacing between the first and second layer is appreciably smaller than the interlayer spacing in subsequent layers below (49). The several ordered surface structures of chemisorbed oxygen have been investigated. The formation of FeO in the surface layer has been detected and the structure of this complicated but important surface layer has been solved (50). Oxygen chemisorption, the initial stages of oxidation and bulk oxide formations have all been studied by electron spectroscopy techniques (51). X-ray photoelectron spectroscopy revealed the various oxidation states of the transition metal ions as a function of oxygen partial pressure and surface concentrations (52). Ultraviolet photoelectron spectroscopy showed drastic changes in the density of states at the Fermi level and changes in work function as oxygen chemisorption and oxidation proceeds. Auger electron spectroscopy clearly detected different Auger peak intensity ratios for the iron and oxygen Auger peaks when FeO, Fe_2O_3 and Fe_3O_4 are present. These Auger peak ratios can then be used to determine the surface stoichiometry of the iron oxides samples. The chemisorption of small molecules, including nitrogen, carbon monoxide, and small hydrocarbons have all been studied on both iron and iron oxide surfaces (53). A detailed discussion of their structure and chemisorption properties are outside the scope of this review.

Since iron oxide Fe_2O_3 has been investigated as a photoelectrode for the photocatalyzed dissociation of water, its surface properties should be discussed in some detail. Recently, detailed examinations of the Fe_2O_3(001) crystal surface using single crystals have been undertaken using low energy electron diffraction (54), X-ray and ultraviolet photoelectron spectroscopies (XPS and UPS), Auger electron spectroscopy and temperature programmed desorption (55). With these techniques, the properties of the clean stoichiometric surface and, the argon ion bombarded non-stoichiometric surface have been studied, in particular, for the absorption and reaction of water on these surfaces. The basal plane (001) of Fe_2O_3 that is stoichiometric is very inert even to water adsorption. From thermal desorption studies, water is observed to desorb between 175 and 200K. The activation energy for desorption is about 12 kcal/mol which is very close to the sublimation energy of ice. These results indicate that water forms ice clusters that are only weakly bound to the (001) Fe_2O_3 plane with minimal chemical interaction. Electron spectroscopy studies of this system confirm this observation.

A more reactive surface can be prepared by introducing surface defects such as oxygen vacancies through argon ion sputtering. Figure 14

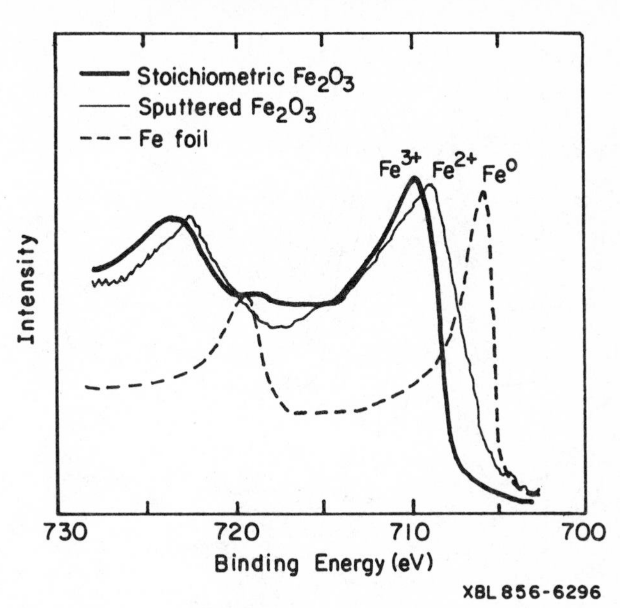

Figure 14.- MgK. X-ray photoelectron spectra of stoichiometric Fe_2O_3, sputtered Fe_2O_3 and clean Fe foil.

illustrates how the presence of reduced Fe^{+2} species is dectected by XPS. The non-stoichiometric iron oxide surface also has a lower work function by about 0.8 V than the stoichiometric iron oxide crystal. Annealing of the surface defects introduced by argon bombardment is difficult. Only when heating about 600K in 5 x 10^{-6} Torr of oxygen will the original stoichiometric iron oxide surface regenerate. The defect species that are created by argon ion bombardment give rise to photoelectron emission above the valence band within the bandgap. This is clearly shown by ultraviolet photoelectron spectroscopy. The iron species responsible for this emission could be either small metallic clusters or Fe^{2+} in various local oxygen vacancy configurations. When water is adsorbed on the defective argon sputtered iron oxide surface, in addition to the adsorption of molecular H_2O, hydrogen evolution is observed upon heating the system in the temperature range between 220 to 750K. It appears that the reduced Fe^{2+} or metallic iron species that are created at the surface adsorb water strongly, react with it, and give rise to the formation of Fe^{3+} ions as well as hydrogen evolution. The UPS spectra of Fig. 15 illustrated how the presence of H_2O and OH species on the sputtered Fe_2O_3 surface is detected. The water induced features (vertical broken line) are visible up to 370 K. The shoulder observed near zero binding energy is due to Fe^{2+} species.

Another way of increasing the bonding strength of water to iron oxi

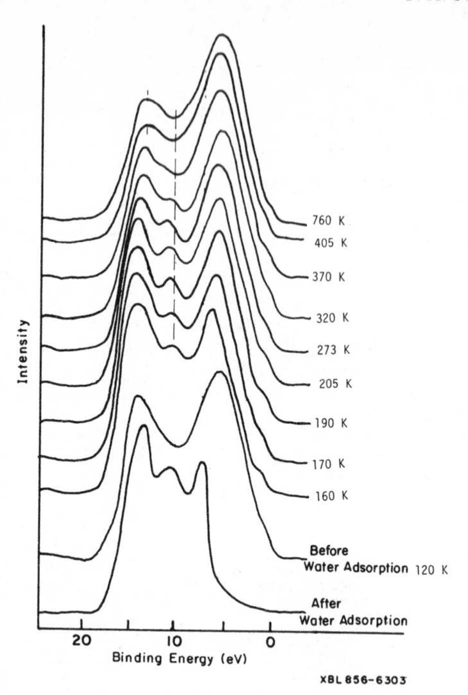

Figure 15.- UPS spectra from the H_2O covered Fe_2O_3 (sputtered) as a function of temperature. The bottom curve corresponds to a thick (20) ice overlayer. The water induced features disappear at 375K approximately. The shoulder near zero binding energy is due to Fe^{+2} defects.

de is by the co-deposition of potassium. When potassium is deposited and oxidized, there is evidence for the formation of a K_2O Fe_2O_3 structure from thermal desorption data. When water and potassium were coadsorbed, not only water desorbs but also there are two broad hydrogen peaks, one at 260 K and the other at 410 K. Potassium binds strongly to the (001) surface of Fe_2O_3 and also accelerates the decomposition of water to produce hydrogen and potassium oxide.

In summary, the stoichiometric basal plane of the corundum phase Fe_2O_3 is extremely inert. Water physisorbs only as ice at low temperatures and it desorbs around 175 K. Argon ion sputtering produces oxygen vacancies that expose Fe^{2+} species on the outermost atomic layer that lower the work function of the sample. These reduced iron species are unstable and disappear rapidly from the surface layer upon heating above 375 K. However, the Fe^{2+} species that can be detected under the surface are stable at temperatures up to 775K for extended periods of time. The surface Fe^{2+} species, when present in the outermost layer, can adsorb water strongly and evolve hydrogen upon heating. Potassium binds strongly to the (001) surface of Fe_2O_3 and when coadsorbed with water it forms KOH that decomposes at 275K and above to hydrogen and potassium oxide (55).

7. THE PHOTOCHEMISTRY OF IRON OXIDE

7.1. Photochemical Studies of Germanium Substituted Fe_2O_3 and Niobium Doped Fe_2O_3 Single Crystals

Iron oxides have recently been investigated as possible electrode materials for the photoelectrolysis of water using sunlight. In an overview, the large adsorption coefficients (56) and the favorable bandgap (2.2 eV) of α-Fe_2O_3 suggests that a large portion of the solar spectrum could be used efficiently in an iron oxide based semiconductor-liquid junction p-n assembly. Most of the studies using n-type Fe_2O_3 photoanodes have focused on thin films (57,61) or polycrystalline sintered disks (58) primarily because of the ease of sample preparation. It was found that iron oxides might be useful if the quantum efficiency of this material could be improved. The origin of the low quantum efficiency of Fe_2O_3 is unclear because the inhomogeneities of thin films or polycrystalline materials make it difficult to separate bulk effects from surface properties. One of the main problems associated with the application of Fe_2O_3 photodiodes is the intrinsic high resistivity of the hematite corundum phase.

Photoelectrolysis processes require high conductivity semiconducting materials and unlike the binary oxides of TiO_2 or WO_3 the corundum phase of iron oxide has a low tolerance for deviations from stoichiometry (48). The purpose of our single crystal studies (59) is to ensure that the low quantum efficiency exhibited by Fe_2O_3 sintered discs is not due to grain boundaries and also to prepare more uniform samples, both in terms of composition and structure. We have prepared germanium and niobium-doped single crystals with well-characterized crystallographic, electrical and photoelectrical properties. We found that germanium substituted Fe_2O_3 crystallized with a corundum structure and it is an extrinsic n-type semiconductor.

Niobium-substituted iron oxide also crystallizes in the corundum structure. It is also an n-type semiconductor and behaves as an electrochemical anode for the photooxidation of water. A room temperature resistivity of 85 ohm cm for these crystals could be achieved. For germanium substituted Fe_2O_3 we could achieve room temperature resistivities of about 5 ohm cm. Magnetic susceptibility studies suggest that the conductivity of these single crystal samples arises from charge compensation resulting from reduction of Fe^{3+} to Fe^{2+} upon the substitution of either germanium or niobium. The photoelectronic properties of these substituted Fe_2O_3 crystals suggest that the low quantum efficiency of Fe_2O_3 photoanodes is related to energy levels near to conduction band of Fe_2O_3. The low quantum efficiency therefore cannot be explained on the basis of such effects as grain boundary recombination because the same quantum efficicncy has been observed with polycrystalline samples as was found with single crystals.

Comparison of the flat band potentials determined by photocurrent measurements with that determined using capacitance techniques implies that an energy level near the conduction band of Fe_2O_3 must be ionized before photoelectrolysis can occur. The ionized energy level may then lead to extensive recombination of photogenerated electron-hole pairs in the space charge layer and therefore lowered quantum efficiencies.

7.2. The Si-doped and Mg-doped Iron Oxide Photochemical Diode

Iron oxide has been studied by Hackerman, et al. (60), that demonstrated desirable properties of this material for water photodissociation. When Fe_2O_3 is used against a platinum counterelectrode photocurrents corresponding to oxygen production are generated for an applied bias of larger than 700 mV. It has been found in our laboratory that doping with silicon reduces the magnitude of the bias and increases the magnitude of anodic photocurrents. This result motivated a systematic study of various dopants in iron oxide which showed that the introduction of magnesium could yield iron oxide electrodes with cathodic behavior. Our recent studies indicate that this cathodic behavior is restricted to the near surface region of the material and it is very much dependent on materials preparation (59). Preparation of the samples required the heating of the mixed magnesium oxide-iron oxide powder to 1400 C followed by rapid quenching in water. The resulting material was a highly heterogeneous magnesium-doped iron oxide also containing phases of magnesium ferrate, $MgFe_2O_4$, and magnesium oxide, MgO. Individual photocurrent vs. voltage characteristics for the silicon-doped and magnesium-doped iron oxide electrodes show the anodic and cathodic behaviour that corresponds to n-type and p-type samples. Individually these electrodes require an external bias to sustain oxygen or hydrogen evolution under illumination when employed against a platinum counterelectrode. However, when connected in a short-circuit configuration the diode assembly assumes an intermediate operating bias of approximately 750 mV, so that O_2 and H_2 production will occur simultaneously without external bias (61). The locations of conduction- and valence-band edges with respect to redox couples (H^+/H_2 and O_2/OH^-) in solution are of critical importance for photodissociate water. To determine the location of these edges on the electrochemical scale, Mott-Schotky measurements were performed in which the capacitance of the space-charge layer was determined using a phase-shift technique. From these studies the energy level diagram that depict band edge locations of doped iron-oxide electrodes relative to hydrogen and oxygen redox couples could be obtained and are as in Fig. 16. Detailed photoelectrochemical studies using this system have been published elsewhere (59). The quantum efficiency as a function of wavelength was determined as well as the catalytic nature of oxygen evolution using oxygen-18 isotope-labeled water (19). The efficiency of the iron oxide diode assembly was low, approximately 0.1%. It appears that the very small thermodynamic driving forces that control the hydrogen evolution and the oxygen evolution is perhaps responsible for the low quantum efficiency. It would be of importance to find other dopants in addition to magnesium that could move the flat band potential further in the cathodic direction so as to provide a greater thermodynamic driving force for the photosplitting of water.

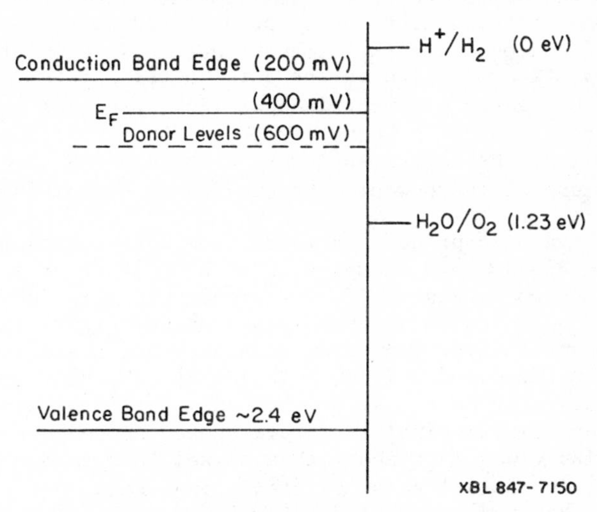

Figure 16.- Hypothetical energy diagram for Fe_2O_3.

7.3. Photochemical Hydrogen Production from a Water Methanol Mixture with Iron Oxide Suspensions

Recent studies by Bard (62), Sakata (63) and Sato (64) have shown that quantum efficiencies for hydrogen production could be obtained from water, water alcohol, and other reaction systems using platinum impregnated semiconductor powders such as TiO_2-Pt or CdS-Pt. Such treatment of the semiconductor particles decreases electron-hole pair recombination and also catalyses hydrogen recombination on the surface. Almost no studies of iron oxide small particle systems for photocatalytic reactions have been reported in comparison to the many studies conducted using TiO_2, $SrTiO_3$, CdS, InP, or ZnO powders (62-65). This is due mainly to the low efficiency exhibited by iron oxides for H_2 photogeneration reactions. We have tested and investigated the photocatalytic hydrogen production abilities of magnesium-doped α-Fe_2O_3 polycrystalline powders and, in particular, the positive effect of spinel phase inclusions in the α-Fe_2O_3 corundum structure for catalytic H_2 production (66). The reactivities of these materials were measured for the photoproduction of CO_2 and H_2 from a liquid phase water methanol 1:1 mixture. The H_2 photoproduction efficiency for this mixture has been shown to be much higher than that from water, which makes it easier to measure H_2 amounts quantitatively. It was found that the existence of spinel phase inclusions in the α-Fe_2O_3 corundum structure enhances the photocatalytic activity, and that the reaction can be regarded as photocatalytic.

Magnesium-doped α-Fe_2O_3 samples were prepared by using a powder mixture of α-Fe_2O_3 and MgO. These materials were mixed using a methanol slurry, then pressed into discs, which were then sintered at temperatures between 1380 and 1425 C for 20 hours. The discs were then quenched in water to room temperature.

The platinum was deposited on the semiconductor powders by mixing appropriate amounts of the semiconductor powders and platinum black. Photoproduction of hydrogen was performed by decomposing water methanol 1:1 mixture. No reaction occurred in the dark in this system. Under illumination H_2 and CO_2 gases were obtained as final products according to the following reaction $H_2O + CH_3OH + \text{light} = 3H_2 + CO_2$. ΔG_{298} is +2.2 kcal/mole. All of the experiments were carried out at room temperature. The structure of the powders was studied by X-ray powder diffraction.

The measured hydrogen production rates from water/methanol mixture using magnesium doped α-Fe_2O_3 samples which had different spinel corundum ratios are shown in Figure 17. The spinel phase is due to the presence of $Fe_{3-x}M_xO_4$ which is a cubic spinel phase. Figure 17 shows the correlations between the hydrogen production rate and the spinel corundum ratio for magnesium doped α-Fe_2O_3 with 10% Pt. The hydrogen production rate depends largely on the spinel corundum ratio. As expected, pure Fe_3O_4, which has a 100% spinel structure showed no photoreactivity. The maximum efficiency was obtained with a spinel to corundum phase ratio of about 0.37. The spinel phase existence in α-Fe_2O_3 may cause a cathodic shift of the conduction band edge of α-Fe_2O_3 while the decreasing resistivity of the catalyst may help the mobility and transfer of photoregenerated carriers. From studies of the hydrogen evolution rate change with photon excitation energy, it was found that this photocatalytic reaction was driven mainly by bandgap (2.2 eV) radiation.

7.4. Catalysis and Photocatalysis by Iron and Iron Oxides

Metallic iron is an excellent catalyst for the synthesis of ammonia from nitrogen and hydrogen. This technologically important structure sensitive reaction has been described in this paper in some detail. Iron oxides can photodissociate water to hydrogen and oxygen and recent studies indicate the photoproduction of ammonia from nitrogen and water at 300K. Both, iron and iron oxides are active catalysts for the hydrogenation of carbon monoxide. This is an important reaction for the production of synthetic fuels and chemicals, alkanes, alkenes and alcohols. Iron oxide is also a good catalyst for the water-gas shift reaction $CO + H_2O \rightarrow CO_2 + H_2$.

It is interesting to speculate the possibility of inorganic photosynthesis using iron and iron oxides as possible catalysts for the process. The photodissociation of water over iron compounds yields hydrogen and oxygen. Oxygen escapes while hydrogen which is most reactive could further react with carbon dioxide and dinitrogen on the iron surface to produce organic molecules and nitrogen containing organic molecules. Once hydrogen is produced its subsequent reactions with either CO_2 or N_2 are thermodynamically downhill, that is, energetically feasible. The only energetically uphill step that requires the input of external energy (solar light) is the photodissociation of water.

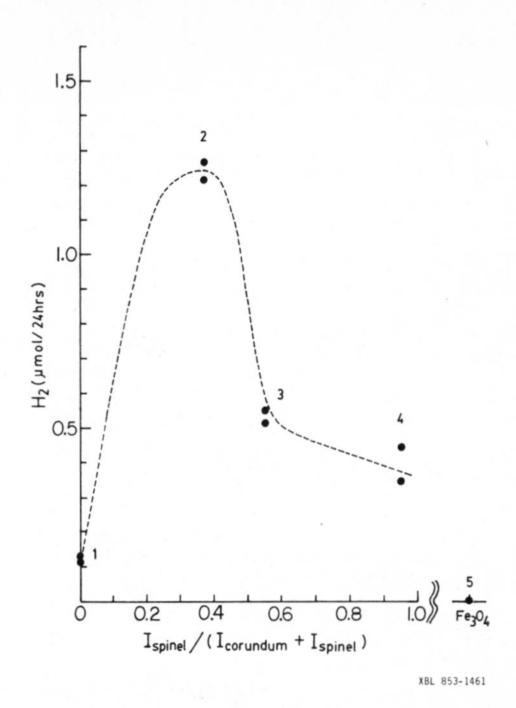

Figure 17.- The dependence of H_2 production from a water methanol mixture (1:1), 5 ml, on the spinel concentration in Mg-doped -Fe_2O_3. All samples have been loaded with 10% Pt.

8. ACKNOWLEDGEMENT

This work was supported by the Director, Office of Energy Research, Office of Basic Energy Sciences, Chemical Science Division of the U.S. Department of Energy under contract number DE-AC03-76SF00098.

9. REFERENCES

(1) J. Berzelius, "Jahres-Bericht über die Fortscri der Physichen Wissenschafter", Tübingen, 1836, p. 243.

(2) S.M. Davis and G.A. Somorjai, "The Chemical Physics of Solid Surfaces and Heterogeneous Catalysis", Vol. 4, D.A. King and D.P. Woodruff, Eds., Elsevier, Amsterdam 1982.

(3) N.D. Spencer and G.A. Somorjai, J. Catal, 74 (1982) 129.

(4) M. Asscher, J. Carrazza, M.H. Khan, K.B. Lewis, and G.A. Somorjai, submitted to J. Catal.

(5) C.T. Campbell, M.T. Paffett, Surf. Sci., 139 (1984) 396.

(6) D.J. Dwyer, J.H. Hardenbergh, J. Catal., 87 (1984) 66.

(7) D.W. Goodman, R.D. Kelley, T.E. Madey, and J.T. Yates, J. Catal., 63 (1980) 226.

(8) R.D. Kelley, D.W. Goodman, Surf. Sci. Lett., 123 (1982) L743.

(9) M. Logan, A.J. Gellman, and G.A. Somorjai, submitted to J. Catal.

(10) J.E. Turner, B.C. Sales, and M.B. Maple, Surf. Sci., 103 (1981) 54.

(11) R.C. Yeates, J.E. Turner, A.J. Gellman and G.A. Somorjai, Surf. Sci., 149 (1985) 175.

(12) F. Zaera and G.A. Somorjai, J. Am. Chem. Soc., 106 (1984) 2288.

(13) A.J. Gellman, M.H. Farias and G.A. Somorjai, J. Catal., 88 (1984) 546.

(14) D.W. Blakely, E. Kozak, B.A. Sexton, and G.A. Somorjai, J. Vac. Sci. Technol., 13 (1976) 1901.

(15) J.A. Dumesic, H. Topsoe, and M. Boudart J. Catal., 37 (1975) 513.

(16) G. Ertl, Catal. Rev. Sci. Eng., 21 (1980) 201.

(17) N.D. Spencer, R.C. Schoonmaker and G.A. Somorjai, J. Catal., 74 (1982) 129.

(18) L.M. Falicov and G.A. Somorjai, submitted to Proc. Nat. Acad. Sci., 1984.

(19) M. Salmeron and G.A. Somorjai, J. Phys. Chem., 86 (1982), 341.

(20) R.J. Koestner, M.A. Van Hove, G.A. Somorjai, J. Phys. Chem., 87 (1983) 203.

(21) L.L. Kesmodel, L.H. Dubois, and G.A. Somorjai, Chem. Phys. Lett., 56 (1978) 267.

(22) L.L. Kesmodel, L.H. Dubois and G.A. Somorjai, J. Phys. Chem., 70 (1979) 2180.

(23) J.E. Demuth, Surf. Sci., 93 (1980) 182.

(24) P. Skinner, M.W. Howard, I.A. Oxton, S.F.A. Kettler, D.B. Powell, and N. Sheppard, J. Chem. Soc., Faraday Trans. 2, 77 (1981) 1203.

(25) B.E. Koel, B.E. Bent, and G.A. Somorjai, Surf. Sci., in press.

(26) F. Zaera, Ph.D. Thesis, Univ. of California, Berkeley, 1984.

(27) A. Wieckowski, S.D. Rosasco, G.N. Salaita, B. Bent, F. Zaera and G.A. Somorjai, J. Am. Chem. Soc., in press.

(28) S. Kolboe, Can. J. Chem., 47 (1969) 352.

(29) A.J. Gellman, M.H. Farias, M. Salmeron, and G.A. Somorjai, Surf. Sci., 136 (1984) 217.

(30) L.J. Clarke, Surf. Sci., 102 (1981) 331.

(31) A. Farkas and L. Farkas, J. Am. Chem. Soc., 60 (1938) 22.

(32) D. Beeck, Rev. Mod. Phys., 17 (1945) 61.

(33) G.C. Bond, Trans. Faraday Soc., 52 (1956) 1235.

(34) V.B. Kazanskii and V.P. Stronin, Kinet. Catal. (Eng. Trasl.), 1 (1960) 517.

(35) T.A. Darling, M.J. Eastlake, and R.L. Moss, J. Catal., 14 (1969)23.

(36) J.C. Schlatter and M. Boudart, J. Catal., 24 (1972) 482.

(37) M. Salmeron, R.J. Gale, and G.A. Somorjai, J. Chem. Phys., 67 (1977) 5324.

(38) S.M. Davis, F. Zaera, and G.A. Somorjai, J. Catal., 85 (1984) 206.

(39) J.E. Crowell, E.L. Garfunkel, and G.A. Somorjai, Surf. Sci., 121 (1982) 303.

(40) G. Blyholder, J. Phys. Chem., 68 (1964) 2772.

(41) J.W.A. Sachtler and G.A. Somorjai, J. Catal., 81, (1983) 77.

(42) J.E. Crowell and G.A. Somorjai, Appl. Surf. Sci., 19 (1984) 73.

(43) J.E. Crowell, W.T. Tysoe and G.A. Somorjai, J. Phys. Chem., 89 (1985) 1598.

(44) See for example, "The Oxide Handbook" Edited by G.V. Samsonov, IFI/ Plenum, N.Y. 1982.

(45) W.H. Strehlow and E.L. Cook, J. Phys. Chem., 2, (1973) 163.

(46) F.P. Koffyberg, K. Dwight and A. Wold, Sol. St. Comm., 30 (1985) 1735.

(47) B.M. Warnes, F.F. Aplan and G. Simkovich, Sol. St. Ionics, 12, (1984), 271.

(48) P. Merchant, R. Collins, K. Dwight, and A. Wold, J. Solid State Chem., 27 (1979) 307.

(49) a) R. Feder, Phys. Status Solidi, 58 (1973) K137;

b) K.O. Legg, F. Jona, D.W. Jepsen, and P.M. Marcus, J. Phys., C10 (1977) 937;

c) H.D. Shih, F. Jona, D.W. Jepsen, P.M. Marcus, Surf. Sci., 104, (1981) 39.

(50) a) C. Leygraf and S. Ekelund, Surf. Sci., 40 (1973) 609;

b) C.F. Brucker and T.N. Rhodin, Surf. Sci., 57 (1975) 523;

c) P.B. Sewell, D.F. Mitchell, and M. Cohen, Surf. Sci., 33 (1972) 535.

(51) G.A. Somorjai and M. Langell, J. Vac. Sci. and Tech., 21(3) (1982) 858.

(52) C.R. Brundle, T.J. Chuang, and K. Wandelt, Surf. Sci. 68 (1977) 459.

(53) G.A. Somorjai, "Chemistry in Two Dimensions: Surfaces", Cornell Univ. Press, 1981.

(54) C. Sanchez, M. Hendewerk, K.D. Sieber, and G.A. Somorjai, J. Sol. State. Chem., in press (1985).

(55) M. Hendewerk, M. Salmeron and G.A. Somorjai, to be published.

(56) L.A. Marasak, R. Messier and W.B. White, J. Phys. Chem. Solids 41, (1980) 981.

(57) a) A.S.N. Murthy and K.S. Reddy, Mat. Res. Bull., 19, (1984) 241;

b) L.-S.R. Yeh and N. Hackerman, J. Electrochem. Soc., 124(6) (1977) 833.

c) J.S. Curran and W. Gissler, J. Electrochem. Soc., 126(1) (1979) 56;

d) K.L. Hardee and A.J. Bard, J. Electrochem. Soc., 124(2) (1977) 215;

e) R.M. Candea, Electrochim. Acta, 26(12) (1983) 1803.

(58) a) R. Shinar and J.H. Kennedy, J. Electrochem. Soc., 130(2) (1983) 392;

b) P. Iwanski, J.S. Curran, W. Gissler, and R. Memming, J. Electrochem. Soc., 128(10) (1981) 2128;

c) K.G. McGregor, M. Calvin, and J.W. Otvos, J. Appl. Phys., 50(1), (1979) 396.

(59) K.D. Sieber, C. Sanchez, J.E. Turner, and G.A. Somorjai, J. Chem. Soc., Faraday Trans. I, in press.

(60) S.M. Wilhelm, K.S. Yun, L.W. Ballenger and N. Hackerman, J. Electrochem. Soc., 126(3) (1979) 419.

(61) G.A. Somorjai and M. Hendewerk, and J.E. Turner, Catal. Rev.-Science.

(62) B. Kraeutler and A.J. Bard, J. Am. Chem. Soc., 100 (1978) 4317.

(63) a) T. Kawai and T. Sakata, J. Chem. Soc., Chem. Commun., 694 (1980);

b) Nature, 256 (1980) 474;

c) Chem. Phys. Lett., 72 (1980) 87.

(64) S. Sato and J.M. White, Chem. Phys. Lett., 72 (1980) 85.

(65) a) D.E. Aspres and A. Heller, J. Phys. Chem., 87 (1983) 4919.

(66) H. Nakanishi, C. Sanchez, M. Hendewerk, and G.A. Somorjai, J. Phys. Chem., submitted February 1985.

BOLTON - Perhaps one of the reasons for the low efficiency of H_2 evolution from H_2O on the doped iron oxide surfaces is that your flat band potential is about the same as the H_2/H^+ couple. However, bandgap energy is internal energy and not Gibbs energy. You will need additional potential to achieve reasonable rates.

SOMORJAI - I agree completely. It is very likely that there is not sufficient thermodynamic driving force to permit efficient H_2 production from water. Experiments are under way to test a variety of iron compounds that contain both Fe^{3+} and Fe^{2+} ions and may have flat band potentials in more favourable positions with respect to the H_2/H^+ couple than the magnesium doped Fe_2O_3.

GRESS - Is it possible that Fe(0) instead of Fe(II) is responsible for the catalytic activity of Fe_2O_3?

SOMORJAI - If metallic or zero valent iron can be photogenerated, it could certainly explain the catalytic behaviour of iron oxide in producing hydrogen or even ammonia. The difficulty is to find proof for the presence of Fe(0) that should have a short lifetime under the reaction conditions and therefore a low steady state concentration under constant illumination. I shall certainly be looking into the different methods that may be used to detect a small concentration of Fe(0) in the background of a large concentration of Fe^{3+}.

McLENDON - A follow-up on Dr. Gress' question: I would expect Fe_2O_3 to be a very poor electrode for H_2 (in essence, a high overvoltage is required to obtain high rates). What information do you have on this point, which would address the alternate mechanism:

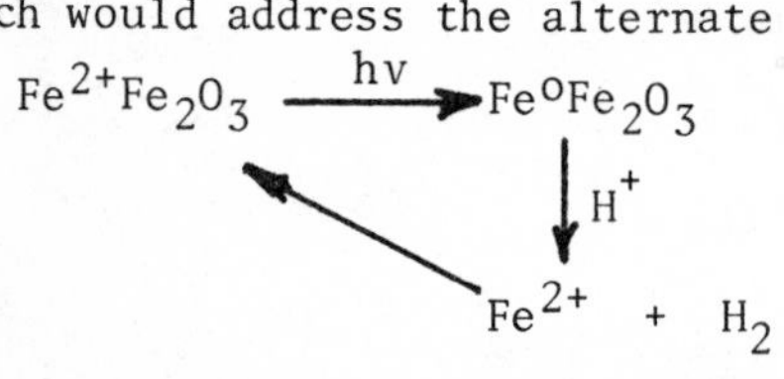

SOMORJAI - There is little doubt that a redox process leading to H_2 evolution from water in the presence of light involves a change in the oxidation state of iron species. The $Fe^{2+} \rightarrow Fe^o$ transition is certainly a possibility. Alternately, surface chemistry studies indicate the reaction of water with Fe^{2+} species on the surface leading to oxidation to Fe^{3+}. Hydrogen is then evolved upon heating.

BRUS - Could you please explain Prof. Falicov's arguement about highly coordinated metal atom such as an adatom. For what class of reactions is the result predicted?

SOMORJAI - The theory is applicable for reactions that occur on transition metal surfaces directly and therefore are likely to be structure sensitive. The theory assumes that rapid catalytic turnover, that requires both breaking and forming strong chemical bonds depends on the availability of degenerate electron vacancy states near the Fermi level. Thus, the activity is proportional to the density of hole states. The calculations indicate that the density of hole states increases with the number of nearest neighbours reaching its highest value for a bulk atom with 12 nearest neighbours. This is why, open, atomically rough surfaces that expose high coordination atomic sites in the second layer to the incoming reactants are such active catalysts.

FRANK - Is it possible to avoid or to reduce the photo-uptake of O_2 in the Fe_2O_3 system?

SOMORJAI - It may be possible to scavenge oxygen by another reactant before it reacts with Fe_2O_3. Since nitrogen produces ammonia more efficiently under illumination, then the rate of hydrogen evolution detectable in the absence of nitrogen may serve this purpose. We should explore the reactions of photoproduced oxygen with a variety of reactants in the presence of Fe_2O_3.

PICHAT - From your experiments, what is your opinion about the origin of the so-called "strong metal-support interaction", namely how do you explain the decrease in the H_2 amount adsorbed (for catalysts with low M content this amount can be nearly nil) and the reversibility of the effect when exposing the catalysts to O_2 at room temperature?

SOMORJAI - The best explanation at present, using the available experimental evidence is the formation of a surface compound between the partially reduced titanium oxide and the transition metal. Hydrogen binds only weakly to the transition metal ions in this compound that gives rise to the so-called strong metal-support interaction. Once reoxidized, there is a phase separation to TiO_2 and the transition metal phase which destroys the effect. However, the surface compound can be regenerated again by subsequent reduction of the TiO_2-metal system in hydrogen.

PHOTOCATALYSIS OVER CLAY SUPPORTS

H.VAN DAMME, F.BERGAYA and D.CHALLAL
C.N.R.S. - C.R.S.O.C.I.
45045 ORLEANS CEDEX
France

ABSTRACT. Clay Minerals can be used in photocatalysis (i) to support heterogeneous redox catalysts, including metals, oxides and sulphides (ii) to anchor, by covalent linkage, homogeneous redox catalysts, and (iii) to intercalate photocatalytic transition metal complexes. The photocatalytic reaction can be performed either at the solid-solution interface in aqueous clay suspensions or at the solid-gas interface on clay powders or clay films. The influence of the clay microstructure on the distribution of molecular species in the reaction medium is of primary importance for the efficiency of the process. The reaction rate can be either considerably enhanced or totally depressed. The examples which are treated include the photo-oxidation of water, the photoreduction of water, the photocleavage of water, and the photoreduction of carbon dioxide by hydrogen (reverse water gas shift reaction).

1. INTRODUCTION

Among all naturally occuring minerals, clays occupy a unique position in chemistry and physical chemistry [1]. This position is partly due to the importance of clays in soil fertility, and partly to their exceptionnaly broad spectrum of applications [2,3]. In this respect, only zeolites can compete with them. Typical and classical applications of clays are paper coating, ceramics, water purification, oil well drilling muds, detergents , pesticide carriers, cosmetics, paints ,.. Most of these applications are ultimately based upon the small size of clay particles, and upon their surface or colloïdal chemistry.

The interest in clays in the field of catalysis [4] is liable to some fluctuation, although the catalytic activity of clays in natural processes such as petroleum genesis or soil chemistry has long been recognized [5]. More than twenty years ago, acid-treated bentonites were extensively used as petroleum-cracking catalysts, but the advent of synthetic zeolites led to the vanishment of this use. Clays are still used for the dimerization and oligomerization of oleic acid. Recent advances in the intercalation chemistry of swelling clays opened new prospects for these materials [6] both for highly selective reactions in

E. Pelizzetti and N. Serpone (eds.), Homogeneous and Heterogeneous Photocatalysis, 479–508.

mild conditions, and for more energetic reactions at higher temperature. In the first type of reactions, the clay is used as an (active) matrix for performing homogeneous catalysis thanks to intercalated metal complexes [7],whereas in the latter type, one takes advantage of the ability of swelling clays to generate porous materials (the so-called pillared clays) analogous to zeolites to perform classical heterogeneous catalysis [8].

As far as photocatalysis is concerned, a first point which should be realized is that clays are not (and will probably never be) photocatalysts by themselves. Indeed, clays are basically insulating materials which do not absorb light in the visible and near UV region of the spectrum, unless they are doped on purpose by transition metal ions such as Cr or Cu. Hence, there is little catalytic activity to be expected upon shining light on a pure clay sample. In fact, the interest in clays in the field of photocatalysis stems from the possibility of using them as an organized medium for running complex photochemical reactions involving molecular species. Clays are a very versatile ensemble of colloïdal solid particles, with different morphologies, surface-electrical and surface-chemical properties which make them suitable as catalyst support and (or) as adsorbant for controlling the position or the movements of molecular species. In addition one should be able to use their own surface chemistry in order to orient the selectivity of a photochemical reaction.

The purpose of this contribution is to review a few typical applications of clays in photocatalysis. We will not consider elementary photophysical and photochemical processes, such as luminescence quenching, energy transfer or electron transfer reactions on clay surfaces, although this is a rapidly growing field [9-12],which raises very fundamental problems on excited states and on surface kinetics. We will rather focus on more complex reactions which involve either a truly photocatalytic step (i.e. a light-induced catalytic reaction in which the catalyst is also the light-absorbing species), or a dark catalytic step, as a part of a light-induced reaction sequence. Only a few examples of such reactions have been published. Most of them take place at the solid-solution interface, and have been developped for the photochemical cleavage of water. One of them will however deal with a reaction (the reduction of CO_2) occuring at the solid-gas interface, while still involving a molecular catalyst. This should be a good (though largely improvable) example of the unique contribution that clays might bring in photocatalysis.

2. STRUCTURE, CLASSIFICATION AND MORPHOLOGY OF CLAYS

2.1. Structure [13]

Before entering photocatalysis, we will briefly recall some basic features of clays which are usefull to understand the variety of possible situations. For geologists and soil scientists, clays are defined as the fine fraction of rocks and soils, with an upper limit for the particle size at 2 μm. It turns out that this purely granulometric definition corresponds closely to a particular class of hydrous silicates with

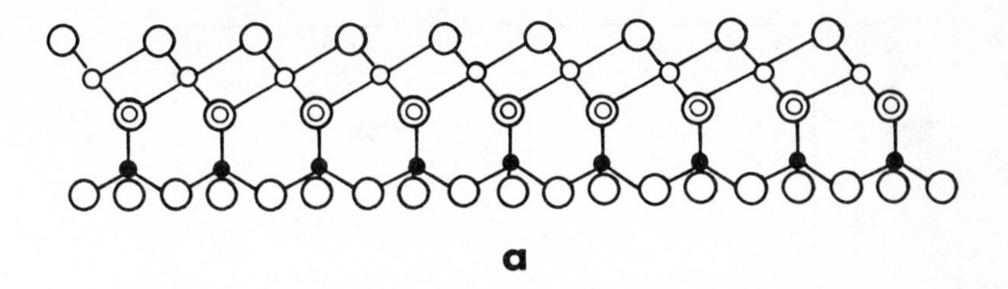

a

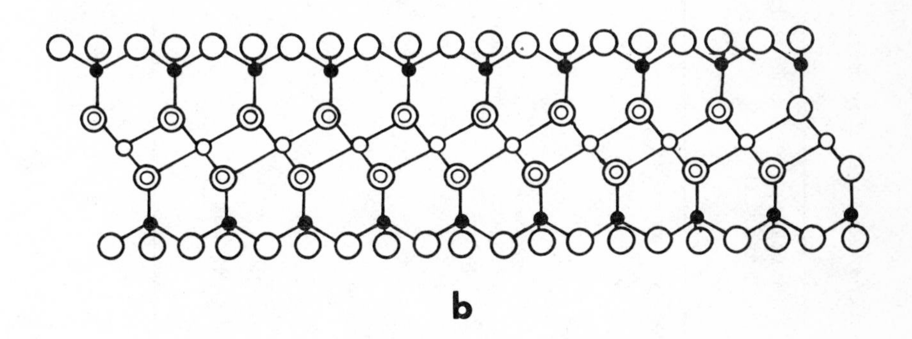

b

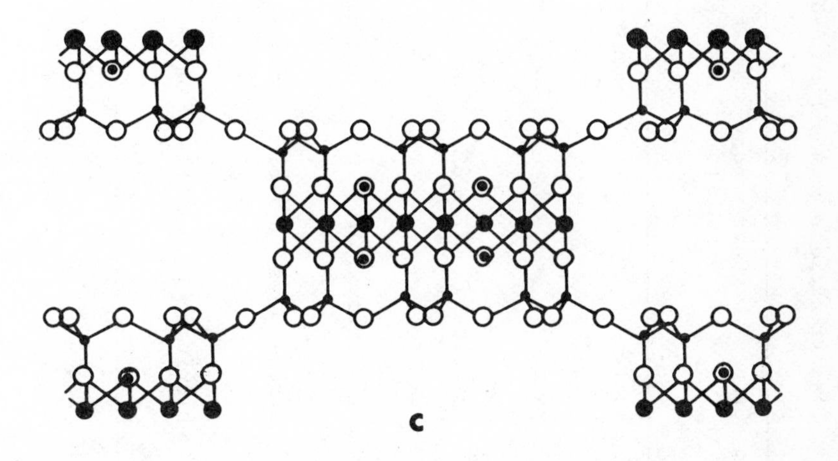

c

Figure 1. Structures of (a) 1:1 layers (kaolinite); (b) 2:1 layers (smectite); (c) inverted 2:1 ribbons (sepiolite).

Table 1. Classification, composition and properties of some clay minerals

Structure type	Group	Charge per formula unit	Behavior in water	Interlayer[a] material	Subgroups	Potential for photo-catalysis
1:1 layer	kaolin	0	non expanding	no		low
2:1 layer	talc	0	non expanding	no		low
2:1 layer	smectite	0.2–06	expanding	Na^{+}, Ca^{2+} water	montmorillonite hectorite-laponite	high
2:1 layer	illite-mica	0.6–1.0	non expanding	K^{+}		low
2:1 inverted ribbons	fibrous clays	variable	non expanding	-	sepiolite palygorskite	high
2:1 layer	chlorites	variable	non expanding	$Mg(OH)_2$		low

[a] in the natural clays

layer structure. The elementary polyhedra of layer silicates are tetrahedra and octahedra of oxygen ions. The coordinating cations, T, in the center of the tetrahedra are essentially Si, Al or Fe^{3+}. Si is by far the most frequent. The coordinating cations, O, in the center of the octahedra are usually Al, Mo, Fe^{3+} and Fe^{2+}. The "backbone" of phyllosilicates is a continuous two-dimensional tetrahedral T_2O_5 sheet, in which the tetrahedra are linked by sharing three corners. The apical oxygen at the fourth corner, usually directed normal to the sheet, forms part of an immediately adjacent octahedral sheet in which octahedra are linked by sharing edges. The unit formed by linking one octahedral sheet to one tetrahedral sheet is named a 1:1 layer (Fig.1a), whereas the unit formed by sandwiching one octahedral sheet between two tetrahedral sheets is named a 2:1 layer (Fig.1b).

1:1 and 2:1 layers are not always neutral. Very often, as a result of ionic substitutions in their lattice, the layers bear a net negative charge which is balanced by interlayer material. This interlayer material can be individual cations, hydrated cations, or an additional octahedral hydroxide layer. The highest charges are found in 2:1 layers.

2.2. Classification and morphology

The layer type, the charge on the layer and the nature of the interlayer material provide the basis for the classification of clay minerals. We will of course not consider here all the varieties of clay minerals common in soils and sediments.An exhaustive compilation can be found in ref.14. A brief description of the major groups is shown in Table 1. Each of them display very distinct and typical structural, morphological, colloïdal and surface-chemical properties.

Kaolinite is a typical 1:1 clay, with zero layer charge. There is no interlayer material in the natural clay, and the elementary layers stack on each other thanks mainly to hydrogen bonds between adjacent layers (Fig.1a). The resulting cristallites are rigid particles. Typical shapes and sizes are 0.5 µm hexagonal platelets, a few hundreds of Å thick. Their specific surface area is rather low : of the order of 20-50 m^2/g. Water does not enter the interlayer space of kaolinites,and kaolinites do not form stable gels in water. As expected from their zero lattice charge, kaolinites have a very small cation exchange capacity, of the order of $2x10^{-5}$ eq./g. When taken together, these properties make kaolinite of little interest for photocatalysis.

Smectites have properties almost opposite to those of kaolins. The layer charge is large, typical 0.5 negative charge per formula unit layer. This corresponds to a cation exchange capacity of the order of 10^{-3} eq/g. i.e. fifty times more than kaolinite. Statistically, each lattice charge covers a surface area of $\sim$ 130 Å^2. In natural smectites,this charge is compensated by hydrated interlayer sodium or calcium cations. Most important, the hydration energy of the interlayer cations allow water to enter the interlayer space (in clays with a lattice charge even larger than in smectites, the attractive interaction becomes too strong to be overcome by hydration forces). This makes the original interlayer cations easily exchangeable by almost any cationic species. The interlayer space is also accessible to a welth of neutral molecules [15]. Morphologically, smectites appear as large (∿ 5 x 10^3 Å) and thin (∿ 10 Å) flexible sheets. The most common members of the family are montmorillonite and hectorite. In montmorillonite, the octahedral cations are Al ions and the layer charge originates from the substitution of some Al^{3+} by Mg^{2+}. In hectorite, the octahedral cations are Mg^{2+} and the lattice charge originates from the substitutions of some Mg^{2+} by Li^+ : a synthetic hectorite is commercialized under the name laponite. In other smectites, like beidellite for instance, the lattice charge originates from substitutions (Si^{4+} by Al^{3+}) in the tetrahedral sheets. Beidellite can also be easily synthesized in the laboratory. Tetrahedrally charged smectites have a larger surface field than octahedrally charged smectites. This is easily understandable since in the latter case the charge is embedded in the center of the layer and is screened by a large number of oxygen atoms. Such a small structural difference can lead to drastically different catalytic activities. Generally speaking, the high surface area of smectites, their large cation exchange capacity, their transparency and the versatility of their intercalation chemistry makes them extremely interesting materials for photocatalysis.

Fibrous clays are another important group of clay minerals. Sepiolite and palygorskite are the most common members of the group. Both have a 2:1 layer structure, like smectites, but the lateral extension of the layers is restricted to six tetrahedra in sepiolite and four tetrahedra in palygorskite, so that the layers become parallepipedic ribbons. Each ribbon is linked to four parallel ribbons by sharing edges. This leads to an alternate ribbon-channel structure (Fig.1c). Morphologically, fibrous clays appear either as rigid needles or as more or less flexible fibers. Typical sepiolite fibers have a diameter of ∿ 0.1 μm and are ∿ 2-3 μm long. The channels contain small ions and water molecules. Though smaller than that of smectites, the cation exchange capacity of sepiolite and palygorskite is not negligible : of the order of 10^{-4} eq/g. This, together with their high surface area (which can be larger than 200 m^2/g) and their aptitude for derivatization (section 5) makes fibrous clays also useful for photocatalysis.

3. CLAY-WATER SYSTEMS - INTERNAL *VS* EXTERNAL SURFACES - PILLARED CLAYS.

Performing and controlling a catalytic or photocatalytic process in an heterogeneous environment clearly implies the knowledge of the texture of the medium. One has to know the chemical nature of the surface sites and the quantitative importance (i.e. the surface area) of the interfacial regions where the reaction takes place. For instance, in zeolite catalysts, the size and the electrical charge of the reactants will determine whether they have access to the internal space of the zeolitic cages (or channels) or not. A similar but more subtle situation occurs with clays, because the texture of the heterogeneous medium is very often dependent upon the reaction conditions (solid-liquid or solid-gas) and the nature of the reactants. In this section, we will briefly consider the various situations which may arise with the two groups of clays which are the most interesting for photocatalysis, namely smectites and fibrous clays (sepiolite).

Let us first consider smectites. Among the best known properties of smectites is their ability to swell and to shrink upon addition or removal of water. For this reason, smectites are also very often referred to as swelling clays [16]. They are also able to form gels and stable colloïdal suspensions. As already pointed out, the first stage of swelling is the hydration of the interlayer cations. This allows a few layers of water molecules to enter the interlayer space. Increasing further the water content of the clay yields first a plastic solid, then a fluid paste, a gel, and finally a diluted colloïdal suspension. In this high water content regime, the swelling of the clay texture ("or microstructure") is essentially an osmotic phenomenon.

Swelling is clearly associated with a statistical increase of the average distance between smectites layers [17,18]. However, even in very diluted colloïdal suspensions, the layers are not totally dissociated [19]. The basic microstructural units in smectite suspensions are aggregates of a few layers more or less parallel to each other, as sketched in Figure 2A. The size of the aggregates is strongly dependent upon the ionic strength of the medium and the nature of the counterions. High ionic strengths floculate the clay, i.e. leads to the growth of the aggregates up to a point where they are too heavy to remain in suspension (this happens, for instance, when a smectite in dispersed in an aqueous solution containing a large amount of EDTA as sacrificial electron donor in water photoreduction experiments). Highly charged and polarizing counterions and, most important, the cationic organic dyes and cationic organometallic complexes often used in photocatalysis also tend to flocculate the smectites in water.

The existence of aggregates considerably modifies the surface area available for chemical reactions. A smectite/water suspension in which all the clay layers would be separated from each other would have a surface area close to 750 m^2/g. In addition, all this surface area would be in direct contact with the aqueous phase and could therefore be considered as entirely "external". As soon as aggregation occurs, a fraction of the total surface becomes part of the interlayer space of the clay, i.e. becomes "internal", and the external surface decreases accordingly. For instance, in a Na^+-hectorite/water suspension containing 10% w/w of solid, the external surface area of the aggregates drops to $\sim$ 70 m^2/g [19]. The same sample exchanged with a tris-bipyridine Ru(II) cation, $Ru(bpy)_3^{2+}$, would have an external surface area still much lower.

Internal surfaces are much less accessible from the solution than external surfaces, even for the water molecules themselves. This is an important factor to take into account in the design of a photocatalytic reaction. For instance, a reaction sequence involving a light-induced electron transfer between a cationic sensitizer, most of which will be on the internal surfaces of the smectite aggregates, and a neutral acceptor molecule in the solution phase, would have only a very limited efficiency. The internal *vs* external segregation would of course be even stronger with a negatively charged acceptor molecule, because of the electrostatic repulsion (depletion layer) between the smectite aggregate and the acceptor molecule.

Further segregation phenomena can take place within the internal space of the aggregates. Indeed, the charge density of the individual

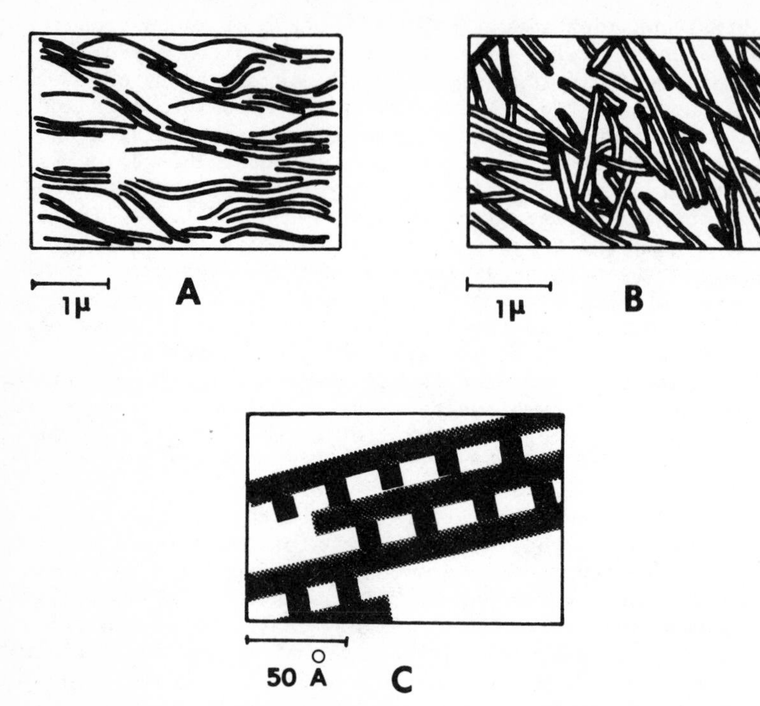

Figure 2. Microstructure of smectite (A), sepiolite (B) and pillared smectite (C).

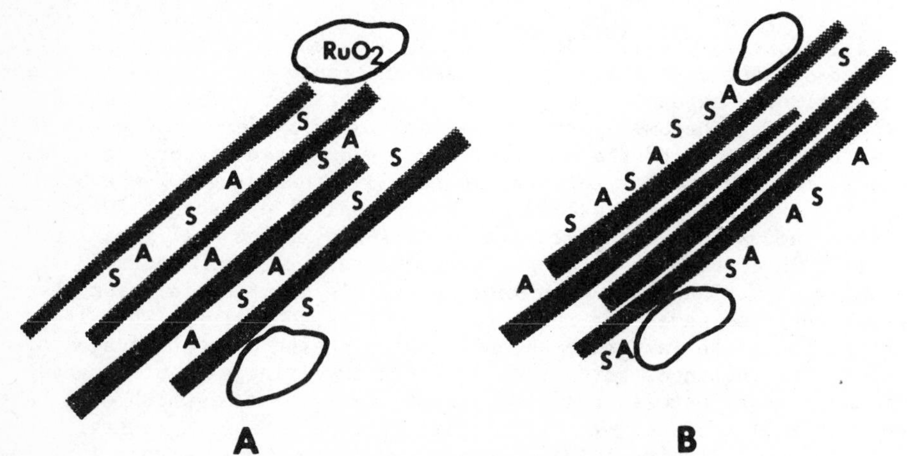

Figure 3. Sketch of the molecular distribution during the photo-oxidation of water in suspensions of untreated (A) or thermally treated (B) smectite.

smectite layers is not homogeneous [20], and this can lead to the preferential location of some compounds on high or low charge density surface. In an aggregate containing two types of cationic reactants, this can impede the encounter of the reactants. This happens, for instance, in hectorite containing co-adsorbed $Ru(bpy)_3^{2+}$ and methylviologen [21,12]. In water, the luminescence of $Ru(bpy)_3^{2+}$ is strongly quenched by methylviologen through a diffusion-limited electron transfer reaction [22].On the opposite, on the hectorite surface, only a negligible quenching effects is observed. Since the redox properties of adsorbed methylviologen are not modified with respect to methylviologen in water, this effect must be ascribed to a segregation phenomenon. Fortunately, segregation does not always occur. For instance, with $Ru(bpy)_3^{2+}$ and $Cr(bpy)_3^{3+}$ co-adsorbed on hectorite, no anomalous behavior is observed [21].

The accessibility problems related to the existence of aggregates and internal surfaces do not arise in suspensions of fibrous clays like sepiolite, because the fibers make an entangled network in which the close contact points between fibers represent only a very small fraction of the total sufrace area of the fibers, as sketched in Figure 2B (it should be pointed out that, although the structure of fibrous clays contains internal channels, the "window" of those channels is too small to allow for the penetration of most molecules of interest in photocatalysis). Hence, almost all the surface of the fibers is available for photocatalysis, and this surface is directly open to the dispersion medium. For instance, in a suspension of Na^+-sepiolite in water (10%, w/w), the surface area in direct contact with the solution is $\sim 250\ m^2/g$ [19].

Running a reaction on a "dry" sepiolite powder, at the gas solid interface, is as easy as in a aqueous dispersion, because the microstructure of the medium is basically the same (an entangled network of fibers with very limited contact points). Smectites behave again very differently. The removal of water collapses most of the interlayer spaces, the size of the connected aggregates grows up to macroscopic sizes, and the surface area accessible to non polar molecules such as nitrogen drops to very small values. Only a very small fraction of the adsorbed species are still able to do chemistry.

This difficulty with smectites can be circumvented by pillaring the clay, i.e. by intercalating the clay with large cations wich keep the individual layers apart, even when water is removed.The most widely used pillaring agents are polynuclear hydroxy aluminum and zirconium cations. The structure of a smectite pillared with hydroxy aluminum cations is shown in Figure 2C.They yield high spacing materials,with a good thermal stability. Alkylammonium ions and bicyclic amine cations have also been used (in the so-called organo-clays). Interestingly, several bulky cations used as sensitizers in photochemistry, like tris-bipyridine or tris-phenanthroline complexes, have also good pillaring properties.

4. PHOTOCATALYSIS OVER CLAY-SUPPORTED Ru METAL, RuO_2, and RuS_2

4.1. Photo-oxidation of water

Oxygen evolution from water is a multi-electron reaction which is notoriously known to be difficult, and it is one of the limiting steps in the photocatalytic cleavage of water. The most active heterogeneous (electro)catalysts are oxides of ruthenium and irridium, and their mixed oxides. Their activity has been established in electrochemical, chemical, and photochemical conditions [23-26].

The first attemps to photo-oxidize water with clay-supported catalysts were performed with hectorite (a smectite)-supported RuO_2 dispersed in water [27]. This catalyst can be prepared in various ways. For instance, one can hydrolyze ruthenium trichloride in an aqueous clay suspension, freeze-dry the mixture, and calcine the product in air at 450 K. One can also exchange the clay with the hexamine ruthenium(III) cation, and calcine the product. The [hectorite-RuO_2] catalyst can be further loaded with a sensitizer, $Ru(bpy)_3^{2+}$, to yields a [hectorite-RuO_2-$Ru(bpy)_3^{2+}$] system, which might be considered as a model for photosystem II particles. The photocatalytic activity of this system in the oxidation of water was tested in sacrificial conditions, with Co(III) chloropentamine, $Co(NH_3)_5Cl^{2+}$, as sacrificial acceptor, according to a now classical reaction scheme [24-25] :

$$S + A \xrightarrow{h\nu} S_{OX} + A_{RED} \tag{1}$$

$$A_{RED} \longrightarrow \text{Decomposition products} \tag{2}$$

$$S_{OX} + OH^- \xrightarrow{Cat_{OX}} S + 1/2\ H_2O + 1/4\ O_2 \tag{3}$$

No oxygen was evolved upon photolysis of the hectorite-based system. This was a frustrating result because very active catalysts for the same reaction were prepared by LEHN et al. [24], by supporting the redox catalyst on faujasite-type zeolites, which are also negatively charged silicates.

In fact the absence of activity in the hectorite-based system can be explained by considering the microstructure (internal and external surfaces) of the system. $Ru(bpy)_3^{2+}$ and $Co(NH_3)_5Cl^{2+}$ are both cations and they are essentially on the internal surfaces of the hectorite aggregates in the suspension. The situation is very different for the RuO_2 catalyst particles. Due to their large size (from 50 to 500 Å, depending on the preparation procedure), they are expelled from the interlayer space, and they are located on the external surfaces of the clay aggregates, as sketched in Figure 3A. This configuration is, of course, very unfavourable since S has little access, if any, to the catalyst. Taking into account the location of the components either on the internal surfaces or on the external surfaces, reactions 1-3 can be written :

$$\underline{\overline{S}} + \underline{\overline{A}} \xrightarrow{h\nu} \underline{\overline{S_{OX}}} + \underline{\overline{A_{RED}}} \tag{4}$$

$$\underline{\overline{A_{RED}}} \longrightarrow \text{decomposition products} \tag{5}$$

$$\underline{\overline{S_{OX}}} + \underline{Cat_{OX}} + H_2O \longrightarrow \text{no reaction} \tag{6}$$

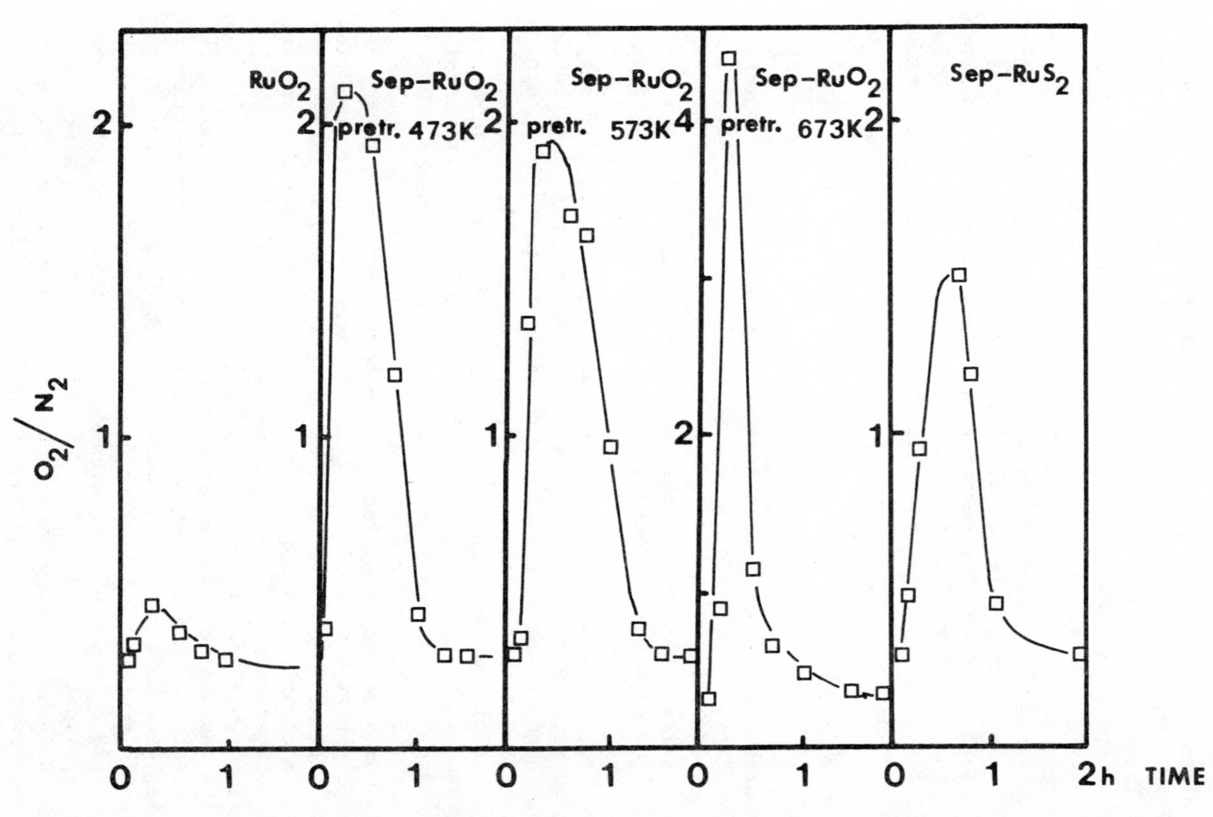

Figure 4. Oxygen to nitrogen ratio in the gas phase during the photo-oxidation of water with various catalysts (50 ml H_2O; 250W tungstan lamp; 0.04 mole $Co(NH_3)_5Cl^{2+}$; $5x10^{-5}$mole $Ru(bpy)_3^{2+}$; pH 4). The amount ofmetal catalyst is the same in each experiment ($8x10^{-5}$m).

In this notation, the species "intercalated" between an upper and a lower line are those which are in the interlayer space of the clay aggregates, whereas the species which are only underlined are those which are on the external surfaces of the aggregates.

This "microstructural" hypothesis was confirmed [27] by pretreating the [hectorite-RuO_2] samples in hydrogen, at 873 K, before adsorbing the sensitizer. This severe thermal treatment irreversibly collapses most of the interlayer spaces and leads a low surface area material (∿ 20 m^2/g) which does no longer swell in water. The $Ru(bpy)_3^{2+}$ ions are forced in this way to remain on the external surfaces of the aggregates, where they can react with water over the catalyst particles. This is sketched in Figure 3B. The situation is now similar to that occuring in zeolite-supported RuO_2 catalysts where, due to their large diameter, the $Ru(bpy)_3^{2+}$ ions cannot penetrate the zeolitic cages (although, as shown by LUNSFORD et al. [28], it can be synthesized within the cages). Using the same notation as above eq.6 becomes :

$$\underline{S_{OX}} + \underline{Cat_{OX}} + H_2O \longrightarrow \underline{S} + \underline{Cat_{OX}} + 1/2\ H_2O + 1/4\ O_2 \qquad (7)$$

It is worth noting that a treatment in hydrogen at 873 K reduces ruthenium to the metallic state. This implies that ruthenium metal particles (or metal particles coated with a few monolayers of oxide) are also active catalysts for the oxidation of water. This is not surprising since Ru metal on carbon electrodes are used in the electrochemical generation of oxygen.

As expected, these accessibility problems do not arise with sepiolite. Its high surface area - all external - and its non negligible cation exchange capacity make it a very interesting support for running a photocatalytic reaction with molecular species at the solid-solution interface. The activity of sepiolite supported-RuO_2 is compared with that of powdered RuO_2 in Figure 4 (In fact, what is shown is not the total gas production but the oxygen to nitrogen ratio in the gaseous products. The reason for this xill be briefly discussed in section 5). In this case, ruthenium dioxide was deposited by decomposing ruthenium tetroxide in a aqueous dispersion of the clay. After deposition, the freeze-dried material was activated in air at various temperatures. The clay-supported catalyst is clearly much more active than the unsupported catalyst. The optimum activation temperature is around 600 K. This is somewhat higher than for RuO_2/Y-zeolite catalysts [24].

Sepiolite can also be used as support for sulphided catalysts like RuS_2, for instance. Such a catalyst can be prepared by impregnating the sepiolite clay with $RuCl_3$, and subsequently sulphiding the catalyst in a 90/10 H_2/H_2S mixture at 700 K, according to the procedure of PECORADO and CHIANELLI [29]. RuS_2/sepiolite proved to be a very acceptable catalyst for water oxidation (Figure 4). This is not surprising since, as an electrode, RuS_2 has already been shown to have a high catalytic activity for oxygen evolution from aqueous solution. As shown in Fig.4, the activity of RuS_2/sepiolite is of the same order of magnitude as that of RuO_2/sepiolite activated at 473 K.

4.2. Photoreduction of water

So far, clay-supported catalysts have been much less developped for the photoreduction of water than for its photo-oxidation, and almost nothing has been published on the subject. Early attempts were devoted to running a classical "sacrificial electron donor - sensitizer - relay - catalyst" system in a clay environment, according to the reaction scheme:

$$S + A \xrightarrow{h\nu} S_{OX} + A_{RED} \quad (8)$$

$$S_{OX} + D_{SAC} \longrightarrow S + D_{SAC}^{OX} \quad (9)$$

$$D_{SAC}^{OX} \longrightarrow \text{Decomposition products} \quad (10)$$

$$A_{RED} + H^+ \xrightarrow{Cat_{RED}} A + 1/2\ H_2 \quad (11)$$

The EDTA-$Ru(bpy)_3^{2+}$-methylviologen-Pt (or RuOx) model system [30] was first examined. In section 3 we already mentionned the segregation problems which arise upon co-adsorption of $Ru(bpy)_3^{2+}$ and MV^{2+} on smectites. As expected, no significant hydrogen evolution was observed with $Ru(bpy)_3^{2+}$ and MV^{2+} adsorbed on a Pt/smectite or RuO_2/smectite catalyst. In order to avoid segregation, sepiolite was used, and hydrogen evolution was indeed observed. Nevertheless, the activity of all the sepiolite-supported catalysts which were tested is lower than that of the classical catalysts such as unsupported platinum metal (generated in sity by reduction of chloroplatinate), platinum metal protected by PVA, sulphided platinum, or powdered RuO_2. A few examples, including Pt/sepiolite, RuO_2/sepiolite are compared to Pt/PVA in Figure §. The drop in activity is very important.

The reason for this low activity has again to be looked for in the microsctucture of the medium. The clay surface is negatively charged, and the redox catalyst is embedded in this surface field. EDTA, the sacrificial electron donor, is also negatively charged above pH . It is therefore repelled from the clay surface. On the other hand, the sensitizer and the relay (acceptor) are both cationic species and they interact attractively with the surface field of the clay. Even without any S-A segregation, this is not a very favourable configuration, for at least two kinetic reasons. First, the clay-solution interfacial region, in which the electrical double layer developps, is depleted in EDTA with respect to the bulk of the solution, whereas it is of course enriched in sensitizer. Hence, reaction 9 should be slower in a clay environment than in a purely homogeneous medium. Secondly, the reorganisation of the double layer, which takes place after the S → A light-induced electron transfer, modifies the structure of the interfacial region in a way which is detrimental to hydrogen evolution. Indeed, the selectivity of the clay surface for highly charged cations should lead to a very narrow Stern-like double layer for S_{OX} and to a more diffuse adsorption layer for A_{RED}. Since hydrogen evolution requires the "encounter" of A_{RED} with the redox catalyst particles which are on the clay

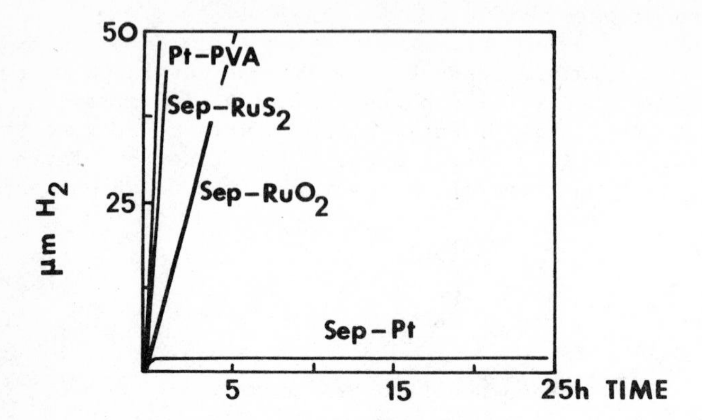

Figure 5. Hydrogen evolution in the sacrificial photoreduction of water with various catalysts (50 ml water; 250W tungsten lamp; $5x10^{-5}$ mole $Ru(bpy)_3^{2+}$; $5x10^{-4}$ mole MV^{2+}; 10^{-2} mole EDTA). The amount of metal catalyst is the same in each experiment ($2x10^{-5}$ MOLE).

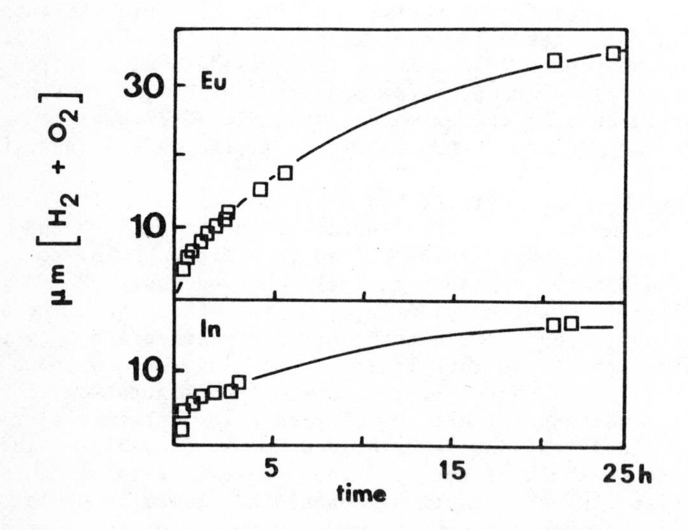

Figure 6. Photocleavage of water with $Al_{0.7}Eu_{0.3}(OH)_3$ and $Al_{0.5}In_{0.5}(OH)_3$ relay colloids in mixed RuO_2-$Ru(bpy)_3^{2+}$-sepiolite hydroxide suspensions.

surface, one expects a detrimental effect on reaction 11 (it should be pointed out that this reorganisation of the double layer is favourably acting in the photocatalytic oxygen evolution system discussed in section 4.1). This can be sketched as follows :

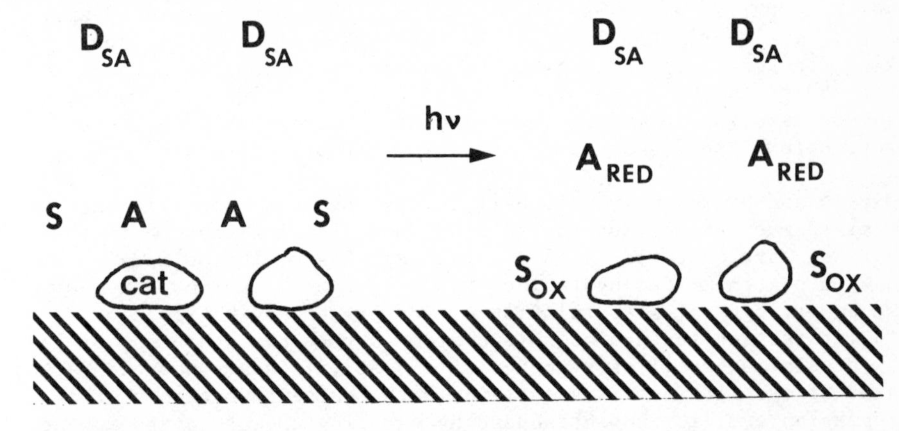

Both points might explain the low efficiency of the overall photoreduction process with sepiolite-supported catalysts and methylviologen as electron relay.

4.3. Photocleavage of water

The conclusion from the results summarized in section 4.1 is that fibrous clays like sepiolite provide an interesting basis for a water oxidation photocatalytic subsystem : [sepiolite - $Ru(bpy)_3^{2+}$- RuO_2]. This subsystem was therefore chosen for being the basis of a complete water photocleavage system, according to the reaction scheme

$$Ru(bpy)_3^{2+} + A \xrightarrow{h\nu} Ru(bpy)_3^{3+} + A_{RED} \quad (12)$$

$$Ru(bpy)_3^{3+} + OH^- \xrightarrow{Cat_{OX}} Ru(bpy)_3^{2+} + 1/2\ H_2O + 1/4\ O_2 \quad (13)$$

$$A_{RED} + H^+ \xrightarrow{Cat_{RED}} A + 1/2\ H_2 \quad (14)$$

The thermodynamic and kinetic difficulties for running an actual molecular system according to this mechanism are notorious [31] . The kinetic difficulties are associated with the stringent necessity to accelerate as much as possible reactions 13 and 14 in order to avoid back electron transfer. Interfacial systems have been proposed very early to solve these problems,and a wide range of situations have been studied. Micelles [32], polyelectrolytes [33], and colloïdal oxides [34] have been shown to promote very efficiently the separation of the redox intermediates by electrostatic interactions, provided the charge of the interface and of the S - A system was varefully chosen. The coupling of the oxygen evolution catalyst with the sensitizer on the sepiolite surface is in this respect very beneficial. In order to achieve an equivalent coupling for the hydrogen evolution reaction, a second colloïdal subsystem was designed [35], in which the support is an aluminum hydroxide, the catalyst is platinum,and the acceptor is a simple transition metal ion with suitable redox potential like Eu^{3+}, In^{3+} or Sm^{3+} doped into the hydroxide particles by co-precipitation. The subsystem can be written $[Al_xA_{1-x}(OH)_3\text{-}Pt]$. Coupling of the two subsystem is easily achieved by mixing the clay suspension with the hydroxide suspension. Indeed,the electric charge on the doped aluminum gel being positive below pH 6, spontaneous association of the two colloïds occurs in this pH range (however, below ∿ pH 3, magnesium ions from the sepiolite are leached out).

Oxygen and hydrogen evolution is indeed observed upon illumination of the mixed suspension, but the turnover numbers with respect to the acceptor ions are quite low (∿ 5 electron/A). Gas evolution is also observed with platinum-free hydroxides [36,37], but the turnover number is even lower. Typical gas evolution curves are shown in Figure 6.

An interesting feature of this system is that the evolution of gases may display an oscillatory behavior [25,36]. It seems now clear that the onset of oscillations is associated with the perturbation of the gas-suspension equilibrium introduced by sampling the gas phase in the closed reactor [36].

Examination of the relay hydroxides in sacrificial conditions,with EDTA as sacrificial donor, shows that hydrogen evolution in these conditions is characterized by turnover numbers with respect to the relay ions of the same order as in cyclic conditions [37]. These low turnover numbers most probably stem from the irreversible character of the A^{3+}/A^{2+} redox couples which are involved, and they raise serious doubts about the future of these hydroxides as relay colloïds in water cleavage systems. Nevertheless, the fact that some hydrogen and some oxygen has been produced is an encouraging point and suggests that the design of the system is worth being developped. The improvements imply the use of faster and more reversible relay colloïds, and a better control of the mixed colloïdal microstructure, which, so far, has been only achieved on a statistical basis.

5. PHOTOCATALYSIS OVER CLAY-ANCHORED CATALYST :PHOTO-OXIDATION OF WATER

One of the classical ways to "heterogenize" homogeneous catalysis is to anchor the catalyst on a mineral support by covalent linkage. Fibrous clays are interesting supports for that purpose. Several methods have been developped to prepare organo-mineral derivatives as rubber reinforcing agents, and the same methods can be applied in photocatalysis. A classical method is to react a magnesian clay, like sepiolite for instance, with a chlorosilane in acidic medium. The mechanism involves elimination of HCl, from the OH groups of the mineral surface and the chlorine of the silane, and formation of Si-O-Si linkages. The reaction has some autocatalytic character since the protons liberated by the reaction generate new OH groups through an hydrolysis reaction

$$\text{-O-Mg-O-} + 2\ H^+ \longrightarrow \text{-OH HO-} + Mg^{2+} \qquad (15)$$

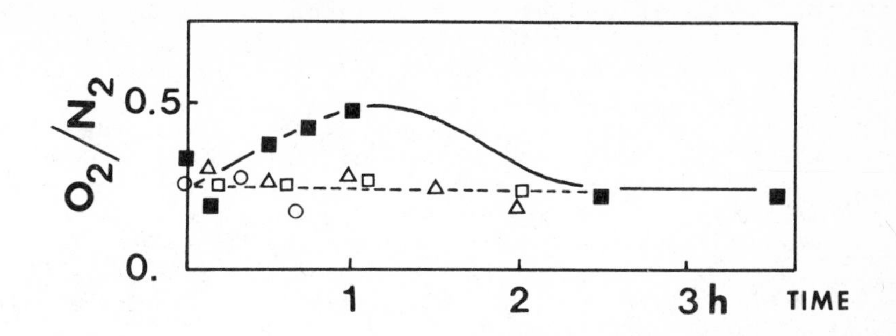

Figure 7. O_2/N_2 ration in the gas phase during the photo-oxidation of water with Os-sepiolite as catalyst (215 mg sepiolite, 7% w/w Os; other conditions as in Fig.4). The pretreatments are as follows: (○) no pretreatment; (■) H_2, 473K, 5 hrs; (△) H_2, 573K, 15 hrs; (□) air, 573K, 15 hrs.

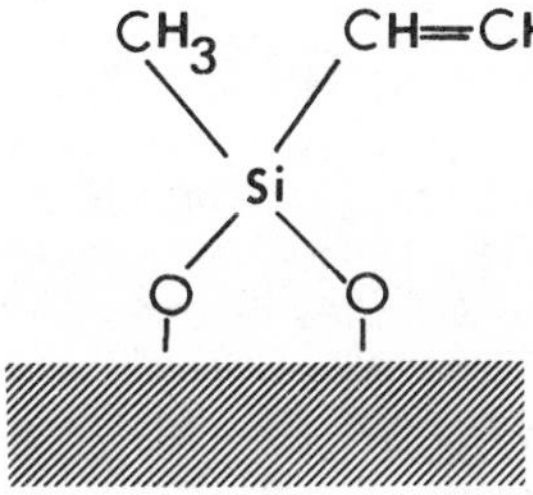

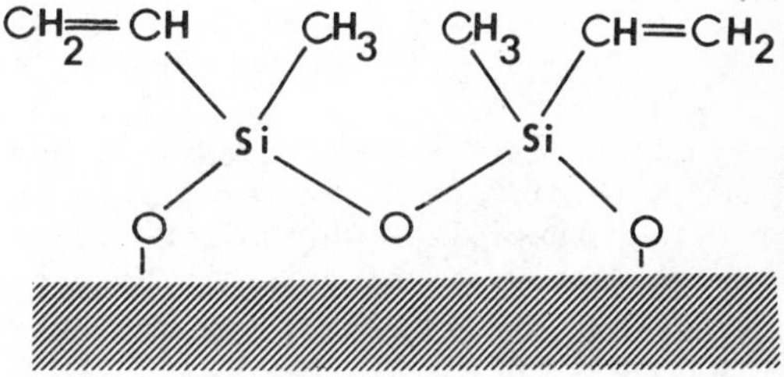

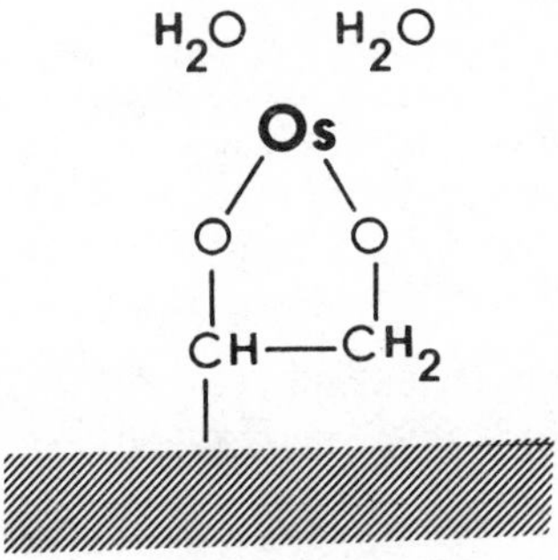

This method has been used to link methyvinylsiloxyl residues to sepiolite [38].As evidenced by laser microprobe mass analysis (LAMMA)[39],the resulting grafted species are either monomeric (1) or polymeric, the dimer (2) being the dominant species.

In an attempt to synthesize a catalyst for the (photo)oxidation of water, osmium tetroxide, OsO_4, was reacted with this material. OsO_4 is indeed easily added to carbon-carbon double bonds in organic solvents. This procedure leads to an ordered distribution of osmium atoms on the sepiolite. Examination of the Os-sepiolite (OsS) by high resolution electron microscopy shows rows of black dots along the fiber axis [40]. Each dot has a diameter of about 10 Å and most probably results from a few (predominantly two) osmium atoms.

OsS was shown by BARRIOS et al.[41] to be able to adsorb dinitrogen at room temperature, after mild reduction in hydrogen at 473 K. It turned out that OsS, after activation in the same conditions, has also some catalytic activity for oxidation of water [42]. Figure 7 shows the results obtained in a sacrificial experiment, with the same sensitizer-sacrificial acceptor couple as in section 4.1 (again, what is shown is not the total oxygen production but the oxygen to nitrogen ratio in the gas phase of the static reactor). One can see that (i) OsS, without further pretreatment, has no detectable activity. In the same conditions, RuO_2/sepiolite is already very active; (ii) activation in air is not effective. In fact, activation in an oxidizing atmosphere burns the surface organic groups and volatilizes the osmium atoms as OsO_4. This is also very different from the behavior of RuO_2/sepiolite, which is improved by a pretreatment in air; (iii) thermal reduction in hydrogen induces some activity, but the treatment has to be moderate. Reduction above 500 K is already too severe. This is again different from the behavior of RuO_2/sepiolite which is converted into an active Ru metal/sepiolite catalyst upon reduction at high temperature (section 4.1). All this shows that the nature of the OsS catalyst is different from that of conventional sepiolite-supported heterogeneous catalysts. Its structure, after activation has still to be determined. Diaquo Os(II)surface species (3) might be involved.

Unfortunately, OsS is not especially active, as compared to supported RuO_2 for instance (Figure 4). Actually, the interest of OsS stems from the fact that it is (probably) the first example of a moleculary dispersed water oxidation catalyst anchored by covalent linkage on a colloïdal support. This opens the way to more complex clay-organic systems mimicking the function of photosystem II reaction centers (with a sensitizer, a catalyst, and an acceptor co-grafted on the same surface).

6. PHOTOCATALYSIS AT THE GAS-SOLID INTERFACE, IN THE INTERLAYER SPACE OF SMECTITES;CARBONYLATION AND REVERSE WATER-GAS SHIFT REACTIONS.

This section will be devoted to (photo)catalysis in what is probably the most unique environment provided by clays : the interlayer space in smectite aggregates. We will focus on a reaction which has not attracted much attention in photocatalysis : the reduction of carbon dioxide by hydrogen, i.e. the reverse water-gas shift (RWGS) reaction,catalyzed by ruthenium complexes.

$$CO_2 + H_2 \longrightarrow CO + H_2O \qquad \Delta G_0 = 6.8 \text{ kcal/mole}$$

As far as energy storage is concerned, this is not an uninteresting reaction.

Three complex cations will be considered as (photo)catalyst precursors : Ru(III) hexamine, $Ru(NH_3)_6^{3+}$:diaquobis-bpy Ru(II),$Ru(bpy)_3(H_2O)_2^{2+}$: and a μ-oxo dimer of Ru(III), $(bpy)_2(H_2O)Ru\text{-}O\text{-}Ru(H_2O)(bpy)_2^{4+}$. These complexes intercalated in smectites are not active catalysts as such.A carbonylation pretreatment is necessary to activate them. We will therefore consider both carbonylation and the RWGS reaction. For the sake of comparison,the purely thermal (dark) behavior will also be described.

The bisbpy complex is particularly interesting since it is known from the work of COLE-HAMILTON and coworkers [43,44] that the monocarbonylated form is an active photocatalyst for the WGS reaction.

6.1. Thermal carbonylation and RWGS reaction

6.1.1. $Ru(NH_3)_6^{3+}$-smectites . The chemistry of hexamine Ru(III) in the intracrystal space of zeolites has been studied extensively by JACOBS and coworkers [45,46]. They showed that activation of a $Ru(NH_3)_6^{3+}$-zeolite of the faujasite type in a CO/H_2O atmosphere at low temperature (423 K < T < 523 K) yields very active WGS catalysts. The catalytic, spectroscopic and thermochemical data show that the active sites are not ruthenium metal particles but, most probably, mononuclear Ru(I) carbonyl species.

$Ru(NH_3)_6^{3+}$ - smectites behave quite similarly to the zeolites. The chemical transformations of the intercalated complex during activation are easily followed by IR spectroscopy on clay film samples [47]. The first step upon room temperature evacuation of the sample is partial hydrolysis of the complex, thanks to the residual interlayer water molecules :

$$\overline{\underline{Ru(NH_3)_6^{3+}}} + H_2O \longrightarrow \overline{\underline{Ru(NH_3)_5OH^{2+}}} + NH_4^+ \qquad (16)$$

Heating the sample at 358 K for one hour in 100 torr CO and 20 torr H_2O is enough to transform the partially hydrolyzed complex into monocarbonyl ruthenium(II) species, which absorb strongly around 1940 cm^{-1}.

$$2\ \overline{\underline{Ru(NH_3)_5OH^{2+}}} + 3\ CO \longrightarrow 2\ \overline{\underline{Ru(NH_3)_5CO^{2+}}} + H_2O + CO_2 \qquad (17)$$

Heating in the same atmosphere at higher temperature leads to the formation of biscarbonylruthenium(II) or, more likely,-ruthenium(I) species, which absorb at 2060 and 2000 cm^{-1}. Their maximum concentration is reached around 413 K and they start decomposing rapidly at 480 K.

Activation of a $Ru(NH_3)_6^{3+}$ - smectite sample in CO/H_2O and in the temperature range where biscarbonyl species are stable (423 < T < 453 K), produces an active low temperature WGS and RWGS catalyst. The major reaction product of the RWGS reaction is CO, as expected, but some methane

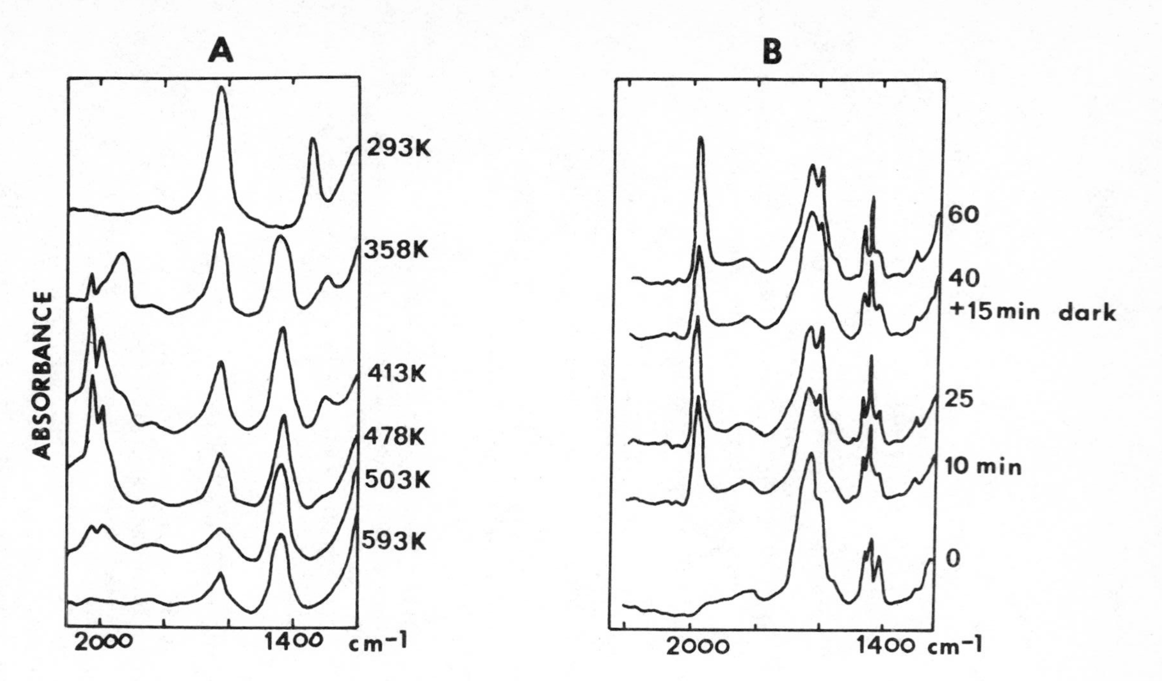

Figure 8. IR spectra of $Ru(bpy)_2(H_2O)_2^{2+}$ - montmorillonite film samples during (A) thermal activation in 100 torr CO and 20 torr H_2O, in the dark, and (B) photochemical activation at room temperature, in the same atmosphere.

is formed above 473 K. Traces of ruthenium metal, which is known to be an excellent methanation catalyst, are probably responsible for this. The CO production rate on a montmorillonite, in 100 torr CO_2 and 100 torr H_2 at 443 K, is of the order of 0.7×10^{-6} CO molecules/s/Ru atom, and the activation energy is $\sim$ 45 Kj/mole.

Little can be said about the mechanism of the reaction. The most significant point is that the catalyst shows appreciable activity for the reduction of CO_2 in a temperature range where the biscarbonyl seems to be the dominant species. Also significant is the non-zero reaction order with respect to carbon dioxide ($\sim$ 1), which shows that CO_2 does not react as a ligand of ruthenium. The mechanism might involve monohydride and formate intermediates, as proposed in zeolites.

6.1.2. $\underline{Ru(bpy)_2(H_2O)_2^{2+}}$ - montmorillonite is also an active catalyst for the RWGS reaction, after appropriate activation. Interestingly, this bis-bpu Ru(II) complex has been shown to have a water-dependent geometry in the interlayer space of smectites [48]. Indeed a "wet" clay, i.e. in a clay containing statistically more than one monolayer of water molecules on its total surface, the trans isomer is the dominant form. On the other hand, in a "dry" clay, i.e. in a clay containing less than about half a monolayer of water on its total surface, the cis isomer was found to be the stable form.

Smectites exchanged with the bispy complex can, to some extent, be considered as pillared clays. Indeed, the basal spacing between the clay layers increases from $\sim$ 5 Å to $\sim$ 9 Å, for clay samples in equilibrium with $\sim$ 15 torr H_2O. This increase is of course favourable to the accessibility of the interlayer space.

The thermal activation of a bis-bpy Ru(II)-montmorillonite in CO yields very rapidly a monocarbonylated complex, characterized by a CO stretching band at 2000 cm^{-1} in the IR spectrum (Figure 8A) :

$$\overline{\underline{Ru(II)(bpy)_2(H_2O)_2^{2+}}} + CO \rightarrow \overline{\underline{Ru(II)(bpy)_2(H_2O)(CO)^{2+}}} + H_2O \quad (18)$$

The reaction starts already at room temperature. The maximum concentration is reached around 413 K. Heating at higher temperature yields a biscarbonyl complex, characterized by a doublet at 2040 and 2090 cm^{-1}. A priori, two reactions might be considered : either carbonylation without reduction of the metal center

$$\overline{\underline{Ru(II)(bpy)_2(H_2O)(CO)^{2+}}} + CO \rightarrow \overline{\underline{Ru(II)(bpy)_2(CO)_2^{2+}}} + H_2O \quad (19)$$

or reductive carbonylation

$$2\ \overline{\underline{Ru(II)(bpy)_2(H_2O)(CO)^{2+}}} + H_2O + 2CO \rightarrow 2\ \overline{\underline{Ru(I)(bpy)_2(CO)_2^{+}}} + H_3O^{+} + CO_2 \quad (20)$$

Although some CO_2 has been detected by mass spectrometry, this is not the proof that reduction of ruthenium takes place, since CO_2 might also be produced by the WGS reaction.

The thermal catalytic activity for the RWGS reaction of the bisby Ru-clay activated at 443 K is somewhat higher than that of the clay prepared from the hexamine precursor. At 443 K, in 100 torr CO_2 and 100 torr H_2, the reaction rate is 1.5×10^{-5} CO molecules/s/Ru atom. The activation energy is only 11 Kj/mole. No methane is formed.

6.1.3. μ-oxo dimer of Ru(III)-montmorillonite undergoes also rapid carbonylation at low temperature. However, in the temperature range 293 - 500 K, only one CO molecule per metal center is fixed. The activity of the carbonylated complex for the RWGS reaction is negligible.

6.2. Photochemical carbonylation and RWGS reaction

In fact this section will be devoted to the bisbpy complex. Indeed, neither the hexamine complex nor the μ-oxo dimer intercalated in smectites show a significant response to visible light (as anticipated for the hexamine complex which has no strong absorption band in the visible).

Irradiation of a bisbpy Ru-smectite with visible light in a CO atmosphere at room temperature leads to the formation of monocarbonyl complexes. The reaction is fast, but it is even faster in a mixture of CO and H_2O. A typical set of IR spectra is shown in Figure 8B. Biscarbonylation is much harder to achieve. Only traces of biscarbonyls are obtained after very long irradiation times.

Though slowly, the reaction is reversible to some extent, in the dark, and in the presence of water. The starting diaquobisbpy complex is regenerated, as easily monitored by the growth of the strong MLCT band of the complex around 480 nm (Figure 9).

Bisbpy Ru-montmorillonite photoactivated to the monocarbonyl state has no detectable activity for the RWGS reaction at room temperature, in the dark. It is also very stable under visible irradiation in vacuum. However, when it is irradiated in a CO_2/H_2 atmosphere, carbon monoxide is produced and the growth of the O.D. at 480 nm shows that the diaquo complex is regenerated (Fig.10). This activity can be assigned to a photochemical RWGS reaction, followed by photoaquation.

$$CO_2 + H_2 \xrightarrow[\overline{\underline{Ru(bpy)_2(H_2O)(CO)}}]{h\nu} CO + H_2O \qquad (21)$$

$$\overline{\underline{Ru(bpy)_2(H_2O)(CO)}} + H_2O \xrightarrow{h\nu} \overline{\underline{Ru(bpy)_2(H_2O)_2}} + CO \qquad (22)$$

The photoaquation of bisbpy carbonyl Ru complexes in solution is a well-known reaction [49]. In our case, it is an undesirable reaction since it produces an inactive species. In order to avoid it and to allow for reaction 21 to proceed for longer times, the water vapor can be trapped at low temperature, but the affinity of the interlayer of smectites for water is so high that this procedure is not effective. Running the photochemical reaction over a sample kept well above room temperature might also be considered.

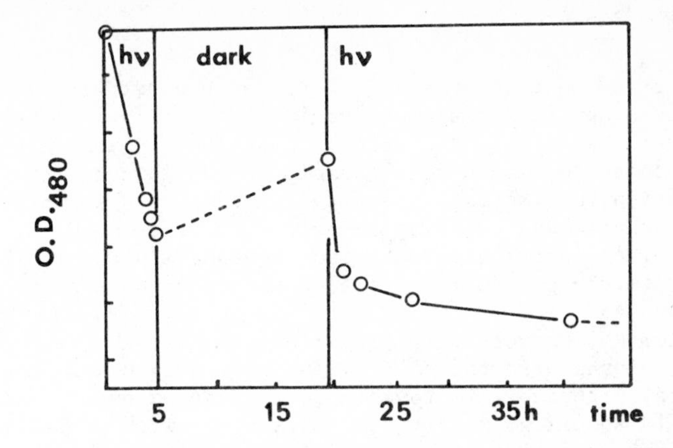

Figure 9. Photochemical activation in CO/H_2O of a $Ru(bpy)_2(H_2O)_2^{2+}$-montmorillonite film sample, monitored by the evolution of the absorption band of the diaquo complex at 480 nm.

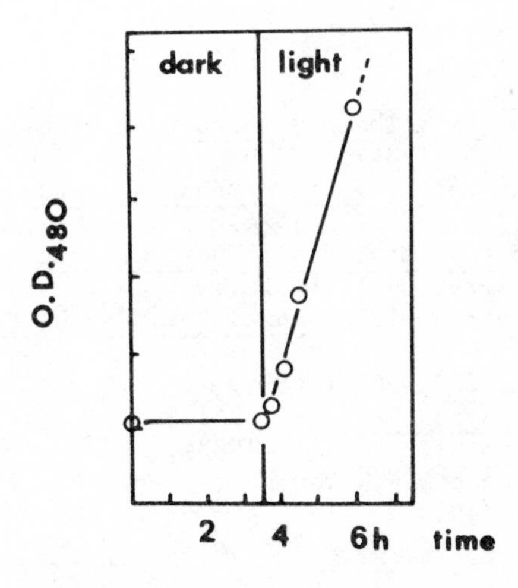

Figure 10. Increase of the optical density at 480 nm, due to $Ru(bpy)_2(H_2O)_2^{2+}$, during the photolysis of $Ru(bpy)_2(CO)_2^{2+}$-montmorillonite in a CO_2/H_2 atmosphere.

7. CONCLUSIONS

The reactions that we have considered are not an exhaustive collection of all the possible applications of clays in photocatalysis. They should merely be considered as a set of currently available typical examples, which may be expected to expand considerably in the future. It seems to us that, in addition to classical uses such as catalysts supports, the most unique role that clays can play in photocatalysis is the structuration of complex electron transfer reaction sequences on the one hand, and the transfer to the gas-solid interface of homogeneous photocatalysis on the other hand.

the photochemical conversion of solar energy through water cleavage has motived most of the efforts, so far. The currently developped clay-based systems are fortunately improvable to a large extent, and the effort in this field should go on for some more time, with probably an outlook towards the reduction of carbon dioxide. On the other hand, simpler photoreactions like the WGS and RWGS reactions are well suited for being performed in the interlayer space of pillared smectites. Due to their simplicity and to the detailed spectroscopic studies that they allow for, this type of reaction involving only one type of intercalated complex will probably draw more attention in the future.

REFERENCES

1. (a)R.E.Grim, Clay Mineralogy, Mc Graw Hill (1968); (b) R.E.Grim, Applied Clay Mineralogy.
2. J.E.Odom, in Clay Minerals : Their Structure Behavior and Use, London, Royal Society, pp.171-190 (1984).
3. W.B.Jepson, in Clay Minerals : Their Structure, Behavior and Use, London, Royal Society, pp.191-212 (1984).
4. R.M.Barrer, F.R.S., in Clay Minerals : Their Structure, Behavior and Use, London, Royal Society, pp.113-132 (1984).
5. J.J.Fripiat and M.I.Cruz-Cumplido, Annu. Rev. Earth. Planet. Sci., 2, 239 (1974).
6. T.J.Pinnavaia, Science, 220, 365 (1983).
7. (a) T.J.Pinnavaia, ACS Symp. Ser., 192, 241 (1982); (b) R.Raythatha and T.J.Pinnavaia, J. Catal., 80, 47 (1983).
8. R.J.Lussier, J.S.Magee and D.E.W.Vaughan, preprints, 7th Canadian Symp. Catal., Edmonton (1980).
9. A.Habti, D.Keravis, P.Levitz and H.Van Damme, J.C.S. Faraday Trans., 2, 80, 67 (1984).
10. R.A.Dellaguardia and J.K.Thomas, J. Phys. Chem., 87, 990 (1983).
11. R.A.Schoonheydt, P.de Pauw, D.Vliers and F.C.de Schrijver, J. Phys. Chem., 88, 5113 (1984).
12. P.K.Ghosh and A.J.Bard, J. Phys. Chem., 88, 5519 (1984).
13. G.Brown, in Clay Minerals : Their Structure, Behavior and Use, London, Royal Society, pp.1-20 (1984).
14. G.W.Brindley and G.Brown, Crystal Structures of Clay Minerals and their X-ray Identification, London, Mineralogical Society (1980).
15. (a) B.K.G.Theng, The Chemistry of Clay-Organic Reactions, Adam Hilger

Ltd, London (1974); (b) G.Lagaly, in Clay Minerals : Their Structure, Behavior and Use, London, Royal Society, pp.95-112 (1984).
16. H.Van Olphen, An Introduction to Clay Colloïd Chemistry, New York (1977).
17. K.Norrish, Disc. Farad. Soc., 18, 120 (1954).
18. C.H.Pons, F.Rousseaux and D.Tchoubar, Clay Min., 16, 23 (1981).
19. J.J.Fripiat, J.Cases, M.François and M.Letellier, J. Coll. Interf. Sci., 89, 378 (1982).
20. (a) M.S.Stul and W.J.Mortier, Clays Clay Miner., 22 , 391 (1974); (b) P.Peigneur, A.Maes, A.Cremers, ibid., 23, 71 (1975); (c) G.Lagaly and A.Weiss, in Proc. Int. Clay Conf., 1975, Mexico, S.W.Bailey Ed., pp.157-172.
21. J.J.Fripiat and H.Van Damme, in "Surface Mobilities on Solid Materials. Fundamental Concepts and Applications", Vu Thien Binh, Ed., Plenum Press, pp.493-526 (1983).
22. C.R.Bock, T.J.Meyer and D.G.Whitten, J. Am. Chem. Soc., 96, 4710 (1974).
23. L.D.Burke, J.O.Murphy, J.F.O Neill and S.Venkatesan, J. Chem. Soc. Faraday Trans. 1, 73, 1659 (1977).
24. J.M.Lehn, J.P.Sauvage and R.Ziessel, Nouv. J. Chim., 3, 423 (1979); 4, 355 (1980); 4, 623 (1980).
25. J.Kiwi and M.Grätzel, Angew. Chem. Int. Ed. Engl., 17, 11 (1978) ; 18, 624 (1979).
26. A.Harriman and A.Mills, J. Chem. Soc. Faraday Trans., 2, 77, 2111 (1981).
27. H.Nijs, H.Van Damme, F.Bergaya, A.Habti and J.J.Fripiat, J. Molec. Catal., 21, 223 (1983).
28. W.de Wilde, G.Peeters and J.H.Lunsford, J. Phys. Chem., 84, 2306 (1980).
29. T.A.Pecorado and R.R.Chianelli, J. Catal., 67, 430 (1981).
30. (a) P.Keller, A.Moradpour, E.Amouyal and H.Kagan, Nouv. J. Chim.,4, 377 (1980); (b) E.Amouyal, P.Keller and A.Moradpour, J. Chem. Soc., Comm., 1019 (1980).
31. A.Harriman, J. Photochem., 25, 33 (1984).
32. P.A.Brugger and M.Grâtzel, J. Chem. Soc., 102, 2461 (1980).
33. D.Meyerstein, J.Rabani, M.S.Matheson and D.Meisel, J. Phys. Chem., 82, 1879 (1978).
34. I.Willner, J.M.Yand, C.Laane, J.W.Otvos and M.Calvin, J. Phys. Chem., 85, 3277 (1981).
35. H.Nijs, J.J.Fripiat and H.Van Damme, J. Phys. Chem., 87, 1279 (1983).
36. F.Bergaya, D.Challal, J.J.Fripiat and H.Van Damme, J. Photochem.,28, 255 (1985).
37. F.Bergaya, D.Challal, J.J.Fripiat and H.Van Damme, Nouv. J. Chim., in press (1985).
38. E.Ruiz-Hitzky and J.J.Fripiat, Clays Clay Miner., 24, 25 (1976).
39. J.K.de Waele, F.C.Adams, B.Casal and E.Ruiz-Hitzky, Microchim. Acta, III, 117 (1985).
40. J.Barrios-Neira, L.Rodrigue and E.Ruiz-Hitzky, J. Microsc., 20, 295 (1974).
41. J.Barrios, G.Poncelet and J.J.Fripiat, J. Catal., 68,362 (1981).
42. B.Casal, F.Bergaya, D.Challal, J.J.Fripiat, E.Ruiz-Hitzky and

H.Van Damme, J. Molec. Catal., in press (1985).
43. D.J.Cole-Hamilton, J.C.S. Chem. Comm., 1213 (1980).
44. D.Choudhury and D.J.Cole-Hamilton, J.C.S. Dalton Trans., 1885 (1982).
45. J.J.Verdonck, R.A.Schoonheydt and P.A.Jacobs, J. Phys. Chem., 85, 2393 (1981) and 87, 684 (1983).
46. P.A.Jacobs, R.Chantillon, P. de Laet, J.Verdonck and M.Tielen, Adv. Chem. Ser., 204, 439 (1983).
47. D.Challal, F.Bergaya and H.Van Damme, Bull. Soc. Chim. Fr., (1985).
48. M.I.Cruz, H.Nijs, J.J.Fripiat and H.Van Damme, J. Chim. Phys., 79, 753 (1982).
49. J.M.Kelly, C.M.O'Connell and J.G.Vos, Inorg. Chim. Acta, 64, L75 (1982).

MEISEL - Do you have an explanation on the molecular level for the segragation that you observe? Is it a cooperativity phenomenon or is it heterogeneity in the clay?

VAN DAMME - The heterogeneity in the clays is certainly playing a role. The charge distribution within the clay layers is known to be inhomogeneous and this has been shown, by X-ray diffraction, to lead to segregation phenomena in the case of simple inorganic ions. In the case of methylviologen and trisbpy-Ru(II), a cooperative effect, possibly related to the different shape of the molecules (the Ru complex is an octahedral, bulky molecule, whereas the methylviologen is a linear, flat molecule) is probably also acting. If surface heterogeneity was the only factor, segregation would be a general phenomenon. This is not the case.

BALZANI - The redox potential of an ion adsorbed on a charged surface should change considerably compared with the redox potential of the same ion in homogeneous solution. I wonder whether this may be an explanation for the different behaviour you observed for oxygen generation when you used the Ru dimer adsorbed on the clay or in homogeneous solution. I would also like to know whether you have any evidence for change in the redox potential upon adsorption on clays (for example, for excited $Ru(bpy)_3^{2+}$).

VAN DAMME - This is an interesting suggestion. Up to now, we have no direct evidence that the redox potential of ions adsorbed on clays is different from the potential in homogeneous solution. This should be best studied by electrochemical methods, on clay-modified electrodes such as those studied in Allen Bard's group. We have however evidence, from X.P.S. measurements, that the binding energy of the electrons of the Ru ion in $Ru(bpy)_3^{2+}$-exchanged clays can be different from that of the pure complex.

LEWIS - Can you address the question of whether any of the $Ru(bpy)_2$-$(OH_2)_2^{2+}$ dimer catalysis of O_2 evolution can be ascribed to the presence of RuO_2?

VAN DAMME - A priori, we cannot discard this possibility. However, there is indirect evidence that the activity cannot be ascribed to RuO_2 impurities. It comes from the fact that the activity critically depends upon the isomer which is used to prepare the catalyst. If RuO_2 impurities were the reason for the observed activity, I would expect the same activity for either cis- or trans- isomers.

FRANK - A comment on N. Lewis' question: We have investigated the spectroelectrochemistry of $(Ru(bpy)_2(H_2O))_2O$ and have found that the oxo-bridged Ru dimer itself, and not RuO_2 impurities, is catalytic for O_2 production from water. The oxo-bridged ruthenium bipyridine dimer is not, however, a very active catalyst for O_2 evolution from water.

VAN DAMME - I agree entirely with this. Still remains the fact that the oxo-bridged dimer is a unique case of a molecule able to catalyze the tetraelectronic charge transfer involved in oxygen evolution from water.

SAKATA - About the dynamical properties of the excited $Ru(bpy)_3^{2+}$. Clay surfaces are thought to have a lot of impurities such as O_2, metal ions like Fe^{3+}, and also defects. Those surface states can function as electron acceptors or energy acceptors. Did you observe fast decay components of luminescence besides slow ones on the clay surface? And can you get some information from those decay properties?

VAN DAMME - We and others have shown that Fe^{3+} is the most important quenching impurity for $Ru(bpy)_3^{2+}$ adsorbed on clays. Because of the restricted mobility of the Ru complex on the surface, one has to consider a whole distribution of local situations, and this leads to a multi-exponential decay, with a single exponential limit at long time. This limit is independent of iron concentration, whereas the amplitude of the multiexponential decay at short time is directly related to the iron concentration. We interpret this in terms of a confined exploration of a heterogeneous medium.

McLENDON - A comment on Prof. Sakata's question: Non-exponential decays are to be expected in a doped system reflecting the distance distribution of donor-acceptor pairs (cf. Thomas/Hopfield and Inokuti + Hirayama). Of course a quantitative fit is difficult, since neither the radical distribution function, nor barrier heights are known in your case.

VAN DAMME - The original expressions derived by Inokuti and Hirayama do not give a good fit in our case. However, as you point out, this could be due to the details of the donor-acceptor correlation function at short distances. In fact, several models could be used. The physically meaningful point is that the motion of the fluorescent probe has to be confined, and this, to some extent could be described in terms of a rigid medium doped with quenchers. However, one has to introduce a cut-off distance for the quenching reaction. Otherwise, one cannot account for the fact that the long-time limit of the decay is independent of quencher concentration.

MANASSEN - Is it possible to anchor dibasic organic molecules between the clay layers? These organic molecules show some resemblance to viologen molecules. Would there be some possibilities in this direction?

VAN DAMME - You are right. The type of material you are talking about (organo layers) has been developed many years ago. The purpose of intercalating the dibasic molecule was to produce a material which would retain an open porosity, even above room temperature, in conditions where, because of dehydration, the interlayer space would normally collapse. Other types of molecular or inorganic properties have been developed since that time, and this led to the whole family of so-called pillared clays. This type of material is clearly of interest for combining photocatalysis with molecular sieving. In the case of clays pillared with dibasic molecules, you probably have also in mind their use as electron relays for the photoreduction of water. This has not been studied so far, but provided their redox potential is in the suitable range, these materials could be investigated in this respect. Thank you for the suggestion.

KISCH - On clean metal surfaces, the activation energy of diffusion is about one tenth of the heat of adsorption. Is there a similar correlation for the movements of $Ru(bpy)_3^{2+}$ on the surface of the clay minerals?

VAN DAMME - I cannot give a detailed answer to your question for several reasons. The first one is that, as far as I know, nobody has measured the heat of adsorption of $Ru(bpy)_3^{2+}$ on clays. The second one is that several workers have shown that the charge density distribution in clays can be very heterogeneous, and this should lead to a broad distribution of adsorption heats. The third one, as I showed briefly, is that on such very heterogeneous surfaces, transport of ions on the surface can no longer be described in terms of homogeneous-type diffusion. One has to switch to a percolation description, with a time-dependent diffusion coefficients and rate constants.

ARCHER - Is there any evidence that any solar-driven photocatalytic process occurs in natural clay deposits?

VAN DAMME - I do not believe that the type of multi-component photocatalysis that I considered in my talk - photoreduction, photooxidation, photocleavage of water, or the WGS reaction - is taking place in natural environments. A priori, one might expect some photoprocesses catalyzed by adsorbed porphyrin compounds, coming from the degradation of green plants. We have shown however that most metalloporphyrins are very rapidly demetallated on clay surfaces. One has also to consider the photocatalytic processes which might be sensitized by the strongly coloured iron-rich clays such as nontronite or glauconite. On the other hand, adsorption on clays is certainly going to modify the photodegradation of pesticides, biological organic matter and wastes.

MUNUERA - Just a short comment on the origin of the oscillation phenomenon in the H_2 evolution you have observed. This phenomenon is well documented for exothermic reactions such as H_2-O_2 recombination on heterogeneous metallic catalysts and critically depends on the experimental conditions (pressure, composition, temperature, etc...). I do not understand why you discard a chemical origin for the phenomenon, though probably perturbed by the gas sampling method. I have observed similar oscillatory H_2 evolution during oxygen photoadsorption from the gas phase on TiO_2/RuO_2, which can be attributed to H_2/O_2 recombination on the RuO_2 active phase.

VAN DAMME - Your observation is very interesting. I believe that the occurrence of oscillations in redox photocatalysis is probably a more general phenomenon than we thought. In our case, I do not entirely discard a chemical origin for the oscillations. There is little doubt that

H_2-O_2 recombination is involved. However, this is not enough to induce oscillations. Since oscillations only develop when the pressure of the gas phase is described by sampling for gas chromatography, we believe that the perturbation of the gas-solution equilibrium, maybe associated with a composition change, is also involved.

PHOTOREDUCTION AND -OXIDATION OF MOLECULAR NITROGEN ON TITANIUM DIOXIDE AND TITANIUM CONTAINING MINERALS

G.N.SCHRAUZER, T.D.GUTH, J.SALEHI, N.STRAMPACH, LIU NAN HUI, and M.R.PALMER

Department of Chemistry, University of California, San Diego
Revelle College,
La Jolla, Calif. 92093

ABSTRACT. Photoexcitation of titanium dioxide produces reactive centers capable of reducing and possibly oxidizing chemisorbed nitrogen. Activity of the photocatalysts is significantly dependent on the source of the titanium oxide and the pretreatment conditions employed. Photocatalysts with the highest activity typically consisted of anatase containing 20-40% of rutile, obtained from iron-doped anatase by a heat-pretreatment. The photocatalytic reactions of nitrogen can occur in nature on the surface of titanium containing desert sands on exposure to sunlight and thus have ecological significance as parts of the nitrogen cycle.

1. INTRODUCTION.

The reduction of molecular nitrogen on illuminated TiO_2 surfaces(1) is perhaps the most interesting of the many photocatalytic reactions that have been described. Our demonstration of nitrogen reduction on titanium containing sands under simulated terrestrial conditions(2,3) suggests that this process has ecological significance and is a part of the natural nitrogen cycle. Evidence for nitrogen photooxidation on TiO_2 and titanium-containing minerals has since also been obtained(3,4). The present account focuses on theoretical, experimental and ecological aspects of these intriguing reactions.

2. NITROGEN PHOTOREDUCTION ON TITANIUM OXIDE SURFACES

2.1. Energetics.

The band-gap of TiO_2 is in the order of 2.9 - 3.2 eV (290 - 335 kJ), corresponding to the energy of near uv light of wavelengths between 390 - 420 nm(4). The gap is sufficiently large to provide energy for the water splitting reaction (eq 1), which, after its discovery by Fujishima and Honda(5,6) is still being extensively studied for potential applications in solar energy conversion systems. The ΔG of eq 2, expressed per equivalent of O_2, suggests that nitrogen reduction is energetically favored over reaction eq 1. However, the energetics of the formation of hydrazine according to eq 3 would be somewhat less favorable.
Moreover, the formation of diazene from nitrogen according to eq 4 must

E. Pelizzetti and N. Serpone (eds.), Homogeneous and Heterogeneous Photocatalysis, 509–520.

also be considered, as it is unlikely that ammonia or hydrazine are formed directly:

$$H_2O + h\nu \xrightarrow{[TiO_2]_s} H_2^{g} + \tfrac{1}{2}\, O_2^{g} \quad (\Delta G^{298°} = 286 \text{ kJ/mol}) \quad (1)$$

$$N_2^{g} + 3\, H_2O^{l} + h\nu \xrightarrow{[TiO_2]_s} 2\, NH_3^{g} + 1.5\, O_2^{g} \quad (\Delta G^{298°} = 766 \text{ kJ/mol}) \quad (2)$$

$$N_2^{g} + 2\, H_2O^{l} + h\nu \xrightarrow{[TiO_2]_s} N_2H_4^{l} + O_2^{g} \quad (\Delta G^{298°} = 626 \text{ kJ/mol}) \quad (3)$$

$$N_2^{g} + H_2O^{l} + h\nu \xrightarrow{[TiO_2]_s} N_2H_2^{g} + \tfrac{1}{2}O_2^{g} \quad (\Delta G^{298°} = 440\text{-}482 \text{ kJ/mol}) \quad (4)$$

The ΔG of formation of diazene in the gas phase is not accurately known; current estimates range from 134 to 196 kJ/mol(7,8). Even assuming the lower value, the formation of gaseous diazene according to eq 4 would be energetically unfavorable. However, as diazene would actually be formed in the chemisorbed state, its ΔG could be diminished by as much as 100-140 kJ/mol, sufficiently to bring the ΔG of reaction eq 4 into the acceptable range.

2.2. Mechanism of Photoexcitation.

Exposure of TiO_2 to near-uv light (390-420 nm) causes the excitation of electrons from the valence band to the conduction band, generating pairs of reducing (e^-) and oxidizing (p^+) centers.
The p^+-centers are short-lived and trapped by surface-OH^- groups to yield chemisorbed OH radicals which terminate with the formation of molecular oxygen.

Electrons in the conduction band are trapped by metal ions in the lattice, giving rise to excited titanium ions in lower oxidation states at which the reduction of the substrates presumably is taking place.

Although the presence of Ti(III) ions on the surface of irradiated rutile has been demonstrated(9), these are not necessarily the nitrogen reducing sites. The Ti(III)-ions could dimerize, form Ti(+II)-centers through disproportionation or the trapping of additional electrons, resulting in sites capable of transferring 2 or more electrons to the chemisorbed nitrogen molecule or to other substrates.

2.3. Photocatalytic Reactions of Acetylenic and Olefinic Substrates.

The photoreduction of acetylene on incompletely outgassed or humidified rutile, mixtures of ethylene, ethane, methane are typically observed(1), at yields depending on details of the experimental conditions and the method of catalyst pretreatment. In addition, traces of propane, butene and of other hydrocarbons are sometimes observed(10). On completely outgassed TiO_2, i.e. in the absence of surface-OH groups, acetylene is cyclotrimerized to benzene, in the light as well as in the dark(10). During the interaction of acetylene with the photocatalyst, surface organotitanium compounds are formed, which, in the presence of surface hydroxyl groups, are protonated to yield olefinic products. Ethylene thus may form from surface-$(CH{=}CH)^{--}$-species(eq 5):

$$C_2H_2 + h\nu \xrightarrow{[TiO_2]_s} [C_2H_2]^{--} \xrightarrow{+2H^+} C_2H_4 \quad (5).$$

A major pathway of methane formation appears to involve product ethylene, which in addition to being photoreduced to ethane, is also undergoing C=C bond cleavage on TiO_2(11). Photoreduction of ethylene on humidified TiO_2 typically afforded a 4:1 mixture of methane and ethane. The formation of higher hydrocarbons from acetylene and ethylene favored if the experiments are conducted on titania surfaces lacking surface hydroxyl groups. Under these conditions, the organotitanium intermediates may undergo methathesis, insertion, isomerization and other reactions instead of protolysis. On completely outgassed TiO_2, ethylene is converted to butene-1 (10).

2.4. Nitrogen Photoreduction.

In principle, the reduction of nitrogen to hydrazine and ammonia could occur by way of 2-, 4- or 6 electron transfer reactions. Inhibition experiments with acetylene favor the 2-electron transfer mechanism: In the photoreduction of acetylene, nitrogen has a strong inhibitory effect only on ethylene production, reaching up to 90% under optimal conditions(1,10). This suggests that N_2 is reduced at the ethylene-producing sites, and that chemisorbed diazene or an equivalent 2-electron reduced species of nitrogen, is the initial product of nitrogen photoreduction. After its formation, diazene, a reactive and shortlived compound, may either decompose either into the elements or by disproportionation (eq 6):

$$N_2 + h\nu \xrightarrow{[TiO_2]_s} [N_2^{--}]_s \xrightarrow{+2H^+} [N_2H_2]_s \begin{cases} \nearrow N_2 + H_2 \\ \searrow \tfrac{1}{2}N_2 + \tfrac{1}{2}N_2H_4 \end{cases} \quad (6)$$

Decomposition into the elements is favored at low stationary concentrations of diazene. However, under optimal reaction conditions, diazene is apparently generated at stationary concentrations high enough for disproportionation into nitrogen and hydrazine, and hydrazine is in turn reduced to ammonia.

Diazene decomposition into the elements is suppressed under optimal conditions of nitrogen photoreduction, especially low irradiation temperatures (30-50°), as virtually no hydrogen is formed. Diazene decomposition appears to be more favored at higher irradiation temperatures, at which hydrogen is formed as a by-product of nitrogen photoreduction. As will be outlined in Section 2.7, a nitrogen-stimulated evolution of hydrogen, attributed to the formation and decomposition of diazene, is observed under certain experimental conditions (see Section 2.7).

2.5 Preparation of Photocatalysts.

Commercial anatase preparations as such are as a rule poor catalysts of nitrogen photoreduction and the water splitting reaction. Some, but not all, are converted into active photocatalysts by impregnating with iron (ferric sulfate) and a thermal pretreatment, resulting in a partial rearrangement into rutile. We obtained our most active samples using powdered anatase of 99.9% purity and particle size 2 micron, prepared by the old wet process (hydrolysis of titanyl sulfate). This type of anatase is no longer readily available, for reasons given below.

Samples of this anatase were impregnated with ferric sulfate, typically corresponding to 0.2 wt.-% in Fe_2O_3, and subjected to a heat treatment, usually from 1-3 hrs at 1000°C(1). Iron accelerates the rearrangement of anatase into rutile(12). The requirement for iron is not absolute however, as active substrata were also obtained with oxides of cobalt or molybdenum(1).

Our best photocatalysts contained from 20-40% rutile (see Fig. 1), as obtained after 1-3 hrs of heating at 1000°, and subsequent storage in humid nitrogen or argon atmospheres prior to the experiments. The newly formed rutile crystals in the most active photocatalysts were very small, their diameter ranged from 0.1 to 0.3 microns. Heating for 5 instead of for 1-3 hrs at 1000° completes the rearrangement and accelerates the growth of the newly formed rutile crystals. This reduces the active surface and greatly diminishes nitrogen reducing activity (Fig 1).

Most currently marketed anatase is no longer prepared by the wet process (hydrolysis of titanyl sulfate), but instead by combustion of $TiCl_4$ in oxygen, or by other vapor phase methods. Such anatase samples produced poor catalysts of nitrogen photoreduction and water splitting. Their small particle size causes them to rapidly rearrange to rutile on heating. The process is accordingly difficult to control and further complicated by accelerated grain growth, which reduces the active surface and also diminishes photocatalytic activity. Some anatase samples furthermore contain proprietary brighteners or other stabilizing additives which render them light-stable; these were also unsuitable for our purposes. To prevent controversy as to the reproducibility of findings, investigators are advised to confer with the manufacturers prior to ordering anatase for their experiments and to carefully optimalize pretreatment conditions.

2.6. Alternate Methods of Photocatalyst Preparation.

V. Augugliaro et al.(13) obtained active photocatalysts by supporting

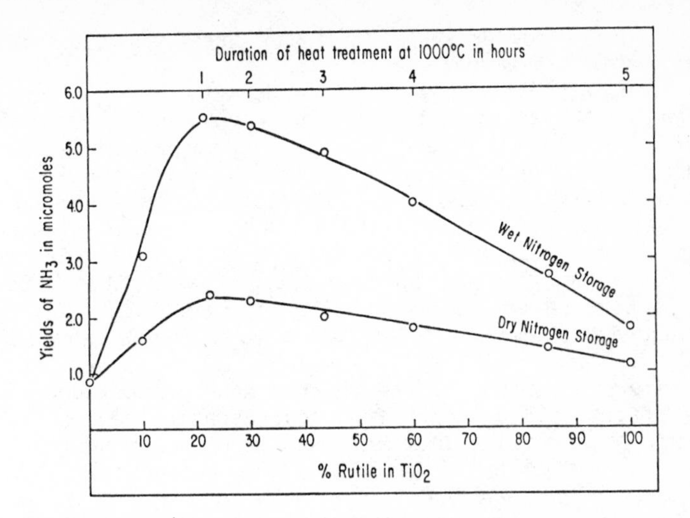

Fig.1. Effect of duration of heat-pretreatment and of storage conditions on photocatalytic activity of TiO_2 containing 0.2 wt.-% of Fe_2O_3, initially consisting of 99% anatase of particle size 2 μ. Sample size: 0.1 g, irradiation temp.: 35°, 500 W Hanovia Hg-arc lamp. Yields measured after 3 hrs. of illumination under 1 atm. of N_2.

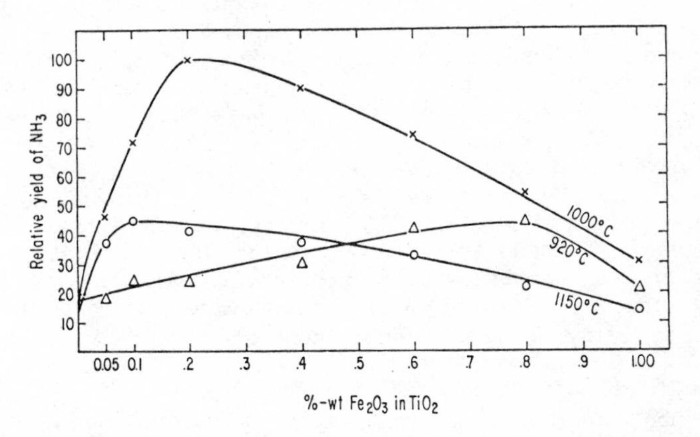

Fig.2. Dependence of the yields of NH_3 on Fe_2O_3 concentration and TiO_2 pretreatment temperature. All samples were heated for 1 hr at stated temperature and stored in H_2O-saturated N_2 atmosphere for 24 hrs prior ot experiment. Sample size: 0.1 g. Other conditions as in legend to Fig. 1.

iron-impregnated rutile powder on alumina for use in a fluidized bed reactor for continuous ammonia production.

Radford and Francis(14), on the other hand, obtained an apparently very active photocatalyst by impregnating powdered anatase with bis(toluene)iron, a highly reactive and temperature sensitive organoiron compound. After evaporation of the solvent, the organoiron compound was thermally decomposed, affording a TiO_2 loaded with elemental iron to approximately 1-2% by mass. Due to the presence of elemental iron in much higher concentrations, this system may be different from ours and may be photoassisted rather than genuinely photocatalytic. According to Radford and Francis, reduction of nitrogen occurs even in aqueous suspensions(14), which we have not seen with our photocatalysts.

2.7. Effects of iron concentration and pretreatment temperature.

Fig.2 shows the effect of increasing iron concentrations on photocatalytic activity. With this particular sample of anatase, the 1 hr heat-treatment at 1000° produced the most active photocatalysts at 0.2 wt.-% of Fe_2O_3. At a pretreatment temperature of 1150°, the maximum photoreducing activity was observed with samples containing only 0.1 wt.-% Fe_2O_3. At the pretreatment temperature of 920°, photocatalytic activity reached a maximum at 0.8 wt.-% Fe_2O_3, indicating a reciprocal relationship between the iron content and the rate of the anatase-rutile conversion.

3.EFFECTS OF NITROGEN ON HYDROGEN PRODUCTION.

Under argon, the humidified photocatalysts promote the water-splitting reaction. With the most active samples, continuous H_2 and O_2 evolution was observed for at least 80 hrs of illumination.

Depending on the catalyst pretreatment temperature, nitrogen may inhibit or stimulate the evolution of hydrogen. Inhibition of hydrogen production by N_2 was observed with catalysts pretreated at temperature between 980 and 1040°, which also exhibited the highest nitrogen photoreducting activity (Fig. 3). However, with catalysts pretreated at 920-960°, only traces of ammonia were formed and a nitrogen-stimulated H_2-production was observed instead. This is attributed to the formation and decomposition of diazene into the elements: The catalysts, due to the low pretreatment temperatures, were still mainly anatase and contained only traces of rutile. In consequence, they showed low photocatalytic activity, and diazene, generated at low stationary concentrations, was lost through decomposition.

4.TEMPERATURE AND PRESSURE DEPENDENCE.

With active photocatalysts, maximum yields of ammonia are observed between 30-50°. The yields decline at higher temperatures, presumably because of diminished N_2 chemisorption and accelerated diazene decomposition.

As a function of nitrogen pressure, the yields of ammonia increase linearly with increasing pN_2 up to about 0.6 atm. Maximum nitrogen binding capacity of the TiO_2 powders was apparently reached at about 1.5 atm, as no further yield increments were observed at higher N_2 pressures (Fig. 4).

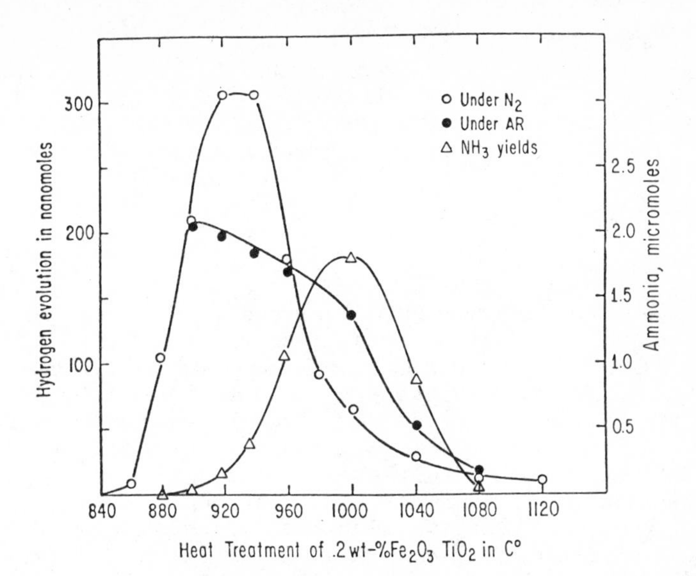

Fig.3. Hydrogen production under Ar and N_2 (1 atm.) as a function of pretreatment temperature. TiO_2 samples were doped with 0.2 wt.-% Fe_2O_3 and after heat treatment were stored in H_2O-saturated Ar atmosphere. Sample size: 0.2 g, irradiation temp.: 80°, 500 W Hanovia Hg-arc lamp. Reaction time: 3 hrs.

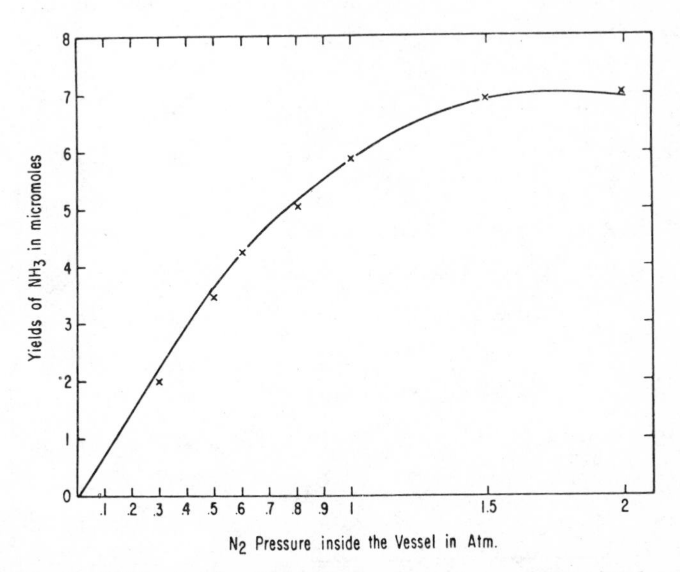

Fig.4. Effect of nitrogen pressure on photocatalytic activity. TiO_2 contained 0.2 wt.-% Fe_2O_3, pretreated at 1000° for 1 hr. Sample size: 0.1 g, irradiation temperature, 35°. Yields measured after 3 hrs of irradiation.

5. NITROGEN PHOTOOXIDATION.

In addition to reduction, oxidation reactions of chemisorbed nitrogen also appear to be promoted on illuminated TiO_2 surfaces. This is because reaction eq 7 is energetically not unfavorable and would be further lowered through chemisorption of product NO:

$$N_2^g + O_2 + h\nu \xrightarrow{[TiO_2]_s} 2\ NO^g\ (\Delta G^{298°} = 181.44\ kJ/mol) \quad (7)$$

In 1979, Bickley and Vishwananthan(15) reported the detection of a chemisorbed nitrogen oxidation product after illuminating H_2O_2-pretreated TiO_2 powders in the presence of nitrogen which produced NO on thermolysis. In similar experiments, we detected the formation of traces of nitrite, but not of nitrate, in aqueous extracts of of TiO_2 substrata after uv-exposure in air. A mixture of nitrite and nitrate was obtained from N_2 photooxidation experiments with H_2O_2-pretreated TiO_2 as the substratum. Although additional studies are necessary it appears that nitrogen photooxidations proceed with low efficiency. As TiO_2 promotes the photooxidation of ammonia by oxygen, nitrogen oxidation products are also formed via initially generated ammonia. Last but not least, nitrite is photoreduced to N_2; from nitrate, no N_2 was generated under similar conditions (TABLE I).

TABLE I

Photooxidation of nitrogen and ammonia in the presence of O_2 and H_2O_2 over 0.2 g of Fe-doped, heat-treated TiO_2. Light source: 500 W Hg-arc, T = 30°, Irrad. Time 6 hrs. Operator: J.Salehi (1982).

No.:	Substrates/Conditions[a]	Gas Phase	Products (μmoles)
1	H_2O	Ar	H_2, O_2
2	H_2O	N_2	NH_3 (1.6),
3	H_2O	Air	NH_3 (1.1), NO_2 (0.4)[b]
4	H_2O, 20 μmoles NO_2^-	Ar	N_2 (5.6)
5	H_2O 200 μmoles NO_2^-	Ar	N_2 (12.4)
6	H_2O 20 μmoles NO_3^-	Ar	No N_2 detected
7	H_2O 200 μmoles NO_3^-	Ar	No N_2 detected
8	20 μL H_2O_2, 30 μmoles NH_3	Ar	NO_2^- (0.5); NO_3^- (2.5)[b]
9	20 μL H_2O_2, 300 μmoles NH_3	Ar	NO_2^- (0.6); NO_3^- (1.4)[b]
10	H_2O, 300 μmoles NH_3	O_2	NO_2^- (0.6); NO_3^- (4.0)[b]
11	20 μL H_2O_2	N_2	NO_2^- (0.4); NO_3^- (2.0)[b]

[a] 0.2 g of TiO_2 photocatalyst, with 20 μL of H_2O.

[b] N_2 not measurable due to presence of large background of O_2.

6. PHOTOCATALYTIC REACTONS OF NITROGEN IN NATURE

In 1979, we demonstrated that sterilized samples of titanium containing sand collected in a California desert photoreduce isotopically labeled molecular nitrogen on exposure to sunlight(2); the experimental design is represented schematically in Fig.5.

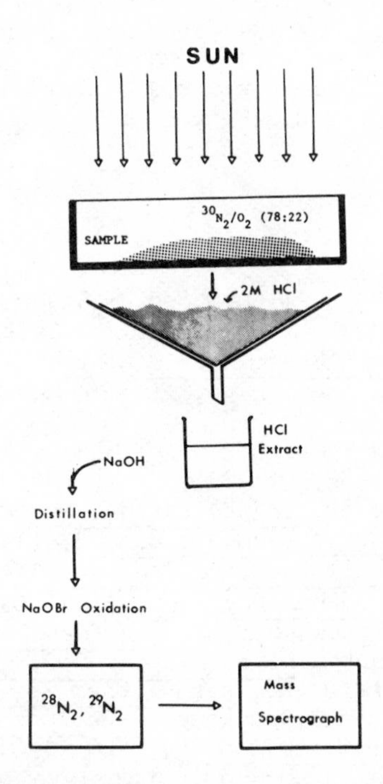

Fig.5. Experimental design of N_2 photofixation experiments with desert sands. Adapted from (2).

Similar experiments were since conducted with samples of sand from various collection sites in California, Kuwait, Egypt, India, China and Saudi Arabia. In addition, the sands were also tested for activity in acetylene photoreduction. A statistically significant direct correlation was observed between the yields of ammonia from nitrogen and ethylene from acetylene ($P<0.01$). The yields of ammonia and acetylene were inversely proportional to the titanium, and inversely proportional to the iron contents of the sands, with $P<0.01$. This indicates that the reactions occur on titanium minerals with a low iron content, i.e. rutile rather than ilmenite. In some of the active sands, the presence of rutile could be demonstrated by x-ray diffractometry. Other samples were photoactive even though rutile could not be detected by x-ray analysis. These, however, contained substantial amounts of sphene ($CaTiO_5$), a mineral known to be susceptible to Ca loss on leaching, resulting in the formation of a layer of TiO_2 on the mineral surface(16).

The sands also exhibited apparent nitrogen photooxidation activity. The yields of ammonia and of nitrogen oxidation products were directly correlated, with $P<0.05$.
Because of the abundance of titanium in the earth's crust, photocatalytic reactions of nitrogen must now be considered as parts of the nitrogen cycle. Conservative estimates suggest that annually about 10 million tons of ammonia are generated in the semi-arid regions of the earth, corresponding to about 10 kg per acre. This is about 10% of the amount of nitrogen fixed biologically and 1/3 of the amount of nitrogen oxidized by atmospheric electric discharges.

REFERENCES

(1) G.N.Schrauzer and T.D.Guth, J.Am.Chem.Soc. 99, 7189 (1977).
(2) G.N.Schrauzer, T.D.Guth, M.R.Palmer and J.Salehi, in: Solar Energy: Chemical Conversion and Storage (R.R.Hautala, R.Bruce King and C.Kutal, eds.), pp. 261-269, The Humana Press, Clifton, N.J., 1979.
(3) G.N.Schrauzer, N.Strampach, Liu Nan Hui, M.R.Palmer and J.Salehi, Proc.Natl.Acad.Sci.USA 80, 3873 (1983).
(4) C.F.Goodeve and J.A.Kitchener, Trans. Faraday Soc. 34, 470 (1938).
(5) A.Fujishima and K.Honda, Nature(London) 238, 37 (1972).
(6) A.Fujishima and K.Honda, Bull.Chem.Soc.Jpn. 44, 1148 (1971).
(7) S.N.Foner and R.L.Hudson, J.Chem.Phys. 68, 3162 (1978).
(8) N.Wiberg, G.Fischer and H.Backhuber, Z.Naturforsch. 34 b, 1385 (1979).
(9) R.I.Bickley and R.K.M.Jayanty, Disc. Faraday Soc. 58, 194 (1974).
(10) A.H.Boonstra and C.A.H.A.Mutsaers, J.Phys.Chem 79, 2025 (1975).
(11) C.Yun, M.Ampo, S.Kodama, and Y.Kubokawa, J.C.S.Chem.Comm 1980, 609.
(12) Y.Iida and S.Ozaki, J.Amer.Ceram.Soc. 44, 120 (1961).
(13) V.Augugliaro, A.Lauricella, L.Rizzuti, M.Schiavello and A.Sclafani, Proc. 3rd World Hydrogen Energy Conf., Tokyo, June 1980.
(14) P.P.Radford and C.G.Francis, J.Chem.Soc.Chem.Comm. 1983, 1520.
(15) R.I.Bickley and V.Vishwanathan, Nature(London) 280, 306 (1979).
(16) G.M.Bancroft, J.B.Metson, S.M.Kanetkar and J.D.Brown, Nature(London) 299, 708 (1982).

MEMMING - What is really the oxidation product at TiO_2 particles when reducing N_2 to NH_3? If it would be O_2, then this result is surprising because illumination of TiO_2 suspensions leads to H_2 formation but O_2 is not found!

SCHRAUZER - Irradiation of TiO_2 in aqueous suspensions may produce surface peroxo-titanium species or H_2O_2 instead of O_2.

FOX - (1) Why is H_2 not incorporated into the NH_3? (2) Why does more severe pretreatment, which gives rutile,not more active for nitrogen fixation?

SCHRAUZER - N_2 photoreduction in our system is not a photo-assisted hydrogenation. Under the reaction conditions, the hydrolysis of surface diazenido or nitrido species is faster than their hydrogenolysis.

KIWI - (1) Does Fe_2O_3 intervene directly on the TiO_2 photocatalysis of fixing N_2? If so, how does it take place? Is it not a sacrificial reaction that uses up the Fe^{3+} in the 10 hours the N_2 fixation proceeds? (2) What is the physical aspect of the 0.2% Fe_2O_3-TiO_2 catalyst? Are there patches of Fe_2O_3 on the surface of TiO_2? Is it a layer of Fe_2O_3 on TiO_2? How big are the clusters of Fe_2O_3 on TiO_2? Is it homogeneously dispersed on TiO_2? (3) 1000°C pretreatment of 1 hour you have reported as fast to evolve NH_3 out of TiO_2. What does 1000°C pretreatment do? Can you quantify the changes taking place with this pretreatment?

SCHRAUZER - There are not detectable patches of Fe_2O_3 on the surface of our photocatalytic TiO_2. The 1000°C pretreatment temperature produced a limited amount of rutile (typically 10-40%) of optimal grain sizes and active surface. Higher heating temperatures, or even prolonged heating at 1000°C (more than 3 hours) caused rutile grain growth with attendant diminution of active surface and photocatalytic activity.

SAKATA - We followed your experiment. We used various semiconductors besides TiO_2 such as CdS and $SrTiO_3$ and various metals on the semiconductor. Interestingly, among various semiconductors and metals, only the combination of TiO_2(anatase) and Fe showed some activity. However, O_2 was not evolved, and the photocatalyst deteriorated after about 10 hours irradiation. How about in your case? In order to improve the efficiency the use of an electrode system would be attractive. Have you tried your TiO_2/Fe_2O_3 as a cathode to reduce nitrogen?

SCHRAUZER - On prolonged irradiation some NH_3 is lost through photo-oxidation. High concentrations of NH_3 may block N_2 binding sites and thus lower efficiency of N_2 reduction. The use of a cathodic potential to influence N_2 photo-reduction has not been attempted as yet. We also have found no activity with CdS or $SrTiO_3$ as substrata of N_2 photo-reduction. With our TiO_2 photocatalysts, O_2 was clearly and reproducibly detected in the expected amounts. Under your conditions, the product O_2 could have remained chemisorbed as $\cdots O_2^-$, or a chemically related species.

HELLER - Photogenerated holes in n-TiO_2 will oxidize an organic adsorbate much more easily than water. Therefore, continuing Prof. Sakata's question, would it not be more likely that ammonia production will be of higher quantum yield when the n-TiO_2 grains are covered in part with an oxidizable organic material?

SCHRAUZER - It would be very difficult to achieve coverage of TiO_2 particles with an oxidizable substance without also covering N_2 binding centers. The reduction of N_2 in my opinion occurs primarily on short-lived excited reduced titanium centers ("Ti^*_{red}") which are not in direct communication with the oxidizing centers. Pre-irradiated TiO_2 (under argon) does not appreciably reduce N_2 on subsequent immediate exposure to N_2 in the dark.

PHOTOASSISTED REDUCTION OF CARBON AND NITROGEN COMPOUNDS WITH SEMICONDUCTORS

M. Halmann and K. Zuckerman
Isotope Department
Weizmann Institute of Science
Rehovot 76100
Israel

ABSTRACT. The conversion of visible light into the stored energy of chemical fuel was studied, using semiconductors as photosensitizers. The reactions investigated were the reduction of carbonate ions to organic compounds and of nitrogen oxyanions to ammonia. Using suspensions of CdS, ZnS-CdS or TiO_2 in 1M KOH and 0.1M Na_2S at 61° C, illuminated with a 150W Xe-lamp, we found that there was a slow release of ammonia, 0.014 micromol h^{-1} cm^{-2} (illuminated area) which was the same whether the mixture was purged with argon or nitrogen gas. Addition of 0.2M potassium nitrate did not change the rate of release of ammonia. However, addition of potassium nitrite caused a marked increase in the rate of production of ammonia. Using 0.13M potassium nitrite in 1M KOH, 0.1M sodium sulfide at 61° C, the production rate of ammonia was 0.97 micromol/(h cm^2) in the presence of TiO_2 (70mg in 70ml reaction solution), 0.60 micromol/(h cm^2) in the presence of CdS (same concentration), and 0.55 micromol/(h cm^2) in the presence of ZnS-CdS (70mg - 50mg mixture). In the absence of sodium sulfide, and if sodium sulfite was substituted for sodium sulfide, the above photoassisted reduction of nitrite ions to ammonia was negligible. Also, there was no reaction in the dark. With CdS - ZnS in 0.5M potassium carbonate - 0.1M sodium sulfide, illumination caused the production of ammonia, methanol and formaldehyde, at rates of 7.1×10^{-9}, 1.5×10^{-9} and 0.2×10^{-9} mol h^{-} cm^{-2}, respectively.

1. INTRODUCTION

The energy and charge transfer processes required for fuel forming reactions may be achieved using semiconductor materials as photocatalysts. In n-type semiconductors, illumination with light more energetic than the bandgap causes an increase in the number of holes in the valence band, while in p-type semiconductors, such illumination increases the number of electrons in the conduction band. Such semiconductors, immersed in solutions containing suitable reduced or oxidized species, can cause the oxidation or reduction of these compounds. Extensive work has been carried out during the last years on the photoassisted reduction of carbon dioxide to organic compounds. This work has been the subject of several reviews [1-5].

E. Pelizzetti and N. Serpone (eds.), Homogeneous and Heterogeneous Photocatalysis, 521–532.

Such reactions are energetically uphill, as shown in the following reactions, in which the free energy change per electron transfer is given by E^o

(1) $CO_2(g) + 2H_2O(l) = HCOOH(aq) + 1/2O_2(g) \quad E^o = 1.428eV$

(2) $CO_2(g) + H_2O(l) = HCHO(g) + O_2(g) \quad E^o = 1.350eV$

(3) $CO_2(g) + 2H_2O(l) = CH_3OH(aq) + 3/2O_2(g) \quad E^o = 1.119eV$

(4) $CO_2(g) + 2H_2O(l) = CH_4(g) + 2O_2(g) \quad E^o = 1.037eV$

The photoreduction of solutions of sodium carbonate to formaldehyde was also described, using aqueous suspensions of TiO_2 [6]. Reduction of dinitrogen to hydrazine and ammonia has been carried out in various photocatalytic systems [7], and is the subject of a previous paper in this Workshop [8]. Nitrogen and sulfur oxides, NO_x and SO_x, and hydrogen sulfide are important pollutants released from fossil fuel burning power stations, internal combustion engines and many mining and industrial operations. There exist thus a considerable need for the conversion of these pollutants into harmless, or even useful products.

2. EXPERIMENTAL

Materials: Potassium hydroxide, potassium nitrite, potassium nitrate, sodium sulfide (Merck, pro analysi), titanium oxide (Degussa P25, anatase), and potassium carbonate (BDH, Anal. grade) were used without purification. For Cd-ZnS mixtures, cadmium sulfide (Fisher Sientific) and zinc sulfide (BDH, Lab. grade) were used. In experiments using cadmium sulfide alone, it was prepared as described [9] by adding a 0.02M sodium sulfide solution (240ml) to 0.05M cadmium nitrate (120ml), filtering, washing with distilled water and drying overnight under vacuum at 65° C.

Light source: A 150W xenon lamp (Oriel) was used, illuminating through a 4 cm long cuvette filled with water to filter out infrared radiation. The light flux incident at the reaction flask was 30-40 mW cm^{-2}, illuminating an area of about 8 cm^2 diameter. The light flux was monitored with a Yellow Springs Instrument Radiometer (Model 65A).

Irradiation procedure: The irradiation flask was a double - walled borosilicate vessel of about 100 ml capacity. Through the outer compartment deionized and thermostatted water was circulated. The flask was fitted with gas inlet and outlet tubes and with magnetic stirring. The gas, argon or nitrogen, was supplied at a constant flow rate (65 ml min^{-1}) and was purified by passing it through a series of wash - bottles containing chromic sulfuric acid (to remove ammonia and organic impurities) and distilled water (to remove acid spray and to saturate the gas with water). The outgoing gas from the reaction flask was led through a series of three small traps, containing each 2-3 ml of 0.1M

HCl, to collect the ammonia produced. For collection of organic products, these traps were kept in an ice bath.

Ammonia analysis was performed by colorimetry using the hypochlorite - phenol method, as previously described [10]. Hydroxylamine was assayed by the color reaction with sulfanilic acid and alpha naphthyl ethylene diamine [11].

Formic acid, formaldehyde, methanol, acetaldehyde and ethanol were analyzed as previously described [34].

3. RESULTS

3.1 Nitrogen compounds:

The first objective was to find out if dinitrogen does undergo photocatalyzed reduction to ammonia in the presence of various semiconductor materials in an alkaline sulfide medium. Results presented in Figure 1 show that ammonia is indeed released from illuminated suspensions of either CdS, CdS-ZnS or TiO_2 in 1M KOH - 0.1M Na_2S. However, the rate of the release of ammonia was unchanged whether the gas bubbling through the medium was argon or nitrogen. This rather small release of ammonia was faster in the presence of TiO_2 or of CdS-ZnS than of CdS. There was no reaction in the dark. This experiment suggested that the medium contained a nitrogenous component, which underwent photoassisted conversion to ammonia. A similar conclusion had been reached previously in an electrochemical study [12]. In an effort to identify this component, 0.20M KNO_3 in 1M KOH - 0.1M Na_2S was illuminated in the presence of CdS (1.0 mg/ml). The rate of release of ammonia was similar to that without added nitrate. Thus, the unknown component was not the nitrate ion. In further experiments, KNO_2 was added to the above KOH - Na_2S solutions with various catalysts. This time there was a dramatic increase in the rate of production of ammonia. Hence, traces of nitrite ion impurities could presumably be the source of the photoinduced production of ammonia in the above experiments in KOH - Na_2S mixtures with CdS or TiO_2. As shown in Figure 2, this production of ammonia is again faster in the presence of TiO_2 than of CdS, and is approximately linear with time during a period of up to about 40 hours. This production of ammonia did not occur in a medium of only 1M KOH and 0.2M KNO_2 with TiO_2 (see Figure 2c). Thus the presence of the Na_2S component is essential for the reaction. If 0.1M sodium sulfite was substituted for the sulfide, the rate of ammonia release was also very small (Figure 2d). Hydroxylamine was not produced in detectable amounts during the photocatalytic reduction of nitrite ions in aqueous KOH - Na_2S in the presence of either TiO_2 or CdS. A comparison of the time course of evolution of ammonia with diferent catalysts and temperatures in 1M KOH - 0.1M Na_2S is given in Figure 3.

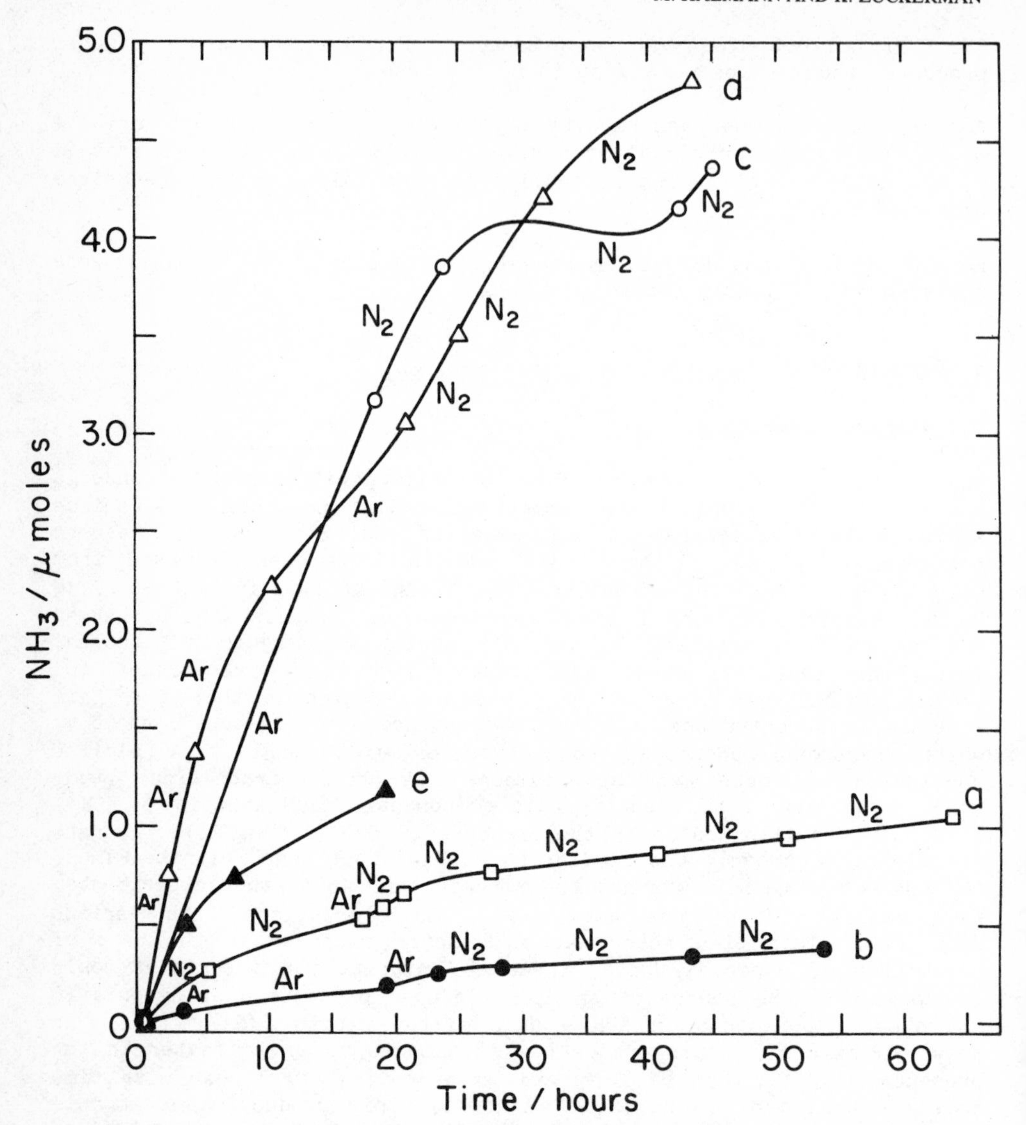

Figure 1. Ammount of ammonia produced (micromoles) as a function of illumination time (hours) in the photocatalytic reduction of nitrogenous impurities in suspensions of powdered semiconductors in aqueous 1M KOH - 0.1M Na_2S. In Run e, the solution contained 0.1M sodium sulfite instead of sodium sulfide. Catalysts: Runs a, b: CdS; c, e: TiO_2; d: CdS-ZnS. Temperatures: Run a, 24° C; Runs b,c, 45° C; Runs d, e, 61° C. Argon or N_2 (as indicated) was bubbled through the reaction flask at 65 ml min^{-1}. Light source: 150W xenon lamp.

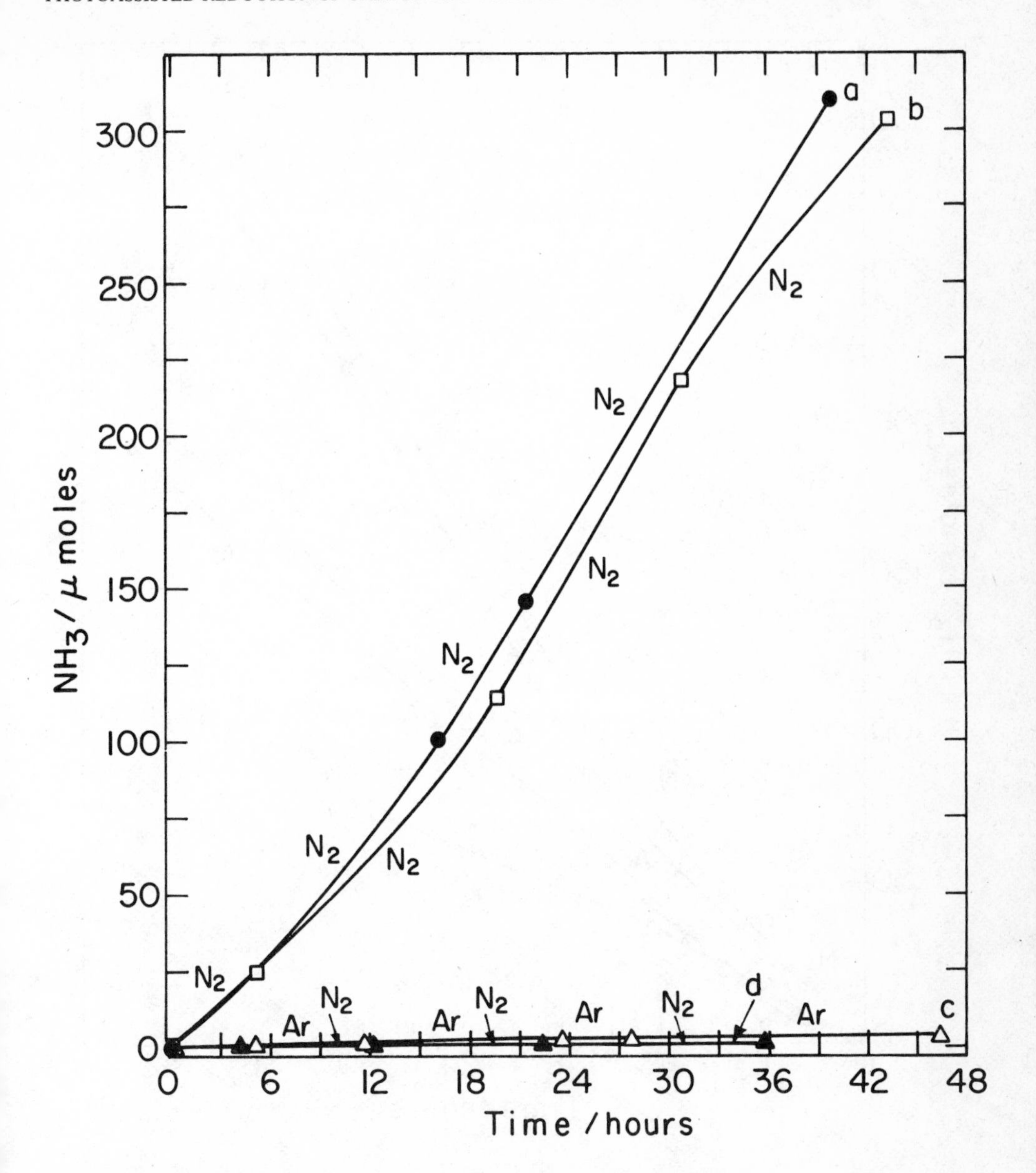

Figure 2. Amount of ammonia produced (micromoles) as a function of illumination time (hours) in the photocatalytic reduction of potassium nitrite (0.12M) on TiO_2 (1.1 mg ml^{-1}) in 1M KOH. Runs a and b, also 0.1M sodium sulfide. Runs c, d, no sodium sulfide. In d, 0.1M sodium sulfite. Temperatures: Run a, c, d 61° C; Run b, 45° C.

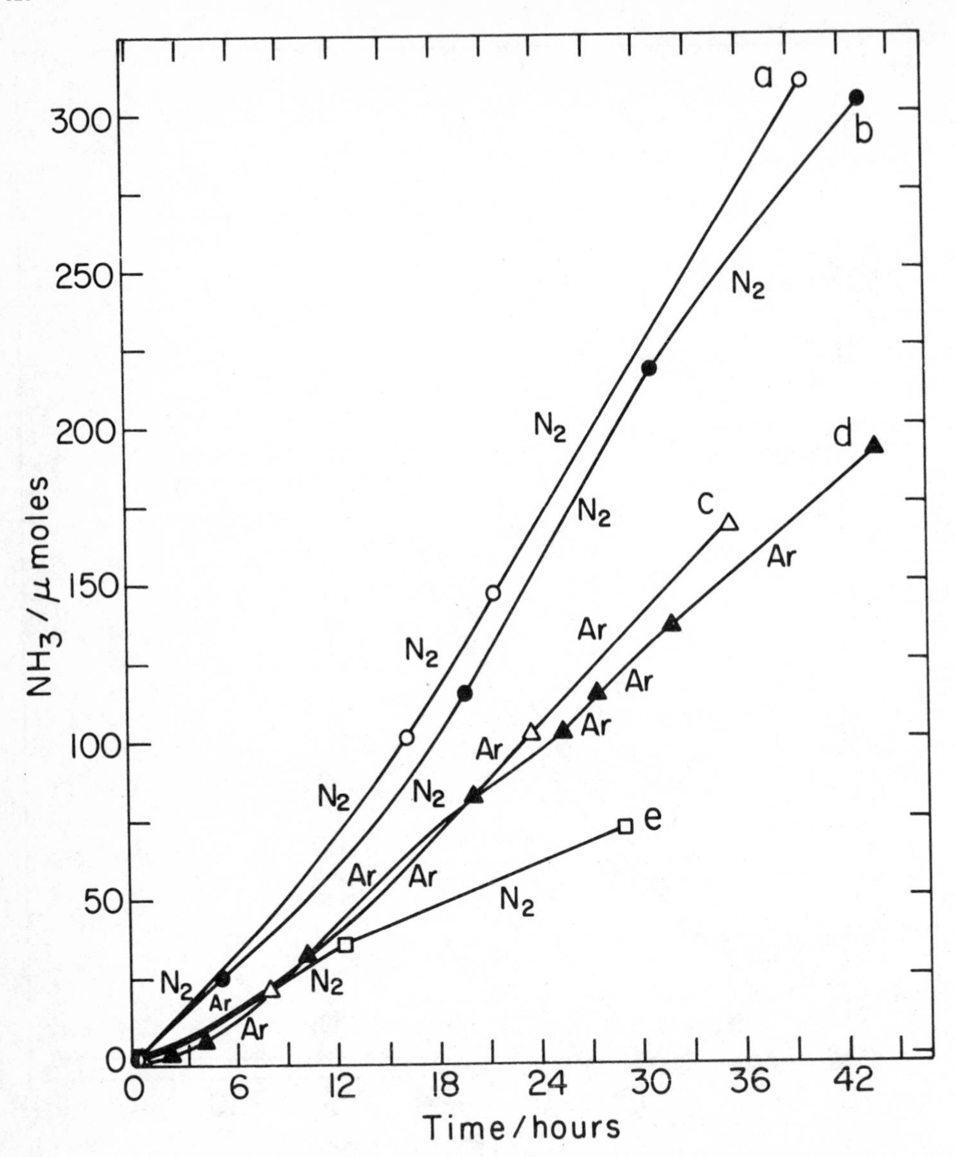

Figure 3. Amount of ammonia produced (micromoles) as a function of illumination time (hours) in the photocatalytic reduction of potassium nitrite (concentration, Runs a, b: 0.13M; c, d: 0.12M; e, 0.23M) in 1M KOH - 0.1M sodium sulfide. Catalysts, a, b: Titanium oxide; c, e: CdS; d, CdS-ZnS. Temperatures, a, c, d: 61° C; b, e: 45° C.

3.2 Carbon compounds:

A summary of previous studies on the heterogeneous photocatalytic reduction of carbon dioxide is presented in Table 1. Most of the previous work was in acid media, with the neutral CO_2 molecule as the reactive species. Only one study was on the photocatalytic reaction of titanium oxide in an alkaline medium, with aqueous sodium carbonate [6]. In Figure 4, we present results for the illumination of suspensions of CdS - ZnS in aqueous 0.5M potassium carbonate - 0.1M sodium sulfide, carried out in an argon atmosphere. Ammonia, methanol and formaldehyde were formed. Integrated production rates (during 34 hours) for ammonia, methanol and formaldehyde were 7.1×10^{-9}, 1.5×10^{-9} and 0.2×10^{-9} mol h^{-1} cm^{-2} (illuminated area), respectively. Here again, as in the above experiments in aqueous potassium hydroxide, the precursor of the ammonia formed must be a nitrogenous impurity. Similar experiments, but with added potassium nitrite, showed a marked increase in the rate of production of ammonia. Concerning the observed production of formaldehyde and methanol, further work will be needed to identify the possible precursors and reaction mechanisms. Formic acid was searched for, but could not be detected.

4. DISCUSSION

According to current concepts on photocatalytic processes induced by illumination of suspensions of small particles of semiconductors, the mechanism involves electron - pair separation. Thus, the particles act as microscopic electrolysis cells, with different sites on the surface serving as the anode and the cathode [13]. While there have been many studies on the electrochemical reduction of the nitrite and nitrate ions in acid media, there have been no investigations on the photocatalytic reduction in alkaline media.

The electroreduction of nitric acid and of the nitrate and nitrite ions in strongly acid solutions has been the subject of several studies using cyclic voltammetry and rotating disk electrodes [14-15]. On copper cathodes, nitrate ions were found to undergo reduction in high yields below pH 2, but to be resistant to reduction above pH 3. Nitrite ions also readily undergo reduction at pH 1 in perchloric acid solution. In acid media, with both nitrate and nitrite ions, the limiting current was proportional to the square root of the rotation rate of the copper disk. Hence the current was limited by the rate of mass transport of the NO_3^- or NO_2^- ions to the copper electrode surface [14]. The electrochemical behavior of nitrogen monoxide, nitrous acid and nitric acid in aqueous solutions was also investigated on platinum metal, using stationary, rotating disc and rotating ring - disk electrodes [15]. In 4-5 M H_2SO_4 solutions, nitrous acid was reduced in three steps, involving the successive formation of nitrogen monoxide, nitrous oxide and hydroxylamine. The results confirm the Abel - Schmid mechanism for the autocatalytic reduction of nitric acid to nitrous acid [16-18]:

TABLE 1.
Photocatalyzed reduction of carbon dioxide and of the carbonate ion to organic compounds in aqueous suspensions or gas - solid reactions with semiconductors.

Catalysts	Medium	Products							Ref.
		CH_4	CO	HCOOH	HCHO	CH_3OH	CH_3CHO	C_2H_5OH	
$SrTiO_3$/Pt	Gas-solid	+							28
Inorg.*	Suspens.				+	+			29
Inorg.	Suspens.				+	+			30
Chlorophyll	Suspens.			+					31
Inorg.	Suspens.				+	+			
	Gas-solid				+	+			32
Inorg.	Suspens.			+	+	+	+	+	33
$SrTiO_3$	Suspens.			+	+	+	+	+	34
Inorg.	Suspens.			+	+	+	+	+	35
TiO_2	Suspens.				+				6
TAM#			+						36
Inorg.+H_2	Gas-solid		+	+	+	+	+		37

* Inorg. = Various inorganic materials;

\# TAM = Tetraaza macrocyclic Co(II)

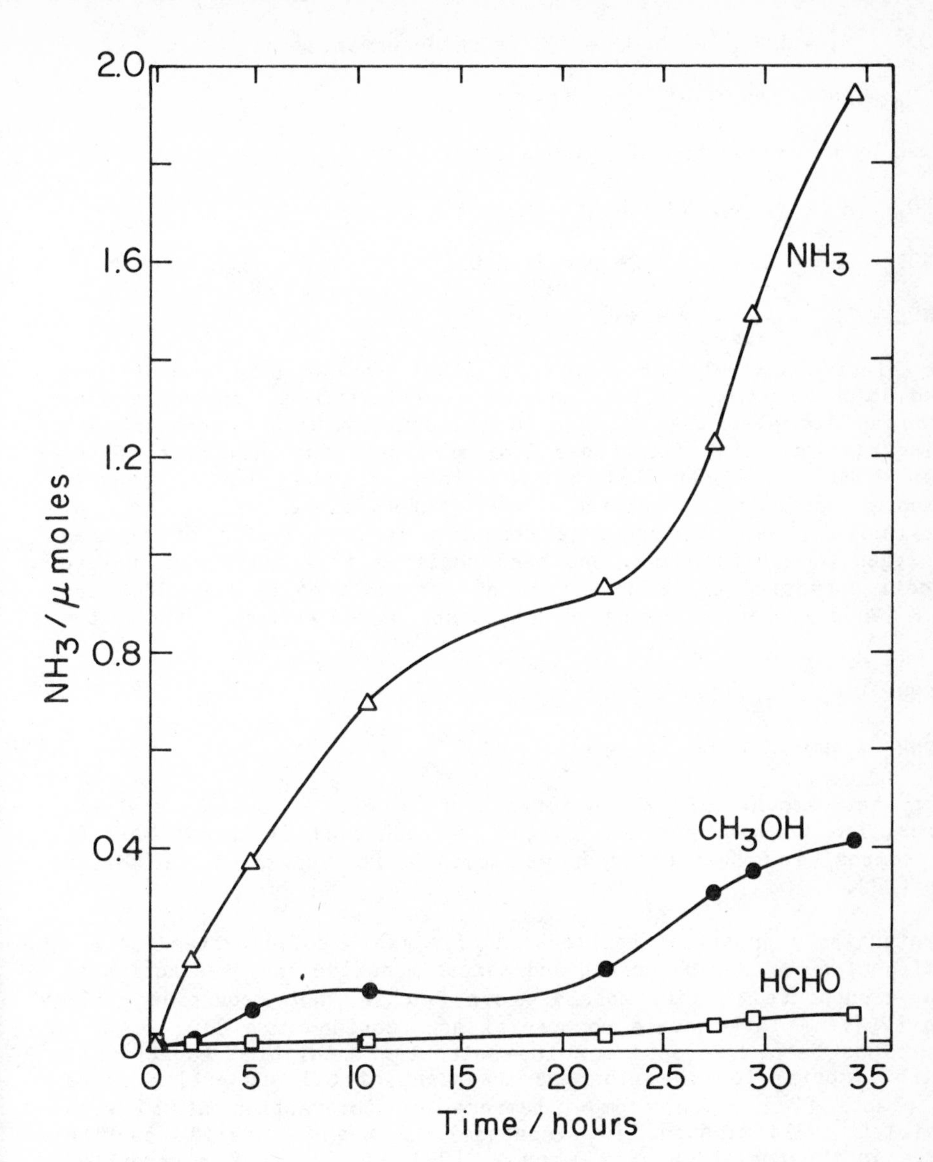

Figure 4. Amounts of ammonia, methanol and formaldehyde formed (micromoles) as a function of time (hours) during the illumination of a suspension of CdS - ZnS (70 and 50 mg) in 0.5M potassium carbonate - 0.1M sodium sulfide (70ml). Carrier gas, argon. Temperature, 61° C.

(5) $NO_3^- + H^+ + HNO_3 \dashrightarrow N_2O_4 + H_2O$ (rate determining)

(6) $N_2O_4 + 2NO + 2H_2O \dashrightarrow 4HNO_2$ (fast)

followed by the reduction of nitrous acid:

(7) $HNO_2 + H^+ + e^- \dashrightarrow NO + H_2O$

(8) $HNO_2 + 2H^+ + 2e^- \dashrightarrow 1/2N_2O + 3/2H_2O$

(9) $HNO_2 + 5H^+ + 4e^- \dashrightarrow H_3NOH^+ + H_2O$

The electrochemical conversion of nitric oxide into ammonia was studied at pH 6, using Fe(II) - complexes such as iron - phenanthroline to trap and complex the NO. Cyclic voltammetry with stationary and rotating platinum disk electrodes indicated that the reduction of the complexed NO was controlled by the rate of the charge transfer reaction, possibly involving the intermediate formation of hydroxylamine [19]. The reverse reaction, the oxidation of ammonia with oxygen to nitric oxide, has been realized in a solid electrolyte fuel cell operating on ammonia fuel at temperatures between 1000 and 1200° K [20-21]. The following two reactions were found to occur at the anode:

(10) $2NH_3 + 5O^{2-} \dashrightarrow 2NO + 10e^-$

(11) $2NH_3 + 3NO \dashrightarrow 5/2N_2 + 3H_2O$

Complete removal of NO by means of a gas - solid catalytic decomposition to nitrogen and oxygen was achieved at about 400° C, using composite catalysts such as Co,La_2O_3,Pt supported on active carbon [22].

A potentially practical application of a gas - solid system for the reduction of NO or NO_2 by carbon and carbon monoxide involves activated carbon - supported alkali metals above 500° K, not requiring noble metals [23]. A large scale commercial application of a TiO_2 - based catalyst has been developed for the selective catalytic reduction of NO_x with ammonia, for the flue gas treatment of oil and coal - fired power plants [24]. The room - temperature interaction of NO with ultraviolet - illuminated TiO_2 (anatase) in a gas - solid reaction resulted in the formation of N_2O and N_2 [25]. The catalytic reduction of NO by CO on TiO_2 - supported rhodium indicated the existence of strong metal - support interactions (SMSI) and dissociative adsorption of NO on Rh/TiO_2 [26].

The photocatalytic reduction of the nitrite ion to ammonia observed in the present work in suspensions of catalysts such as TiO_2, CdS and CdS-ZnS differs from the electrochemical reduction in acid media in the failure to detect hydroxylamine - which is an important product and probable intermediate in acid solutions. Coprecipitated ZnS - CdS has previously been used for the photoassisted production of hydrogen with visible light [27]. In the present work, a simple mixture of ZnS and CdS was shown to provide efficient photoreduction of the nitrite ion to ammonia, and also in small yields reduction of aqueous carbonate solutions. The generally accepted mechanism for the reduction of carbon dioxide in acid media involves electron capture as the primary step,

(12) $CO_2 + e^- \rightarrow CO_2^-$

producing the CO_2^- radical anion as an intermediate. In alkaline media, with aqueous carbonate, a tentative suggestion for the mechanism may involve as the primary step hydrogen atom abstraction by a hydroxyl radical from the bicarbonate ion,

(13) $OH + HCO_3^- \rightarrow H_2O + CO_3$

This may be followed by decomposition to O_2^- and CO. Carbon monoxide may react with OH^- to produce the formate ion.

5. ACKNOWLEDGEMENT

This study was supported in part by a grant from the Adler Foundation for Space Research, administered by the Israel Academy of Sciences and Humanities.

6. REFERENCES

1. G. J. F. Chittenden and A. W. Schwartz, Biosystems, 14 (1981) 15.
2. R. P. A. Sneeden, J. Molec. Catal., 17 (1982) 349.
3. M. Halmann, in Energy Resources Through Photochemistry and Catalysis, M. Gratzel, Ed., Academic Press, New York, 1983, p. 507.
4. R. Ziessel, Nouv. J. Chim., 7 (1983) 613.
5. M. Ulman, B. Aurian-Blajeni and M. Halmann, CHEMTECH 14 (1984) 235.
6. K. Chandrasekaran and J. K. Thomas, Chem. Phys. Lett., 99 (1983) 7.
7. P. P. Radford and C. G. Francis, J. Chem. Soc., Chem. Commun. (1983) 1520.
8. G. N. Schrauzer, "Selective activation of nitrogen", this Volume.
9. M. Serpone, E. Borgarello, M. Barbeni and E. Pelizzetti, Inorg. Chim. Acta, 90, 191, 1984.
10. S. Grayer and M. Halmann, J. Electroanal. Chem., 170 (1984) 363.
11. F. Fiadeiro, L. Solarzano and J. D. H. Strickland, Limn. Oceanogr., 12 (1967) 555.
12. M. Halmann, J. Electroanal. Chem., 181 (1984) 307.
13. H. Gerischer, J. Phys. Chem., 88 (1984) 6096.
14. D. Pletcher and Z. Poorabedi, Electrochim. Acta, 24 (1979) 1253.
15. A. Katagiri, M. Maeda, T. Yamaguchi, Z. Ogumi, Z. Takehara and S. Yoshizawa, Nippon Kagaku Kaishi, (8) (1984) 1221.
16. E. Abel and H. Schmid, Z. physik. Chem., 132 (1928) 55.
18. E. Abel, H. Schmid and E. Romer, Z. physik. Chem., 148 (1930) 337.

18. G. Schmid and M. A. Lobeck, Ber. Bunsenges. Phys. Chem., 73 (1969) 189.
19. K. Ogura and H. Ishikawa, J. Chem. Soc., Farad. Trans. I., 80 (1984) 2243.
20. C. G. Vayenas and R. D. Farr, Science, 208 (1980) 593.
21. C. T. Sigal and C. G. Vayenas, Solid State Ionics, 5 (1981) 567.
22. T. Inui, T. Otowa and Y. Takegami, J. Chem. Soc., Chem. Commun., (1980) 94.
23. F. Kapteijn, A. J. C. Mierop, G. Abbel and J. A. Moulin, J. Chem. Soc., Chem. Commun., (1984) 1085.
24. S. Matsuda and A. Kato, Appl. Catal., 8 (1984) 3175.
25. H. Courbon and P. Pichat, J. Chem. Soc., Farad. Trans. I, 80 (1984) 1085.
26. V. Rives-Arnau and G. Munuera, Appl. Surface Science, 6 (1980) 1.
27. N. Kakuta, K. H. Park, M. F. Finlayson, A. Ueno, A. J. Bard, A. Campion, M. A. Fox, S. E. Weber and J. M. White, J. Phys. Chem., 89 (1985) 732.
28. C. Hemminger, R. Carr and G. A. Somorjai, Chem. Phys. Lett., 57 (1978) 100.
29. T. Inoue, A. Fujishima, S. Konishi and K. Honda, Nature, 277 (1979) 637.
30. M. Halmann and B. Aurian-Blajeni, 2nd European Community Photovoltaic Solar Energy Conference, Berlin (West), D. Reidel Publ., 1979, p. 682.
31. D. R. Fruge, G. D. Fong and K. K. Fong, J. Amer. Chem. Soc., 101 (1979) 3694.
32. B. Aurian-Blajeni, M. Halmann and J. Manassen, Solar Energy, 25 (1980) 165.
33. M. Ulman, B. Aurian-Blajeni and M. Halmann, Israel J. Chem., 22 (1982) 177.
34. M. Ulman, A. H. A. Tinnemans, A. Mackor, B. Aurian-Blajeni and M. Halmann, Internat. J. Solar Energy, 1 (1982) 213.
35. M. Halmann, M. Ulman and B. Aurian-Blajeni, Solar Energy, 31 (1983) 429.
36. A. H. Tinnemans, T. P. M. Koster, D. H. M. W. Thewissen and A. Mackor, Rec. Trav. Chim. Pays Bas, 103 (1984) 288.

ADSORPTION AND DESORPTION PROCESSES IN PHOTOCATALYSIS

Pierre Pichat
Ecole Centrale de Lyon
Equipe CNRS Photocatalyse*
B.P. 163, 69131 Ecully Cedex,
France

ABSTRACT. In this review, the following topics are discussed: (i)effect of illumination on the adsorption equilibrium at semiconductor surfaces (photoadsorption and photodesorption) particularly for the system gaseous oxygen-semiconductor oxides, with a special development on the interest of oxygen isotope exchanges to assess the lability of adsorbed and surface oxygen atoms; (ii) nature of adsorption sites and of adsorption species, competition between reactants and products derived from kinetic studies of photocatalytic reactions, from ESR spectra and from laser-flash excitation of semiconductor colloids coupled with time-resolved spectroscopies; (iii)evidence for hydrogen spillover between the two components of group VIII metal/titanium dioxide photocatalysts.

INTRODUCTION

Adsorption and desorption phenomena, and the related concepts of surface sites or surface arrangements, play an important role in heterogeneous photocatalysis since they are included in the sequence of events which lead from gaseous and/or liquid reactants to gaseous and/or liquid products over solids. In addition, the photons which are used to activate the catalyst can modify the adsorption equilibrium as compared with the situation in the dark, and, furthermore, can increase the lability of surface ions.

Because of the extent of the present topic, this review is voluntarily limited to some aspects. To complete the information,recent reviews on photo-induced processes at gas-solid interfaces are recommended (1-3). General articles on photocatalytic reactions provide a useful background knowledge (4-9).

In the first part, we discuss studies dealing with the effects of band-gap illumination on the adsorption equilibrium between a gas (most often oxygen) and a semiconductor (most often an oxide). The second part is more specifically devoted to investigations aiming to determine

* J.E. CNRS 4594

E. Pelizzetti and N. Serpone (eds.), Homogeneous and Heterogeneous Photocatalysis, 533–554.

the adsorbed species,the adsorption sites and the charge transfers involved in photocatalytic reactions by kinetics or various physical methods. In the third part, for reactions including hydrogen as a reactant or as a product, experiments are presented, showing the existence of hydrogen spillover in adsorbed phase between the two components of bifunctional metal/semiconductor oxide photocatalysts.

Creation of electron-hole pairs in a semiconductor, mainly by band-to-band transitions, is at the basis of heterogeneous photocatalysis if one considers the general and simple definition given by Morrison (1) :"(heterogeneous) photocatalysis is the increase in a chemical reaction rate produced by optical excitation of the solid where the solid remains chemically unchanged". In terms of energy levels, the charge carriers thus generated can interact with so-called surface states. These surface states can be due to surface impurities or to surface groups or to adsorbed species. This emphasizes the importance of adsorption for the fate of the photoproduced charges, and accordingly for photocatalysis. The nature of the surface states determines their capture cross section - their reactivity - towards holes or electrons. It ensues that a knowledge, as accurate as possible, of the surface (impurities, defects, groups) and of the adsorbed species is essential.

The importance of recombination, in other words, the mean life-time of the photoproduced electron-hole pairs, depends (i) upon the size of the absorbing particle (ratio of its diameter to the thickness of the space charge layer)(3), and (ii) upon the nature and number of the surface states and, in particular, upon the reactants in contact with the semiconductor. Not only the energy levels should be considered, but also the kinetics of the reactions of the photoproduced charges with the various surface states. This means that models of the semiconductor surfaces based on electron energy bands or on active sites should together be considered(2).

Photoproduced charges can recombine via interaction with an occupied or unoccupied surface state:

$$A^{n+} \xrightarrow{p^+} A^{(n+1)+} \xrightarrow{e^-} A^{n+} \qquad (1)$$

where, for instance, A^{n+} can be an OH^- group, in the case of a semiconductor oxide. They can also be captured by different surface entities so that a net chemical reaction can take place. A^{n+} symbolizes any occupied surface state with which a hole reacts; in fact, it appears that the interactions reported concern species carrying one negative charge.

$$A^{n+} + p^+ \rightarrow A^{(n+1)+}$$

$$B^{(n+1)+} + e^- \rightarrow B^{n+}$$

$$\overline{A^{n+} + B^{(n+1)+} \rightarrow A^{(n+1)+} + B^{n+}} \qquad (2)$$

A photocatalytic reaction thus occurs if at least one reactant is at the origin of species A or B of Eq.(2), or contribute to their renewal (viz. the conversion of OH groups to OD groups by D_2). For instance, in the case of the dehydrogenation of alcohols, we have proposed (10) that the alcohol molecules, dissociatively adsorbed, react as follows:

$$>CHOH \rightarrow >CHO^- + H^+ \quad (3)$$
$$>CHO^- + p^+ \rightarrow >C = O + H^\circ \quad (4)$$
$$H^+ + e^- \rightarrow H^\circ \quad (5)$$

where $>CHO^-$ and H^+ play the role of A and B of Eq.(2), respectively.

1. PHOTOADSORPTION AND PHOTODESORPTION

Band-gap illumination by increasing the number of charge carriers (in particular, the relative increase in the minority charges carriers can be sonsiderable) can change the adsorption equilibrium of certain compounds. In a simple theory model the species involved are those which carry a charge; in the case of large adsorbed species, configurations with high localized partial charge can also be affected. Photosorption processes resulting from a direct interaction of adsorbates with photons will not be considered here.

As several parameters intervene, it is not possible to predict whether the net effect of illumination will be photoadsorption or photodesorption. Besides the gas pressure and temperature,these parameters include (i) the adsorption equilibrium in the dark (at low temperature and pressure, the ionosorption can be irreversible), (ii) the number and nature of surface states - which in turn depend on the history of the sample (preparation, pretreatment), and the interaction rates of these surface states with the photoproduced charges. Theoretical considerations are developed in refs 1,11,12.

1.1. Photoadsorption

Because most studies have dealt with the photoadsorption of gaseous oxygen on semiconductor oxides, this paper is also principally limited to this system.

1.1.1. Extent of photoadsorption. The amounts of oxygen molecules which can be photoadsorbed on semiconductor oxides are low. They, of course, depend on the pressure. Maxima of the order of 0.3 molecules nm^{-2} have been reported for the O_2- TiO_2 system for pressures of a few hundreds Pa (13-16). Somewhat lower amounts have been indicated for oxygen photoadsorbed on ZnO (17) or on SnO_2 (18).

1.1.2 Effect of surface OH groups.For TiO_2, oxygen photoadsorption is decreased by thermal removal of water and hydroxyl groups as shown by detailed studies (13,14,19,20). A beneficial effect of OH groups on O_2 photoadsorption has also been mentioned in the case of BeO (21). The positive effect of the OH groups upon O_2 photoadsorption is attributed to hole trapping by these groups (13,14,19,20,22,23). The resulting OH° radicals combine, producing H_2O_2, which in turn can decompose to form O_2

$$OH^- + p^+ \rightarrow OH^\circ \quad (6)$$
$$2\ OH^\circ \rightarrow H_2O_2 \quad (7)$$
$$H_2O_2 \rightarrow H_2O + 1/2\ O_2 \quad (8)$$

However, the resulting O_2 photodesorption cannot be observed under oxygen pressures of the order of 100 P_a(20). The formation of HO_2^- and of OH_2° entities has also been proposed; the latter radicals would arise from

$$O_2^- + H^+ \rightarrow O_2H^\circ \quad (9)$$

at 77K and/or from

$$OH^\circ + H_2O_2 \rightarrow O_2H^\circ + H_2O \quad (10)$$

These proposals were supported by studies of H_2O_2 photodecomposition over TiO_2(23,24). See also paragraph 1.1.4.

For fully hydrated TiO_2 samples, water photodesorption was reported to accompany O_2 photoadsorption (20); this desorption was assumed to originate from a weakening of the bond between adsorbed water and titanium cations when their charge is decreased from 4 + to 3 + because of the band-gap illumination and subsequent electron trapping. This is an example of a light-induced desorption of species which were not ionosorbed.

1.1.3. Photoadsorption kinetics. Several authors found a Roguinski-Zeldovitch-Elovich law

$$P_o - P_t = A \ln\ (1 + t/t_o) \quad (11)$$

for several systems : O_2 on TiO_2 (17,25), ZnO (17), SnO_2 (18); H_2 and CH_4 dissociative adsorptions on TiO_2 (17). This law is consistent with a linear relationship between the amount of negatively charged adsorbed oxygen and the surface potential (25). For O_2 photoadsorption on TiO_2, different laws have been observed. For hydroxylated surfaces a first order kinetics was followed

$$\ln P_t = \ln P_o - k_a t \quad (12)$$

whereas, for dehydroxylated surfaces, the photoadsorption rate obeyed a diffusion law

$$P_o - P_t = k_{df} t^{1/2} \quad (13)$$

interpreted as reflecting the diffusion of the photoproduced excitons towards the residual OH groups (20). Adjustement of the kinetic curves to the Elovich law would have required a too high t_o parameter. Presumably, the differences reported arise from various samples, surface treatments and oxygen pressure domains.

1.1.4. Photoadsorbed oxygen species. From ESR spectra recorded at 77 K with TiO_2 samples which have been exposed to gaseous O_2, the existence of O_2^-, O_3^-, O_3^{3-} and O_2^{2-} species was deduced (22,26-28). Only the former one was stable at 300 K (22). It is also the less reactive species (27-30). Equations showing the ways of formation of these species and their reciprocal transformation have been proposed. For instance, O_3^- species are supposed to be produced by the interaction of O_2 molecules with O^- ions-radicals.

Although they are indirect, electrical methods can yield information on the nature of oxygen adsorbed species on illuminated semiconductors.

A detailed study (25,31) of the TiO_2 - O_2 system under UV-illumination using the vibrating capacitor method, which yields the variations of the work function of the powder semiconductor, has led to the following conclusions:

(i) two types of ionosorbed oxygen can explain the results; they are tentatively attributed to O_2^- and to O^-;

(ii) simultaneous photodesorption and photoadsorption of oxygen occur; the photodesorption prevails at the very beginning;

(iii) qualitatively similar results are found for rutile and nouporous anatase specimens of various specific areas;

(iv) the amount of photoadsorbed oxygen depends on the number of available free electrons as shown by the effects of small quantities of homogeneously introduced dopes (Nb^{5+} or V^{5+}, and Ga^{3+} have opposite effects); higher quantities of V^{5+} decrease the oxygen photoadsorption, presumably because of increase charge recombination (32,33).

In the author's group, the effect of various gases on the photoconductance σ of powder n-type semiconductor oxides has been investigated (34-36). Electrophilic gases, such as O_2 (34-36) and NO (36), decrease σ of preevacuated samples. In addition, from the slope of the σ variations as a function of O_2 or NO pressure in a log-log plot, the nature of the oxygen or nitric oxide species controlling the adsorption equilibrium between the semiconductor free electrons and the gas (for given illumination, temperature and pressure range) can be deduced. Nevertheless, this implies assumptions on the ways of formation of these oxygen species from gaseous O_2. For example, the predominance of

$$O_2(g) \rightleftarrows O_2(ads) \tag{14}$$

$$O_2(ads) + e^- \rightleftarrows O_2^{\cdot-} (ads) \tag{15}$$

should give rise to a - 1 slope, while a - 1/2 slope can result from

$$1/2\ O_2(g) \rightleftarrows O(ads) \tag{16}$$

$$O(ads) + e^- \rightleftarrows O^{\cdot-} (ads) \tag{17}$$

or from Eq (14) and

$$O_2(ads) + 2e^- \rightarrow O_2^{2-}(ads) \rightarrow 2\ O^{\cdot-}(ads) \tag{18}$$

However, in this latter case, σ would be proportional to the square root of the flux power, whereas Eqs (16) and (17) correspond to a linear σ-flux power relationship (35). Therefore the effect of light intensity should allow one to discriminate between these mechanisms of formation of O^- species. Similarly, for NO adsorption over illuminated TiO_2, an inverse dependence of σ upon NO pressure was found (36) and interpreted as indicating that each NO molecule captures one semiconductor free electron. However, data of other technics are required to determine whether NO^- or NO_2^- species (where the second oxygen atom comes from the TiO_2 surface) or any other species corresponding to the capture of one electron per NO molecule are thus formed. Also, note that the fact that one species governs the adsorption equilibrium do not necessarily imply that other species are not present, but can mean that they saturate the surface.

Moreover, if the semiconductor samples previously evacuated under illumination to remove the labile oxygen species under these conditions, are exposed to organic compounds involved in oxidations, as reactants or products, the effect on the photoconductance indicates whether or not these compounds compete with O_2 or NO for electron capture (34). However, the absence of competition for electrons does not exclude adsorption competition for the same surface sites.

In an attempt to determine the active oxygen species in hydrocarbon photocatalytic oxidations in gas phase, we have carried out simultaneous measurements of σ and of the photocatalytic activity A in a specially designed cell. For instance, in the case of isobutane (IS) oxidation over TiO_2, the following relations were found (37):

$$\sigma = k_\sigma \, P_{O_2}^{-1} \, P_{IS}^{0} \quad (19), \qquad A = k_A \, P_{O_2}^{0} \, P_{IS}^{0.35} \quad (20)$$

The independence of σ on IS pressure indicates that IS does not capture nor release electrons, whereas the fractional kinetic order 0.35 shows that IS reacts in an adsorbed phase, since this value is very close to the apparent order of adsorption 0.3 found for the surface coverage in IS according to a Langmuir model in the pressure range investigated (13-60 kPa). The σ oxygen pressure relationship is a hint that $O_2^{\cdot -}$ species control the adsorption equilibrium for the pressures chosen. Since A is unaffected by oxygen pressure, it is deduced that the active oxygen species are associated with $O^{\cdot -}$ ion-radicals whose adsorption sites are saturated.

Results relative to adsorbed oxygen species on semiconductor oxides suspended in liquid water are scarce apart from studies dealing with the formation of $O_2^{\cdot -}$, OH° and O_2H° species over TiO_2 pigments (38,39). Sustained splitting of liquid water over TiO_2 (40,41) or CdS (40,42,43) particles supporting Pt and RuO_2 has been reported, although periodical removal of the H_2-O_2 mixture formed was required to maintain the rate (41). However, it was indicated afterwards that O_2 was not evolved when using a closed photoreactor (44). As the formation of a monolayer of $O_2^{\cdot -}$ species cannot explain the missing amount of O_2, the formation of μ-peroxo bridges has been proposed (44):

1.1.5. Effect of co-adsorbates. We will mention two examples concerning the influence of co-adsorbed gases on oxygen photoadsorption.

The effect of CO_2 co-adsorption on powder TiO_2 was investigated at 77K (45). Whereas for hydroxylated samples no influence was found, important decreases in the ESR signals corresponding to $O_2^{\cdot -}$ and O_3^- species were observed for dehydroxylated samples. This phenomenon is related to a modification of the surface by CO_2 adsorption in the second case.

By constrast, in the case of SnO_2, preadsorbed CO produces a substantial increase in the amount of O_2 photoadsorbed under sub-bandgap irradiation and furthermore the sensitivity threshold is displaced to longer wavelenths (46). In short, these effects arise from a modification of the surface states in the band gap because of exposure of SnO_2 to CO which namely forms carbonate species.

In both cases, the presence of gases capable of reacting with the oxide results in surface changes which, as expected, affect the photoadsorption of oxygen (cf. the effect of the degree of hydroxylation).

1.2. Photodesorption

1.2.1. Definition. For a n-type semiconductor, intrinsic photodesorption results from the interaction of holes, produced by band-gap illumination, with negatively charged adsorbed species, generating neutral molecules which are thermally released in the ambient. For several semiconductor oxides, except TiO_2, photodesorption seems to have been observed at energies lower than the band gap (47). This has been ascribed to extrinsic mechanisms. In particular, holes can be produced by light excitation of electrons from acceptor levels to the conduction band with subsequent thermal transition of electrons from the valence band to the acceptor levels. Also, optical excitation of electrons from adsorbed species to the conduction band can cause photodesorption; in that case, the desorption optical threshold should depend on the nature of the adsorbed species (48).

1.2.2. Photodesorption, thermodesorption and photolysis. Distinction between real photodesorption and thermodesorption arising from sample heating by the illumination is not easy. Local raises in temperature can be caused by the recombination of charges produced by band-gap irradiation. The dependence of the desorption rate upon light intensity should be discriminating: linear for photodesorption, non-linear for thermodesorption (intervention of Boltzmann-type factors, $-E_a/T$) (2,48, 49). Also, the response time to pulsed illumination can allow a distinction, since photoeffects should be faster (2). This implies experimental devices with sufficient sensitivity and speed to accurately analyze the desorbed molecules. Unfortunately, in earlier studies, photoadsorption and photodesorption were often evaluated only from changes in the gas pressure over the illuminated solid.

Also, photodesorption should be distinguished from photolysis (1,2,50). For instable semiconductor oxides oxygen evolution can stem from

$$O_s^{2-} + 2\ p^+ \rightarrow 1/2\ O_2 \qquad (21)$$

associated with the corresponding electron-consuming reaction; for instance, in the case of ZnO (2,50,51):

$$Zn^{2+} + e^- \rightarrow Zn^+_{interstitial} \qquad (22)$$

However, in the presence of gaseous oxygen, the cross-section for hole capture of oxygen ionosorbed species, such as O_2^-, being much greater than that of O_s^{2-} (50), photodesorption should predominate over photolysis.

1.2.3. Nature of photodesorbed species. Contradictory results have been reported. Evolution of CO_2 on illuminating ZnO (52-54), TiO_2 (53,55,56), niobium oxide (57), $SrTiO_3$ (58), CdS (48), has been observed by mass spectrometry. In general terms, this phenomenon can be explained by a reaction between adsorbed oxygen and carbon-containing adsorbed species or impurities, since the photocatalytic oxidation of a variety of organic compounds is well established (4-9). Indeed, Auger spectra showed a marked decline in surface carbon after prolonged UV-illumination of ZnO, for instance (52b). Also, it has been found in the author's laboratory that the photodesorption of CO_2 from TiO_2, as determined from mass spectra, was considerably decreased by a overnight pretreatment of the powder in O_2 at 723 K. Oxygen photodesorption studies should be carried out with metal oxide surfaces whose carbon contamination has been eliminated as much as possible (2). The existence of a chemisorbed "CO_2^-" complex, formed by oxygen adsorption onto a carbon impurity atom, has been suggested, although it was recognized that the nature of the carbon and the carbon complexes are not well understood (59). At least direct formation of such a CO_2^- complex seems not possible on TiO_2, since it was found that the photoconductance of this material is unaffected by exposure to CO_2 (34,36). Furthermore, CO_2^- is a paramagnetic species and no corresponding ESR signal has ever been observed for TiO_2 or ZnO exposed to O_2 or CO_2 (29).

Direct observation by mass spectrometry of oxygen photodesorption is not an easy task. Complete removal of carbon-containing surface species is difficult as known from infrared spectroscopy and from other modern surface analysis methods. Inevitable presence of a peak at the mass-to-charge ratio 16 in the mass spectrometer background is an impediment for evaluating low amounts of photodesorbed oxygen. The best experimental procedure seems to be that in which the sample is exposed to a continuous flow of $^{16}O_2$ or $^{18}O_2$ at very low pressure (2,60,61). Under these conditions, an increase in O_2 mass peak has been found during the first three seconds of continuous illumination, with a low flux power, of powder TiO_2 (60). More sophisticated experiments yielding time-profiles of mass-resolved changes allowed one to conclude to the phodesorption of $^{18}O_2$ from TiO_2, V_2O_5, Cr_2O_3, Fe_3O_4 and ZnO powders after a 50 µs UV-pulse; however, in this latter case, heating effects might have been produced by the high power of the flash (2,61).

The following observations constitute indirect evidence of O_2 photodesorption: (i) photodesorption is related to previous exposure to O_2 (exposures to CO_2 or CO are inefficient); (ii) the conductivity increase resulting from illumination under vacuum corresponds to the decrease found when adding an electrophilic gas such as oxygen. On the other hand, the observation of NO photodesorption from alumina (62) can also be considered as an argument in favour of the possibility of oxygen photodesorption from clean surfaces. The photodesorption of nitric oxide probably results from the neutralization of NO negatively adsorbed species; the photoadsorption of NO by electron capture has been shown for TiO_2 as mentioned in the section on photoadsorption (36).

1.3. Use of oxygen isotope exchanges to access the light-induced oxygen hability.

Isotopic exchanges of gaseous $^{18}O_2$ (62-64) and $N^{18}O$ (66) with surface and adsorbed oxygen atoms can be used to show the lability of these oxygen atoms at room temperature under band-gap illumination. By contrast, in the dark, temperatures largely above the ambient are generally required to allow such isotopic heteroexchanges with equivalent rates.

Note that over illuminated powder-TiO_2, the initial rate of exchange is much faster with $N^{18}O$ than with $^{18}O_2$, which can be interpreted in terms of relative stabilities of NO_2 or O_3 surface species if an associative mechanism of the exchange is assumed.

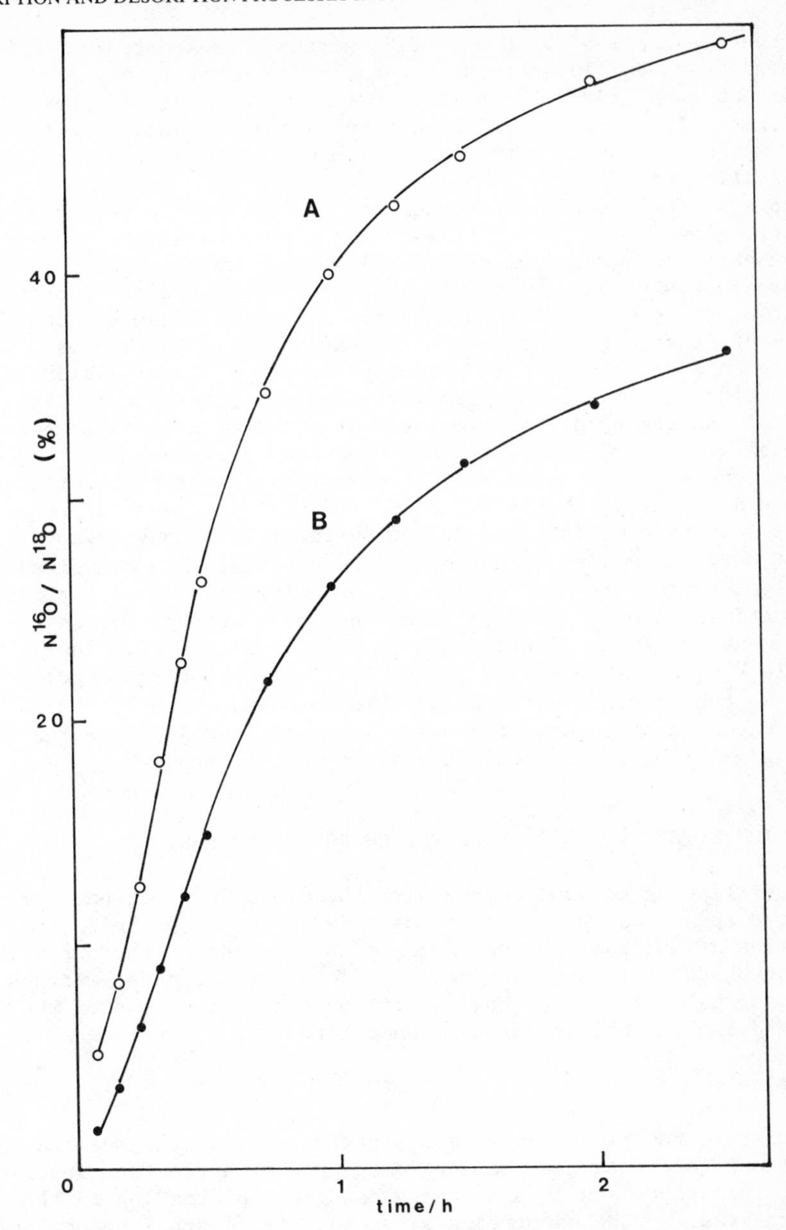

Figure 1. Variations in the isotopic composition of gas phase NO as a function of illumination time over (A) preoxidised (treatment in O_2 at 723K, evacuation at 423K) or (B) prereduced (treatment in H_2 at 723K, evacuation at 723 K) powder TiO_2 (Degussa, P-25). Conditions: ca.495 Pa $N^{18}O$ (98.5% isotopic purity); 100 mg photocatalyst (fixed bed); illumination with a Philips HPK 125 W UV lamp coupled with a 300-410 nm filter, flux power ca.3.85 mW cm^{-2}. See ref. 65.

Instantaneous exchange of $N^{18}O$ with TiO_2 surfaces prereduced in H_2 at 723 K showed that UV-illumination renders labile surface oxygen atoms which have not been eliminated by this pretreatment (Fig. 1). However, a higher initial rate for preoxidised samples (Fig.1) indicated that adsorbed oxygen atoms take part in the exchange, as expected from their greater lability (66).

Moreover, the isotopic heteroexchange of oxygen is a very sensitive photoinduced gas-solid reaction to access the photocatalytic oxidizing properties of semiconductor oxides. For instance, when comparing a TiO_2 sample homogeneously doped with 0.85 at % Cr^{3+} cations with undoped titania of nearly equal surface area, the rates of various photocatalytic oxidations of organic compounds were decreased from ∿ 25 to ∿ 85 times, whereas the initial rate of oxygen isotope exchange under illumination was ∿ 1000 times lower (32-33). Similarly, whereas the photocatalytic oxidation of isobutane in gas phase over a series of anatase samples varied by a factor of about 3, the photoinduced heteroexchange of oxygen over the same specimens was affected by a factor of about 45(63).

On the other hand, it has been inferred that, at least in some cases, the same type of labile adsorbed/surface atomic oxygen species which take part in $^{18}O_2$ or $N^{18}O$ isotopic exchange are also involved in photocatalytic oxidations over semiconductor oxides since : (i) the same orders of activities were found in both types of reactions for several TiO_2 specimens as mentioned above (63), as well as for SnO_2, ZnO, ZrO_2 samples (a V_2O_5 specimen was inactive) (64), (ii) these exchanges were inhibited in the presence of an oxidizable compound (63,65). The role of O_2 or NO in the photocatalytic oxidation reactions is to replenish the coverage of TiO_2 in removable atomic oxygen species.

2. ADSORPTION SITES. ADSORBED SPECIES. CHARGE TRANSFERS.

Although overlapping to some degree with the preceding section, this part is more specifically devoted to methods directed to the determination of the role of the adsorbed species and of the surface features in photocatalytic reactions. As in catalysis, these methods are essential to obtain a rational basis for the improvement of the activity and the selectivity of semiconductors.

2.1. Kinetics

As for catalytic reactions, experiments performed with various reactant concentrations (or pressures) can indicate whether a reactant intervenes in the adsorbed state and in some cases can give information on the adsorbed species, on the adsorption sites (use of poisons) and on the adsorption competition between the reactants and the products.

Some examples are presented in the following paragraphs. For instance, if the reciprocal of the reaction rate r is proportional to the reciprocal of the concentration C (or pressure) of one reactant, this can be accounted for by $r = k\theta$, where θ is the surface coverage by this reactant given by the Langmuir expression $\theta = KC/(1 + KC)$. Such a relationship has been found for the oxidation of iodide ions (66) and of oxalic acid (67) in aqueous solution over TiO_2. By contrast, a linear relationship between $1/r$ and $C^{-1/2}$ indicates that r depends on a species arising from the dissociative adsorption of the reactant considered. The dehydrogenation of alcohols over Pt/TiO_2 provides an example (10). For the oxidation of NH_3 over TiO_2 (68), a plot of the reciprocal of the formation rate of N_2 against the reciprocal of NH_3 pressure yielded non-parallel straight lines for various O_2 pressures. This allowed one to conclude that the oxidation occurs via a Langmuir-Hinshelwood mechanism, oxygen and ammonia being adsorbed on different sites.

Langmuir forms for the degradation rates of various chlorinated or brominated organic pollutants over UV-illuminated TiO_2 aqueous slurries have been proposed (69-71). Moreover, the kinetic equations can explain the degree of inhibition by chloride (or bromine) ions and allow one to suggest hypotheses concerning the type of adsorption and the nature (acidic or basic) of the adsorption sites. For instance, trichloethylene would be adsorbed on Ti sites through its pi-electrons, whereas dichloroacetaldehyde - an intermediate in the degradation of trichloroethylene - would be dissociatively chemisorbed by attachment of the α - hydrogen to a surface O^{2-} ion and attraction of the other moiety to a Ti site (69).

Rate expressions, consistent with the effects of the pressures of all reactants, have been derived for the gas phase oxidation to acetone of 2 - methyl - 2 - butyl - alcohol and isobutane over UV-illuminated powder TiO_2 (72). These expressions are based on the Langmuir model of adsorption. Although the organic compounds do not compete with oxygen for electron capture (34,36), the kinetics analysis assumes the same adsorption sites for all reactants (72). It is concluded that the oxygen intervenes as an adsorbed monoatomic species which transforms adsorbed isobutane into an a surface tert-butanol intermediate. For the further transformation of this intermediate or for the oxidation of the adsorbed alcohol considered, an expression assuming a two-site dehydration as the rate-determining step is in agreement with the experimental data.

These kinetic experiments offer the advantage of being simple and to be performed with unexpensive apparatuses. However, the indications they yield are general and physical methods are necessary to obtain a more detailed picture of the phenomena occurring in the adsorbed phase.

2.2. ESR spectra

The intense use of ESR spectroscopy to determine the nature of some oxygen adsorbed species and the formation of OH° radicals, has been evoked in section (1.1.). The detection of an electron transfer from UV-illuminated TiO_2 (grafted on porous Vycor glass) to adsorbed N_2O at 77 K was also made by ESR; the N_2O^- ion-radicals thus formed were unstable: they disappeared when the temperature was raised to 300 K or they were replaced by adsorbed O_2^- species when O_2 was admitted at 77 K (73). ESR has also been successfully employed to show the formation of methyl or triphenylmethyl radicals during the decarboxylation of acetic acid and triphenylacetic acid, respectively, over TiO_2 or

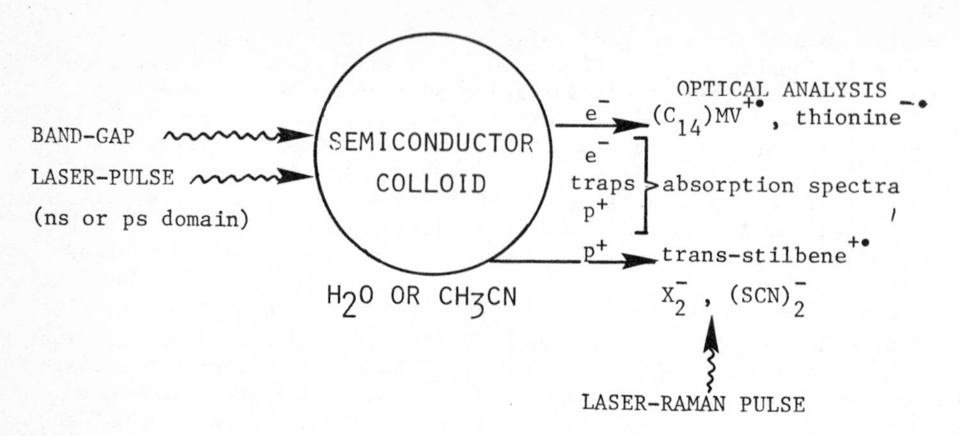

Figure 2. Principle of investigations on the transfers of photoproduced charges and on the detection of adsorbed ions-radicals by laser-flash excitation coupled with time-resolved optical analysis. For references, see text.

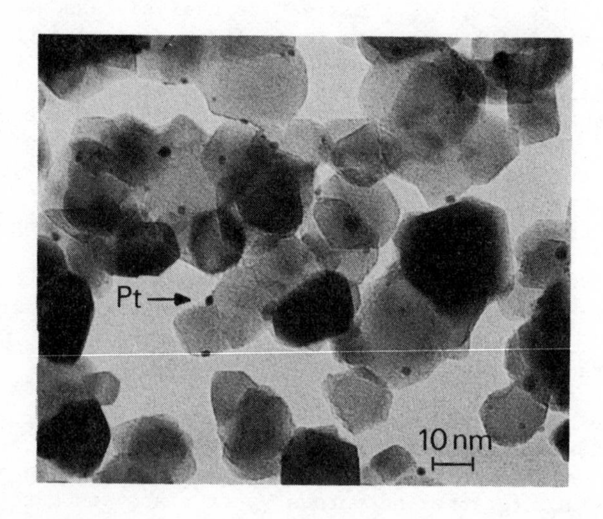

Figure 3. Transmission electron micrograph of a 1 wt% Pt/TiO_2 sample showing the Pt particles on top of the TiO_2 grains.

platinized TiO_2(74). In the second case, the radical ESR signal was observed directly, whereas methyl radicals were trapped with a nitrone.

2.3. Laser-flash excitation coupled with fast optical analysis

This method (Fig.2) has allowed considerable progresses in the understanding of charge transfers between an excited semiconductor colloid and adsorbed reactants.

For instance, a 475-nm absorption band has been assigned to trapped holes at the TiO_2 surface as a result of a laser flash in the presence of an adsorbed electron scavenger (75,76). The time decay of this band has been monitored for various concentrations of sodium citrate, acetic acid or ethanol in the Pt-TiO_2 sol (76). Initially sodium citrate was the most effective, whereas, after a while, ethanol become the most effective. This has been interpreted in terms of competition between surface coverage by each reactant and ease of oxidation of each reactant. This shows that the charge transfer concerns adsorbed species.

The transfer of holes from excited colloidal TiO_2 to reductants such as halide ions and thiocyanide ions in aqueous solution was followed by the absorption spectra of the resulting X_2^- and $(SCN)_2^-$ species (40,77). The transfer was achieved within the ∿ 10 ns flash which means that only the adsorbed species participate. These types of studies have been extended to non-aqueous solutions (78,79). Over TiO_2 in acetonitrile the absorption spectra showed the fast transformation of trans-stilbene into the corresponding radical-cation and the slower transfer of electrons to methylviolen cations(78). Also, other results concern colloidal CdS. Very recently, experiments in the pico-second domain have been carried out (80). The time decay of a 600-nm absorption, attributed to electrons trapped, presumably at Ti^{3+} sites, was followed 20 ps after the 30 ps excitation pulse of colloidal TiO_2 in deaerated water at pH 2.7. A recombination coefficient of the electron-hole pairs of 4×10^{-11} cm^3 s^{-1} was calculated.

The use of time-resolved Raman spectrocopy yields more detailed information on the nature of the species involved in charge transfer or resulting from it (81,82). For instance, hole or electron transfers were studied from excited TiO_2 or CdS colloids in aqueous solutions to SCN^-, MV^{2+} (methylviologen) and C_{14} MV^{2+} ions. The bands of the Raman spectrum of $MV^{+\cdot}$ on CdS showing a growth of the Langmuir type as a function of the initial MV^{2+} concentration, it was concluded that the transfer involves adsorbed MV^{2+} ions; moreover, calibrations indicated that at pH10 one cation was adsorbed per CdS particle (∿ 4-4.5 nm dia.). In addition, it was inferred that the adsorbed ions retain their first water layers and are adsorbed either electrostatically or by an hydrophobic chain ($C_{14}MV^{2+}$), since the Raman spectra of the resulting $(SCN)_2^-$, $MV^{+\cdot}$ or $C_{14}MV^{+\cdot}$ ions were indistinguishable from those of the same species in aqueous solutions (81). An example of a reverse transfer of electron between a dye and a semiconductor further illustrates the interest of time-resolved Raman spectroscopy. In the case of colloidal TiO_2 at pH3, resonance Raman spectra of adsorbed, semioxidized eosin Y have been obtained following optical excitation of the dye below the semiconductor band gap. In addition, these spectra indicated a protonation of the positively charged adsorbed species in an environment less polar than that corresponding to pure water, which gave an estimation of the perturbation brought by the colloid surface (82).

3. HYDROGEN SPILLOVER ON M/TiO_2 PHOTOCATALYSTS

Photocatalysts of this type are used in photocatalytic reactions involving H_2 either as a reactant or as a product. In the case of Pt and of some other noble metals, they can be prepared in the form of small particles (about 2 nm dia.) on top of the TiO_2 grains (Fig.3) by impregnation with a salt of the desired metal and reduction in H_2 at high temperature (10,84,93).

The following paragraphs describe and discuss experiments showing the existence of hydrogen spillover (migration of hydrogen species) between the two components of these bifunctional photocatalysts.

Isotopic exchange between unlabelled cyclopentane and deuterium (CDIE) (83), as well as the dehydrogenation of various primary or secondary, saturated (10,84-95) or unsaturated (96,97) alcohols (DHA), require the presence of a deposit of a group VIII transition metal on a semiconductor oxide, such as TiO_2, to be photocatalytic(fig.4).Over the naked semiconductor, the initial reaction rate is much lower and the reaction stops after a certain period of time (10,83). Studies of these reactions over a Pt/TiO_2 sample at different temperatures show that below $\sim$ 260 K (CDIE) (83[b]) or $\sim$ 310 K (DHA) (84) the desorption of H_2 or HD is the rate-limiting step. The value of the heat of hydrogen reversible adsorption derived from these studies is identical for both reactions; furthermore, it corresponds to the value measured by microcalorimetry and is equal to that reported for Pt/Al_2O_3 catalysts, which shows that H_2 (or HD) evolves from the metal and not from the semiconductor. In this particular case, simple experiments - studies of the variations of the photocatalytic activity against temperature and comparison with values of adsorption heats - have allowed one to determine the nature of the desorption sites of one of the products.

At temperatures $\geqslant$ 273 K, CDIE produces polydeuterated cyclopentane molecules in the absence of UV light. On the contrary, photocatalytic CDIE at temperatures $\leqslant$ 263 K (to avoid the thermally catalyzed exchange) initially yields only C_5H_9D (83). It can be inferred that the reaction in the dark takes place on the metal particles where the residence time of the chemisorbed cyclopentane molecules enables a multiple exchange, whereas, under illumination and at lower temperatures, the exchange occurs on TiO_2 where cyclopentane is weakly adsorbed and accordingly does not stay on the surface long enough to exchange more than one hydrogen atom for each adsorption act. Other experiments have shown that OH/OD groups of the TiO_2 surface are involved in this exchange (4,83). On neat anatase the OH groups resulting from the isotopic exchange of OD groups with cyclopentane cannot exchange with gaseous deuterium at the temperatures used (98). By contrast, the OH-D_2 exchange can occur in the presence of a deposited group VIII metal, which again implies a spillover of some deuterium atoms, dissociatively adsorbed on the metal particles, to the semiconductor.

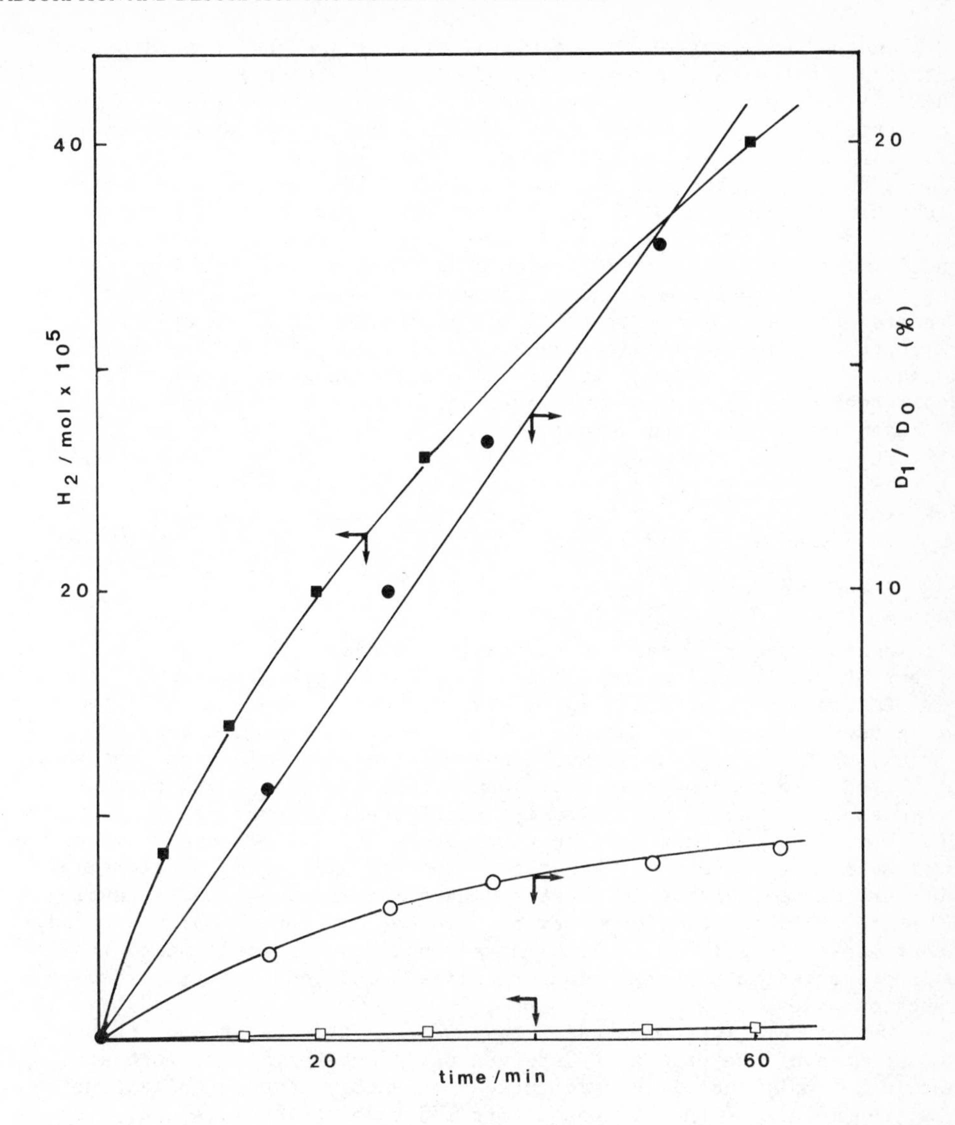

Figure 4. Effect of Pt deposited on TiO_2 (Degussa P-25) for the propanol-1 dehydrogenation in liquid phase at 298K and for the cyclopentane-deuterium isotopic exchange in gas phase at 263K. Open symbols: naked TiO_2. Solid symbols: 0.5 wt% Pt/TiO_2. D_o=C_5H_{10}; D_1=C_5H_9D. Conditions: 70 mg photocatalyst in 20 cm^3 propanol-1 or 100 mg photocatalyst pretreated in D_2 at 573K exposed to a total pressure of 2.65 kPa (94%D_2). Quantum yields:ca.0.2 for the dehydrogenation;ca.0.018 for the exchange. See refs 83a,84.

Another evidence of hydrogen spillover is brought forward by the effect on CDIE of a pretreatment of Pt/TiO_2 catalysts in D_2 at 773K(83[b]). This pretreatment causes a large decrease in D_2 chemisorption on the Pt particles (cf. the so-called "strong metal-support interaction") and also markedly affects the CDIE (Fig.5).

Finally, the increased conductivity (99,100) and photoconductance (100,101) of evacuated M/TiO_2 catalysts when these samples are exposed to H_2 can also be interpreted in terms of H_2 spillover. Qualitatively, these increases are consistent with a decrease in the work function of the metal which has chemisorbed hydrogen (102-104), so that the charge transfer between the non-preoxidized TiO_2 to the metal in order to align their Fermi levels is reduced. In addition, the half-order dependence of the conductivity or photoconductance on H_2 pressure is consistent with a hydrogen spillover accompanied by a release of electrons as schematized below:

$$Pt_s - H + O^{2-} \rightleftarrows Pt_s + OH^- + e^- \quad (23)$$

In conclusion, the reversible transfer of adsorbed hydrogen atoms (or protons) on M/TiO_2 photocatalysts seems a well established phenomenon and is supported by other studies (93).

CONCLUDING REMARKS

(i) Numerous experimental and theoretical studies have considered the effect of band-gap illumination on the adsorption equilibrium at the gas-semiconductor particles interface, whereas those about the liquid-semiconductor particles interface are scarce. As a result the photoadsorption of gaseous oxygen on semiconductor oxides and the role of the surface OH groups is well documented. The characteristics of photodesorption are not so clearly established; however, photodesorption does not seem very important in photocatalysis. (ii) To study adsorption over solid photocatalysts, kinetics, isotopic exchanges and classical physical methods are of interest and have increased our understanding of photocatalytic reactions. New in situ methods for the liquid-solid interface are highly desired. Great advances can be anticipated from pulsed excitation of semiconductor colloids coupled with time-resolved spectroscopies.

More generally, the studies reviewed in this paper show that a description of the processes intervening in heterogeneous photocatalysis should not only include considerations on energy levels (collective electron models of the semiconductors and redox potentials) but also an approach of the surface features.

ACKNOWLEDGEMENTS

The author is indebted to his CNRS collaborators Dr. J.-M. Herrmann, Mr H. Courbon, Mr J. Disdier, Mrs M.-N. Mozzanega whose contributions appear in the list of references. For these contributions, several financial supports from the CNRS interdisciplinary research program on energy and raw materials (PIRSEM) are gratefully acknowledged.

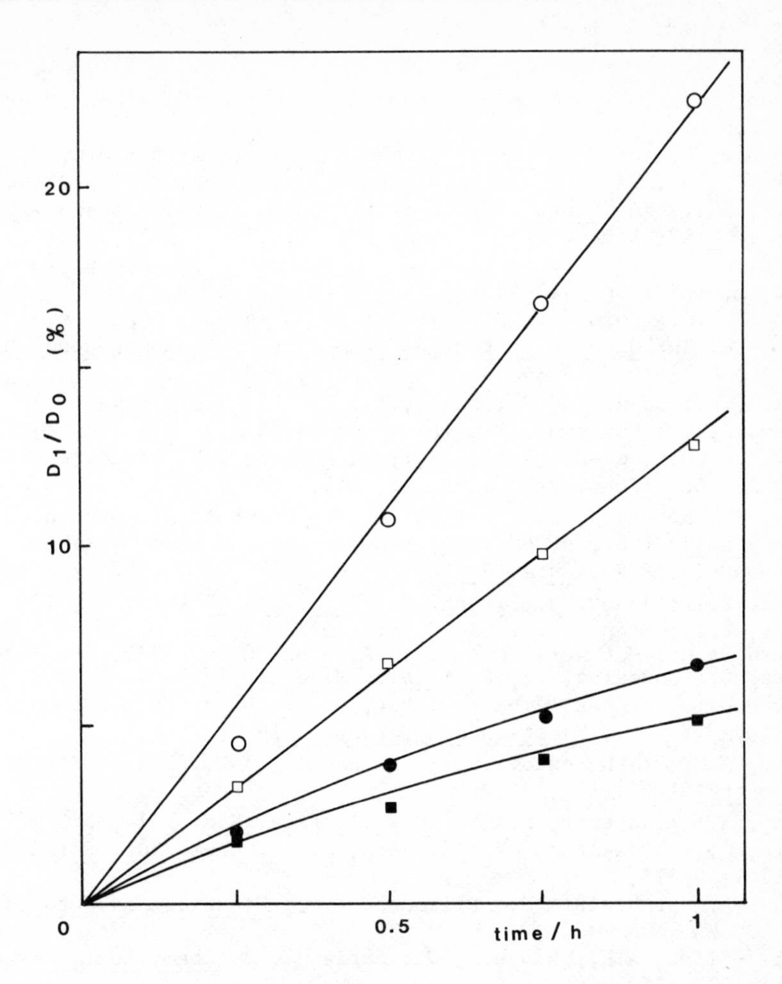

Figure 5. Variations of the ratio C_5H_9D/C_5H_{10} in gas phase in the presence of D_2 as a function of illumination time at 260K over Pt/TiO_2 (Degussa, P-25) powder samples containing 5 (circles) or 10 (squares) wt% Pt pretreated in D_2 at 573 K (open symbols) or 773K (solid symbols). Conditions: 2.13 k Pa total pressure (molecular ratio deuterium/cyclopentane: 3); 100 mg photocatalyst (fixed bed); illumination with a Philips HPK 125 W UV lamp coupled with a 300-410 nm filter, flux power ca. 3.5 mW cm^{-2}. See ref. 83.

REFERENCES

1. Morrison, S.R., "The Chemical Physics of Surfaces", Plenum Press, 1977, Chap.9.
2. Cunningham, J., in "Comprehensive Chemical Kinetics", Bamford, C.H.; Tipper, C.F.H., Eds.; Elsevier, 1984; Vol. 19; Chap. 3, pp 291-427.
3. Bickley, R.I., in "Photoelectrochemistry, Photocatalysis and Photoreactors"; Schiavello, M., Ed; Reidel, 1985; pp 379-388; 491-502.
4. Pichat, P.; Ibid.; pp 425-455.
5. Teichner, S.J.; Formenti, M.; Ibid.; pp 457-489.
6. Pichat, P., in "Organic Phototransformations in Nonhomogeneous Media"; Fox M.A., Ed; A.C.S. Symp. Ser. 1985, 278, pp 21-42.
7. Tokumaru, K.; Sakuragi, H.; Kanno, T.; Oguchi, T.; Misawa, H.; Schimamura, Y.; Kuriyama, Y., Ibid.; pp 43-55.
8. Fox, M.A.; Chen, C.-C.; Park K.-H.; Yornathan, J.N., Ibid.; pp 69-78.
9. Fox, M.A., Acc. Chem. Res. 1983, 16, 314.
10. Pichat, P.; Herrmann, J.-M.; Disdier, J.; Courbon, H.; Mozzanega, M.-N., Nouv. J. Chim. 1981, 5, 627.
11. Vol'kenshtein, F.F., Russ. J. Phys. Chem. 1974, 48, 1555.
12. Gersten, J.I.; Janow, R.; Tzoar, N., Phys. Rev. B 1975, 11, 1267.
13. Bickley, R.I.; Stone, F.S., J. Catal. 1973, 31, 389.
14. Boonstra, A.H.; Mutsaers, C.A.H.A., J. Phys. Chem. 1975, 79, 1694.
15. Munuera, G.; Gonzalez, F., Rev. Chim. Minérale 1967, 4, 207.
16. Courbon, H.; Herrmann, J.-M.; Pichat, P., J. Phys. Chem.1984, 88, 5210.
17. Solonitsyn, Yu. P., Kinetics and Catalysis 1966, 7, 424.
18. Volodin, A.M.; Zakharenko, V.S.; Cherkashin, A.E., React. Kin et Catal. Lett. 1981, 18, 321.
19. Bickley, R.I.; Jayanty, R.K.M., Disc. Faraday Soc. 1974, 58, 194.
20. Munuera, G.; Rives-Arnau, V.; Saucedo, A., J. Chem. Soc., Faraday Trans.I 1979, 75, 736.
21. Basov, L-L.; Drobinin A.N.; Filimonov V.N., Kinetics and Catalysis 1982, 23, 85.
22. Gonzalez-Elipe, A.R.; Munuera, G.; Soria J., J. Chem. Soc., Faraday Trans.I 1979, 75, 748.
23. Gonzalez-Elipe, A.R.; Munuera, G.; Sanz J.; Soria, J., J. Chem. Soc., Faraday Trans.I 1980, 76, 1535.
24. Bickley, R.I.; Vishwanathan, V., Nature 1979, 280, 306.
25. Bourasseau, S.; Martin, J.-R.; Juillet, F.; Teichner, S.J., J. Chim. Phys. 1974, 71, 122.
26. Meriaudeau, P.; Védrine J.C., J. Chem. Soc., Faraday Trans.I 1976, 72, 472.
27. Che, M.; Tench, A.J., Adv. Catal. 1982, 31, 77.
28. Che, M.; Tench, A.J., Adv. Catal. 1983, 32, 1.
29. Lunsford, J., Catal. Rev. 1973, 8, 135.
30. Bielanski, A.; Haber, J., Catal. Rev. 1979, 19, 1.
31. Bourasseau, S.; Martin, J.-R.; Juillet F.; Teichner, S.J., J. Chim. Phys. 1973, 70, 1467; 1472; 1974, 71, 1017.
32. Herrmann, J.-M.; Disdier, J.; Pichat, P., Chem. Phys. Lett. 1984, 108, 618.
33. Pichat, P.; Herrmann, J.-M.; Disdier, J.; Mozzanega, M.-N.;

Courbon, H., in "Catalysis on the Energy Scene", Studies in Surf. Sci. Catal., 19; Kaliaguine, S., Mahay, A., Eds; Elsevier, 1984, pp 319-326.
34. Herrmann, J.-M.; Disdier, J.; Pichat, P., Proc. 7th Int. Vacuum Congr. 3 rd Int. Conf. Solid Surfaces; Dobrozemski, R., et al, Eds; F.Berger Söhne, Horn (Austria), 1977, Vol.II, pp 951-954.
35. Herrmann, J.-M.; Disdier, J.; Pichat, P., J. Chem. Soc., Faraday Trans.I 1981, 77, 2815.
36. Pichat, P.; Herrmann, J.-M.; Courbon, H.; Disdier, J.; Mozzanega, M.-N. Canad. J. Chem. Eng. 1982, 60, 27.
37. Herrmann, J.-M.; Disdier, J.; Mozzanega, M.-N.; Pichat, P., J. Catal. 1979, 60, 369.
38. Völz, H.G.; Kaempf, G.; Fitzky, H.G.; Klaeren, A. in "Photodegradation and Photostabilization of Coatings"; Pappas, S.P.; Winslow, F.H., Eds.; A.C.S. Symp. Ser. 1981, 151, pp 163-182 and references therein.
39. Ceresa, E.M.; Burlamacchi, L.; Visca, M., J. Mater. Sci. 1983, 18, 289.
40. Grätzel, M., Pure Appl. Chem. 1982, 54, 2369; Biochimica Biophysica Acta 1982, 683, 221; Chem. Future 1983, pp 171-8.
41. Borgarello, E.; Kiwi, J.; Pelizzetti, E.; Visca, M.; Grätzel, M., Nature 1981, 281, 158; J. Am. Chem. Soc. 1981, 103, 6324.
42. Kalyanasundaram, K.; Borgarello, E.; Grätzel, M.; Helv. Chim. Acta 1981, 64, 362.
43. Kalyanasundaram, K.; Borgarello, E.; Duonghong D.; Grätzel, M., Angew. Chem. Int. Ed. Engl. 1981, 20, 987.
44. Yesodharan E.; Grätzel, M., Helv. Chim. Acta 1983, 66, 2145.
45. Gonzalez-Elipe, A.R.; Munuera, G.; Soria, J., React. Kinet. Catal. Lett. 1981, 18, 367.
46. Zakharenko, V.S.; Cherkashin, A.E.; Ibid., 1983, 23, 131.
47. Kornblit, L.; Ignatiev, A., Surf. Sci. 1984, 136, L.57 and references therein.
48. Baidyaroy, S.; Bottoms, W.R.; Mark, P., Surf. Sci. 1971, 28, 517.
49. Genequand, P., Surf. Sci. 1971, 25, 643.
50. Grade, M. in "Current Topics in Materials Sci."; Kaldis, E., Ed.; North Holland Publ. Co., 1981, 7, pp 339-363.
51. Bonasewicz, P.; Hirschwald, W., Ibid., pp 410-431.
52. [a]Shapira, Y.; Cox, S.M.; Lichtman, D., Surf. Sci. 1975, 50, 503; [b]1976, 54, 43.
53. Lichtman, D.; Shapira, Y., J. Nucl. Mater. 1976, 63, 184.
54. Shapira, Y.; Mc Quistan, R.B.; Lichtman, D., Phys. Rev. B 1977, 15, 2163.
55. Cox, S.M.; Lichtman, D., Surf. Sci. 1976, 54, 675.
56. Van Hieu, N.; Lichtman, D., Surf. Sci. 1981, 103, 535.
57. Lichtman, D.; Lin, T., Proc. 7th Int. Vacuum Congr. 3 rd Int. Conf. Solid Surfaces; Dobrozemski, R., et al., Eds; F. Ferger (Söhne, Horn (Austria), 1977, Vol.II, pp 1277-1280.
58. Van Hieu, N.; Lichtman, D., J. Catal. 1982, 73, 329.
59. Lichtman, D.; Shapira, Y.; CRC Crit. Rev. Solid St.Mater.Sci. 1978, 8, 93.
60. Courbon, H.; Formenti, M.; Juillet, F.; Lisachenko, A.A.; Martin, J.-R., Teichner, S.J., Kinetics and Catalysis, 1973, 14, 84.

61. Cunningham, J.; Doyle, B.; Morrissey, D.J.; Samman, N., Proc. 6th Int. Congr. Catal.; Bond, G.C.; Wells, P.B., Eds.; Chem. Soc., London, 1976, Vol.2, pp 1093-1101.
62. Ryabchuk, V.K.; Basov, L.L.; Lisachenko, A.A.; Vilesov, F.I., Sov. Phys. Tech. Phys. 1974, 18, 1349.
63. Courbon, H.; Formenti, M.; Pichat, P., J. Phys. Chem. 1977, 81, 550.
64. Courbon, H.; Pichat, P., C.R. Acad. Sci. 1977, 285C, 171.
65. Courbon, H.; Pichat, P., J. Chem. Soc., Faraday Trans.I 1984, 80,3175.
66. Herrmann, J.-M.; Pichat, P., Ibid, 1980, 76, 1138.
67. Herrmann, J.-M.; Mozzanega, M.-N.; Pichat, P., J. Photochem. 1983, 22, 333.
68. Mozzanega, H.; Herrmann, J.-M.; Pichat, P., J. Phys. Chem. 1979, 83, 2251.
69. Pruden, A.L.; Ollis, D.F., J. Catal. 1983, 82, 404.
70. Hsiao, C.-Y.; Lee, C.-L.; Ollis, D.F. J.Catal. 1983, 82, 418. Ollis, D.F.; Hsiao, C.-Y.; Budiman, L.; Lee, C.-L., J. Catal. 1984, 88, 89.
71. Nguyen, T.; Ollis, D.F., J. Phys. Chem. 1984, 88, 3386.
72. Pruden-Childs, A.; Ollis, D.F., J. Catal. 1981, 67, 35.
73. Anpo, M.; Aikawa, N.; Kubokawa, Y., J. Chem. Soc., Chem. Commun.1984, 644.
74. Kraeutler, B.; Jaeger, C.D.; Bard, A.J., J. Am. Chem. Soc. 1978, 4903.
75. Bahnemann, D.; Henglein, A.; Lilie, J.; Spanhel, L., J. Phys. Chem. 1984, 88, 709.
76. Bahnemann, D.; Henglein, A.; Spanhel, L. Faraday Discuss. Chem. Soc. 1984, 88, 151; Henglein, A.,Pure Appl. Chem. 1984, 56, 1215.
77. Moser, J.; Grätzel, M., Helv. Chim. Acta 1982, 65, 1436.
78. Fox, M.A.; Lindig, B.; Chen, C.-C., J. Am. Chem. Soc. 1982, 104,5828.
79. Kamat, P.V., J. Photochem. 1985, 28, 513.
80. Moser, J.; Grätzel, M.; Serpone, N.; Sharma, D.K., J. Am. Chem. Soc., in press.
81. Rossetti, R.; Beck, S.M.; Brus, L.E., J. Am. Chem. Soc. 1984, 106, 980.
82. Rossetti, R.; Brus, L.E., J. Am. Chem. Soc. 1984, 106, 4336.
83. [a] Courbon, H.; Herrmann, J.-M.; Pichat, P., J. Catal. 1981, 72, 129; [b] Id., in press.
84. Pichat, P.; Mozzanega, M.-N.; Disdier, J.; Herrmann, J.-M., Nouv. J. Chim. 1982, 6, 559.
85. Teratini, S.; Nakamichi, J.; Taya, K.; Tanaka, K., Bull. Chem. Soc. Jpn 1982, 55, 1688.
86. Domen, K.; Naito, S.; Onishi, T.; Tamaru, K., Chem. Lett. 1982, 555.
87. Muradov, N.Z.; Buzhutin, Y.; Bezugalya, A.G.; Rustamov, M.I., Russ. J. Phys. Chem.1982, 56, 1082.
88. Borgarello, E.; Pelizzetti, E.; Chim. Ind. 1983, 65, 474.
89. Oosawa, Y., Chem. Lett. 1983,577.
90. Taniguchi, Y.; Yoneyama, H.; Tamura, H., Chem. Lett. 1983, 269.
91. Prahov, L.T.; Disdier, J.; Herrmann, J.-M.; Pichat, P., Int.J.Hydrogen Energy 1984, 9, 397.
92. Matsumura, M.; Hiramoto, M.; Iehara, T.; Tsubomura, H., J.Phys.Chem. 1984, 88, 248.

93. Aït-Ichou, I.; Formenti, M.; Teichner, S.J., in "Spillover of Adsorbed Species", Studies in Surf. Sci. Catal., 17; Pajonk, G.M.; Teichner, S.J.; Germain, J.E., Eds; Elsevier, 1983, pp 63-75.
94. Id., in "Catalysis on the Energy Scene", Studies in Surf. Sci. Catal., 19; Kaliaguine, S.; Mahay, A., Eds; Elsevier, 1984, pp 297-307.
95. Aït-Ichou, I.; Formenti, M.; Pommier, B.; Teichner, S.J., J. Catal. 1985, 91, 293.
96. Pichat, P.; Disdier, J.; Mozzanega, M.-N.; Herrmann, J.-M., in Proc. 8th Int. Congr. Catal.; Verlag Chemie, Dechema; 1984, Vol. III, pp 487-498.
97. Hussein, F.H.; Pattenden, G.; Rudham, R.; Russel, J.J., Tetrahedron Lett. 1984, 25, 3363.
98. Primet, M.; Pichat, P.; Mathieu, M.-V., J. Phys. Chem. 1971, 75, 1216.
99. Herrmann, J.-M.; Pichat, P., J. Catal. 1982, 78, 425.
100. Id., in "Spillover of Adsorbed Species", Studies in Surf. Sci. Catal., 17; Pajonk, G.M.; Teichner, S.J.; Germain, J.E., Eds; Elsevier, 1983, pp 77-87.
101. Disdier, J.; Herrmann, J.-M.; Pichat, P. J. Chem. Soc., Faraday Trans. I 1983, 79, 651.
102. Yamamoto, N.; Tonomura, S.; Matsuoka, T.; Tsubomura, H., Surf. Sci. 1980, 92, 400.
103. Hope, G.A.; Bard, A.J., J. Phys. Chem. 1983, 87, 1979.
104. Aspnes, D.E.; Heller, A., J. Phys. Chem., 1983, 87, 4919.

ARCHER - HREELS of oxygen molecularly adsorbed on metals usually shows an O-O vibration frequency corresponding to transfer of a non-integral number of electrons from the metal to the adsorbed O_2, i.e. to formation of $M^{x+}O_2^{x-}$ surface compounds in which the electron distribution in the $M-O_2$ molecular orbital corresponds to non-integral x. You have discussed ESR data for O_2 molecularly adsorbed on semiconductors in terms only of integral x and formation of adsorbed O_2^-. Could you comment on this apparent difference in interpretation? Presumably, ESR of $M^{x+}.O_2^{x-}$ with non-integral x would detect the same total number of spins as the ESR of $M^+.O_2^-$ but the g factor would be affected?

PICHAT - As ESR detects paramagnetic species, it implies transfers of integral numbers of electrons between the adsorbent and an individual adsorbed species. For oxide catalysts exposed to O_2, the existence of O_2^- species has been reported by many authors. However, a detailed analysis of the ESR spectra - in particular of those obtained by using O-17 enriched O_2 - can give information on the environment of these species on the surface which depends inter alia on the pretreatment. Transfers of non-integral numbers of electrons certainly occur in many cases and other spectroscopic techniques are appropriate to detect them. On the other hand, different adsorbed species are expected on metals and oxides.

PHOTO-INDUCED PROCESSES AT TITANIUM DIOXIDE SURFACES

R.I. Bickley, R.K.M. Jayanty and V. Vishwanathan,
School of Chemistry, University of Bradford,
Bradford, BD7 1DP, U.K.
J.A. Navio,
Dept^{o} de Quimica Inorganica, Universidad de Sevilla, Espana.

ABSTRACT. Evidence is presented to demonstrate the participation of surface hydroxyl groups in the photosorption of oxygen on hydroxylated rutile (TiO_2) surfaces under the stimulation of near u.v. radiation ($310nm < \lambda$). Species resembling adsorbed hydrogen peroxide are produced which have the capability of interacting with dinitrogen in a consecutive thermal process.

Introduction

Interest in the uses of titanium dioxide in photocatalysis and in photoelectrochemistry has become widespread, since the photoelectrochemical experiments of Fujishima and Honda (1) demonstrated the photoelectrolysis of water into its constituent elements. As testimony to the efforts which have been made subsequently to design a satisfactory system which might utilise solar energy to drive the electrolysis of water, and moreover to the dominant use of TiO_2 and related mixed oxides, reference need be made only to the published proceedings of successive International Conferences on the Photochemical Conversion and Storage of Solar Energy. (2)

In the context of research on Heterogeneous Photocatalysis recent reviews of the subject by Formenti and Teichner (3), and by Bickley (4), and the published proceedings of the N.A.T.O. A.S.I. on Photoelectrochemistry, Photocatalysis and Photoreactors (5) gives an excellent means of identifying the common factors in these popular but difficult areas of research.

In this presentation evidence will be given to emphasise the importance of hydroxylated surfaces of titanium dioxide (rutile) in some aspects of heterogeneous photocatalysis. It should be stressed that the presence of adsorbed water is not essential for TiO_2 to behave as a photocatalyst but that in situations in which it is present it can exert a dominant effect. Of particular interest are those photochemical processes which may be possible in the natural environment of the Earth's surface and which may be stimulated by direct solar radiation. Examples of such processes are:- the photosorption of oxygen (6), the photo-synthesis of hydrogen peroxide (7), the

E. Pelizzetti and N. Serpone (eds.), Homogeneous and Heterogeneous Photocatalysis, 555–565.

oxidative photo-fixation of atmospheric dinitrogen (8) and the photo-oxidative degradation of numerous organic molecules to form CO_2 (9).

Experimental

The experiments, which are described, were performed in a vacuum system which was capable of attaining ultimate low pressures of ca. 10^{-4} Pa. Measurements of adsorption were made in a small constant volume section of the apparatus (240 cm^3) which contained the adsorption cell and two pressure gauges; a Pirani gauge and a pressure transducer (Bell and Howell Ltd. 0-35 mb A). The constant volume system could be connected to a pair of mercury filled gas burettes (300 cm^3) by which experiments at constant pressure could be achieved. Gas analyses were made by releasing the contents of the adsorption system through a calibrated needle valve into a magnetic deflection mass-spectrometer (Vacuum Generators, Micromass 2A) which was capable of detecting molecules with relative masses of 2, 3, 4, 12-60 and 48-240.

The adsorption cell, into which about 0.5 - 1g. of the TiO_2 specimen was usually placed, was constructed of clear silica tubing ($\emptyset$=30 mm) to which a silica side-arm and two Vitreosil clear silica optical windows were fused. Irradiation of the surface of the powdered specimens was effected from above by a 500W medium pressure mercury arc contained within a water cooled pyrex glass condenser.

Analyses of the strongly adsorbed surface species were made using temperature programmed desorption (t.p.d.) linked to continuous monitoring of the gas phase using the mass-spectrometer in its scanning mode of operation.

The materials used in this work were as follows:-

1) All gases were supplied in pyrex glass bulbs (1 dm^3) and were designated as spectroscopically pure by the manufacturers, British Oxygen (Gases) Ltd.
2) The TiO_2 was supplied by Tioxide International Ltd. according to the following specifications; a) a low area rutile, ex sulphate of 4.2 m^2g^{-1} containing Si 300 ppm, Fe 3 ppm, Al_2O_3 13 ppm, K_2O 20 ppm and CaO 13 ppm
   b) a higher area material, ex chloride hydrolysis of 20.2 m^2g^{-1} containing Si 100 ppm, CaO 60 ppm, Fe 6 ppm, K_2O 18 ppm, Zno 23 ppm and Al_2O_3 33 ppm
3) H_2O_2(l) was supplied by BDH Ltd. ; 30 volume.
4) H_2O(l) was obtained from a laboratory deioniser with a conductivity of 1.3×10^{-4} Ω^{-1} m^{-1}

Results

Figure 1 depicts the photosorption of O_2 upon rutile (TiO_2) surfaces. The results show the extent of the photoadsorption during continuous illumination, and the kinetic profiles approximate to first order behaviour. Most significantly is demonstrated the effect of progress-

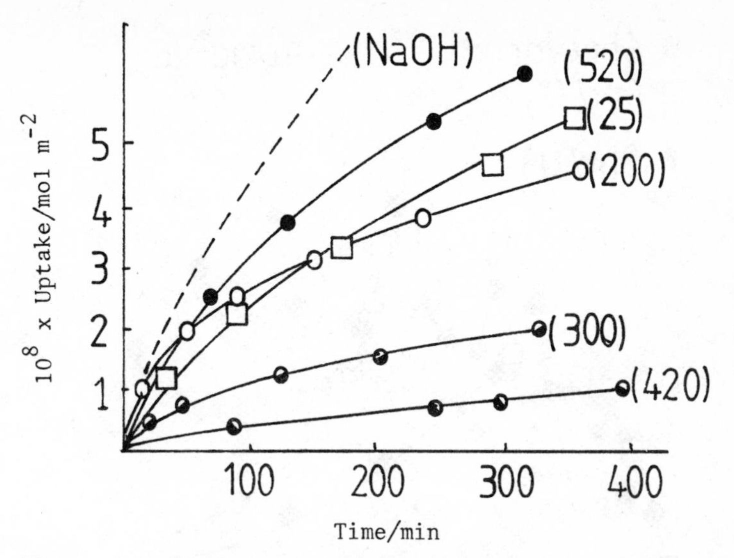

Figure 1. O_2 photosorption on TiO_2 as a function of the outgassing temperature.

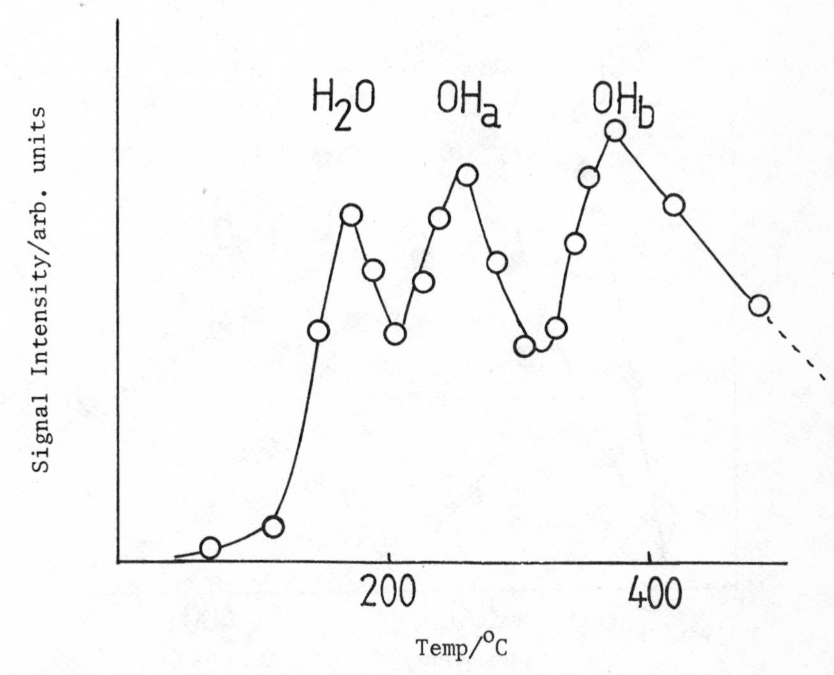

Figure 2. T.p.d of water from the surface of hydroxylated TiO_2.

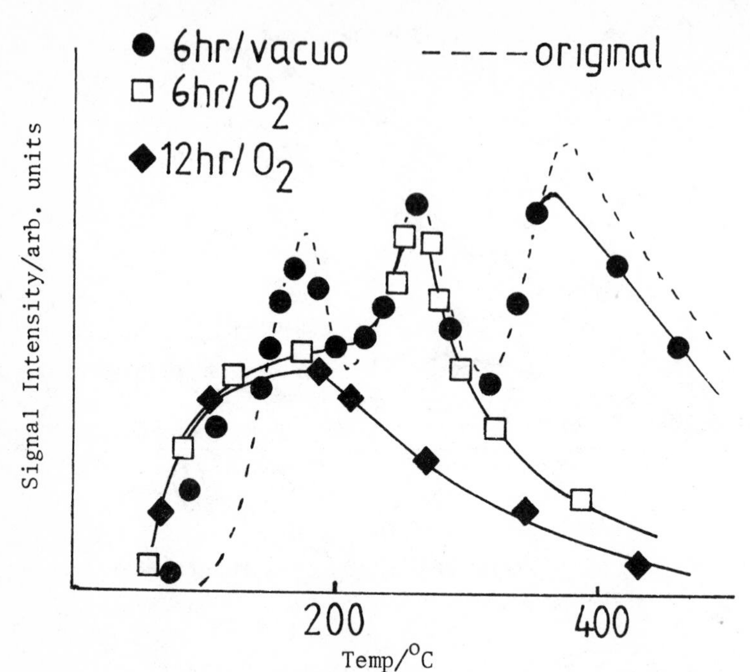

Figure 3. T.p.d of water from the surface of hydroxylated TiO_2 after irradiation with u.v. light

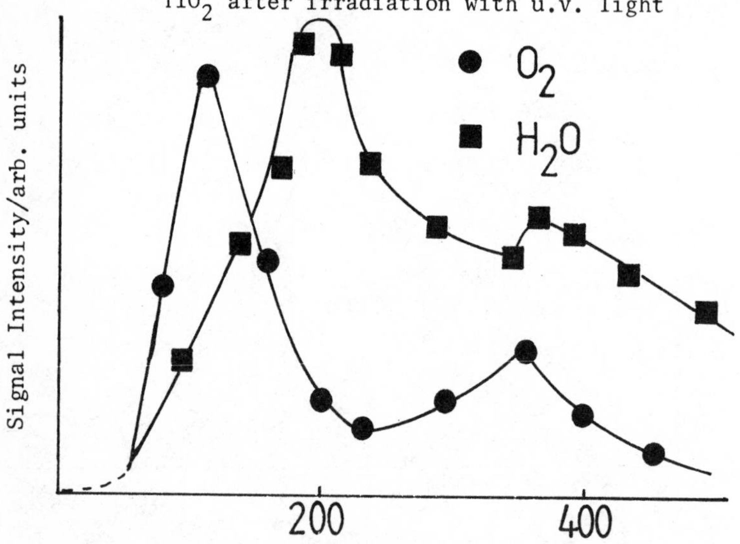

Figure 4. T.p.d from the surface of TiO_2 containing preadsorbed H_2O_2.

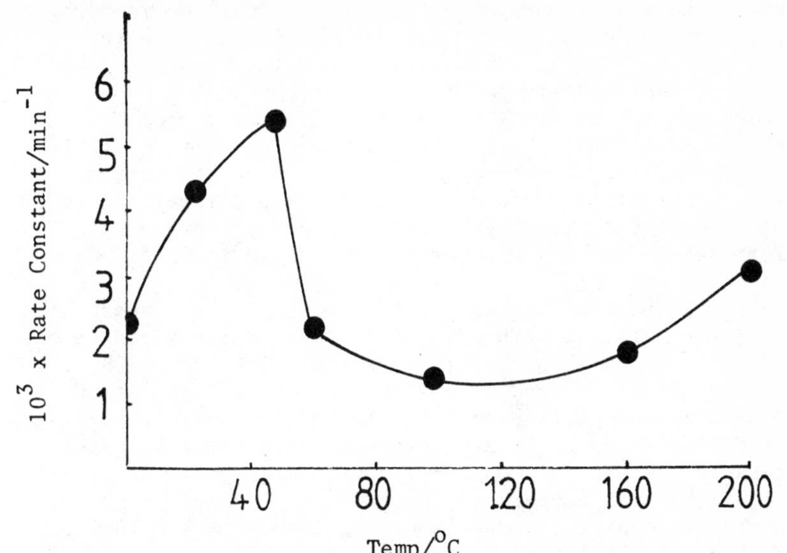

Figure 5. Variation of the First Order Rate Constant for O_2 photosorption

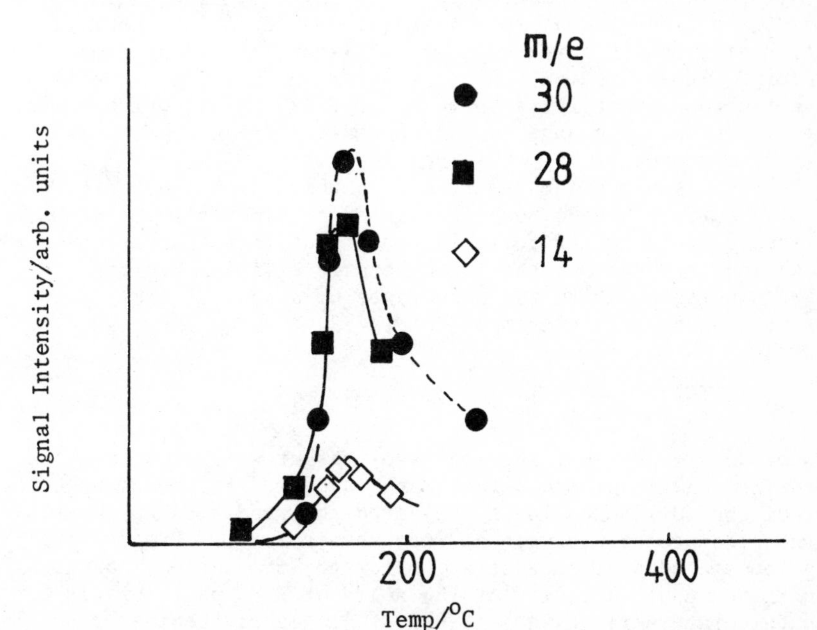

Figure 6. Desorption of nitrogen containing species from the surface of hydroxylated TiO_2 after u.v. irradiation and exposure to N_2(g).

ively increasing the temperature of outgassing of the specimen prior to the study of photosorption at room temperature (25°C). The interpretation of these thermally induced variations in activity are best understood by referring additionally to Fig. 2 in which the t.p.d. profiles of water on the rutile surface is presented.

Figure 2 indicates that the surface of TiO_2 can be effectively dehydroxylated by treating the specimen at 550°. Following the reoxidation of the specimen in pure O_2(g), the original water desorption profile can be regenerated by a thermal pretreatment in water vapour (20 Torr) at 300°C, or by immersing the solid in liquid water at ∿100°C.

Figure 3 represents the t.p.d. profiles of rutile surfaces which have been irradiated at 25°C with u.v. light under various conditions a) in vacuo, b) in O_2 (0.1 Torr) for 6 hrs and c) O_2 for 12 hrs. Significant changes are observed as a result of these irradiation treatments, the original profile of the unirradiated surface showing a progressive erosion of the high temperature water desorption peaks, and the appearance of a new region due to water at temperatures lower than those at which water appears in the original desorption profile. This low temperature peak for water is accompanied by a simultaneous release of dioxygen as indicated by the existence of a sharp peak for $m/e^+ = 32$ over a fairly narrow range of temperature.

The basis for interpreting Figure 3 is provided in Figure 4, in which t.p.d. profiles are presented for a titanium dioxide surface containing H_2O_2 preadsorbed at room temperature and for the thermal decomposition of titanium peroxide. In addition, the temperature dependence of O_2 photosorption, which is shown in Figure 5, reveals a progressive increase in activity up to about 323K, beyond which temperature there is an abrupt change to a much smaller value which then progressively increases at higher temperatures.

Finally results are presented in Fig. 6 which show the t.p.d. profile of a TiO_2 surface having been irradiated previously in moist N_2, or having been exposed to air following an irradiation in moist oxygen. In either of these conditions the t.p.d. characteristics depicted in Figure 4 are suppressed, with the appearance of a peak at $m/e^+ = 30$ in the corresponding range of temperature.

Discussion

It is evident from the foregoing results that water adsorbed upon the surface of titanium dioxide exerts a marked influence upon the reactivity of the solid when it is subjected to irradiation by near u.v. light in the presence of dioxygen at room temperature. Progressively increasing temperatures of outgassing, prior to irradiation, causes a diminution in the photo-activity of the solid at 25°C particularly within the temperature range 200 - 420°C. At higher temperatures (520°C), there is an enhancement of the activity, which is considered to be related to a slight change of colour of the specimen, a bluish appearance, which is indicative of thermal reduction of the TiO_2 to $TiO_{2\,(1-x)}$. Subsequent reoxidation of the specimen restores the solid to its stoichiometric condition and virtually destroys the capacity of the solid for oxygen photoadsorption. In all of the experiments the maximum nett adsorption of oxygen corresponds to a surface coverage of less than 1% of a monolayer.

From Figure 2, it is clear that the marked decrease in activity cannot be related to the first water desorption peak, which is regarded as originating from non-dissociatively adsorbed molecular water (10) since its coverage will be extensively reduced by prolonged outgassing at 200°C without there being a corresponding decrease in photosorption activity. Above 200°C, the t.p.d. profile indicates that the second and third water desorption peaks are progressively decreased, and this decrease appears to correlate with the decrease in activity of the solid for oxygen photoadsorption. Water released from the surface of the TiO_2 at the higher temperatures arises from the condensation of surface hydroxyl groups (OH_a and OH_b), there being two types which differ in their basicities:-

$$OH_a + OH_a \longrightarrow H_2O(g) + O^{=} + V_{\ddot{o}}$$

$$OH_b + OH_b \longrightarrow H_2O(g) + O^{=} + V_{\ddot{o}}$$

Since the majority of the photoadsorption activity can be destroyed by outgassing, and by subsequent reoxidation, it is considered that the loss of the hydroxyl groups brings about this change and points strongly to their involvement in the photosorption process. Complementary support for the involvement of hydroxyl species is provided by the marked enhancement of the photosorption activity by the addition, at 25°C, of a superficial layer of NaOH(s) to the TiO_2 (Fig. 1). The enhancement in the activity of the reduced speciman $TiO_{2(1-x)}$ is regarded as being associated with the presence of Ti^{3+} centres in the solid. When destroyed by reoxidation, these centres are also no longer able to participate in the photoprocesses and the activity disappears.

T.p.d. profiles which have been made after prolonged irradiation of the TiO_2 specimens (Figure 3) provide further clues to the manner in which OH groups are involved in the photoprocess. Six hours of irradiation in vacuo lead to only minor changes in the shape of the original t.p.d. profile of water; a slight erosion of the third peak (OH_b), and the formation of a small shoulder to the initial peak (H_2O(ads)). The presence of O_2(g) in an otherwise identical experiment resulted in O_2 photosorption, the complete disappearance of the peak due OH_b, and the creation of a marked shoulder to the leading edge of the original peak due to $H_2O_{(ads)}$. Twelve hours of irradiation caused also a significant decrease in the second peak, (OH_a).

During the t.p.d. profiles after irradiation, the shoulder to the initial water peak is accompanied by the simultaneous release of dioxygen over a very limited range of temperature. The concurrent release of H_2O and O_2 is strongly suggestive of the heterogeneous thermal decomposition of hydrogen peroxide and implies that H_2O_2 is a product during the photoadsorption process. Similar desorption characteristics are confirmed (Fig. 4) for a TiO_2 surface containing adsorbed H_2O_2, and for the thermal decomposition of titanium peroxide (11).

Evidence is provided which clearly shows the involvement of surface OH groups in the photosorption of O_2, there being a nett adsorption of <1% of the total surface sites ($\sim 5 \times 10^{18} m^{-2}$). Most significantly the entire populations of OH groups are destroyed. The significance of this observation becomes clear only when attention is given to the manner in which the powdered solid is presented to the light. The specimen, in its cell, forms a loose layer of material of approximately 0.5-1cm in thickness into which the light can penetrate to the depth of a few particles. For such quantities of solid the adsorption activity (rate) is independent of the mass, whereas for smaller masses e.g.<0.2g the rate of photosorption is proportional to the mass of solid. Thus it appears that the rate of photoadsorption is controlled by the number of photons which strike the surface but that areas of the surface which are unirradiated undergo changes as a consequence of the irradiation.

In attempting to reconcile the nett quantity of oxygen molecules which are photoadsorbed ($\sim 2 \times 10^{16} m^{-2}$), with the quantities of OH groups which are transformed ($7 \times 10^{17} m^{-2}$), it is concluded that some $O_2(g)$ may be released from the solid as a result of the photodecomposition of adsorbed H_2O_2. Evidence to support this suggestion comes from Munuera (12) and from Boonstra (13) both of whom have demonstrated the intrinsic instability of adsorbed H_2O_2 on TiO_2 surfaces under u.v. irradiation. Additionally there is little evidence for the release of H_2O_2 molecules in experiments which have been conducted in the liquid phase (14).

In the absence of $O_2(g)$ the conversion of OH groups is slow, and O_2 appears to act as an electron scavenger to minimise hole-electron recombination, in order that the holes can be trapped by the OH centres. Hydrogen peroxide is then formed by the combination of two OH radicals, but the decomposition of the H_2O_2 molecules may also occur through a mechanism which involves OH radicals abstracting hydrogen from H_2O_2. It raises the question of whether O_2 ($^1\Delta g$) can be formed via such processes on rutile surfaces, in contrast to anatase surfaces (15).

Finally if free OH radicals are the origin of the H_2O_2, then their presence might have been expected to stimulate the equilibration of a non-equilibrium mixture of H_2 and D_2 at room temperature. Recently this experiment has been performed (16) but it was found to have little or no influence upon the rate of equilibration of the H_2/D_2 mixture, although there was evidence of powerful oxidising species having been formed which were capable of fixing atmospheric dinitrogen (Figure 6). The inference to be drawn from this experiment is that whereas OH radicals in their electronic ground states ($^2\pi$) are capable of combining to form H_2O_2, they are insufficiently energetic to dissociate the dihydrogen bond to produce an H-atom or presumably to diffuse across the surface by interaction with water molecules. It would seem in these circumstances that water molecules must act mainly as a source of OH centres through the act of dissociative chemisorption.

Reaction Scheme.

$H_2O(g) \rightleftharpoons H_2O(ads)$	Non Dissociative Adsorption of Water
$H_2O(ads) + O^= + V_{\ddot{O}} \rightleftharpoons OH_a^- + OH_b^-$	Dissociative Adsorption of Water
$TiO_2 + h\nu\ (E > Eg) \longrightarrow h^+ \sim e^-$	Exciton Formation
$h^+ \sim e^- \longrightarrow h^+_{vb} + e^-_{cb}$	Charge Separation in the Space Charge Layer
$h^+_{vb} + OH^-_b \longrightarrow \dot{O}H$	Hole Trapping
$h^+ \sim e^- + OH^-_b \longrightarrow \dot{O}H + e^-_{cb}$	
$h^+_{vb}\ Ti^{3+}_{lattice} \longrightarrow Ti^{4+}_{lattice}$	
$e^-_{cb} + O_2(g) \longrightarrow O_2^-\ (ads.)$	Electron Capture
$e^-_{cb} + Ti^{4+}_{lattice} \longrightarrow Ti^{3+}_{lattice}$	
$H\dot{O} + \dot{O}H \longrightarrow H_2O_2(ads.)$	Formation of Hydrogen Peroxide
$H_2O_2 + \dot{O}H \longrightarrow H\dot{O}_2 + H_2O$	
$H\dot{O}_2 + H\dot{O}_2 \longrightarrow H_2O_2 + O_2(g)$	Decomposition of Hydrogen Peroxide

---

$H_2 + \dot{O}H \not\longrightarrow H + H_2O$	No H_2/D_2 Exchange
$H_2O + OH \longrightarrow H\dot{O} + H_2O$	Possible means of Radical Transport
$N_2 + O_2^{*} + 2e^- \longrightarrow 2\ N_2O_2^{2-}{}_{ads}$	Dinitrogen Fixation

Conclusion

Oxygen photoadsorption on hydroxylated surfaces of TiO_2 occurs with the formation of adsorbed hydrogen peroxide at temperatures below $\sim$ 323K. Above this temperature adsorbed H_2O_2 species are thermally unstable but photoadsorption of O_2 continues albeit by some alternative mechanism. Some of the oxidising species which are produced are sufficiently reactive to be able to fix dinitrogen which is released during t.p.d. as NO(g) and N_2(g). The hydroxyl radicals from which the H_2O_2 originates are insufficiently energetic to dissociate H_2.

Acknowledgements

The authors wish to acknowledge the financial support of the following bodies without which this work would not have been possible: The Spanish Ministry of Education, The British Council and Tioxide International Ltd. The principal author (RIB) wishes to thank also many friends in various countries for their encouragement during the course of this work.

References

1. A. Fujishima and K. Honda, Nature, 1972, 238, 37
2. "Photochemical Conversion and Storage of Solar Energy" 4th Int. Conf. Jerusalem 1982, Ed. J. Rabani, Weizmann Press 3rd Int. Conf. Boulder 1980, Ed. J.S. Connolly Acad.Press
3. M. Formenti and S.J. Teichner, Specialist Periodical Reports of the Royal Society of Chemistry (London). Catalysis, 1978, 2, 87-106
4. R.I. Bickley, Specialist Periodical Reports of the Royal Society of Chemistry (London). Catalysis, 1982, 5, 308-332
5. Photoelectrochemistry, Photocatalysis and Photoreactors. Fundamentals and Developments. Ed. M. Schiavello, NATO ASI Series C, 1984, Vol. 146
6. R.I. Bickley and F.S. Stone, J.Catalysis 1983, 31, 389-398
7. H.G. Voltz, G. Kampf and H.G. Fitzky 10th Fatipec Congress, 1970 107-113
8. R.I. Bickley and V. Vishwanathan, Nature 1979, 280, 306
9. E. Pelizzetti in Ref. 5 373-375
10. G.D. Parfitt, Progress in Surface and Membrane Science, 1976, 11, 181-226
11. R.I. Bickley and J.A. Navio (unpublished results)

12a G. Munuera, V. Rives-Arnau and A. Saucedo, J.C.S. Faraday I 1979, 75, 736

12b A. Gonzalez Elipé, G. Munuera, and J. Soria, J.C.S. Faraday I, 1979, 75, 748

13. A. Boonstra and C.A.H.A. Mutsaers, J.Phys.Chem. 1975, 79, 1940
14. R.B. Cundall, R. Rudham and M.S. Salim, J.C.S. Faraday I, 1976, 72, 1642

15. G. Munuera, J.A. Navio and V. Rives Arnau, J.C.S. Faraday I, 1981, 11, 2747
16. R.I. Bickley, J.A. Navio and C. Payne (unpublished results)

PHOTOLYSIS OF ADSORBED PHASES: ETHANOIC ACID-INSULATORS AND SEMICONDUCTORS SYSTEMS. INFLUENCE OF ACID-BASE PROPERTIES.

V. Augugliaro, L. Palmisano, M. Schiavello, and A. Sclafani.
Istituto di Ingegneria Chimica, University of Palermo, Italy.

ABSTRACT. The paper reports the results of experiments in which the photodecomposition of adsorbed ethanoic acid (and in few cases of adsorbed acetone and ethanol) was carried out by using several silica gel samples, whose properties were modified by adding oxidic dopants as γ-Al_2O_3, which is acidic in character, and MgO, which is mainly basic. Also the related system MgO-Al_2O_3 was investigated. Few IR spectra on selected pure specimens and on specimens containing adsorbed ethanoic acid were recorded with the aim of following the variation of the surfaces-adsorbate interaction with the variation of the acid-base properties. It was also of interest to compare the photoreactivity of insulators and semiconductors compounds in order to see the influence between the acid-base properties and the semiconducting properties. Therefore the photoreactivity of a series of oxides was measured and the results are discussed within this framework.

1. INTRODUCTION

Photocatalysis by semiconductors is becoming an overwhelming area of interest for its potentiality in solar energy conversion and storage (1,2).

Whatever the regime used, gas-solid or gas-solid-liquid, the main steps which can be envisaged in a photocatalytic process are the followings:

- photogeneration of electron-hole pairs and their trapping by suitable reducible-oxidable species to avoid recombination;
- chemical transformation of these species into suitable products and their separation (desorption) from the reacting media;
- **reconstruction of the photocatalysts initial conditions.**

E. Pelizzetti and N. Serpone (eds.), Homogeneous and Heterogeneous Photocatalysis, 567–580.

An ever increasing number of photoreactions have been carried out using a great variety of semiconductors (1,2). All of such reactions are characterized by low yields, both for the energy and chemical conversion. The reasons for these low conversions were attributed to the difficulty of efficiently performing one or more of the above described steps.

It must be outlined, however, that, in principle and for suitable adsorbates, a different reaction mechanism may take place in addition or not to that above described. This other mechanism may appear when the reacting molecules have or develop upon adsorption suitable chromophore groups. The adsorbate species containing such groups undergo chemical transformations under suitable light irradiation, and therefore the photocatalysts have in these instances only the role of adsorbents.

This process is well known as photolysis of adsorbate species and it was studied mainly for investigating spectroscopic features of adsorbate species (3,4). The photolytic process, of course, may occur with any type of solids in principle, including also semiconductors: the only condition is the formation of species containing chromophore groups and the use of suitable light irradiation. When semiconductors solids are used, the redox reaction mechanisms, specific for the semiconductors, may mix up with photolytic processes.

Recently it was reported a study (5) in which the photodecomposition of ethanoic acid adsorbed onto silica gels, carried out in a flow system, was attributed to the presence of isolated species containing the $\gtrdot$C=O group.

These species were esthers not having hydrogen bonding with the surroundings. These silyl-esther species were identified by IR spectra and were found partly reversible upon outgassing. Therefore it appears that the photoreactivity of silica gel is linked to the nature of the species adsorbed on its surface. The amount and the type of these species are affected in broad terms by the chemical properties of the adsorbents. Among these the acid base properties play, of course, a relevant role.

The present review reports the results of experiments in which the photodecomposition of adsorbed ethanoic acid (and in few cases, of adsorbed acetone and ethanol) was carried out by using several silica gel samples, whose properties were modified by adding oxidic dopants as γ-Al_2O_3, which is acidic in character, and MgO, which is mainly basic. Also the related system MgO-Al_2O_3 was investigated. In this way it is possible to compare the photoreactivity of several related surfaces for which the acid-base properties vary over a broad range: **and therefore correlations can be sought.**

Few IR spectra on selected pure specimens and on specimens containing adsorbed ethanoic acid were recorded with the aim of following the variation of the surfaces-adsorbate interaction with the variation of the acid-base properties.

It was also of interest to compare the photoreactivity of insulators and semiconductors compounds in order to see the relation, for this reaction, between the acid-base properties and the semiconducting properties. Therefore the photoreactivity of a series of oxides was measured and the results are discussed within this framework.

2. EXPERIMENTAL

2.1. Apparatus

All catalytic experiments have been performed in a flow apparatus using a fixed bed flat reactor, whose internal dimensions (width, thickness and height) were 3.7, 0.2 and 9 cm, respectively. The photoreactor was vertically mounted inside a thermostated chamber and irradiated on one side by a 1000 W Hg-Xe lamp (Hanovia L 5173). The approximate distance between the reactor and the lamp was of 50 cm. A He flow (99.5 % purity), at a constant rate of 0.1 cm^3s^{-1} at standard conditions, was bubbled in a bottle containing glacial ethanoic acid or acetone or ethanol when necessary (Carlo Erba reagents grade).

The gaseous mixture from the bubbling bottle was then fed to the photoreactor. Finally, the gas from the reactor outlet was sent to a gaschromatograph (Varian, Vista 6500), equipped with FID and TCD detectors for continuously monitoring the mixture composition. A Porapak QS column, 3 m in length, was used to perform the separation of gases. The temperatures of the saturator and of the photoreactor had the values of 40 and 50 °C, respectively.

The run procedure was as follows: the photoreactor was filled with the catalyst (1.5 g) and time was allowed in order to reach thermal equilibrium in all the system. Then, the feeding of the gaseous mixture and the irradiation of the reactor started at the same instant and this time has been considered the run zero time. Analyses of gas leaving the reactor were performed a few minutes after the start and then every about three hours. The runs lasted several hours and they were stopped when no changes in catalytic activity were observed.

A more rapid procedure which yielded the same results and which consisted in using specimens previously saturated with the organic molecules was also adopted.

Details on the apparatus and on the procedures can be found elsewhere (5).

2.2. Catalysts Preparation and Characterization

The description of the several silica gel specimens used can be found elsewhere (5).

The catalysts with variable ratio of the two components were prepared in these ways. The binary samples of SiO_2-MgO and Al_2O_3-MgO were prepared by impregnating SiO_2 or Al_2O_3, respectively, with a $Mg(NO_3)_2$ solution and then by performing a thermal decomposition. The binary samples of SiO_2-Al_2O_3 were prepared by a coprecipitation method at pH=5.5. In order to correctly compare the reactivity of these binary catalysts with that of pure components, the previous ways were also used for preparing pure SiO_2, Al_2O_3, and MgO. All the samples were fired at 400 °C in air for 24 h. Some of them, checked by X-ray diffractometry, were found amorphous. Other details will be published soon (6).

Surface area measurements were performed by a dinamic BET method using dinitrogen as adsorbate and a Micromeritics Flowsorb 2300 apparatus.

Reflectance spectra, in the range 600-300 nm, performed on several specimens containing, or not, adsorbed organic molecules, were recorded by a spectrophotometer Varian DMS 90.

An IR spectrophotometer (Perkin Elmer, mod. 580) has been used to perform spectra in the range 4000-1200 cm^{-1} on a disk obtained by pressing the powders at a pressure of $1.3 \cdot 10^9$ Pa. The cell containing the sample was equipped with CsI window and was connected to a high vacuum system equipped with suitable traps.

3. RESULTS AND DISCUSSION

3.1. SiO_2 Systems

The results obtained on silica gels, which are relevant for this review, can be summarized as follows (5):

- the adsorbed ethanoic acid was decomposed to CO_2 and to CH_4 in the ratio 2:1 under radiation whose wavelength was between 300 and 400 nm;
- traces of products as C_2H_6 and CH_3OH were also detected; no

reactivity was detected when the light was turned off;

- the photoactivity is a consequence of the interaction between silica gel surface and ethanoic acid molecule and it cannot be attributed to contaminants or to impurities present on silica specimens;
- adsorbed acetone was also photodecomposed, although to a lesser extent than ethanoic acid, while ethanol was not affected at all;
- an absorption from about 400 nm to lower wavelengths was observed over silica gel specimens saturated with ethanoic acid or with acetone, while the reflectance spectrum of silica gel specimen saturated with ethanol was the same of that of pure silica;
- infrared measurements, carried out on pure silica and on silica containing various amounts of adsorbed ethanoic acid, allowed to identify a free esther containing a $\geq C=O$ group which is not hydrogen bridged. It is likely that this species is responsible for the photoproduction of CO_2 and CH_4.

The above results allow to make few straightforward considerations trying to understand why ethanoic acid (and acetone) molecules undergo photodecomposition when adsorbed on silica gel surfaces.

The reflectance spectra clearly indicate the presence of species able to absorb light starting from 400 nm when ethanoic acid is adsorbed on silica gel surfaces.

The IR spectra allow to identify, among several adsorbed species, an esther species, free from hydrogen bonds. It has been shown, on spectroscopic grounds, that these species can undergo the transitions $n-\pi^*$ or $\pi-\pi^*$ (and therefore they can be photodecomposed) in the range of wavelengths 400-300 nm. The above transitions occur at about 180 nm in the gas phase in aprotic media. Therefore, the adsorbed phase is decomposed at lower energy in comparison with gas and liquid phases. Silica behaves as electron donor towards the adsorbed ethanoic acid species, causing the increase in energy of the ground states and therefore the decrease in energy of the above mentioned transitions, $n-\pi^*$ and $\pi-\pi^*$.

It seems thus that the interaction between silica gel surfaces and organic molecules, leading to a species containing $\geq C=O$ chromophore group, is the reason for the photodecomposition of adsorbed ethanoic acid.

After the establishment of this essential point, the question to be answered is related to the factors affecting the interaction.

In the previous paper (5) it was advanced the hypothesis that the acid-base properties of the silica surfaces could determine the level of photoreactivity since the chemisorption process between ethanoic acid and silica gel is a typical acid-base reaction. Moreover by

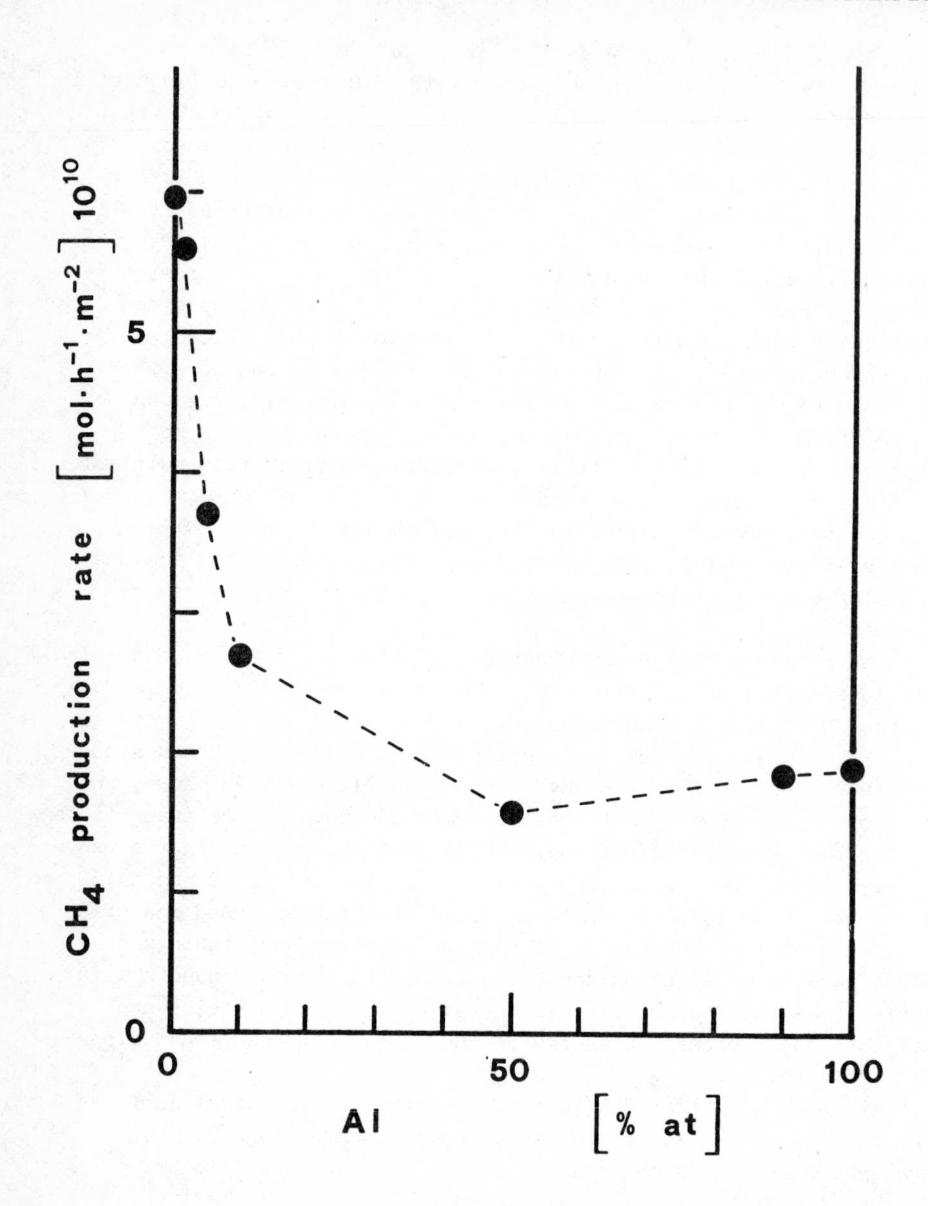

Fig. 1 - Experimental results of methane production rate for the Al_2O_3-SiO_2 system.

changing the acid-base properties of silica samples the redox properties are affected as well as the energy level of the free esther species, responsible for the photoreactivity.

To this aim two silica gel specimens, one treated for one week with HF solution (pH=3) and another with NaOH solution (pH=10), were prepared and tested in the same conditions as previous pure silica gel specimens (7). It resulted the following decreasing order of photoreactivity: NaOH treated SiO_2, not treated SiO_2, and HF treated SiO_2.

As expected, the acidic SiO_2 gel specimen resulted relevantly less active than the basic specimen. From these and previous results the indication that the Brønsted acidity is detrimental for establishing high photoreactivity is rather evident. One reason is of course the likely formation of hydrogen bridged esther species.

Other experiments were performed having in mind to test the photoreactivity of acidic specimens having mainly Lewis character and specimens having mainly basic properties. To this aim specimens of Al_2O_3-SiO_2 and MgO-SiO_2 as well MgO-Al_2O_3 at various ratios of the two components were prepared and tested in the same conditions as the other specimens.

3.2. Al_2O_3-SiO_2 System

The experimental results are reported in Fig. 1. The photoreactivity is expressed as $mol \cdot h^{-1} \cdot m^{-2}$, which can be considered a heuristic parameter for comparison within similar solids when the radiation conditions are maintained constant.

The Al_2O_3 surface is considered acidic with prevailing Lewis character (8). Therefore, according to that previously reported (5), the interaction between ethanoic acid and Al_2O_3 electron acceptor centres (Lewis acid sites) would lead to esther species energetically more stabilized than those formed on SiO_2. This behaviour is similar to that invoked to occur on the Lewis sites of α-Fe_2O_3, when ethanoic acid is adsorbed (9). As a consequence the photodecomposition is less easy and a decrease of photoreactivity should be observed. As one may note from the results reported in Fig. 1, this is the case.

Moreover the addition of Al_2O_3 to SiO_2 should result in a decrease of the photoreactivity, more or less pronounced according to the Al_2O_3 content.

The pertinent literature reports that the Al_2O_3-SiO_2 system, at a content of about 20% of Al_2O_3, shows a maximum of the total acidity, that is the Brønsted acidity of SiO_2 plus the Lewis acidity of Al_2O_3 (10,11,12). The reported photoreactivity results reproduce this trend.

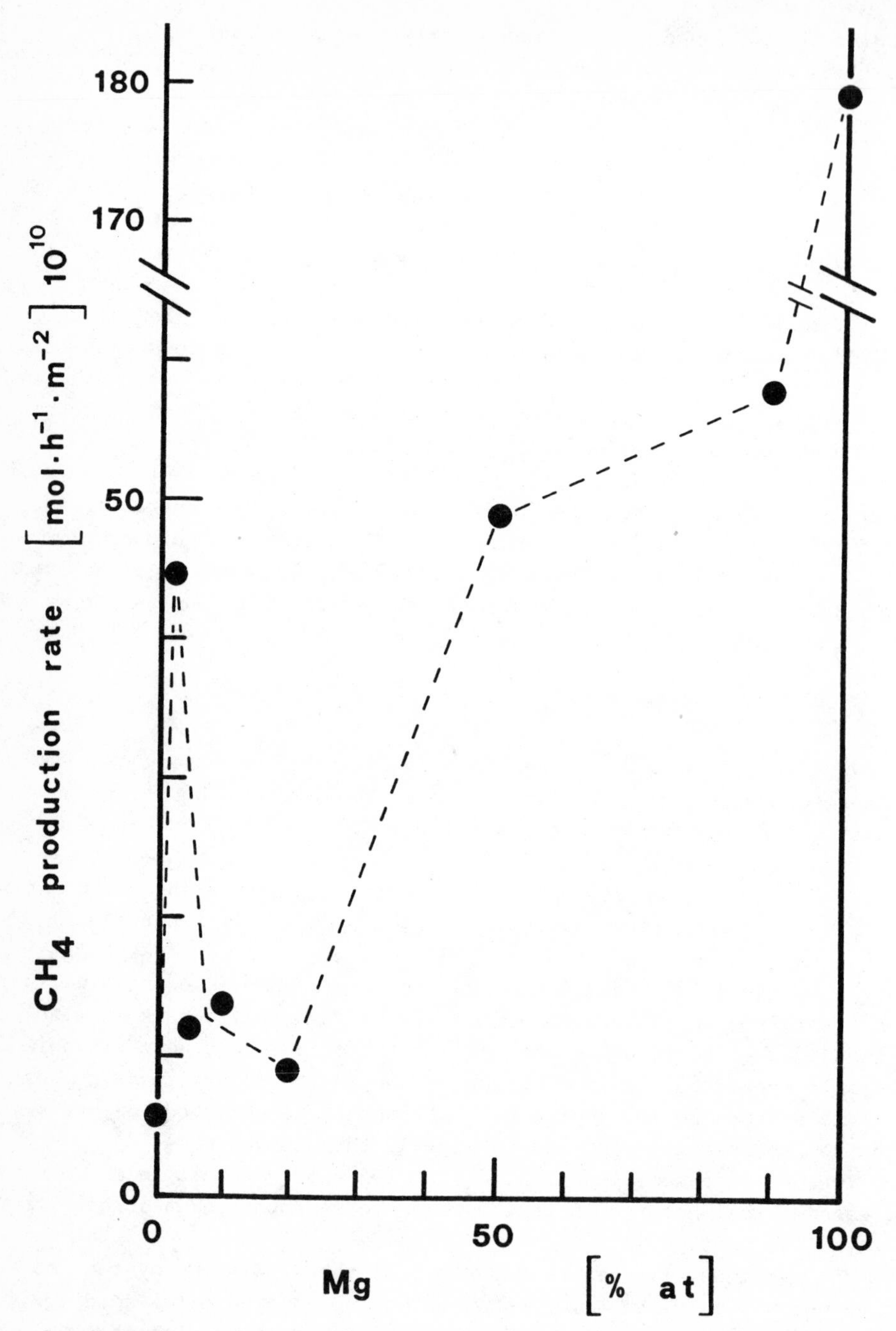

Fig. 2 - Experimental results of methane production rate for the $MgO-SiO_2$ system.

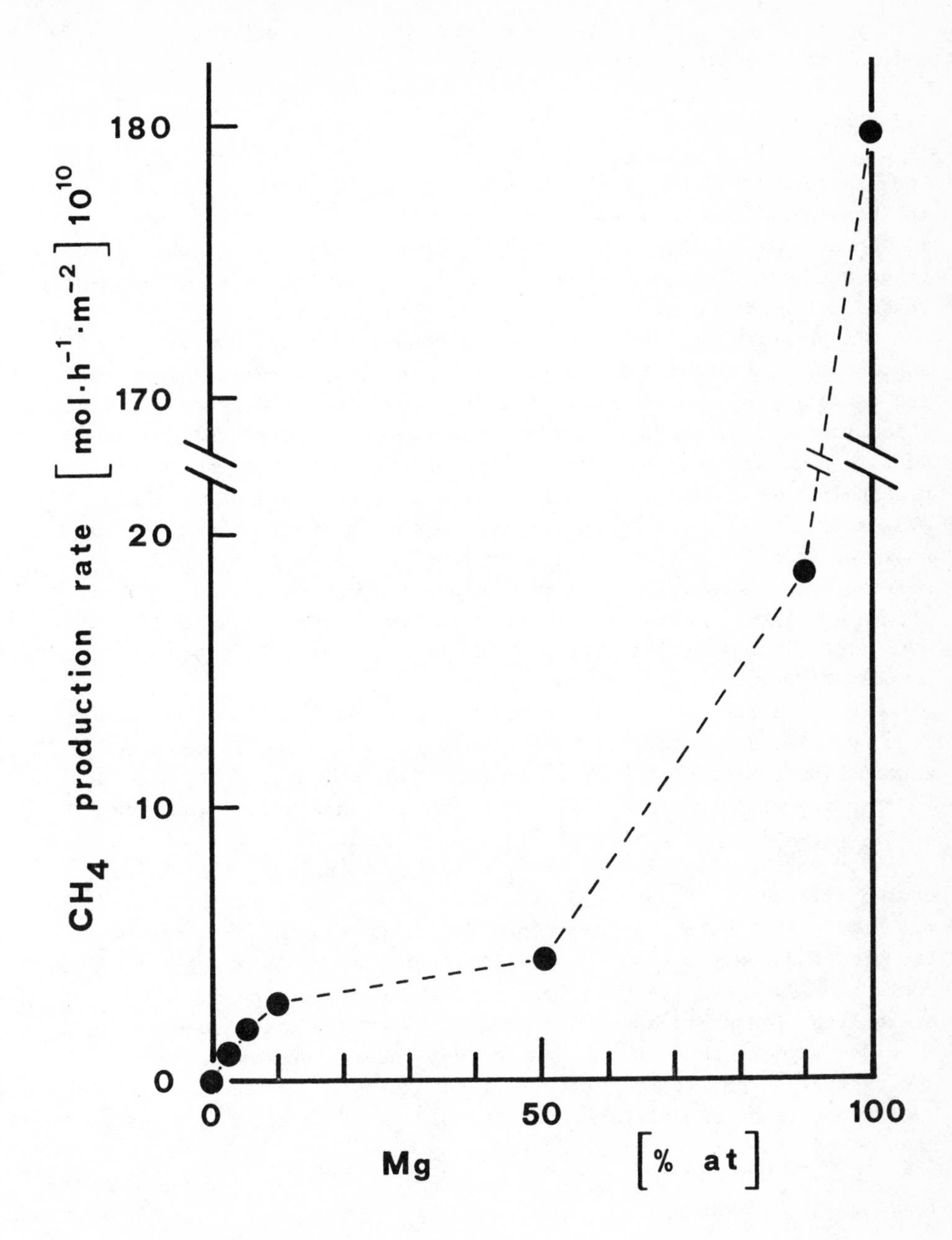

Fig. 3 - Experimental results of methane production rate for the $MgO-Al_2O_3$ system.

By concluding it can be inferred that also the Lewis acidity is detrimental for the studied reaction.

3.3. $MgO-SiO_2$ System

The experimental results are reported in Fig.2. By taking into account the considerations made about the energetics of the esther species, it is foreseeable an increase of the electron density of the chromophore group due to the electron-donor capacity of the MgO base. Consequently the effect on the photoreactivity is beneficial.

The results confirm the previous considerations. However, it is worth outlining the following point: it is relatively easy to predict the variation of photoreactivity within a series of related solids in which the acid-base properties are varying in a controlled and known way. This means that the trend of the energy content of species responsible for the level of photoreactivity can be qualitatively evaluated. In a series of solids of various chemical nature, although known as mainly acids or bases, the level of photoreactivity is difficult or impossible to predict since the nature and strength of the interaction, leading to the photoactive species, cannot be evaluated. Therefore the type and the strength of this interaction can be only deducted 'a posteriori'.

For instance, as it will be reported later on, the CaO compound does not exhibit photoreactivity at all. It can be explained by considering that CaO is more basic than MgO and, therefore, the interaction produces ionic species which are unreactive. The same behaviour was found when $MgAc_2 \cdot 4H_2O$ was tested.

For the $MgO-SiO_2$ system the reactivity increases with the MgO content. It has been reported by several Authors (13,14,15) that the $MgO-SiO_2$ system exhibits acidic properties due to the presence of Bronsted and Lewis sites. The acidity increases up to about a 30 % content of MgO. At higher MgO content, the basic nature of MgO is prevailing. The photoreactivity results reproduce this trend.

One maynote that at 1 % MgO content the photoreactivity is unexpectedly high. It is likely that by adding a small amount of the base there is a neutralization of the SiO_2 surface acidity as it happens for the SiO_2 sample treated with NaOH.

3.4. $MgO-Al_2O_3$ System

The experimental results are reported in Fig. 3. The considerations previously made for $Al_2O_3-SiO_2$ and $MgO-SiO_2$ systems are also valid for explaining the photoreactivity of this system.

It has been reported that the addition of MgO to Al_2O_3 determines a continuous increase of the basic properties of the system (16,17), that is a decrease of the strength of the Lewis acidic sites (18).

Such behaviour is clearly reflected in the photoreactivity results.

3.5. Insulators and Semiconductors Oxides

The experimental results are reported in Tab. 1 in which the oxides order is that of decreasing reactivity. Few considerations can be drawn from the observation of these results. A basic semiconductor, as ZnO, is more active than an acidic semiconductor, as TiO_2, while the basic insulator MgO is sensibly more active of both. Acidic semiconductors as Fe_2O_3, WO_3, etc. are almost completely inactive.

From this findings the following main consideration can be drawn. The acid-base properties are more effective than the semiconducting ones in determining the level of photoreactivity. Of course this behaviour is not generalized but it is limited to the ethanoic acid photodecomposition, which occurs, as clearly shown, via the formation of a species containing a chromophore group. The interaction between this species and the surroundings affects the level of reactivity. On this ground the semiconducting properties seem less relevant because they do not affect the formation of the species containing the chromophore group.

It is worth noting that the bulk $MgAc_2 \cdot 4H_2O$ is not at all active, indicating that, when an adsorbed ionic species is present, the photoreactivity decreases or disappears completely.

4. CONCLUSIONS

Few conclusions can be drawn from the results obtained in this work:

- an adsorbed species, containing chromophore groups, can undergo chemical transformation upon irradiation with suitable radiation;
- such species can be adsorbed both over insulators and over semiconductors materials; over these latter materials the radiation may stimulate both the solid and the adsorbate species, leading, in principle, to both the photolytic and to the semiconducting mechanism for the photochemical reaction. For the insulators only the photolytic process is possible;
- for the systems studied in this work (ethanoic acid adsorbed over silica, over modified silica, and over a series of insulators and semiconductors oxides) it appears that the acid-base properties are

TABLE 1. Experimental results of CH_4 and CO_2 production rates obtained by using insulators and semiconductors oxides.

PHOTOCATALYST (Source)	SURFACE AREA	RUN TIME	CH_4	CO_2
	$\frac{m^2}{g}$	h	$\frac{mol}{h \cdot m^2}\ 10^{10}$	$\frac{mol}{h \cdot m^2}\ 10^{10}$
MgO (Carlo Erba)	10.1	21	342	455
MgO (home prepared)	2.5	25	179.4	922.6
MgO (JMC Specpure)	10.3	25	117	176
ZnO (Höechst)	4	68	84	244
TiO_2 (home prepared)	110	38	35	59
TiO_2 (BDH)	10.5	73	18	370
SiO_2 (BDH, pH=10)	260	26	14	29.7
SiO_2 (BDH)	240	33	7	14
γ-Al_2O_3 (Akzo Chemie)	160	56	5.4	21
SiO_2 (BDH, pH=3)	256	38	2.3	11
WO_3 (Carlo Erba)	15	29	1.7	17
MnO_2 (Carlo Erba)	-	27	traces	traces
Cu_2O (Höechst)	-	15	-	-
MoO_3 (Ventron)	-	15	-	-
MoO_2 (Ventron)	-	20	-	-
Cr_2O_3 (Höechst)	-	23	-	-
Fe_2O_3 (Carlo Erba)	-	20	-	-
$MgAc_2 \cdot 4H_2O$ (Sigma)	-	46	-	-
CaO (Carlo Erba)	-	19	-	-
PbO (Carlo Erba)	-	15	-	-

essential in determining the type and the energetics of the adsorbed species and consequentely the level of photoreactivity. A basic and insulator oxide as MgO resulted the more active among all the solids studied, indicating that the acid-base interaction, rather than the semiconducting properties, is the key factor for this photoprocess. Of course there is an optimum in the strength of the acid-base interaction, as it is shown by the complete inactivity of the more basic CaO.

ACKNOWLEDGEMENTS

This work was supported by CNR (Roma) and MPI (Roma). The Authors whish to thank prof. S. Coluccia (Dept. of Physical Chemistry, University of Torino) for recording the IR spectra and prof. R. Maggiore (Dept. of Chemistry, University of Catania) for most of the surface area measurements.

REFERENCES

1. Grätzel, M. (Ed) 1983, "Energy Resources through Photochemistry and Catalysis" Academic Press, New York.
2. Schiavello, M. (Ed) 1985, "Photoelectrochemistry, Photocatalysis and Photoreactors. Fundamentals and Developments" D. Reidel Publishing Co., Dordrecht.
3. Terenin, A. 1964, Adv. Catal. 15, 227.
4. Nicholls, C. H., Leemakers, P. A. 1971, Adv. Photochem. 8, 315.
5. Augugliaro, V., Palmisano, L., Schiavello, M., Sclafani, A. 1985, J. Catal. in press.
6. Sclafani, A., Coluccia, S., Palmisano, L., Augugliaro, V. 1985, to be published.
7. Augugliaro, V., Palmisano, L., Schiavello, M., Sclafani, A. 1985, Proc. 5th Italian-Czechoslovak Symposium on Catalysis, Bechine (Czechoslovachia), September 16-20, 4.
8. Parry, E. P. 1963, J. Catal. 2, 371.
9. Lorenzelli, V., Busca, G., Sheppard, N. 1980, J. Catal. 66, 28.
10. Yamagata, n., Owada, Y., Okazaki, S., Tanabe, K. 1977, J. Catal. 47, 358.
11. Benesi, H. A., Winquist, B. h. c. 1978, Adv. Catal. 27, 97.
12. Tanabe, K. 1970, "Solid Acids and Bases" Academic Press, New York.

13. Karachiev, L. G., Barachevski, V. A. 1964, Kinetika i Kataliz. 5, 630.
14. Kermarec, M., Briend-Faure, M., Delafosse, D. 1974, J. Chem. Soc. Far. Trans. 70, 2180.
15. Dzis'ko, V. A., Borisova, M. S., Karakchiev, L. G., Makarov, A. D., Kotsarenko, N. C., Zusman, R. J., Khripin, L. A. 1969, Kinetika i Kataliz. 6, 1033.
16. Miyata, S., Kumura, T., Hattori, H., Tanabe, K. 1971, Nippon Kagaku Zasshi 92, 514.
17. Lercher, J. A., Colombier, C., Vinek, H., Noller, H. 1985, "Catalysis by Acids and Bases" Imelik B. et al (Eds.) Elsevier Amsterdam.
18. Cordischi, D., Indovina, V., Occhiuzzi, M. 1984, "Applicazioni industriali della chimica fisica delle superfici" CNR (Ed), 55.

CADMIUM SULFIDE PHOTOCATALYZED HYDROGEN PRODUCTION FROM AQUEOUS SOLUTIONS OF SULFITE

Michio Matsumura, Yukinari Saho and Hiroshi Tsubomura
Laboratory for Chemical Conversion of Solar Energy and
Department of Chemistry, Faculty of Engineering Science
Osaka University, Toyonaka, Osaka 560
Japan

ABSTRACT. Hydrogen, sulfate, and dithionate are photocatalitically produced in good yield from aqueous solutions of sulfite by using platinum loaded cadmium sulfide powder as reported previously. The reaction rate has been greatly enhanced by the heat treatment of the catalyst, the refined method of platinum loading, the choice of pH of the solution, and temperature. Further details of the experimental conditins of the photocatalytic reactions have been pursued, and reported in this paper.

The photocatsalyst's properties have been discussed on the basis of photoelectrochemical studies of cadmium sulfide electrodes (either single crystal or sinter) and the platinum electrode. These results have led to a further understanding of the reactivity of sulfite ion on cadmium sulfide, the stability of the photocatalyst, temperature dependence of the phototocatalytic reaction, etc. The current-voltage characteristics of CdS-Pt diodes have also been measured to deepen the understanding ot the properties of the electrical contact at the CdS-Pt interface. The barrier hight at the CdS-Pt interface has been discussed for several cases, and the temperature effect has been explained thereof.

INTRODUCTION

Photocatalytic reactions of cadmium sulfide have been actively studied for the past five years. Hydrogen production was investigated using sulfides [1-5] or organic sacrificial donors such as methanol [6] and cysteine[7]. We reported hydrogen production under visible light from aqueous solutions of sulfite, SO_3^{2-}, by use of platinum loaded CdS powder, the sulfite ion being converted into sulfate, SO_4^{2-}, and dithionate, $S_2O_6^{2-}$,ions [8,9]. Independently, Bühler et al. and Tamaru et al. have reported hydrogen production from sulfite by the CdS photocatalyst [10,11]. This reaction is valuable because sulfate and dithionate, as well as hydrogen, are produced from an industrial nuisance, sulfur dioxide.

E. Pelizzetti and N. Serpone (eds.), Homogeneous and Heterogeneous Photocatalysis, 581–591.

We found that the photocatalytic hydrogen production can be best carried out at pH 8, where H_2, SO_4^{2-} and $S_2O_6^{2-}$are produced at the mole ratio of 1:0.57:0.43. The reaction is thought to proceed in the following way, with e and h denoting electron and hole respectively:

$$CdS + h \longrightarrow e + h$$

$$H^+ + 2e \longrightarrow H_2$$

$$SO_3^{2-} + H_2O + 2h \longrightarrow SO_4^{2-} + 2H^+$$

$$2SO_3^{2-} + 2h \longrightarrow S_2O_6^{2-}$$

We have succeeded to improve largely the efficiency of the reactions by investigating the effect of the CdS crystal structure [9]. The commercial products of the CdS powder have two crystal structures: hexagonal (wurtzite) and cubic (zinc blende), and the former shows much higher efficiency than the latter. When a CdS specinem contains both structures, the hydrogen production rate increases with heat treatment, mainly due to the conversion of the cubic structure into hexagonal.

In the initial stage of our work, the photocatalyst was prepared by grinding the CdS powder together with a small amount of platinum powder in an agate mortar. This method was known to be effective for the case of TiO_2 photocatalyst. In the case of CdS, however the catalytic efficiency decreased with the grinding time, presumably because the CdS microcrystals are softer than TiO_2 and liable to be damaged by the grinding. In support of this view, the reflection spectra of CdS showed increased absorption in the long wavelength region and the photoemission intensity decreased very much by grinding. The best method of the platinum loading so far attained is to shake a mixture of CdS and 2 wt % of Pt in a glass vessel for ca. 1 hr.

With these improvements and at an optimal pH range of the solution, the rate of hydrogen production under simulated solar radiation (AM1, 1.00 kWm^{-2}) reached 0.50 mol $m^{-2}h^{-1}$ at 70°C [9], 3 times higher than that of our initial value [8] and much higher than those reported by other groups [10,11]. In the present paper, the results of further work for the improvement of the efficiencies, together with relevant electrochemical studies, are presented.

EXPERIMENTAL

CdS powder specimens obtained from Fruuchi Chemical Co. and Kojundo Chemical Laboratory were used in this work. X-ray diffraction measurements showed that they had mainly hexagonal (wurtzite) structure and cubic (zinc blende) structure, respectively. The platinum loading of the CdS powder was carried out by shaking the CdS powder together with platinum powder (Japan-Engelhard) in a glass vessel for 1 h. The photocatalyst prepared by this method showed high efficiency and good reproducibility.

The photocatalytic reaction was performed in a 100-mL flask, containing the Pt-loaded CdS powder (250 mg) and 20 ml of an aqueous solution of Na_2SO_3 buffered by adding boric acid and sodium hydroxide. After the air in the flask was removed by repeated freeze-pump-thaw cycles, the flask was illuminated from the bottom by a 500-W high-pressure mercury lamp combined with a uv cutoff filter (Toshiba, L39), whose transmittance at 390 nm was 50 %. The amount of hydrogen produced was measured by introducing it into a vacuum line provided with an oil manometer.

The CdS electrodes were fabricated using either single crystal wafers (Teikokutsushin Co.) having the (1000) face or sintered disks prepared by heating compressed CdS powder at 750°C under nitrogen flow for 4 h. Electrochemical measurements were performed by use of a three-electrode cell provided with a CdS electrode, a Pt counterelectrode, and a saturated calomel electrode (SCE). During the measurements, the solution was bubbled with high purity nitrogen gas and stirred magnetically.

RESULTS

1. Effect of Heat Treatments

As stated previously, the content of the crystallites having the hexagonal structure increased with heat treatment and the catalytic efficiency increased in parallel with that [9]. Even the efficiency of the powder mostly consisting of the hexagonal increased sometimes with heating. This is presumably due to the mending of lattice defects in the crystallites. In such a case, the photoluminescence efficiency also increased with heating.

2. Method of Platinum Loading

As platinum photodeposition was employed frequently for the preparation of the CdS photocatalyst, we tried platinum loading by illuminating a nitrogen bubbled solution of suspended CdS containing potassium hexachloroplatinate(IV) and sodium sulfite. The efficiency of hydrogen production by use of this photocatalyst thus prepared was not good, ca. 60 % of that carried out by use of a photocatalyst prepared by the shaking method.

3. Temperature

The rate of hydrogen production increased with the temperature of the solution as shown in Fig. 1. Temperature enhanced photocatalytic reactivity has been reported by several groups [10, 11], and has been attributed to enhancements of the diffusion rate of reactants, etc. In the present case, however, the main reason for the temperature effect is the enhanced electron transfer rate from CdS to platinum as

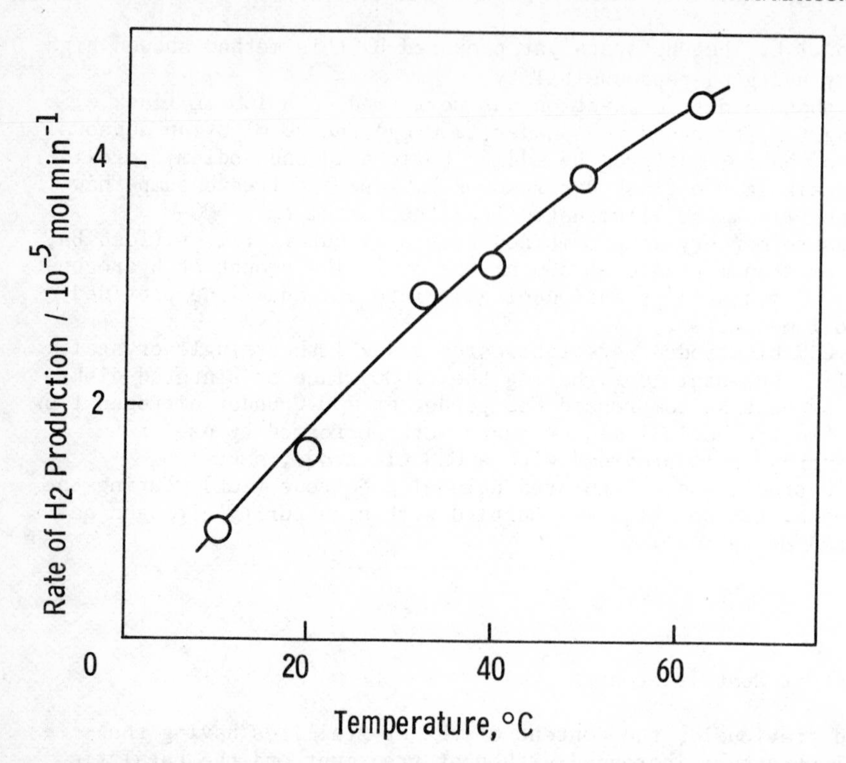

Fig. 1. Temperature dependence of the rate of photocatalytic hydrogen production from 1.0 mol dm^{-3} Na_2SO_3 (pH 8.7) using CdS powder (Furuuch Chem.) heat treated at 800 °C and loaded with 2 wt % Pt powder.

clarified by the electrochemical studies (see later section).

4. Effect of Added Chemicals

The hydrogen evolution was decreased by addition of iodide, hexacyanoferrate(II) or other reducing agents, because they compete with the sulfite ion in reactions with the hole and their reduced forms inhibit hydrogen evolution. Inclusion of oxygen also decreased the efficiency, presumably because it inhibits the H^+ reduction. Removal of oxygen can be done not only by pumping or N_2 bubbling but also by reaction with sulfite, converting it to sulfate. The presence of cadmium ion also retarded the reaction by precipitating Cd metal on Pt.

5. Photocatalyst Stability

As reported in our previous paper [8], no change in the weight of the photocatalyst nor in the catalytic activity was observed after illumination for 65 hours during which the solution was renewed many times so as to maintain the sulfite concentration and the pH. A total of 1.12 L H_2 was produced. No change in the ESCA profile of the CdS surface was observed. However, after prolonged reaction, in which 3.2 L H2 was obtained, the rate of reaction decreased to ca. 70 % of the initial.

6. Electrochemical Studies

The flat-band potential of a single crystal CdS electrode has been determined as -0.85 V vs. SCE from the Mott-Schottky plots in a buffer solution of pH 7, while that in the same solution containing 0.1 mol dm^{-3} Na_2SO_3 was -0.95 V, only 0.1 V more negative. On the other hand, the current-potential curve of a CdS electrode in the sulfite solution shifted by ca. 0.4 V to negative compared with that in a solution without sulfite (Fig. 2). These results indicate that the oxidation of sulfite on CdS is spontaneous, and it does not need any significant activation energy, while the anodic process (corrosion of CdS) in a solution without sulfite needs a fairly high activation energy. In Fig. 2 is also shown the hydrogen evolution current for a platinum electrode in the same buffer solution. The very small overlap in potentials between this curve and that for the anodic corrosion current indicates that CdS is fairly stable for illumination in aqueous solutions even in the absence of sulfite. The results in Fig. 2 also show that with the presence of sulfite a fairly high current flows when a CdS electrode and a Pt electrode are short-circuited. This in turn indicates that a Pt-loaded CdS can work as a good photocatalyst for the hydrogen evolution in the sacrificial oxidation of sulfite.

When a CdS sinter electrode was illuminated by a solar simulator (Wacom Co., AM 1, 100 mW cm^{-2}), saturated photocurrent density of 6.9 mA cm^{-2} was observed at potentials higher than 0.2 V vs. SCE. This current density corresponds to the rate of hydrogen production of 1.29 mol $m^{-2}h^{-1}$. The observed rate of hydrogen evolution by the photocatalytic reaction was 0.5 mol $m^{-2}h^{-1}$ under illumination by the same solar simulator at 70°C, 39 % of that calculated from the saturated photocurrent density. This is in good agreement with the observed apparent quantum efficiency of the reaction, i.e., 35 % at 60°C [9].

Both the current-potential curve of the photoanodic oxidation of sulfite on CdS and that of the hydrogen evolution current on Pt showed little change by temperature. On the other hand, the hydrogen evolving cathodic current on a Pt-loaded CdS sinter electrode changed with temperature as shown by solid lines in Fig. 2. In this experiment, the Pt-loading was carried out by shaking the CdS sinter electrode in platinum black, until the surface became gray. As the hydrogen evolution on Pt is well known to be a reaction of very small activation energy, it is concluded that the electron transfer from CdS to Pt in the Pt-loaded CdS catalyst has a considerable potential barrier.

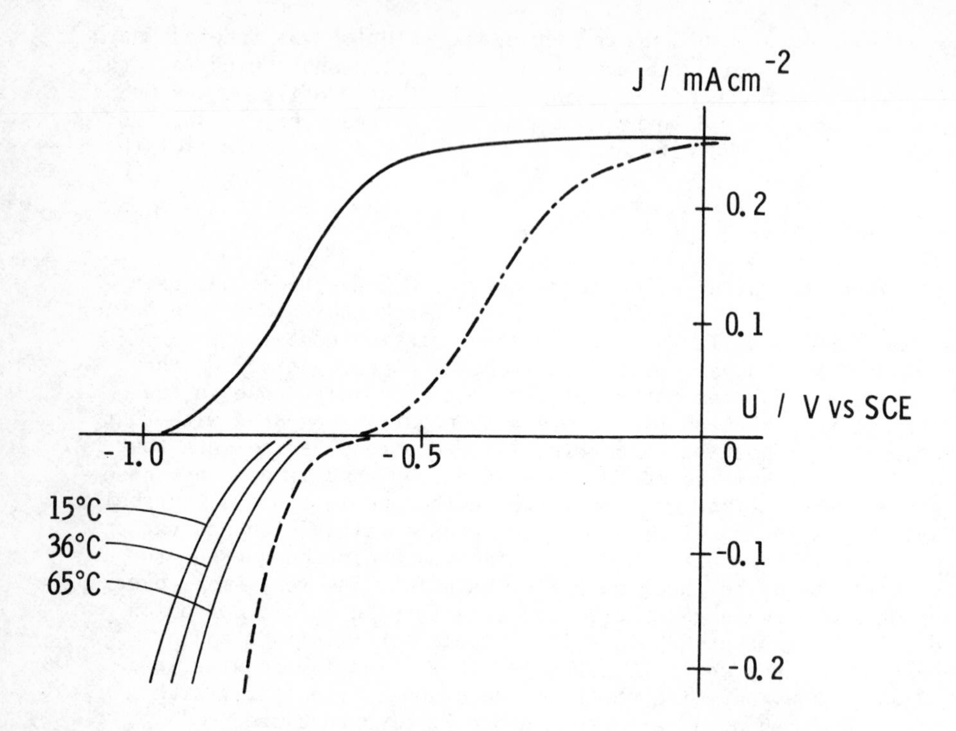

Fig. 2. Current-potential curves for a CdS single crystal electrode measured under illumination in 0.5 mol dm^{-3} Na_2SO_3 (————) and in the solution without Na_2SO_3 (– – – –), and that for a Pt electrode in 0.5 mol dm^{-3} Na_2SO_3 (— — —). The solutions were buffered at pH 8.2 and illumination was made by a 6-W tungsten-halogen lamp. The solid lines (————) show the current-potential curves for the platinum-loaded CdS sinter electrode measured in the dark at various temperatures in the solution without Na_2SO_3 (pH 8.5).

DISCUSSION

The current in a metal-semiconductor junction diode is generally given by the following equations:

$$J = J_o \{\exp(eU/kT) - 1\}$$

$$J_o = A^* T^2 \exp(- e\phi_B/kT)$$

where U is a voltage applied to the diode, A^* a Richardson Constant, and ϕ_B the barrier between metal and semiconductor. J_o corresponds to the exchange current density of hydrogen evolution, and can be determined by extraporating the Tafel line to the thermodynamic hydrogen evolution potential (-0.74 V vs. SCE) using the data such as shown in Fig. 2. Plot of J_o thus obtained with the inverse of temperature is made as shown by Fig. 3, from which ϕ_B is obtained to be 0.17 ± 0.04 eV. Based on this value, the potential energy diagram of CdS/Pt/solution junctions under hydrogen evolution is shown in Fig. 4 (A). On the other hand, the potential energy diagram of a bare CdS-solution interface may be given by Fig. 4 (B), where the difference between the bottom of the conduction band and the flat-band potential U_{fb}, determined before as - 0.95 V vs. SCE, is estimated to be 0.1 V. From this diagram, it can be understood why Pt-loaded CdS acts as a good photocatalyst for hydrogen evolution from sulfite;

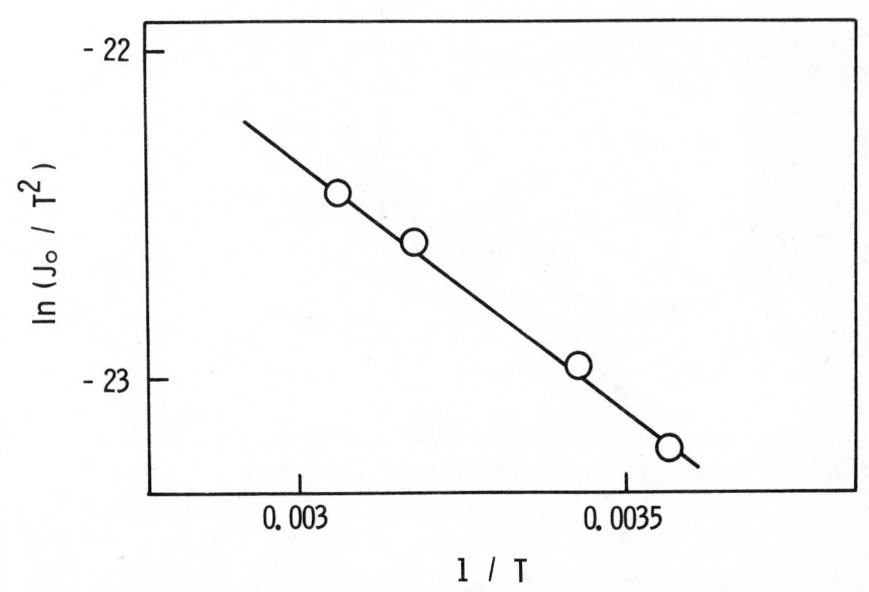

Fig. 3. Temperature dependence of the exchange current density (J_o) for hydrogen evolution at a platinum loaded CdS sinter electrode at pH 8.5.

namely, the photoexcited holes preferentially move toward the bare part of the CdS surface due to the stronger electric field toward the naked CdS-solution interface than the platinum covered surface and react with sulfite, while the electrons can transfer to the H^+ ion via platinum through a relatively small barrier.

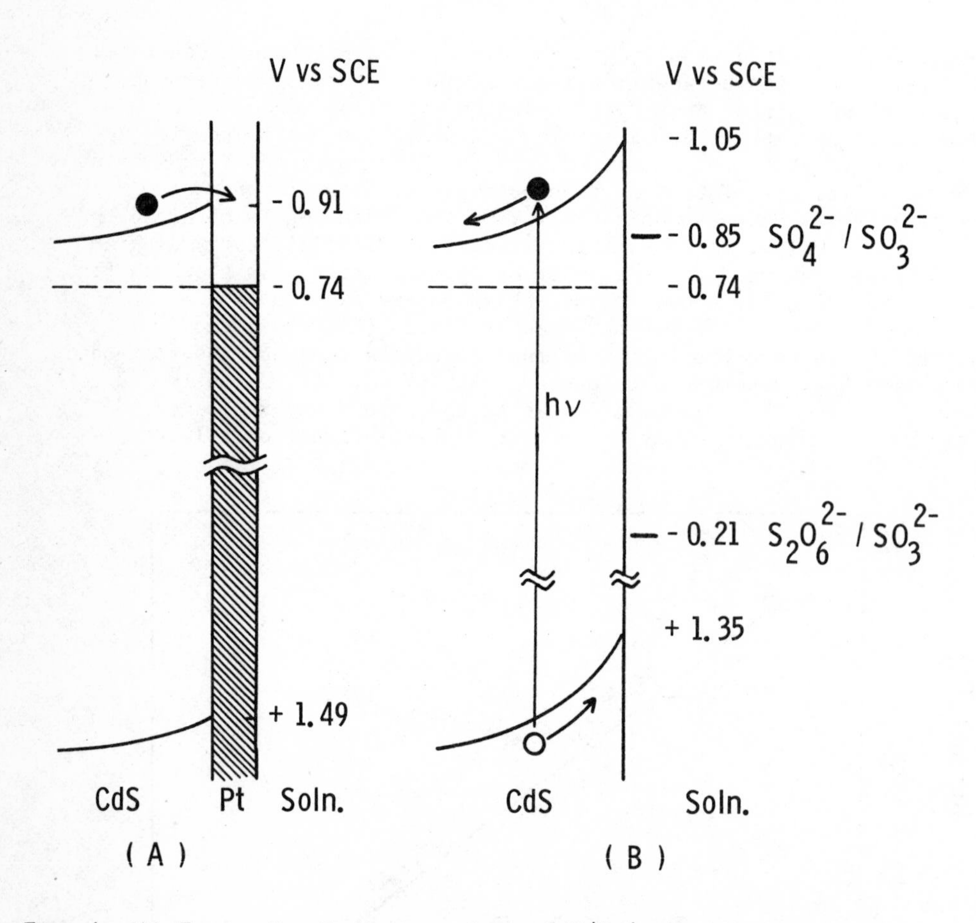

Fig. 4. (A) Energy level diagram of the CdS/Pt/solution junctions at pH 8.5. (B) The same for the bare CdS/solution junction.

The barrier height between an n-type semiconductor, such as TiO_2 or ZnO, and metal, such as Pt or Pd, changes with the atmosphere [12,13]. This was explained by assuming that the work function of the metal is high when it is in the air owing to oxygen chemisorption, but becomes low by the presence of hydrogen because it reacts with adsorbed oxygen and removes it [12]. To ascertain the presence of a similar mechanism in the case of CdS, we measured the current-potential curves of a CdS single crystal-Pt junction diode prepared by electron-beam depositing Pt (5 nm) on CdS. The rectifying behavior of the diode changed into ohmic in hydrogen (Fig. 5). This result is in contradistinction with that of Aspnes and Heller, who reported that ca. 1 eV barrier remained even in the hydrogen atmosphere [13]. For the case of a CdS-Pt diode made by coating Pt black suspended in methanol on sintered CdS and drying, we found that the i-v curve in hydrogen still showed some rectifying property as shown by the broken line in Fig. 6, like in the case of the Pt loaded CdS sinter electrode (Fig. 2). These results indicate that the barrier hight between CdS and Pt largely depends on the method of Pt loading, and a higher efficiency of the photocatalytic reaction might be obtained by searching for a new method of metal loading on CdS, leading to a smaller barrier hight between the semiconductor and the metal.

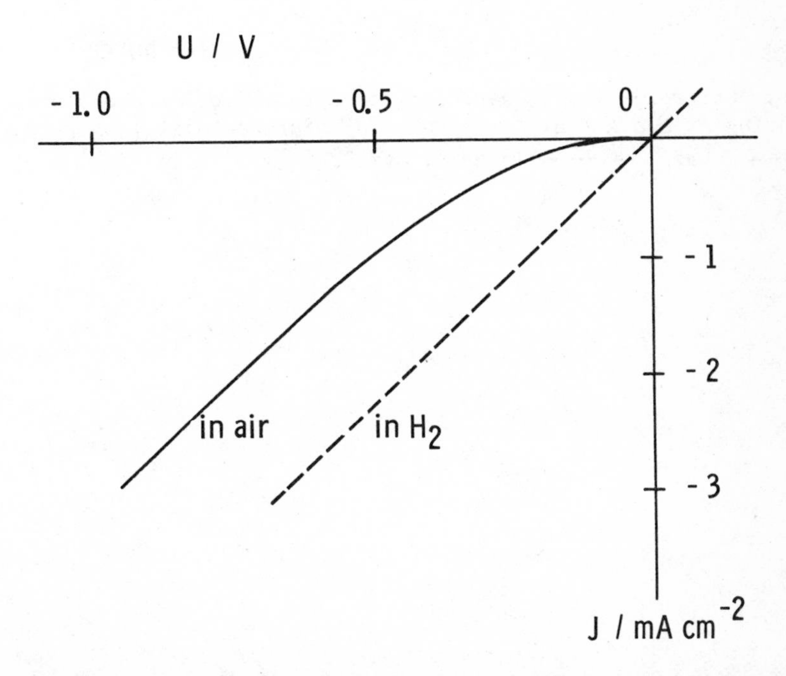

Fig. 5. Current-voltage characteristics for a CdS-Pt diode prepared by depositiing Pt (5 nm) on a CdS single crystal wafer by an electron beam evaporator; ——— in air, — — — in H_2 (1 atm.).

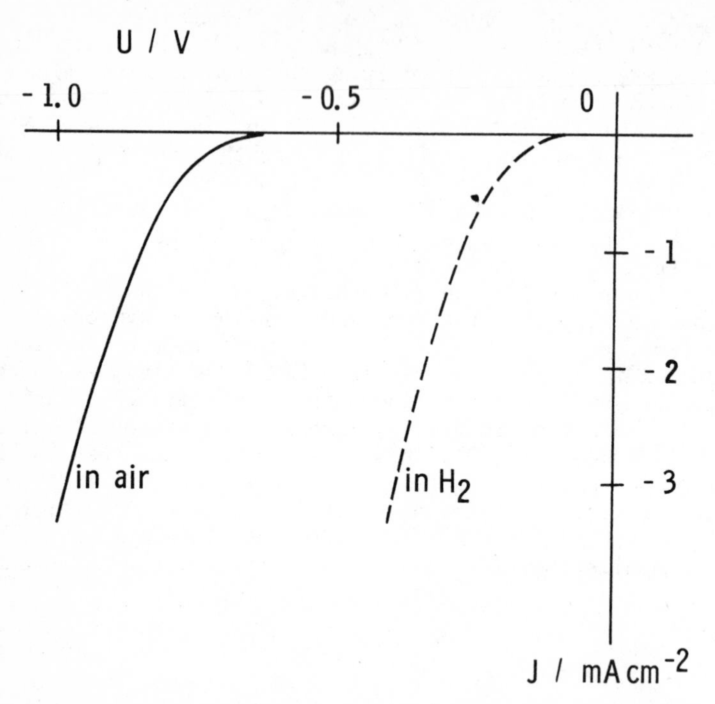

Fig. 6. Current-voltage characteristics for a CdS-Pt diode prepared by coating Pt black (ca. 1 mm) on a CdS sintered disk (see text); ——— in air, — — — in H_2 (1 atm.).

REFERENCES

1) E. Borgarello, K. Kalyanasundaram, M. Grätzel, E. Pelizzeti, Helv. Chim. Acta, 65, 243 (1982).
2) N. Serpone, E. Borgarello, M. Grätzel, J. Chem. Soc. Chem. Commun, 1982, 342.
3) D. J. Meissner, R. Memming, B. Kastening, Chem. Phys. Lett., 94, 34 (1983).
4) A. W. -H. Mau, C. -B. Huang, N, Kakuta, A. J. Bard, A. Campion, M. A. Fox, J. M. White, S. E. Webber, J. Am. Chem. Soc., 106, 6537 (1984).
5) D. H. M. W. Thewissen, A. H. A. Tinnemans, M. Eeuwhorst-Reiten, Nouv. J. Chim., 7, 191 (1983).
6) M. Matsumura, M. Hiramoto, T. Iehara, H. Tsubomura, J. Phys. Chem., 88, 248 (1984).
7) J. R. Darwent, G. Porter, J. Chem. Soc., Chem. Commun., 1981, 145.
8) M. Matsumura, Y. Saho, H. Tsubomura, J. Phys. Chem. 87, 3807 (1983).
9) M. Matsumura, Y. Saho, H. Tsubomura, J. Phys. Chem. 89, 1327 (1985).
10) N. Bühler, K. Meier, J-F. Reber, J. Phys. Chem., 88, 3261 (1984).
11) T. Aruga, K. Domen, S. Naito, T. Onishi, K. Tamaru, Chem. Lett., 1983, 1037.
12) N. Yamamoto, S. Tonomura, T. Matsuoka, H. Tsubomura, J. Appl. Phys., 52, 6230 (1981).
13) D. E. Aspnes, A. Heller, J. Phys. Chem., 87, 4919 (1983).

ENVIRONMENTAL PHOTOCHEMISTRY OF CHLORINATED AROMATICS IN AQUEOUS MEDIA. A REVIEW OF DATA

D. Cesareo, A. di Domenico, S. Marchini, L. Passerini and M.L. Tosato

Istituto Superiore di Sanità
Department of Comparative Toxicology and Ecotoxicology
Viale Regina Elena, 299
00161 Roma
Italy

ABSTRACT. A review of data relative to the photochemistry in aqueous media of chlorinated benzenes (PCBzs), phenols (PCPs), naphthalenes (PCNs), dibenzofurans (PCDFs), and dibenzo-*p*-dioxins (PCDDs) is presented. Analysis of data is focussed on direct, sensitized, and catalysis-assisted photoprocesses occurring under environmentally relevant conditions. The purpose of the review is to provide a basis for preliminary assessment of potential effectiveness of sunlight photochemistry as detoxication mechanism in natural water bodies for the above chlorinated aromatics which, due to their potential high exposure and toxic effects (see Appendix), are among those pollutants which have priority for hazard assessment.

INTRODUCTION

Increasing concern as to the environmental burden of a large variety of man-made chemicals that may irreversibly affect natural biological equilibria, and the parallel enactment, in a number of Countries, of chemical control legislations have, in recent years, stimulated research activities in rather new areas. Sunlight photochemistry is one example. Sunlight-induced (photo)chemical processes may in fact play some role in determining the fate of environmental chemicals and be a possible route for their removal.

In the water compartment, for instance, these processes may provide a unique sink for biodegradation-resistant, non-hydrolyzable chemicals that find their way to the sunlit surface layers of natural waters, as a consequence of natural transport processes following uncontrolled release from various point sources, and/or of local discharge of industrial or other wastes.

State-of-art understanding of photochemical processes in natural waters has been recently reviewed. Their variety and complexity,

E. Pelizzetti and N. Serpone (eds.), Homogeneous and Heterogeneous Photocatalysis, 593–627.

and their potential relevance as removal factors for broad classes of chemical pollutants have been highlighted, and an exhaustive bibliography given (Zafiriou et al., 1984).

In the present work, laboratory and field data relevant to assessing the environmental photochemistry in water of chlorinated aromatics - namely, PCBzs, PCPs, PCNs, PCDFs, and PCDDs - have been assembled and analyzed. Rather extensive photochemistry studies have been performed in recent years on these chemicals which meet both requirements for being potentially hazardous to the environment: in fact, they elicit severe toxic effects to aquatic life at very low concentrations and have the potential for high exposure due to large production volume and/or highly dispersive uses, resistance to both microbial and hydrolytic degradation, and potential for bioaccumulation.

The purpose of the review is to provide a basis for comparative analysis of the effectiveness of photochemical processes as detoxication mechanisms in water systems.

SELECTION OF DATA

Data relative to experiments carried out outside the environmentally relevant wavelengths and in non-aqueous solvents have been considered only when deemed useful or in the absence of pertinent data. Furthermore, data have been selected taking into account that natural waters contain numerous chemicals (both natural and xenobiotic) so that rates and products of environmental phototransformations cannot be predicted by direct photolysis evaluation only. Indirect interactions by energy transfer from an excited photosensitizer, which basically acts as a catalyst, may occur, as well as secondary reactions between the substrates and photolytically or photochemically-generated reactive species (radicals, singlet oxigen, ozone, photoconductive metal oxides, bases, etc.). Therefore, data referring to homogeneous and heterogeneous photocatalysis have been particularly considered whenever available, since photocatalysis, in addition to being always likely to occur, represents the only possible photochemical path for chemicals which, in water, are substantially transparent to sunlight radiations, i.e. their molar extinction coefficients are lower than 1 M^{-1} cm^{-1} above 290 nm (ECETOC, 1981). The fact that light may interact in different ways with chemicals in solution, or present as suspended particles, or sorbed onto particulate matter surface has also been taken into account.

In addition to rates of phototransformation, identities, relative quantities, and potential hazard of photoproducts have been particularly addressed to, whenever this information was available.

Preliminary comparative analysis of effectiveness of different photochemical processes as detoxication routes has been tentatively carried out considering data relative to classes of congeners and obtained under as far as possible equivalent experimental conditions.

In the Appendix, available physico-chemical properties and other data concerning potential partitioning among environmental compartments and persistency of the examined chlorinated aromatics, as well as available data relative to their aquatic toxicities, are reported. The mentioned fate- and effects-related data are basic elements for preliminary hazard assessment, and provide the rationale for the need of exploring photochemical processes as detoxication routes for chlorinated aromatic.

ANALYSIS OF DATA

Polychlorinated Benzenes

Information relevant to environmental photochemistry was available for each of the 12 PCBzs. A selection of data is reported in Table I. For each PCBz, in each considered process, relevant experimental condition, amount of substrate loss, photoproducts and their relative quantities are reported as available. Photoproducts have been tentatively assigned to three groups: dechlorinated congeners, other products (including isomers and a variety of divers organics, different from polychlorinated biphenyls (PCBs) and for which the chlorine content and not the identity is given), and PCBs. These groups may be assumed as indicators of processes leading to degraded, equivalent or largely unknown, and potentially upgraded hazard, respectively.

Analysis of data (some of which not reported in Table I) will be performed in two steps. Firstly the photochemical behaviour of PCBzs will be separately examined by groups, from mono- to hexachlorobenzene, according to chlorine content. Secondly a comparative analysis of photochemical behaviour of PCBzs, also in relation to the number of chlorine atoms, will be performed making reference to data drawn from experiments of direct, sensitized, and heterogeneous catalysis-mediated photoprocesses.

Monochlorobenzene (MCBz). Direct non-negligible dechlorination of MCBz was observed only upon irradiation (λ<290 nm) of its cyclohexane or methanol solutions. Under the same conditions, the formation of condensation products (byphenyls and other complex products between substrate and sensitizer) was observed in the presence of amines acting as sensitizer (Arnold and Wong, 1977). Quantitative direct photohydrolysis of MCBz in dilute aqueous solution into phenol was reported to occur whatever the pH (1-12) and wavelengths (254 or 300 nm). The quantum yields of phenol formation were 0.1 and 0.2 at 254 and 300 nm, respectively (Tissot et al., 1983).

Upon irradiation (λ >300 nm) of MCBz in a 0.1% TiO_2-water slurry (Table I), benzene or phenol formation was not observed. Under these heterogeneous phase catalysis conditions, degradation was rather slow. The primary photoproducts were chlorophenols, and these were subsequently dechlorinated (probably under homogeneous conditions) to give the corresponding hydroquinones and/or catechols which, in turn, were readily converted to benzoquinones. PCBs (dichlorinated) were formed after longer exposure times (Ollis et al., 1984).

TABLE I. Phototransformations of PCBzs: substrate loss and products in direct, sensitized and heterogeneous catalysis mediated photoprocesses (λ >285) in various solvents (matrix)

SUBSTRATE	EXPERIMENTAL CONDITIONS			SUBSTRATE LOSS	PHOTOPRODUCTS[a)]				
	IRRADIATION		MATRIX		CONGENERS		OTHERS		PCBs
	λ	time	solvents sensitizer catalyst		mono- dechlorinated	di- dechlorinated	divers	isomers	
	nm	b)		%	%	%	(%)	%	%
MCBz	300	-	H_2O	100	-		phenol(100)		-
	>300	2h	H_2O-TiO_2(.1%)	20	-		Cl_1(15)		-
	"	10h	" " "	-	-		-		Cl_2
1,2-DCBz	>300	short	H_2O-TiO_2(.1%)	-	-	-	Cl_2-Cl_1	-	-
	"	long	" " "	-	-	-	-	-	Cl_2
1,4-DCBz	300	5m	H_2O-TiO_2(1%)	50	-	-	-	-	-
	"	-	H_2O-clays[c)]	0					
1,2,3-TrCBz	>285	51h	H_2O-CH_3CN(1:1)	40	54	-	-	-	-
	"	7h	MeOH-O_2	96	74	-	-	-	-
	"	5h	MeOH-acetone[d)]	53	81	0.6	-	-	-
1,2,3-TrCBz	>285	51h	H_2O-CH_3CN(1:1)	28	38	-	-	9	-
	"	7h	MeOH-O_2	63	85	5	-	-	-
	"	1.5h	MeOH-acetone[d)]	25	99	-	-	-	-
1,3,5-TrCBz	>285	51h	H_2O-CH_3CN(1:i)	39	23	-	-	10	Cl_5(5)
	"	22h	MeOH-O_2	31	96	-	-	-	-
	"	3h	MeOH-acetone[d)]	72	37	3	-	-	Cl_5-Cl_4(8)

TABLE I. (continued)

1,2,3,4-	>285	40h	H_2O-CH_3CN(1:1)	56	35	32	Cl_6-Cl_2	3	Cl_7
TCBz	"	4h	" " " + acetone[e]	78	42	7	Cl_4-Cl_3	-	Cl_7-Cl_5(12)
1,2,3,5-	>285	36h	H_2O-CH_3CN(1:1)	48	36	4	Cl_5(1.5)	6	Cl_7-Cl_5(3)
TCBz	"	1.5h	" " " + acetone[e]	64	60	2	Cl_4-Cl_2	-	Cl_7-Cl_5(13)
1,2,4,5-	>285	8h	H_2O-CH_3CN(1:1.5)	98	28	9	Cl_5	2	Cl_7-Cl_5(3)
TCBz	"	4h	" " " + acetone[e]	90	25	12	Cl_3		Cl_7-Cl_4(7)
PeCBz	>285	24h	H_2O-CH_3CN(1:1.5)	41	20	21[f]	Cl_2(tr)	-	-
	"	4h	" " " + acetone[e]	54	75	8[f]	-	-	Cl_9-Cl_7(4)
	sunlight	7d	MeOH	0					
HxCBz	>285	8h	H_2O-CH_3CN(1:9)	34	77	3	-	-	-
	"	16h	" " " + acetone[e]	29	71	7	-	-	-
	sunlight	15d	MeOH	62	5	-	Cl_5-Cl_4(84)	-	-

a) these are grouped as: congeners, containing one or two chlorine atoms less than the parent compound; others, including isomers and divers, the latter being various chlorinated organics (different from congeners and PCBs) containing the listed maximum and minimum number of Cl-atoms; and PCBs which are usually mixtures of polychlorinated biphenyls whose maximum and minimum number of Cl-atoms are listed. The percentages of photoproducts refer to substrate loss.

b) times are in minutes (m), hours (h), or days (d).

c) this include various natural clays, sediments, or Ti containing ores.

d) acetone was 5.53 mM.

e) acetone was 0.553 M.

f) including tri-dechlorinated congeners, i.e. DCBzs, whose amounts were about 7% in the unsensitized photolysis, and less tha 1% in the sensitized one.

Dichlorobenzenes (DCBzs). Direct dechlorination of 1,2-, 1,3-, 1,4-DCBz to MCBz was reported following irradiation (λ<290 nm) of the respective methanol solutions (Parlar and Korte, 1979; Mansour et al., 1980). The relatively high dechlorination yields were proportional to the respective decomposition quantum yields, and were practically unmodified in aerated as well as deaerated methanol, methanol-dioxan, and methanol added with sensitizers. Much lower yields were obtained in water-methanol mixtures. 1,4-DCBz was by far the more resistant isomer to dechlorination.

Irradiation (λ>300 nm) of 1,2-DCBz in 0.1% TiO_2-water slurry (Table I) gave the same results as with MCBz under equivalent conditions (Ollis et al., 1984). Heterogeneous phase photooxidation to chlorophenols was followed by dechlorination, and PCBs were formed after prolonged exposure to the light source.

1,4-DCBz in a 1% TiO_2-water slurry was also rapidly degraded upon irradiation at λ = 300 nm. No photoreaction was, viceversa, observed if the semiconductor powder was substituted with Ti-containing ores, or different kinds of clays, or suspended natural sediments (Oliver et al., 1979).

Trichlorobenzenes (TrCBzs). All results reported in Table I were obtained using wavelengths above 285 nm in aqueous media or in methanol added with quenchers (O_2 and isoprene) or sensitizer (acetone) (Choudhry et al., 1979) which are substances that may be assumed to mimic natural components of environmental water bodies.

In water-acetonitrile (1:1) solution, all TrCBzs (5-6x10^{-3} M) underwent phototranformation at similar rates decreasing in the order 1,2,4->1,3,5->1,2,3-TrCBz. Approximately 50% of the converted substrate was not accounted for in terms of identified photoproducts. Among identified photoproducts, DCBzs, formed by reductive dechlorination, were the only products of 1,2,4-TrCBz, and the preponderant products among those of 1,2,3- and 1,3,5-TrCBz, which also formed isomers. PCBs (mainly pentachlorinated), were detected only among reaction products of 1,3,5-TrCBz.

In the methanol solutions, either deaerated or added with quenchers or sensitizer, the disappearance of TrCBzs was much faster than in aqueous media. The following tabulation of half-lives (in hours) of TrCBzs, calculated assuming pseudo-first order loss, may allow some analysis of specific effects of quenchers and sensitizer on reaction rates:

TrCBz	Water-acetonitrile	Methanol (deaerated with N_2)	Methanol-(oxygenated with air)	Methanol-isoprene (8.7E-3 M)	Methanol-acetone (5.5E-3 M)
1,2,4-	70	3	1.5	2.5	4.5
1,3,5-	72	22	41	31	1.5
1,2,3-	108	14	15	18	3.5

No very significant changes were observed in rates, and in relative rates, of TrCBzs loss in the deaerated, aerated, or added with quenchers, methanol (in which rates decrease in the order 1,2,4->1,2,3->1,3,5-TrCBz). Viceversa, rates and relative rates of loss are modified in the presence of the sensitizer: the process is generally faster and rates of loss decrease in the order 1,3,5->1,2,3->1,2,4-TrCBz, which is opposite in respect to what was found above.

In the methanol solutions, unidentified products were a small percentage. Among identified photoproducts, DCBzs accounted for a very high percentage of the converted substrates, chlorine rearrangement was never observed, and PCBs were formed only in the sensitized reaction of 1,3,5-TrCBz.

Tetrachlorobenzenes (TCBzs). All the results reported in Table I refer to experiments carried out irradiating at λ >285 nm the TCBz isomers (approximately 10^{-3} M) in water-acetonitrile (1:1, except for 1,2,4,5-TCBz: 1:1.5) mixtures in the absence or presence of 0.55-M acetone, added as sensitizer (Choudhry and Hutzinger, 1984). Therefore, the results may allow to compare the effects of direct and indirect photolysis on TCBzs degradation. The comparison will be performed on the basis of the reaction rates (calculated assuming pseudo-first order loss of substrates) and of other parameters that may be useful indicators of the relative capability of the two processes as hazard removal factors: these are, in addition to the quantities of PCBs which are formed, the ratio between the percentage of mono- and di-dechlorinated congeners (a relevant parameter inasmuch as toxicity is directly dipendent on the number of chlorine atoms), and the percentage of unidentified products which indicates uncontrolled (de)toxication paths. These parameters are listed in Table II which will be examined later.

In the unsensitized reaction, the 1,2,4,5-isomer was by far the more labile (its half-life was 1.5 h, while half-lives were 33 and 38 h for 1,2,3,4- and 1,2,3,5-TCBz, respectively). Reductive dechlorination to both TrCBzs and DCBzs was the major reaction path as it accounted for 80 to 95% of the identified reaction products. The ratios TrCBz/DCBz are 1, 3, and 9 for the 1,2,3,4-, 1,2,4,5-, and 1,2,3,5-isomer, respectively. Small quantities of isomers, of divers chlorinated organics, and of PCBs (mainly heptachlorinated) were formed in any case.

In the sensitized reaction, the rates of disappearance of the three isomers were approximately the same. Strangely enough the sensitizer did not affect the half life of 1,2,4,5-TCBz, whereas the half lives of 1,2,3,4- and 1,2,3,5-TCBz became about 16 and 38 times reduced, respectively, in the presence of acetone. 25 to 50% of photoproducts were not identified. The identified photoproducts pattern was largely maintained in presence of sensitizer, as dechlorinated congeners were the predominant photoproducts. However, the ratios TrCBzs/DCBzs were increased (except for 1,2,4,5-TCBz, see Table II), isomers were not formed, the "divers" chlorinated organics included reaction products with acetone, and the PCB (mainly heptachlorinate) yield was at least twice as much.

TABLE II. Phototranformation parameters (rates, dechlorination ratio(R), % PCBs, and % of unknown photoproducts) of Chlorobenzenes under unsensitized and sensitized photolysis conditions.

SUBSTRATE	UNSENSITIZED REACTION (water-acetonitrile)				SENSITIZED REACTION (water-acetonitrile + acetone[a)])			
	t/2 h	mono-deCl / di-deCl R[b)]	PCBs %	unknown photoproducts %	t/2 h	mono-deCl / di-deCl R[b)]	PCBs %	unknown photoproducts %
HxCBz	13	25	-	20	32	10	-	22
PeCBz	31	1	-	59	3.5	9	4	13
TCBz								
1,2,3,4-	33	1	tr.	33	2	6	12	39
1,2,3,5-	38	9	3	57	1	30	13	25
1,2,4,5-	1.4	3	3	48	1.2	2	7	50
					(methanol + acetone[a)])			
TrCBz								
1,2,4-	70	-	-	46	4.5	-	-	19
1,2,3-	108	-	-	53	3.5	-	-	1
1,3,5-	72	12	5	62	1.5	-	8	52

a) acetone concentrations were 0.55M in the water-acetonitrile solutions, and $5.5x10^{-3}$M in the methanol solution.

b) R represents the ratio between the percentages of the mono-dechlorinated congeners and the di- (or di- + tri-) dechlorinated congeners.

Pentachlorobenzene (PeCBz). Phototransformation of PeCBz was studied under the same conditions as TCBzs by the same authors (Choudhry and Hutzinger, 1984).

In water-acetonitrile (1:1.5) solution, the half-life of PeCBz (approximately 10^{-3} M) was about 32 hours. 60% of the converted PeCBz was accounted for by the dechlorinated congeners TCBzs, TrCBzs, and DCBzs, whose ratio (TCBzs/TrCBzs + DCBzs) was approximately 1. No other photoproducts were identified.

The sensitized reaction was 9 times faster: less than 13% of transformed PeCBz was unidentified. The major path was again dechlorination to congeners whose ratio (TCBz/TrCBz + DCBz) was about 9. In addition to this, 4% of PCBs (mainly nonachlorinated) was formed.

In a different experiment, PeCBz dissolved in methanol (or hexane), despite the fact that its absorption spectrum overlaps to some extent sunlight wavelengths, was found unchanged after a 7-day exposure to sunlight (Crosby and Hamadmad, 1971).

Hexachlorobenzene (HxCBz). Phototransformation of HxCBz was studied (see Table I) in water-acetonitrile with and without added acetone, under the same conditions as for TCBzs and PeCBz (Choudhry and Hutzinger, 1984).

In water-acetonitrile (1:9) solution, the half-life of HxCBz (approximately 10^{-3} M) was about 13 hours. Only dechlorinated congeners (PeCBz and TCBzs, whose ratio was 25) were identified among photoproducts and they accounted for 80% of transformed HxCBz.

Strangely enough, the acetone-sensitized reaction was found slower (HxCBz half-life, 32 hours), whereas photoproducts were the same and accounted for the same percentage of transformed parent compound. Only the relative quantities of PeCBz and TCBz were different: the ratio PeCBz/TCBz was 10.

In a different experiment, HxCBz in a methanol solution, showed, contrary to PeCBz, a 60% loss after 15 days of exposure to sunlight. 84% of the protoproducts were identified as penta- and tetrachlorinated aromatics resulting from reaction between substrate and solvent, whereas PeCBz was 5%, as reported by Plimmer and Klingebiel (1976). These authors found the same photoproducts after 30-minute irradiation with higher energy frequencies (λ = 260 nm) of a methanolic suspension of HxCBz.

Comparative analysis of PCBzs photodegradation in different processes. The majority of available data refer to direct, sensitized, and catalysis-mediated processes. For the purpose of comparing the effectiveness of these processes as detoxication routes for PCBzs it would be useful to have, for each process, a set of homogeneous data on each PCBz or, at least, on one PCBz for each group of isomers. This is because toxicity is directly dependent on the number of chlorine atoms. However, for no process such a set is available. Very few inhomogeneous data refer to direct photolysis in water; homogeneous data are available for direct (unsensitized) photolysis in water-acetonitrile mixture of tri- to hexa-chlorobenzenes, for indirect (sensitized) photolysis of tetra- to hexa-chlorobenzene in the presence of acetone, and for TiO_2 catalyzed photooxidation of MCBz, 1,2- and 1,4-DCBz.

Direct photolysis in water. Among various processes, this is the most enviromentally significant. One finding, although single, seems relevant: at λ = 300 nm, MCBz, which should be completely transparent (estimated logε = -2 at 300 nm), was found to photohydrolyze quantitatively into phenol. Therefore, it is not unreasonable to think that MCBz congeners, as far as they occur in the solution phase, may undergo degradation through the same dechlorination mechanism whose rates and effectiveness are, however, unpredictable.

Sensitized versus unsensitized photolysis. Relevant parameters useful for assessing relative propensity to photodetoxication of variously chlorinated PCBzs in the two processes are collected in Table II. These parameters, already introduced in previous sections are: relative half-lives ($t_{\frac{1}{2}}$), ratios between the mono-dechlorinated and the di- (or di- + tri-) dechlorinated congeners (R), percentage of PCBs formed, and percentage of unidentified photoproducts. The choice of these parameters was made in such a way that, very rudimentally speaking, the lower the figures the better the process from the point of view of detoxication.

In the unsensitized reaction, HxCBz is the less persistent PCBz (with the unique exception of 1,2,4,5-TCBz). The photolability apparently decreases along with decreasing chlorine content. PCBs are absent among PeCBz and HxCBz photoproducts; they do not exceed 3 and 5% of photoproducts formed from TCBz and TrCBz, respectively. The ratio R is particularly high with HxCBz (which is practically dechlorinated to PeCBz only) whereas for other PCBzs the dechlorination process involve parent compounds and their primary photodechlorinated products. The percentage of unknown photoproducts is rather high except for HxCBz.

In the sensitized, reaction HxCBz is the more persistent PCBz. Excluding TrCBzs from consideration (as they were examined in methanol-acetone solutions), the photolability apparently increases along with decreasing number of chlorine atoms. PCBs, whose content may be as high as 13%, are present among photoproducts of each group of isomers, except among those of HxCBz. The ratios R are rather high showing that mono-dechlorination prevails largely on di-dechlorination. The quantity of unknown photoproducts is equivalent to that observed in the unsensitized reaction.

The comparison among rates and products in the sensitized and unsensitized reactions shows that:

- the reactions are faster under sensitized conditions; however, the most hazardous chemical among those under examination is degraded slower;
- the proportion of mono- to di-dechlorinated congeners is generally higher in the sensitized reaction (only HxCBz makes exception);
- the quantities of PCBs formed are always lower in the absence of sensitizer;
- the proportion of unknown vs. identified photoproducts is more or less equivalent in the two processes.

Preliminary conclusions may be as follows: acetone may mimic impurities present in natural waters. However, as a sensitizer, it does not necessarily improve detoxication as, although substrate loss is generally faster in presence of acetone, HxCBz makes a relevant exception. Furthermore, the fact that both PCB formation is enhanced up to more than twice as much, and mono-dechlorinated congeners generally prevail on the di- or tri-dechlorinated ones, indicates that detoxication is less efficient in the presence of sensitizers.

TiO_2-catalyzed phototransformation. Homogeneous data refer to MCBz and 1,2-DCBz tested in a 0.1% TiO_2-water slurry. The few qualitative data show that both substrate loss and dechlorination are relatively slow, and that mineralization does not occur. Furthermore multiple photoproducts, including PCBs, are formed; therefore TiO_2 particles do not seem a good candidate material for decontamination of MCBz or DCBzs containing waters, although much higher degradation rates were observed with 1,4-DCBz dissolved in water containing 1% TiO_2.

Polychlorinated Phenols

The photochemistry of few of the 19 congeners, all of which meet the necessary conditions for direct photolysis in water, has been studied. They are 4-chlorophenol (4-MCP), 2,4-dichlorophenol (2,4-DCP), and pentachlorophenol (PeCP). Available data refer to direct photolysis in water and to TiO_2-mediated phototransformation.

Direct photolysis in water. 4-MCP was reported by Crosby and Wong (1973) to undergo very slow decompomposition in water solution (UV light, pH = 8, about 100% loss after 130 hours irradiation; half-life: 40-50 hours) attributed to photonucleic substitution of chlorine by hydroxide ion. PeCP was found by the same authors (Wong and Crosby, 1978) to undergo faster decompostion under equivalent conditions (UV light, pH = 7.3, 100% loss after about 20 hours irradiation; half-life: 3-4 hours). Primary photonucleic substitution products of PeCP were tetra- and trichlorinated phenols. Additional identified photoproducts were chlorinated diols and dichloromaleic acid (see Table III). The latter was the unique identifiable product when the test compound loss was almost complete; this finding suggested the possibility of complete mineralization of PeCP under sunlight, following the ring opening step. Recently, direct photolysis was found the major sink for PeCP in experimental ponds (Crossland and Wolff, 1984). Furthermore, for PeCP and 2,4-DCP the yearly average photolytic half-life when dissolved in the top millimeters of natural waters, was calculated to be about 6 and 3 minutes respectively (Lemaire et al., 1985), based on measured quantum yields, molar extinction coefficients, and computed total absorbed light (Zepp and Cline, 1977).

The formation of chlorinated dibenzo-*p*-dioxins (CDDs), from PCPs in water solution was also monitored. In the riboflavine sensitized photolysis of 2,4-DCP none was found, although small amounts of dimeric structures (tetrachlorinated phenoxyl phenols) were identified among photoproducts (Plimmer and Klingebiel, 1971). OCDD, octachlorinated dioxin, was reported to form, by Stehl et al., 1973, in very small amounts from pentachlorophenol (e.g. 0.03% OCDD was found after 5 days exposure to sunlight of a pentachlorophenate solution when the loss of the parent compound was about 50%).

TABLE III- Direct photolysis and heterogeneous photocatalytic degradation of 4-chlorophenol(4-MCP) and pentachlorophenol (PeCP)*.

SUBSTRATE	IRRADIATION		DIRECT PHOTOLYSIS (H_2O solution)			HETEROGENEOUS PHOTOCATALYTIC DEGRADATION (.2% TiO_2 - H_2O suspension)		
	λ nm	time h	pH[a)]	SUBSTRATE LOSS %	PRODUCTS	pH[a)]	SUBSTRATE LOSS %	PRODUCTS
4-MCP	⩾340	1	5.6	: 0	-	4.5	40	$HCl + CO_2$
	"	"				12.0	70	"
	"	22	5.6	<10	-	4.5	98[b)]	"
PeCP	⩾310	1	3.0	3[c)]	-	3.0	94[d)e)]	$HCl + CO_2$
	⩾330	1	10.5	6	-	10.5	99	"
	>290	÷3	7.3	÷50[f)]	Cl_4-Cl_3 phenols and diols Cl_2 maleic acid			

*Barbeni et al., 1984 and 1985, except for PeCP (λ >290 nm), Wong and Crosby, 1978.
a) initial pH (except for pH 7.3: borate - phosphate buffer); b) this was found negligible in degassed TiO_2 slurry or in TiO_2-CH_3CN slurry and <20% if TiO_2 was substituted with SiO_2; c) loss was found 44% at pH 10.5; d) including a minor loss of PeCP adsorbed on TiO_2 (adsorbtion on TiO_2 was 20% after 2 hours equilibration in the dark at pH 3.0 and was negligible at pH 10.5);e) this became 50% in the degassed dispersion and 20% when TiO_2 was substituted with Al_2O_3; f) this became 100% after 20 hours (100% loss of PeCP was also observed after 10 days exposure of the same solution to sunlight),whereas at pH 3.3, 50% loss was only observed after about 100 hours illumination (with same light).

TiO_2-catalyzed phototransformation. The role of heterogeneous photocatalysis on PCPs degradation was also investigated. The efficiency of TiO_2 particles on photodegradation of 4-MCP and PeCP in water solution was found very high (Barbeni et al., 1984, 1985) provided sufficient aeration was available for the photooxidation process to occur. A selection of data relative to this kind of experiments and the corresponding data obtained in the direct photolysis experiments is reported in Table III.

4-MCP ($5x10^{-4}$ M) dissolved in a .2% TiO_2-water slurry was found to undergo, upon illumination (λ>340 nm), quantitative mineralization to HCl and CO_2; the initial half-life was 35 minutes at pH 12, and 80 minutes at pH 4.5.

PeCP ($4x10^{-5}$ M), under the same experimental conditions, was almost completely mineralized at higher rate. Its half-life was 20 and 15 minutes as unionized and ionized species, respectively. The same experiments carried out in the absence of the semiconductor powder showed negligible photodegradation of both compounds. Much lower mineralization yields were obtained when TiO_2 was substituted with SiO_2, Al_2O_3 or other metal oxides. Sunlight had equivalent effects as artificial light on the degradation of PeCP and 4-MCP dissolved in TiO_2-water suspensions.

Comparative analysis of PCP direct and catalyzed photodegradation. The following general consideration may be drawn from available data.

Direct photolysis under sunlight is an important route for PCPs dechlorination. PeCP is degraded faster then 4-MCP, and its dechlorinated products are readily converted into secondary products of increasingly lower clorine content. Therefore direct photolysis, although some evidence of formation of PCDDs in very small quantities is available, can be considered an effective route for detoxication of PCPs present in the photic zone.

Photomineralization of PCPs in water in sunlight may be artificially promoted by TiO_2 suspended particles. This process may potentially be applied for treatment of PCPs containing waste waters, provided that the process is conducted under controlled conditions that prevent eventual formation of toxic by-products.

Polychlorinated Naphthalenes

In the UV range, PCNs in hexane exhibit their most intense ($\log\varepsilon$ = 4-6) absorption band at wavelengths shorter than 290 nm. In general, introducing chlorine atoms into the naphthalene nucleus determines a bathochromic shift of the band, so that monochloronaphthalenes have the absorption maximum at approximately 220 nm, whereas that of the octachloroderivative is at 275 nm.

All PCNs show other, much less intense absorption bands at longer wavelengths, and characterized by the presence of several maxima. These bands overlap the UV section of the solar spectrum to an extent depending on each individual compound spectral features so that in most cases sunlight of high energy is available to some extent directly to the chemicals. Both the intensity and the position of such bands are affected by the number of chlorine atoms and their substitution pattern.

A limited amount of data is available on PCN photodegradation: the following information has been drawn from Ruzo et al. (1975a,b). Data reported in Tables IV and V refer to eleven congeners, with substitution degree up to four chlorine atoms, and to their direct or catalyzed (sensitized) photolysis in homogeneous phase. No information was provided in the original papers as to the total amounts of parent compounds lost.

Mono- and dichloronaphthalenes (Table IV) were irradiated at ∿30°C in either methanol, cyclohexane, or acetonitrile-water (4:1) by means of a UV lamp with a maximum output at 300 nm (cutoff, ∿285 nm).

In the acetonitrile-water mixture, monochloronaphthalenes were converted chiefly into chlorobinaphthyls (PCBNs), while naphthols, arising from the photosubstitution of chlorine, and small amounts of hydroxylated dimers were also identified. However, in the presence of oxygen, naphthol was the preponderant product. It was also observed that 1-monochloronaphthalene disappearance quantum yield increased eightfold in the presence of the electron-donor triethylamine.

As visible from Table IV, additional data for different media are available. They show that while major photochemical pathways are substantially the same, the relative photoproduct yields may change remarkably. In particular, monochloronaphthalenes (MCNs) in water tend to form mainly binaphthyls, whereas naphthalene is a major product in organic solvents unless a sensitizer is added (as for 2-MCN).

Results collected in Table V complement previous ones (Table IV) and allow some considerations on the relative photoreactivity of a number of selected PCNs irradiated under equal experimental conditions. It may be observed that the dechlorination/dimerization ratios cover a wide range of values, which indicates marked substituent effects. In general, dechlorination is favored with PCNs that have adjacent (vicinal or peri) chlorine atoms, while unhindered PCNs yield mostly dimers. Relative reaction rates are greater for the former type of PCNs, with the somewhat expected exception of 1,2,3,4-tetrachloronaphthalene (1,2,3,4-TCN).

Sunlight irradiations of a number of PCNs as solid films on quartz surface gave only insoluble polymeric material.

No data in water media seem to be available. However, by analogy with similar compounds, the experimental evidence in organic solvents suggests that sunlight might promote phototransformation of substrates also in aqueous systems yielding analogous photoproducts. These processes, however, might not entail an effective detoxication as PCNs with intermediate chlorosubstitution level are generally more toxic than the higher congeners (see Appendix). Furthermore, no information is available on toxicity of PCBNs and other photoproducts observed.

TABLE IV

Distribution of photoproducts and quantum yields of mono- and dichloronaphthalenes (derived from Ruzo et al., 1975a)

Substrate	Solvent	Reaction quantum yield(a)	Dechlorination(b)	Binaphthyls(b)	Other(b,c)	
1-MCN	Methanol(d)	0.005	74	25		<1
1-MCN	Methanol-oxygen	0.002	76	23		<1
1-MCN	Methanol-hydrobromic acid		88	12		
1-MCN	Cyclohexane		88	12		
1-MCN	Acetonitrile-water		<1	94	5	<1
2-MCN	Methanol(d)	0.007	58	38	4	
2-MCN	Methanol-benzophenone	0.007	2	97	1	
2-MCN	Cyclohexane		72	28		
2-MCN	Acetonitrile-water		2	94	4	
1,2-DCN	Methanol	0.012	32(e)	66	2	
1,2-DCN	Methanol-benzophenone	0.014	28	68	4	

(a) Degassed solutions, unless differently stated. 20-60 h irradiations.
(b) Yields were estimated as percentage of total product formation by comparison with standard concentrations of naphthalene and binaphthyl.
(c) Substitution (left column) and chlorination (right column) products.
(d) The material balance on naphthyl residues was >95% in early stages of the reaction (6 h).
(e) Approximately equal amounts of 1- and 2-MCN.

TABLE V

Polychloronaphthalene degradation in methanol(a) under UV irradiation (derived from Ruzo et al., 1975b)

Substrate	Irradiation time (hours)	Relative reaction rate	Dechlorination products (%)	(Type)	Other(b)
1,8-DCN	40	56.0	86	1-MCN	TrCBN
1,2-DCN	40	10.1	25	1-, 2-MCN (1:1)	TrCBN
1,3,5,8-TCN	120	9.0	80	1,4,6-TrCN	HCBN
1-MCN	40	8.6	28		MCBN
2,7-DCN	120	8.3	6	2-MCN	TrCBN
2,3-DCN(c)	120	8.0	25	2-MCN	TrCBN
1,4-DCN	120	6.3	10	1-MCN	TrCBN
2-MCN	120	5.9	15		MCBN
1,5-DCN	120	4.0	2	1-MCN	TrCBN
1,3,5,7-TCN	120	2.3	1	1,3,7-TrCN	HCBN
1,2,3,4-TCN	120	1.0	80	Mixture of various congeners	HCBN

(a) In the presence of air. Degassed solutions containing benzophenone (0.15 M) exhibited the same reaction rates. All photolyses carried out at $\sim$30°C.
(b) In addition to chlorobinaphthyls, small amounts of methoxylated naphthalenes (<2%) and methoxylated binaphthyls were observed.
(c) Methanolic 2,3-DCN irradiated by sunlight showed a similar photodegradation pattern.

Polychlorinated Dibenzofurans and Dibenzodioxins

An extensive review of PCDF and PCDD photochemistry is available from the literature (Choudhry and Hutzinger, 1982). What follows is a concise summary of literature data on the photochemical behavior of PCDFs and PCDDs in liquid media. For PCDDs, results of photodegradation studies in complex matrices have also been reported due to their environmental interest.

Polychlorodibenzofurans. A number of pertinent data on PCDF photochemistry are reported in Table VI.

Sunlamp irradiation of 2-chloro- and 2,8-dichlorodibenzofuran (2-MCDF and 2,8-DCDF) in water determined a scant loss of both compounds (Crosby and Moilanen, 1973). Moreover, 2-MCDF in purified methanol was also scarcely affected by light. Sunlamp irradiation of 2,8-DCDF in highly purified methanol - or acetone-added purified methanol - yielded photodegradation to 2-MCDF with slow but similar reaction rates. However, photolysis occurred at a remarkably faster rate when utilizing a laboratory-grade solvent or after addition of 4,4'-dichlorobenzophenone as a sensitizer to the high purity methanol (Crosby and Moilanen, 1973; Crosby et al., 1973). The conclusion was drawn that medium impurities can drastically alter photodecomposition rates, and that the environmental sensitization of PCDF photolysis seems quite plausible.

UV-lamp irradiation of 2,8-DCDF in methanol or hexane solutions resulted in a rapid loss of the original substrate (Hutzinger et al., 1973): decomposition was faster in methanol than in hexane. The same may be said for octachlorodibenzofuran (OCDF). However, in the same solvent both chemicals decomposed with similar reaction rates. No other products but those formed by progressive dechlorination of original substrates were found to be present.

Although not strictly related to the subject of this paper, it is worthwhile to mention that Buser (1976) obtained mixtures of various PCDF congeners by reductive dechlorination of OCDF in solution exposed to γ-ray or UV light from a mercury lamp.

It may be pointed out that 2,8-DCDF and OCDF exposed to sunlight as thin film on quartz underwent a progressive loss of chlorine atoms, although in the case of DCDF a trichloroderivative was also detected. In both cases, parent compound disappearance was limited (Hutzinger et al., 1973).

Polychlorodibenzo-p-dioxins. Owing to the interest in the spontaneous - or, eventually, man-induced - reduction of the environmental burden of these compounds, their photodecomposition in various matrices has been investigated by several authors. It is not surprising that the congener which has received most attention is the 2,3,7,8-tetrachloroderivative (often abbreviated as TCDD or "dioxin"): in fact, TCDD was a normal trace-level impurity of phytotoxic industrial products massively employed during the Vietnam war, but has also been the toxicant responsible for highly hazardous environmental poisonings as a consequence of a number of accidents - among which, that of Seveso (Milan, Italy) in July 1976 (see references cited hereafter). A few examples follow relative to TCDD and octachlorodibenzodioxin (OCDD) in complex media.

TABLE VI

Photochemistry of MCDF, DCDF, and OCDF in water, methanol and hexane

Substrate	Solvent	Irradiation Light	Time (hours)	Substrate loss (%)	Products
2-MCDF(a)	Aqueous suspension	Sunlamp		Low	
2-MCDF(a)	Methanol (at 30°C)	Sunlamp	>125	<10	
2,8-DCDF(a,b)	Aqueous suspension	Sunlamp		Low	2-MCDF
2,8-DCDF(c)	Methanol	Sunlamp	48	>95	2-MCDF
2,8-DCDF(a,c)	High purity methanol, or acetone-added purified methanol	Sunlamp	∿112	58-83	2-MCDF
2,8-DCDF(a,c)	Purified methanol, DCBPh(f) added after 90 h	Sunlamp	∿112	100	
2,8-DCDF(d)	Methanol	UV lamp (310 nm)	<1	∿100	2-MCDF(e)
2,8-DCDF(d)	Hexane	UV lamp (310 nm)	6	>90	2-MCDF(e)
OCDF(d)	Methanol	UV lamp (310 nm)	<1	100	TCDF, PeCDF, HxCDF, HCDF. Some DCDF and TrCDF(e)
OCDF(d)	Hexane	UV lamp (310 nm)	2	>90	TCDF, PeCDF, HxCDF, HCDF. Some DCDF and TrCDF(e)

(a) From Crosby and Moilanen (1973).
(b) 2,8-DCDF absorption maximum in methanol located at ∿ 290 nm. Absorption negligible at wavelengths >320 nm (Crosby et al., 1973).
(c) From Crosby et al. (1973).
(d) From Hutzinger et al. (1973).
(e) Unidentified polymeric products when irradiation ∿ 20 h.
(f) 4,4'-Dichlorobenzophenone.

TCDD exposed to sunlight or other light sources, on glass, marble, or wet or dry soil exhibited only a negligible decay, or none at all (Crosby et al., 1971; Allegrini et al., 1977; Plimmer, 1978). However, a decay was detected following treatment of the contaminated medium with hydrogen-donor solvents, such as ethyl oleate-xylene mixtures, with or without emulsified water (Allegrini et al., 1977; Bertoni et al., 1978). TCDD on interior or exterior wall surfaces, sprayed with the same mixture and UV-irradiated, was completely decomposed (Allegrini et al., 1977).

TCDD also disappeared to a variable extent from silica gel, alumina, and aluminium, with or without ethyl oleate addition (Gebefügi et al., 1977; Allegrini et al., 1977).

TCDD loss was also observed when the compound was in the presence of those phytotoxic industrial compounds to which is production-related, and exposed to sunlight (Crosby and Wong, 1977; Nash and Beall, 1980).

In the experiments described above, aside from determining the amount of substrate disappearance, no other satisfactory determination - qualitative and/or quantitative - was carried out (this holds also for reactions reported in Table VII). However, in one case (Plimmer, 1978), sunlight irradiation of TCDD on silica was accompanied by a modest substrate loss and the appearance of unidentified polar compounds.

OCDD, in a few liquid and uncommon solid media, was reported to undergo photodegradation under both natural and artificial sunlight (Arsenault, 1976; Lamparski et al., 1980).

Irradiation with a UV mercury lamp (254 nm) of PCDDs in various hydrocarbon solvents always resulted in extensive loss of the parent substrate and formation of several related congeners with lower chlorosubstitution degree (Buser, 1976, 1979; Desideri et al., 1979). Conversion yields could be optimized so as to employ such method to obtain reference substances for analytical uses (Buser, 1976, 1979).

Table VII exhibits selected data on PCDD photochemistry in liquid media. It may be observed that in few cases only reaction products have been investigated, and always on a merely qualitative basis: a quantitative determination of photoproducts, which would account satisfactorily for the amount of substrate loss, is normally absent.

All PCDDs shown degraded in organic media under various irradiation conditions all of which, however, provide some energy in the solar spectrum UV section capable of inducing direct photolysis. Aside from eventual impurities present in the matrix, no sensitizer appears to have been added to modify photoreaction rates and yields.

As experimental conditions are generally diverse, quantitative comparison between findings of different authors may not be carried out. However, one striking feature appears recurring in various experiments: UV-induced loss of chlorine atoms in solubilized higher PCDD congeners seems to occur preferentially from lateral (2-, 3-, 7-, or 8-) positions flanked on both sides by adjacent chlorines (Buser, 1976, 1979; Dobbs and Grant, 1979; Nestrick et al., 1980). It was also shown that the _peri_ positions (1-, 4-, 6-, or 9-) lost chlorine at a rate slower than the lateral positions. Therefore, 2,3,7,8-TCDD was predicted as likely

TABLE VII

Photochemistry of DCDD, TrCDD, TCDD, HxCDD, HCDD, and OCDD in various organic solvents and in aqueous media

Substrate	Solvent	Irradiation Light	Time (hours)	Substrate loss (%)	Products
2,7-DCDD(a)	Methanol(b)	UV lamp	6	~70	
2,7-DCDD(c)	Octanol, or isooctane	Sunlamp	0.67	50	
2,3,7-TrCDD(d)	Hexane	UV lamp (313 nm)	1.17	61(e)	
2,3,7,8-TCDD(a)	Aqueous suspension	UV lamp		Negligible	
2,3,7,8-TCDD(f)	Water, CPC(g) added	UV lamp (254-356 nm)	4	~90	
2,3,7,8-TCDD(f)	Water, SDS(h) added	UV lamp (254-356 nm)	8	>90	
2,3,7,8-TCDD(d)	Water-acetonitrile (75:25)	UV lamp (313 nm)	24	62(e)	
2,3,7,8-TCDD(d)	Water-acetonitrile (75:25)	Sunlight	~26	49(e)	
2,3,7,8-TCDD(a)	Methanol(b)	UV lamp	24	100	2,3,7-TrCDD and DCDD
2,3,7,8-TCDD(f)	Methanol	UV lamp (254-356 nm)	18	100(i)	
2,3,7,8-TCDD(a)	Methanol	Sunlight	<8(j)	100	
2,3,7,8-TCDD(d)	Hexane	UV lamp (313 nm)	4	66(e)	
2,3,7,8-TCDD(c)	Isooctane	Sunlamp	0.67-3(k)	50	No TCDD after 24-h irradiation
1,2,4,6,7,9,-HxCDD(1)	Hexane	Sunlight	47	50	
1,2,3,6,7,9-HxCDD(1)	Hexane	Sunlight	17	50	
1,2;3,7,8,9-HxCDD(1)	Hexane	Sunlight	5.4	50	
1,2,3,4,6,7,9-HCDD(1)	Hexane	Sunlight	28	50	
1,2,3,4,6,7,8-HCDD(1)	Hexane	Sunlight	11	50	

(over)

(Table VII, continued)

OCDD(a)	Methanol(b)	UV lamp	24	10	Series of PCDD's of decreasing chlorine content
OCDD(c)	Octanol, or isooctane	Sunlamp	18-20	6- ∿20	
OCDD(l)	Hexane	Sunlight	16	50	<10% of degraded OCDD converted into less chlorinated compounds (e.g. HCDD)(m)

(a) Crosby et al. (1971).
(b) Irradiation also carried out in ethanol with similar results.
(c) Stehl et al. (1973).
(d) Mill et al. (1983).
(e) Disappearance quantum yields, Φ : Φ(TrCDD) = 0.22; Φ(TCDD, in hexane) = 0.049; Φ(TCDD, in water-acetonitrile) = 0.0022; Φ(TCDD, sunlight) = 0.00068.
(f) Botrê et al. (1978).
(g) 1-Hexadecylpyridinium chloride.
(h) Sodium dodecyl sulfate.
(i) Estimated by these authors.
(j) Production of yellow gum after 36-h irradiation.
(k) Light source placed either 0.5 or 1.0 m, respectively, from sample.
(l) Dobbs and Grant (1979).
(m) Also compounds with longer GC retention time were observed.

being the most photolabile of all PCDDs (Dobbs and Grant, 1979): this was later confirmed for the TCDD subgroup and a few other higher PCDDs including OCDD (Nestrick et al., 1980).

Methanolic or ethanolic solutions of 2,7-dichlorodibenzodioxin (2,7-DCDD), 2,3,7,8-TCDD, and OCDD were exposed to UV light in sealed borosilicate glass containers (Crosby et al., 1971). TCDD was also exposed to natural sunlight either in sealed tubes or in open beakers: photolysis decay patterns were similar with both light sources. Rate of decay decreased with increasing number of chlorosubstitution. The disappearance of TCDD and OCDD was accompanied by the appearance of chlorinated dioxins with decreasing chlorine content: for instance, 2,3,7-TrCDD and a DCDD were identified as photoproducts of TCDD decay.

All decays were found to follow first-order kinetics (see various authors and especially Dobbs and Grant, 1979, and Desideri et al., 1979).

It can be pointed out that negligible loss of 2,3,7,8-TCDD was determined when the substrate was UV-irradiated as an aqueous suspension (Crosby et al., 1971). Addition to the water medium of a surfactant (Tween-80) and benzene produced some decay of the compound when the light source was a sunlamp (Plimmer et al., 1973).

As TCDD is so scarcely soluble in water, addition of a surfactant seems to be necessary in order to obtain an increase of the substrate solubility and a reasonable stabilization of the system as due to the surfactant micellar action. The study by Botrè et al. (1978) shows clearly that TCDD photodecomposition in such media may be complete and occurs at a rate faster than that in methanol under equal irradiation conditions. According to the authors, the reason for such an increase in the decay rate is not completely understood as yet, but the experimental evidence seems to support the existence of a stabilizing interaction via $\pi-\pi$ orbitals between the pyridin ring and the aromatic nucleus of TCDD, allowing some energy to be transferred with consequent enhancement of the toxicant photodecomposition rate.

TCDD went through UV irradiation substantially unchanged when distributed on dry or wet soil, when in the form of a thin dry film on a glass dish, or even when suspended in distilled water.

Photochemical data - although absent for the majority of these compounds, very few for some of them, or many but inhomogeneous as for TCDD - show that sunlight may be capable of inducing phototransformation of PCDFs and PCDDs. Experimental evidence shows that an important pathway is reductive dechlorination to congeners with lower chlorine content in both direct and sensitized (catalyzed) photolysis. Whether dechlorination be a way to detoxication may not be assessed. In fact, in addition to the number of chlorine atoms, at least another factor interplays in determining actual toxicity (see Appendix). Furthermore, rates of substrate loss as well as those of formation of daughter-compounds have been investigated to a very limited extent, or are absent.

CONCLUSIVE REMARKS

This review suggests that sunlight photochemistry may be an important factor in the elimination of chlorinated aromatics from the upper layers of natural water bodies. This may occur by means of direct, sensitized, and catalysis-mediated processes in homogeneous and heterogeneous phases.

It has also been shown that sunlight may be utilized for waste water detoxication under controlled conditions.

Environmental significance of available data cannot be readily evaluated: in fact, data are largely lacking - few of them being comparable - and rarely obtained in water systems and natural sunlight.

Nevertheless, as none of the chemicals dealt with in this paper - substantially non-hydrolyzable and generally very resistant to biodegradation - have proved to be unreactive towards environmentally available UV wavelengths, sunlight photochemistry should not be underestimated as a natural removal factor. However, for the purpose of scoring photodegradability and including the constants of photodegradation processes among the inputs of predictive models of environmental fate, adequate and agreed upon experimental protocols - also different from those for direct photolysis studies - should be recommended.

APPENDIX

ELEMENTS OF HAZARD ASSESSMENT

Prediction of the fate and toxic effects of chemicals in the environment is essential to the assessment of their potential hazard. The fate of a chemical depends upon the characteristics of the receiving environment and upon the chemical's properties, that regulate its tranfer (partitioning) processes between environmental compartments and its transformation processes. Toxic effects depend on the nature of both the chemical and the biological receptors.

Probable fate in the water compartment can therefore be tentatively predicted making reference to elementary physico-chemical properties such as water solubility (Sw), vapor pressure (VP) etc.; to partition constants such as the air/water Henry's constants (H), the n-octanol/water partition coefficients (log(P)) etc.; and to kinetic constants, or related data, that refer to transformation processes such as hydrolysis, biodegradation and sunlight induced transformations.

Probable adverse effects to the aquatic life can be predicted based upon a number of biological assays that monitor different toxic endpoints. Very often toxic endpoints, such as statistically determined LC50 (lethal concentrations 50%, i.e. the concentration that kills 50% of the exposed organisms) after appropriate exposure times, are used for predictions.

Fate-related properties and effects-related toxic endpoints are therefore basic elements for hazard assessment. Some are reported in the following for chlorinated benzenes and phenols; fewer data, as available, are reported for compounds belonging to the other considered classes of chlorinated aromatics, i.e. chlorinated naphtalenes, dibenzofurans and dibenzo-*p*-dioxins. Some general consideration concerning probable aquatic fate and adverse effects to aquatic life will be drawn from this information.

Polychlorinated benzenes. Some physico-chemical properties, partitioning and persistency data for each of the 12 PCBzs are listed in Table A. As a general trend, all properties, with the exception of the Henry's constant, increase (or decrease) as a function of the number of chlorine atoms, although non-negligible differences in some physico-chemical properties may be observed among isomers.

PCBzs have low water solubilities and vapour pressures which decrease with increasing chlorine content; the resulting Henry's constants (in atm $m^3 mol^{-1}$) are generally higher than 10^{-3}, so that all PCBzs are likely to volatilize from water (Mackay and Wolkoff, 1973; see also footnote next page). All PCBzs have high affinity for lipophilic materials which increases with chlorination as shown by the n-octanol/water partition coefficients (log(P)) and by the bioconcentration factors (BCF); bioaccumulation was demonstrated to occur, at high degree, for PCBzs with more than two chlorine atoms (Kitano, 1984). Based on the log(P) values, sorption on soil and sediments can also be predicted to increase with molecular weight.

TABLE A. Environmental fate-related physico-chemical properties, partitioning and persistency data for chlorinated benzenes (PCBzs).

SUBSTRATE	MP* °C	BP* °C	Sw a) $mol.m^{-3}$ (25°C)	VP* atm. (25°C)	Log P*b)	H c) $atm.m^3.mol^{-1}$	BIOACCUM d) Log BCF	BIOACCUM d) +/-	BIOD e)	λ max f) nm
MCBz	-45	132	446	$1.55x10^{-2}$	2.84	$3.47x10^{-3}$		-	-	272
1,2-DCBz	-17	180	0.93	$1.97x10^{-3}$	3.38	$2.12x10^{-3}$	3.80	-		277
1,3-DCBz	-25	173	0.97	$2.49x10^{-3}$	3.38	$2.56x10^{-3}$	4.02	-		278
1,4-DCBz	53	174	0.44	$1.31x10^{-3}$	3.37	$2.98x10^{-3}$	3.96		+	280
1,2,3-TrCBz	53	218	0.10	$9.18x10^{-5}$	4.19	$9.27x10^{-4}$	4.47	+	-	280
1,2,4-TrCBz	17	214	0.17	$3.82x10^{-4}$	4.20	$2.25x10^{-3}$	4.56	+	-	287
1,3,5-Tr	63	208	0.033	$1.97x10^{-4}$	4.26	$5.97x10^{-3}$	4.67	+	-	281
1,2,3,4-TCBz	46	255	0.027	$5.23x10^{-5}$	4.99	$1.94x10^{-3}$	5.13	+		291
1,2,3,5-TCBz	51	246	0.024	$9.17x10^{-5}$	4.99	$3.82x10^{-3}$				291
1,2,4,5-TCBz	139	248	0.002	$6.60x10^{-5}$	4.80	$3.10x10^{-2}$	5.17	+		294
PeCBz	85	276	$5x10^{-4}$	-	5.63	-	5.36	+		298
HxCBz	229	319	$7x10^{-5}$	$2.20x10^{-8}$	6.37	-	5.37	++	-	301

*Values from Sadagopa Ramanujam et al.,1983; a) values from Banerjee,1984, except for PeCBz and HxCBz taken from *; b) P = n-octanol/water partition coefficient; values for Log P differing among them up to about 2 logarithmic units, are reported in the literature: this may depend on the experimental procedure or on the regression equation used for calculation of Log P; c) Henry's constants calculated as the ratio between VP and Sw (Lyman et al., 1982) d) potential for bioaccumulation; this is expressed as the BioConcentration Factor (BCF)(from Oliver and Niimi, 1983) or using the symbols -, +, or ++ which indicate non(low)bioaccumulative, accumulative, or highly accumulative chemicals respectively (Kitano, 1984); e) -, non-biodegradable; +, biodegradable (Kitano,1984); f) lowest energy electronic transitions in ethanol, except for PeCP (isooctane); log ε (max) are between 2 and 3 (Weast,1976-77)

Hydrolysis of aryl halides is extremely unlikely especially at environmentally relevant pHs. Biodegradation was reported to occur for 1,4-DCBz (Kitano, 1984); for other PCBzs biodegradation might occur under special conditions but at very low rates. As to photodegradation, direct photolysis could only occur for the heaviest term of the series whose uv absorption bands overlap the UV section of sunlight spectrum.

From the above, volatilization, sorption, and bioaccumulation are competing partitioning processes; their relative rates will determine the fate of PCBzs. Should volatilization prevail, atmospheric photodegradation would regulate the final fate of PCBzs. Should sorption and/or bioaccumulation prevail, degradation by aquatic microorganisms would be expected to regulate the disappearance of PCBzs. Since this process is expected to be extremely slow, PCBzs should tend to accumulate in natural water systems unless phototransformation processes may provide an additional removal route.

Some data concerning the toxicity of PCBzs to selected aquatic organisms are reported in Table B. Toxicities are given as LC50, in ppm, for the tested species (lowest LC50 values correspond to highest toxicity). It can be seen that lethal concentrations are relatively high for all PCBzs, and decrease markedly with increasing chlorination. For HxCBz the only available lethal concentration data is 0.3 ppm. Lethal concentrations for PeCBz range between 0.2 and 5.3 ppm whereas for MCBz they range between 19 and 86 ppm.

Polychlorinated Phenols. Some fate- and toxic effects-related properties relevant to the assessment of the potential hazard in the aqueous environmental compartment are listed in Tables C and D, respectively, for some PCPs (among the 19 congeners) that are representative of various degrees of chlorination of the phenolic ring, and for which enough data were available.

As with PCBzs, the physico-chemical properties, as well as the partitioning and persistency data of PCPs largely depend on the number of chlorine atoms. 4-MCP, for example, is the most soluble PCP and the less bioaccumulative. It is 20,000 times more soluble, and 200 times less bioaccumulative than PeCP (BCF values were obtained by means of the

---

The following mass distribution fractions (%) (and the relative concentrations) were calculated for 1,2,4-TrCBz and HxCBz by means of a model developed by Yoshida, 1983.

	Air	Water	Soil	Sediment	Biota	Suspended sediments
1,2,4-TrCB	99.16 (0.103)	0.43 (1.00)	0.07 (89.1)	0.34 (178)	1.1×10^{-4} (491)	3.8×10^{-4} (178)
HxCBz	95.1 (2.5×10^{-2})	2.63 (1.00)	0.40 (78.3)	1.86 (157)	1.0×10^{-2} (8.6×10^{3})	2.1×10^{-3} (157)

TABLE B- Aquatic toxicities of PCBzs: LC_{50} values (in ppm) in different organisms following exposure times of 96h or 14 days.

SUBSTRATE	DAPHNIA MAGNA[a]	FATHEAD MINNOW[a]	BLUEGILL[b]	POECILIA RETICULATA[b]
	96h	96h	96h	14d
MCBz	86	29	20	19
1,2-DCBz	2.5	57	10	5.9
1,3- "	28	10	5	7.4
1,4- "	11	4	4.3	4
1,2,4-TrCBz	-	-	-	2.4
1,2,3- "	50	2.9	3.4	2.4
1,3,5- "	-	-	-	3.3
1,2,3,4-TCBz	-	1	-	0.8
1,2,3,5- "	9.7	-	6.4	0.8
1,2,4,5- "	-	-	0.15	0.3
PeCBz	5.3	-	0.25	0.2
HxCBz	-	-	-	0.3

a) Sadagopa Ramanujam, 1982; b) Verschueren, 1983.

TABLE C. Environmental fate related physico-chemical properties, partitioning and persistency data for some chlorinated phenols (PCPs)

SUBSTRATE	MP* °C	BP* °C	Sw* $mol.m^{-3}$ (25°C)	VP* atm. (20°C)	Log P*b)	H c) $atm.m^3mol^{-1}$	pK_a*	BIOACCUM d) LogBCF	BIOACCUM d) +/-	BIODEGR. e) (relative rate)	λmax. nm f)
4-MCP	40	219	211	$1.3x10^{-4}$	2.40	$6x10^{-7}$	9.18	1.0	-	76,000	298
2,4-DCP	43	210	27.6	$1.3x10^{-4}$	3.18	$4.7x10^{-6}$	7.68	1.53	-	530	304
2,4,5-TrCP	68	245	6.1		3.72		7.43	1.79	-	20	310
2,4,6-TrCP	69	246	4.6		3.83		7.42	1.78	-		311
2,3,4,6-TCP	69		0.4		4.10		5.38	1.97	-	1 g)	
2,3,5,6-TCP	115				5.05		5.48	2.68	-	1 g)	
PeCP	190	310	0.07	$1.4.10^{-7}$	5.43	$2x10^{-6}$	4.92		-		320

*Values from Sadagopa Ramanujan et al.,1983; b)P=n-octanol/water partition coefficient; a number of different values, experimental or calculated, are found in the literature for Log P of PCPs: these may differ among them up to about 1 logarithmic unit depending on the experimental procedure or the regression equation used for calculation; c)Henry's constants (VP/Sw) calculated neglecting thet Sw and VP refer to different temperatures since the order of magnitude of H is of interest; d)potential for bioaccumulation; this is expressed as Log BCF(BioConcentration Factor)(Values from Verschueren, 1983) or with the symbols + or - indicanting confirmed accumulative, and confirmed non(low)accumulative (Kitano, 1984); e)relative rates of biodegradation (Banerjee et al.,1984); f)lowest energy electronic transition in 0.1N NaOH: the Log (max) values are from 3 and 4 (Weast, 1976-77); g)the reported value refer to the isomer 2,3,4,5-TCP and might be somewhat different for the other two isomers.

TABLE D. Aquatic toxicities of PCPs containing 1 to 5 chlorine atoms: LC 50 values (in ppm) to different organisms following exposure times of 24, 48, or 96 hours

SUBSTRATE	DAPHNIA MAGNA[a]		FATHEAD MINNOW[b]	GOLDFISH[b]	BLUEGILL[c]	POECILIA RETICULATA[b]
	24h	48h	96h	24h	96h	24h
4-MCP	8.8	4.1		9.0	3.8	
2,4-DCP		2.6	8.2	7.8	2.0	4.2
2,4,5-TrCP	3.8	2.7		1.7	0.5	
2,4,6-TrCP	15.0	6.0		10.0	0.3	
2,3,4,6-TCP		0.2		0.7	0.1	
2,3,5,6-TCP	2.5	0.6			0.2	1.4
PeCP	1.5	0.3	0.2	0.3	0.1	0.4

a) LeBlanc, 1980;
b) Verschueren,1983
c) Buccafusco et al, 1981

regression equation: log(BCF) = 0.76xlog(P) - 0.23, derived by Veith, 1980); and its biodegradation is 76,000 times faster than that of TCP. Whereas the poorly chlorinated PCPs are associated with relatively low persistence in water (due to their fast biodegradation), and low propensity to bioaccumulate in biota and water sediments (due to their low log(P)), the fully chlorinated phenol, PeCP, which is also the most widely industrially used PCP, is resistant to biodegradation and tends to accumulate in sediments and biota; furthermore PeCP is often used as the water soluble sodium salt: this fact provides for dilution but at the same time favours widespread contamination of water bodies and sediments. Direct photolysis occurs to different exent for PCPs that absorb above 290 nm; photolysis probably consists in the photonucleic substitution of a chlorine atom with an OH group.

Since volatilization is unlikely for PCPs (see H values), they mainly distribute in water where biodegradation and photolysis should be the two major fate determining processes; considering the position of the uv absorption bands of PCPs, biodegradation may be easily predicted the dominant process for practically all PCPs, except for PeCP for which photolysis and biodegradation should be competing processes.

All PCPs are rather toxic to aquatic organisms and their toxicities increase regularly with the number of chlorine atoms (Table D): in fact, PeCP is, almost for all tested species, the compound associated with the lowest LD50. The PeCP concentration that killed 50% of the exposed organisms was found to range from 0.1 to 1.5 ppm, whereas for 4-MCP LC50 values range between 3.8 and 9 ppm.

Polychlorinated Naphthalenes. PCNs have physical and chemical properties similar to those of polychlorobiphenyls (PCBs) and are manufactured, as complex mixtures, for analogous uses on industrial scale (Brinkman and Reymer, 1976).

The PCN family is made of 75 different congeners, for many of which a satisfactory chemico-physical characterization is lacking. From what is known, the great majority of PCNs melt above room temperature, with the octachloroderivative melting as high as ~200°C. Melting points of commercial mixtures are generally in the range 65-140°C, although there are Halowaxes melting at the extreme values of approximately -33 and 185°C.

PCNs and PCN mixtures have specific gravities comprised between >1 and 2, and are substantially insoluble in water (Brinkman and Reymer, 1976). They exhibit a high degree of chemical stability even at temperatures up to their boiling range (>200°C), are resistant to biodegradation and very bioaccumulative (Kitano, 1984). Similarly to PCBs, PCNs are widespread and persistent pollutants.

PCNs - especially those containing five and six chlorine atoms - have been found to be very toxic to man and several animal species (Brinkman and Reymer, 1976).

Polychlorinated Dibenzofurans and Dibenzodioxins. Several references are available on the subject. Most of the information reported hereafter has been drawn from Taylor (1979), Moore et al. (1979), National Research Council of Canada (1981), Choudhry and Hutzinger (1982), and Rappe (1984).

None of the chemicals belonging to either family are produced by the industry for commercial purposes, aside from the very small amounts utilized for special purposes. However, PCDFs (135 congeners) and PCDDs (75 congeners) have widespread diffusion in the environment - although normally at very low and variable concentration levels (ppb range, or below) - as they are trace-contaminants of PCPs and their salts and derivatives which have found very extensive and dispersive uses since the 1930's.

Furthermore, PCDFs and PCDDs can be formed in combustion processes when the proper precursors (polychlorinated biphenyls and phenols, respectively, and polychlorinated benzenes) are present. For instance, industrial and municipal incinerators appear to contribute to a major extent to the presence of PCDFs and PCDDs in the environment.

PCDFs and PCDDs may also originate from higher congeners by light-induced progressive loss of chlorine, as will be discussed later. However, they may also arise in the environment from diffuse photolabile chloroaromatics acting as precursors. Conversion of PCBs and PCDPEs to PCDF's has been studied by Crosby and Moilanen (1973), Crosby et al. (1973), Norström et al. (1976, 1977), and Choudhry et al. (1977a,b). Similarly, irradiation of polychlorophenoxy phenols in methanol yielded a mixture of various products including PCDDs (Nilsson et al., 1974).

Undoubtedly there is a remarkable lack of information on the chemico-physical properties of these halogenated compounds, and only few congeners, of the PCDD family in particular, have been characterized to some degree. Characterization is often by analogy with similar substances (e.g. PCBs and chlorinated pesticides).

Congeners with four chlorine atoms or more can be expected to have a very low water solubility (<1 ppb), decreasing with increasing level of chlorosubstitution. Lipophilicity, biomagnification, octanol/water partition coefficient, and soil sorption partition coefficient (K_{oc}), instead, appear to increase with increasing chlorosubstitution. A few data for 2,3,7,8-TCDD are: MP = 306°C; VP = $<1x10^{-6}$ mmHg (25°C); P = 1.4-$19x10^{6}$ (25°C); K_{oc} = 9.9-$33x10^{5}$; Sw = 0.2 ppb (25°C).

PCDFs and PCDDs are generally characterized by a high stability. Therefore, they are presumed - as found experimentally in a few cases - to have a lasting environmental persistence which, combined with their constant input into the environment, could at length lead to a general accumulation of such compounds in both the biotic and abiotic media.

Several PCDFs and PCDDs - and especially a number of those with four, five, and six chlorine atoms - possess a high toxicity and, indeed, the 2,3,7,8-tetrachlorodibenzodioxin (often abbreviated to TCDD or "dioxin") is commonly considered as the most toxic man-made chemical.

Although toxic properties are related to the degree of chlorination, positional isomerism plays a critical role as well to the extent that one should expect toxic potency to vary greatly even within the same isomer group. By far, 2,3,7,8-TCDD is the congener which has received most attention.

In general, PCDFs and PCDDs with low chlorosubstitution degree (mono-, di-, and trichloroderivatives) are considered of much less concern for the health of man and the environment.

REFERENCES

Allegrini, I., Bertoni, G., Brocco, D., Liberti, A., and Possanzini, M. (1977): Decontaminazione mediante Radiazioni Ultraviolette da Inquinamento da 2,3,7,8-Tetraclorodibenzodiossina, Chim. Ind. 59, 541-544.

Arnold, D.R., and Wong, P.C. (1977): The Photochemistry of Chloroaromatic Compounds. Is "π-Chlorobenzene" an Intermediate?, J. Am. Chem. Soc. 99, 3361-3366.

Arsenault, R.D. (1976): Pentachlorophenol and Contained Chlorinated Dibenzodioxins in the Environment. A Study of Environmental Fate, Stability, and Significance when Used in Wood Preservation. In: Proceedings of the American Wood-Preservation Association 72, 122-148.

Banerjee, S. (1984): Solubility of Organic Mixtures in Water, Environ. Sci. Technol. 18, 587-591.

Banerjee, S., Howard, P.H., Rosenberg, A.M., Dombrowski, A.E., Sikka, H., and Tullis, D.L. (1984): Development of a General Kinetic Model for Biodegradation and Its Application to Chlorophenols and Related Compounds, Environ. Sci. Technol. 18, 416-422.

Barbeni, M., Pramauro, E., Pelizzetti, E., Borgarello, E., Grätzel, M., and Serpone, N. (1984): Photodegradation of 4-Chlorophenol Catalyzed by Titanium Dioxide Particles, Nouv. J. Chim. 8, 547-550.

Barbeni, M., Pramauro, E., Pelizzetti, E., Borgarello, E., and Serpone, N. (1985): Photodegradation of Pentachlorophenol Catalyzed by Semiconductor Particles, Chemosphere 14, 195-208.

Bertoni, G., Brocco, D., Di Palo, V., Liberti, A., Possanzini, M., and Bruner, F. (1978): Gas Chromatographic Determination of 2,3,7,8-Tetrachlorodibenzodioxin in the Experimental Decontamination of Seveso Soil by Ultraviolet Radiation, Anal. Chem. 50, 732-735.

Botrè, C., Memoli, A., and Alhaique, F. (1978): TCDD Solubilization and Photodecomposition in Aqueous Solutions, Environ. Sci. Technol. 12, 335-336.

Brinkman, U.A.Th., and Reymer, H.G.M. (1976): Polychlorinated Naphthalenes, J. Chromatogr. 127, 203-243.

Buccafusco, R.J., Ells, S.J., and LeBlanc, G.A. (1981): Acute Toxicity of Priority Pollutants to Bluegill, Bull. Environ. Contam. Toxicol. 26, 446-471.

Buser, H.-R. (1976): Preparation of Qualitative Standard Mixtures of Polychlorinated Dibenzo-p-dioxins and Dibenzofurans by Ultraviolet and γ-Irradiation of the Octachloro Compounds, J. Chromatogr. 129, 303-307.

Buser, H.-R. (1979): Formation and Identification of Tetra- and Pentachlorodibenzo-p-dioxins from Photolysis of Two Isomeric Hexachlorodibenzo-p-dioxins, Chemosphere 8, 251-257.

Choudhry, G.G., Sundström, G., van der Wielen, F.W.M., and Hutzinger, O. (1977a): Synthesis of Chlorinated Dibenzofurans by Photolysis of Chlorinated Diphenyl Ethers in Acetone Solution, Chemosphere 6, 327-332.

Choudhry, G.G., Sundström, G., Ruzo, L.O., and Hutzinger, O. (1977b): Photochemistry of Chlorinated Diphenyl Ethers, J. Agric. Food Chem. 25, 1371-1376.

Choudhry, G.G., Roof, A.A.M., and Hutzinger, O. (1979): Photochemistry of Halogenated Benzene Derivatives. I. Trichlorobenzenes: Reductive Dechlorination, Isomerization and Formation of Polychlorobiphenyls, Tetrahedron Lett. 22, 2059-2062.

Choudhry, G.G., and Hutzinger, O. (1982): Photochemical Formation and Degradation of Polychlorinated Dibenzofurans and Dibenzo-p-dioxins, Residue Rev. 84, 113-161.

Choudhry, G.G., and Hutzinger, O. (1984): Acetone-Sensitized and Nonsensitized Photolyses of Tetra-, Penta-, and Hexachlorobenzenes in Acetonitrile-Water Mixtures: Photoisomerization and Formation of Several Products Including Polychlorobiphenyls, Environ. Sci. Technol. 18, 235-241.

Crosby, D.G., and Hamadmad, N. (1971): The Photoreduction of Pentachlorobenzenes, J. Agric. Food Chem. 19, 1171-1174.

Crosby, D.G., Wong, A.S., Plimmer, J.R., and Woolson, E.A. (1971): Photodecomposition of Chlorinated Dibenzo-p-dioxins, Science 173, 748-749.

Crosby, D.G., and Moilanen, K.W. (1973): Photodecomposition of Chlorinated Biphenyls and Dibenzofurans, Bull. Environ. Contam. Toxicol. 10, 372-377.

Crosby, D.G., Moilanen, K.W., and Wong, A.S. (1973): Environmental Generation and Degradation of Dibenzodioxins and Dibenzofurans, Environ. Health Perspect. 5, 259-266.

Crosby, D.G., and Wong, A.S. (1973): Photodecomposition of p-Chlorophenoxyacetic Acid, J. Agric. Food Chem. 21, 1049-1052.

Crosby, D.G., and Wong, A.S. (1977): Environmental Degradation of 2,3,7,8-Tetrachlorodibenzo-p-dioxin (TCDD), Science 195, 1337-1338.

Crossland, N.O., and Wolff, C.J. (1984): In Aquatic Toxicology (in press). Cited by Zafirou et al. (1984).

Desideri, A., di Domenico, A., Vanzati, R., Tancioni, P., and Di Muccio, A. (1979): Photolysis of 2,3,7,8-Tetrachlorodibenzo-p-dioxin (TCDD) in Iso-octane, Hexane, and Cyclohexane, Boll. Chim. Farm. 118, 274-281.

Dobbs, A.J., and Grant, C. (1979): Photolysis of Highly Chlorinated Dibenzo-p-dioxins by Sunlight, Nature 278, 163-165.

ECETOC (1981): An Assessment of Test Methods for Photodegradation of Chemicals in the Environment - Technical Report No. 3. European Chemical Industry, Ecology and Toxicology Centre (Brussels, Belgium).

Gebefügi, I., Baumann, R., and Korte, F. (1977): Photochemischer Abbau von 2,3,7,8-Tetrachlordibenzo-p-dioxin (TCDD) unter simulierten Umweltbedingungen, Naturwissenschaften 64, 486-487.

Hutzinger, O., Safe, S., Wentzell, B.R., and Zitko, V. (1973): Photochemical Degradation of Di- and Octachlorodibenzofuran, Environ. Health Perspect. 5, 267-271.

Kitano, M. (1984): Safety Examination of Existing Chemicals in Japan. In: Proceedings of the International Workshop on Recent National and International Approaches to the Control of Existing Chemicals, Rome (Italy), October 1984 (in press).

Lamparski, L.L., Stehl, R.H., and Johnson, R.L. (1980): Photolysis of Pentachlorophenol-Treated Wood. Chlorinated Dibenzo-p-dioxin Formation, Environ. Sci. Technol. 14, 196-200.

LeBlanc, G.A. (1980): Acute Toxicity of Priority Pollutants to Water Flea (Daphnia magna), Bull. Environ. Contam. Toxicol. 24, 684-691.

Lemaire, J., Guth, J.A., Klais, O., Leahey, J., Merz, W., Philp, J., Wilmes, R., and Wolff, C.J.M. (1985): Ring Test of a Method for Assessing the Phototransformation of Chemicals in Water, Chemosphere 14, 53-77.

Lyman, W.J., Reehl, W.F., and Rosenblatt, D.H., Eds. (1982): Handbook of Chemical Property Estimation Methods. McGraw Hill Co.

Mackay, D., and Wolkoff, A.W. (1973): Rates of Evaporation of Low Solubility Contaminants from Water Bodies to Atmosphere, Environ. Sci. Tecnol. 7, 611-614.

Mansour, V.M., Wawrik, S., Parlar, H., and Korte, F. (1980): Synthese Monosubstituierter Chlorbenzole durch Photoinduzierte Dechlorierung, Chemiker-Zeitung 104, 339-340.

Mill, T., Drossman, H., Podoll, T., Jaber, H., Spanggord, R., and Combs, D. (1983): Tetrachlorodibenzodioxin: Rates of Photolysis and Volatilization in the Environment. Personal communication.

Moore, J.A., McConnell, E.E., Dalgard, D.W., and Harris, M.W. (1979): Comparative Toxicity of Three Halogenated Dibenzofurans in Guinea Pigs, Mice, and Rhesus Monkeys, Ann. N.Y. Acad. Sci. 320, 151-163.

Nash, R.G., and Beall, M.L., Jr. (1980): Distribution of Silvex, 2,4-D, and TCDD Applied to Turf in Chambers and Field Plots, J. Agric. Food Chem. 28, 614-623.

National Research Council of Canada (1981): Polychlorinated Dibenzo-p-dioxins - Criteria for Their Effects on Man and His Environment. Publication NRCC No. 18574 of the Environmental Secretariat (Ottawa, Canada).

Nestrick, T.J., Lamparski, L.L., and Townsend, D.I. (1980): Identification of Tetrachlorodibenzo-p-dioxin Isomers at the 1-ng Level by Photolytic Degradation and Pattern Recognition Techniques, Anal. Chem. 52, 1865-1874.

Nilsson, C.-A., Andersson, K., Rappe, C., and Westermark, S.-O. (1974): Chromatographic Evidence for the Formation of Chlorodioxins from Chloro-2-phenoxyphenols, J. Chromatogr. 96, 137-147.

Norström, A., Andersson, K., and Rappe, C. (1976): Formation of Chlorodibenzofurans by Irradiation of Chlorinated Diphenyl Ethers, Chemosphere 5, 21-24.

Noström, A., Andersson K., and Rappe, C. (1977): Studies on the Formation of Chlorodibenzofurans by Irradiation or Pyrolysis of Chlo rinated Diphenyl Ethers, Chemosphere 6, 241-248.

Oliver, B.G., Cosgrove, E.G., and Carey, J.H. (1979): Effect of Suspended Sediments on the Photolysis of Organics in Water, Environ. Sci. Technol. 13, 1075-1077.

Oliver, B.J., and Niimi, A.J. (1983): Bioconcentration of Chlorobenzenes from Water by Rainbow Trout: Correlations with Partition Coefficients and Environmental Residues, Environ. Sci. Technol. 17, 287-291.

Ollis, D.F., Hsiao, C.-Y., Budiman, L., and Lee, C.-L. (1984): Heterogeneous Photoassisted Catalysis: Conversions of Perchloroethylene, Dichloroethane, Chloroacetic Acids, and Chlorobenzenes, J. Catal. 88, 89-96.

Parlar, H., and Korte, F. (1979): The Significance of Quantum Yield during Determination of Environmental Photochemical Degradability of Organic Compounds, Chemosphere 8, 797-807.

Plimmer, J.R., and Klingebiel, U.I. (1971): Riboflavin Photosensitized Oxidation of 2,4-Dichlorophenol: Assessment of Possible Chlorinated Dioxin Formation, Science 174, 407-408.

Plimmer, J.R., Klingebiel, U.I., Crosby, D.G., and Wong, A.S. (1973): Photochemistry of Dibenzo-p-dioxins, Adv. Chem. Ser. 120, 44-54.

Plimmer, J.R., and Klingebiel, U.I. (1976): Photolysis of Hexachlorobenzene, J. Agric. Food Chem. 24, 721-723.

Plimmer, J.R. (1978): Photolysis of TCDD and Trifluralin on Silica and Soil, Bull. Environ. Contam. Toxicol. 20, 87-92.

Rappe, C. (1984): Analysis of Polychlorinated Dioxins and Furans, Environ. Sci. Technol. 18, 78-90A.

Ruzo, L.O., Bunce, N.J., and Safe, S. (1975a): Photoreactions of Simple Halonaphthalenes in Solution, Can. J. Chem. 53, 688-693.

Ruzo, L.O., Bunce, N.J., and Safe, S. (1975b): Photodegradation of Polychloronaphthalenes in Methanol Solution, Bull. Environ. Contam. Toxicol. 14, 341-345.

Sadagopa Ramanujam, V.M., Trieff, N.M., and Harper, B.L. (1983): Structure-Based Classification, Correlation and Toxicity Ranking of Carcinogens and Other Toxic Compounds. US EPA contract No. 68-03-3125, ECAO (Cincinnati, Ohio 45268).

Stehl, R.H., Papenfuss, R.R., Bredeweg, R.A., and Roberts, R.W. (1973): The Stability of Pentachlorophenol and Chlorinated Dioxins to Sunlight, Heat, and Combustion, Adv. Chem. Ser. 120, 119-125.

Taylor, J.S. (1979): Environmental Chloracne: Update and Overview, Ann. N.Y. Acad. Sci. 320, 295-307.

Tissot, A., Boule, P., and Lemaire, J. (1983): Photochimie et Environnement. V. Photohydrolyse du Monochlorobenzene en Solution Aqueuse Diluée, Chemosphere 12, 859-872.

Veith, G.D. (1980): State-of-the-Art Report on Structure-Activity Methods Development. US EPA report for the Office of Toxic Substances prepared by Environmental Research Laboratory (Duluth, Minnesota 55804).

Verschueren, K. (1983): Handbook of Environmental Data on Organic Chemicals, 2nd ed., Van Nostrand Reinhold Company (NY).

Weast, R.C. (1976-77): Handbook of Chemistry and Physics, 57th ed. CRC Press, Inc. (Boca Raton, Florida).

Wong, A.S., and Crosby, D.G. (1978): Photolysis of Pentachlorophenol in Water. In: Pentachlorophenol: Chemistry, Pharmacology and Environmental Toxicology, pp. 19-39. Ranga Rao, K., Ed., Plenum Press (New York, NY).

Yoshida, K., Shigeoka, T., and Yamauchi, F. (1983): Non-Steady-State Equilibrium Model for the Preliminary Prediction of the Fate of Chemicals in the Environment, Ecotoxicol. Environ. Safety 7, 179-190.

Zafiriou, O.C., Joussot-Dubien, J., Zepp, R.G., and Zika, R.G. (1984): Photochemistry of Natural Waters, Environ. Sci. Technol. 18, 358-371A.

Zepp, R.G., and Cline, D.M. (1977): Rates of Direct Photolysis in the Aquatic Environment, Environ. Sci. Technol. 11, 359-366.

PHOTODEGRADATION OF WASTES AND POLLUTANTS IN AQUATIC ENVIRONMENT

Barry G. Oliver and John H. Carey
Environmental Contaminants Division
National Water Research Institute
Canada Centre for Inland Waters
P.O. Box 5050
Burlington, Ontario, Canada L7R 4A6

ABSTRACT. Photocatalytic processes have considerable potential to contribute to the degradation of pollutants in the aquatic environment and in wastewater treatment. The sunlight irradiation of humic substances in natural waters has been shown to produce several reactant species including singlet oxygen, superoxide, hydroxyl radical and hydrogen peroxide. These species can react with organic pollutants to cause oxidation and/or dechlorination of these chemicals. Biological components of natural waters such as algae can contain a significant fraction of certain contaminants and studies have shown that algae can photocatalyze the breakdown of several persistent organic compounds. Heterogeneous photolysis using semiconductors such as titanium dioxide (anatase) has been shown to break down several organic and inorganic pollutants. Some mechanistic considerations as well as practical applications of this method for wastewater treatment are discussed.

1. PHOTODEGRADATION IN THE AQUEOUS ENVIRONMENT

The direct and indirect photolysis of environmental contaminants represents an important degradative pathway for certain contaminants in the aqueous environment[1,2]. For example for 3-trifluoromethyl-4-nitrophenol (TFM), a lampricide used in streams flowing into the Great Lakes, direct photolysis is the only known decomposition route[3]. The photochemical half-life of TFM in the aquatic environment is about three days and the photolysis by-products have been shown to be non-toxic to aquatic organisms[3]. Other examples of the importance of direct aquatic photolysis have been demonstrated for the herbicide 2,4-D (2,4-dichlorophenoxyacetic acid)[4], for the wood preservative pentachlorophenol[5], and for primary aromatic amines[6]. But these compounds represent only a small fraction of environmental contaminants and are fairly unique since they exhibit significant light absorption in the solar spectral region. Many environmental pollutants are poor absorbers of solar light - either

E. Pelizzetti and N. Serpone (eds.), Homogeneous and Heterogeneous Photocatalysis, 629–650.

exhibiting no absorbance or weak absorbance in the 290-320 non-region. For these chemicals indirect photosensitized and photo-catalyzed reactons must be evaluated to assess the importance of photolysis as a degradative pathway.

1.1 Mass Balance and Phase Partitioning Considerations

Preliminary mass balance calculations and, where possible, an evaluation of the phases in which the chemicals of interest are found in a particular river/lake system are useful first steps in assessing the potential significance of photolysis and designing appropriate photochemical studies. An illustration of this type of preliminary analysis for several chlorinated contaminants in the Niagara River/Lake Ontario system follows. Figure 1 shows the major potential loss mechanisms for chemicals in the lake. Chemicals could be: 1) absorbed to settling particles and wind up in bottom sediments, 2) discharged via the St. Lawrence River, 3) lost from the water column by volatilization, 4) photolyzed and/or biodegraded. A preliminary mass balance calculation for Lake Ontario for some chlorobenzenes, CB's, polychlorinated biphenyls, PCB's, and mirex for which the Niagara River is thought to be the main source is shown in Table I[7,8]. The bulk (>80%) of the CB's and PCB's

TABLE I Proportion of CB, PCB and Mirex Input to Lake Ontario from the Niagara River Lost by Various Processes

Chemical	% * Sedimenting	% St.** Lawrence River	% Unaccounted For
1,2,4-Trichlorobenzene	1	3	96
1,2,3,4-Tetrachlorobenzene	2	2	96
Pentachlorobenzene	4	3	93
Hexachlorobenzene	15	5	80
2,5,2'-Trichlorobiphenyl	5	5	90
2,5,2',5'-Tetrachlorobiphenyl	10	4	86
2,4,5,2',5'-Pentachlorobiphenyl	9	3	88
2,4,5,2',4',5'-Hexachlorobiphenyl	11	6	83
Mirex	70	18	12

* From Reference 7.
** From Reference 8.

entering Lake Ontario from the Niagara River are unaccounted for in Table I and are probably lost from the lake by volatilization. Biodegradation for the more highly chlorinated compounds is not likely to be significant[9]. Recent laboratory studies[10,11] have known that CB's and PCB's can be dechloriated photochemically, so photolysis may provide an alternative loss mechanism to volatilization for these chemicals in the lake. Mass balance calculations for mirex, on the other hand, show that most of this chemical is

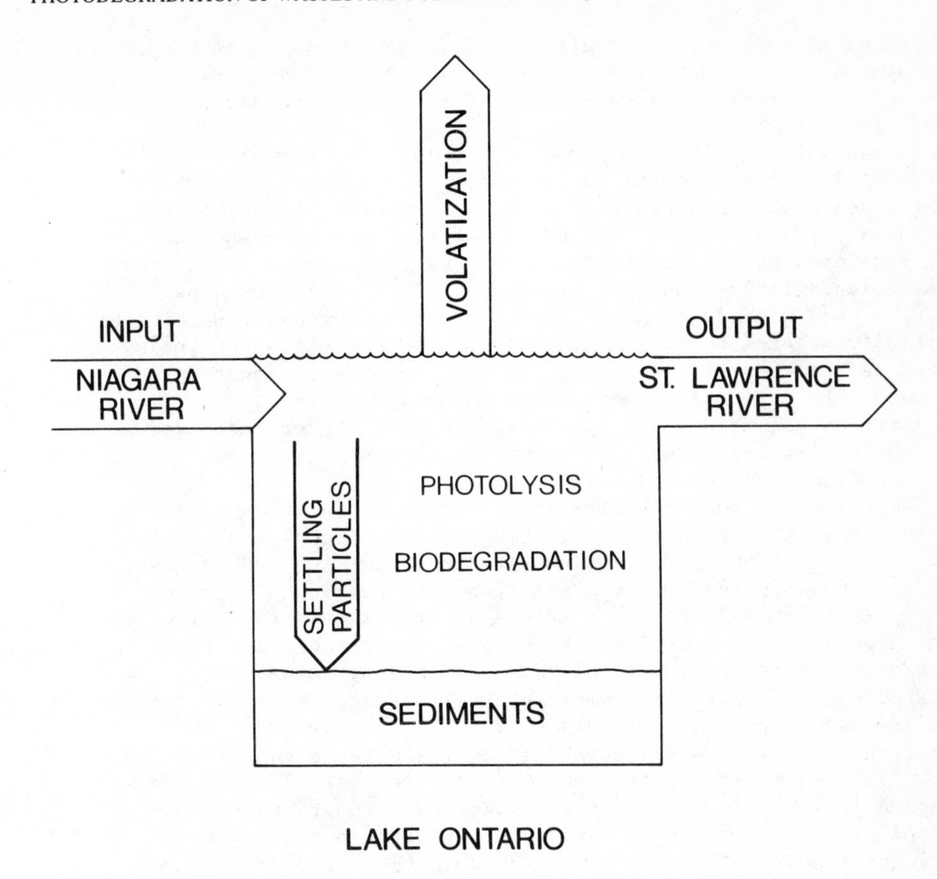

Figure 1 Block diagram of major loss processes for Lake Ontario.

sequestered to settling particulates and likely winds up in bottom sediments. Because Lake Ontario's depth averages 86 meters, photolysis of mirex can largely be ruled out as a significant degradative pathway.

In general, the lower a compound's solubility in water, the greater will be its tendency to partition into the sediment phase. In the absence of field concentration measurements, relationships using physical and chemical properties of the substance such as those developed by Karickhoff[12] can be used to estimate the degree of partitioning into suspended sediments. These relationships should be used with caution, however, since the degree of partitioning is also related to suspended sediment concentration and sediment organic carbon content[12]. In the case of the Niagara River during the course of these measurements, the suspended sediment concentration was about 5 mg/L. In rivers with higher suspended sediment levels, a greater portion of the chemicals will be partitioned into the sediments.

The phases in which chemicals are found in the study area are also an essential element in evaluating degradation pathways. A crude estimate of the partitioning between phases in the Niagara River/Lake Ontario system was accomplished by passing a large volume of water (600 liters) through a series of fine sieves to collect the larger algae and then through a high speed centrifuge to collect the remaining suspended sediment. The centrifuged water was then extracted with a 200 liter stainless steel extraction device[13] and the algae and suspended sediments were soxhlet extracted. The organic extracts were analyzed using capillary gas chromatography[14]. The results of this analysis for several CB's and PCB's are shown in Table II. Most of the lower CB's (up to pentachlorobenzene) are found mainly in the aqueous phase. Significant fractions of HCB and the PCB's are found in the suspended sediment and algae phases. Molar concentrations are extremely low in the

TABLE II Phase Partitioning of CB's and PCB's in the Niagara River Plume (Average of 4 Samples, September, 1984)

Chemical*	% Aqueous	% Suspended Sediment	% Algae
1,2,4-Trichlorobenzene	96	3	1
1,2,3,4-Tetrachlorobenzene	97	2	1
Pentachlorobenzene	90	8	2
Hexachlorobenzene	67	28	5
2,5,2'-Trichlorobiphenyl	86	8	6
2,5,2',5'-Tetrachlorobiphenyl	64	17	19
2,4,5,2',5'-Pentachlorobiphenyl	47	23	30

* 2,4,5,2',4,',5'-Hexachlorobiphenyl and Mirex were not included in this table because they were not detected in every phase.

aqueous phase 10^{-11} to 10^{-12} M/L and four orders of magnitude higher 10^{-7} to 10^{-8} M/kg in the suspended sediment and algae phases. Based on the data in Tables I and II, photolysis could be important for tri-, tetra- and pentachlorobenzenes in the aqueous phase (or in the atmosphere as a result of volatilization). For hexachlorobenzene and the PCB's, photolysis in algae and on suspended sediments should also be considered. In the Lake Ontario system, because of the extreme depth of the water (86 m) and the high settling velocity of the particulates (≈ 1 m/day), association with suspended sediments effectively removes the chemical from the photic zone. However, since photolysis in suspended sediments is possible in shallower water bodies, discussion of photolysis of chemicals on suspended sediments will be included here for completeness.

1.2 Atmospheric Photolysis

Because of the surprising importance of volatilization of chemicals from certain waterbodies, photolysis in the atmosphere should be examined as a possible major route of decomposition of persistent organics. Direct vapour phase photolysis in pesticide drift during application has been demonstrated to occur[15]. However, very little is known about the atmospheric photolysis of chemicals such as chlorobenzenes and PCB's. PCB's are known to be present globally in air samples[16], so long range transport is important. PCB's are also found in significant concentrations in precipitation[17], so some fraction of volatilized material is scavenged by airborne particles and returned to the earth via rain. Hydrogen peroxide and OH° and HO_2° radicals are known to be photochemically generated in the atmosphere and in clouds[18,19]. Free radicals such as hydroxyl radicals have been shown to produce stepwise dechloriantion of PCB's in aqueous solutions[20], so these reactions may also be important in the atmosphere.

Although there are 209 PCB congeners, most environmental data to date report total PCB concentrations. Improved capillary gas chromatographic methods have made it possible to quantify most individual PCB congeners[21]. A combination approach of (1) laboratory experiments on the sensitized photolysis of major individual PCB congeners present in commercial PCB mixtures (Aroclors) with emphasis on identifying partially dechlorinated products and (2) quantification of atmospheric and precipitation samples for individual PCB congeners should readily show whether or not significant atmospheric photolysis of PCB's is occurring. It is probable that some PCB's will be photochemically dechlorinated to PCB products which are rare or non-existent in the commercial Aroclor mixtures. Finding these rare PCB's in significant quantities in atmospheric and precipitation samples would then indicate that photolysis of PCB's was occurring. Should these initial experiments prove fruitful, detailed work on the photodegradation of PCB's, CB's and other persistent contaminants in simulated atmospheres containing fog and clouds are recommended to evaluate decomposition rates.

1.3 Indirect Photolysis in the Aqueous Phase

Colored organic material is present in significant concentrations in most natural waters. The major class of materials present is humic substances, HS, of which fulvic acid is the most prevalent[22]. Humic substances, which are yellow/brown in color, have no distinct light absorption bands. They exhibit high light absorption in the ultraviolet region and gradually decreasing absorption throughout the visible. The sunlight irradiation of oxygen-containing colored humic waters has been demonstrated to produce singlet oxygen likely by the following series of reactions[23,24,25].

$$ {}^{1}HS \xrightarrow{\text{light}} {}^{1}HS^{*} \rightarrow {}^{3}HS^{*} \quad (1)$$

$$ {}^{3}HS^{*} + {}^{3}O_2 \rightarrow {}^{1}HS + {}^{1}O_2 \quad (2)$$

Singlet oxygen is much more reactive than ground state oxygen. Because it reacts rapidly with 2,5-dimethylfuran (DMF) to produce cis-1,2-diacetylethylene[26], the DMF reaction is usually used as a probe to show the presence of $^{1}O_2$. The irradiation of several natural waters from the United States was shown by Zepp _et al._[23] to produce between 10^{-12} and 10^{-13} molar singlet oxygen.

A sharp decrease in the DMF reaction rate has been observed after addition of known singlet oxygen quenchers. DABCO (1,4-diazabicyclo [2.2.2] octane)[23] and sodium azide[24]. This provides further evidence for singlet oxygen formation in these irradiations. Singlet oxygen can rapidly oxidize furans, sulfides and electron-rich olefins so pollutants with the required structure may be efficiently degraded by this mechanism. But, singlet oxygen also reacts with biological substrates such as histidine and α chymotrypin[24]. Reactions with biota in natural waters would, therefore, compete with pollutant degradative reactions. Thus the rate of contaminant breakdown by $^{1}O_2$ reactions should be greater in colored oligotrophic as compared with colored eutrophic waterbodies.

Another reactant species produced in sunlight irradiated natural waters is superoxide, $O_2^{-\circ}$[27]. The presence of superoxide was demonstrated by irradiation of a natural water containing hydroxylamine. The hydroxylamine is converted to nitrite by reaction with superoxide and the reaction was shown to be inhibited by superoxide dismutase (SOD)[27]. Recently, laser flash kinetic spectroscopic measurements have shown that the likely precursor of superoxide is the hydrated electron[28]. This electron is produced either through direct photoionization of the humic substance or from ionization of the excited singlet state[28].

$$ HS \xrightarrow{\text{light}} HS^{+\circ} + e^{-}_{(aq)} \quad (3)$$

$$ HS \xrightarrow{\text{light}} {}^{1}HS^{*} \rightarrow HS^{+\circ} + e^{-}_{(aq)} \quad (4)$$

The hydrated electrons quickly react with oxygen to produce superoxide.

$$e^-_{(aq)} + O_2 \rightarrow O_2^{-\circ} \qquad (5)$$

Superoxide has been shown to be capable of dehalogenating some pesticides[29] and undergoes a variety of oxidation and reduction reactions with other organic and inorganic substrates[30]. But its major fate in natural waters is probably disproportionation to hydrogen peroxide.

$$2O_2^{-\circ} + 2H^+ \rightarrow H_2O_2 + O_2 \qquad (6)$$

Hydrogen peroxide has been shown to be produced photochemically in irradiated surface and groundwaters[31,32]. Hydrogen peroxide is a fairly strong oxidant. It is also easily photolized to produce the extremely strong and indiscriminant oxidant hydroxyl radical.

$$H_2O_2 \xrightarrow{\text{light}} 2OH^\circ \qquad (7)$$

Hydroxyl radicals as well as alkylperoxy radicals have been shown to be present in sunlight irradiated natural waters[33] and have been shown to be capable of breaking down several persistent pesticides[34].

Zepp _et al._[35] have recently demonstrated that energy transfer reactions can occur in natural waters. The direct triplet energy transfer from humic substances to cis and trans-1,3-pentadiene resulting in subsequent isomerization of the acceptor has been observed.

$$HS \xrightarrow{\text{light}} {}^1HS^* \rightarrow {}^3HS^* \qquad (8)$$

$${}^3HS^* + cP \rightarrow T + HS \qquad (9)$$

$${}^3HS^* + tP \rightarrow T + HS \qquad (10)$$

$$T \rightarrow cP \text{ or } tP \qquad (11)$$

Charge transfer reactions between metals, such as iron and copper, and organic ligands such as EDTA and NTA have been shown to occur at solar wavelengths[36-40]. Charge transfer reactions such as these may be involved in the photobleaching observed in humic waters[41] since environmental humics contain large quantities of complexed iron and some copper. Definitive studies should be carried out to estimate the relative importance of charge transfer compared to energy transfer reactions in natural waters.

There is some controversy whether the surface microlayer, a layer about 100 microns thick on the water surface, should be treated as bulk water or as a separate entity. The quantity of chemicals in the microlayer can vary dramatically over time[42]. The microlayer is probably important when a pollutant has been discharged to the system in a water-insoluble solvent which has a lower density than water - for example oil. Several studies have shown that photosensitized reactions do occur in oil surface films and that the reaction rates are closer to those observed in hydrocarbon solvents than to water[43,44,45]. Except under special conditions such as those encountered after a large chemical spill to the aquatic environment, photolysis in the microlayer probably does not represent a significant pathway for pollutant degradation.

Zepp and Cline[46] have presented detailed methods, using literature solar flux measurements at various latitudes, for estimating half-lives of direct photolysis for several compounds. Similar techniques are needed for computing half-lives of pollutants degraded by the indirect photolytic processes described above.

1.4 Photolysis on Suspended Sediments

As mentioned earlier, contaminants adsorbed to suspended sediments may be removed from the photic zone by settling. However, in rivers and in shallow lakes where sediment resuspension is an important process, suspended sediments can influence photolysis in several ways. They can reduce photolysis rates by competitively absorbing sunlight[47] or by quenching excited states of absorbed chemicals which normally photolyze[48]. They can enhance photolysis rates by increasing the diffuseness of incoming light[49]. Adsorption to sediments may shift light absorption bands of chemicals into the solar spectral region where photolysis could occur[50]. But the major potential indirect of photocatalytic effect of suspended sediments is probably via the semiconductor-type mechanism which will be discussed in detail latter. The decomposition of 1,4-dichlorobenzene in irradiated TiO_2 (anatase phase) slurries has been observed in the laboratory[47]. TiO_2 is found in the environment but mainly in the less photoreactive rutile phase. The decomposition of carboxylic acids has been observed in irradiated slurries of the mineral goethite (iron oxy-hydroxide)[51]. Although there are some instances of the formation of pure metal oxide and sulfide precipitates in natural waters, most suspended sediments in the aqueous environment are coated with an organic film. These films would strongly inhibit semiconductor-like reactions. Since the organic coatings on the particles contain humic substances, indirect photochemical generation of singlet oxygen and/or superoxide may occur with irradiation. However, because oxygen would have to be adsorbed or in close proximity to the particulates for these reactions to occur, the quantum yields would probably be much lower than for solution phase humics. Thus, the importance of indirect and photocatalyzed reactions on suspended sediments in the aqueous environment has yet to be demonstrated.

1.5 Photolysis in Algae

Algae may contain an important fraction of persistent organic chemicals in certain water systems. Association with algae would function to keep the contaminant in the photic zone for a longer period and this could enhance direct photolysis rates. Indirect, sensitized, photolysis of several nonionic organic chemicals has been observed by Zepp and Schlotzhauer[52]. For example, methyl parathion and parathion photoreacted 390 times more rapidly when sorbed to algae than in distilled water, and aniline and m-toluidine reacted over 12,000 times faster[52]. The most critical step in algal photolysis seems to be adsorption of the chemical by the algae[53]. Chemicals which had higher octanol-water partition coefficients and partitioned into the algae to a greater extent, showed the greatest reaction rate enhancement[53]. Some preliminary mechanistic studies by Zepp *et al.*[53] have shown that hydrogen peroxide does not appear involved in algal sensitized photoreactions.

It is of interest to note that the mode of activity of many herbicides, especially the dinitroaniline family, involves scavenging of electrons from the photosynthetic electron transfer system of the target plant. The fate of the resulting herbicide radicals has not been examined. Photochemistry in aquatic biological systems would appear to be an important area for future research.

2 APPLICATION OF PHOTOLYSIS TO WASTEWATER TREATMENT

2.1 Homogeneous Photolysis

Ultraviolet light at 254 nm has excellent germicidal properties so it has been used as a point-of-use water treatment disinfectant device and as an alternative to chlorination for disinfection of treated municipal wastewater[54, 55, 56]. But direct photolysis of organic and inorganic wastes with ultraviolet, U.V. light is not an efficient process so it is usually used in combination with sensitizers for wastewater treatment. The combination U.V. and hydrogen peroxide has been shown to destroy organics in surface[57] and wastewater[58]. The UV/ozone combination has also been used in water and wastewater treatment[59]. There are two main problems with the combination treatment methods above: 1) the high cost of generating 254 nm light and 2) the high cost of continual addition of the photochemical reactants which are consumed in the reaction. These factors make these processes expensive so, except for special applications where no other methods are available, they are not likely to be widely used.

2.2 Heterogeneous Photocatalysis

The use of sensitizers which use sunlight and are not consumed during the treatment process should provide a more cost-effective wastewater treatment process. Some work has been done on dyes chemically bonded to porous polymer beads[60,61] but the most promising approach is the use of semiconductors. Photoredox processes can occur at semiconductor interfaces as a result of the absorption of "bandgap" radiation, preferably in the solar spectral region. The immediate products of this excitation are an electron in the conduction band and an electron vacancy or "hole" left in the valence band. Both these reactive species

$$\text{Semiconductor} \xrightarrow{\text{light}} e^-_{CB} + h^+_{VB} \tag{12}$$

can migrate to the solid/solution interface and lead to charge transfer across the interface and redox reactions in solution. In order for these reactions to be significant, both the charge separation and the charge transfer processes must be efficient enough to compete with electron-hole recombination.

We tested a group of seven photostable titanium-containing semiconductors to find out which had the highest photoreactivity for organic oxidation[62]. Methanol and isopropanol were chosen as susbstrates for oxidation since they react with a wide variety of oxygen containing radicals, including those expected to result from the photooxidation of water on the metal oxide surface. In order to suppress chain reactions, ferric perchlorate was included at a concentration large enough to react with all alcohol radicals formed. Alcohol concentrations were chosen from studies of iron photochemistry[63,64] and were large enough to scavenge all hydroxyl and hydroperoxyl radicals in solution. The presence of iron allowed the solutions to be degassed which removed the possible complication of peroxyl radical formation by reaction of the alcohol radicals with dissolved oxygen. Thus the degree of alcohol oxidation can be expected to be a true measure of primary photoefficiency for each catalyst. The results of these irradiations are listed in Table III. The quantum efficiency ϕ, also called the quantum yield, is defined as

$$\phi = \frac{\text{number of molecules reacting or formed per unit time}}{\text{number of quanta absorbed per unit time}} \tag{13}$$

The results clearly show anatase to be superior to all semiconductors tested by an order of magnitude, so further studies were limited to this semiconductor. Irradiation of this semiconductor with light of wavelength less than 400 nm produces the electron/hole pair.

Table III Primary Quantum Efficiencies (ϕ) for Products of the Irradiation of Semiconductor Slurries

Semiconductor	$\phi Fe(II)$	ϕCH_2O)
TiO_2 (anatase)	.334	.076
TiO_2 (rutile)	.020	.005
$ZnTiO_3$	.010	.0001
$MgTiO_3$	.034	.0006
$CaTiO_3$	.030	.0006
$BaTiO_3$	.012	.0001
$SrTiO_3$	.020	.0005

Solution composition:
$[HClO_4]=0.5M$; $[CH_3OH]=0.494M$; $[(CH_3)_2CHOH] = 0.260M$; $[Fe^{3+}] = 0.01M$

As mentioned above, for the semiconductor pathway to be important, both electrons and holes must be efficiently separated and both must reach the particle surface and react. This requirement partially explains why many pure mineral substances that absorb visible light are not efficient photocatalysts. In many of these substances, visible light excites electrons from impurity levels located within the bandgap to the conduction band. The holes are not mobile and accumulate in the crystal until recombination. The rate of recombination, which depends on concentrations of both electrons and holes, becomes the predominant process. Even if both charge carriers reach the surface, recombination may still predominate if both species are not consumed in appropriate fast reactions. This kinetic requirement implies that the most effective reaction partners will be already present at the surface when the charge carriers arrive. The thermodynamic factors affecting redox efficiencies across semiconductor-solution interfaces have also been extensively studied and reviewed[65]. Efficient charge transfer may occur only if there is sufficient quantum mechanical overlap between the redox level in solution and the energy level of the charge carrier at the interface. Taken together, these two requirements lead to the conclusion that the most efficient processes will occur for reactions of the solvent or of adsorbed species of appropriate redox potential. Surface states may function as 'storage' traps if their redox potentials permit interfacial electron transfer.

In order for efficient photocatalysis to occur, both oxidation and reduction reactions must occur at the interface. From a practical standpoint, it is useful to consider each process separately. Consider first the conduction band electrons. They will be consumed in reactions resulting in reduction of species on the solution side of the interface. Because it occupies most of the sites at the interface, the first acceptor that must be considered is the solvent itself. The possibility of reducing water to hydrogen has stimulated a great deal of research interest because of the potential application to convert sunlight to a useable fuel. Unfortunately, water is not that easy to reduce since there are no known naturally occurring substances that can catalyze this reaction without an externally applied potential. A more promising electron acceptor is the oxygen molecule. The generation of superoxide from the one-electron reduction of the oxygen molecule has been shown to occur in humic waters[27]. Superoxide is unstable in aqueous solution, undergoing rapid disproportionation to hydrogen peroxide and oxygen. Many studies of photocatalysis at semiconductor surfaces, including the photolysis of PCBs in anatase slurries, have noted that in the absence of oxygen, no reaction could be observed[20]. Superoxide is a good nucleophile that may become involved in reactions with organic contaminants such as the dechlorination of chlorinated aromatics. In addition to this general route of electron scavenging, if there are electron acceptors present at the surface with the right energy levels, they may also be reduced.

Reactions of valence band holes in TiO_2 with the solvent water have recently been reported by Carey and Langford[66]. They irradiated 0.5% anatase aqueous slurries which were 0.1 M in perchloric acid, 0.01 M in ferric perchlorate, and 0.2 - 1.7 M in tertiary butanol or 2 to 7 M in methanol. Previous studies[63] have shown that t-butanol reacts only with hydroxyl radicals while methanol reacts with hydroxyl radicals and directly with other free radicals in the system. The primary quantum yields of iron (II) in these systems were 0.17 for t-butanol and 0.33 for methanol. These results show that about half the "holes" produce hydroxyl radicals in the solution while the other half produce weaker oxidants. Comparable scavenging studies with ethylene glycol by the same authors[66], similar to those conducted by Carey et al.[67], showed that equal quantities of the products acetaldehyde (the hydrogen abstraction product from hydroxyl radicals) and formaldehyde (the direct reaction product) were produced in yields >0.1. These results can be rationalized from surface studies of anatase. Two types of OH group exist on an hydroxylated anatase surface - one bound to a single titanium and the other bound to two titanium atoms. Hole migration to the singly bound hydroxyl likely provides the excited state that produces solution hydroxyl radicals. Hole formation at the doubly bound hydroxyl forms an excited state that leads to a weaker oxidizing radical in solution. The two types of hydroxyl groups are about equally abundant on the anatase surface in good agreement with the quantum yield studies.

The irradiation of o, m, and p-chlorobenzoic acids in anatase slurries produced chloride with an average quantum yield of 0.035. The yield of salicylic acid from irradiation of anatase slurries of benzoic acid was 0.017. These results indicate that dechlorination of these aromatics, probably via hydroxyl radical attack, is the preferential process. The photodechlorination of PCB's in anatase slurries has been reported[20]. We irradiated a saturated aqueous solution (30 ppb) of the commercial PCB mixture Aroclor 1254 containing 0.5% anatase with 350 nm light for 1 hour. The effect of the irradiated, unirradiated and distilled water control solutions on algae were measured by monitoring the optical density at 540 nm. The curves in Figure 2 show that the growth of the algae in the irradiated and distilled water control were similar, whereas, the unirradiated Arochlor solution suppressed the algal growth. Recently, Hidaka *et al.*[68] and Barbeni *et al.*[69] have demonstrated partial photodecomposition of sodium dodecylbenzene sulfonate and the total decomposition of 4-chlorophenol in anatase slurries.

These data show that heterogeneous photolysis could be used to detoxify wastewater. It is probably not necessary for the photochemical treatment to completely decompose the organics but it can be used to convert them to biodegradable compounds[62]. The treatment would be combined with a second stage biological reactor, such as the activated sludge process, to provide the final conversion to CO_2. The most promising types of wastewater for potential heterogeneous photochemical treatment are low volume, fairly concentrated industrial wastewaters containing organics which are not biodegradable. Another potential application is the treatment of leachates from chemical dump sites. In North America there are thousands of dump sites which contain large quantities of chlorinated solvents and chlorinated aromatics such as PCB's. These dump sites are currently polluting groundwater and surface waters throughtout the continent. One current method of alleviating this problem is to pump the leachate from below the dump to the surface for treatment. This contaminated water may prove amenable to combined photochemical/biological treatment.

Inorganic pollutants such as cyanide have also been shown to degrade via semiconductor photodegradation processes[70]. Figure 3 shows a plot of CN^- concentration versus irradiation time at 350 nm for a KCN solution containing 0.5% anatase. The likely reaction mechanism for cyanide disappearance is:

$$CN^- + 2OH^- + 2h^+ \rightarrow CNO^- + H_2O \qquad (13)$$

Cyanide is a problem in gold mine effluents since it is used to extract the gold from the ore. The mine wastewater also contains cyanide complexes such as ferricyanide which are difficult to break down with chemical treatment. Wastewater containing these cyanide complexes can be completely nontoxic to biota when first discharged to the aquatic environment. But it can exhibit a phototoxic effect downstream when sunlight photolyzes the complexes and releases free cyanide. Fish kills have been observed several miles downstream of some gold mine discharges. The TiO_2 photolysis process has been shown to be capable of destroying complexed "cyanide" as well as "free" cyanide[62]. Some recent studies explored the economics of this process for treatment of gold mill effluents[71].

In summary, the treatment of wastewater containing both organic and inorganic pollutants using heterogeneous photocatalysis would seem to have considerable potential. More studies are required to elucidate the reaction mechanisms, to design treatment units and to examine the economics of the process.

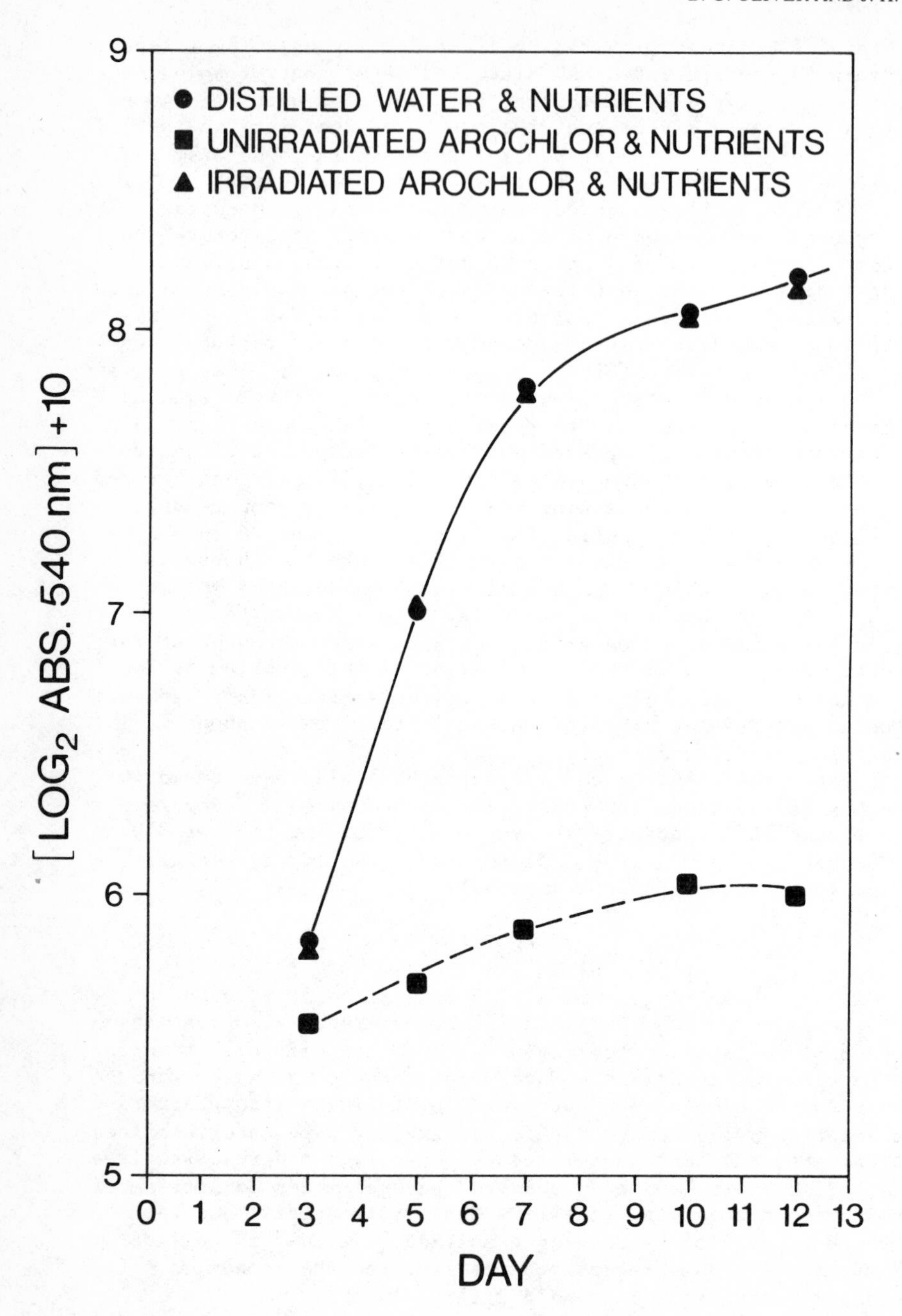

Figure 2 The effect of PCB's and PCB photoproducts on the growth of *Scenedesmus quadricauda*.

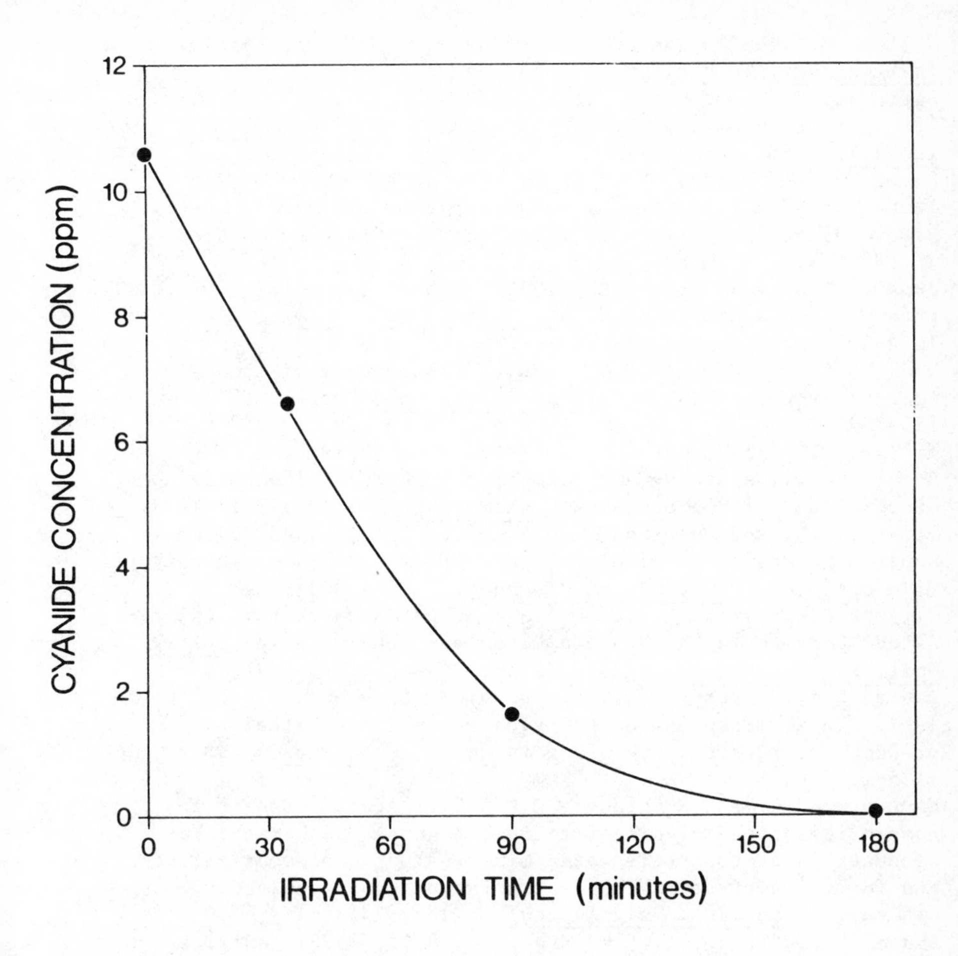

Figure 3 Irradiation of a 0.5% TiO_2 slurry of a 10.6 ppm KCN, 0.01 M NaOH solution with 350 nm light.

REFERENCES

1. Miller, S. 1983. Photochemistry of natural water systems. Environ. Sci. Technol. **17** : 568A-570A.
2. Zafiriou, O.C. Joussot-Dubien, J., Zepp, R.G., Zika, R.G. 1984. Photochemistry of natural waters. Environ. Sci. Technol. **18** : 358A-371A.
3. Carey, J.H. and Fox, M.E. 1981. Photodegradation of the lampricide 3-trifluoromethyl-4-nitrophenol (TFM). 1. Pathway of the direct photolysis in solution. J. Great Lakes Res. **7** : 234-241.
4. Crosby, D.G. and Wong, A.S. 1973. Photodecomposition of p-chlorophenoxyacetic acid. J. Agric. Food. Chem. **21** : 1049-1052.
5. Wong, A.S. and Crosby, D.G. 1981. Photodecomposition of pentachlorophenol in water. J. Agric. Food Chem. **29** : 125-130.
6. Miller, G.C. and Crosby, D.G. 1983. Photooxidation of 4-chloroaniline and N-(4-chlorophenyl)-benzenesulfonamide to nitroso- and nitro- products. Chemosphere. **12** : 1217-1227.
7. Oliver, B.G. and Charlton, M.N. 1984. Chlorinated contaminants on settling particulates in the Niagara River vicinity of Lake Ontario. Environ. Sci. Technol. **18** : 903-908.
8. Oliver, B.G. 1984. Distribution and pathways of some chlorinated benzenes in the Niagara River and Lake Ontario. Water Poll. Res. J. Canada. **19** : 47-58.
9. David, E.M., Murray, H.E., Liehr, J.G., and Powers, E.L. 1981. Basic microbial degradation rates and chemical byproducts of selected organic compounds. Water Res. **15** : 1125-1127.
10. Choudhry, G.G. and Hutzinger, O. 1984. Acetone-sensitized and nonsensitized photolysis of tetra-, penta-, and hexachlorobenzenes in acetonitrile-water mixtures: photoisomerization and formation of several products including polychlorobiphenyls. Environ. Sci. Technol. **18** : 235-241.
11. Bunce, N.J., Kumar, Y., and Brownlee, B.G. 1978. An assessment of the impact of solar degradation of polychlorinated biphenyls in the aquatic environment. Chemosphere. **7** : 155-164.
12. Karickhoff, S.W. 1981. Semi-empirical estimation of sorption of hydrophobic pollutants on natural sediments and soils. Chemosphere. **10** : 833-846.
13. McCrea, R.C. 1982. Development of an aqueous phase liquid-liquid extractor (APLE). Interim Report. Inland Waters Directorate, Ontario Region, Water Quality Branch, Burlington, Ontario.
14. Oliver, B.G. and Nicol, K.D. 1982. Gas chromatographic determination of chlorobenzenes and other chlorinated hydrocarbons in environmental samples using fused silica capillary columns. Chromatographia. **16** : 336-340.

15. Woodrow, J.E., Crosby, D.G., and Seiber, J.N. 1983. Vapor-phase photochemistry of pesticides. Residue Rev. **85** : 111-125.
16. Atlas, E. and Giam, L.S. 1981. Global transport of organic pollutants: ambient concentrations in the remote marine atmosphere. Science. **211** : 163-165.
17. Swain, W. 1978. Chlorinated organic residues in fish, water and precipitation from the vicinity of Isle Royale, Lake Superior. J. Great Lakes Res. **4** : 398-407.
18. Chameides, W.L. and Davis, D.D. 1982. The free radical chemistry of cloud droplets and its impact upon the composition of rain. J. Geophys. Res. **87** : 4863-4877.
19. Zika, R., Saltzman, E., Chameides, W.L., and Davis, D.D. 1982. H_2O_2 levels in rainwater collected in south Florida and the Bahama Islands. J. Geophys. Res. **87** : 5015-5017.
20. Carey, J.H., Lawrence, J., and Tosine, H.M. 1976. Photodechlorination of PCB's in the presence of titanium dioxide in aqueous suspensions. Bull. Environ. Contam. Toxicol. **16** : 697-701.
21. Mullin, M.D., Pochini, C.M., McCrindle, S., Romkes, M., Safe, S.H., and Safe, L.M. 1984. High-resolution PCB analysis: synthesis and chromatographic properties of all 209 PCB congeners. Environ. Sci. Technol. **18** : 468-476.
22. Thurman, E.M. and Malcolm, R.L. 1981. Preparative isolation of aquatic humic substances. Environ. Sci. Technol. **15** : 463-466.
23. Zepp, R.G., Wolfe, N.L., Baughman, G.L., and Hollis, R.C. 1977. Singlet oxygen in natural waters. Nature. **267** : 421-423.
24. Baxter, R.M. and Carey, J.H. 1982. Reactions of singlet oxygen in humic waters. Freshwat. Biol. **12** : 285-292.
25. Wolff, C.J.M., Halmans, M.T.H., and Van der Heijde, H.B. 1981. The formation of singlet oxygen in surface water. Chemosphere. **10** : 59-62.
26. Foote, C.S., Wuesthoff, M.T., Wexler, S., Burstain, I.G., Denny, R., Schenck, G.O., and Schulte-Elte, K.H. 1967. Photosensitized oxygenations of alkylsubstituted furans. Tetrahedron. **23** : 2583-2599.
27. Baxter, R.M. and Carey, J.H. 1983. Evidence for the photochemical generation of superoxide ion in humic waters. Nature. **306** : 575-576.
28. Fischer, A., Kliger, D., Winterle, J., and Mill, T. 1985. Primary photochemical processes in photolysis mediated by humic substances. Presented before the Division of Environmental Chemistry, American Chemical Society, Miami, Florida.
29. Dureja, P., Casida, J.E., and Ruzo, L.O. 1982. Superoxide-mediated dehydrohalogenation reactions of the pyrethroid permethrin and other chlorinated pesticides. Tetrahedron Lett. **23** : 5003-5004.
30. Lee-Ruff, E. 1977. The organic chemistry of superoxide. Chem. Soc. Rev. **6** : 195-214.

31. Draper, W.M. and Crosby, D.G. 1983. Photochemical generation of hydrogen peroxide in natural waters. Arch. Environ. Contam. Toxicol. **12** : 121-126.
32. Cooper, W.J. and Zika, R.G. 1983. Photochemical formation of hydrogen peroxide in surface and ground waters exposed to sunlight. Science. **220** : 711-712.
33. Mill, T., Hendry, D.G., and Richardson, H. 1980. Free-radical oxidants in natural waters. Science. **207** : 886-887.
34. Draper, W.M. and Crosby, D.G. 1984. Solar photooxidation of pesticides in dilute hydrogen peroxide. J. Agric. Food Chem. **32** : 231-237.
35. Zepp, R.G., Scholtzhauer, P.F., and Sink, R.M. 1985. Photosensitized transformation involving electronic energy transfer in natural waters: role of humic substances. Environ. Sci. Technol. **19** : 74-81.
36. Trott, T., Henwood, R.W., and Langford, C.H. 1972. Sunlight photochemistry of ferric nitrilotriacetate complexes. Environ. Sci. Technol. **6** : 367-368.
37. Langford, C.H., Wingham, M., and Sastri, V.S. 1973. Ligand photooxidation in copper (II) complexes of nitrilotriacetic acid: implications for natural waters. Environ. Sci. Technol. **7** : 820-822.
38. Carey, J.H. and Langford, C.H. 1973. Photodecomposition of Fe (III) aminopolycarboxylates. Can. J. Chem. **51** : 3665-3670.
39. Stolzberg, R.J. and Hume, D.N. 1975. Rapid formation of iminodiacetate from photochemical degradation of Fe (III) nitrilotriacetate solutions. Environ. Sci. Technol. **9** : 654-656.
40. Lockhart, H.B. Jr. and Blakeley, R.V. 1975. Aerobic photodegradation of Fe (III)-(ethylenedinitrilo) tetraacetate (ferric EDTA). Environ. Sci. Technol. **9** : 1035-1038.
41. Miles, C.J. and Brezonik, P.L. 1981. Oxygen consumption in humic-colored waters by a photochemical ferrous-ferric catalytic cycle. Environ. Sci. Technol. **15** : 1089-1095.
42. Maguire, R.J. Kuntz, K.W., and Hale, E.J. 1983. Chlorinated hydrocarbons in the surface microlayer of the Niagara River. J. Great Lakes Res. **9** : 281-286.
43. Larson, R.A., Bott, T.L., Hunt, L.L., and Rogenmuser, K. 1979. Photooxidation products of a fuel oil and their antimicrobial activity. Environ. Sci. Technol. **13** : 965-969.
44. Patel, J.R., Politzer, I.R., Griffin, G.W., and Laseter, J.L. 1978. Mass spectra of the oxygenated products generated from phenanthrene under simulated environmental conditions. Biomed. Mass Spectros. **5** : 664-670.
45. Larson, R.A. and Rounds, S. 1985. Photochemistry in aquatic surface microlayers. Presented before the Division of Environmental Chemistry, American Chemical Society, Miami, Florida.
46. Zepp, R.G. and Cline, D.M. 1977. Rates of direct photolysis in aquatic environment. Environ. Sci. Technol. **11** : 359-366.

47. Oliver, B.G., Cosgrove, E.G., and Carey, J.H. 1979. Effect of suspended sediments on the photolysis of organics in water. Environ. Sci. Technol. **13** : 1075-1077.
48. Zepp, R.G. and Schlotzhauer, P.F. 1981. Effects of equilibrium time on photoreactivity of the pollutant DDE sorbed on natural sediments. Chemosphere. **10** : 453-460.
49. Miller, G.C. and Zepp, R.G. 1979. Effects of suspended sediments on photolysis rates of dissolved pollutants. Water Res. **13** : 453-459.
50. Leermakers, P.A. and Thomas, H.T. 1965. Electronic spectra and photochemistry of adsorbed organic molecules. 1. Spectra of ketones on silica gel. J. Amer. Chem. Soc. **87** : 1620-1622.
51. Goldberg, M.C. 1985. Aquatic photolysis of oxy-organic compounds adsorbed on goethite. Presented before the Division of Environmental Chemistry, American Chemical Society, Miami, Florida.
52. Zepp, R.G. and Schlotzhauer, P.F. 1983. Influence of algae on photolysis rates of chemicals in water. Environ. Sci. Technol. **17** : 462-468.
53. Zepp, R.G., Schlotzhauer, P.F., and Gordon, J.A. 1985. Photobiological transformations of xenobiotics in natural waters: role of microalgae. Presented before the Division of Environmental Chemistry, American Chemical Society, Miami, Florida.
54. Oliver, B.G. and Cosgrove, E.G. 1975. The disinfection of sewage treatment plant effluents using ultraviolet light. Can. J. Chem. Eng. **53** : 170-174.
55. Oliver, B.G. and Carey, J.H. 1976. Ultraviolet disinfection: an alternative to chlorination. J. Water Poll. Control Fed. **48** : 2619-2624.
56. Whitby, G.E., Palmateer, G., Cook, W.G., Maarschalkerweerd, J., Huber, D., and Flood, K. 1984. Ultraviolet disinfection of secondary effluent. J. Water Poll. Control Fed. **56** : 844-850.
57. Malaiyandi, M., Sadar, M.H., Lee, P., and O'Grady, R. 1980. Removal of organics in water using hydrogen peroxide in presence of ultraviolet light. Water Res. **14** : 1131-1135.
58. Koubek, E. 1975. Photochemically induced oxidation of refractory organics with hydrogen peroxide. Ind. Eng. Chem. Process Des. Dev. **14** : 348-350.
59. Peyton, G.R. and Glaze, W.H. 1985. The mechanism of photolytic ozonation. Presented before the Division of Environmental Chemistry, American Chemical Society, Miami, Florida.
60. Nilsson, R. and Kearns, D.R. 1974. Some useful heterogeneous systems for photosensitized generation of singlet oxygen. Photochem. Phtobiol. **19** : 181-184.
61. Seely, G.R. and Hart, R.L. 1977. Preparation of stained alginate beads for photosensitized oxidation of organic pollutants. Environ. Sci. Technol. **11** : 623-625.

62. Carey, J.H. and Oliver, B.G. 1980. The photochemical treatment of wastewater by ultraviolet irradiation of semiconductors. Water Poll. Res. J. of Canada. **15** : 157-185.
63. Langford, C.H. and Carey, J.H. 1975. The charge transfer photochemistry of the hexaaquoiron (III) ion, the chloropentaaquo (III) ion, and the μ-dihydroxo dimer explored with tertbutyl alcohol scavenging. Can. J. Chem. **53** : 2430-2435.
64. Carey, J.H. and Langford, C.H. 1975. Outer sphere oxidations of alcohols and formic acid by charge transfer excited states of iron (III) species. Can. J. Chem. **53** : 2436-2440.
65. Gerischer, H. 1977. On the stability of semiconductor electrodes against photodecomposition. J. Electroanal. Chem. **82** : 133-143.
66. Carey, J.H. and Langford, C.H. 1985. Photocatalysis by inorganic components of natural waters. Presented before the American Chemical Society, Miami, Florida.
67. Carey, J.H., Cosgrove, E.G., and Oliver, B.G. 1977. The photolysis of hexaaqroiron (III) perchlorate in the presence of ethylene glycol. Can. J. Chem. **55** : 625-629.
68. Hidaka, H., Kubota, H., Grätzel, M., Serpone, N., and Pelizzetti, E. 1985. Photodegradation of surfactants. I: Degradation of sodium dodecylbenzene sulfonate in aqueous semiconductor dispersions. Nouv. J. Chimie. In press.
69. Barbeni, M., Pramauro, E., Pelizzetti, E., Borgarello, E., Grätzel, M., and Serpone, N. 1985. Photodegradation of 4-chlorphenol catalyzed by titanium dioxide particles. Nouv. J. Chimie. In press.
70. Frank, S.N. and Bard, A.J. 1977. Semiconductor electrodes. II. Photoassisted oxidations and photoelectrosynthesis at polycrystalline TiO_2 electrodes. J. Amer. Chem. Soc. **99** : 4667-4675.
71. Carey, J.H. and Zaidi, S.A. 1985. The use of low pressure mercury lamps for the photodecomposition of iron cyanide in gold mill effluents. Can. J. Water Poll. Res. In press.

FOX - (1) Would not TiO_2 be coated by humic acid which would inhibit photoactivity? (2) Is photocatalysis competitive with filtering (e.g., activated carbon) or ion exchange columns?

OLIVER - (1) This would probably happen eventually but usually the water to be treated would contain low concentrations of natural organics. Some type of heat treatment would probably be required to restore the photoactivity of the semiconductor after some period of use for waste water treatment. (2) Adsorption and ion-exchange processes remove the contaminant from the water but do not destroy it. For example, if you adsorb PCB's on activated carbon, at some stage you would have to heat-treat the carbon to destroy the PCB's. If this is not done under proper conditions, you could wind up with a more toxic product (dibenzodioxins or dibenzofurans). The photocatalytic process dechlorinates the PCB's and renders them non-toxic.

MUNUERA - You presented some results using ethanol and t-butanol. The yield was about half for the second alcohol. I would like to draw your attention to the fact that maximum coverage for t-butanol on anatase is about half of that of ethanol. The different yield can be alternatively explained on this basis without assuming any change of mechanism.

OLIVER - I still feel our results indicate two reactive sites on TiO_2. Two products, formaldehyde and acetaldehyde, are generated when ethylene glycol is used as the scavenger. This result indicates that two types of free radicals are produced at the TiO_2 surface by reactions similar to those found in the irradiation of hexaaquoiron(III) complexes.

KIWI - Why has not Fe_2O_3 been used more to destroy the wastes of water in circumstances where it absorbs visible light up to 570 nm in contrast to the much used TiO_2 which only absorbs light up to 400 nm?

OLIVER - The use of other semiconductors with lower bandgap energies than TiO_2 should be explored, since they may be useful for waste treatment for certain chemicals. This research area may provide useful future results, but the redox chemistry of the reactions in the wastewater must also be carefully considered.

McLENDON - (1) How does one minimize competitive reactions of

$$TiO_2 + \text{humates} \longrightarrow (\text{humates})_{ox}$$

(2) How do you deal with the problem of $R + TiO_2/O_2 \longrightarrow R_{ox}$ where R_{ox} is an ultimate carcinogen. Such reactions are very well known in biology. Indeed most aromatic heterocycles owe their toxicity to oxygenation by cyt P450.

OLIVER - (1) This competition would be difficult to deal with. A prerequisite for water treatable by the TiO_2 process would likely be the presence of very low concentrations of naturally occurring organics. (2) The major process we observe in using TiO_2 for treating chlorinated organics is reductive dechlorination. In most instances, direct and indirect photolysis yields products that are less toxic than parent material. In the small percentage of cases where more toxic products are produced by photolysis, this method would not be useful for treatment.

SOMORJAI - Have minimum concentrations been established for some or most of the chemicals that pollute water that are (i) life threatening or (ii) biologically harmful? Has any thought been given to concentrating the chemicals that are to be removed by adsorption to improve on the economy of cleaning processes? Adsorption on active carbon or on zeolites come to mind.

OLIVER - (1) A considerable amount of work is conducted on feeding rats and mice chemicals to see how toxic and carcinogenic they are. This work is very expensive, so only a small percentage of compounds have been tested. For chemicals that have been tested, limits for food and water have been established. However, little is known about the health effects of exposure to hundreds of chemicals at low concentration (synergestic and antagonistic effects). (2) As a basic environmental rule, chemicals that are persistent in the environment should not be discharged to the environment. These chemicals should be recovered at the source and destroyed. If such chemicals are released, they become dispersed throughout the ecosystem and usually become a problem. Discharge limits can be set for chemicals that are biodegradable or photodegradable since the rate of disappearance from the ecosystem can be predicted.

HETEROGENEOUS PHOTOCATALYSIS FOR WATER PURIFICATION: PROSPECTS AND PROBLEMS

D. F. Ollis
Chemical Engineering Department
North Carolina State University
Raleigh, NC 27695-7905, USA

ABSTRACT. Heterogeneous photocatalysis has demonstrated the complete wet oxidation (mineralization) of a number of common contaminants of water supplies, especially halogenated alkanes, alkenes, and aromatics. We review these recent results, including influence of reactant concentration, product inhibition, intermediate formation and disappearance, synthetic vs. solar illumination, and kinetic rate forms, and discuss prospects and problems for future essays.

1. INTRODUCTION

Contamination of surface and well water supplies by halogenated organic compounds has been noted with increasing frequency in many populated areas over the last decade. Such contamination may arise from direct industrial discharges, from leakage from landfill or other land disposal sites or from underground solvent storage tanks. Chlorinated hydrocarbons are also produced during conventional drinking water chlorination procedures. As TiO_2 was early on identified as a photocatalyst with dechlorination activity, it has been examined as a potential photocatalyst for water purification by destruction of halogenated hydrocarbon contaminants.

The earliest reports of photon-driven degradation activity of TiO_2 included conversions of interest for water contaminant oxidations: oxidation of CO (1,2), of cyanide (3,4), of sulfite (4), decarboxylations of carboxylic acids (5,6). Further, chloride release from chlorinated hydrocarbons was noted in the presence of illuminated TiO_2 for chlorinated biphenyls (7), and p-dichlorobenzene (8), Illuminated ZnO powders led to demonstrated chloride transfer to ethane from tetrahalomethanes (CCl_4, $CFCl_3$, and CF_2Cl_2) (9) and dechlorination of CF_2Cl_2 and $CFCl_3$ (10).

An early review of reaction data associated with illuminated semiconductors established that a reasonable number of reports had concerned truly photocatalyzed reactions, as judged by the (arbitrary) criterion of demonstrating more than 100 monolayer equivalents of reactant conversion without noticeable change of the illuminated solid (11).

E. Pelizzetti and N. Serpone (eds.), Homogeneous and Heterogeneous Photocatalysis, 651–656.

2. REACTANT CONVERSIONS (12)

2.1 Halomethanes

Aqueous solutions of 10-100 ppm of CH_2Cl_2 (13), $CHCl_3$ (14), or CCl_4 (13) may be completely mineralized to HCl and CO_2. Results with $CHCl_3$, the most common product of water chlorination practices (15-17) and a suspected carcinogen (18), were supportive of a potential water treatment procedure. Upon sufficient exposure of 10-100 ppm solutions of chloroform to illuminated TiO_2, no reactant remained (within detection limits of 1 ppm), no other halocarbon by-products were noted, and a 100% chlorine recovery as (H)Cl was demonstrated, as was the presence of CO_2 (14). Further, this conversion process was shown to be photocatalytic by the previously mentioned criterion (11).

The bromomethanes CH_2Br_2 and $CHBr_3$ are completely mineralized by photocatalysis with TiO_2. The relative reactivities over TiO_2, seen from the corresponding rate constants, k, in a Langmuir-Hinshelwood rate form (Table 1), are: $CHBr_3 > CHCl_3 > CH_2Br_2 > CH_2Cl_2 >> CCl_4$.

2.2 Haloethanes, haloethylenes, and haloacetic acids

Two-carbon species, especially the industrial solvents trichloroethylene (TCE) and perchloroethylene (PCE) as well as dichloroethane and chloroacetic acids, have been noted in halocarbon contaminated waters (19-21). The first three contaminants were completely mineralized to HCl and CO_2 upon sufficient contact with illuminated TiO_2, as were dichloro- and monochloroacetic acids (21-23). Intermediates were identified in two cases: Trichloroethylene yielded dichloroacetaldehyde, and dichloroethane produced traces of vinyl chloride. These intermediates were subsequently completely degraded. The importance of such intermediates is discussed below in section 3.

Both 1,2-dibromoethane (also known as ethylene dibromide or EDB) and its (1,1) isomer were completely degraded, as demonstrated by 100% recovery of initial bromine as bromide ion when reactant had dropped to an undetectable level (25). EDB produced a trace intermediate, vinyl bromide, which was subsequently degraded.

2.3 Chloroaromatic Conversions

Monochlorobenzene was not mineralized in our reactor with illuminated TiO_2. GC/MS study of reacted solutions indicated instead the conversion of monochlorobenzene to other aromatics, in the sequence chlorobenzene to chlorophenol to hydroquinone to quinone (22). The first two steps appear to be heterogeneously photocatalyzed; the last step may have homogeneous as well as heterogeneous paths.

Other investigators have noted the mineralization of haloaromatics over illuminated TiO_2. 4-chlorophenol conversion was shown by Barbeni, Pramauro, Pelizzetti, Borgarello, Gratzel, and Serpone (25), and

Matthews (26) has found chlorobenzene mineralization. The apparent discrepancy between our study (22) and others (25-26) may be due to catalyst or dissolved oxygen levels. The complete degradation of two chloroaromatics has, however, been demonstrated by these last two papers (25-26).

The above results demonstrate the complete mineralization of a variety of haloalkanes, haloalkenes, and haloaromatics, including all of the most common water supply halocarbon contaminants reported in Europe and the USA.

3. INTERMEDIATES

Any water treatment process must yield an effluent stream with (i) the initial contaminant level is greatly reduced and (ii) no appreciable resultant product of a hazardous or objectionable nature. Both (1,2) dibromoethane (EDB) and (1,2) dibromoethane conversions above were found to yield trace levels of vinyl halide, which in each case was subsequently rapidly degraded. The appearance of a known carcinogen (vinyl chloride) as a trace intermediate provides a challenge for future work, it is possible that higher dissolved oxygen levels or other catalysts may reduce these trace contaminants to acceptable levels.

Similarly, our identification via GC/MS of chlorophenol from chlorobenzene indicates the potential production of a compound with objectionable taste. The later report of 4-chlorophenol degradation (25) suggests that satisfactory destruction of such distasteful intermediates (of chlorobenzene conversion) may be achievable photocatalytically.

4. INHIBITORS

Rate inhibition by haloacid products (HCl, HBr) was noticed at 20-50 ppm. While these effects must be included in a kinetic analysis of data (or avoided by use of initial rate data), these observed levels of inhibition should be insignificant in an actual water source purification application, as surface and well water contamination levels (and mandated US maximum allowable levels) are typically in the 10-50 ppb range. Thus, photocatalytic treatment of a marginal supply for drinking water would not be expected to generate an inhibitory level of haloacid product.

5. PLATINUM METALLIZATION

Howard (27) has examined the influence of platinum loading and subsequent hydrogen pretreatment on the rate of chloroform conversion. Platinum loadings at several tenths of a percent, pretreated by 480°C reduction in hydrogen, yielded increased rates vs. untreated TiO_2, and 3.7% platinum/TiO_2 gave approximately a 50% rate reduction. This behavior closely parallels that noted for other Pt/TiO_2 conversions, including those noted by Pichat (28) for methanol and 1-propanol dehydrogenation, cyclopentane-denterium isotopic exchange, and oxygen isotope heteroexchange.

However, virtually the same rate enhancement was obtained by hydrogen reduction at 480°C of unplatinized TiO_2 (27). Thus, for halocarbon conversion, platinization at low loadings (<1%) per se does not lead to rate enhancement (vs. identically treated TiO_2), and higher metal loadings are deleterious to activity.

6. SOLAR vs. SYNTHETIC ILLUMINATION

While the ground level solar spectrum contains only the order of 1% near-UV photons (photons of energy sufficient for TiO_2 photo-excitation), the complete mineralization of a 50 ppm solution of TCE in 2-3 hours was demonstrated in a quartz recirculating reactor with 0.3% TiO_2 (just sufficient for opacity) (29). Interestingly, the rate was approximately constant from 3 hours after sunrise to about 3 hours before sunset, apparently reflecting the nearly constant scattered solar "blue" flux which has about a 10% variation over this interval (30).

7. KINETICS

All of the halocarbon degradations examined provided initial rate data which could be described by a simple Langmuir-Hinshelwood model, equation (1):

$$\text{rate} = r = \frac{k \cdot K\ (\text{reactant})}{1 + K\ (\text{reactant})} \tag{1}$$

Plots of reciprocal rate vs. reciprocal reactant concentration were linear, as required by the reciprocal of equation (1):

$$\frac{1}{r} = \frac{1}{k\ K\ (\text{reactant})} + \frac{1}{k} \tag{2}$$

The rate parameters k and K for the various reactants are given in Table 1.

These studies were typically carried out at 10-100 ppm reactant. At the levels of 10-50 ppb which characterize drinking water standards for many halocarbons, the apparent reaction rate would be linear (assuming no other reaction paths appear at lower concentrations). Thus the apparent rate would be first order, given by equation (3):

$$\text{rate} = r = k\ K\ (\text{reactant}) \tag{3}$$

From this equation, future challenges are evident. The rate is first order in reactant; thus conversions of trace contaminants will be

Table 1 Rate Parameters for Halocarbon Mineralization (12)

Reactant	k (ppm/min-g(catalyst))	K (ppm^{-1})
CH_2Cl_2	1.6	0.02
$CHCl_3$	4.4	0.003
CCl_4	0.18	0.005
CH_2Br_2	4.1	0.02
$CHBr_3$	6.2	0.01
$Cl_2C{=}CClH$	830	0.01
$Cl_2C{=}CCl_2$	6.8	0.02
$H_2ClCCClH_2$	1.1	0.01
$CHBr_2CH_3$	3.9	0.02
CH_2BrCH_2Br	2.2	0.02
$H_2ClCCOOH$	5.5	0.002
HCl_2CCOOH	8.5	0.003
Cl_3CCOOH	∿ 0	-

slow. Further research will need to address means for increasing catalyst activity (k) and reactant binding (K), and thus catalyst composition and pretreatment.

8. CONCLUSIONS

Heterogeneous photocatalysis has demonstrated the ability to completely destroy a number of halogenated alkanes, alkenes, and aromatics which are commonly found in halocarbons contaminated water supplies. Further improvements, needed to increase the application potential of this technique, must address dissolved oxygen (or other oxidant) levels, catalyst activity and reactant binding, elimination of undesirable intermediates, and solar vs. synthetic sources.

9. REFERENCES

1. Murphys, W. R., Veerkamp, R. F., and Leland, T. W., J. Catalysis, 43, 304 (1976).
2. Juillet, F., LeComte, F., Mozzenega, H., Teichner, S. J., Thevenet, A., and Vergnon, P., Faraday Symp., Chem. Soc., 7, 57 (1973).
3. Frank, S. W., and Bard, A. J., J. Amer. Chem. Soc., 99, 303 (1977).
4. Frank. S. W., and Bard, A. J., J. Phys. Chem., 81, (15), 1484 (1977).
5. Krautler, B., and Bard, A. J., J. Amer. Chem. Soc., 100, 2239 (1978).

6. Krautler, B. and Bard, A. J., J. Amer. Chem. Soc., 100, 5985 (1978).
7. Corey, J. H., Lawrence, J., and Tosine, H. M., Bull. Environ. Contam. Toxicol., 16, 697 (1976).
8. Oliver, B. G., Cosgrove, E. G., and Corey, J. H., Environ. Sci. Tech., 13, (9), 1075 (1979).
9. Ausloos, P., Rebbert, R. F., and Glasgow, L., J. Res. NBS, 82, 1 (1977).
10. Filby, W. G., Mintos, M., and Gusten, H., Ber. Bunsenges Phys. Chem., 85, 189 (1981).
11. Childs, L. P., and Ollis, D. F., J. Catal., 66, 383 (1980).
12. Ollis, D., Environ. Sci. Technol., 19, 480 (1985).
13. Hsiao, C.-Y., Lee, C.-L. and Ollis, D. F., J. Catal., 82, 418 (1983).
14. Pruden, A. L. and Ollis, D. F., Environ. Sci. Tech., 17, 628 (1983).
15. Rook, J. J., Water Treatment Exam., 23, 234 (1974).
16. Bellar, T. A., Lichtenberg, J. J., and Kramer, R. D., J. Am. Water Works Ass'n., 66, (12), 703 (1974).
17. Oliver, B. G., and Lawrence, J., J. Am. Water Works Ass'n., 71. 161, (1979).
18. Ann. Report on Carcinogenesis of Chloroform, Nat'l. Cancer Institute, Bethesda, MD (3/1/76).
19. Brass, H. J., J. Am. Water Works Ass'n., 74, 107 (1982).
20. Symons, J. M., et al., J. Am. Water Works Ass'n., 67, 634 (1975).
21. Pruden, A. L., and Ollis, D. F., J. Catal., 82, 404 (1983).
22. Ollis, D. F., Hsiao, C.-Y., Budiman, L., and Lee, C.-L., J. Catal., 88, 96 (1984).
23. Gauron, M., MS Thesis, Univ. Calif., Davis, CA, 1983.
24. Nguyen, T. and Ollis, D. F., J. Phys. Chem., 88, 3386 (1984).
25. Barbeni, M., Pramauro, E., Pelizzetti, E., Borgarello, E., Gratzel, M., and Serpone, N., Noveau Journal de Chimie, 8, (8/9), 547, 1984.
26. Matthews, R., J. Catalysis, (in press).
27. Howard, T., MS Thesis, Univ. of Calif, Davis, CA, 1984.
28. Pichat, P., in Photoelectrochemistry, Photocatalysis, and Photoreactors: Fundamentals and Developments, M. Schiavello (ed.), (NATO ASI Series C, Vol. 146), D. Reidel Publishing Co., Dordrecht, Holland, 1985.
29. Ahmed, S., and Ollis, D. F., Solar Energy, 32, 597 (1984).
30. Meinel, A. B. and Meinel, M. P., Applied Solar Energy, Addison-Wesley Publishing Co., Reading, MA, p. 63, 1976.

A GAS RESEARCH INSTITUTE PERSPECTIVE ON INORGANIC SYNTHESIS OF GASEOUS FUELS

K. Krist and R. V. Serauskas
Gas Research Institute
8600 W. Bryn Mawr Ave.
Chicago, IL 60631
U. S. A.

ABSTRACT. The incentives, concepts and status for research on light-driven synthesis of gaseous fuels from inorganic materials are discussed from the viewpoint of a basic research program at the Gas Research Institute. The investigation of these systems has an important role to play in the effective development of new energy supply technologies. However, significant technical advances will be required for their implementation. The key issues for discrete and integrated photovoltaic-electrochemical systems, for which hydrogen production is technically feasible, concern cost, relationship to solar-electric power distribution, and extension to methane production. The key issues for electrolyte junction and photocatalytic concepts concern the ability to demonstrate technical feasibility and the characterization of low-cost photochemical reactors.

1. INCENTIVES

The synthesis of gaseous fuels from common, well-defined inorganic materials is being investigated by the Basic Research Department of the Gas Research Institute as a possible long-term substitute for natural gas. Solar radiation among other primary energy sources may be used to drive fuel-formation from these energy poor substances.

Without new supplements, a decline of natural gas production in the lower 48 states to about 40% of demand by 2010 is forecast.[1] Although other imported gas options are available, and promising developments are occurring in enhanced conventional and unconventional natural gas supply technologies, new approaches to gas supply need to be investigated because of uncertainty about either the feasibility, cost, resource size, or environmental effects of various alternatives. The possibility that a future economically attractive inorganic-based method might expand the market for gaseous fuels or enable gaseous fuel production to compete with the advent of inexpensive solar-electric power production also cannot be excluded, although many advances would be required.

E. Pelizzetti and N. Serpone (eds.), Homogeneous and Heterogeneous Photocatalysis, 657–671.

Certain inorganic materials such as H_2O, CO_2, and N_2 could provide for essentially inexhaustible supplies of gaseous fuels. H_2S and mineral carbonates could also be very significant sources. Unlike coal and biomass as synthetic gas feedstocks, these materials are simple and well-defined. Such properties may help to significantly improve fuel production rates for a given process. For example, the rate of methane production in unoptimized stirred fermentors grown on H_2/CO_2 are 3-200 times faster than the rates in digestion systems for farm wastes, sugar beet wastes, municipal refuse, and cow manure.[2] These desirable feedstock properties also greatly facilitate fundamental mechanistic research. The relative ease with which fundamental studies can be performed and the new spectroscopic and surface science techniques that are becoming available to study well-defined fuel reaction mechanisms, may lead to significant performance improvements.

While the environmental impact of these systems (for example, due to wastes associated with the fabrication of solar grade materials) needs further evaluation, the clean feedstocks and capability for remote location suggest that they may be relatively free from the environmental or resource management issues associated with other synthetic gas technologies. Acid rain, particulate emissions, and atmospheric CO_2 might be minimized, and possible competition with agriculture and aquaculture for arable land and fresh water avoided.

Apart from advancing the possible development of practical fuel gas production technology, this new catalytic fuel chemistry research has crosscutting significance for a number of gas technologies. To a significant degree, the science associated with the catalysis of small molecule, multiequivalent redox reactions of inorganic synthesis applies to organic gas synthesis reactions as well as fuel cell, catalytic combustion, and certain gas separations reactions. Two important areas of overlap concern catalyst development and characterization of interfaces at which reactions are occurring. Moreover, because detailed mechanistic research can be performed, inorganic systems serve as useful models of gas synthesis. The catalytic chemistry in solar-driven systems is also relevant to systems driven by other primary energy sources.

Applications of inorganic synthesis that might involve acceptable infrastructure changes include: direct use of natural gas blended with H_2 - about 10% may be added to existing pipelines without unduly compromising gas quality;[3] chemical feedstock, transportation, or fuel cell applications of H_2; and methane production. Approaches to methane production might include: using H_2 and O_2 derived from water-splitting in other synthetic gasification technologies;[4] hydrogenation of natural or industrial exhaust sources of concentrated CO_2 in thermal, biological, or electrochemical processes; direct reduction of water and CO_2 to methane; active upgrading of subquality gas through processes for converting N_2, CO_2, and H_2S to useful products; and O_2 by-product use for O_2-enhanced combustion, or sale as a commodity.

Various alternatives might be considered for using H_2 and O_2 produced in pure, separated form from hydro- or solar-driven water-splitting processes in coal or biomass gasification, or to upgrade residual oils. In coal gasification, for example, hydrogenation of the coal feedstock, the CO intermediate, or excess CO_2 by-product may be possible.[5] Such processes might utilize O_2 for the gasification reactor, in O_2-enhanced combustion, or for sale as a commodity; and reduce the amount of CO_2 produced.

2. CONCEPTS

To date the research community has emphasized the study of water-splitting by various "quantum" and thermal concepts. GRI is attempting to complement this effort and reflect industry priority for methane fuel by emphasizing novel CO_2 reduction processes for the generation of methane or methane precursors along with research on water-splitting. Research on other inorganic fuel synthesis reactions of H_2S and N_2 and certain aspects of natural photosynthesis are also reviewed continually.

The organization of the concepts for GRI's program is outlined in Table I. In the quantum photo- and electrocatalytic approaches, absorbed light energy is channelled into the fuel evolution redox reaction in successful competition with its dissipation as heat. Photocatalytic systems have photosensitizers that are distributed throughout a support or in solution, and effect primary charge separation by molecular electron transfer redox steps. Primary charge separation in electrocatalytic systems, by contrast, involves electronic conduction in semiconductor materials.

Three types of photocatalytic concepts are distinguished through differences in the photosensitization process. For photoredox reactions, the photosensitizers are transition metal complexes or dye molecules dissolved in solution. For molecular assemblies, a set of linked molecules, usually ordered relative to each other, is involved in photosensitization. The assembly may involve the species that mediate charge separation or the support for these species, and can serve to improve the charge separation process or protect it from photodegradation. The assemblies may be monolayers, multilayers, micelles, vesicles, or biomimetic. Biomimetic systems mimic the primary photophysical and photochemical processes in natural photosynthesis. They can involve "synthetic reaction centers" as well as artificial or extracted membranes. Biological systems, on the other hand, refer to certain living organisms such as cyanobacteria or green algae. Under proper conditions these organisms can utilize photosynthesis for water-splitting instead of carbohydrate production. These systems are distinct from biomass conversion which require harvesting for consumption or further conversion.

TABLE I: Concept Elements for Inorganic Synthesis of Gaseous Fuels

Pathway* End Points →		Splitting of Water $H_2O \rightarrow H_2$	Aqueous Reduction of CO_2 $CO_2+H^+ \rightarrow CH_4$ or CH_4 precursors	Hydrogenation of CO_2 $CO_2+H_2 \rightarrow CH_4$
Pathway Routes*	Method of Primary Charge Separation	Systems	Systems	Systems
(PC) Photocatalytic/ (Photochemical)	Molecular Electron Transfer Steps in Condensed Media	Photoredox Rxns	Photoredox Rxns	
		Molecular Assemblies		
		Biological Systems		
(EC) Electrocatalytic/ (Electrochemical)	Electrolyte Junctions	Dispersed Systems		
		Photoelectrosynthetic Cells	Photoelectrosynthetic Cells	
	Solid State Junctions	Integrated Photovoltaic Electrolysis		
		Discrete Photovoltaic Electrolysis	Discrete Photovoltaic Electrochemical	
(TC) Thermocatalytic/ (Thermochemical)	Thermal Bond Breaking	Single-step Reaction		Direct Reaction
		Cyclic Reaction		

*Pathways include "quantum" and "thermal" routes to products.

Four types of light-driven electrocatalytic processes can be distinguished. Dispersed systems and photoelectrosynthetic cells involve semiconductor-electrolyte junctions for charge separation instead of the solid-state junctions of photovoltaic systems. In dispersed systems semiconductor particles or colloids are distributed in a support or in solution, while the electrolyte junction in photoelectrosynthetic cells is at a macroscopic electrode. In integrated photovoltaic-electrochemical systems fuel is generated on the photovoltaic device surface. These three systems generate gaseous fuels directly at the site of charge separation. Discrete photovoltaic-electrochemical systems involve an electrical network that physically separates fuel synthesis from charge separation.

Photoelectrosynthetic cells, and integrated and discrete photovoltaic systems are able to spatially separate gaseous fuel production from co-products on a practical scale while product separation for the remaining quantum concepts is a research issue.

Hybrid thermal/electrochemical systems may be a useful approach for discrete photovoltaic systems because the electrochemical step can be heated separately, lowering the thermodynamic and kinetic barriers for the fuel reaction. The utility of other hybrid quantum systems may be limited because of rapid recombination associated with light-induced charge separation at high temperatures. Thermocatalytic approaches may be carried out directly in single-step reactions or by means of chemical cycles that result in fuel synthesis.

3. STATUS

3.1 Photo- and Electrocatalytic Concepts

Table II surveys the current performance of inorganic synthesis systems in terms of the types of products that can be formed selectively, efficiency, durability, and cost. GRI currently supports research related to each of the solar quantum concepts. Rough estimates of performance requirements for proof of concept of these approaches include: 10 - 15% efficiencies, 20 year durabilities with minimal degradation in performance, and small photoconversion materials cost relative to the cost of array support structures.[7]

Apart from issues of technical feasibility, the quantum concepts face significant process design and cost issues due to the distributed character of solar radiation. High balance of system costs (cost for everything but the photochemical materials or photovoltaic modules) may be associated with the distribution of reactors or photovoltaic modules over large areas and, concurrently, with transporting the reagents and product gases within the system.

Preliminary studies vary considerably in their assessment of the potential for competitive costs in centralized solar fuel-producing facilities. Although life cycle costs in some solar systems may eventually be competitive, the extent to which inexpensive construction materials and modular approaches can reduce capital costs and pay-back periods needs to be carefully evaluated. Obtaining reasonable capital costs is particularly important in view of the significant possibility that lower-cost future natural gas supply options will be developed through research on improved gas exploration and deep drilling techniques; and the development of tight sands, Devonian shale, and coal seam unconventional sources.

Some potential for competitive inorganic synthesis costs exists. For example, assuming the achievement of certain reasonable performance targets, one study suggests that average net energy yields (energy derived per unit energy input) of photovoltaic systems (for electric power production) may exceed those of other future energy technology options.[8] Minimal moving parts, and virtually unmanned, automated operation may limit operating and maintenance costs. Recycle of solar conversion materials may improve net energy yields.

Comparisons of land use among synthetic gas approaches do not exclude the possibility of a least-cost inorganic synthesis option. Solar energy converted to fuel at efficiencies of 15%, would require an area about one fifth the state of Arizona (assuming 50% ground coverage by the array) for a 20 quadrillion Btu gas demand. Furthermore, the system could be located in remote areas. By contrast, the practical efficiencies of 0.3% typical of processes for which biomass gasification is a step could require much greater arable land areas.[9] Land area for coal gasification, (related to the plant site and the average mining area exposed during the life of the plant), has been estimated to be within an order of magnitude of that for inorganic synthesis.[10]

3.1.1 Discrete and Integrated Electrochemical Concepts. Discrete electrolytic and, in a marginal sense, integrated electrolytic processes for water-splitting are technically feasible. The long-term character of these systems is due to costs that are roughly an order of magnitude too high. However, photovoltaic costs are being lowered through the development of cheaper single-crystal, polycrystalline, and thin film materials; efficiencies are being raised through the use of tandem cell and light-concentrator arrangements; and some progress is also being made in reducing balance of systems costs.[11]

The discrete systems may be used for load management of electrical power distribution, for the supply of gas pipelines, or for a combination of these applications. The potential for gas supply applications of these concepts requires further evaluation. In future photovoltaic systems, gas production may cost more than electric power because of the electrolysis step. This penalty and the end-use value may favor direct use of electric power over gas production.

However, various production factors may enable certain economically competitive gas supply roles for photovoltaic systems. These include:

- the ability to avoid an inverter by use of direct current;
- the possibility of employing low voltage and a simpler fault detection system in a gas producing photovoltaic array;
- the possibility of developing novel array materials and designs to reduce gas production costs relative to electric power;
- the minimal load management requirement;
- the production of a pure, valuable O_2 by-product;
- and the possibility of supplemental uses that maintain a least-cost gaseous fuel product (such as blending H_2 in pipelines, coupling to synthetic gasification, or upgrading subquality gas).

Certain transport and end-use factors may also favor the prospects for these systems. These include:

- the transport advantage of gas over long distances;
- finding applications for H_2 involving minimal or economically acceptable infrastructure changes, and for which the form value of H_2 is increased relative to electricity.

Market niches relative to electric power distribution may also develop for processes driven by hydroelectric or nuclear primary energy.

In view of their technical feasibility, gradually decreasing costs, and close relationship with electric power production, GRI is evaluating two gas-supply strategies for these systems. The first concerns the design and characterization of novel systems that can compete with solar-electric power distribution either on the basis of a centralized production facility or a total gas transport system application dedicated to particular end-uses. Production aspects such as the distribution of electrolyzers within an array, operating electrolyzers at elevated temperatures with solar thermal or natural gas energy, and the possibility of increasing light concentration ratios through the use of integrated photovoltaic-electrochemical arrays, need to be considered. The second strategy concerns the extension of these concepts to methane production through the electrochemical reduction of aqueous CO_2 or the thermal hydrogenation of CO_2 by electrolytic H_2.

3.1.2 Electrolyte-Junction and Photocatalytic Concepts Because of their low efficiencies and poor durabilities, the remaining quantum concepts are essentially unproven. However, these concepts involve exciting research areas such as natural and artificial photosynthesis, semiconductor-electrolyte interfacial behavior, and the mechanisms of inorganic redox reactions. These areas are being widely pursued in the scientific community using a variety of elegant technical approaches. They also have particular crosscutting significance because fuel evolution must successfully compete with the rapid auxiliary and reverse reactions of which photogenerated species are capable. Catalytic research for satisfying this constraint may lead to improvements in small molecule redox reaction catalysis beneficial to gas utilization and supply technologies.

Ultimately, these approaches might be more selective for methane formation, simpler, more efficient, and less costly than discrete and integrated photovoltaic-electrochemical systems, and might enable gaseous fuels to be generated more cheaply than electricity from sunlight. On the other hand, reducing the cost of fabrication and transporting reagents in photochemical reactor arrays may require significant engineering advances.[12] Because of their unproven, long-term character, little design and cost analysis of the physical structures of photochemical reactor systems has taken place. Although it is premature to try to analyze these systems in detail, a better understanding of the potential for constructing low-cost array structures for these concepts is needed.

The following steps summarize technical issues for these quantum systems:
- efficient absorption of solar energy;
- minimization of reverse reactions of photogenerated species before fuel evolution;
- selective catalysis of gas evolution, particularly for O_2 evolution;
- limitation of light-induced degradation, corrosion and other side reactions;
- involvement of low-cost materials;
- loss of efficiency, yield, or stability upon scale-up;
- adapting and matching different elements of a particular solar conversion systems for optimal operation of a complete system.

The key strategies being pursued by GRI for investigating these concepts include:
- minimization of reverse reactions through the use of such effects as interfacial gradients and molecular activation barriers to electron transfer;
- *in situ* molecular level characterization of the solid-liquid interface during fuel evolution;
- catalysis of multiequivalent gas evolution reactions;
- molecular engineering of highly absorbing, complete, efficient, stable, low-cost systems;
- design and characterization of low-cost photochemical reactors.

3.2 Thermocatalytic Concepts

Experimental research on direct thermal methods and thermochemical cycles for water-splitting are not supported in GRI's program currently, although they have been supported previously.[13] Calculations for known direct water-splitting processes suggest that, if the processes occur at practical rates instead of the infinitely slow rates of an ideal, reversible process, the efficiencies can be very low or negative.[14] The many proposed thermochemical cycles for water-splitting have the following types of significant difficulties:
- difficult materials separations;
- contamination of chemical reactions;
- destructive corrosion and containment problems;
- make-up of chemical reactant losses;

TABLE II Status of Inorganic Synthesis

System	Selective Products Formed[a]	Laboratory Light-to-Chemical Efficiency	Durability	Cost
Non Biological Photocatalytic	H_2	below 0.1%	Minutes to weeks	NA
Biological Photocatalytic	H_2	10% at low light intensities under special anaerobic conditions	Weeks or longer helped by organism repair mechanisms	NA
	carbohydrate	1-5%	Years	
Dispersed	H_2	about 0.5%	Minutes to weeks	NA
Photelectrosynthetic Cells	H_2	Up to 1% /with potentially cheap materials	Minutes to months	NA
	HCOOH CO CH_3OH	NA	NA	
Integrated PV-Electrolysis	H_2	3% /amorphous silicon 8% /single crystal	Possibly good	High
Discrete PV-Electrochemical	H_2	9% /single crystal silicon	Years	High
	$CO + H_2$ HCOOH CH_3OH CH_4 (30%)	NA	NA	
Photovoltaic-High Temperature Electrolysis	H_2	possibly 20-30%[b]	Months	High
Solar Thermal Electrolysis	H_2	12%	Years	High
Thermochemical	H_2	about -5 to +5%/ direct splitting[c]	NA	High

(a) For aqueous solutions. (b) Based on informal discussion.
(c) Calculated according to Reference 14.

- kinetic limitations due to slow forward, or fast "back" or side reactions;
- heat transfer inefficiencies;
- environmental damage due to losses.

In the future, however, concepts for more efficient thermochemical processes may be developed. For example, new strategies might involve the use of non-energy-intensive methods for separating high temperature intermediates or products before equilibrium is reached so that efficiency-compromising rapid quenching processes are not required. Such strategies might be applied to either H_2 or CH_4 production. GRI is continuing to monitor this area for significant new ideas.

4. RESULTS

Despite the long-term, difficult nature of this research, a number of significant advances have been achieved thus far by GRI-supported project area researchers. These results and the high quality of the research efforts underway are helping to sustain gas industry commitment. A few results are highlighted below. Not discussed here are important fundamental advances that are providing the basis for future performance improvements. These concern areas such as energy and electron-transfer dynamics in light-absorbing systems, elucidation of reaction mechanisms, molecular-level characterization of the chemical, physical, and electronic nature of catalytic interfaces and membranes, analysis of solvent reorganization barriers to electron transfer kinetics, the reactivity of surface immobilized catalysts, and the synthesis of new "artificial reaction centers" and catalysts.

SRI International has achieved the first electrochemical reduction of CO_2 and water to significant amounts of methane. Faradaic yields of 30 percent at 61° C and -0.55 V (sce) were obtained at current densities of about 100 mA in 0.2M Na_2SO_4 using specially prepared ruthenium electrodes.[15] Valuable by-products, methanol and CO were also formed. Methanol has also been selectively generated at similar current densities and overpotentials at low bandgap, single-crystal semiconductor, and molybdenum electrodes; and significant progress has been made in understanding the mechanisms involved.[16] The role of impurities, electrode corrosion, and electrode roughness, the loss of product selectivity with increasing current density, and improving methane selectivity are among many issues that still need to be addressed. However, the research is stimulating interest in converting aqueous CO_2 to highly reduced methane or methane precursor products.

Another investigation of electrochemical aqueous CO_2 reduction at Stanford University has led to state-of-the-art rates of production of CO and H_2 using Co phthalocyanine deposited pyrolytic graphite or carbon cloth electrodes.[17] Reduction was achieved within 300 mV of the thermodynamic CO_2/CO redox potential with turnover rates exceeding 100 s^{-1}; and essentially the only carbon-containing product was CO.

At Oak Ridge National Laboratory, instrumentation for detecting transmitted light and preserving algal activity in a laboratory flow apparatus has led to the measurement of equivalent solar-to-chemical conversion efficiencies of 10% for the decomposition of water to H_2 and O_2 by green algae.[18] This efficiency compares to the 2-3% efficiencies previously obtained in this system, and is believed to be the highest efficiency reported for living photosynthetic systems. The high efficiency relative to the production of carbohydrate in plant systems or H_2 by cyanobacteria, is related to the simpler, non ATP dependent, anaerobic photosynthesis in green algae. Although obtained under artificial conditions, the result suggests long-term potential for high practical efficiencies.

As part of a project to develop artificial photosynthetic systems, Stanford University researchers have produced human myoglobin by genetic engineering techniques.[19] The myoglobin protein is being used as a host to form complexes with chlorophyll-like molecules. The result will enable the study of a key problem in these systems, the effect of protein medium on the electron transfer mechanisms in water-splitting. The complementary DNA for myoglobin was cloned, and myoglobin was expressed in E. coli and reconstituted to produce large (gram) quantitites. Individual bases of the DNA have been modified to produce site-specific changes in the amino acids of the myoglobin. The result is a significant biotechnology advance because very few functional proteins of higher organisms have been produced successfully at high levels in bacteria.

Research at the Ecole Polytechnic Federale-Lausanne has led to several significant advances in the use of stable wide bandgap semiconductors for water-splitting.[20] TiO_2 surfaces have been modified by photochemical derivatization and chemisorption using tris(2,2'-bipyridyl-4,4'dicarboxylate) Ru(II) to improve photosensitization. Monochromatic photocurrent yields based on incident light of 36% and 44% have been obtained respectively, a considerable improvement over the efficiency of 1.5% obtained previously with tris(2,2'-bipyridyl) Ru(II). An increase in overall light-to-chemical efficiency from 0.1% to 0.7% has also been obtained for water-splitting in suspensions of Pt-loaded titania particles containing Ba^{2+} ions.[20] Scavenging of peroxide by Ba^{2+} decreases back reactions. Cyclic performance of the system can be achieved by decomposing BaO_2 at 200-300 °C.

Other results obtained thus far in GRI's program concern:

- photochemical H_2 generation with porphyrin complexes; synthesis of Group VII and VIII transition metal complex catalysts for CO_2 reduction; artificial photosynthesis with photosystem II chlorophyll species; the development of factor analysis algorithms for resolving multicomponent spectra in photosynthetic systems;[21]
- the construction of novel semiconductor/membrane systems for H_2 generation; the synthesis of novel "molecular semiconductors" for H_2 generation; H_2 generation at alkaline p-InP photocathodes;

CO_2 reduction to CO at CdTe photocathodes; the characterization of oxide semiconductor photoelectrodes;[22]
- the development of computer algorithms for analyzing the finite-rate performance of inorganic synthesis concepts.[14]

Crosscutting developments are also beginning to occur within the program. In situ studies of fuel evolution at the solid-liquid interface are leading to the refinement of new techniques needed for fundamental gas-technology studies, such as novel Fourier transform infra-red and Raman spectroscopic methods, and steady state and transient space charge spectroscopic techniques. Advances in understanding the reactivity of immobilized catalysts and in methods for the chemical derivatization of surfaces have occurred that have general implications for small molecule reactions. GRI's interest in photovoltaic devices in this area has led to an evaluation of thermophotovoltaic concepts for converting radiant energy derived from clean gaseous fuel combustion to electrical power for cogeneration applications. Such systems are without moving parts, and have potential for low-cost operation that is more efficient than comparable thermoelectric or thermionic devices.

5. CONCLUSION

GRI is attempting to conduct a limited but sustained program for stimulating and monitoring long-term research for inorganic-based gas synthesis because of its potentially strong significance for future gas technologies. The outset of this program has been designated for exploring a number of different concepts for fuel generation. The key issues for discrete and integrated photovoltaic-electrochemical systems, for which hydrogen production is technically feasible, concern cost, relationship to solar-electric power distribution, and extension to methane production. The key issues for electrolyte junction and photocatalytic concepts concern the ability to demonstrate technical feasibility and the characterization of low-cost photochemical reactors. Decisions for further modifications to the program are planned for 1987. These will be based upon previous program developments and the requirements of the regulated gas industry and gas ratepayer.

6. REFERENCES

1. T. J. Woods, 'The Long-Term Trends in Lower-48 Gas Supply and Prices: The 1984 GRI Baseline Projection of U. S. Energy Supply and Demand, 1983-2010', Gas Research Insights (1985).

2. L. Daniels, Personal Communication, Department of Microbiology, The University of Iowa (1984).

3. (a) J. D'Acierno and M. Beller, 'Analysis of the Use of Hydrogen as a Supplement to Natural Gas', Office of Advanced Energy Systems Research, Division of Energy Storage, DOE, Publication No.'s DE82006844 and BNL-51442, September, (1981).

(b) Public Service Electric and Gas Co., 'Blending of Hydrogen in Natural Gas Distribution Systems', NTIS Report Nos. CONS-2925-1, CONS-2925-2, CONS-2925-3, June 1, 1976 - April 30, 1978.

4. W. Donitz and E. Erdle, 'High-Temperature Electrolysis of Water Vapor - Status of Development and Perspectives for Application', Int. J. Hydrogen Energy, 10, 291 (1985).

5. A. Flowers and H. Meyer, 'Prospects for Hydrogen Utilization in the Long-Term: A Gas Industry View', Paper presented at the Second International Symposium on Hydrogen Energy, Beijing, China (1985).

6. For particles that are very small or non-crystalline, semiconductor conduction band properties may not be present so that the dispersed systems are essentially photocatalytic in character.

7. Solar Photovoltaic Energy Conversion, H. Ehrenreich, ed., American Physical Society, N. Y. (1979).

8. C. J. Cleveland, R. Costanza, C. A. S. Hall, R. Kaufmann, 'Energy and the U. S. Economy: A Biophysical Perspective', Science, 225, 890 (1984).

9. (a) E. S. Lipinsky, D. M. Jenkins, B. A. Young, and W. J. Sheppard, 'Review of the Potential for Biomass Resources and Conversion Technology', Final Report for Contract No. 5083-511-0768 prepared by Battelle Columbus Laboratories for the Gas Research Institute, (1983).

(b) R. M. Gifford, J. H. Thorne, W. D. Hitz, and R. T. Giaquinta, 'Crop Productivity and Photoassimilate Partitioning', Science, 225, 801 (1984).

(c) H. L. Chum and M. Baizer, The Electrochemistry of Biomass and Derived Materials, ACS Monograph 183, American Chemical Society, (1985).

10. V. Smil, 'On Energy and Land', American Scientist, 72, 15 (1984).

11. R. J. Sprafka and W. J. D. Escher, 'Study of Systems and Technology for Liquid Hydrogen Production Independent of Fossil Fuels', Final Report to J. F. Kennedy Space Center, NASA Procurement Office, Kennedy Space Center, FL 32899, #N84-22768, August, 1983.

12. J. R. Biddle, D. B. Peterson, and T. Fujita, 'Solar Photochemical Process Engineering for Production of Fuels and Chemicals', Jet Propulsion Laboratory Report for DOE, JPL Publication 84-31, (1984).

13. R. H. Carty, M. M. Mazumder, J. D. Schrieber, J. B. Pangborn, Final Report for Contract No. 5014-323-0117 prepared by the Institute of Gas Technology for the Gas Research Institute, (1981).

14. J. W. Warner and R. S. Berry, "Hydrogen Separation and the Direct High-Temperature Splitting of Water," submitted to Int. J. Hydrogen Energy, (1985).

15. K. W. Frese and S. Leach, "Electrochemical Reduction of Carbon Dioxide to Methane, Methanol and CO on Ru Electrodes, " J. Electrochem. Soc. 132, 259 (1985).

16. "Reduction of CO_2 on n-GaAs Electrodes and Selective Methanol Synthesis," J. Electrochem. Soc. 131, 2518 (1984).

17. C. M. Lieber and N. S. Lewis, 'Catalytic Reduction of CO_2 at Carbon Electrodes Modified with Cobalt Phthalocyanine', Extended Abstracts for the Electrochemical Society Meeting, Toronto, Ontario, Canada, Abstract No. 649, (1985).

18. E. Greenbaum, 'Photosynthetic Water-Splitting' Quarterly Report for GRI Contract No. 5083-260-0880, March-May, 1985.

19. R. Varadarajan, A. Szabo and S. G. Boxer, "Cloning, Expression in E. Coli, and Reconstitution of Human Myoglobin", Biochemistry, accepted (1985).

20. (a) J. Desilvestro, M. Gratzel, L. Kavan, J. Moser, J. Augustynski, 'Highly Efficient Sensitization of Titanium Dioxide' J. Am. Chem. Soc., 107, 2988, (1985).

(b) L. Kavan, M. Gratzel, and J. Augustynski, 'Photochemical Surface Derivatization of Titanium Dioxide Electrodes; Discovery of a Highly Efficient and Stable Semiconductor/Dye System for Visible Light Energy Conversion', Submitted to Nature, (1985).

(c)B. Gu, J. Kiwi, and M. Gratzel, 'Photochemical Water Cleavage in Suspensions of Pt-Loaded Titania Particles with 0.7% Overall Light to Chemical Conversion Efficiency, Nouveau Journal de Chimie, in press, (1985).

21. (a) J. A. Shelnutt, 'Photoreduction of Methylviologen by Dihydroxytin(IV) Uroporphyrin' J. Am. Chem. Soc., 105, 7179 (1983).

(b) B. P. Sullivan and T. J. Meyer, 'Organometallic Trifluoromethanesulfanato Complexes of Re (I) and Os (II) as Synthetic Intermediates. Electrochemical Properties of Possible Catalysts for CO_2 reduction' Submitted (1984).

(c) M. S. Showell, and F. K. Fong, 'Elementary Reconstitution of the Water Splitting Light Reaction in Photosynthesis. 3. Photoelectrochemical Properties of Chlorophyll', J. Am. Chem. Soc., 104, 2773 (1982).

(d) M. A. Marchiarullo and R. T. Ross, 'Resolution of Component Spectra for Spinach Chloroplasts and Green Algae by Means of Factor Analysis', Biochimica et Biophysica Acta, 83, 52 (1985).

22. (a) M. Krishnan, J. R. White, M. A. Fox, A. J. Bard, 'Integrated Chemical Systems: Photocatalysis at Semiconductors Incorporated into Polymer (Nafion)/Mediator Systems', J. Am. Chem. Soc., 105, 7002 (1983).

(b) V. Houlding and A. Frank, 'Cooperative Excited State Behavior in Pt(II) Magnus-Type Double Salt Materials. Active and Inactive Photosensitizers for H_2 Production in Aqueous Suspension', Submitted to Inorg. Chem. (1984).

(c) P. G. P. Ang and A. F. Sammells, "Hydrogen Evolution at p-InP Photocathodes in Alkaline Electrolyte," J. Electrochem. Soc., 131 1464 (1984)

(d) I. Taniguchi, B. Aurian-Blajeni and J. O'M. Bockris, 'Photo-Aided Reduction of Carbon Dioxide to Carbon Monoxide' J. Electroanal. Chem., 157, 179 (1983).

(e) B. Odekirk and J. S. Blakemore, 'Ceramic $SrTiO_3$ Photoanodes: Enhancement of Photoactivity Through Donor-Doping,' J. Electrochem. Soc., 130, 321 (1983).

PUTTING PHOTOCATALYSIS TO WORK

Enrico Borgarello,* Nick Serpone,* Massimo Barbeni,** Claudio Minero,** Ezio Pelizzetti,** and Edmondo Pramauro**
*Department of Chemistry, Concordia University, Montréal, Québec, CANADA H3G 1M8
**Dipartimento di Chimica Analitica, Università di Torino, 10125 Torino, ITALIA

ABSTRACT. This paper emphasizes the practical aspects of photocatalysis as applied to practical problems of some concern and employs semiconductor dispersions of CdS and TiO_2. It presents our recent contributions to (i) the photocleavage of hydrogen sulfide, (ii) the treatment of wastes, and (iii) the photoreduction of and recovery of metals on TiO_2. It further stresses the point that our present knowledge of photocatalysis and photocatalytic processes can be and must be applied to resolving environmental problems caused by discharge of toxic materials into the aquatic and atmospheric ecosystems. The problems are many; they must be addressed.

1. INTRODUCTION

In recent years, experiments with semiconductor powders, coupled or uncoupled to suitable redox catalysts, have demonstrated the feasibility of carrying out several reactions induced by light [1]. Reactions involving hydrogen production, synthetic processes, treatment of wastes, and refining and recovery of metals can benefit from the performance of semiconductor particulate materials. In the near future, the cleavage of hydrogen sulfide [1], the dehydrogenation of alcohols [1], and the decarboxylation of carboxylic acids [1,2] (products of fermentation processes or of controlled degradation of cellulose or other biomass) will prove useful in the manufacturing of chemicals, clean fuels, and will, without doubt, contribute to the solution of environmental problems [1b]. The degradation of contaminants and the potential of photocatalytic processes in the treatment of waste waters have been discussed extensively [3], and the time for the implementation of these processes is immediate. Also, an attractive and practical implementation of photocatalysis concerns the recovery and refining of noble metals, as for example, gold, platinum, palladium, rhodium, and silver from industrial water effluents and dilute solutions [4]. Recent results from our laboratories have shown the selective recovery of one specific metal over several others in solution by an appropriate choice of experimental conditions [4d]. Most of our work has been performed using CdS or TiO_2 semiconduc-

E. Pelizzetti and N. Serpone (eds.), Homogeneous and Heterogeneous Photocatalysis, 673–689.

tor dispersions. These were chosen on the basis of chemical and physical properties, stability, availability, and cost [5].

The aim of this paper is to illustrate our recent contributions to the search for practical applications of photocatalytic processes. Specifically, we have focussed our attention to the improvement of existing catalysts for a given process, to the design of new catalysts and devices to facilitate the separation of products, reactants, and catalysts, and to the extension of our knowledge of semiconductors to new photochemical reactions. Another paper in these proceedings demonstrates another aspect of our current work which concerns fundamental studies into the primary events that occur following light excitation of semiconductor colloidal sols.

2. PHOTOCLEAVAGE OF HYDROGEN SULFIDE

Hydrogen sulfide is one of the major constituent of natural gas [5,6]. It is the principal product of the hydrodesulfurization of petroleum, coal, and it is present in many other energy sources. Oxidation of sulfur during the combustion process must be avoided to prevent large increases of sulfur oxides (SO_2, SO_3,...) in the atmospheric environment and thereby generate severe environmental problems (acid rain, smog, corrosion,...) [6c]. Classically, H_2S is removed by the Claus process [7] and is transformed into water and sulfur. A recent estimation [6] indicates that over 10 million metric tons of H_2S are produced each year; if all the hydrogen were recovered from H_2S, nearly 8 x 10^9 m^3/year of hydrogen could be produced [6,8].

The most studied routes to recover H_2 from hydrogen sulfide are the thermochemical processes in which metal disulfides act as catalysts [6a,b]. Unfortunately, the high temperature required causes sintering of the catalysts with subsequent loss of catalytic activity [9]. An alternative route utilizes an electrochemical device for the recovery of hydrogen [10]. Some years ago, several workers showed that H_2 can be photogenerated when H_2S is bubbled through an alkaline solution containing CdS dispersions [11]. The photocleavage reaction (equation 1) has

$$H_2S \xrightarrow{h\nu} H_2 + S \tag{1}$$

been studied in detail and several improvements were reported with most of the studies concerning the preparation of efficient catalysts[11,12]. In this regard, we have recently shown that H_2 evolution occurs with a quantum efficiency of 0.45 $\pm$ 0.05 and an overall energy conversion efficiency close to 3% for a Rh-loaded cadmium sulfide catalyst in an alkaline solution of H_2S and SO_2 (i.e., SO_3^{2-} ions) [13]. Table I summarizes the rates of hydrogen evolution for different semiconductor dispersions and under the same experimental conditions (0.1 M Na_2S/Na_2SO_3 in 1 M NaOH). It is noteworthy that the performance of the CdS/Rh catalyst is improved from the one observed in the recently proposed interparticle electron transfer mechanism that employs two coupled semiconductors [12g,h]. This suggests a very efficient charge separation in this new Rh-loaded CdS catalyst. The high quantum efficiency obtained and the facile preparation of the catalyst show that this device can achieve

TABLE I.- Rates of hydrogen evolution for various semiconductor dispersions: 10 mg of CdS (Fluka-1), 0.1 M S^{2-}/SO_3^{2-} in 1 M NaOH, 5-ml samples, Hg/Xe lamp operated at 900 watts, 405 nm cutoff filter, argon-purged for ca. 15 min.[a]

Dispersion	$r(H_2)$, ml/h
CdS	0.27 ± 0.03
CdS + TiO_2/0.5 wt.% RuO_2	2.84 ± 0.30
CdS + 0.04 wt.% Rh^{3+}	3.34 ± 0.35
CdS + 0.2 wt.% Rh^{3+}	7.30 ± 0.70
CdS + 0.4 wt.% Rh^{3+}	5.30 ± 0.50
CdS + 2 wt.% Rh^{3+}	5.13 ± 0.55

[a] Reference 13.

practical yields and must be considered in practical applications when considering the usefulness of hydrogen as a fuel along with the concomitant environmental advantage of removing H_2S and SO_2 from the atmospheric ecosystem.

What is also remarkable about this Rh-loaded CdS catalyst is that the hydrogen evolution rate is independent of the presence of air in the system. This is depicted in Figure 1, where the temporal hydrogen evolution is plotted for argon-purged and air-equilibrated samples. This discovery is very important, especially in the design of a reactor for the removal of H_2S and SO_2 from the environment anf from energy sources containing H_2S since, in this case, the relative cost of the device would be minimized. Moreover, two principal items must be considered in any practical application of photocatalytic processes:(a) the long term stability of the catalyst, and (b) the separation of the products from the reactants and from the catalytic slurry.

In our laboratories, we do not dispose of photochemical reactors designed to study the long term stability of the catalysts over several months of performance, but studies in this direction are in progress at the Ciba-Geigy Laboratories [11d]. Rather, we have focussed our efforts to the separation of products, reactants, and catalyst(s) at the end of the process.

The complete cleavage of hydrogen sulfide in alkaline solution in the presence of SO_2 and CdS leads to the formation of H_2 and thiosulfate $S_2O_3^{2-}$ in a 1:1 stoichiometric ratio [11b,c]. While H_2 is found in the gas phase, $S_2O_3^{2-}$ ions are in solution. Separation of products by centrifugation is not the best means since small colloidal particles of CdS stay in solution [13]. Several interesting suggestions have been proposed by different groups on this problem [14]. Our approach was to immobilize a CdS-based catalyst on a fixed polymeric material [15]; among the different polymeric materials tested, preference was given to a polycarbonate plate because of its stability under the drastic experimental conditions (0.1 M Na_2S/Na_2SO_3, 1 M NaOH, 70°C, continuous flow of H_2), using artificial light (irradiation with a 450-watt or a 900-

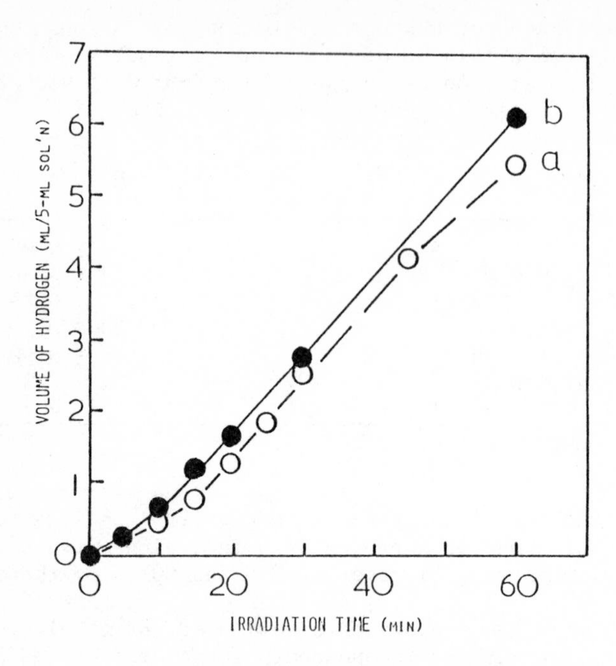

Figure 1.- Volume of hydrogen formation (STP) as a function of irradiation time for the CdS/0.2 wt.% Rh^0 dispersion containing S^{2-}/SO_3^{2-} (0.1 M); solution (a) was argon-purged, solution (b) was air-equilibrated. Other conditions were 10 mg of catalyst, 5-ml samples, 405-nm cut-off filter, temperature ca. 35°C. From reference 13.

watt Xe lamp) or direct sunlight. The CdS/RuO_2 catalyst was supported on the polycarbonate plate by a silicon-based adhesive [15]. Tests have been carried out over a 2-year period by exposure of this device to natural sunlight and to simulated AM1 sunlight radiation. Under concentrated sunlight irradiation (500 times AM1), stirring was provided by thermal convection; in all other experiments, no stirring was used. The reproducibility of the activity of different CdS/RuO_2-polycarbonate plates was very good [15], provided that the same CdS/RuO_2 was employed for different preparations. In a long term irradiation experiment, in the initial two hours, the hydrogen evolution rate decreases following loss of some of the powder from the plate. After this, the rate of H_2 formation stays constant over several weeks of irradiation (see Figure 2). The quantum efficiency of hydrogen generation was ca. 10% for this device, and the overall energy conversion efficiency (using the caloric value of hydrogen) was 0.5%. This value can easily be improved on optimizing the photochemical reactor [1b], and stirring the solution. Most of our efforts are presently directed at studying better techniques at fixing the catalyst on the polycarbonate plate and on other suitable substrates.

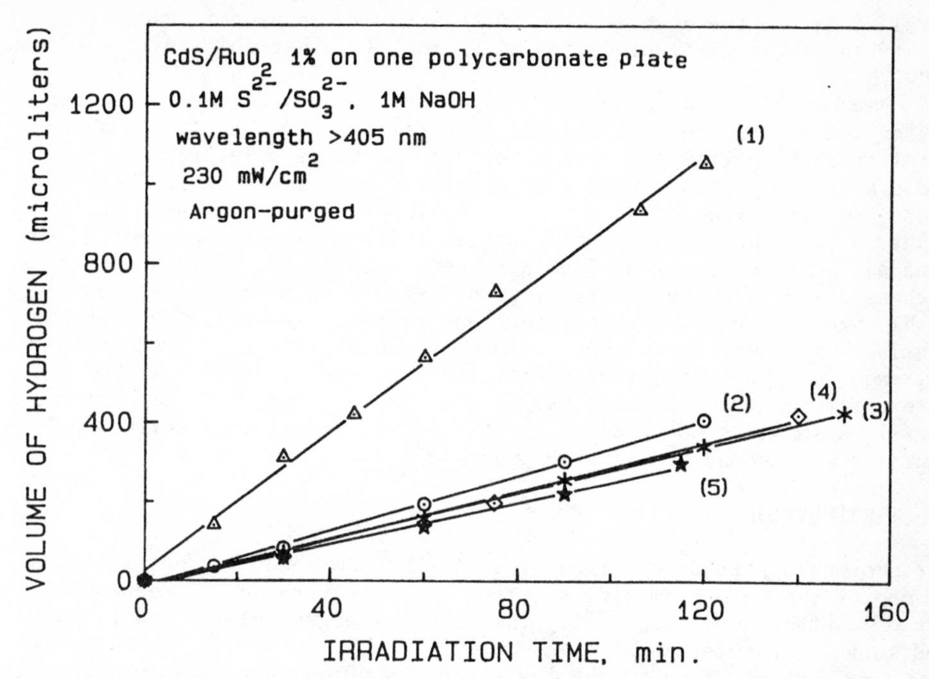

Figure 2.- Volume of hydrogen formed (STP) as a function of irradiation time in the photocleavage of hydrogen sulfide. The experimental conditions are noted in this figure. The same prepared catalyst supported on one plate was used for all the runs shown. From reference 15.

Another interesting reaction, presently under active investigation in our laboratories, is the removal of CN^- ions from waste waters by coupling the process to the photochemical cleavage of H_2S in alkaline media. It is well known that polysulfide ions, S_n^{2-}, can react thermally with CN^- to form the thiocyanate ion, SCN^- according to reaction 2 [16]:

$$S_2^{2-} + CN^- \longrightarrow SCN^- + S^{2-} \tag{2}$$

Irradiation of a CdS/Rh catalytic slurry containing 0.1 M Na_2S and 0.1 M NaCN in 1 M NaOH solution leads to the evolution of H_2 and to the formation of SCN^- [17]. The quantum efficiency of this process is ca. 0.4 depending on the CdS preparation [12i]. The cyanide ion is a rather frequent pollutant occurring, for example, in rinse waters following steel surface hardening treatments, electroplating, gold extraction, metal cleaning, and in mining processes, to mention but a few. Its removal is currently carried out by chemical oxidation (with ozone and chlorine) or by electrochemical means at some considerable cost [18]. Bard and coworkers [19] have demonstrated the conversion of CN^- ions to CNO^- in TiO_2 dispersions. The advantage of our method for the photochemical removal of CN^- is the high quantum efficiency of the process and, more significant, the final product SCN^- is much less harmful (toxicity) than CN^- [20].

Because of its wide possible application, both for hydrogen production and for the removal of contaminants (H_2S, SO_2, CN^-), the photocatalytic H_2S splitting reaction could prove to be a leading process in the practical application of photocatalytic processes. Several suggestions have already been put forth on the feasibility of a large scale plant for the photocleavage of hydrogen sulfide [1b,21]. Presently, in Canada, the recovery of H_2 from H_2S (some natural gas fields contain as much as 90% of hydrogen sulfide) is not viewed as an economical and viable process inasmuch as numerous and enormous sources of energy are available at lower costs than would be photocatalytically produced H_2. Instead, the photocatalytic splitting of hydrogen sulfide will be of interest to Canada, to Italy, as well as other countries, for environmental reasons, as SO_2 and CN^- constitute at present very serious pollutants in waters and in the atmosphere [22].

3. TREATMENT OF WASTES

The photocatalytic degradation of wastes by semiconductor particulates is one of the more promising solution to the environmental decomposition of contaminants, especially the class of organochlorinated compounds and surfactant molecules.

The contamination by chemicals containing halogen atoms presents a serious environmental problem owing to the toxicity [23] and to the widespread utilization of these contaminants as pesticides and insecticides. The subject is acute and immediate both in Europe [24] and in North America [25], where contamination by aromatic chlorinated organic compounds is increasing at an alarming rate. According to the U.S. Environmental Protection Agency estimate of 1980 [25c], some 57 million metric tons of non-radioactive hazardous wastes were generated by manufacturing industries in the United States alone. The large use of chlorinated compounds such as pentachlorophenol (PCP) and others as preservatives in woods, paints, drilling muds, photographic emulsions, hides and leathers, and textiles is one of the major causes of this environmental problem. Chlorinated phenols are also used as antimicrobials in industrial cooling systems and in pulp and paper mill systems, as herbicides and insecticides in agriculture, and have also found use in health care and veterinary products and in disinfectants [26]. The environmental contamination by surfactant molecules also poses a severe ecological problem (e.g., in Japan) which is aggravated by the fact that biodegradation of these compounds is often too slow and too inefficient. For example, in the case of sodium dodecylbenzenesulfonate (DBS), biodecomposition requires a period of two days; branched isomers of DBS are not biodegraded even after a week of exposure to bacteria [27].

The photocatalytic degradation of aliphatic and aromatic halogeno compounds and surfactants was recently achieved with TiO_2 powders [3b, 28]. Table 2 reports the half-life for the total photodegradation of several contaminants (see Figure 3) studied in TiO_2 slurries and under exposure to simulated sunlight (AM1). The process leads to the total

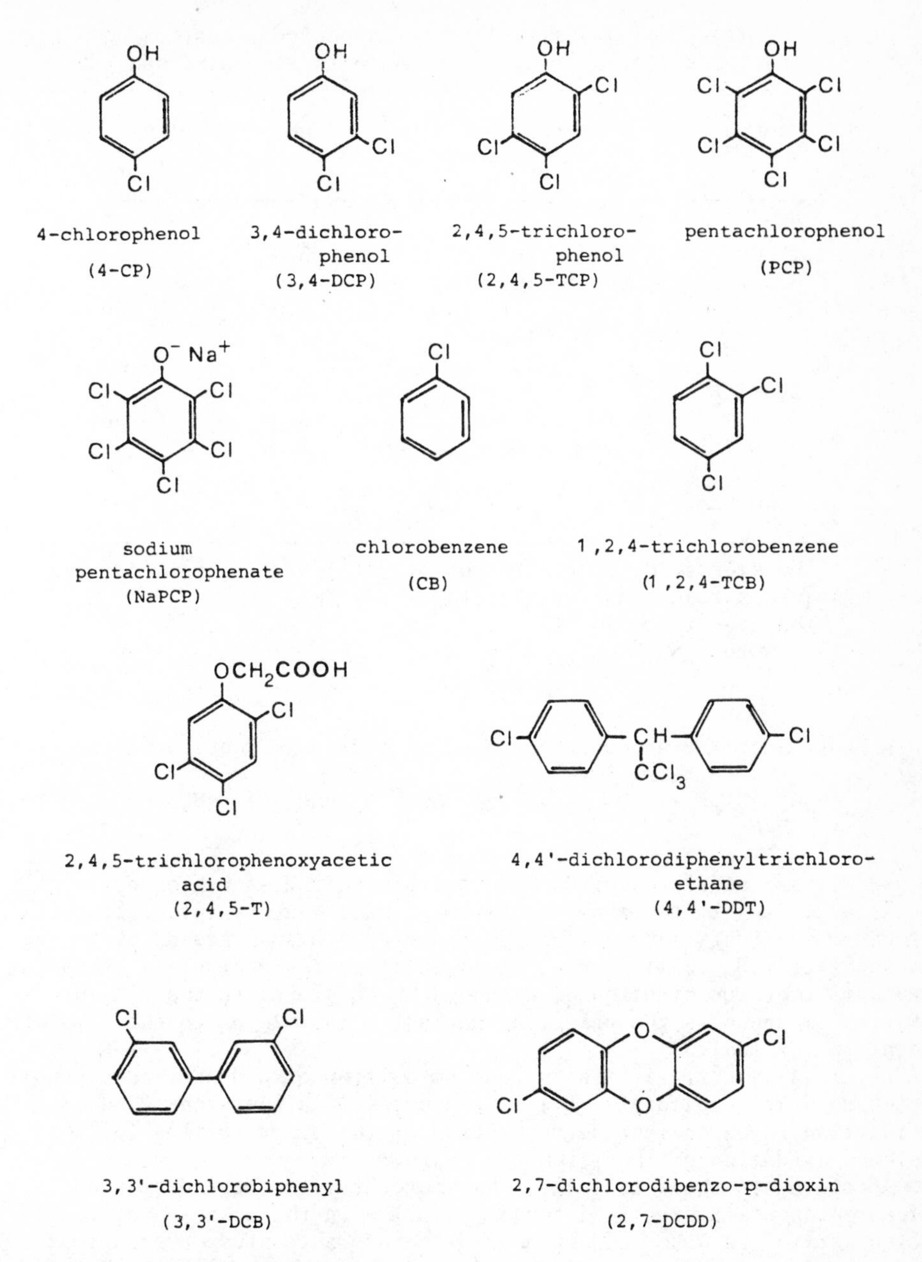

Figure 3.- Contaminants that have been investigated in our laboratories. From reference 3b.

TABLE II.- Half-lives for the total photodegradation of contaminants assisted by TiO_2 on exposure to simulated sunlight[a]

Compound	Concentration (ppm)	pH	$t_{1/2}$ (min)
4-CP[b]	6	3.0	14
3,4-DCP	18	3.0	45
2,4,5-TCP	20	3.0	55
PCP[b]	12	3.0	20
NaPCP	12	10.5	15
CB	45	2.5	90
1,2,4-TCB	10	3.0	24
2,4,5-T	32	3.0	40
4,4'-DDT[c]	1	3.0	46
3,3'-DCB[c]	1	3.0	10
2,7-DCDD[c]	0.2	3.0	46

[a] Reference 3b; concentration of catalyst, 2.0 g/l; aerated aqueous solutions; wavelength > 330 nm.
[b] Wavelength > 310 nm.
[c] Adsorbed on TiO_2.

mineralization of the aromatic compound according to reaction 3. Both

$$C_nH_mO_zCl_y \; + \; xO_2 \xrightarrow{h\nu} nCO_2 \; + \; yHCl \; + \; wH_2O \qquad (3)$$

HCl and CO_2 are always observed in quantitative and stoichiometric amounts according to the above equation [3,28]. What is interesting is that under sunlight, the degradation reaction 3 also proceeds at a significant rate [28a, 29]; work along these lines is in progress. Thus far it appears that substitution of Cl atoms by OH groups in the case of aromatic compounds is the principal mechanism that leads to the complete mineralization [29].

Figure 4 illustrates the photodecomposition of n-dodecylbenzenesulfonate and benzenesulfonate (BS) in aqueous TiO_2 dispersions [28d,e]. The reaction involves fast decomposition of the aromatic ring followed by slower oxidation of the aliphatic chain.

According to these examples, the photocatalytic degradation of wastes represents a practical route to follow in the clean-up of contaminated waters. We have been intrigued therefore to study a practical device where waste waters could be treated photochemically with TiO_2 and the resulting aqueous system easilty separated from the catalyst. In Figure 5 we represent such a simple device that consists of a column filled with TiO_2-coated glass beads; irradiation was done with a Hg/Xe lamp. The water containing the wastes is passed through the column and subsequently illuminated; the mineralization takes place in the column,

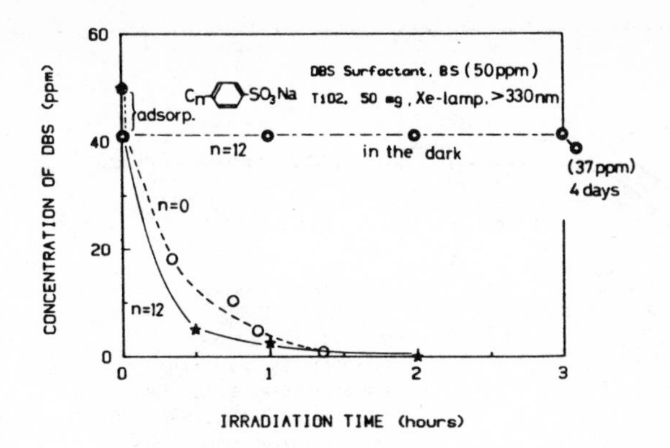

Figure 4.- Photodecomposition of n-dodecylbenzenesulfonate (DBS) and benzenesulfonate (BS) in aqueous TiO_2 dispersions. The amount of DBS as calculated from its absorption at 224 nm is plotted as a function of irradiation time. Conditions: 50 mg of TiO_2, dispersed in 25 ml of neutral water, concentration of substrate is 50 mg/l. From reference 28d.

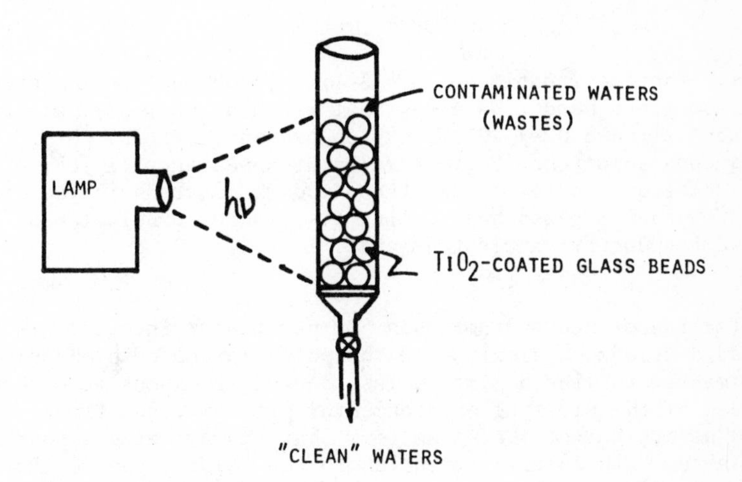

Figure 5.- Column-type reactor. From reference 30.

after which the water containing CO_2 and HCl is recovered. Preliminary experiments using 2,4,5-trichlorophenol (2,4,5-TCP) have shown the feasibility of this device [30]. In Figure 6, we report the disappearance of the 2,4,5-TCP as a function of time in a 10-cm column filled with 4-mm glass beads coated with TiO_2 and irradiated with a 900-watt Hg/Xe lamp. Our present efforts are also focussed at studying the complete

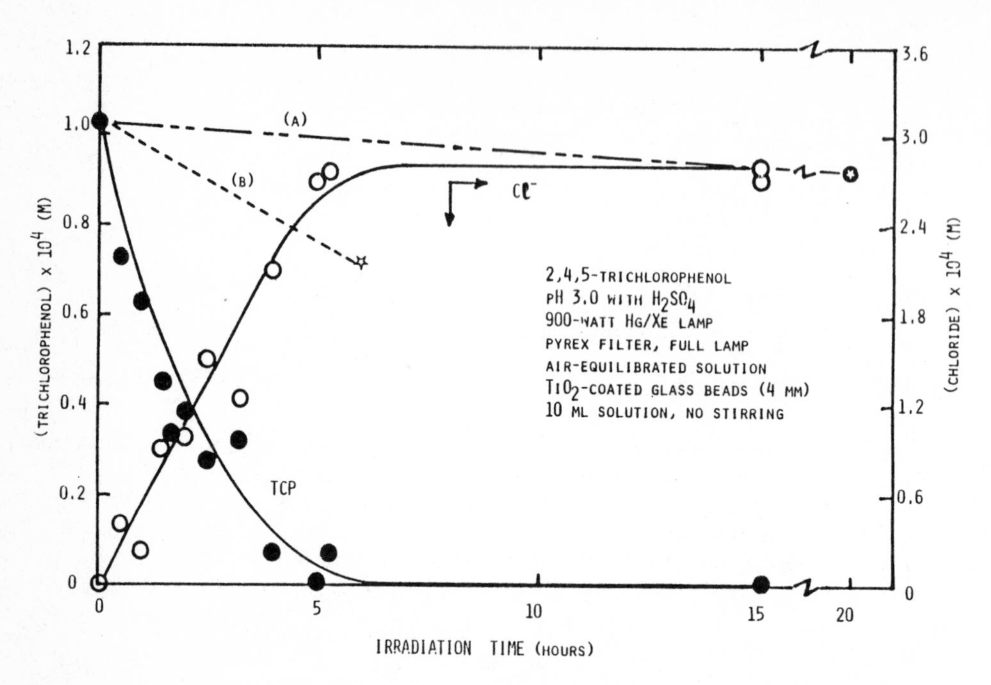

Figure 6.- Photodegradation of 2,4,5-trichlorophenol in the presence of TiO_2-coated glass beads. Wavelength of irradiation > 330 nm; initial TCP concentration 1.02 x 10^{-4}M; oxygen present; initial pH 3.0; unbuffered aqueous solutions. Solid circles:disappearance of TCP; empty circles:chloride ion formation. Line A: dark reaction in the presence of the TiO_2-coated glass beads. Line B: irradiated solution at > 330 nm but no semiconductor catalyst present.

mineralization of new contaminants. Of particular interest are the chlorinated dibenzo-p-dioxins and the polychlorinated biphenyls in their natural matrix to find a simple, fast and yet inexpensive method to resolve some of the pressing environmental problems. Additional work is directed at evaluating the parameters that characterize a possible practical reactor with special attention to the column-type reactor depicted in Figure 5, but optimizing the absorption of light by the TiO_2 catalyst.

4. PHOTOREDUCTION OF AND RECOVERY OF METALS ON TiO_2

Many reasons have led us to investigate the reduction of metals on TiO_2 particles irradiated with ultraviolet light. First, noble metals, as well as others [4], can easily be photoreduced on the surface of TiO_2 particles and thus are removed from solution. Application of this photoreduction process to the recovery and refining of metals in wastes or in mixtures [31], was obvious as this might afford an efficient and facile route to concentrate such noble metals as Pt, Au, Pd, Rh, and Ag from very diluted and dirty solutions. A second and equally important consequence of the photoreduction of metals on TiO_2 is the possibility of preparing metal redox catalysts supported on TiO_2 [32]. A third aspect of this photocatalytic metal reduction on TiO_2 involves the possibility of eliminating trace metal contaminants from solutions (for example, As(III), V(V), and Hg(I)) [33].

Figure 7 depicts the experimental results of the reduction of Au^{3+} on TiO_2 and on WO_3 using, respectively, ultraviolet and visible light [4c]. We have shown that it is possible to find conditions such that gold(III) can be reduced selectively in the presence of Cu(II), Ni(II), and Zn(II) without any interference from these metals [4c]. This result

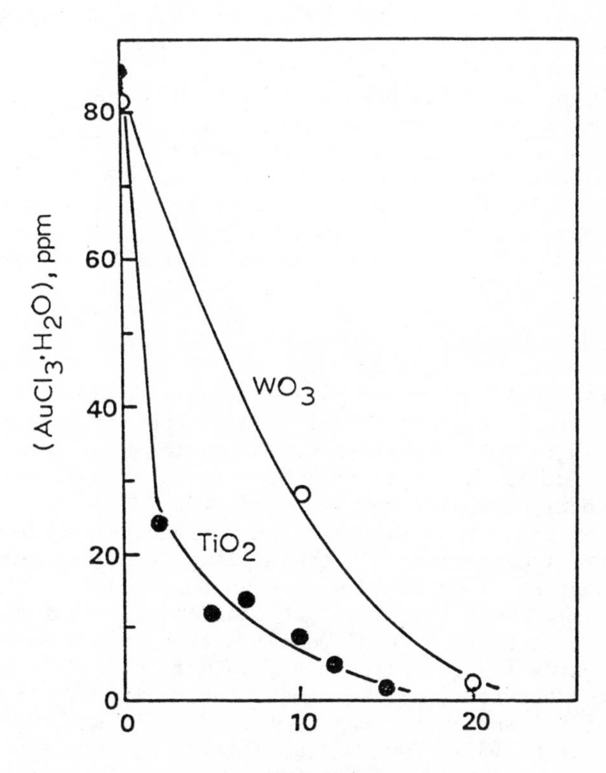

Figure 7.- Plot showing variations in the concentration of Au(III) in the aqueous phase as a function of irradiation time of a TiO_2 or WO_3 suspension. Conditions: TiO_2 100 mg (UV light) full lamp; WO_3 100 mg (visible light > 405 nm); pH 3.1 (natural pH); air-equilibrated or Ar-purged suspensions show identical rates of disappearance of Au(III) within experimental error. From reference 4c.

has opened up the possibility of employing photocatalytic processes to recover Au, Pt, Pd, and Rh, even when mixed with other metals. Under appropriate conditions, a few ppm of the noble metal(s) can be reduced selectively even when several hundred ppm of other metals are present (Fe^{3+}, Cu^{2+}, Ni^{2+}, Zn^{2+}) [31]. Our findings lead us to extend the technique at the selective photochemical deposition of either Au or Pt or Rh or Pd on TiO_2 when all are present in the same solution [31]. The separation is based on pH variations and on the presence of O_2 during the reduction process. The rate of reduction on TiO_2 for the four metals in solution follows the order Au $\gg$ Pt $\sim$ Pd $>$ Rh in argon-purged solutions; the order is Au $\gg$ Pt $\sim$ Pd, and Rh is not reduced in air-equilibrated systems at pH 3.0 [31].

Another aspect of some concern is the recovery of the noble metal from the surface of the TiO_2 particles, once the reduction has occurred. We have found that metal-coated TiO_2 (by the photoreduction process) can easily be demetallized by treating the powder with aqua regia. Atomic absorption analyses showed that the metal is recovered quantitatively and that no dissolution of TiO_2 had occurred [4c] confirming the strong stability of TiO_2 powder (TiO_2 Degussa P25) in HCl and HNO_3 solutions. We are also pursuing metal recovery with the type of reactor containing TiO_2-coated glass beads (Figure 5) and the prospects are exciting, especially when large volumes of solution have to be treated to recover or remove noble metals. Experiments are in progress to elucidate all the appropriate parameters. It is also worth pointing out that after approximately 100 uses of the glass beads of Figure 5, there appears to be no loss of activity by the TiO_2.

5. FINAL REMARKS

In addition to the applications of photocatalysis noted in this paper, other photocatalytic processes on semiconductor powders are presently being investigated in our laboratories. The photodehydrogenation of alcohols (aliphatic) has been one area of recent interest [34], as was the CO_2 photoreduction on differently doped TiO_2 [35].

The oxidation of various thiosalts is yet another field where photocatalysis might be applied. In the milling of sulfide ores, part of the sulfide content is oxidized by oxygen to thiosalts ($S_2O_3^{2-}$, $S_3O_6^{2-}$, $S_4O_6^{2-}$,...). These thiosalts, at concentrations of a few hundred ppm can cause serious environmental contamination in the receiving water streams through oxidation to sulfuric acid, thereby leading to an increase in the water acidity [36]. We have noted earlier how $S_2O_3^{2-}$ ions can be reduced to S^{2-} and SO_3^{2-} ions in TiO_2 dispersions irradiated with ultraviolet light [37]. Some of our efforts, in the near future, will address this question. The literature suggests the process to be feasible [38].

The applications treated in this paper constitute valid and economical solutions to some present and pressing problems, particularly those regarding contamination of the environment.

6. ACKNOWLEDGEMENTS

Our work has benefited from generous support from the Natural Sciences and Engineering Research Council of Canada, and from the Consiglio Nazionale delle Ricerche (Roma) through its "Progetto Finalizzato, Chimica Fine e Secondaria". We are grateful to these agencies for their continuing support.

7. REFERENCES

[1] a) M. Gratzel, Ed., Energy Resources through Photochemistry and Catalysis, Academic Press, New York, 1983.

b) M. Schiavello, Ed., Photoelectrochemistry, Photocatalysis and Photoreactors, Reidel Publ. Co., Dordrecht, Holland, 1985.

c) E. Pelizzetti, M. Barbeni, E. Pramauro, W. Erbs, E. Borgarello, M.A. Jamieson, and N. Serpone, Quimica Nova (Brasil), in press.

[2] B. Krautler and A.J. Bard, J.Am.Chem.Soc., 100, 2239 (1978); ibid., 100, 5985 (1978).

[3] a) R. W. Matthews, Nature, 11, 23 (1984).

b) E. Pelizzetti, M. Barbeni, E. Pramauro, N. Serpone, E. Borgarello M.A. Jamieson, and H. Hidaka, Chim. Ind. (Milano), in press; and references therein.

[4] a) H. Reiche, W.W. Dunn, and A.J. Bard, J.Phys.Chem., 83, 2248 (1979).

b) J.J. Curran, J. Domenich, N. Jaffrezic-Renault, and R. Philippe, J.Phys.Chem., 89, 957 (1985).

c) E. Borgarello, R. Harris, and N. Serpone, Nouv.J.Chim., in press (1985).

d) E. Borgarello, N. Serpone, and E. Pelizzetti, manuscript in preparation.

[5] J.F. Reber, in reference 1b.

[6] a) T.N. Veziroglu, W.D. VanVorst, and J.H. Kelly, Eds., Hydrogen Energy Progress IV, Pergamon Press, New York, 1982.

b) T.N. Veziroglu and A. Taylor, Hydrogen Energy Progress V, Pergamon Press, New York, 1984.

c) E. Pelizzetti, Energia e Materie Prime, 17, 31 (1981).

[7] A. Kohl and F. Riensfeld, Gas Purification, Gulf Publishing Co., Houston, Texas, 1979, p. 410.

[8] R.W. Bartlet, D. Cubicciotti, D.H. Hildebrand, D.D. McDonald, K. Jemran, and M.E.D. Raymont, Preliminary Evaluation of Processes for Recovering Hydrogen from Hydrogen Sulfide, Final Report, SRI Project No. 8030, J.P.L. Contract No. 955272, SRI International, Menlo Park, California, 1981.

[9] a) H. Kiuchi, K. Funaki, Y. Nakai, T. Tanaka, in reference 6a, p. 543.

b) Y. Krishnan and D.H. Hildebrand, in reference 6a, p. 829.

c) T. Kamyana, M. Dokyia, M. Fushishiga, H. Yoko, K. Fukuda, Intern. J.Hydrogen Energy, 1, 5 (1983).

[10] D.W. Kalina and E.T. Maas, Jr., Intern.J.Hydrogen Energy, 10, 157 (1984).

[11] a) E. Borgarello and M. Gratzel, in reference 6a, p. 739.

b) E. Borgarello, W. Erbs, M. Gratzel, E. Pelizzetti, Nouv.J.Chim., 7, 195 (1983).

c) D.H.M.W. Thewissen, K. Timmer, M. Eenwhorst-Reinten, A.H.A. Tinnemans, and A. Mackor, Nouv.J.Chim., 7, 191 (1983).

d) N. Buhler, K. Meier, and J.F. Reber, J.Phys.Chem., 88, 3261 (1984).

[12] a) E. Borgarello, K. Kalyanasundaram, M. Gratzel, and E. Pelizzetti Helv.Chim.Acta, 65, 243 (1982).

b) E. Borgarello, N. Serpone, M. Gratzel, and E. Pelizzetti, Intern J.Hydrogen Energy, in press (1985).

c) D.H. Thewissen, E.A. Zouven-Assink, K. Timmer, A.H.A. Tinnemans, and A. Mackor, J.Chem.Soc.Chem.Commun., 941 (1984).

d) T. Rajh, O. Micic, Bull.Soc.Chim.Beograd, 48, 335 (1983).

e) E. Borgarello, N. Serpone, M. Gratzel, and E. Pelizzetti, Inorg. Chim.Acta, submitted.

f) J.F. Reber and K. Meier, J.Phys.Chem., 88, 5903 (1984).

g) N. Serpone, E. Borgarello, and M. Gratzel, J.Chem.Soc.Chem. Commun., 342 (1984).

h) M. Barbeni, E. Pelizzetti, E. Borgarello, N. Serpone, M. Gratzel L. Balducci, and M. Visca, Intern.J.Hydrogen Energy, 10, 249 (1985).

i) N. Serpone, E. Borgarello, M. Barbeni, and E. Pelizzetti, Inorg. Chim.Acta, 90, 191 (1984).

[13] E. Borgarello, N. Serpone, E. Pelizzetti, and M. Barbeni, Nouv.J. Chim., submitted.

[14] a) Y.M. Tricot and J.H. Fendler, J.Am.Chem.Soc., 106, 2475 (1984).

b) D. Meissner, R. Memming, and B. Kastening, Chem.Phys.Letters, 96, 34 (1984).

c) J.P. Kuczynski, B.H. Milosavjevic, and J.K. Thomas, J.Phys.Chem. 88, 980 (1984).

d) M. Krishnan, J.R. White, M.A. Fox, and A.J. Bard, J.Am.Chem.Soc. 105, 7002 (1983).

e) A.N.H. Mau, C.H. Huang, N. Kakuta, A.J. Bard, A. Campion, M.A. Fox, J.R. White, and S.E. Weber, J.Am.Chem.Soc., 106, 6537 (1984).

f) N. Kakuta, J.R. White, A. Campion, A.J. Bard, M.A. Fox, and S. E. Webber, J.Phys.Chem., 89, 48 (1985).

[15] E. Borgarello, N. Serpone, P. Liska, W. Erbs, M. Gratzel, and E. Pelizzetti, Gazz.Chim.Ital., submitted.

[16] E Schulek, Z.anal.Chem., 65, 352 (1925).

[17] E. Borgarello, Rita Terzian, N. Serpone, E. Pelizzetti, and M. Barbeni, submitted for publication.

[18] a) S.P. Tucker and G.A. Carson, Environ.Sci.Technol., 19, 215 (1985).

b) A.D. Mehrkam, Met.Prog., 108, 103 (1975).

c) K.F. Cherry, Plating Waste Treatment, Ann Arbor Science, Ann Arbor, Michigan, 1982.

[19] S.N. Frank and A.J. Bard, J.Am.Chem.Soc., 99, 203 (1977);J.Phys. Chem., 81, 1484 (1977).

[20] H.E. Christensen and E.J. Fairchild, Eds., Registry of Toxic Effects of Chemical Substances, U.S. Department of Health, Education and Welfare, Rockville, Maryland, June 1976.

[21] D.A. Armstrong, Energy Processing (Canada), July 1984, p. 2.

[22] See various issues of Chem.& Eng. News, 1984.

[23] a) D.Y. Lai, J.Environ.Sci.Health, C2, 135 (1984).

b) C.R. Pearson, in Handbook of Environmental Chemistry, O. Hutzinger, Ed., Springer-Verlag, Berlin, vol.3, part B, pp 86-116, 1982.

c) Y.R. Harvey, W.Y. Steinhauer, and J.M. Teal, Science, 180, 643 (1973).

[24] G.U. Fortunati, Symposium on Chlorinated Dioxins and Dibenzofurans in the Environment, American Chemical Society Annual Meeting, Miami, Florida, 1985.

[25] a) National Geographic, 167, 318 (1985).

b) See various recent issues of Chem.& Eng. News, 1984, 1985.

[26] a) C. Rappe, in Handbook of Environmental Chemistry, O. Hutzinger, Ed., Springer-Verlag, Berlin, vol.3, part A, pp. 157-180, 1980.

b) R.A. Bailey, H.M. Clarke, J.P. Ferris, S. Krause, and B.L. Strong, Chemistry of the Environment, Academic Press, New York, 1978.

c) J. Josephson, Environ.Sci.Technol., 17, 124A (1983).

d) P.A. Jones, Chlorophenols and their Impurities in the Canadian Environment, Report EPS 3-EC-81-2, Environmental Impact Control Directorate, Environment Canada, March 1981; see also the 1983 supplement, Report EPS 3-EP-84-3, March 1984.

[27] a) R.D. Swisher, J.Am.Oil Chem.Soc., 40, 648 (1963).

b) R.D. Swisher, J.Water Poll.Control.Fed., 35, 1557 (1963).

c) R.D. Swisher, J.Water Poll.Control.Fed., 35, 877 (1963).

[28] a) D.F. Ollis, Environ.Sci.Technol., 19, 480 (1985).

b) M. Barbeni, E. Pramauro, E. Pelizzetti, E. Borgarello, N. Serpone, and M. Gratzel, Nouv.J.Chim., 8, 547 (1984).

c) M. Barbeni, E. Pramauro, E. Pelizzetti, E. Borgarello, and N. Serpone, Chemosphere, 14, 195 (1985).

d) H. Hidaka, K. Kubota, M. Gratzel, N. Serpone, and E. Pelizzetti, Nouv.J.Chim., 9, 67 (1985).

e) I. Izumi and J. Kyokane, Kenkyu Kiyo-Nara Kogyo Koto Senmon Yakko, 19, 43 (1983); CA 101:173458n(1984).

[29] M. Barbeni, M. Vincenti, E. Pelizzetti, E. Borgarello, and N. Serpone, work in progress.

[30] N. Serpone, E. Borgarello, P. Cahill, M. Barbeni, and E. Pelizzetti work in progress.

[31] N. Serpone, E. Borgarello, R. Harris, G. Emo, and E. Pelizzetti, to be submitted for publication.

[32] a) B. Krautler and A.J. Bard, J.Am.Chem.Soc., 100, 4317 (1978).

b) E. Borgarello, J. Kiwi, E. Pelizzetti, M. Visca, and M. Gratzel, J.Am.Chem.Soc., 103, 6324 (1981).

[33] a) A. Kudo, S. Miyahara, and D.R. Miller, Progr.Wat.Tech., 12, 509 (1980).

b) R. Batti, R. Magnaval, and E. Lanzola, Chemosphere, 4, 13 (1975).

[34] N. Serpone, E. Borgarello, E. Pelizzetti, and M. Barbeni, Chim. Ind.(Milano), in press (1985).

[35] M. Halmann, V. Katzir, E. Borgarello, and J. Kiwi, Sol.Energy Mat., 10, 85 (1984).

[36] M. Wasserlauf and J.E. Dutrizac, The chemistry, generation, and treatment of thiosalts in milling effluents - A non-critical summary of CANMET investigation 1976-1982, Division Report MRP/MSC 82-28(R), CANMET, Energy, Mines and Resouces Canada, 1982.

[37] E. Borgarello, J. DeSilvestro, M. Gratzel, and E. Pelizzetti, Helv. Chim.Acta, 66, 1827 (1983).

[38] Y. Matsumoto, H. Nagai, and E. Sato, J.Phys.Chem., 86, 4664 (1982).

PANEL DISCUSSION on the SELECTIVE ACTIVATION AND CONVERSION OF MOLECULES

Co-Chairmen: D. Meisel and D. F. Ollis

Many of the studies presented during the Workshop have demonstrated that a variety of molecules can be selectively activated to preferentially proceed in a desired pathway provided some physico-chemical principles are applied in the design of the photosensitizer/substrate/catalyst system. We would like therefore to highlight in the following discussion recent advances and future directions that should be pursued if the field of photocatalysis in the broadest sense is to benefit through selective activation. In particular, we would like to emphasize those directions which may lead to advancement of applicable technologies.

For the purpose of this discussion we explore the points raised above from two different angles. The first will be from the point of view of the reactants. We ask ourselves how we can selectively activate a particular type of molecule in the presence of a variety of other molecules which may include the medium reaction products, sensitizer, etc... The second angle concerns how we can convert the activated species through a particular pathway when several are available. In other words, how can we be kinetically specific? Once we answer these questions, we may also wonder which pathways should we pursue and toward what end?

It seems quite clear that selectivity in excitation poses no particular problem. We can be extremely selective in the excitation down to the vibronic level or to the selective isotopic level. The problems that we do face are on the catalytic side or in coupling the excitation state with the catalytic step. We should therefore concentrate on these aspects.

A. Selective Activation of Reactants

Several methods to obtain selectivity in activation of reactants were presented in the Workshop. These may roughly be grouped into the following four categories.

1. Surface site selectivity
2. Geometrically constrained environment (zeolites, clays,..)
3. Selectivity by molecular assemblies (micelles, vesicles, membranes, etc...)
4. Selectivity through quantum size effects (very small particles).

We shall discuss them in the above order.

Surface site selectivity is typically observed at the solid-gas surface as has been shown by the work of Somorjai, Pichat, Teichner, Bickley, and others. However, surface site selectivity has also been observed by Amouyal at the solid-water interface and by Fox and Kisch on other solid-liquid systems. What are therefore the prospects of this type of **activation?**

E. Pelizzetti and N. Serpone (eds.), Homogeneous and Heterogeneous Photocatalysis, 691–698.

TEICHNER - I would like to draw your attention to the fact that the question of selectivity may be strongly medium dependent. We have seen several cases in which reactions that proceed smoothly and selectively at the solid-gas interface either lose their selectivity or change their course when performed at the solid-liquid interface.

McLENDON - As an innocent bystander, it seems to me that many examples of selectivity were presented. To mention only a few, Fox's work showed new chemical pathways which were not reproduced by conventional electrochemical pathways. Amouyal showed that the competitive reactions of H_2 evolution and hydrogenation can be controlled, and Somorjai, Schrauzer, and Bickley showed how photocatalysis can reduce N_2 to NH_3 even under unusual conditions (H_2O, O_2, ...).

FOX - Site selectivity for olefin oxygenation may reflect different mechanisms in the gas phase and liquid phase. Dielectric changes of solvent certainly influence the rate of electron/hole recombination, so that reaction kinetic differences may be amplified by altering liquid phase electrolyte composition. The mechanism may also change as a function of the oxidation potential of the adsorbed substrate; for example, contrasting modes of oxygenation for cyclohexene and styrenes show the first may involve free radical chain oxygenation, while the second may involve quantum-controlled ($\emptyset \lesssim 1$) electron exchange.

MUNUERA - Let me draw your attention to the possible role of adsorption displacement equilibria in controlling selectivity in systems running at room temperature. I will try to illustrate it using two simple examples such as photo-dehydrogenation of isopropanol and water cleavage on Pt/TiO_2 either in gas/solid or liquid/solid phases.

In the first case, the formed acetone is displaced from the TiO_2 surface by new alcohol molecules from the gas or liquid phase in a 1:1 ratio so that 100% oxidation to acetone is achieved. In the case of water cleavage according to:

$$2H_2O \xrightarrow[Pt/TiO_2]{h\nu} H_2 + H_2O_2$$

unfortunately H_2O_2 is much more tightly adsorbed on the TiO_2 surface vs. H_2O so that it cannot be displaced by H_2O molecules from the gas or liquid phase and the reactions become at the end poisoned. In fact, introduction of another component, a solvent, made even more complex the adsorption and displacement equilibria at the TiO_2 surface.

SAKATA - In order to achieve a high selectivity of a reaction making use of catalytic processes is essentially important. In this sense, understanding of catalytic properties of semiconductor surface is necessary. As Prof. Fox kindly introduced in her lecture, lactic acid is decomposed into H_2 and pyruvic acid on CdS and ZnS surface, while it is decomposed into H_2, CO_2, and acetaldehyde on TiO_2, $SrTiO_3$, and MoS_2. Although the mechanism controlling the reaction paths is not clear, it seems that the catalytic properties of the semiconductor surfaces influence this reaction. Since the production of a keto acid from a hydroxy acid is related to amino acid synthesis, investigation of this kind of phenomena would be very important. In addition, as Dr. Krist remarked, making a new catalytic surface by adding a catalyst such as metals and organometallic compounds would be important to achieve a selective reaction. Finally, relating to Dr. Heller's remarks, I would like to point out the importance of exploiting new semiconductors besides CdS which can work with visible light.

A second kind of selectivity is introduced by geometrical constraints imposed on the molecule by the physical arrangement of host catalysts. This type of selectivity has been observed by N. Turro (Columbia University) in the photodecomposition of dibenzyl ketones in zeolites. Host-guest complexes are another type of geometrically constraining environment where selectivity has been observed. In this meeting, H. VanDamme has presented his work in clays where similar selectivity can be expected.

VAN DAMME - Generally speaking, one may expect two types of selectivities in zeolites, clays, and pillared clays: (i) selectivities due to geometrical constraints, associated with window or pore sizes. It is quite obvious however that these selectivities (molecular sieving, shape selectivities) should only appear for rather large molecules, of dimension of the order of,or larger than the pore size. (ii) selectivities due to the chemical properties of the surface. For instance, strongly acidic sites, such as those present in clays and zeolites, will clearly favour the reaction paths involving protonated species.

FOX - While chemical reactivity of the surface may certainly affect selectivity, I would like to indicate that our present level of knowledge is such that we don't even know the effect of impurities on such systems, in particular in solid-liquid systems. These might be impurities in the electrolyte or in the electrode material, surface states or bulk imperfections.

KRIST - Two examples of GRI supported work indicate some promise for selective formation of highly reduced products of aqueous CO_2 reduction:
1. Research by K. Frese at SRI International on the electrochemical reduction of aqueous CO_2. At GaAs (111) phase single crystal electrodes, methanol can be produced with Faradaic yields above 70% with current densities of approximately 100 uA/cm^2. Methanol is only observed in a small range of current densities and slightly acidic pH values, and an impurity may be involved in the mechanism . However, the reaction appears to be catalytic with large turnover numbers. Preliminary research on specifically prepared Ru electrodes has also resulted in CH_4 formation. More work is needed to establish whether this reaction is catalytic.
2. Research by N. Lewis at Stanford University on Co-phthalocyanine modified electrodes has resulted in the selective reduction of aqueous CO_2 to CO and H_2 with high turnover numbers.

The third type, selectivity provided by molecular organisates has been extensively studied in particular in respect to charge separation and inhibition of the back reaction. Could they further contribute towards selective activation?

FENDLER - Organized assemblies do contribute to selectivity by altering ground and excited state pK's, oxidation and reduction potentials, reaction rates and product distributions. Many things can be considered as organized assemblies: even the addition of hexametaphosphate to colloid semiconductors, the stacking of porphyrins, or the use of surfactants, membranes, clays, zeolites, etc... Due care needs to be exercised however in selecting the most appropriate systems for a given application. Selectivity by application of magnetic fields (external or intrinsic) should also be added to the list. Indeed there are substantial magnetic effects on the early steps of photosynthesis. Indeed, these magnetic effects are optimal in conjunction with organized assemblies.

LEWIS - Organized assemblies are indeed necessary, and I cannot think of a much more organized system than a perfect semiconductor interface with a spatially dependent Coulomb potential (band bending) used to achieve electron/hole separation in a defined direction. We in fact thus agree. I would like to further suggest that immobilization is not useful when an ineffective catalyst is immobilized, and that an effective catalyst need not be immobilized to yield the desired redox products. Immobilization has several advantages, including the opportunity to achieve unique chemistry not available in conventional environments, and these opportunities should be explored.

Perhaps the newest and most exciting opportunities arise through the long sought but only recently discovered quantum size effects in very small particles. In fact, O. Micic has already presented to us the first observations of chemical selectivity resulting from such effects. Perhaps Louis Brus, who gave us the theoretical basis to these effects, could provide us with some of his ideas regarding their possible utilization in redox chemistry.

BRUS - Semiconductor clusters having effective band gaps larger than the bulk value, should also have size dependent redox potentials for both hole and electron. The principal reason is that carriers necessarily acquire quantum localization energy in physically small clusters. This extra energy makes the electron a better reducing agent, and the hole a better oxiding agent. The shift ΔV can be simply estimated by $\Delta V = \hbar^2\pi^2/2em^*R^2$, where m* is the carrier effective mass and R is the cluster radius. A more elaborate treatment is given in the reference: J.Chem.Phys., 79, 5566 (1983).

Selectivity in photoredox processes can then be achieved by choosing particle size to give the required redox potential. Clear experimental evidence that smaller crystallites can carry out higher energy redox processes than large crystallites has been given in this Workshop by Olga Micic and co-workers.

The kinetic rates of charge transfer, across the cluster

interfaces to adsorbed molecules, should be enhanced in small crystallites. This occurs because the wavefunction (e.g., charge density) is pushed towards the surface, and away from the interior of the cluster, in smaller clusters as shown in the above reference.

A negative aspect of these increased hole and electron energies in small clusters is an increased probability of lattice dissolution by these redox agents. This probability is increased by the fact that the lattice binding energy per atom is lowered in small clusters. Perhaps strong surface binding agents, such as hexametaphosphate as used by Henglein and co-workers, will lower dissolution rates.

NOZIK - In addition to the features of size quantization summarized by Brus, I would also add the possibility of achieving high thermodynamic conversion efficiency (~66%) by utilizing hot electrons and hot holes in quantized systems. We have seen that charge transfer can occur out of excited (higher lying) quantum states in the quantum wells. In order to achieve the high conversion efficiency possible with hot carriers, one might expect to have to utilize the hot carriers at their individual quantized levels. This would be a very difficult problem, requiring electron and hole acceptors at each level. However, another possible approach is to equilibrate the hot electrons and holes each among themselves to create a hot carrier pool, with a high effective temperature and affect charge transfer at one energy at this pool average. This internal equilibration occurs in femtoseconds. The down side of quantization is the enhanced probability for photocorrosion by the higher energy carriers. The connection between superlattices and small particles is very straightforward in that the former represents quantization effects in one dimension while the latter represents the same basic effects in three dimensions. We have seen enhanced photochemistry from both types of systems.

HELLER - An opportunity - yet unexplored - may exist in the expansion of photocatalysis to small band gap semiconductors. Infrared photons are much less expensive than ultraviolet photons. We may apply two catalysts to the semiconductor surface, one adsorbing an 'oxidizer', the other adsorbing a 'reducer'. Upon illumination, we should see the clear reduction of the oxidizer by the reducer. Stability need not be an issue if two stable metals are used to form the semiconductor contacts, which will act as unidirectional gates, determining the direction of the reaction [example (Ru) and (PtSi) islands on silicon particles].

The discussion above clearly provides the rationale for fundamental research which is still required in order to enable us to better design in a predictive way photocatalytic systems. We now turn to the other angle of the same problem: kinetic selectivity.

B. Kinetic Selectivity in Activation of Molecules

By kinetic selectivity we mean approaches to achieving passage of an activated, adsorbed intermediate down one of several possible paths by virtue of a kinetically fast pathway for the desired reaction and kinetically slow rates for accomplishment of other reactions also possible (at some conditions on the same catalyst).

Examples of kinetic selectivity abound in the presentations heard already, as well as those summarized below by Halman and Pichat. We may list them:

1. Reduction of CO_2 (to HCOOH, CH_2O, CH_3OH, or CH_4) or its conversion via the reverse water gas shift reaction.

HALMAN - Selectivity in CO_2 photoreduction up to now has varied tremendously in different laboratories. Somorjai and co-workers, about 7 years ago, reduced CO_2 and H_2O in a gas-solid interface with $Pt/SrTiO_3$ to CH_4. K. Frese and co-workers at SRI International have electrochemically reduced CO_2 on n-GaAs and n-GaP (in the dark) or p-GaAs and p-GaP (under illumination) with high efficiency to methanol. On metal electrodes, CO_2 is mainly reduced to HCOOH.

2. The reduction of N_2 may potentially pass through the sequence of species dissolved (or equivalently adsorbed) of diazene (N_2H_2), hydrazine (N_2H_4), ammonia, as discussed earlier this week by Schrauzer using $V(OH)_n$ catalyst. An additional complication to this serial kinetic sequence is the spontaneous and also base catalyzed conversion of diazene into (N_2, H_2) in dilute solutions and to products including (N_2, N_2H_4) in more concentrated solutions.

3. Sequential metal ion removal from multicomponent aqueous solutions involves selectivity, although whether thermodynamic or kinetic over illuminated TiO_2 based catalysts remains to be seen. Borgarello, Serpone, Pelizzetti et al. reported earlier examples of quantitative recovery of gold (full report to appear in the Nouveau Journal de Chimie). Potential applications in precious metal recovery would depend strongly on selectivity (observed) and on ease of TiO_2 recovery and regeneration, evidently accomplished with aqua regia, without TiO_2 damage.

Hydrocarbon and oxygenate conversions have long provided numerous examples of kinetic selectivity, as Dr. Pichat's following discussion shows.

PICHAT - 4. Oxidations have been observed with gaseous compounds (dissolved in 'inert solvents') and liquid compounds (alkanes, alkenes, alcohols, and other classes of compounds, cf. Dr. Fox's lecture) over neat semiconductor oxides, the best results being obtained over TiO_2 specimens.
Alkanes, alkenes and alcohols are converted to aldehydes or ketones; however, acids can be progressively formed if the primary products are left in the reaction medium. The selectivity can be high in some cases (for instance for the trans formation of gaseous or liquid alkyltoluenes into alkylbenzaldehydes) and different from that found with other modes of activation, but undesired cleavages can occur. In sol-

utions, high chemical yields have been obtained. Unfortunately, quantum yields are too often not reported. They appear to vary considerably with the oxidizable molecule and the photocatalyst.
Regarding mechanism, results indicate that the oxygen active species are probably dissociated. Few attempts have been made to determine how they incorporate into the organic molecules. For gas phase oxidations, the reaction path

alkane ⟶ alkene ⟶ alcohol ⟶ >C=O

and for solutions, the formation of radical ions have been proposed.

PICHAT - 5. Dehydrogenations at ambient conditions are uphill reactions which have been observed with gaseous (dissolved) and liquid alcohols, saturated or unsaturated (such as terpene alcohols) over bifunctional photocatalysts constituted of group VIII transition metals (including Ni) deposited on semiconductor oxides. A double bond is partly hydrogenated only if conjugated (in the case of terpene alcohols). The quantum yields can be high.
The mechanism is based upon the dissociative adsorption of the alcohols on the basic sites of the semiconductor and on hydrogen spillover between the two components of the photocatalyst. This class of photocatalytic reactions should be extended to other types of molecules.

Applications of photocatalytic oxidations present several advantages with regard to other mild oxidation methods: use of air, cheap semiconductors (such as TiO_2), ease of separation of products and photocatalyst no polluting residuals (compare with the use of MnO_2 + H_2SO_4, for example). However, in every particular case, the problems of selectivity (as in catalysis) and quantum yields are yet to be resolved. Photocatalysts are to be improved. Designof special reactors, such as "falling-film" or more exactly "falling-suspension" devices, should be envisaged to avoid over-oxidation. In the absence of oxygen (dehydrogenations) this latter problem does not exist of course. As a final remark, I would say that efforts should be made to convince organic chemists in industry to use photons and a catalyst at the same time, since they are often already reluctant to use photons.

6. Another selective conversion involves amino acid syntheses, presented later in this Workshop by Sakata, where both amination of β-hydroxy acids and β-keto acids, and hydrogenation of imino acids, must be accomplished without concurrent decarboxylations. The importance of CdS rather than TiO_2 as a selective catalyst here is noteworthy.

7. For environmental conversions involving contaminated waters this Workshop has demonstrated from the presentation of

Borgarello, Serpone, Pelizzetti et al., and of Ollis that most simple one and two-carbon halogenated derivatives of alkanes, alkenes, and acetic acids, and haloaromatics (including two and three ring compounds) may be completely mineralized to CO_2 and HX (X = Cl, Br). However, it appears that intermediate formation is not uncommon (trichloroethylene $\rightarrow$ dichloroacetaldehyde $\rightarrow$ CO_2; 1,2-dihaloethanes $\rightarrow$ vinyl halide (trace) $\rightarrow$ CO_2; and chlorobenzene $\rightarrow$ (chlorophenol, hydroquinones and quinones) $\rightarrow$ CO_2). In all cases, continual complete conversion to CO_2 and HX appears achievable provided sufficient dissolved oxygen level is maintained.

A final selectivity topic not yet fully explored involves the full description of adsorbate interactions with holes and electrons. In other words, the selective fate of adsorbates is coupled with the selective fate of the solid state-originated photoproduced holes and electrons.

In sum, both selective activation and selective conversions have now been demonstrated for several major kinds of chemical conversions, and the quality of understanding in most instances is such as to strongly suggest future directions of inquiry. In some instances, the situation remains unclear and a consensus is yet to be discerned, suggesting very strongly a need for standard conditions and independent repetition of experiments. The range of potential applications for the reactions cited above includes fuel production (CH_4, N_2H_4), fertilizers (NH_3), solvents (oxygenates), and intermediate chemicals (amino acids, ascorbic acid), and water purification and decontamination. Thus selectivity challenges span virtually the full gamut of heterogeneous photocatalyzed conversions.

PANEL DISCUSSION on SENSITIZATION and IMMOBILIZATION of CATALYSTS on VARIOUS SUPPORTS

Co-Chairmen: J. Fendler and J. Bolton

Sensitization - Before describing the characteristics of sensitizers, let us first present a definition of a redox catalyst: it is a substance which can accept or donate several electrons ($n \leq 8$) such that each electron is removed or added at approximately the same potential. When fully charged, the catalyst should be capable of driving a multi-electron redox reaction in a concerted process with few, if any intermediate steps.

For example, a good oxygen catalyst should be oxidized in four one-electron steps with the oxidation of water to oxygen occurring only when the last electron is removed. The ease of removing the last electron should not be much more difficult than removing the first electron.

The sensitizer is the substance which absorbs light and then transfers an electron into or out of the redox catalyst. If the catalyst is a semiconductor, the sensitizer should absorb at wavelengths longer than λ_{bg}, the band gap wavelength of the semiconductor. For optimal efficiency, a sensitizer should possess the following characteristics:

(i) Absorption into the red and near-IR regions of the solar spectrum.
(ii) High quantum yield for charge injection into the catalyst.
(iii) High absorption cross section.
(iv) High stability ($\leq 10^{-6}$ quantum yield for degradative side reactions).
(v) Good redox match to the band gap of the semiconductor catalyst.

Some of outstanding problems yet to be solved are:

(a) How to get a good quantum yield for charge injection for more than a monolayer.
(b) What is the role of the solvent in mediating electron transfer at the interface.
(c) How important is the nature of the semiconductor other than the band edge energies (i.e., doping levels, defects, surface treatment, etc...).

Immobilization of semiconductors and catalysts is an essential requirement of efficiency. It allows the in situ generation or incorporation of controllable sized uniform particles and provides a means for their stabilization. Immobilization also leads to a micro-environment which alters oxidation and reduction potentials and may facilitate charge separation.

E. Pelizzetti and N. Serpone (eds.), Homogeneous and Heterogeneous Photocatalysis, 699–701.

Catalysts and semiconductors have been immobilized fruitfully in (or on) surfactants, aqueous and reversed micelles, microemulsions, vesicles and polymerized vesicles, silicates, clays, zeolites, polyelectrolytes and polymeric membranes. In the absence of additives, no colloidal particles, in the order of 50-5000 A, remain monodisperse for any lengths of time. Advantage should be taken of the differences between the available methods of immobilization for a given application. Properties of the different immobilizers should be well understood prior to their utilization. Well conceived semiconductor immobilization will also aid selective doping and deposition of oxidation and reduction catalysts at destined sites.

Immobilization also aids the development and stabilization of ultra-small colloidal particles. Particles in the 10-50 A range - the modern equivalent of the world of neglected dimensions - have unique properties. Quantum size effects transition from individual molecules to cooperative units and semiconductors occur in this range. Photocataysis at these transitions may well lead to new chemistries.

Our accomplishments and future directions should be highlighted in the following discussions.

TSUBOMURA - To overcome the dilemma existing in dye sensitization of semiconductor electrodes, namely high efficiency of monolayer coverage leading to low absorptivity, we prepared porous ZnO sinter electrodes which adsorbed rose bengal or some other dyes on a monomolecular basis but to a large amount in the total. The photocurrents and ultimate efficiency of energy conversion of PEC cells using such porous electrodes improved dramatically. We obtained photocurrents of the order of 5 mA and the efficiency of monochromatic conversion of 2 to 3%.

However, after many attempts, the quantum efficiency of the monochromatic photocurrent remained as low as ca. 30%. No further improvement was successful. It is most probable that the quantum efficiency of charge injection from excited dye to semiconductor conduction band is almost unity, but after injection, the electron is trapped by the surface states and returns back to the dye, thus wasting the photon energy. We feel, at the moment, that it will be very difficult to prepare more efficient PEC systems by use of dye-sensitized semiconductor electrodes.

FOX - I think we are being pessimistic in overlooking many well-developed multi-electron systems: e.g., Whitten, Wilner, many multi-electron events on metal complexes. Many such systems are currently being developed.

The need for vectorial electron transfer systems persists.

NOZIK - I would like to point out a promising area of research for colloidal and/or particulate semiconductor systems. This involves developing two-photon structures, similar to what has been shown in macrocrystals, that consist of n-type and p-type semiconductor regions coupled through an ohmic contact, such that the majority carriers combine and leave the minority electron in the p-type region and minority holes in the n-type regions to drive a much higher energy reaction. These systems are analogous to biological photosynthesis in the use of 2 photons per net e^-/h^+ pair used in the chemistry, and will increase the upper theoretical thermodynamic efficiency from about 30% to about 40%. Although the 2-photon system was demonstrated for large crystals, it has yet to be demonstrated for small particles.

HELLER - Photoelectrolytic processes may lead some day to energetically uphill processes that are economical. They may also shed light on and lead to the improvement of a small group of industrially important catalysts. In general, photons from artificial sources are expensive and can be used economically only when the product is expensive. With these, we shall not produce bulk chemicals. With sunlight, using small band gap semiconductors, we may produce even bulk chemicals. In energetically downhill (i.e., photocatalytic) reactions, we are unlikely to use photons to drive industrial processes, because even the solar photons cost money and because the plants cannot be run continuously. Nevertheless, with the understanding of the catalyst-semiconductor systems that is resulting from our research, we shall be able to develop new unidirectional catalysts. These will have two sites on which two different molecules ("A" and "B") will be adsorbed. One site will exclusively oxidize "A" to "C", the other exclusively reduce "B" to "D" with the net reaction being:

$$A + B \longrightarrow C + D$$

These correspond to spontaneous microanodic and microcathodic processes with the semiconductor's junctions acting as unidirectional gates.

MEMMING - Concerning sensitization, it has been studied in detail at extended electrodes. True sensitization was only found with adsorbed molecules. The quantum efficiency can be nearly unity. Since a monolayer absorbs only about 1% of the incident light, the efficiency is extremely low. Honda has shown that an increase of the number of layers does not lead to higher efficiencies because of recombination effects. Therefore, there is not much chance to sensitize processes at extended electrodes at higher efficiencies. The situation may be very different for particles loaded with a sensitizer because photons can be absorbed by many sensitizer molecules. However, it has not been proven yet whether a sensitizer molecule being oxidized in the excited state is capable of oxidizing water or some other system.

KRIST - Some of the many issues where further attention is needed in photocatalytic systems for gas evolution include:

(1) molecular or interfacial barriers to back reactions.
(2) catalytic, small molecule multi-equivalent redox chemistry.
(3) detailed, in situ characterization of the solid-electrolyte interface during fuel evolution.
(4) molecular engineering of systems that satisfy selectivity, stability, efficiency, and cost requirements.
(5) careful control and characterization of defects and impurities in both electrodes and electrolytes.

PANEL DISCUSSION on ELECTROCATALYSIS AND PHOTOCATALYSIS

Co-Chairmen: G. McLendon and R. Memming

SERPONE - Before I give the floor to our co-chairmen, I would like this meeting to address two important questions raised by Janos Fendler on Tuesday.

(i) What have we accomplished (in the last decade)?

(ii) Where do we go from Maratea?

McLENDON - Introductory Remarks

It seems to me that much has been accomplished in the last ten years, but much remains to be done. My simple view of this Workshop is that (at least) four major areas can serve as a focus for further development. These include:

1. New chemical reactions involving new pathways and specificity. Elegant work in this area was reported by Halman, Fox, Schrauzer, Somorjai, Teichner, Whitten, Kisch, Sakata. Further, such reactions may be further modified in unusual environments as pointed out by Fendler, VanDamme, Bickley, and others.

2. Detoxification/Environmental Processes. Ollis, Borgarello, Serpone, Oliver, and others noted that detoxification may occur in the near term since the problems can be dealt with based on ongoing results in photocatalyses. By dealing with local "crises" cost concerns are of less immediate concern. Plating can also be obtained, as pointed out by Borgarello.

3. "Nontraditional" Applications. For example, electronics, optics,... Heller has pointed out some unusual applications of photoelectrochemistry already in use at Bell Laboratories. Many of the materials made for photocayalysis may prove quite important in other fields, and such studies should not be naively yielded to physicists or engineers.

4. Finally, in the longer terms the area of photochemical fuel generation remains quite promising. Significant advances have been made in the areas of multielectron redox processes, understanding of solution redox chemistry, and so forth. In photovoltaics, a far more mature field, ($\geq$ 40 years old) the future is now. Nonetheless, continuing adv-

E. Pelizzetti and N. Serpone (eds.), Homogeneous and Heterogeneous Photocatalysis, 703–706.

ances suggest fuel production remains a viable alternative over ≳10+ year timeframe.

Some Problems/Projections

The lectures at this Workshop have clearly brought out some of the problems in the area of photocatalysis. A better understanding of intrinsic structural and kinetic properties are required in order to rationally optimize photocatalytic systems. Detailed analysis of (interfacial) redox reactions and competing non-radiative processes will be required for optimizing the efficiency of liquid junction semiconductors. In this context, both fundamental and exploratory studies are required. Significant progress also is required in the structural chemistry of semiconductors. Some questions include: (a) can special catalytic sites be created at the surface to promote selective chemistry? (b) what is the chemical nature of energy-dissipating 'surface states', and how can they be modified and controlled? The synthesis of new photoactive materials and catalysts is an area of obvious interest. Important advances are occurring in such areas as graded semiconductor junctions and quantized semiconductor materials among others. The relative strengths of colloid semiconductors versus bulk semiconductors for various different applications should be reviewed critically.

Finally, in the area of fuel production, best case engineering studies would be helpful in comparing various strategies. For the long term future of the gas industry, a number of possible lessons emerge from this Workshop. First, it is clear that major progress has been made over the past decade in photoelectrochemistry and photocatalysis. Direct technological applications are already occuring, and major advances are anticipated in several areas of particular interests to the gas industry. Several (by no means all) pertinent examples include:

1. Multielectron catalysis, which is central to fuel cell technology.

2. Controlled synthesis of materials for photoconversion, which might improve fuel production to an economically attractive level.

3. Upgrading of currently uneconomic gas (eg. by catalytic H_2S removal).

4. Photochemical water splitting or CO_2 reduction either directly or via photoelectrolysis, which although involving intermediate electrical power, does not entail capital intensive power grids and load leveling as in the electrical power industry.

Finally, the field of gas production from inorganic precursors is by no means closed and serves as an important hedge for the twenty-first century. It is clear that such investments are long term. A useful benchmark is provided by photovoltaic devices, which after forty years of R & D are now reaching economic viability.

FENDLER - Have we lost sight of how nature converts solar energy to chemical energy? As we go back for further inspiration to improve our current microchemical and photocatalytic semiconductor systems.

McLENDON - While I am not a photosynthesis expert, I have carried out extensive work on protein electron transfer reactions and two points emerge from such studies. First, as John Hopfield points out, semiconductors and photosynthesis are already surprisingly similar: both involve electron hopping over a series of spatially fixed positions to provide vectorial charge separation.

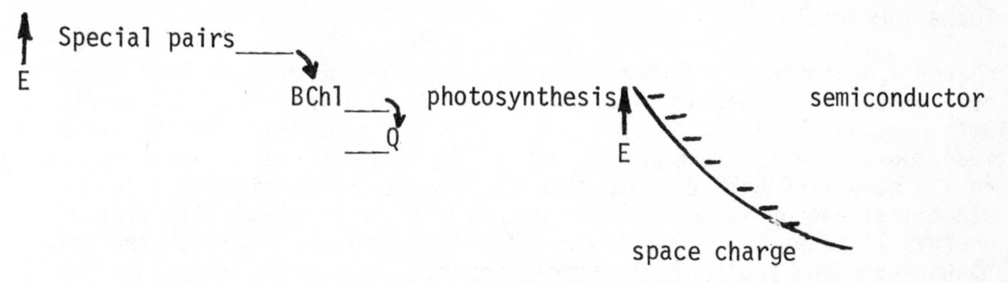

Second, proteins do not appear on careful examination to use any unusual mechanisms either in terms of reorganization energies or electronic coupling; that is, rates are not increased in special ways to be emulated. The key to a successful organism is to balance many processes under special limiting conditions (pH 7, 37°C,...) which need not be imposed on the chemist. From these perspectives, I think it quite possible that in the limited preview of photoelectrochemistry, we might do a bit better than nature.

BICKLEY - The question of O_2 evolution from photocatalysis is raised through the indirect information which we have obtained concerning the net adsorption of O_2 and the erosion of 'OH' groups from hydroxylated TiO_2 surfaces. The net production of O_2 represents something of the order of ca.$\leq$5% of the total activity which is suggested. Moreover, the question of providing oxidizable species in a natural environment is not only presented by organic materials but also by N_2 which does appear from our work to be non-dissociatively oxidized to $(NO^-)_2$ and which may then further participate in another step whereby ammonia is formed.

MANASSEN - I would like to bring up the enginnering question. We are starting to hear more about oxygen evolution instead of hydrogen evolution. Would it be advisable to make oxygen and a reduced substance, which can be led towards an electrochemical cell, where it will act as a depolariser for water electrolysis, forming hydrogen at a low voltage in a central location?

OLLIS - Prior research in practical gas/solid heterogeneous catalysis and in fundamental surface science was characterized by high surface area powders and low index plane single crystals. As a common ground appeared, work with powders and sinters began to include multiple characterizations of the catalyst surface (IR spectroscopy, x-ray diffraction, specific chemisorptive titrations, Auger and photoelectron spectroscopy) and work with single crystals added examination of high index crystal planes, addition of additives, and again elemental and spectroscopic surface characterizations. This fruitful history suggests strongly that photoelectrochemical and photocatalytic systems should repeat major aspects of this sympathetically collaborative history.

McLENDON - I would like to address this question to the experts: To what extent can or should photoelectrochemistry compete with conventional solid state photovoltaic devices, and to what extent is this desirable focus for future work?

HELLER - Photosynthetic ("photocatalytic") fuel generation fascinates our community because it allows to demonstrate our ability to compete with photosynthesis in green plants and micro-organisms. "Photocatalytic" fuel generation for common terrestrial applications will, however, lag in its economics behind photovoltaics. The reason is simple: a joule of electrical energy is more valuable than a joule of chemically stored energy. It takes 2.5 - 3.0 joules of energy stored in oxidizable fuels to generate one joule of electrical energy.

Nevertheless, there is an abundance of economically valid opportunities for the utilization of sunlight in the synthesis of chemicals. I shall mention three: First, I see an opportunity in the energetically downhill photosynthesis of dilute aqueous calcium and magnesium nitrate solutions from air (nitrogen and oxygen), rainwater and limestone (calcium and magnesium carbonates). Such solutions may be used in situ, without concentration, in agricultural applications. Second, we should also explore the in situ synthesis of harmless materials whose synthesis now involves the separation and storage of dangerous chemicals. Thus, in the preparation of polyvinyl chloride now involves vinyl chloride, a carcinogenic precursor. Photochemical preparation of the monomer and its in situ photochemical polymerization may provide an attractive path to this material. Third, based on the present level of understanding of the photoelectrochemistry and photocatalysis of non-colloidal, large grained n-TiO_2 (anatase or rutile) and n-Fe_2O_3, we could design long-lasting, non-flaking, non-chalking paints. While the direct photochemical degradation of the organic binders in paints is well suppressed, the pigment-mediated degradation, associated with direct and indirect oxidation of the organic binders by holes photogenerated in the semiconducting oxides, is poorly controlled. Industry's current approach, that of encapsulating the pigment particles in silicon dioxide, does not work well, because it is difficult to completely coat every particle by exposure to either silicon tetrachloride or to a silicon alkoxide and by baking in humid air.

The dollar volume of sales of nitrogen fertilizers, of polyvinyl chloride and of white and red-pigmented paints greatly exceeds that of the photovoltaic industry. Thus, there are opportunities in photocatalysis that are socially, scientifically and economically significant.

FINAL REMARKS

The workshop, with its informal format and the Panel Discussion Sessions, provided an important forum for exchange of ideas and presentation of the latest 'state-of-the-art' research in many of the intriguing aspects of Homogeneous and Heterogeneous Photocatalysis. The proceedings of the three Panel Discussions on (i) Sensitization and Immobilization of Catalysts on Various Supports, on (ii) Selective Activation and Conversion of Molecules, and on (iii) Electrocatalysis and Photocatalysis are reported in this book.

Several areas of future research directions and focus were noted in both fundamental studies and in applied practical investigations. Indeed, some of the research in the applied area is at the stage of technological transfer to industry. The consensus emerged that much has been accomplished in the last decade, yet much remains to be done. The tools and techniques are available to forge ahead in our quests for an understanding of known and yet to be uncovered processes. The examples are many and were explored in the discussions. It remains for the researchers to pursue the several excellent proposals developed at this NATO Workshop.

Lastly, but not least, it remains for us to express our sincere thanks to the Scientific Affairs Division of the North Atlantic Treaty Organization for their generous support and sponsorship to hold the Workshop. Our co-sponsors, the Gas Research Institute (Chicago), the Consiglio Nazionale delle Ricerche (Roma), and Concordia University (Montreal) were most supportive in providing additional funding. Our many thanks to the Plenary Speakers and to the participants for making the Workshop a highly scientific success.

Ezio Pelizzetti — Italia

Nick Serpone — Canada

AUTHOR INDEX

SUBJECT INDEX